MAJOR
GLACIAL
INTERVALS

SOME MAJOR MOUNTAIN-BUILDING EPISODES

WESTERN
NORTH AMERICA

Eastern America–
Northern Europe

ASIA

Himalayan

Alpine

Laramide

Sevier

Nevadan

Sonoma

Alleghenian-
Hercynian

Antler

Acadian-
Caledonian

Taconic

Symbols for maps, stratigraphic sections, and cross sections

Gravel or conglomerate

Limestone or other carbonate rock

Sand or sandstone

Evaporite deposit

Silt or siltstone

Extrusive igneous rock

Clay, mud, shale, or
mudstone

Mud/sand or mudstone/
sandstone mixture

Crystalline
(igneous or metamorphic)
rock

B
A
Surface position (map view) of a thrust fault; block A has been
thrust over block B

Fault; arrows show relative movements of rocks on opposite sides
of the fault

Unconformity in cross section

EARTH AND LIFE THROUGH TIME

EARTH AND LIFE THROUGH TIME

STEVEN M. STANLEY   *Johns Hopkins University*

W. H. FREEMAN AND COMPANY   *New York*

TO NELL

Illustration credits: Cover image, Bill Ratcliffe; Part I,
© Tom McHugh 1977; Part II, N. K. Huber, U.S. Geological Survey;
Part V, © Tom McHugh 1976, American Museum of Natural History;
Part VI, Alison F. Richard.

Book and cover design by Lisa Douglis

Library of Congress Cataloguing in Publication Data

Stanley, Steven M.
    Earth and life through time.

    Includes bibliographies and index.
      1. Historical geology.   I. Title.
QE28.3.S73 1985 551.7 84-21098
ISBN   0-7167-1677-1

Printed in the United States of America

1 2 3 4 5 6 7 8 9 0 KP 3 2 1 0 8 9 8 7 6 5

Contents

PREFACE   ix

CHAPTER 1   **Introduction to the Earth, Life, and Geologic Time**

The principle of uniformitarianism   1
Life on earth   5
Movements of the earth   13
Fundamental principles of geology   17
Geologic time   18
Growth of the geologic time scale   22
Fields of geologic science   23
This book: An overview   23
Chapter summary / Exercises / Additional reading

PART ONE   THE ENVIRONMENTAL SETTING

CHAPTER 2   Environments and Life

Land and sea   29
Principles of ecology   30
The atmosphere   34
The terrestrial realm   39
The marine realm   45
BOX 2-1 Coral reefs in the tropics   56
Chapter summary / Exercises / Additional reading

CHAPTER 3   Nonmarine Sedimentary Environments

Soil environments   62
Lakes as depositional environments   65
Glacial environments   65
Deserts and arid basins   69

River systems   74
Chapter summary / Exercises / Additional reading

CHAPTER 4   Marine Sedimentary Environments

Deltas   83
The barrier island–lagoon complex   86
Organic reefs   90
Carbonate platforms   95
Marginal marine sabkas   103
Submarine slopes and turbidites   104
Pelagic sediments   108
Chapter summary / Exercises / Additional reading

PART TWO   THE DIMENSION OF TIME

CHAPTER 5   Correlation and Dating of the Rock Record

Index fossils and zones   115
BOX 5-1 Graptolites: Index fossils of the Paleozoic   117
Radioactivity and absolute ages   119
Fossils versus radioactivity:
   The accuracy of correlation   121
Time-parallel surfaces in rocks   123
The units of stratigraphy   130
Chapter summary / Exercises / Additional reading

CHAPTER 6   Evolution and the Fossil Record

Charles Darwin's contribution   137
Genes, DNA, and chromosomes   143

BOX 6-1 Mendel's telltale ratios   144

BOX 6-2 DNA, the genetic code   146

Populations, species, and speciation   147

Extinction   147

Convergence and iterative evolution   151

Evolutionary trends   152

BOX 6-3 The generation of large-scale trends   163

Chapter summary / Exercises / Additional reading

PART THREE   MOVEMENTS OF THE EARTH

CHAPTER 7   Plate Tectonics

The history of opinion about continental drift   169

The rise of plate tectonics   182

Clues to ancient plate movements   195

Chapter summary / Exercises / Additional reading

CHAPTER 8   Mountain Building

Plate tectonics and orogenesis   205

The anatomy of a mountain chain   206

Foredeep deposition   207

The mechanism of deformation   208

The youthful Alps   209

BOX 8-1 Geosynclines   210

The lofty Himalayas   215

The Andes: Mountain building
    without continental collision   217

The Appalachians: An ancient mountain system   219

BOX 8-2 Where was Gondwanaland
    in Late Carboniferous time?   234

Chapter summary / Exercises / Additional reading

PART FOUR   THE PRECAMBRIAN WORLD

CHAPTER 9   The Archean Eon
of Precambrian Time

Ages of the universe and the planet Earth   241

The planets   242

Origin of the solar system and Earth   245

How the Earth and its fluids
    became concentrically layered   248

The great meteorite shower   251

Archean rocks   253

Archean tectonics   259

Large cratons appear   259

Archean life   262

Chapter summary / Exercises / Additional reading

CHAPTER 10   Global Events
of the Proterozoic

A modern style of orogeny   273

Proterozoic glacial deposits   276

Atmospheric oxygen   279

Life of the Proterozoic Eon   283

Chapter summary / Exercises / Additional reading

CHAPTER 11   Proterozoic Cratons:
Foundations of
the Modern World

North America, Greenland, and
    northern Britain as Laurentia   298

The early history of Gondwanaland   311

Eurasia: A composite landmass   317

Chapter summary / Exercises / Additional reading

PART FIVE   THE PALEOZOIC ERA

CHAPTER 12:   The Early Paleozoic World

Life   325

Paleogeography of the Cambrian world   343

Paleogeography of the Ordovician world   346

Regional examples   351

Chapter summary / Exercises / Additional reading

CHAPTER 13   The Middle Paleozoic World

Life   360

BOX 13-1 A life cycle typical of
    a middle Paleozoic land plant   372

BOX 13-2 A life cycle of a seed plant,
    which does not require a moist setting   375

Paleogeography   381

Regional examples   383

Chapter summary / Exercises / Additional reading

CHAPTER 14   The Late Paleozoic World

Life   398
Paleogeography   417
Regional examples   423
BOX 14-1 Pennsylvanian cyclothems
    in central North America   428
Chapter summary / Exercises / Additional reading

PART SIX   THE MESOZOIC ERA

CHAPTER 15   The Early Mesozoic Era

Life in the oceans: A new biota   444
Terrestrial life   452
BOX 15-1 Who were the dinosaurs?   458
Paleogeography   461
Regional examples   466
Chapter summary / Exercises / Additional reading

CHAPTER 16   The Cretaceous World

Life   483
Paleogeography   501
BOX 16-1 The early mammals   502
Regional examples   513
Chapter summary / Exercises / Additional reading

PART SEVEN   THE CENOZOIC ERA

CHAPTER 17   The Paleogene World

Worldwide events   528
Regional events   544
Chapter summary / Exercises / Additional reading

CHAPTER 18   The Neogene World

Worldwide events   560
Regional events   589
Chapter summary / Exercises / Additional reading

EPILOGUE   616
APPENDIX I   Minerals and Rocks   618
**APPENDIX II   Deformation Structures
                   in Rocks**   639
**APPENDIX III   Classification of
                   Major Fossil Groups**   648
APPENDIX IV   Stratigraphic Stages   656
GLOSSARY   659
INDEX   672

Preface

Major developments in the study of earth history during the past few years have, for the first time, made it possible to integrate descriptions of such diverse elements of our planet's evolution as earth movements, climatic and oceanographic changes, and the transformation of life by evolution and extinction. I have written this book in the conviction that the physical history and the biological history of the earth are inextricably intertwined. Not only has the evolutionary play been performed in the ecological theater, to borrow G. Evelyn Hutchinson's metaphor, but the players themselves, albeit unwittingly, have changed the stage upon which they have acted.

This integrated view of earth history has led me to write a text which can serve as a self-contained introduction to both the biological and the geological aspects of our planet's evolution. Unlike many other historical geology texts, *Earth and Life Through Time* requires no previous knowledge of physical geology in order to be fully comprehended; yet it may also be easily adapted for use in courses with different emphases and requirements for enrollment, and can therefore be used by majors and nonmajors alike, by students who have had a previous course in physical geology, and by those who are being introduced to geologic techniques and concepts for the first time.

Broadly, the first portion of the text (Chapters 1 through 8) provides facts and principles necessary to understand the second (Chapters 9 through 18), which is a chronological review of the earth and its biota. Unlike comparable segments in the traditional earth history text, however, the introductory chapters are not merely a summary of relevant topics in physical geology. They also contain much additional material that is essential for understanding modern concepts of earth history and that, to date, has been insufficiently covered in the available texts. Chapter 2, for example, presents an overview of ecological principles and the nature of our planet's ecosystem. I hope that the material here will prove generally enlightening to students, quite apart from its immediate purpose of serving as a basis for understanding life and environments of the past. Chapters 3 and 4 are also without counterpart in previous historical geology texts. They review a modest number of important nonmarine and marine depositional settings that account for the bulk of the sedimentary record. I have placed Chapter 5, which introduces correlation and other stratigraphic procedures and principles, ahead of the chapter on evolution (Chapter 6). This ordering may run counter to the inclinations of some teachers. My rationale is that understanding the elements of biostratigraphy requires knowledge of faunal and floral succession but not knowledge of the nature of evolution. Furthermore, only students who first know something about correlation and radiometric dating will be able to appreciate what the fossil record can tell us about rates and patterns of evolution and extinction.

To introduce plate tectonics (Chapter 7), I have employed a largely historical approach in order to give students the opportunity both to share in the excitement that this new paradigm has generated and to learn how scientific theories germinate and develop. The final "background" chapter (Chapter 8), also a new feature for a book of this type, reviews mountain-building processes through analysis of four case studies: the Alps, the Himalayas, the Andes, and the Appalachians. I later refer to these examples in the second part of the book, where each has a place in the chronological sequence. Additional basic information concerning minerals, rocks, and deformational structures in the earth's crust is summarized in Appendixes I and II.

Throughout the book I have relied heavily on figures

and figure legends to introduce students to ancient animals and plants. In Chapters 12 through 18, this approach provides a chronological picture of life on earth. Complementing this perspective, Appendix III places important fossil groups in a taxonomic context; alongside nearly every taxon listed is a cross-reference to one or more figures in the text illustrating one or more members of the taxon. Study of Appendix III and the figures (and figure legends) cited within it will provide the student with a general picture of ancient life.

The broad coverage of this book is designed to give teachers the freedom to lecture only on specific topics that they find appealing. Believing that students are most comfortably immersed in history by degrees, I have organized the chapters on Phanerozoic time so that the general features of each interval (its biological and paleogeographic developments) appear first, and more localized aspects (selected regional events) follow. This arrangement also allows for flexibility in the use of the book: Instructors whose courses are dominated by students not majoring in geology may wish to omit some or all of the "Regional Events" sections of the final seven chapters—or these instructors might recommend that students merely scan those portions of the book; Chapter 11, which describes Proterozoic cratons, might also be omitted or scanned. Such courses would then focus on global events, paleogeography, and the history of life.

Because of its comprehensiveness and flexibility, *Earth and Life Through Time* can also serve as a text for courses on the History of Life. In this role, the book will surely not be assigned in its entirety. Chapters 7 and 8, covering plate tectonics and mountain building, might be scanned rather than read; other expendable sections would include parts of Chapter 9 (which covers Archean time), all of Chapter 11 (which describes Proterozoic cratons), and the "Regional Events" sections of later chapters.

In writing this book I have relied on the criticisms and suggestions of many experts and capable critics. None of these persons has seen a final version of anything he or she has reviewed, so, in the proper tradition, I accept blame for any errors, but I also express much gratitude to all:

Tony Arnold, *Florida State University*

Stanley M. Awramik, *University of California, Santa Barbara*

William B. N. Berry, *University of California, Berkeley*

Harvey Blatt, *University of Oklahoma*

Bruce Bolt, *University of California, Berkeley*

Scott Brande, *University of Alabama*

Allan Cain, *University of Rhode Island*

Barry Cameron, *Acadia University*

James Collinson, *Ohio State University*

Allan Cox, *Stanford University*

Roger Cuffey, *Pennsylvania State University*

Kenneth Deffeyes, *Princeton University*

Lawrence A. Hardie, *Johns Hopkins University*

Rose Mary Harvey, *Milliken Middle School*

Edward Hay, *De Anza College*

Leo J. Hickey, *Yale University*

Allan Hunt, *University of Vermont*

Andrew H. Knoll, *Harvard University*

Peter Kresan, *University of Arizona*

Léo Laporte, *University of California, Santa Cruz*

David Lumsden, *Lumsden, Memphis State University*

George R. McGhee, *Rutgers University*

William Meyers, *State University of New York, Stony Brook*

Eldridge Moores, *University of California, Davis*

Everett C. Olsson, *University of California, Los Angeles*

John H. Ostrum, *Yale University*

A. R. Palmer, *Geological Society of America*

Charles Pitrat, *University of Massachusetts*

Reuben J. Ross, *Colorado School of Mines*

Charles A. Ross, *Gulf Oil Exploration and Production Corporation*

Claude Spinosa, *Boise State University*

Michael Tevesz, *Cleveland State University*

Roger D. K. Thomas, *Franklin and Marshall College*

Charles P. Thornton, *Pennsylvania State University*

Eugene Tynan, *University of Rhode Island*

John A. Van Couvering, *American Museum of Natural History*

James Valentine, *University of California, Santa Barbara*

Jamie Webb, *California State University, Dominguez Hills*

Jack A. Wolfe, *United States Geological Survey*

In the course of preparing the manuscript and developing the extensive illustration program, I was assisted by able persons too numerous to name separately here. Some of the outstanding contributions nevertheless merit special acknowledgment: Anne Boersma, J. M. Pulsford, and Don Baird, for making exceptional efforts in providing illustrative material; Gregory Paul, whose beautiful drawings speak for themselves; Kate Francis and Susan Lubonovich, for their excellent typing; and Robert T. Bakker, for enriching my intellectual life during the past ten years and enlightening me about many things described in the following pages.

I also thank the many people at W. H. Freeman and Company who have aided in this project. John H. Staples has long been a friend and valued advisor, Moira Lerner has been the best and most cheerful developmental editor anyone could wish for, Georgia Lee Hadler coordinated the long transition from manuscript to bound book with great skill, and Lisa Douglis designed what I think is an exceptionally beautiful book.

Finally, I offer apologies and appreciation to my wife, Nell, for tolerating my horrendous daily schedule for more years at work on this book than I wish to count.

I would be grateful if readers would notify me of errors or offer suggestions for improvement, preferably with literature citations where these would be helpful.

Steven M. Stanley

EARTH AND LIFE THROUGH TIME

CHAPTER 1

Introduction to the Earth, Life, and Geologic Time

Few people recognize, as they travel down a highway or hike along a mountain trail, that the rocks they see around them have rich and varied histories. Unless they are geologists, they have probably not been trained to identify a particular cliff as rock formed on a tidal flat that once fringed a primordial sea, to read in a hillside's ancient rocks the history of a primitive forest buried by a fiery volcanic eruption, or to decipher clues in lowland rocks telling of a lofty mountain chain that once stood where the land is now flat. Geologists can do these things because they have at their service a wide variety of information gathered during the two centuries that the modern science of geology has existed. The goal of this book is to introduce enough of those geologic facts and principles to give the student an understanding of the general history of our planet and its life. The chapters that follow will describe how the physical world assumed its present form and where the inhabitants of the modern world came from. They will also reveal the procedures through which geologists have assembled this information. Students of earth history inevitably discover that the perspective this knowledge provides changes their perception of themselves and of the land and life around them.

The San Andreas Fault in California is a great break within the earth's crust separating large segments, or plates, of the crust. The Pacific plate is on the left, and the North American plate is on the right. As indicated by the arrows, the Pacific plate periodically slides northwestward in relation to the North American plate. Movement along this fault led to the San Francisco earthquake of 1906. Many faults transect California, although movement has not occurred along most of these during recorded history. (*R. E. Wallace, U.S. Geological Survey.*)

Knowledge of the earth's history can also be of great practical value. For example, geologists have learned to locate petroleum reservoirs by ascertaining where the porous rocks of these reservoirs tend to form in relation to other bodies of rock. Early in this century, vast stores of natural gas and petroleum were discovered trapped in rocks that lay adjacent to buried pillars of salt in the Gulf Coast region of the United States. It is now known that natural gas, petroleum, and sulfur are trapped in rocks above salt domes. By studying waves of earth movement produced by artificial explosions, geologists can locate salt domes deep below the surface and guide petroleum companies in their extraction of valuable resources from the rock. Geologists have also helped discover many additional kinds of petroleum reservoirs, as well as coal deposits, ore deposits, and other natural resources buried within the earth. Occurrences of some of these resources will be analyzed in the chapters that follow. Before we launch into our detailed examination of the history of the earth and its life, however, an introduction to some of the basic facts and unifying concepts of geology is in order. This chapter will be devoted to laying that groundwork.

THE PRINCIPLE OF UNIFORMITARIANISM

Fundamental to the modern science of geology is the principle of **uniformitarianism**, which is the belief that there are inviolable laws of nature that have not changed in the course of time. Of course uniformitarianism applies not only to geology but to all scientific disciplines—physicists, for example, invoke the principle of uniformitarianism when they assume that the results of an experiment will be applicable to events that take place a day, a year, or a century after the experiment is conducted—but geologists hold the principle of uniformitarianism in particularly great esteem because, as we shall see, it was the widespread adoption of uniformitarianism during the first half of the nineteenth century that signaled the beginning of the modern science of geology.

Actualism: The present as the key to the past

The principle of uniformitarianism governs geologists' interpretations of even the most ancient rocks on earth. It is in the present, however, that many geologic processes are discovered and analyzed, and the application of these analyses to ancient rocks in accordance with the principle of uniformitarianism is sometimes called **actualism**. As an example, when we see ripples on the surface of an ancient rock composed of hardened sand (sandstone), we assume that these ripples formed in the same manner that similar ripples develop today—under the influence of certain kinds of water movement or wind. Similarly, when we encounter ancient rocks that closely resemble those forming today from volcanic eruptions of molten rock in Hawaii, we assume that the ancient rocks are also of volcanic origin.

Commonly, actualism is described by the phrase, "The present is the key to the past." This idea is only partly true, however. Although it is universally agreed that natural laws have not varied in the course of geologic time, not all past events have been duplicated in recent times. Many researchers, for example, believe that the impact of very large meteorites may explain certain past events, such as the extinction of the dinosaurs 65 million years ago. They can calculate that the impact of a huge meteorite—one something like 10 kilometers (6 miles) in diameter—if it were to land in the ocean, would produce a great tidal wave, but because we have never observed the arrival of such a large meteorite, we do not know exactly what else would happen. It has been suggested that the fine dust injected into the upper atmosphere might block the sun's rays from the earth for many days. As we will learn in Chapter 16, there is some evidence to support this contention, but because we cannot observe the consequences of such an event today, the idea is difficult to verify. In other words, in this case actualism cannot be applied.

Similarly, geologists have found that certain types of rocks cannot be observed in the process of forming today. When this is the case, geologists usually assume that (1) the rocks in question formed under conditions that no longer exist; (2) the conditions responsible for the formation of these rocks still exist, but at such great depths beneath the earth's surface that we cannot observe them; or (3) the conditions exist today but produce the rocks only after a long interval of geologic time. Many iron ore deposits more than 2 billion years old, for example, are of types that cannot be found in the process of forming today. It is believed that when the iron ore formed, chemical conditions on earth differed from those of the present world, and furthermore that the rocks underwent slow alteration after they were formed. The existence of these iron ore deposits does not necessarily negate the principle of uniformitarianism, inasmuch as there is no evidence that natural laws were broken, but it does present geologists with a problem that cannot be solved by application of the principle of actualism, since they will not have the opportunity to interpret the rocks by studying their development during a modern human lifetime.

In an attempt to address some of these problems, geologists have learned to form rocks in the laboratory by duplicating the conditions that prevail at great depths within the earth. To accomplish this, they expose simple chemical components to temperatures many times higher—and to pressures many times greater—than those of the earth's surface. Such experiments indicate the range of conditions under which a particular type of rock could have formed in nature. In conducting these experiments, geologists are, in a sense, expanding the domain of actualism by using as a model not only what is happening in nature today, but also what happens under artificial conditions and may have happened under natural conditions of the past.

The uniformitarian view of rocks

Until the early nineteenth century, many natural scientists subscribed to a concept known as **catastrophism**. According to this idea, floods caused by supernatural forces formed most of the rocks visible at the earth's surface. Late in the eighteenth century, Abraham Gottlob Werner, an influential German professor of Mineralogy, claimed that most rocks formed as a result of the precipitation of minerals from a vast sea that periodically flooded and retreated from the earth's surface. These ideas were entirely speculative; they were unsupported by evidence from the physical world.

Not long after Werner published his ideas, James Hutton, a Scottish farmer, established the foundations of uniformitarianism by writing about the origins of rocks in

Scotland. Hutton concluded that rocks formed as a result of a variety of processes presently operating at or near the earth's surface—processes such as volcanic activity and the accumulation of grains of sand and clay under the influence of gravity. It was only after extensive debate that Hutton's interpretation of the origins of rocks was generally accepted by the scientific community. Once established, however, uniformitarianism soon dominated the science of geology, gaining almost total acceptance after Charles Lyell, an Englishman, popularized it in the 1830s in a three-volume book entitled *Principles of Geology*. Let us briefly examine the uniformitarian view of how rocks form.

Rocks consist of interlocking or attached grains that are typically composed of single minerals. A **mineral** is a naturally occurring inorganic solid element or compound with a particular chemical composition or range of compositions and a characteristic internal structure. Quartz, which forms most grains of sand, is probably the most familiar and widely recognized mineral; other minerals form the materials we call mica, clay, and asbestos. Rocky surfaces that stand exposed and are readily accessible for study are generally designated as **outcrops** or **exposures**, although some geologists restrict the first term to rocks laid bare by natural processes and the second to rocks exposed by human activities such as quarrying or road building. Scientists also have access to rocks that are not visible in outcrops or exposures. Well drilling and mining, for ex-

ample, allow geologists to sample rocks that are still buried beneath the earth's surface.

On the basis of modes of origin, many of which can be seen operating today, early uniformitarian geologists, led by Hutton and Lyell, came to recognize three basic types of rocks: igneous, sedimentary, and metamorphic. **Igneous rocks**, which form by the cooling of molten material to thepoint at which it hardens, or freezes (much as ice forms when water freezes in a refrigerator), are composed of interlocking grains, each consisting of a particular mineral. The most familiar igneous rock to the nongeologist is granite. The molten material, or **magma**, that turns into igneous rocks comes from great depths within the earth, where temperatures are very high. This material may reach the earth's surface through cracks and fissures in the crust and then cool to form **extrusive**, or **volcanic**, igneous rock (Figure 1-1), or it may cool and harden within the earth to form **intrusive** igneous rock (Figure 1-2). Igneous rock that solidifies deep within the earth is sometimes uplifted by subsequent earth movements along with surrounding rock and eventually exposed at the earth's surface by **erosion**, which is the group of processes that loosen rock and move pieces of loosened rock downhill.

Sedimentary rocks form from **sediments**, which are materials deposited at the earth's surface by water, ice, or air. Most sediments are accumulations of distinct mineral grains. Some of these grains are products of weathering

FIGURE 1-1 Mount St. Helens in a peaceful state (*left*) and erupting in 1980 (*right*). The cone of the volcano is itself formed of volcanic igneous rock extruded from the volcano. (*U.S. Geological Survey.*)

FIGURE 1-2   Igneous rock that formed by cooling of magma within the earth. The mineral grains of this type of igneous rock are larger than those of volcanic igneous rock, giving the rock a coarser texture. The hammerhead rests against an inclusion—a piece of older rock that was trapped within the magma and "frozen" into it as it cooled. (*J. J. Witkind, U.S. Geological Survey.*)

(i.e., decay and breakup) of older rocks, while others result from the chemical precipitation of minerals from water. Grains of sediment seldom become mutually attached to form a hard rock until long after they have accumulated. The two important agents of this rock-forming process, which is known as **lithification**, are compaction of sediment under

the influence of gravity and cementation of grains by the precipitation of mineral cement from solutions that flow between the grains. Lithification is a form of **diagenesis**, which is the full set of processes, including solution, that alter sediments at low temperatures after burial. Alteration at high temperatures constitutes metamorphism, which will be described shortly.

Most igneous rocks consist of silicate minerals (see Appendix I) and so do most sedimentary particles, or **clasts**, derived from them. Sedimentary rocks formed primarily of silicate minerals are thus known as **siliciclastic** rocks, and these are the most abundant sedimentary rocks of the earth's crust.

Sediments usually accumulate during episodes of deposition, each of which forms a tabular unit known as a **stratum** (plural, **strata**). Strata tend to remain distinct from one another even after lithification because the grains of adjacent beds usually differ in size or composition. Because of their differences, the contacting surfaces of the strata usually adhere to each other only weakly, and sedimentary rocks often flake or fracture along these surfaces. The result is that sedimentary rocks exposed at the earth's surface often have a steplike configuration when viewed from the side (Figure 1-3). **Stratification** is the word used to describe the layered character of sedimentary rocks. **Bedding** is stratification in which layers exceed one centimeter (~0.4 inches) in thickness, and **lamination** is stratification on a finer scale.

Metamorphic rocks form by the alteration, or **meta-**

FIGURE 1-3   Horizontal bedding in sedimentary rocks of the Bob Marshall Wilderness Area in Montana. Erosion of these rocks has produced a steplike outcrop pattern. Sediment loosened by erosion forms a sloping deposit at the foot of the cliff. (*M. R. Mudger, U.S. Geological Survey.*)

morphism, of rocks within the earth under conditions of high temperature and pressure. By definition, metamorphism alters rocks without turning them to liquid. If the temperature becomes high enough to melt a rock and the molten rock subsequently cools to form a new solid rock, this new rock is by definition igneous rather than metamorphic. Metamorphism produces minerals and textures that differ from those of the original rock and that are characteristically arrayed in parallel wavy layers (Figure 1-4).

Geologists also classify rocks into units called **formations**. Each formation consists of a body of rocks of a particular type that formed in a particular way—for example, a body of granite, of sandstone, or of alternating layers of sandstone and shale. A formation is formally named, usually for a geographic feature such as a town or river where it is well exposed. Smaller rock units called **members** are recognized within some formations. Similarly, some formations are united to form larger units termed **groups**, and some groups, in turn, are combined into **supergroups**.

More about the nature and origin of minerals and the three basic types of rocks can be found in Appendix I.

FIGURE 1-4   Metamorphic rock showing parallelism of mineral grains that formed at high pressure and high temperature. (*M. H. Staatz, U.S. Geological Survey.*)

blueprint is the cell's built-in ability to duplicate itself so that a replica can be passed on to another cell or to an entirely new organism.

LIFE ON EARTH

The organisms that have inhabited the earth in the course of geologic time have left a partial record in rock of their presence and their activities. This record reveals that life has changed dramatically during the history of the earth and that its transformation has been intimately associated with changes in physical conditions on earth —in climates or in the position of continents, for example.

It is not easy to provide a precise definition of **life**, but two attributes that are generally regarded as essential to life are the capacity for self-replication and the capacity for self-regulation. Viruses are simple entities that can replicate themselves (or reproduce) but they do not regulate themselves—that is to say, they do not employ raw materials from the environment to sustain orderly, internal chemical reactions. Thus, viruses are not considered to be living things. On the earth today, all entities that are self-replicating and self-regulating also share the quality of being cellular, which means that they consist of one or more discrete units called cells. A living **cell** is a module that includes a number of distinct features, including apparatuses that facilitate certain chemical reactions. The chemical "blueprint" for a cell's operation is coded into the chemical structure of the DNA molecule, which we will examine more closely in Chapter 6. An essential feature of this

Taxonomic groups

Until well into the nineteenth century, scientists divided all living things into two categories: the animal kingdom and the plant kingdom. As various forms of life came to be better understood, however, these distinctions became increasingly difficult to maintain. Today five **kingdoms** are recognized—the Monera, Protoctista, Fungi, Animalia, and Plantae. They are illustrated in Figure 1-5.

A more detailed classification of many forms of life can be found in Appendix III of this book. As examination of this appendix will indicate, each of the five kingdoms is divided into numerous subordinate groups. These kingdoms and their subordinate groups are known as **taxa** or **taxonomic groups**, and the study of the composition and relationships of these groups is known as **taxonomy**. Taxa within kingdoms range from the broad category known as the **phylum** (plural, **phyla**) to the narrowest category, the **species** (Table 1-1). The basic categories of higher taxa— the kingdom, phylum, **class, order, family,** and **genus** (plural, **genera**) are sometimes supplemented by categories such as the subfamily and the superfamily. Names of genera are printed in italics, as are species designations. Actually, the name of the species consists of two words, the first of which is the name of the genus to which the species belongs.

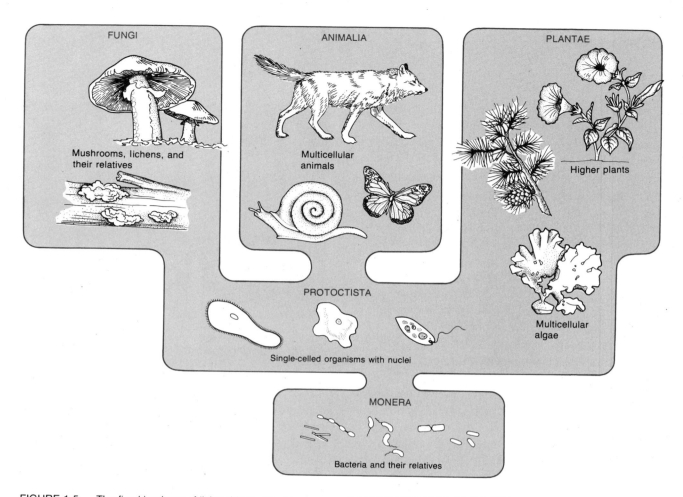

FUNGI

Mushrooms, lichens, and their relatives

ANIMALIA

Multicellular animals

PLANTAE

Higher plants

Multicellular algae

PROTOCTISTA

Single-celled organisms with nuclei

MONERA

Bacteria and their relatives

FIGURE 1-5   The five kingdoms of living things. The Monera include simple forms whose cells lack the internal organization represented by subcellular bodies such as nuclei and chromosomes. Some experts divide the Monera into two kingdoms andhence recognize six kingdoms altogether. The Protoctista include single-celled organisms that possess nuclei and chromosomes; animal-like protists eat other organisms, while plantlike protists manufacture their own food. Red, green, and brown algae are multicellular, but their cells are not differentiated into tissues of discrete cell types; some experts classify these algae as protists, while others classify them as members of the Plantae. Plantae are multicellular organisms that manufacture their own food, and animals are multicellular organisms that ingest food and digest it within their bodies. Fungi include mushrooms and molds; they absorb food from their environment.

Figure 1-6 further illustrates how humans are classified within the order Primates of the class Mammalia. In general, the narrower the taxonomic category, the greater the degree of biological similarity. Humans and gorillas, for example, have enough in common to be assigned to the same superfamily, but humans and gorillas are grouped with monkeys only within the broader suborder category. Within this suborder, monkeys are placed in a superfamily of their own. Often one or a small number of biological features serve to distinguish one higher taxon from other closely related taxa of the same rank. Dinosaurs, for ex-

Table 1-1   Major taxonomic categories within a kingdom, as illustrated by the classification of humans

Between these categories, intermediate ones (e.g., superorders, suborders, superfamilies, and subfamilies) are sometimes recognized.

Kingdom: Animalia
Phylum: Chordata
Class: Mammalia
Order: Primates
Family: Hominidae
Genus: *Homo*
Species: *Homo sapiens*

PRIMATES

SUBORDER ANTHROPOIDEA

SUPERFAMILY HOMINOIDEA

FAMILY
Hylobatidae

Hylobates (gibbon)

FAMILY
Hominidae

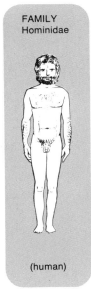

(human)

FAMILY
Pongidae

Gorilla
(gorilla)

Pan
(chimp)

Pongo
(orangutan)

SUPERFAMILY CERCOPITHECOIDEA

Old World
monkeys

SUPERFAMILY CEBOIDEA

New World monkeys

SUBORDER PROSIMII

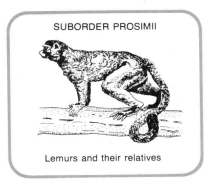

Lemurs and their relatives

FIGURE 1-6   The taxonomic position of the human genus *Homo* within the order Primates. There are four other genera in the superfamily Hominoidea: three ape genera of the family Pongidae and one genus of the family Hylobatidae.

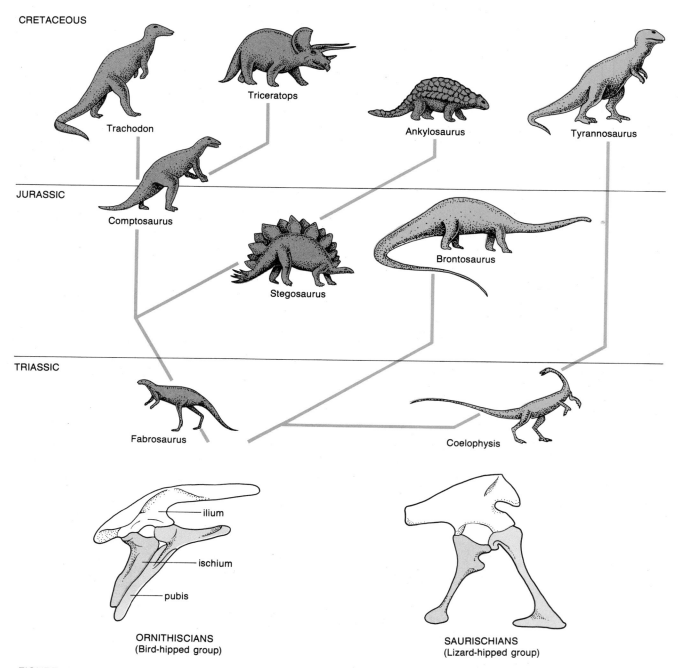

CRETACEOUS

Trachodon

Triceratops

Ankylosaurus

Tyrannosaurus

JURASSIC

Comptosaurus

Stegosaurus

Brontosaurus

TRIASSIC

Fabrosaurus

Coelophysis

ilium

ischium

pubis

ORNITHISCIANS
(Bird-hipped group)

SAURISCHIANS
(Lizard-hipped group)

FIGURE 1-7   The division of the dinosaurs into two orders: the ornithischian, or "bird-hipped," dinosaurs and the saurischian, or "lizard-hipped," dinosaurs (in color). Of the three large pelvic bones, the pubic bone in these two groups differs the most both in shape and in position. (*After E. H. Colbert, Evolution of the Vertebrates, John Wiley & Sons, Inc., New York, 1980.*)

ample, are divided into two orders on the basis of pelvic structure (Figure 1-7).

Although the grouping of species into higher taxa is largely a subjective matter, species are generally considered to be natural entities that are defined in terms of reproductive isolation. Essentially, each species represents a group of actually or potentially interbreeding individuals that do not interbreed with other interbreeding groups. In Chapter 6, we will examine how new species originate from previously existing species by the process of evolution.

Fossils

Most of our knowledge about life of past intervals of geologic time is derived from fossils, which are actual remains or tangible signs of ancient life that have been preserved in rocks. Because few fossils can survive the high temperatures at which igneous and metamorphic rocks form, almost all fossils are found in sedimentary rocks. Of the five kingdoms, only the fungi are poorly represented in the fossil record. Fossils of the other four kingdoms depict many aspects of the history of life on earth—a history that, as will be discussed in Chapter 6, encompasses the evolution of living things.

It is impossible to decide exactly when something becomes a fossil. A shell that has been covered by sand for just 5 or 6 years or a cow's skull that has been buried in the floodplain of a river for a similar length of time would not ordinarily be regarded as a fossil. On the other hand, a beetle specimen found in 18,000-year-old ice age deposits is certainly a fossil even though it may represent a species that is still living. Sometimes the term *subfossil* is applied to the remains of an organism that is only a few years or a few tens or hundreds of years old. The term **fossil** is usually restricted to tangible remains or signs of ancient organisms that died thousands or millions of years ago.

The most readily preserved features of animals are the structures that are informally described as "hard parts"— the teeth and bones of vertebrate animals and comparable solid structures of invertebrate animals (Figure 1-8). Many

FIGURE 1-8   Fossilized hard parts of organisms. *A.* A slice of a fossil tree trunk permineralized with silica (scale 5 centimeters, or ~ 2 inches). *B.* The skeleton of a Mesozoic sea urchin (phylum Echinodermata) about 5 centimeters (~ 2 inches) long. This skeleton closely resembles those of living sea urchins that burrow in the sea floor. *C.* A Mesozoic ammonite shell (phylum Mollusca) about 7 centimeters (~ 2.8 inches) in diameter; ammonites were squidlike swimmers that fed on other ocean animals. *D.* Teeth in jaw fragments of early Cenozoic mammals. (*Photographs by the author.*)

groups of invertebrates lack skeletons and therefore have poor fossil records or no fossil records at all. Some invertebrate animals, on the other hand, have internal skeletons embedded in soft tissue; among these are the sea urchins. Others have protective external skeletons; among these are the bivalve and gastropod mollusks, whose tissues are housed inside skeletons popularly known as "seashells." Hard parts are often preserved with only a modest amount of diagenetic alteration, but at times they are completely replaced by minerals that are unrelated to the original skeletal material.

Although plants do not have skeletal structures per se, their cells have rigid walls of cellulose. As a consequence, woody tissue and even many leaves are much more likely to be preserved than the flesh of animals. After plants are buried in sediment, the spaces left inside the cell walls of woody tissue are often replaced by inorganic materials—most commonly by finely crystalline quartz, known as chert. This filling process, which produces the fossils that are often called "petrified wood," is known as **permineralization** (Figure 1-8). Permineralization is not restricted to woody plant tissue. Porous animal skeletons, such as the bones of vertebrates, also become permineralized.

FIGURE 1-9   Fossil molds of Paleozoic brachiopods. These natural molds (M) are in sandstone. The shells that the molds represent were dissolved away by groundwater flowing through the rock. Artificial casts have been made by pouring rubber (R) over the surface that bears the molds. Shells (S) resembling the ones that formed the molds have been placed on the rock. The largest shell and mold are about 5 centimeters (~ 2 inches) across. (*Photograph by the author.*)

FIGURE 1-10   Fossils that are impressions, with traces of carbon produced by the carbonization of original tissues. *A.* A butterfly. *B.* A palm tree about 2.2 meters (~ 7 feet) tall from Italy. Both are of Cenozoic age. (*A. Courtesy of F. M. Carpenter. B. Courtesy of Leo F. Hickey, Smithsonian Institution.*)

FIGURE 1-11   Trace fossils. *A*. Sedimentary filling of an early Mesozoic dinosaur footprint about 10 centimeters (~ 4 inches) long. *B*. The track of an unknown marine animal of Paleozoic age; this animal meandered over the sediment searching for food. (*A. Courtesy of Donald Baird. B. Photograph by the author.*)

Sometimes solutions percolating through rock or sediment dissolve fossil skeletons, leaving a space within the rock that is a three-dimensional negative imprint of the organic structure. This type of imprint, which is called a mold, can be artificially filled with wax, clay, or liquid rubber to produce a replica of the original object (Figure 1-9).

Fossils called **impressions** might be viewed as squashed molds. Impressions usually preserve in flattened form the outlines and some of the surface features of soft or semi-hard organisms such as insects or leaves (Figure 1-10). A residue of carbon remains on the surface of some impressions after other compounds have been lost by the escape of liquids and gasses. This process of carbon concentration is known as **carbonization**.

Tracks, trails, burrows, and other marks left by animal activity are known as **trace fossils** (Figure 1-11). Because they are the direct results of activity, trace fossils can reveal a great deal about the behavior of extinct animals—although the animal that made a particular kind of trace cannot always be identified with certainty.

Fleshy parts of animals, or "soft parts," are occasionally found in the fossil record, but only in sediments that date back a few millions or tens of millions of years. The fossil rhino shown in Figure 1-12 was protected from deteriora-

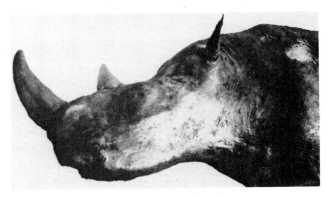

FIGURE 1-12   A fossil rhinoceros found embedded in a Pleistocene asphalt deposit in Spain. This animal represents an extinct species.

FIGURE 1-13   Some remarkable fossils. *A.* A small fish that died while trying to swallow another fish, apparently having overestimated its capacity; the outlines of the fishes are visible as dark films of organic matter in the fine-grained sediment (Eocene of Wyoming). *B.* Crinoids (sea lilies) attached to a fossil tree trunk. The skeletons of these animals consist of many small elements that escaped being torn apart by scavenging animals after the crinoids died because the crinoids and the waterlogged tree trunk were buried on a sea floor where there was too little oxygen for scavengers to live (mid-Mesozoic of Germany). *C.* A small fossil cephalopod (a mollusk related to an octopus) in which the outline of the tentacles is visible; this unusual fossil was preserved in a hard, rounded structure composed of minerals that grew around the animal in soft sediment (late Paleozoic of Illinois). (*A. Museum of Natural History, Princeton University. B. Courtesy of B. Hauff. C. Courtesy of E. S. Richardson.*)

tion by burial in asphalt—oily, tarlike material derived from the partial decay of ancient plants. The deposit most famous for preservation of soft parts is the Eocene Geiseltal of West Germany, which is more than 40 million years old. In the dense Geiseltal sediments, which are rich in oily plant debris, the skin and blood vessels of long-extinct frogs can still be studied, fossil leaves are still green, and insects retain their iridescent color. Protection from oxygen is the secret for survival of soft tissue. This is most likely to occur when organisms are buried in fine-grained, dense sediment, especially if oily water-repellent organic matter is also present.

Some unusual fossils, though not representing well-preserved soft parts, nonetheless reveal outlines of soft parts (Figure 1-13). Most of these fossils were preserved under special conditions such as the absence of oxygen, rapid burial in fine-grained sediment, or protection in a nodular structure called a **concretion**, formed by diagenetic alteration of sediment. Although the origin of concretions is not fully understood, those that form around fossils seem to have resulted from localized diagenesis caused in some way by the emission of chemical compounds from the fossilized organism.

Less spectacular fossils—especially hard parts and molds—are much more common in sedimentary rocks than most people realize. They are especially abundant in sedimentary rocks that were formed in the ocean, where animals with skeletons abound. To appreciate how common fossils are, one need only recognize that many limestones consist largely of fossil skeletons of marine organisms.

Although fossils occur with great frequency in many sedimentary rocks, it is important to recognize that most species of animals and plants have never been discovered in the fossil record. Rare species and those that lack skeletons are especially unlikely to be found in fossilized form. Even most species with skeletons have left no permanent fossil record. A variety of processes destroy skeletons. Animals that scavenge on carcasses, for example, may splinter bones in the scavenging process. Also, many bones, teeth, and shells are abraded beyond recognition when they are transported by moving water before burial. Even after burial, many fossils fail to survive diagenesis, metamorphism, and erosion of the sedimentary rocks in which they are embedded. Finally, many fossil species remain entombed in rocks that have never been exposed at the earth's surface or sampled by drilling operations.

Organisms often lose their identity in contributing to **fossil fuels**, which are condensed and altered forms of organic matter that can be burned to supply energy for human use. **Coal** is a metamorphic rock that serves as a fossil fuel. Coal forms when fossil plants are carbonized under high temperatures and pressures that drive out many compounds in liquid and gaseous states. Most low-grade coals are full of carbonized fossil plants that can still be identified as particular genera or even species. **Peat** is the term used for sediments composed of plant debris that has not yet metamorphosed to form coal.

In Chapter 2, we will learn some of the ways in which fossils reveal information about ancient environments; in Chapter 5, we will consider how fossils are employed to date rocks; and in Chapter 6, we will examine how fossils document the evolution of life.

MOVEMENTS OF THE EARTH

Rocks are not motionless. They not only move but also break, and they can even change shape without breaking. When buried deep within the earth under conditions of great pressure, rocks can bend like soft metal or, like bread dough or modeling clay, be squeezed into new shapes. Most deformed rocks that we now see at the earth's surface underwent deformation when buried far below the surface (Figure 1-14A).

FIGURE 1-14   Deformational structures in rocks. *A*. Folded sedimentary rocks in the Rocky Mountains of Montana; folding occurred when the Rockies were forming. *B*. The Hanning Bay fault scarp, which formed during the Alaskan earthquake of 1964 and along which rocks underwent about 4 meters (~13 feet) of displacement. (*A. M. R. Mudge, U.S. Geological Survey. B. G. Plafker, Science 148:1675–1687.*)

A surface along which rocks of any kind have broken and moved past one another is called a **fault**. Faults have formed during the present century, and some of these have intersected the surface of the earth, where earth movements along them have left visible scars (Figure 1-14*B*). Movements of the earth, which are often associated with faulting, are partially responsible for the origin of mountains, although buildup of volcanic igneous rocks has also contributed to the growth of most mountain chains. Movement along faults is sporadic. Sometimes after years of quiescence, rocks will suddenly move several meters along a fault. Uplift or subsidence (depression) of land, which may or may not be accompanied by faulting, can also be sporadic or gradual. Sudden uplift of local areas and subsidence of others accompanied the 1964 earthquake in Alaska (Figure 1-15). Appendix II of this book reviews the nature of folds, faults, and other structures resulting from the deformation of rocks and illustrates how these and other features of rocks enable us to reconstruct geologic events of the past.

Most of our knowledge about the structure of the earth's interior derives from the study of oscillatory movements called **seismic waves**, which travel through the earth as a consequence of natural or artificial disturbances. An earthquake is an example of a natural seismic disturbance that results from the sudden movement of one portion of the earth against another along a fault. Artificial explosions at the earth's surface also produce seismic waves.

Primary and secondary waves are two types of seismic waves that have provided particularly valuable insight into the nature of the earth's interior. Primary waves, which are also called *P* waves, propagate changes in volume as portions of the earth are alternately compressed and expanded. Secondary waves, or *S* waves, shake the material from side to side perpendicular to their direction of movement (Figure 1-16). Primary waves are so designated because they travel faster than secondary waves.

An earthquake always begins at a **focus**, which is a place within the earth where rocks move against other rocks along a fault, producing both *P* waves and *S* waves. Earthquake foci lie within the earth's mantle and crust, far from the center of the earth—but the waves that foci emit often pass great distances through the earth to emerge at the surface, where they can be detected with machines called **seismographs**. Geophysicists can evaluate the nature of the earth's internal structure by recording at many locations the arrival times of *P* waves and *S* waves of an earthquake.

The study of seismic waves reveals that the materials that form the central part of the earth are much more dense than those near the earth's surface. The density gradient from the surface to the center of the earth is not a gradual one, however; instead, the planet is divided into several discrete concentric layers (Figure 1-17). At the earth's center is the **core**, whose solid, spherical inner portion and liquid outer portion are thought to consist primarily of iron. Forming a thick envelope around the outer core is the **mantle**, a complex body of less dense rocky material. Finally, capping the mantle is the **crust**, which consists of still less dense rocky material. As will be discussed in Chapter 9, the density gradient from the core to the crust developed early in the earth's history, when molten materials of low density came to float on materials of higher density.

There are several ways in which the study of seismic waves has revealed the aspects of the earth's interior just described and others as well. For example, when earth-

FIGURE 1-15   Uplift of the sea floor (white area) during the Alaskan earthquake of 1964. While this area rose above the surface of the sea, other areas sank. (*U.S. Geological Survey.*)

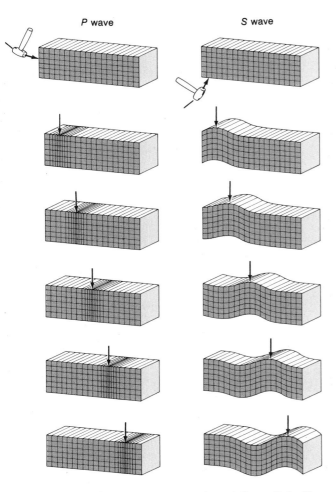

FIGURE 1-16   Seismic waves move through the earth in different ways. A *P* wave moves by alternately compressing and expanding blocks of material. An *S* wave moves by shaking blocks of material up and down and deforming them temporarily. (*After S. Uyeda, The New View of the Earth, W. H. Freeman and Company, New York, 1978.*)

quake waves reach a boundary between two concentric layers within the earth, they are usually both reflected from the boundary and transmitted through it, just as light striking the surface of a body of water is partially reflected and partially transmitted. *S* waves, however, do not penetrate the earth's outer core, and since it is known that liquids cannot transmit *S* waves, this strongly suggests that the outer core is made of liquid. It is also known that both *P* waves and *S* waves travel more rapidly through material of high density than through material of low density. Changes in wave velocity have thus revealed that the earth increases in density with depth and, further, that this increase is not gradual. The passage of seismic waves from the rocks of

the crust to the denser rocks of the mantle, for example, is signaled by an abrupt increase in velocity known as the **Mohorovičić discontinuity**, or **Moho**, for short (Figure 1-18). Because continental crust is much thicker than the crust beneath the oceans, the Moho dips downward beneath the continents.

The rocks that form oceanic crust are the type known as **mafic**—a label whose first three letters indicate that these dark rocks are rich in magnesium (Mg) and iron (Fe). Mafic rocks are much less common in continental crust than are lighter colored, less dense rocks that are described as **felsic**—an adjective derived from the first three letters of feldspar, which is the most common mineral of continental crust. In comparison to mafic rocks, felsic rocks are rich in silicon and aluminum and poor in the heavier elements magnesium and iron. Rocks of the mantle are even richer in magnesium and iron than is the oceanic crust—hence their great density—and they are known as **ultramafic** rocks.

Figure 1-18 illustrates that continental surfaces not only stand above the surface of the oceanic crust but also extend farther down into the mantle than oceanic crust. It also shows that continental crust beneath a mountain range extends even farther down into the mantle than that located elsewhere. **Isostatic adjustment**, or the upward or downward movement that keeps crust in gravitational equilibrium as it floats on the mantle, is responsible for this phenom-

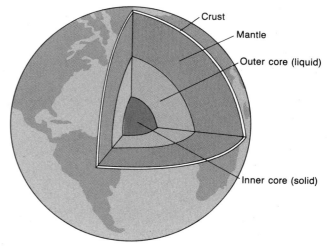

FIGURE 1-17   Zonation of the earth's interior. The crust, which includes continents at the surface of the earth, rests on the mantle. The mantle, in turn, rests on the core. The outer core is liquid, but the inner core is solid. (*After W. J. Kauffman, Planets and Moons, W. H. Freeman and Company, New York, 1979.*)

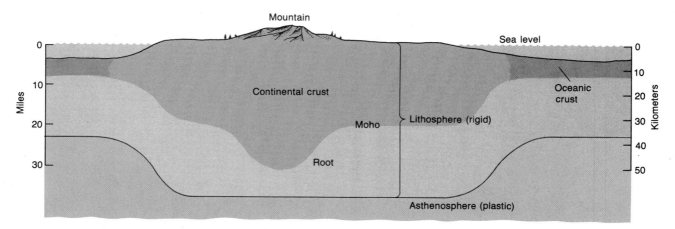

FIGURE 1-18   The structure of the upper part of the earth. Continental crust is much thicker than oceanic crust, and it is especially thick where there are mountains beneath which lie deep roots. The Moho separates the crust from the mantle. The crust and the upper mantle together represent the rigid lithosphere.

enon. In effect, the root beneath a mountain acts to balance the mountain (Figure 1-19).

Although the crust and the upper mantle are separated by a difference in density, they are firmly attached to one another, forming a rigid layer known as the **lithosphere**. Below the lithosphere is the **asthenosphere**, which is also known as the "low-velocity zone" of the mantle because it has been found that seismic waves slow down as they pass through it. This property tells us that the asthenosphere is composed of partially molten rock—slushlike material consisting of solid particles with liquid occupying the spaces in between. Although the asthenosphere represents no more than 6 percent of the thickness of the mantle, the mobility of this layer allows the overlying lithosphere to

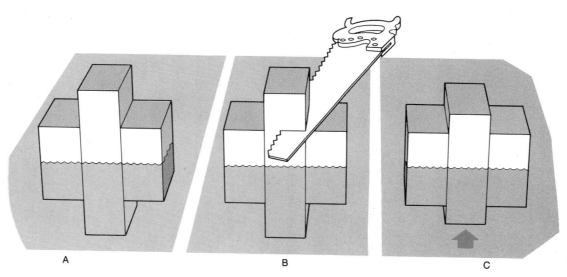

FIGURE 1-19   The principle of isostasy, illustrated by blocks of wood in water. A. Three blocks of wood float adjacent to one another. Because the wood is half as dense as the water, each block lies half above the surface of the water and half below. This means that the weight of a block of wood is equivalent to the weight of the volume of water that it displaces. When the top of the center block is cut off (B), the weight of this block no longer balances the weight of the displaced water. As a result, the block bobs upward until it is balanced once again, lying half above and half below the water (C). The central block is like the part of a continent where a large mountain is present—the low-density elevated crust must be balanced by a root.

move. The lithosphere does not move as a unit, however; instead it is divided into **plates** that move in relation to one another. Some plates carry continents with them as they move, while others carry only oceanic crust.

Some plates, such as the one that includes the continent of Asia, are enormous, while others, such as the one that forms the floor of the Caribbean Sea, constitute only a minute fraction of the skin of the earth. Plates move over the surface of the earth about as rapidly as your fingernails grow. Slow as this rate may seem, the progress of plates over millions of years has been considerable. Many have moved about 500 or 1000 kilometers (~300 to 600 miles) in 10 million years. Since the early 1960s, it has been recognized that many earth movements can be attributed to the motions of plates. Movement of the edge of one plate over the edge of another is in fact a major source of mountain building. We will learn more about plate movements in Chapter 7 and about mountain building in Chapter 8.

FUNDAMENTAL PRINCIPLES OF GEOLOGY

In keeping with the principle of uniformitarianism, processes operating on and within the earth follow physical and chemical laws. Several principles derived from physical laws are fundamental to our method of learning about the earth's history—a method that relies heavily on information gained from studying the relative positions of bodies of rock.

In the seventeenth century, Nicolaus Steno, a Danish physician living in Florence, Italy, formulated three axioms for interpreting stratified rocks. The first is the **principle of superposition**, which states that in an undisturbed sequence of strata, the oldest strata lie at the bottom and successively higher strata are progressively younger. In other words, in an uninterrupted sequence of strata, each bed is younger than the one below and older than the one above. This, of course, is a simple consequence of the law of gravity, as was Steno's second principle, the **principle of original horizontality**, which states that all strata are horizontal when they form. As it turns out, this principle requires some modification. We now recognize that some sediments, such as those of a sand dune (Figure 1-20), accumulate on sloping surfaces, forming strata that lie parallel to the surface of deposition. Sediments seldom accumulate at an angle greater than 45° to the horizontal,

FIGURE 1-20 Sloping beds of a modern dune in southwest Africa. The surface of a dune, visible at the top of the picture, has sloping sides along which sand accumulates. A trench cut across the crest of the dune, shows how the beds slope because they were deposited along the sides of the dune. Note how beds deposited on each side overlap other beds deposited on the other. The scale is in inches. (*E. D. McKee, U.S. Geological Survey.*)

however, so a reasonable restatement of Steno's second principle would be that almost all strata are initially more nearly horizontal than vertical. Steno's third principle is the **principle of original lateral continuity**, which he invoked to explain the occurrence on opposite sides of a valley (or some other intervening feature of the landscape) of similar rocks that seemed previously to have been connected. Steno was, in effect, pointing out that strata originally have continuous tabular shapes, "pinching out" laterally to a thickness of zero or abutting against the walls of the natural basin in which they formed. The original continuity of a stratum can later be broken either by erosion, as in the development of a river valley, or by faulting.

Complex sequences of events can be "read" from geometric relationships more complicated than those dealt with in Steno's principles. **Intrusive relationships**, for example, reveal the relative ages of rocks. When liquid rock penetrates preexisting solid rock, it eats its way into the solid rock or pushes part of the rock aside before cooling to form a body of solid igneous rock. Thus, the invading igneous rock is always younger than the rock that it intrudes (Figure 1-21).

Faults, of course, are also younger than the rocks that they transect, and when the plane of one fault is offset by the plane of another, the fault that is offset is older (Figure 1-22). This is known as the **principle of crosscutting relationships**.

Finally, when fragments of one body of rock are found within a second body of rock, the second is always younger than the first. The second body of rock may be either a sedimentary rock in which some particles have come from another body of rock or an igneous body that incorporated pieces of surrounding rock before cooling (Figure 1-2). This relationship, which is known as the **principle of components**, can be stated more succinctly: A body of rock is younger than another body of rock from which any of its components are derived.

The simple principles outlined above have enabled geologists to establish the relative ages of most bodies of rock that lie adjacent to one another at the earth's surface.

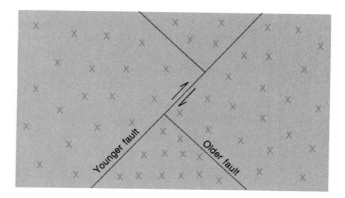

FIGURE 1-22   Two faults of different ages. In accordance with the principle of crosscutting relationships, the fault that is offset by the other is the older of the two.

When the rocks being compared lie at great distances from one another, however, other methods must be employed. Fossils provide one valuable means of establishing the relative ages of rocks that lie far apart. William "Strata" Smith, a British surveyor, noted late in the eighteenth century that fossils are not randomly distributed in rocks. After studying large areas of England and Wales, Smith found that fossils in sedimentary rocks occurred in a particular vertical order ("vertical" in terms of the succession of one layer above another). To the surprise of less experienced observers, Smith could predict the vertical ordering of fossils in areas he had never visited. We now recognize that this ordering, known as **fossil succession**, reflects the sequence of organic evolution—the natural appearance and disappearance of species through time. Although Smith, like nearly all of his contemporaries, did not believe in evolution, he was able to use his knowledge of fossil succession to determine where isolated outcrops of sedimentary rocks fitted into the general sequence of strata in England and Wales.

GEOLOGIC TIME

Although the principles outlined in the previous section allow us to establish the **relative ages** of many bodies of rock, they do not permit us to determine the actual ages of rocks measured in millions of years. As we will see in Chapter 3, some sedimentary beds are produced annually, like the rings in a tree trunk. Unless the latest of a contin-

FIGURE 1-21   Narrow veinlike bodies of igneous rock that pass through lighter-colored rock. In accordance with the principle of intrusive relationships, the veinlike bodies are younger than the rocks they pass through. (*W. R. Hansen, U.S. Geological Survey.*)

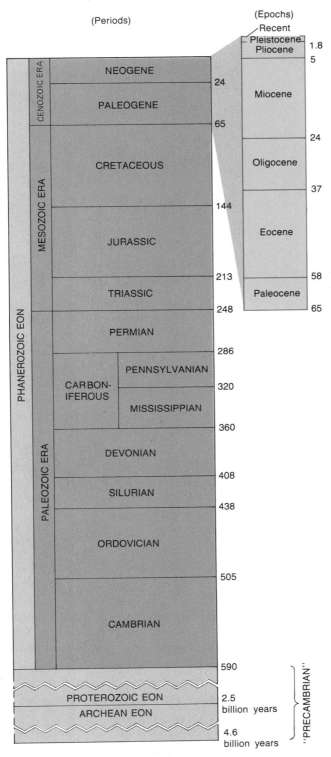

(Periods)   (Epochs)

FIGURE 1-23   The geologic time scale. The numbers on the right give the ages of the boundaries between periods and epochs in millions of years. The Recent Epoch (the past 10,000 years or so) is also known as the Holocene.

uous sequence of annual beds is presently forming, however, it is impossible to count backward in order to determine precisely how many years ago an older bed formed. In other words, if a sequence of this type formed long ago, we cannot tell the actual ages of its beds. Fortunately, there are "geologic clocks" that provide us with a means of approximating the actual ages of ancient rocks. These take the form of naturally occurring **radioactive** materials that decay into other materials at known rates. By measuring the amount of decay in a radioactive material that has been decaying since it became part of a rock, we can estimate the age of the rock. We will learn more about this technique in Chapter 5.

Dating of radioactive materials reveals that some rocks on earth are at least 3.7 or 3.8 billion years old. During the last century, long before the discovery of radioactivity, it became apparent that very old sedimentary rocks contain no identifiable fossils. Beginning with these rocks and examining progressively younger rocks in any region, early geologists discovered that fossils became abundant at a certain level. This level became the boundary at which all of geologic time was divided into two major intervals. The oldest rocks with conspicuous fossils were designated as **Cambrian** in age, and still older rocks became known as **Precambrian** rocks. Today the Precambrian designation is still used informally, but the Precambrian interval is formally divided into the **Archean Eon** and the **Proterozoic Eon**, with the boundary between these two eons placed at 2.5 billion years ago. Subsequent geologic time, from Cambrian on, constitutes the **Phanerozoic Eon**, meaning the "interval of well-displayed life."

Phanerozoic time is divided into three primary intervals, or eras, which the history of life on earth serves to define. The earliest is the **Paleozoic Era**, or the "interval of old life." This is followed by the "interval of middle life," or the **Mesozoic Era**, which is commonly called the Age of Dinosaurs, and by the "interval of modern life," or the **Cenozoic Era**, which is informally designated as the Age of Mammals. Figure 1-23 depicts these eras and the intervals within them known as the geologic **periods**.

Figure 1-23 also illustrates when each period began and ended, as determined by radioactive materials in rocks whose ages approximate period boundaries. Note that the Phanerozoic interval began about 590 million years ago. A human lifetime is so short in comparison to this figure that geologic time seems too vast for us to comprehend; experience does not permit us to extrapolate from the time scale we are familiar with, measured in seconds, minutes,

FIGURE 1-24   Photographs of Bowknot Bend along the Green River in Utah in 1871 (*A*) and almost 100 years later, in 1968 (*B*). Note that during this interval, which is long by human standards, the rocks suffered very little destruction by erosion. (*E. O. Beaman and H. G. Stephens, U.S. Geological Survey.*)

hours, days, and years, to a scale suitable for geologic time. Geologists therefore use a separate scale when they think about geologic time—one in which the units are millions of years. An analogy with a year provides a useful illustration of this geologic time scale. If the Phanerozoic interval of time were compressed into a year, we would find animals with backbones crawling up onto the land for the first time in mid-April, dinosaurs inheriting the earth in early July but then suddenly dying out in late October, and humans appearing on earth within two hours of midnight on New Year's Eve!

Even during the nineteenth century, before there was any means of measuring the actual ages of rocks, geologists were aware that the earth was hundreds of millions of years old. They could see this by observing the slowness of the geologic processes by which rocks are formed and then destroyed (Figure 1-24).

James Hutton, in presenting his view of the earth's history, recognized that one of the greatest differences between uniformitarianism and catastrophism lay in the uniformitarian assumption that the earth was very old—that the sequence of rocks at the earth's surface could not have formed by a few brief upheavals or floods but must instead have formed slowly by processes operating over vast stretches of time. Hutton wrote that, in viewing the earth on these terms, he could envision "no vestige of a beginning, no prospect of an end."

Between 1865 and the beginning of the twentieth century, the physicist Lord Kelvin and his followers issued a challenge to uniformitarianism, arguing that the earth was only about 20 million, or perhaps 30 or 40 million, years old. Lord Kelvin and his followers based their assertion on calculations of the rate at which the earth could be expected to cool from an initially high temperature of formation. Their assumption was that the earth had formed in a molten state and had been cooling rapidly ever since. The fact that the earth's interior was still very hot (temperatures could be seen to rise as a person descended through a mine shaft), seemed to indicate that the earth could not be very old.

It was not until the discovery of naturally occurring radioactive material that these physicists' calculations were disproven. It is now known that radioactive decay releases heat, which has prevented the earth from cooling rapidly since its origin. Thus, the interior of the planet has remained quite hot over more than 4 billion years. The energy of this heat moves the huge plates of rock, some of which carry continents, over the earth's surface.

Dating of rocks by means of radioactive materials has revealed that many major geologic events span millions of years, but in the context of geologic time, these events are of relatively brief duration. We now know, for example, that the Himalayas, the tallest mountains on earth, formed within the past 15 million years or so, but this period of time represents less than one-third of 1 percent of the earth's history. Destructive processes have also yielded enormous changes within a tiny fraction of the earth's lifetime. Mountains that were the precursors of the Rockies in western North America were leveled just a few million years after they formed, and most of the Grand Canyon of Arizona was cut by erosion within just the past 2 or 3 million years. We will examine these events in greater detail in later chapters of this book.

Partly because of periodic earth movements that elevate depositional basins, rocks are not deposited continuously anywhere. As a result, one or more breaks interrupt any local record of deposition. An **unconformity** is a surface between a group of sedimentary strata and the rocks beneath them; it represents an interval of time during which erosion occurred rather than deposition. Because rock deformation is episodic, only rocks of a certain age will have been affected by a particular deformational episode. The surface between the deformed and undeformed rocks represents one kind of unconformity: When a group of rocks has been tilted and eroded and younger rocks have been deposited on top of them, the eroded surface is said to represent an **angular unconformity** (Figures 1-25 and

FIGURE 1-25   Angular unconformity near Socorro, New Mexico. The upper beds were deposited after the lower ones had been tilted and eroded. There is therefore a sizable gap in time between the deposition of the upper and lower sets of beds. (*R. H. Chapman, U.S. Geological Survey.*)

Angular unconformity	Disconformity	Nonconformity

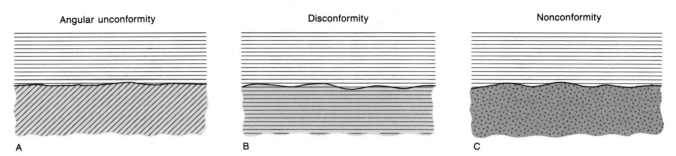

A   B   C

FIGURE 1-26   Diagrams of an angular unconformity, a disconformity, and a nonconformity. An angular unconformity (A) separates tilted beds below from flat-lying beds above. A disconformity (B) separates flat-lying beds below from other flat-

lying beds above, but the upper beds rest upon an erosion surface that developed after the lower beds were deposited. A nonconformity (C) separates flat-lying beds from crystalline (igneous or metamorphic) rocks.

1-26A). Other unconformities are less dramatic. Sometimes the beds below an eroded surface are undisturbed, and only the irregular surface between groups of beds reveals a past episode of erosion. This kind of unconformity is called a **disconformity** (Figure 1-26B). The label **nonconformity** is sometimes employed for an unconformity in which bedded rocks rest on an eroded surface of crystalline rocks (Figure 1-26C).

GROWTH OF THE GEOLOGIC TIME SCALE

In the nineteenth century, when the geologic periods were first distinguished as unique intervals of time, geologists did not know even approximately how long ago each period had begun or ended. Each was simply defined as the undetermined interval of time that a body of rock called a geologic **system** represented. The Cambrian Period, for example, was simply an interval of time that corresponded to those rocks that were designated as the Cambrian System. Geologists at this time did not have the means to study the entire sequence of rocks on earth, from the most ancient to the most modern, in the order in which they occurred in the geologic record. Instead, they were limited to a piecemeal study of whatever promising rock sequences were accessible to them. Thus the Cretaceous System was formally designated in France in 1822, whereas the much older Cambrian and Silurian systems did not gain formal recognition until 1835. System after system was

added up and down the sequence until all the Phanerozoic rocks of Europe were included.

The study of early Paleozoic rocks illustrates how new systems were designated to fill gaps between others that had been recognized earlier. In 1835, Adam Sedgwick and Roderick Murchison presented a joint paper in which they established the Cambrian and Silurian systems, primarily on the basis of geologic studies in Wales. The term *Cambrian* was derived from the Roman name for Wales (Cambria), while *Silurian* was named for the Silures, an ancient tribe that had once inhabited Wales. Murchison defined the Silurian System primarily on the basis of its fossils, and Sedgwick noted that the Cambrian, which is the oldest Phanerozoic system, was less fossiliferous; he recognized, as we do today, that during Cambrian time life on earth was in a primitive stage of development. The very sparseness of fossils in the Cambrian System, however, led Murchison to question if that system was definitely different from the Silurian, and it was not until 1850 that geologists recognized a suite of Cambrian fossils distinct from those of the Silurian. In 1879, Charles Lapworth, a Scottish schoolmaster, showed that in many areas of the world the lower part of the Paleozoic interval, which includes the Cambrian and Silurian systems, actually displayed a succession of three distinctive groups of fossils. He suggested that the label *Cambrian System* be retained for the rocks containing the oldest group of fossils and that the label *Silurian System* be applied to the rocks containing the youngest group. For the intervening rocks, with their own distinctive fossils, Lapworth proposed the name *Ordovician*, for the Ordovices, an ancient Welsh tribe that was the last in Britain to submit to Roman domination.

Lapworth's threefold division of the lower portion of the Paleozoic interval is still accepted, although today we have a more refined view of the formal boundaries between the three systems that he recognized.

As we will learn in greater detail in Chapter 12, fossils contained in Ordovician rocks differ substantially from those of Cambrian rocks because a major extinction of life occurred at the time when the youngest Cambrian rocks were being deposited. Similarly, a major ice age culminated when the youngest Ordovician rocks accumulated, and many groups of Ordovician animals suddenly disappeared. Although not all system boundaries were placed at such meaningful junctures in the geologic record, most geologic systems do contain highly distinctive groups of fossils. Some systems are also characterized by distinctive bodies of rocks produced by forms of life that were uniquely abundant during brief intervals of geologic time. In many parts of the world, for example, the Pennsylvanian System is characterized by numerous coal beds whose origin is traceable to trees of particular types that inhabited vast swampy areas during this period; the logs of these trees were buried in swamps and subsequently turned to coal. Never before or since have swampy forests been so widespread or coal deposits so extensively produced.

The subdivision of rock sequences and geologic time did not cease with the designation of systems and periods. With increased knowledge of the geologic systems and of the periods that these systems represented, nineteenth century geologists subdivided these units into smaller ones, which will be discussed in Chapter 5.

FIELDS OF GEOLOGIC SCIENCE

No geologist can be expert in all of the subjects introduced in this chapter. Like scientists in other fields, each is a specialist in one or a few branches of the larger discipline. The following is a brief outline of the branches of geologic science.

The study of the compositions and origins of rocks is known as **petrology**, from the Greek word *petros*, meaning rock. Within this field are three primary subdivisions: igneous petrology, metamorphic petrology, and sedimentary petrology, or **sedimentology**.

Stratigraphy is the study of the relationships of strata in time and space. Most strata are sedimentary, but some are of volcanic origin. Included in the study of stratigraphy are the use of fossils and radioactive materials to date rocks and the identification of boundaries between geologic systems throughout the world.

Structural geology is the branch of geology concerned with the ways in which rocks bend, break, and flow under stress. **Tectonics** is the subdivision of structural geology concerned with large-scale features such as mountains, and **plate tectonics** is the study of the movements and interactions of lithospheric plates.

Paleontology is the study of ancient life, including its distribution in time and space and its relationship to environments of the past. The primary raw materials of paleontology are fossils. **Vertebrate paleontology, invertebrate paleontology**, and **paleobotany** represent, respectively, the study of ancient animals with backbones, ancient animals without backbones, and ancient plants. Some paleontologists specialize in the study of microscopic fossils of animals or plants, and their field is known as **micropaleontology**. Microfossils are especially important for dating rocks in the search for petroleum because they are so minuscule as to be abundant even in the small chips of rock obtained when geologists drill for oil at great depths within the earth.

During recent decades, emphasis has been placed on the study of processes that have shaped the earth and its life, rather than on the description and classification of rocks and fossils, and this has led to another kind of classification of earth science. The fields that focus on physical, chemical, and biological processes are known, respectively, as **geophysics, geochemistry**, and **geobiology,** or more commonly, **paleobiology**. Geophysicists concern themselves with such things as the movements of plates over the earth's surface, the motions of rock deep within the earth, the behavior of the earth as a giant bar magnet, and the nature of earthquakes; geochemists investigate the ways in which chemical reactions produce rocks and alter them; and paleobiologists study the ways in which animals and plants have evolved, functioned within their environment, and become extinct.

THIS BOOK: AN OVERVIEW

In the second half of this book, we will take a journey through geologic time, beginning in the Precambrian interval and traveling to the present. On this voyage you will come to appreciate that many different kinds of events in nature are interrelated. This history of life, for example, can be understood only in the context of climatic changes and migrations of continents on mobile plates. Mountain building, though a phenomenon consisting of vertical movement, usually results from the horizontal motions and interactions of plates at the earth's surface, and both mountain building and continental migration affect climatic conditions. Interrelationships like these will be recurring themes of this book.

We should not begin our journey through time, however, without first examining the materials that form the earth and the life that inhabits this planet. Thus, the next seven chapters will introduce these subjects and the principles by which their history is reconstructed. The general topics to be covered, in order of appearance, are the relationships of living things to their natural environment; the ways in which sediments accumulate in particular environments; the methods by which ancient rocks are dated; the pattern and process of biological evolution; the nature of plate movements; and the origin of mountains. Mastery of the elements of these topics will provide a firm foundation for understanding the kaleidoscopic changes by which the earth and its inhabitants have attained their present configurations.

CHAPTER SUMMARY

1.   The principle of uniformitarianism, which is fundamental to natural science, asserts that the laws of nature do not vary in the course of time.

2.   Actualism is the uniformitarian procedure whereby events and processes of the geologic past are interpreted in light of events and processes observed in the modern world.

3.   A mineral is an inorganic element or compound that is characterized not only by its chemical composition but also by its internal structure. Rocks are aggregates of mineral grains.

4.   Igneous rocks form by the cooling of liquid rock; sedimentary rocks are layered rocks that accumulate under the influence of gravity; and metamorphic rocks form by the alteration of preexisting rocks at high temperatures and pressures.

5.   Living things are grouped into species, which include individuals that can or do interbreed; species are grouped into categories called higher taxa.

6.   Fossils are remains or tangible signs of ancient life found within rocks.

7.   Rocks break under pressure and move along faults; deep within the earth they also bend and flow.

8.   Earthquakes and artificial explosions create seismic waves that reveal much about the structure of the earth's interior.

9.   The earth is divided into concentric layers. A central core of high density is surrounded by a less dense mantle, which is blanketed by a still less dense crust.

10.   The parts of the earth's crust that form continents are thicker and less dense than the parts that lie beneath the oceans.

11.   The crust and upper mantle constitute the rigid asthenosphere, which is divided into discrete plates that move laterally over a partially molten zone of the mantle.

12.   The relative ages of rocks that come into contact with one another can often be determined by the principles of superposition, original horizontality, lateral continuity, intrusive relationships, crosscutting relationships, and components.

13.   Changes in life on earth during the course of geologic time are reflected in the rock record. The fossil succession observed in the rock record reveals the relative ages of rocks in different regions.

14.   The decay of naturally occurring radioactive materials reveals the actual ages (in years) of some rocks.

15.   The scale that is employed to divide the rock record into units representing discrete intervals of geologic time was developed in Europe during the nineteenth century.

EXERCISES

1.   Give general examples of the use of actualism to interpret ancient rocks.

2.   In which of the three basic kinds of rocks do nearly all fossils occur? Why?

3.   Name and describe as many modes of fossil preservation as you can.

4.   What evidence is there that the outer core of the earth is liquid?

5.   What is isostasy and how does it explain why mountains have roots?

6. What kinds of features distinguish one system of geologic time from another?

7. What are the major fields of geologic science, and which areas of science do they cover?

ADDITIONAL READING

Berry, W. B. N., *Growth of a Prehistoric Time Scale*, W. H. Freeman and Company, New York, 1968.

Clark, S. P., *Structure of the Earth,* Prentice-Hall, Inc., Englewood Cliffs, New Jersey, 1971.

Decker, R., and B. Decker, *Volcanoes*, W. H. Freeman and Company, New York, 1981.

Middlemiss, F. A., *Fossils*, George Allen & Unwin, Ltd., London, 1976.

Press, F., and R. Siever, *Earth,* W. H. Freeman and Company, New York, 1981.

PART ONE

The Environmental Setting

Environments at the earth's surface have changed continuously in the course of geologic time. Sedimentary rocks and the fossils they contain reflect the nature of these environments and thus provide clues that allow geologists to unravel the history of environmental change. These clues make it possible to reconstruct both local structures and broad geographic features of the past.

The most fundamental division of environments on earth—one that is 'reflected' in the distribution of both biological and sedimentary features—is the division between land and sea. Among the nonmarine environments that can be recognized through the study of rocks and fossils are deserts, rivers, lakes, forests, and grasslands; the marine environments include river deltas, lagoons, sandy beaches, reefs built by plants and animals, and deep-sea floors. Within both the terrestrial realm and the marine realm climatic conditions exert strong control over the geographic distribution of plants and animals. Other physical factors, as well as interactions between organisms, determine the distribution of life on a smaller scale.

CHAPTER 2

Environments
and Life

Species are able to live only in environments where they can find food, tolerate physical and chemical conditions, and avoid natural enemies. These were requirements for life of the past just as they are requirements for life of the modern world. Climate is the environmental factor that, on a global scale, exerts the most profound control over the distribution of species, influencing conditions not only on land but also in bodies of water. Temperature and other climatic conditions strongly affect the distribution of plant species, and plant species, in turn, influence the distribution of many animal species. In this chapter, therefore, we will pay close attention to the mechanisms that create the prevailing weather patterns on the earth today. Our major focus, however, will be an overall examination of the relationships between living things and their environments.

If the earth had undergone little change in the course of geologic time, the information contained in this chapter could be applied to ancient fossils and rocks in a simple and direct manner. But the earth has in fact changed dramatically over the course of its history: The earth's materials—including living matter—have changed continually both in composition and in location. Life, of course, has undergone vast evolutionary change, and many conspicuous physical features of the modern world—including the

polar ice caps of Greenland and Antarctica and the ice-cold body of water that now forms the deep ocean—were not present 100 million years ago. Although this situation does not challenge the principle of uniformitarianism (page 1), inasmuch as natural laws are not violated, it does mean that one must take changing environmental conditions into account in interpreting the rock record. What we learn in this chapter will serve as a starting point for exploring environments and life of ancient times later on in this book, starting with the planet's origins and moving forward until we reach the present. In other words, we will ultimately explain the historical development of many of the patterns that will be discussed in this chapter.

LAND AND SEA

One means of gaining an understanding of environments on earth is to examine the configuration of the planet's surface. You will recall that the earth's crust is divided into the thin, dense oceanic crust and the thicker, less dense continental crust—a distinction that accounts for the earth's external shape, in which continental surfaces stand above the sea floor. A **hypsometric curve** illustrates what proportions of the earth's surface lie at various altitudes above sea level and what proportions lie at various depths below sea level (Figure 2-1). This curve shows that more than 70 percent of the earth's surface is covered by oceans and that most of the land above sea level is less than 1 kilometer (~0.62 miles) above sea level. Mountainous terrane, in other words, is relatively uncommon. **Continental shelves**, which are the flooded margins of continents, form only a small proportion of the land beneath the sea; most of the

View of a savannah in Kenya. A savannah is a grassy plain that supports few trees and is generally populated by large animals. The animals in the foreground of this scene are topis, members of the antelope family. (© *Carol First.*)

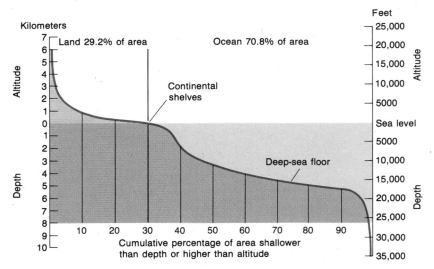

FIGURE 2-1   The hypsometric curve, which represents the surface of the earth. The curve shows the relative amounts of land and sea floor that lie at various elevations with respect to sea level. The plot is cumulative, depicting the total percentage of land that lies above each depth or altitude. About 70 percent of the earth's surface lies below sea level, and most of this forms the deep-sea floor.

ocean bottom stands between 3 and 6 kilometers (~2 to 4 miles) below sea level, forming the deep sea floor.

Those environments on or close to the earth's surface that are inhabited by life are called **habitats**. Nearly all habitats can be classified as terrestrial or aquatic. Aquatic habitats are further divided into freshwater habitats (e.g., lakes, rivers, and streams) and marine habitats (e.g., those within oceans and seas). Birds, bats, and insects use the atmosphere above the ground as a part-time habitat, but virtually all of these creatures also conduct some activities such as feeding, sleeping, and reproduction, at the surface of the earth. Later in this chapter, we will examine the nature of the various habitats on earth together with the forms of life that occupy them. First, however, we will examine some of the principles that govern the occurrence of species within habitats in general.

PRINCIPLES OF ECOLOGY

Ecology is the study of the factors that govern the distribution and abundance of organisms in natural environments. Some of these factors are conditions of the physical environment, while others are modes of interaction between species.

A species' position in its environment

The way in which a species relates to its environment defines the **ecologic niche** of that species. The niche requirements of a species include particular nutrients or food resources and particular physical and chemical conditions. Some species have much broader niches than others. Before human interference, for example, the species that includes grizzlies and brown bears ranged over most of Europe, Asia, and western North America, eating everything from deer and rodents to fishes, insects, and berries. The sloth bear, in contrast, has a narrow niche. It is restricted to Southeast Asia, feeding mainly on insects, for which its peglike teeth are specialized, and on fruits. Similar contrasts in niche breadth characterize many other closely related species.

We speak of the way in which a species lives within its niche as a **life habit** or **mode of life**. Included in a species' mode of life are the method by which it obtains nutrients or food, the manner in which it reproduces, and the way in which it stations itself within the environment or moves about.

Every species is restricted in its natural occurrence by environmental conditions known as **limiting factors**. Among the most important of these factors are physical and chemical conditions. On land, for example, most ferns live only under moist conditions, while some other plants, such as cactuses, require dry habitats. In the ocean, the salt content of the water is an important limiting factor. As an example, few species of the marine phylum Echinodermata, which includes starfishes and sea urchins, can live in lagoons or bays where normal ocean water is diluted by freshwater from rivers.

Almost every species shares part of its environment with other species. Thus, for many species, **competition** with other species—or the process in which two or more species vie for an environmental resource that is in limited supply—constitutes a limiting factor as well. Among the resources for which species commonly compete are food and living space. Often, two species that live in similar ways cannot coexist in an environment because one species competes more effectively, thereby excluding the other species. Plants that grow in soil, for example, often compete for water and nutrients in that soil; as a result, plant species living close together will often have roots that penetrate the soil to different depths (Figure 2-2).

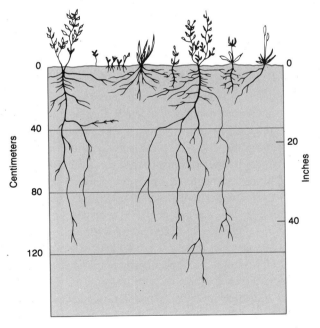

FIGURE 2-2   Differences in the niches of coexisting species of plants. The roots of different species occupy different depth zones of the soil and thus avoid competing for water and nutrients. (*After H. Walter, Vegetation of the Earth in Relation to Climate and Eco-Physiological Conditions, Springer-Verlag, Stuttgart, 1973.*)

Predation, or the eating of one species by another, is also a limiting factor, and one that often reduces the abundance of the victim species. An especially effective predator can prevent another species from occupying a habitat altogether.

Communities of organisms

Populations of several species living together in a habitat form an **ecologic community**. In most ecologic communities, some species feed on others. The foundation of such systems are organisms called **producers**, which are plants or plantlike organisms that manufacture their own food from raw materials in the environment. In contrast, animals and animal-like organisms, or **consumers**, feed on other organisms. Consumers that feed on producers are known as **herbivores**, and consumers that feed on other consumers are known as **carnivores**. Terrestrial herbivores include such diverse groups as rabbits, cows, pigeons, garden slugs, and leaf-chewing insects. Terrestrial carnivores include weasels, foxes, lions, and ladybugs.

The organisms of a community and the physical environment that they occupy constitute an **ecosystem**. Ecosystems come in all sizes, and some include many communities. The earth and all the forms of life that inhabit it represent an ecosystem, but so does a tiny droplet of water that is inhabited by only a few microscopic organisms. Obviously, then, large ecosystems can be divided into many smaller ecosystems, and the size of the ecosystem that is treated in a particular ecologic study will depend on the type of research that is being conducted. The animals of an ecosystem are collectively referred to as a **fauna** and the plants as a **flora**. A flora and a fauna living together constitute a **biota**.

One of the most important attributes of an ecosystem is the flow of energy and materials through it. When herbivores eat plants, they incorporate into their own tissue part of the food that these plants have synthesized. Carnivores assimilate the tissue of herbivores in much the same way. In most ecosystems, carnivores that eat herbivores are eaten in turn by other carnivores; in fact, several levels of carnivores are often present in an ecosystem. An entire sequence of this kind, from producer to top carnivore, constitutes a **food chain**. Because most carnivores feed on animals smaller than themselves, the body size of carnivores often increases toward the top of a food chain (Figure 2-3). Exceptions to this rule include weasels, which often kill rabbits that are larger than themselves.

FIGURE 2-3   A sequence of species forming a food chain within a stream. Single-celled algae are fed upon by nymphs (i.e., the juvenile stage) of mayfly insects. These, in turn, are eaten by sunfishes, which are preyed upon by large-mouthed bass. Otters eat the bass. (Organisms are not drawn to scale.)

Simple food chains—sequences in which a single species occupies each level—are uncommon in nature. Most ecosystems are characterized by **food webs** in which several species occupy each level. Most species below the top carnivore level serve as food for more than one consumer species. Similarly, most consumer species feed on more than one kind of prey (Figure 2-4).

Parasites and scavengers add further complexity to ecosystems. **Parasites** are organisms that derive nutrition from others without killing their victims, and **scavengers** feed on organisms that are already dead. A flea that feeds on the blood of a dog, for example, is a parasite, as is a tapeworm that lives within a human. In contrast, a vulture is a scavenger, as is a maggot, which feeds on dead flesh.

Although material flows from one level of a food web to the next, it does not stop at the highest level. In fact, materials are cycled through the ecosystem continuously with single-celled bacteria completing the cycle (Figure 2-5). Some of these bacteria decompose dead animals and plants of all types into simple chemical compounds, while others transform decomposed material, liberating nutrients that are reused by plants.

The term **diversity** is used to describe the number of species that live together within a community. Diversity can be measured in several ways, some of which take into account the relative abundance of species, but the simplest measure of diversity is nothing more than a count of the number of species present. Sometimes researchers measure the diversity of an entire community, while at other times they may measure only one group of species, such as green plants or herbivores.

Most habitats that present physical difficulties for life support communities of low total diversity. Because plants require water to make food, for example, deserts contain fewer species of plants than do moist tropical forests. Only a few types of plants, such as cactuses, can sustain themselves in desert environments.

Predation is another factor that influences community

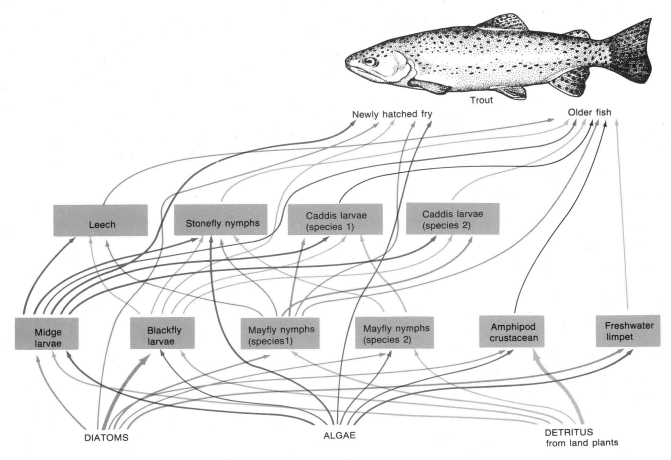

FIGURE 2-4   The relatively complex food web of a trout stream. Arrows show the flow of food, with the thickness of the arrows indicating the rate of flow from one kind of organism to another. There are few if any plankton in the moving water; instead, algae on the stream bottom and plant detritus from the land form the base of the food web. (*After W. D. Russell-Hunter, Aquatic Productivity: An Introduction to Some Basic Aspects of Biological Oceanography and Limnology, The Macmillan Company, New York, 1970.*)

diversity. One or more highly effective predators can, in fact, cause the diversity of a community to remain at an exceedingly low level by preventing all but one or a few species from surviving. Goats, which eat nearly all kinds of vegetation and are therefore very effective predators on plants, provide an excellent example of this phenomenon. Let us imagine that goats within an enclosed area prefer ground vegetation to the more abundant leafy vegetation that grows on trees. The leafy vegetation can sustain several goats even after all of the ground vegetation has been eaten. Thus, the goats will first eat up all of the ground vegetation, and they will keep removing it if seeds are blown in and germinate to produce new seedlings—but they will not starve because of the abundant supply of leafy vegetation. Therefore, these goats will create an environment in which ground vegetation is absent.

On the other hand, certain types of predators actually work to enhance the diversity of plant species within a

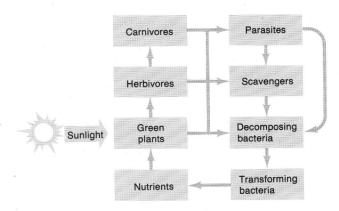

FIGURE 2-5   The cycle of materials through an ecosystem. Herbivores consume plants and are consumed, in turn, by carnivores. Parasites and scavengers derive nutrition from all of these groups—parasites by feeding on them without killing them and scavengers by feeding on them after they are dead. Some bacteria decompose animals and plants after they die, while others transform the decayed material to nutrients, which are then used by plants to make food; thus, the cycle is completed.

community. If the goats in the aforementioned enclosure were replaced by cattle, for example, the cattle, grazing less effectively, might actually cause the diversity of ground-plant species to increase. How could they do this? To understand the mechanism, we must imagine what would happen in the absence of predation. With no grazing pressure, one or a small number of plant species might take over the land area by competing so effectively for water and nutrients that they would exclude weaker species. Now imagine what would happen if we were to add to such an ecosystem animals that remove plants less efficiently than do goats but efficiently enough to continually produce small bare areas of ground. Species of plants that were weak competitors could then invade the bare areas if their seeds happened to arrive from outside the enclosure. Even if the superior competitors soon crowded out these new colonizers, the latter would constantly invade new bare areas— and the result would be a community in which many plant species coexisted. In a similar manner, carnivorous animals can allow many species of herbivorous animals to coexist by preventing the species of superior competitive ability from becoming abundant enough to exclude other species.

Certain types of physical disturbances can also prevent strongly competitive species from excluding less competitive species. Storm waves, for example, may tear animals and plants from rocky shores, leaving bare surfaces for the invasion of species that are weak competitors. Ecologists apply the label **opportunistic** to species that specialize in invading newly vacated habitats, other examples of which are land cleared by fire or new shore areas formed along rivers that change course at flood stage. Populations of opportunistic species seldom survive for long periods of time in the face of better competitors, but because opportunistic species tend to be good invaders, while some of their populations are disappearing from one area, others simultaneously become established elsewhere. The plants that we call weeds are excellent examples of opportunistic species. In gardens and lawns, many types of weeds come and go in the course of a few seasons.

Biogeography

The distribution and abundance of organisms on a broad geographic scale is studied within the biological field known as **biogeography**. The limiting factor that plays the largest role here is temperature, which tends to decrease from the equator toward the poles. Many species are restricted to polar regions, others to tropical regions near the equator,

and still others to latitudes in between, where temperatures are intermediate.

Most large, widespread taxonomic groups of animals and plants include more tropical species than species adapted to cold climates. In addition, tropical communities are, on the average, more diverse than communities of high latitudes; in general, species increase in number toward the equator. This would appear to relate to the fact that animals and plants cannot easily survive in the cold habitats that characterize high latitudes.

Temperature, however, is not the only control of biogeographic patterns of occurrence, as evidenced by the fact that most species will not be found in all the habitats that meet their particular ecologic requirements. The dispersal of most species is also restricted by barriers, the most obvious of which are land barriers for aquatic forms of life and water barriers for terrestrial forms of life. Of course, these barriers change with time, shifting the geographic distribution of species accordingly. *Mammuthus*, the genus that includes the extinct members of the elephant family known as mammoths, evolved in Africa about 5 million years ago during the early part of the Pliocene Epoch. Blocked by northern oceans, mammoths were unable to migrate to North America until the Pleistocene Epoch. During the Pleistocene Epoch, however, large volumes of water were locked up on the land as glaciers (or ice sheets) causing sea level to fall throughout the world. Consequently, a land bridge was formed between Siberia and Alaska, allowing mammoths to invade the Americas, where they survived until several thousand years ago (Figure 2-6).

The survival of mammoths in the Americas represents an interesting biogeographic phenomenon, the development of a **relict distribution**, or the localized occurrence of a taxonomic group after it has died out throughout most of the geographic area that it has occupied. By late in the Pleistocene Epoch, mammoths had disappeared from most of the area that they had previously occupied and remained in only a relatively small area of North America.

THE ATMOSPHERE

The **atmosphere**, or the envelope of gases that surrounds the earth, serves life primarily as a reservoir of chemical compounds that are used within living systems. The atmosphere has no outer boundaries; instead, it grades into interplanetary space. The dense part of the atmosphere,

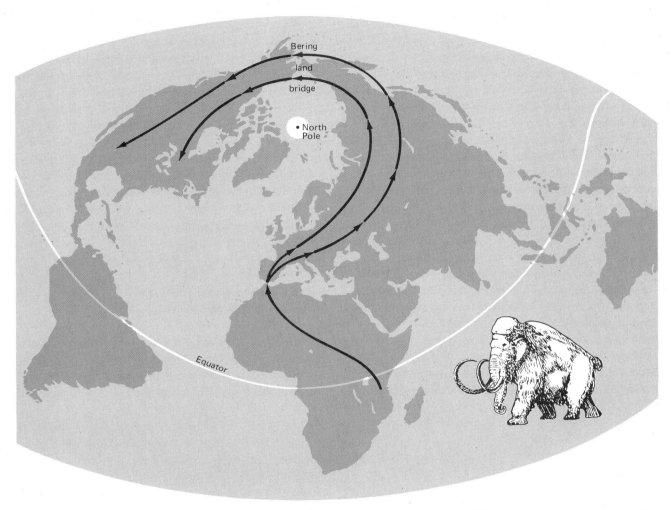

FIGURE 2-6   Changes in the geographic distributions of mammoths, extinct members of the elephant family. Mammoths evolved in Africa during the Pliocene Epoch and spread to Eurasia. Later, during the Pleistocene interval, when sea level was lowered, the Bering land bridge formed between Eurasia and North America. (*After V. J. Maglio, Amer. Philos. Soc. Trans. 63:1–149, 1973.*)

however, forms a thin envelope around the earth, so that more than 97 percent of the mass of the atmosphere lies within 30 kilometers (~19 miles) of the earth's surface. (To place this figure in perspective, note that it resembles the average thickness of the earth's continental crust.)

It is interesting to consider why few forms of life colonize the earth's atmosphere. One problem for life here is the weakness of physical support. Cells and tissue cannot float indefinitely in air, which as a gas has a relatively low density. Another problem is that water is necessary for life, and water is present in the air only as water vapor or as droplets of liquid. Because large, persistent bodies of water such as rivers, lakes, and oceans cannot form in the atmosphere, airborne organisms cannot remain immersed in water or even rely on its presence in large quantities.

Chemical composition of the atmosphere

Nitrogen is the most abundant chemical component of the atmosphere, making up about 78 percent of the total volume of atmospheric gas. It is an important component of proteins, which stimulate chemical reactions and serve as important building blocks in all living things. Second in abundance is oxygen, which forms about 21 percent of the volume of the atmosphere. Both nitrogen and oxygen exist in the atmosphere as molecules that consist of two atoms (N_2 and O_2); this is why we speak of them as compounds rather than as elements.

Atmospheric nitrogen and oxygen are maintained at consistent levels by being cycled through living organisms continuously and returned to the air. Figure 2-7 illustrates the **global oxygen cycle**. Most oxygen enters the atmosphere from plants, which produce it through **photosynthesis**—a process in which water and carbon dioxide are converted to sugar and oxygen. The green compound chlorophyll acts as a catalyst in this process, while the energy that is necessary for the conversion is derived from the sun. A smaller amount of oxygen comes from the upper atmosphere, where sunlight breaks down water vapor (H_2O) into oxygen and hydrogen.

Carbon dioxide (CO_2), from which plants produce oxygen, is contributed to the atmosphere by animal respiration and by the burning of fossil fuels. Carbon dioxide forms only about three hundredths of one percent of the atmospheric volume, but in recent history its abundance has been much more variable than that of nitrogen and oxygen. Since 1900, the burning of coal, oil, and wood has increased the volume of carbon dioxide in the atmosphere by about 10 percent. If this buildup continues, it may eventually cause the earth's atmosphere to warm up through the so-called greenhouse effect. This term refers to the manner in which solar radiation of short wavelength passes through the glass of a greenhouse and then accumulates in the form of heat, which does not easily escape back through the glass—the same mechanism by which the interior of an automobile heats up on a sunny day. In the atmosphere, carbon dioxide acts in the same manner as the glass of a greenhouse, allowing solar radiation to pass to the earth's surface and then preventing much of the resulting heat from escaping from the lower atmosphere. If carbon dioxide continues to accumulate in the atmosphere as a result of the burning of fossil fuels, global temperatures may eventually increase, and local climates may then change in ways that could greatly affect the patterns of life on earth.

Temperatures and circulation in the atmosphere

Both the atmosphere and the ocean consist of liquid or gas that is in constant motion in relation to the earth beneath them, and most of the energy that produces this motion comes from the sun. Thus, solar radiation is responsible

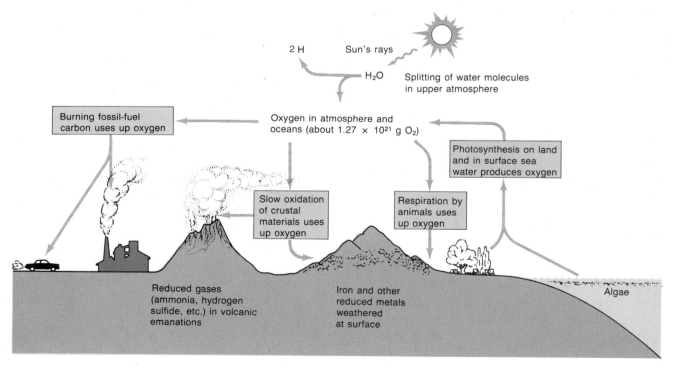

FIGURE 2-7   The cycle of oxygen through the atmosphere. The most rapid input of oxygen is from the photosynthesis of producers on the land and in the sea. Sunlight decomposes water in the upper atmosphere to produce oxygen at a lower rate. (*After F. Press and R. Siever, Earth, W. H. Freeman and Company, New York, 1982.*)

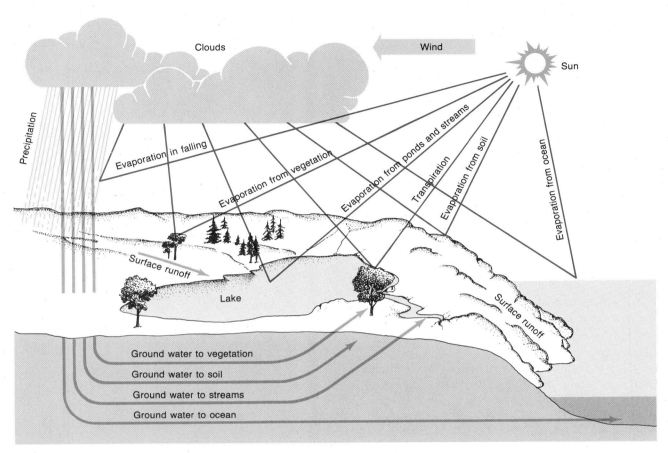

FIGURE 2-8   The global hydrological cycle. The movement of water from plants into the atmosphere is called transpiration.

for much of what takes place in the atmosphere and in the ocean.

When solar energy reaches the earth or the ocean, a good deal of it is absorbed and turned into heat energy. The amount of solar radiation that is absorbed at the earth's surface varies from place to place according to the percentage of solar radiation reflected from the earth's surface, a factor known as the **albedo**. Less heat is generated by sunlight where there is ice than where there is water, soil, or vegetation because ice reflects more radiation. The albedo ranges from 6 to 10 percent for the ocean; from 5 to 30 percent for forests, grassy surfaces, or bare soil; and from 45 to 95 percent for ice or snow.

Heat from the sun causes water at the earth's surface to evaporate and to enter the atmosphere as water vapor (Figure 2-8). Under the cool conditions that characterize high altitudes, water vapor condenses to form clouds. Precipitation from clouds carries water back to earth, where it

evaporates immediately or after moving through the ground or flowing through streams and rivers. Some water reaches the atmosphere again by **transpiration**, or emission from plants that have collected the water with their roots. Evaporation from these sites of accumulation completes the **global hydrological cycle**, or the cycle of water movement.

The movement of water and water vapor through the hydrological cycle involves only a few of the many types of fluid movements driven by solar energy near the earth's surface. Oceans and lakes are warmed by the absorption of solar radiation, and the atmosphere is warmed primarily by heat that rises from the land. This warming produces fluid movement. Much of the transfer of heat from place to place in the ocean or atmosphere occurs by **convection**, which results from the expansion of a liquid or gas that is warmed. Expansion reduces the density of a fluid, causing it to rise through cooler and therefore denser fluid. Thus, if a kettle of water is heated from below, the warmed water

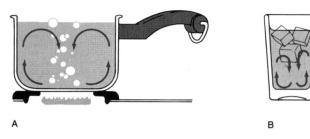

FIGURE 2-9   Convection in liquid. *A.* Water in a pan heated from below turns over as the warm water rises from the bottom and the cooler water at the top descends. *B.* Cooling of water by ice at the top of a glass results in a similar convective motion.

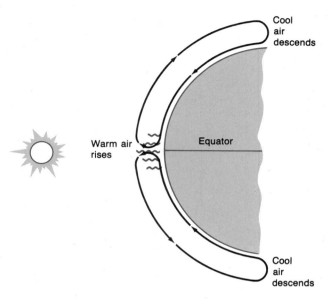

FIGURE 2-10   Diagram of atmospheric circulation for an imaginary earth that receives sunlight equally on all sides from a source rotating in the plane of the equator. One convection cell occupies the Northern Hemisphere and another the Southern Hemisphere.

near the bottom rises through the cooler water above, in other words, the water in the kettle "turns over," continuing to do so as cooler water sinks and is warmed. Similarly, water in a glass with floating ice cools at the top and undergoes convective turnover (Figure 2-9).

Convection operating in conjunction with other forces produces major movements within the earth's atmosphere. Because sunlight strikes the earth's surface at a low angle near the poles, the poles are much cooler than the equatorial region, where the sun's rays impinge on the surface more directly. The atmosphere also becomes cooler with altitude above the earth's surface, from which its heat is derived.

To understand the actual pattern of atmospheric motion, let us consider what might happen if the earth did not rotate but were heated evenly on all sides by a sun that rotated in the plane of the earth's equator. This imaginary system would produce the kind of atmospheric convection shown in Figure 2-10, in which warm air rises near the equator while cool air sinks toward the earth near the poles. The real system, in which the earth rotates, produces a much more complicated pattern of atmospheric circulation (Figure 2-11). To understand this pattern, it is necessary to take into account the **Coriolis effect**.

The Coriolis effect is most easily envisioned for an air current moving from a pole of the earth toward the equator (Figure 2-12). While such an air current is flowing toward the equator, the earth is rotating eastward. Thus, instead of moving directly along a line of longitude toward the equator, an air current will bend to the west with respect to landmarks on the earth's surface and will continue to do so because the surface speed of the earth increases toward the equator. More generally, the Coriolis effect causes an

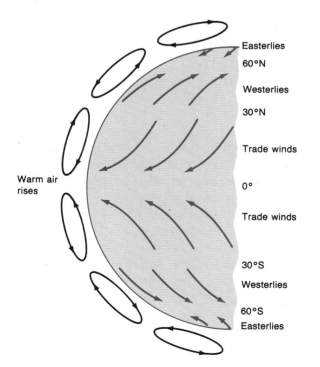

FIGURE 2-11   The major gyres of the earth's atmosphere. The lower segments of these gyres represent the major wind systems, labeled on the right.

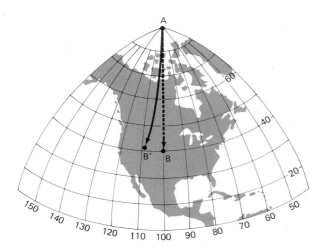

FIGURE 2-12   Illustration of the Coriolis effect for an air current that flows from the North Pole of the earth toward the equator. While the current moves, the earth rotates from west to east, and the current bends to the right with respect to the earth's surface. All currents of air and water in the Northern Hemisphere tend to bend toward the right.

air current flowing in any direction in the Northern Hemisphere to bend toward the right. Similarly, it bends all air currents in the Southern Hemisphere toward the left.

The presence of the Coriolis effect prevents the earth's atmosphere from circulating in the simple pattern shown in Figure 2-10. To understand why, consider what happens near the equator. Although air in this region rises from the warm surface of the earth and spreads in a general way to the north and south, the air that spreads poleward from the equator at high elevations does not move *directly* to the north and south because the Coriolis effect deflects it toward the west. As a result of this deflection, the air piles up north and south of the equator more rapidly than it can escape toward the poles, producing belts of high atmospheric pressure between about 20 and 30° north and south of the equator. This high pressure bears down on the earth's surface, pushing winds toward the north and south. These winds are also deflected by the Coriolis effect. Thus, the winds that flow toward the equator, which are known as **trade winds** (Figure 2-11), move diagonally westward. Note that the trade winds replace the warm air that rises at the equator, completing a cycle, or gyre, of airflow on either side of the equator.

Another gyre is positioned poleward of the trade winds. Here, **westerlies** flow toward the northeast. Like the trades, these winds originate from the high-pressure system where air piles up between 20 and 30° from the equator. Still

farther from the equator in each hemisphere is yet another gyre, which originates where cool air descends near the pole.

THE TERRESTRIAL REALM

At the present time, the continents of the world stand at relatively high elevations above sea level. Thus, broader expanses of land exist today than were present during most of Phanerozoic time. Another unusual feature of the modern world is that temperature gradients between the poles and the equator are unusually steep. While tropical conditions prevail near the equator, where temperatures are warm, ice sheets cover large areas of land near the north and south poles, where the average summer temperature is well below freezing. This is significant in that climatic conditions, as previously noted, have a profound effect on the distribution of organisms on the land. Climates will therefore be our first consideration in this discussion of the distribution of life on the broad continental surfaces of the modern world.

Climates and vegetation

It is remarkable how closely the distribution of terrestrial vegetation corresponds to the geographic pattern of climates. This correspondence, coupled with the fact that plants are the dominant producers of the food web and thus strongly affect the distribution and abundance of animals, makes climate an especially significant factor in terrestrial ecology. Plants, in fact, serve not only as sources of food but also as habitats for many animals; numerous insects, for example, spend their entire lives on certain types of trees, and insects include at least half of all species of organisms living in the world today.

Near the equator, terrestrial climates are not only very warm but often very moist as well. This is attributable to the fact that air in this region rises after being heated at the earth's surface, cooling as it does so (Figure 2-11). Because cool air cannot hold as much water vapor as warm air, moisture is lost in the form of rain. In South America and Africa, the only large continents with equatorial regions (Figure 2-13), the warm, moist conditions that characterize **tropical rain forests** allow many different kinds of plants to thrive (Figure 2-14), forming what are informally called jungles. These plants provide food and shelter for a variety of animals.

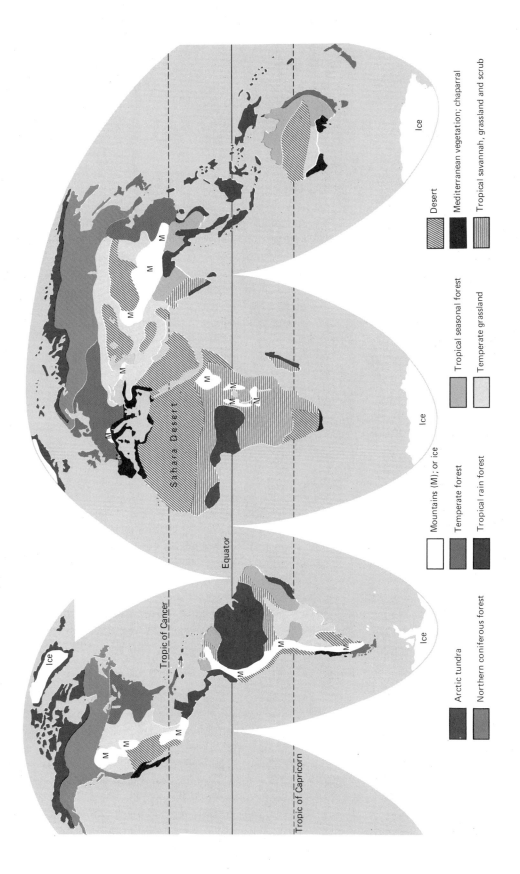

FIGURE 2-13   The major terrestrial communities on earth. Each kind of community is characterized by a particular association of plants adapted to particular climatic conditions. (After C. B. Cox and P. D. Moore, Biogeography: An Ecological and Evolutionary Approach, John Wiley & Sons, Inc., New York, 1980.)

Tropical climates are defined as those in which the average air temperature ranges from 18 to 20°C (~64 to 68°F) or higher. These climates are usually found at latitudes within 30° of the equator. You will recall that between 20 and 30° north and south of the equator, the air that rises from the equator piles up and, after cooling, descends to form trade winds (Figure 2-11). The air of these winds is dry, having dropped much of its moisture at high altitudes as rain. As a result, the trade winds drop little rain; instead, they pick up moisture from the surface of the earth, leaving deserts on broad continental areas that lie about 30° from the equator. The Sahara is the largest of these deserts, but broad deserts also occupy southwestern North America, southern Africa, and central Australia (Figure 2-13).

Most **deserts** receive fewer than 25 centimeters (~10 inches) of rain per year. Because only a few types of plants can live under such conditions, the desert environment is

FIGURE 2-15   Desert along the southern edge of the Colorado Plateau, near the Grand Canyon. The spiny plant in the center is known as Spanish bayonet. (*Courtesy of C. B. Hunt.*)

FIGURE 2-14   The Amazon rainforest near Iquitos, Peru. Many species of plants form the understory beneath a canopy of tall trees. (*Courtesy of Parks Canada. Photograph by M. J. Balick.*)

characterized by sand and bare rock rather than by dense vegetation (Figure 2-15). Some desert plants, such as cactuses, have the capacity to store water, and nearly all have small leaves to conserve against loss of water by evaporation. Few species of animals can survive in the desert environment, and a large percentage of those that do are nocturnal; many are small rodents that find refuge from the hot sun by remaining in burrows during the day.

Deserts also form at latitudes where moisture is more plentiful, but only where there is little moist air. The Sierra Nevada Mountains near the west coast of the United States, for example, leave what is known as a **rain shadow** to their east. Winds rising from the west over the Sierra Nevada cool and lose their moisture rapidly (Figure 2-16), so even the highest elevations of the mountain range receive less rain than the lower western slopes, and very little rain is shed on the eastern side. There, in the rain shadow, lies the Great Basin, a desert that includes most of the state of Nevada. Deserts also exist at high latitudes in central Asia, but only in regions so far from oceans that they receive little moisture (Figure 2-13).

Savannahs and **grasslands** form in areas where rainfall is sufficient for grasses to thrive, but not sufficient to allow the growth of forests. Many savannahs and grasslands are, in fact, positioned between dry deserts and wet woodlands. The Great Plains of the United States are an example, as are the savannahs of Africa, which are noted

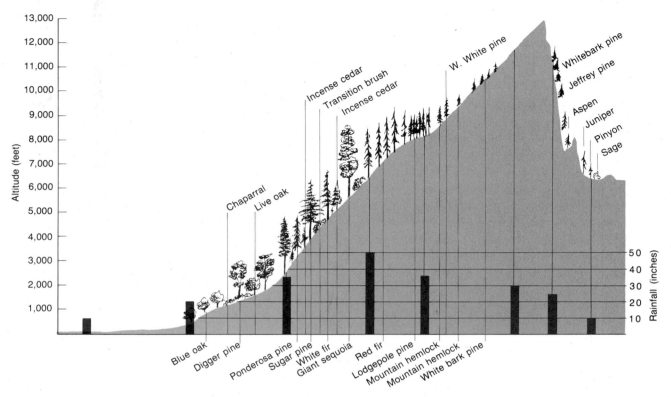

FIGURE 2-16 Changes in the flora of the Sierra Nevada Mountains with altitude. Moist air rises as it approaches the mountains from the west and then cools, dropping its moisture in the form of rain. Rainfall decreases toward the top of the mountain range, and little moisture is left by the time the air moves to the east of the range. Here, in the rain shadow of the mountains, an arid climate prevails. (*U.S. Department of Agriculture.*)

for their large animals (page 28). Most of the herbivores found in savannahs and grasslands—including bisons, antelopes, zebras, and wildebeests—are relatively large animals that graze on grasses and have considerable stamina with which to flee from carnivores such as jackals, lions, and cheetahs. The majority of carnivores in these regions are also large animals that have the ability to capture the large grazers. Many savannahs and grasslands also support scattered trees, and these habitats intergrade with open woodlands, where more trees are present.

In sharp contrast to the rich biotas of tropical rain forests, warm savannahs, and woodlands are the meager biotas located near the north and south poles. Today, large ice caps cover Greenland and Antarctica, the two large continents of polar regions. Such ice caps have been absent from the earth in past times, when no large continents have occupied polar regions or when the polar regions have been warmer. The ice caps of Greenland and Antarctica are now so heavy that they actually depress the continental

crust (Figure 2-17). These ice caps represent continental glaciers.

Glaciers are among the most impressive physical structures on earth. They are not simply masses of ice, they are masses of ice in motion (Figure 2-18). Glaciers form from snow that accumulates until it is so thick that the pressure of its weight recrystallizes the individual flakes into a solid mass. Glaciers slide downhill, but they also spread over horizontal surfaces by means of internal deformation. In so doing, their behavior resembles the flowing of any solid material near its melting point. Glaciers form not only at high latitudes but also at high elevations near the equator, where the atmosphere and earth are cool; they occupy mountain valleys, through which they flow downhill and often melt at the warmer temperatures closer to sea level. Near the poles, glaciers may reach the sea before they melt. Large chunks of ice then break from their terminal portions and float off in the form of icebergs.

Nearly all of Antarctica is covered by ice, but no other

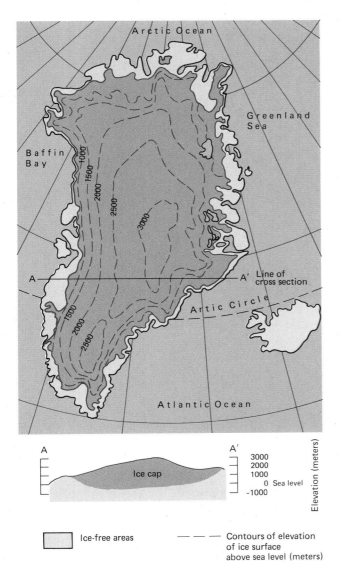

A' Line of cross section

A   A'
                                          3000
                                          2000
            Ice cap   1000
                                             0 Sea level
                                          -1000

Ice-free areas   ——— Contours of elevation
                              of ice surface
                              above sea level (meters)

FIGURE 2-17   The thick glacial ice cap of Greenland, which depresses the continental crust. The broken lines on the map indicate ice thickness. (*After F. Press and R. Siever, Earth, W. H. Freeman and Company, New York, 1982.*)

tall trees, but rather plants that need little moisture, such as mosses, sedges, lichens (associations of algae and fungi), and low-growing trees and shrubs. A broad belt of tundra stretches across the northern margins of North America and Eurasia, supporting a low diversity of animal life (Figure 2-19). Rodents and snowshoe hares are present in tundras today, and the dominant herbivores of larger size are the caribou, reindeer, and musk ox. Foxes and wolves are the primary hunters.

South of the tundra in the Northern Hemisphere, in areas where there is sufficient moisture, forests rather than deserts or grasslands are found. The cold regions adjacent to tundra are cloaked in **evergreen coniferous forests**, where trees such as spruce, pine, and fir dominate. In areas with short summers, these trees are successful because of their ability to conduct photosynthesis (i.e., to make food) year-round. The diversity of animals in cold evergreen forests is not so great. To the south, in slightly warmer climates with longer summers, **temperate forests** replace evergreen forests. Some evergreen trees can be found in such forests, but deciduous trees such as maples, oaks, and beeches are usually present in greater abundance. Ground animals are more diverse in temperate forests than in cold evergreen forests, and birds are especially well represented.

Mediterranean climates, which are characterized by dry summers and wet winters, often prevail along coasts

large continent in the Southern Hemisphere includes large areas that experience very cold conditions year-round. In the Northern Hemisphere and in many tropical areas above the tree line, however, there is a type of subarctic community known as **tundra**. Tundra exists in areas where a layer of soil beneath the surface remains frozen even though air temperatures rise above freezing during the summer. Under these conditions, water is never available in abundance. The dominant plants in tundras are not grasses and

FIGURE 2-18   Hubbard Glacier in Alaska. This glacier flows through mountain valleys to the sea, where pieces of it break off to form icebergs that float off into the ocean. (*U.S. Geological Survey.*)

FIGURE 2-19   Tundra with caribou on Hazen Plateau, Elles-   *Photograph by I. K. MacNeil.*)
mere Island, in northern Canada. (*Courtesy of Parks Canada.*

that lie about 40° from the equator. During the summer, the land in these areas is warmer than the ocean, causing moist air coming off the ocean to be warmed over the land and to retain its moisture. In the winter, the land is cooler than the ocean, causing moist sea air to cool over the land and drop its moisture. This type of climate characterizes the Mediterranean region as well as much of California and southeastern Australia. Mediterranean climates support chaparral vegetation, which consists primarily of shrubby plants with waxy leaves that retain moisture during summer droughts. Such climates have attracted large human populations, which have altered them greatly by decimating the diverse faunas of native animals.

Fossil plants as indicators of climate.

Because plants are so sensitive to environmental conditions, the fossils of many plants can be used to interpret climatic conditions of the past. The **cycads**, for example, are an ancient group of plants that are now found growing only in tropical and subtropical settings (Figure 2-20). Because this phenomenon seems to reflect a fundamental physiological limitation of the group, it is assumed that fossil cycads also lived in warm climates.

Flowering plants, which include not only plants with conspicuous flowers but also hardwood trees (e.g., maples, beeches, and oaks) as well as grasses and their relatives, provide a valuable means of interpreting climates of the past 80 or 90 million years, the interval during which they have been abundant on earth. The thick, waxy leaves found on some hardwood plants, for example, help these plants retain moisture and thus serve to indicate that fossil hardwood plants lived in warm climates. Leaf margins provide an even more useful means of assessing temperatures of the past. Leaves with smooth rather than toothed margins are especially common in the tropics. In fact, if a large flora occupies a region that is characterized by moderate or abundant precipitation, the percentage of plant species with smooth margins provides a remarkably good measure of the average annual temperature of that region (Figure 2-21). As we will see in Chapters 16, 17, and 18, this relationship has revealed a considerable amount of information about temperatures of the Late Cretaceous and Cenozoic intervals of geologic time.

Inasmuch as certain groups of animals are also restricted to warm regions, some animal fossils also serve as indicators of warm climates. Reptiles, which do not maintain constant warm body temperatures, exemplify animals that cannot live in very cold climates.

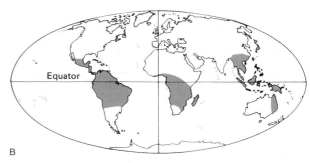

FIGURE 2-20   *A.* A living cycad. Cycads are plants that were especially common during the Mesozoic Era. Today, few cycads are found outside the tropics, and it appears that ancient cycads were also restricted to warm climates. *B.* The distribution of cycads today. (*After C. B. Cox and P. D. Moore, Biogeography: An Ecological and Evolutionary Approach, John Wiley & Sons, Inc., New York, 1980.*)

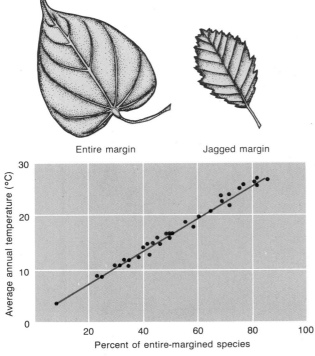

FIGURE 2-21   Relation between climate and leaf shapes of flowering plants. In the modern world there is a close correlation between the average annual temperature of a region and the percentage of plant species with entire (or smooth) margins. (*After J. A. Wolfe, American Scientist, 66:994–1003, 1978.*)

THE MARINE REALM

The ocean floor is a vast basin in which most sediments have accumulated over the course of the earth's history. For this reason, and because so many types of organisms live in the ocean, the sea floor is also where most species of the fossil record have been preserved.

Water movements

The major ocean currents at the earth's surface owe their existence primarily to large-scale winds. The trade winds blow toward the equator from the northeast and southeast (Figure 2-11), pushing equatorial water westward to form the north and south **equatorial currents** (Figure 2-22). These currents pile water up on the western sides of the major ocean basins, where some of the water flows backward under the influence of gravity as **equatorial countercurrents**.

Because equatorial currents are also affected by the Coriolis effect, they are deflected toward the poles as they approach the western boundaries of ocean basins. (Recall that the Coriolis effect bends a current in the Northern Hemisphere toward the right, or clockwise, and bends a current in the Southern Hemisphere toward the left, or counterclockwise.) At the same time, the trade winds drive water along the western margins of continents toward the

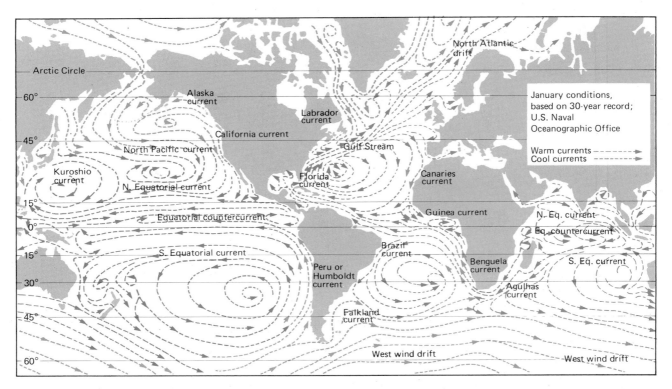

FIGURE 2-22   Major surface currents of the ocean. Note that large gyres north of the equator move clockwise, while those south of the equator move counterclockwise. (*After P. R. Ehr-* *lich, A. H. Ehrlich, and J. P. Holdren, Ecoscience: Population, Resources, and Environment, W. H. Freeman and Company, New York, 1977.*)

equator. The result of these movements for each major ocean is a full clockwise gyre of rotation north of the equator and a counterclockwise gyre south of the equator. The Indian Ocean, most of which lies south of the equator, has only a counterclockwise gyre, while the Gulf Stream, a famous segment of a gyre, carries warm water from low latitudes across the North Atlantic to warm the shores of Great Britain, where, in the southwest, palm trees survive more than 50° north of the equator. An eastern segment of the Pacific gyre has the opposite effect, bringing cool water to the coast of California.

In the Southern Hemisphere, the southern segments of the three gyres are known as **west-wind drifts**. Strengthened by westerly winds (Figure 2-11), these drifts join to form the Antarctic **circumpolar current** (Figure 2-23). In the Northern Hemisphere, the land masses of North America and Eurasia inhibit the development of a comparable circumpolar current.

Near the poles, water that is dense because it is frigid sinks to great depths. When this water reaches the deep-sea floor, it spreads toward the equator. Antarctic water that descends in this manner is slightly colder than the

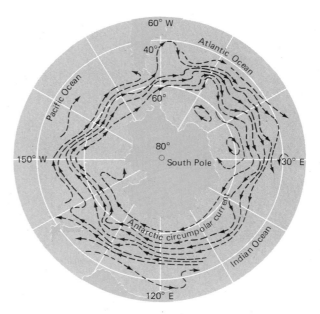

FIGURE 2-23   The eastward-flowing current around Antarctica. Note in Figure 2-22 how this current is formed by the large counterclockwise gyres of southern oceans. (*After A. N. Strahler, The Earth Sciences, Harper & Row, New York, 1971.*)

FIGURE 2-24 Waves on the surface of the ocean produced by winds traveling at 65 kilometers (~39 miles) per hour. These waves are 5 meters (~17 feet) high. (*Atmospheric Environment Service—Environment Canada.*)

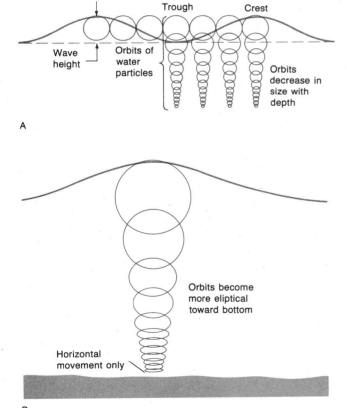

A

B

FIGURE 2-25 Wave motion in the sea. *A*. A wave form at the surface of the water results from the orbital movements of particles of water. Each particle of water rotates in the same orbit while the wave moves. *B*. As a wave moves into shallow water, the lower orbits flatten out. (*After F. Press and R. Siever, Earth, W. H. Freeman and Company, New York, 1982.*)

Arctic water from the north, and it thus hugs the bottom of the sea, flowing well into the Northern Hemisphere. Above this water, which remains at near-freezing temperatures, slightly warmer Arctic water flows southward. These currents supply the deep sea with oxygen from the earth's atmosphere near the poles. This oxygen permits a wide variety of animals to live here despite the freezing temperatures.

Waves represent yet another important form of water movement (Figure 2-24). **Surface waves** result from the circular movement of water particles under the influence of the wind. Because this movement decreases with depth, wave motion has no effect below a certain water depth, often several meters below the surface, depending on the wave size. Waves that are far from shore form swells that lack crests. As a wave approaches the shore, however, the orbits of water movement flatten out until they reach the sea bottom, where there is no room for vertical movement. At this stage, only horizontal, back-and-forth movement is possible (Figure 2-25). Very close to shore, the sea floor interferes with wave movement to such an extent that waves steepen and break (Figure 2-26).

The belt along a beach where waves break is known as the **surf zone**. In this zone, breaking waves move particles of sand shoreward. After a wave breaks, water spreads toward the beach and slides back down the sloping front of the beach. When waves approach a shore obliquely, they move sand obliquely landward, but some of the sand then slides backward in a direction perpendicular to the shoreline, resulting in a zigzag movement of sand grains along the shoreline. This movement, which is known as **longshore drift**, causes large volumes of sand to move parallel to many coastlines.

Tides, which also cause major movements of water in the oceans, result from the rotation of the solid earth be-

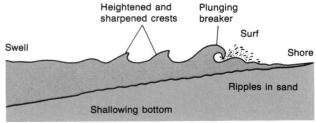

FIGURE 2-26 Waves breaking as they approach a beach. Offshore, the waves take the form of swells; when the orbital motion of a wave below the sea surface becomes obstructed by the sea floor, the upper part of the wave spills over toward the beach. (*After F. Press and R. Siever, Earth, W. H. Freeman and Company, New York, 1982.*)

neath bulges of water on the oceans. These bulges result primarily from the gravitational attraction of the moon (Figure 2-27). As tides approach a coast, they often produce strong currents. A tide flows toward a coast and then ebbs again within a few hours as the earth rotates, moving the coast away from the tidal bulge. As tides ebb and flow at the edge of the sea, they move the shoreline back and forth across an **intertidal zone**. Landward of this zone, there is often a belt that is flooded only during storms or at times when strong onshore winds coincide with high tide; this area, which is usually dry, is known as a **supratidal zone**.

The depth of the sea

Water depth in the sea varies from the thickness of a film of sea water at the shoreline to more than 10 kilometers (~6 miles) in the deep sea. Earlier in this chapter, in the analysis of the hypsometric curve, it was noted that large areas of sea floor lie between 3 and 6 kilometers (~2 to 4 miles) below sea level (Figure 2-1). These areas constitute the **abyssal plain**.

More details of the configuration of the sea floor are shown in Figure 2-28. A continental shelf is nothing more than the submarine extension of a continental land mass. The **continental margin** marks the edge of the shelf; seaward of it, the continent pinches out along the **continental slope**. Near the base of the continental slope, continental crust gives way to oceanic crust. Just seaward of this junc-

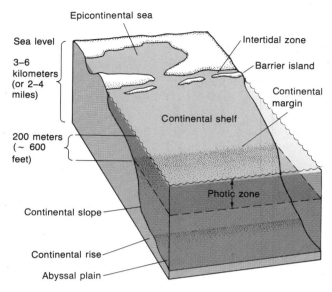

FIGURE 2-28   Diagram illustrating the aquatic environments at the edge of a continent. When the water is clear, the continental shelf—which is the submerged margin of a continent—usually lies within the photic zone, where there is enough sunlight to permit photosynthesis.

ture is the **continental rise**, which, as will be explained more fully in the next chapter, is a prism consisting of sediment that has been transported down the continental slope. Beyond the continental rise lies the previously mentioned abyssal plain, which is the surface of a layer of sediment resting on oceanic crust. When we speak of the **deep-sea floor**, we are usually referring to the region below the continental margin; the area above the deep-sea floor is often referred to as the **oceanic realm**.

Along the margin of the sea, **barrier islands** of sand heaped up by waves often parallel the shoreline. In the protection of these elongate islands are relatively quiet lagoons or bays. **Marshes**, which are formed by low-growing plants that inhabit the intertidal zone, fringe the margins of these ponded bodies of water (Figure 2-29). Here their remains accumulate as **peat**, which, if buried under the proper conditions of temperature and pressure, can turn to coal. In places, the sea spreads farther inland over a continent, forming a broad, semi-isolated **epicontinental sea**. At the present time, the seas happen to stand lower in relation to continental surfaces than they have at most times during the past 600 million years. For this reason, continental seas are not well developed; Hudson Bay in eastern Canada is perhaps the best modern example. As

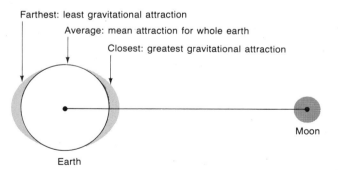

FIGURE 2-27   The origin of tides in the ocean. The moon exerts a gravitational attraction on the oceans that is strongest closest to the moon, causing the ocean to bulge out here. The weakest attraction is on the side farthest from the moon, and thus the ocean bulges out here as well. The solid earth rotates beneath these tidal bulges, causing them to move in relation to the solid earth. (*After F. Press and R. Siever, Earth, W. H. Freeman and Company, New York, 1982.*)

FIGURE 2-29   A marsh formed along the southern part of San Francisco Bay by grasses that tolerate submergence in salt water. Such marshes fringe many of the world's major lagoons. As the dead grass accumulates, it forms peat, which may eventually become coal. (*U.S. Department of the Interior.*)

we journey through Phanerozoic time in later chapters of this book, we will examine many ancient epicontinental seas and the life that they harbored.

How is life of the ocean related to water depth? Depth itself is probably a limiting factor for only a few species, but some significant limiting factors such as light (in the case of plants) and temperature are often closely related to water depth. The upper layer of the ocean, where enough light penetrates the water to permit plants to conduct photosynthesis, is known as the **photic zone**. Although the base of this zone varies from place to place depending on water clarity, it usually lies at a depth that is greater than 100 meters (~300 feet) and less than 200 meters (~600 feet) below sea level. It happens that 200 meters is also the approximate depth of the continental margin in most areas.

For life of the sea floor, the most profound environmental change associated with depth takes place along the margin of the ocean. At the transition between the intertidal zone and the adjacent **subtidal** sea floor, which is never exposed to the air, the biota of the intertidal zone must endure large, often rapid fluctuations in environmental conditions. At some latitudes in this zone, hot, dry conditions prevail at low tide during the summer, and yet winter chills drop temperatures below freezing. In the surf zone, where waves break along a sandy beach, the constant movement of the sand permits only a few species to survive. Species that live in this zone are exceptionally mobile and can quickly reestablish themselves in the sand in the event that they are dislodged by a wave.

Unusual conditions also characterize the deep sea, but here the environment is more stable. As we have seen, the frigid water that descends in polar regions covers the abyssal plain with waters whose temperature approaches the freezing point. In fact, cool waters characterize the entire ocean at depths below about 500 meters (~1600 feet), and these waters form an environment known as the **psychrosphere**, which is inhabited by unique groups of species that are adapted to cold conditions.

In certain groups of bottom-dwelling marine organisms, different species tend to live at different depths in the ocean. The **foraminifers**, a group of single-celled, amoebalike creatures with internal skeletons, are one such group. They

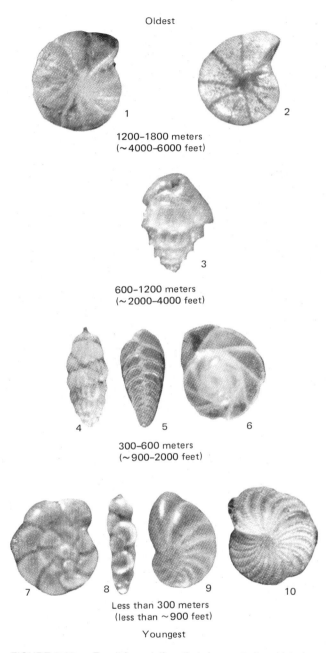

Oldest

1   2

1200–1800 meters
(~4000–6000 feet)

3

600–1200 meters
(~2000–4000 feet)

4   5   6

300–600 meters
(~900–2000 feet)

7   8   9   10

Less than 300 meters
(less than ~900 feet)

Youngest

FIGURE 2-30   Fossil foraminifers that demonstrate a historical trend of shallowing of the marine waters in the Los Angeles
and Ventura basins of Southern California. All the species
shown here are still alive today and live on the sea floor. Pliocene foraminifers found in sediments of the California basins
are deep-water species, and the older of these represent even
deeper water than do the younger ones. The youngest species
of Pleistocene age represent depths ranging from about 300
meters to low tide. Thus, waters of the basin became shallower
with time. (1) *Cibicites mckannai*, (2) *Nonion pompilioides*,
(3) *Bulimina subacuminata*, (4) *Uvigerina peregrina*, (5) *Boli-
vina argentea*, (6) *Epistomella pacifica*, (7) *Ammonia beccarii*,
(8) *Uvigerina juncea*, (9) *Nonion grateloupi*, (10) *Elphidium
crispum*. (*Courtesy of M. L. Natland.*)

have left an excellent fossil record that has provided considerable information about the water depths at which ancient sediments were deposited (Figure 2-30). It is not known
why different species of foraminifers tend to live at different
depths; it may be that these tiny organisms are sensitive
to water pressure, which increases with depth.

Marine life habits and food webs

Most of the photosynthesis that takes place in the ocean
is conducted by single-celled floating algae. For this reason, these algae are considered by many to be the most
important producers at the base of the food web. Organisms that float in water are known as **plankton**, and plantlike organisms that belong to this group constitute the
phytoplankton. The most important groups of phytoplankton are **dinoflagellates, diatoms,** and (in warm regions)
coccolithophores. The general characteristics of these
groups of algae are shown in Figure 2-31. All three groups
have extensive fossil records—the diatoms and coccolithophores because they have hard skeletons and the dinoflagellates because they have durable cell walls.

Feeding on the phytoplankton are floating animals known
as the **zooplankton**, among which are small shrimplike
crustaceans (see Appendix III) and other animals that spend
their full lives afloat. Also included in this group, however,
are the floating larvae of some invertebrate species that
spend their adult lives on the sea floor. These larvae allow
the sea floor species to disperse over large areas of the
ocean; after spending time in the plankton, they settle to
the sea bottom and develop into adult animals. Some members of the zooplankton are carnivores that feed on other
zooplankton.

Although many types of plankton have the capacity to
swim, planktonic species move through the ocean primarily
by drifting passively along with the water in which they live.
Animals that move through the water primarily by swimming
are termed **nekton**. The most important of these are fishes.
Both the plankton and the nekton include not only herbivores, which feed on phytoplankton, but also carnivores,
which feed on other animals. Planktonic and nektonic organisms together constitute **pelagic life**, or life that exists
above the sea floor.

Both immobile and mobile organisms also populate the
sea floor, and these organisms are known as **benthos**, or
benthic life. The sea floor itself is often referred to as the
substratum. Substrata may be hard, as is the case when
they are formed of rock, but are more often composed of

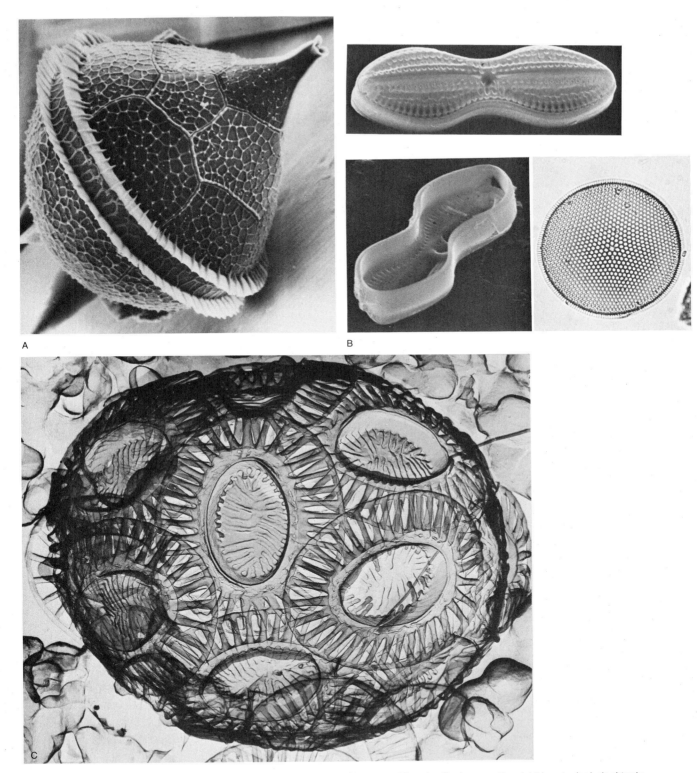

FIGURE 2-31   The major types of single-celled algae in modern oceans. *A.* The dinoflagellate *Protoperidinium steinii* (×2900). Dinoflagellates have two whiplike flagella, one at one end of the cell and the other wrapped around the middle in the troughlike "girdle." *B.* Two kinds of diatoms: The elongate form on the top is a bottom-dwelling species (×800); the lower picture shows the pillboxlike structure of the silica skeleton of all diatoms. The circular form on the right is a typical planktonic species (×512). *C.* The coccolithophore *Emiliania huxleyi* (×1800); the shieldlike structures on the surface of the cell are coccoliths of calcium carbonate. (*A. Courtesy of J. R. Allen. B. Courtesy of H. Tappan. C. Courtesy of A. McIntyre and A. W. H. Bé.*)

soft substances such as loose sediment. Some benthic organisms live on top of the substratum, while others live within it. Animals that dig into or move through soft sediment are usually termed **burrowers**; those that excavate living space in hard substrata are called **borers**.

Just as producers and consumers float in the water, both nutritional groups are found on the sea floor. Benthic producers include certain kinds of single-celled algae as well as multicellular plants. Because they require light, these

FIGURE 2-32   Life habits of animals of the sea floor. (For more biological details about these groups, see Appendix III.) Arrows show movements of water currents into and out of the animals, bringing oxygen and sometimes food, and removing waste. *A.* A snail (gastropod) grazing on algae growing on a rocky surface. *B.* A sea urchin (echinoid echinoderm), also grazing on the rocky surface. *C.* A starfish (echinoderm) feeding on a mussel. *D.* A suspension-feeding bivalve mollusk attached to the rocky surface. *E.* A sponge, which is also a suspension feeder. *F.* A crab feeding as a carnivore. *G.* A cockle, which is a bivalve mollusk that burrows in sediment and obtains food as a suspension feeder. *H.* An annelid worm that lives in the sediment in a tube, through which it pumps water for suspension feeding. *I.* A deposit-feeding bivalve mollusk that uses a long, fleshy siphon to draw in sediment from which it extracts organic matter. *J.* A deposit-feeding worm that lives within a cone-shaped tube of sand grains, removing food from the sediment. *K.* A deposit-feeding sea cucumber (echinoderm) that passes large amounts of sediment through its gut.

photosynthetic forms must live on or close to the surface of the substratum.

Most benthic herbivores fall into three categories (Figure 2-32). One group, known as **grazers**, feed on plant-like forms, especially algae growing on hard surfaces. Most grazers crawl over the sea floor, but some are fishes, which belong to the nekton. The second group of benthic herbivores, known as **suspension feeders**, strain phytoplankton and plant debris from the water. The third group, known as **deposit feeders**, consume sediment and digest organic matter mixed in with the mineral grains. Most deposit feeders live in or on muddy sediment rather than on clean sand because such sediment is richer in organic matter.

Bottom-dwelling carnivores of modern seas include crabs and starfishes (Figure 2-32) as well as several kinds of snails and worms. In addition, many fishes, which are nektonic in habit, swim close to the sea bottom and feed on bottom-dwelling animals.

Figure 2-33 depicts the basic features of the marine food web. The phytoplankton occupy the photic zone above both the continental shelves and the deep sea. Joining them as producers on continental shelves, where the photic zone reaches the sea floor, are bottom-dwelling plants. The high concentration of phytoplankton in the photic zone causes zooplankton to be concentrated there as well. Some herbivorous zooplankton can also be found at greater depths, where they feed on the algal cells and plant debris that rain slowly down from the photic zone.

Different kinds of fishes occupy different depth zones in the ocean. Some, like herring, feed on zooplankton, and these are joined by the great baleen whales, which strain zooplankton through a sievelike bony structure. Other fishes, including nearly all sharks, are carnivorous, as are many kinds of whales. Carnivores are found at all depths of the ocean, although some species are restricted to narrow depth zones.

Many different kinds of benthos are found on the shallow sea floors of the photic zone. In this zone, most of the food at the base of the food web comes from phytoplankton, although some also derives from benthic plants. In the deep sea, however, where suspended food is scarce, there are only a few kinds of benthic suspension feeders. Here most herbivores are deposit feeders, and these are joined by many types of carnivorous fishes. In fact, numerous species live along the cold, dark abyssal plain. Organic debris arrives here at a very slow rate, however, settling from shallow depths; hence, the density of animals is low. Thus, a survey of 1000 square meters of the deep-sea floor might

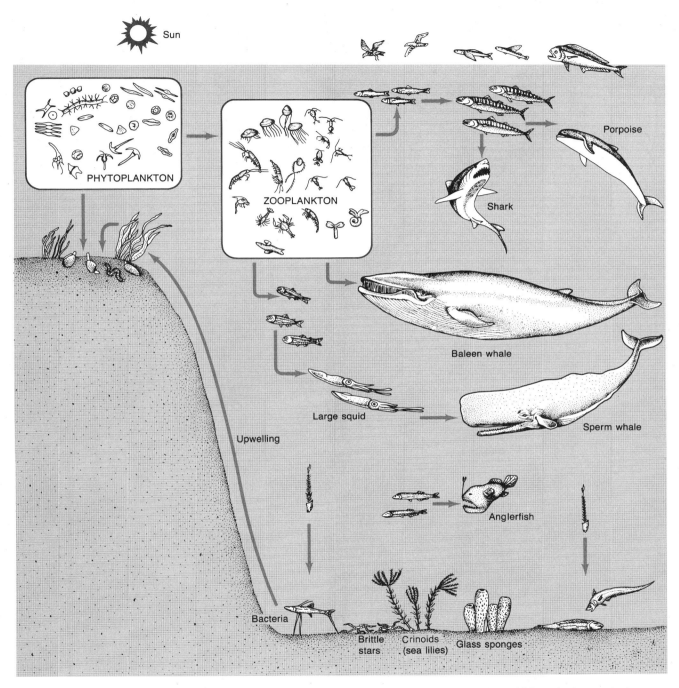

FIGURE 2-33   The food web in the ocean (the various forms of life are not drawn to scale.) Phytoplankton occupy the photic zone of the ocean, and thus most zooplankton, which feed on phytoplankton, also live here. On continental shelves, especially near the shore, bottom-dwelling plants also contribute food to the marine ecosystem. Most species of large carnivores are fishes. Whales are warm-blooded mammals that include carnivorous porpoises and sperm whales, which feed on large ani- mals, as well as baleen whales, which strain tiny zooplankton from the water. As the amount of plant material diminishes with depth, the abundance of animal life diminishes as well. A few suspension feeders, such as sponges and crinoids (sea lilies), live in the deep sea floor, but most herbivores here are deposit feeders. Bacteria in the deep sea turn dead organic matter into nutrients that upwelling currents carry to the surface for use by phytoplankton and other photosynthetic life.

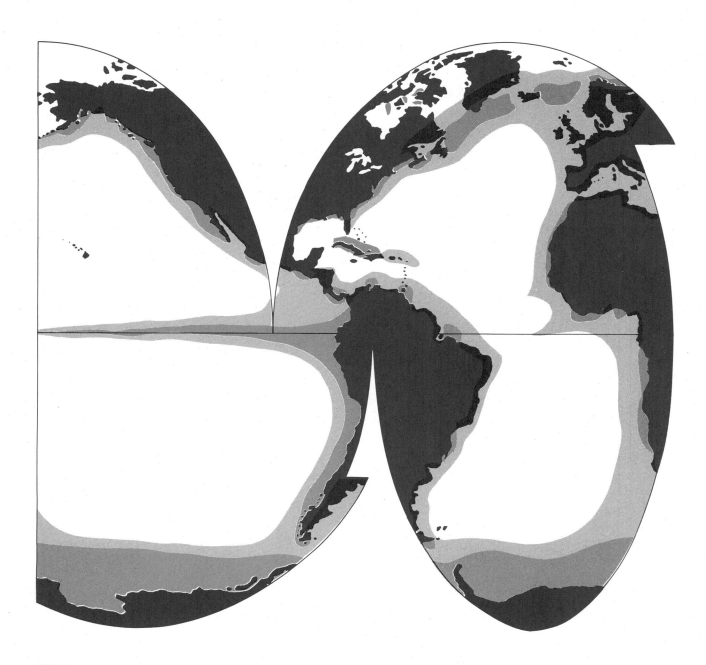

FIGURE 2-34   Upwelling in the Atlantic and western Pacific oceans, where surface currents draw water away from the coast. Dark shading shows strong upwelling; light shading shows weaker upwelling. (*From J. D. Isaacs, "The Nature of Oceanic Life." Copyright © 1969 by Scientific American, Inc. All rights reserved.*)

uncover dozens of species, each represented by a small number of individuals.

Bacteria occur throughout the ocean but are concentrated in the deep sea, where organic debris accumulates. Some bacteria decompose this debris, while others transform some of the products of decay into simple nutrient compounds of nitrogen and phosphorus. Phytoplankton use these compounds to make food, thereby cycling the materials back through the ecosystem. A crucial step in this recycling is the physical process known as **upwelling**, which consists of the movement of cold water upward from the deep sea to the photic zone. Upwelling tends to occur along the margins of continents (Figure 2-34), where the large oceanic gyres drag water away from the continents; this water is replaced by upwelling of water from the deep sea. Upwelling often brings nutrients to the photic zone in large quantities, producing an unusually rich growth of phytoplankton. The phytoplankton support large populations of zooplankton, which, in turn, support large populations of fishes.

Marine temperature and biogeography

In the marine realm, as in the terrestrial realm, temperature plays a major role in determining the geographic distribution of species. Planktonic life exemplifies this pattern. The coccolithophores, for example, live primarily in warm waters; individual species occur in narrow latitudinal belts where

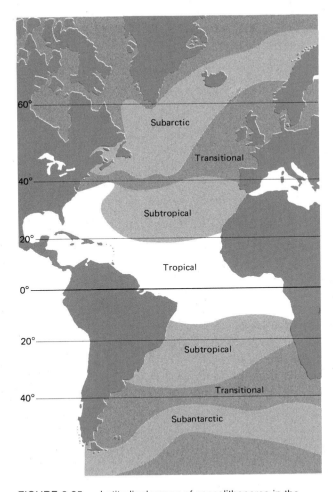

FIGURE 2-35   Latitudinal zones of coccolithopores in the modern ocean. Although these members of the phytoplankton are found in cold waters, fewer species occur here than in warm-water zones. (*After A. McIntyre and A. W. H. Bé, Deep Sea Research, 14:561–597, 1967.*)

water temperatures remain within certain limits (Figure 2-35). In contrast, most species of planktonic diatoms live in cool waters at high latitudes.

The geographic distributions of species of the sea floor are also limited by temperature, and some large groups of animals and plants are restricted to certain latitudes. The reef-building corals represent one of the most important groups of this type, forming massive reefs that are largely restricted to the tropics. Reefs form their own fossil records as they grow. Beneath their living surface they consist of the dead remains of the animals and plants responsible for their construction (Box 2-1). Other kinds of limestones are also largely tropical in distribution (see Appendix I).

Oxygen isotopes of fossil shells have been used in efforts to determine the temperatures at which animals lived millions of years ago. Oxygen occurs naturally in two isotopic forms, or forms of the element that have different atomic weights. Oxygen 18 has two more neutrons than the more common form, oxygen 16, and is thus the heavier **isotope**. The two isotopes have the same chemical properties, but marine organisms that secrete shells incorporate the isotopes in slightly different proportions, depending on the temperature of the environment. As temperature decreases, the percentage of oxygen 18 increases. A major difficulty encountered in attempts to analyze ancient ocean temperatures by means of oxygen isotopes is that most ancient shells have suffered chemical alteration after burial. As a result, temperature estimates based on oxygen isotopes are often inaccurate.

Salinity as a limiting factor

The saltiness of natural water is called **salinity**. Oceanic sea water contains about 35 parts of salt per thousand parts of water, or is said to have a salinity of 35 parts per thousand. Salinities of 30 to 40 parts per thousand are regarded as representing the normal range for sea water; water of lower salinity is labeled **brackish**, while water of higher salinity is termed **hypersaline**.

Brackish and hypersaline conditions are most commonly found in bays and lagoons along the margin of the ocean. Brackish salinities result from an influx of water from rivers into bays or lagoons that are partially isolated from the open ocean. Hypersaline conditions also develop in bays and lagoons, but only in those whose waters experience high rates of evaporation—which is usually in hot, arid climates.

Brackish and hypersaline waters typically undergo frequent salinity changes resulting from changes in rainfall

Box 2-1

Coral reefs in the tropics

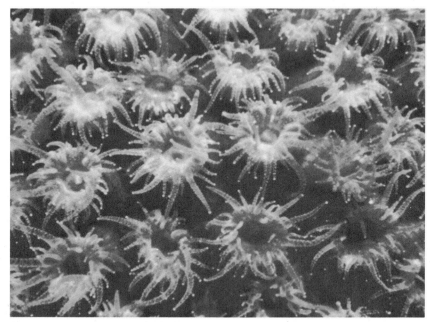

Polyps of a reef-building coral colony. White spots on the tips of the tentacles represent stinging cells. The polyps are about 2 centimeters in diameter. (*Photograph courtesy of P. Dunstan.*)

Living coral reefs, inhabited by numerous, varied, and highly colorful animals and plants, are among the most beautiful biotic communities on earth. In many shallow tropical seas they are readily viewed with a swim mask and snorkel. Fossil reefs are porous structures that serve as important traps for petroleum.

The framework of a modern reef is built by corals, skeleton-secreting animals that belong to the primitive animal phylum **Coelenterata** (sometimes called Cnidaria)—a group that also includes the jellyfishes. Coelenterates are saclike animals with tentacles. Most of them derive nutrition from smaller animals that they capture from the water by means of stinging cells on their tentacles.

Reef-building corals are colonial animals that grow as clusters of connected individuals, or **polyps**. A colony forms from a single original polyp that develops from a larva. The original polyp gives rise to a colony by budding off additional polyps which, in turn, bud off others. Each polyp secretes a cup of calcium carbonate, and the adjacent cups are fused to form a large composite skeleton. There are also species of corals that

do not form reefs; most of these are solitary rather than colonial animals and consist of single polyps. Colonial growth allows corals to form massive reefs, which are limestone structures that stand above the surrounding sea floor. Some other types of reef dwellers join corals in contributing skeletons to the solid reef structure.

Aiding colonial corals in forming their large colonies are single-celled algae that live and multiply in the coral tissues. The corals supplement their diet by digesting some of these algae. Before they are digested, however, algae cells remove carbon dioxide from the corals for use in photosynthesis. In this way, the algae facilitate the corals' secretion of their calcium carbonate skeleton. Because the algae require sunlight for photosynthesis, coral reefs are restricted to moderately shallow depths in the ocean.

Reefs thrive in areas of strong water movement, which brings corals zooplankton in large quantities; in addition, they require relatively clear water because murky, sediment-laden water shields their symbiotic algae from sunlight and hinders their ability to feed on small floating animals. Warm water is also essential for coral reef growth, apparently because it facilitates the secretion of calcium carbonate. Few reefs grow at locations more than 30° north or south of the equator, although some individual coral colonies survive at slightly higher latitudes. Coral reefs thrive only where the water temperature seldom falls below 18°C (~64°F); in other words, they are tropical structures. In Chapter 4, we will examine the kinds of limestone deposits that coral reefs form and the types of reefs that form in various settings.

Modern corals belong to the group of coelenterates known as **hexacorals**, or **Scleractinia**. These animals evolved near the beginning of the Mesozoic Era. In Chapters 12 through 16 we will learn about extinct groups of organisms that built reefs before the hexacorals evolved or that competed with hexacorals for space on reefs during the Mesozoic Era. ☐

Portion of a coral reef near North Rock, Bermuda. One large "brain coral" is visible in the foreground, and another of a different species can be seen to the left of center. The branching organisms are not plants but sea whips and fans, which, like corals, are colonial members of the phylum Coelenterata. The scale is 50 centimeters (~20 inches) long (*Courtesy of P. Garrett.*)

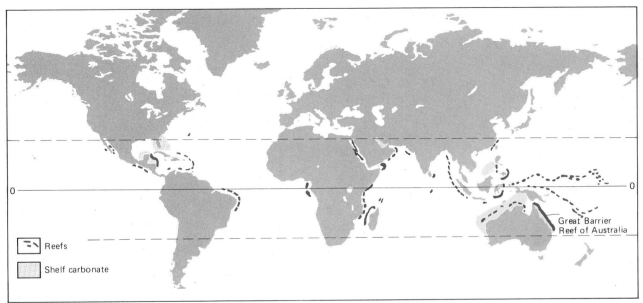

Map of coral reefs and major areas of carbonate sedimentation in the modern world. Carbonate sediments, like reefs, are largely restricted to the tropics.

and in evaporation rates. The tissues of most animals have salt contents similar to the salinity of normal sea water. Marine animals therefore tend to find it difficult to move into habitats characterized by abnormal and fluctuating salinities. It is hardly surprising that most bays and lagoons contain few species of animals in comparison to normal marine habitats.

Freshwater environments

The difficulty of living in water of low salinity is a major reason the faunas of rivers and lakes are not very diverse. Rivers and lakes are freshwater habitats, or habitats whose salinities remain below 0.5 percent. Most freshwater animals must possess ways of excreting excess water that enters their body tissues from the environment. Phytoplankton and zooplankton similar to those of the ocean also inhabit lakes, but in reduced variety. Because of their motion, streams and rivers do not sustain a complex planktonic community equivalent to that of the ocean. As exemplified by the food web of a trout stream (Figure 2-4), most producers occupying rivers live on the bottom, and a large proportion of consumers are immature growth stages of terrestrial insects. Both lakes and rivers differ from the ocean in their small variety of animal species of larger body size. Unfortunately, too, relatively few kinds of these animals are readily preserved as fossils, with fishes and shelled mollusks constituting the primary exceptions.

As we will see in later chapters, most major groups of aquatic life evolved in the ocean, and only secondarily did some invade less hospitable freshwater environments.

CHAPTER SUMMARY

1.   The way in which a species interacts with its environment defines the ecologic niche of that species.

2.   The distribution and abundance of any species are governed by a number of limiting factors. These include the availability of food, the nature of physical and chemical conditions in the environment, and the occurrence of other species that are potential predators or competitors.

3.   Communities are groups of coexisting species that form food webs. Plants are at the base of most food webs, and these are fed upon by herbivores, which, in turn, are fed upon by carnivores.

4.   Bacteria decompose dead organisms and transform the products of decay into compounds that serve as plant nutrients. Thus, materials are cycled through the ecosystem continuously.

5.   Physical barriers to dispersal and changes in environmental temperature are the most important factors that limit geographic distributions of species; many more species exist in warm climates than in cold climates.

6.   Ice caps exist near the earth's poles on Greenland and Iceland.

7.   Tropical rain forests develop near the equator, where warm air rises and loses its moisture. The dried air descends north and south of the equator and circles back toward the equator as trade winds. In many areas that lie between 20 and 30° from the equator, trade winds produce deserts and dry grasslands.

8.   Because land plants are highly sensitive to climatic conditions, fossil land plants are useful indicators of ancient climates.

9.   In large oceans, prevailing winds and the Coriolis effect create huge gyres of water movement.

10.   The dominant food-producers in the ocean are photosynthetic single-celled algae that float at shallow depths.

11.   The deep sea is cold because its waters come from near the north and south poles, where frigid water sinks to great depths and flows toward the equator. Because little food reaches the deep sea, animals are sparsely distributed here, but large numbers of species are present.

12.   In bays and lagoons near the margin of the ocean, the salinity of water is likely to differ from that of normal sea water and is also likely to fluctuate greatly; relatively few species are able to live in these environments.

13.   Freshwater environments, such as lakes and rivers usually harbor fewer species than do normal marine environments because life in freshwater poses physiological problems for many kinds of animals.

EXERCISES

1.   Sometimes the species of a community are described as forming a food chain. Why is it usually more appropriate to speak instead of a food web?

2. Which terrestrial and marine environments characteristically contain few species? Explain why each of these environments is populated in this way.

3. How do the main kinds of producers (photosynthesizers) in the ocean differ in mode of life from those on the land?

4. What does the shape of the hypsometric curve tell us about the distribution of sea floor environments?

5. Rain forests and coral reefs are sometimes compared, inasmuch as both are communities of high species diversity. Why are both communities restricted to the tropics?

ADDITIONAL READING

Cox, C. B., and P. D. Moore, *Biogeography: An Ecological and Evolutionary Approach*, John Wiley & Sons, Inc., New York, 1980.

Gross, M. G., *Oceanography: A View of the Earth*, Prentice-Hall, Inc., Englewood Cliffs, New Jersey, 1976.

Kormondy, E. J., *Concepts of Ecology*, Prentice-Hall, Inc., Englewood Cliffs, New Jersey, 1976.

————, "The Biosphere," *Scientific American*, September 1970.

Laporte, L. F., *Ancient Environments*, Prentice-Hall, Inc., Englewood Cliffs, New Jersey, 1979.

Miller, A., and J. C. Thompson, *Elements of Meteorology*, Charles E. Merrill Books, Inc., Columbus, Ohio, 1983.

Turekian, K. K., *Oceans*, Prentice-Hall, Inc., Englewood Cliffs, New Jersey, 1976.

CHAPTER 3
Nonmarine
Sedimentary
Environments

here are many lessons to be learned from the depositional settings of ancient sedimentary rocks. The most general reason for studying these settings is to reconstruct geography of the past, that is, **paleogeography**. The goal is not only to learn about the distribution of land and sea at a particular time, but also to identify and reconstruct more localized environmental features, such as deserts, lakes, river valleys, lagoons, and submarine continental margins. In most instances, for example, geologists can learn not only where a river valley was located, but also what kind of river occupied that valley—perhaps one formed by many small, intertwining channels choked with bars of gravel and sand, or one that flowed along a single broad, winding channel. Frequently, geologists can also "read" from the sedimentary record whether the terrane that once bordered an ancient river was a dry, sparsely vegetated plain or a swamp densely populated by water-loving trees and undergrowth.

The identification of ancient sedimentary environments also provides geologists with a framework within which to interpret life of the past. Although we can learn some aspects of how an animal or plant species lived by studying the configuration of its fossil remains alone, a fuller understanding of that species can come only when its habitat is taken into consideration. The largest dinosaurs provide an illustration. It has been suggested that these animals were too big to be fully terrestrial and therefore must have spent much of their time in water, like hippopotamuses. As will be discussed more fully in Chapter 15, the fossil record of these giant creatures contradicts this idea: Their bones are frequently found in nonaquatic environments.

The study of ancient sedimentary environments is also of considerable practical value in that many sedimentary deposits occur in conjunction with important natural resources or are themselves natural resources. By understanding the environmental relationships of sedimentary rocks, economic geologists can often predict where these resources will be found beneath the earth's surface. Coal is an example of a resource whose location can be predicted on this basis, as are petroleum and natural gas, which tend to accumulate in porous rocks such as clean sands deposited along ancient shorelines or rivers and in ancient limestone reefs constructed by corals or coral-like organisms.

Before geologists can interpret the origins of ancient sedimentary rocks, however, they must first understand the patterns and processes by which sediment deposition takes place in the modern world. Unfortunately, the products of this deposition are often obscured and thus cannot be observed unless geologists excavate, tunnel, or core in order to observe them. **Coring** is a technique that can be used to examine deposits occupying the center of lakes, lagoons, or deep oceans. A tube is driven into the bottom of a lake or sea and subsequently withdrawn. The core of sediment thus extracted can then be studied to determine the sequence of sediment deposition at that site. Coring at several locations provides a three-dimensional picture of sedimentary deposits. Similarly, geologists who wish to examine a meandering river's depositional record must either dig one or more pits in the valley floor adjacent to the channel or sample that floor by means of coring. (The

Sand dunes near Stovepipe Wells, California. The wind produces the rippled surface on the dunes. (*George Grant, U.S. Department of the Interior.*)

reason for studying the sediments *adjacent* to a river will become clear in the course of this chapter.)

Some sedimentary features provide highly reliable information about the nature of the environments in which they formed. The tiny spherical grains of calcium carbonate known as **ooids**, for example, are found in cross-bedded rocks that almost always represent shallow tropical sea floors agitated by waves. Many other sedimentary features, however, offer little more than ambiguous testimony about their environments of deposition. Most coal deposits, for example, represent swamps choked with vegetation—but such swamps are typical of both the banks of rivers and the shores of marine lagoons. Geologists must therefore consider the nature of the beds that lie above or below coal deposits; if there are burrowed sandy muds with fossils

of marine animals above or below a coal bed, it is possible that a marine lagoon, not a river, lay adjacent to the swamp or marsh in which the coal formed (Figure 2-29).

In this chapter, we will begin our review of depositional environments and their characteristic sediments with examples from nonaquatic settings, and we will then move to freshwater systems. In the next chapter, our focus will shift to the marine realm.

SOIL ENVIRONMENTS

Soil can be defined as loose sediment that has accumulated in contact with the atmosphere rather than under water. Soil rests either on sediment of different character-

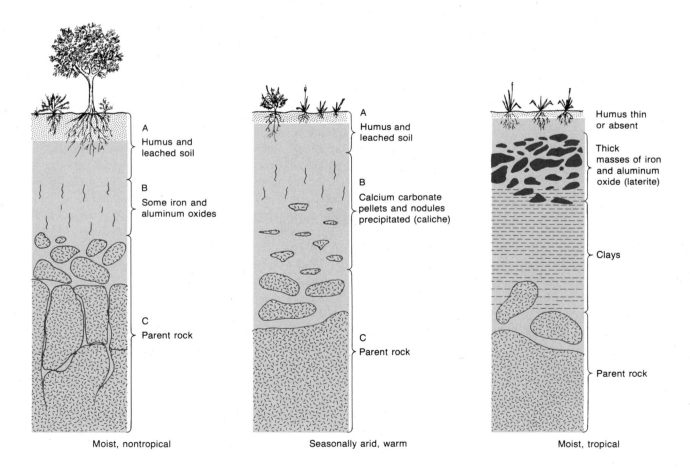

FIGURE 3-1   Soil formation in three kinds of climates. In a moist, nontropical climate, humus-rich soil near the surface (A-horizon) grades into the subsoil (B-horizon), which contains little organic matter, and this grades downward into decaying parent rock (C-horizon). In a seasonally arid, warm climate, cal-cium carbonate precipitates as caliche in the B-horizon. In a moist, tropical climate, humus as well as many silicate minerals are destroyed, leaving a residue of iron and aluminum oxide, la-terite. (*After F. Press and R. Siever, Earth, W. H. Freeman and Company, New York 1982.*)

istics or on rock, and it develops some of its properties at the site where it first accumulated. Soil also serves as a medium for the growth of plants by supplying essential nutrients and by providing a base for the physical support of stalks and stems. Soils form in a variety of environments throughout the world—in tropical rain forests, in arid regions, and even on mountain tops. Moreover, layers of diagenetically altered soil can often be found buried within thick sequences of ancient sediment. Recognition of these ancient soils is of great geologic significance in that they can be used to define the configuration of ancient landscapes. Certain soils also offer important evidence about the climatic conditions under which they formed.

How soils form

Soils develop in part by the weathering processes described in Appendix I. Most soils consist of three intergrading layers known as **horizons**. The **A-horizon**, which is commonly termed **topsoil**, consists primarily of sand and clay mixed with organic matter called **humus**, formed largely from the decay of leaves, woody tissues, and other plant materials. Humus, which gives topsoil its dark color, is derived from plants and, in turn, supplies nutrients for other plants; thus, it occupies an important position in the cycling of materials through the plants of terrestrial ecosystems. Desert soils are poor in humus because vegetation in such environments is sparse. Below the A-horizon of a soil is the **B-horizon**, which contains less humus, and this is underlain by the **C-horizon**, which consists of slightly altered and broken bedrock mixed with some sand and clay.

Different types of soils form under different climatic conditions (Figure 3-1). In temperate climates such as those of the northern United States and central Europe some relatively insoluble oxides of iron and aluminum accumulate in the B-horizon, whereas in warm climates that are dry part of the year, calcium carbonate accumulates in the B-horizon, forming what are known as **caliche** nodules, or **calcrete** (Figure 3-2). The accumulation of caliche nodules results from the evaporation of groundwater under hot, dry conditions.

In moist tropical climates the A-horizon is often unrecognizable because of the high temperatures and abundance of water that characterize such environments. Here, warm waters percolate through the soil, destroying humus by oxidizing it. Silicate minerals in such areas also break down and disappear quickly, leaving the soil rich in aluminum oxides as well as in iron oxides, which give it a

A

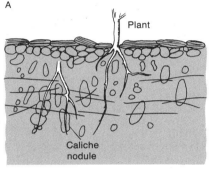

B

FIGURE 3-2   White caliche nodules in an ancient soil of Late Triassic or Early Jurassic age. The soil is now a mudstone bed within the New Haven Formation of Connecticut. As shown in the drawing, the nodules encrusted plant roots in places. The roots are now represented in the ancient rock by cylindrical sediment fillings, some of which pass through caliche nodules. (*From J. F. Hubert, Geology 5:302–304, 1977.*)

rusty red color. Tropical soils of this type are known as **laterites**.

Ancient soils

Ancient buried soils can be exceedingly difficult to recognize or to interpret. One reason for this is that diagenesis often alters the chemical components of soils beyond recognition. One place where ancient soil *is* likely to be found, however, is beneath an unconformity. At such sites, soils will have formed while the rocks below the unconformity were exposed at the earth's surface—that is to say, before these rocks were buried beneath younger sediments. Unfortunately, the alteration of rocks beneath an unconformity does not always signal the presence of an ancient soil. Because unconformities are discontinuities along which water may tend to migrate, these rocks may undergo chemical diagenesis not unlike the chemical weathering that alters soils at the earth's surface.

There are, however, a variety of features that can facilitate the identification of ancient soils. Soil that formed on a rocky surface, for example, can usually be identified by the loose rock fragments of its C-horizon (Figure 3-1). Caliche is also easy to recognize in the fossil record, although it forms only in warm, seasonally arid regions. Figure 3-2 shows an ancient caliche that formed about 200 million years ago in the Connecticut Valley north of New Haven. The abundance of caliche in this area has led geologists to conclude that the climate here was probably warm and semiarid with an annual rainfall of 10 to 50 cencentimeters (~4 to 20 inches). Cylindrical sediment fillings of spaces once occupied by roots indicate the presence of plants in the ancient soil. Nodules of caliche surround some of these tubes, making their presence more apparent than it would otherwise be.

Sediment fillings of root tubes provide one of the best available means of identifying ancient soils—including those in the Badlands National Monument of South Dakota, where a succession of about 90 soil horizons of Oligocene age have been identified. Even more spectacular are the buried forests of Eocene age at Yellowstone National Park in Wyoming where one can see not only the fillings of root tubes, but actual roots and standing tree trunks permineralized with silica. As will be described in Chapter 17, there is a succession of more than 20 such forests in this area. Each was killed by a volcanic eruption, and the next grew on soils that developed on top of the new layer of volcanic rocks.

Plant fossils, however, are not the only biological clues to the identity of ancient soils. Burrows made by animals such as insects and rodents are also diagnostic features. Certainly the most unusual of these are the structures known as "devil's corkscrews," which are actually burrows that beavers of an extinct species dug with their teeth in the Oligocene and Miocene soils of Nebraska (Figure 3-3). Skeletons of these beavers have been found in the burrows, and scratches on the burrow walls match their front teeth! The fact that these animals lived as far as 10 meters (~33 feet) below ground level indicates that the level of standing water in the ancient soil barely stood above this depth; had it been higher, the beavers would have drowned.

The nature of soils has changed in the course of the

A

B

FIGURE 3-3   Fossil burrows made by beavers in soils more than 20 million years old in Nebraska. Unlike modern beavers, these animals did not live near water, but instead lived in grasslands, where they formed colonies like those of prairie dogs. The burrows extend downward as far as 10 meters (~33 feet) into the soil, and approximately equal numbers spiral clockwise and counterclockwise. Tooth marks on the burrow walls were dug with the beavers' front teeth, as shown in the drawing. (A. Carnegie Museum of Natural History. B. After L. D. Martin and D. K. Bennett, Palaeogeography, Palaeoclimatology, Palaeoecology 22:173–193, 1977.)

earth's history. One major step in soil evolution was taken with the origin of large, woody plants and forests during the Devonian Period. Only from this time on has humus accumulated at a rapid rate on land. An important environmental change that took place even earlier, however, was the development of a reservoir of free oxygen in the atmosphere, which occurred during Precambrian time. Before this time, oxygen was scarce, and weathering was therefore slow and often incomplete. The result, as we will discuss in Chapter 10, was that numerous grains of unoxidized and weakly oxidized minerals accumulated at the earth's surface.

LAKES AS DEPOSITIONAL ENVIRONMENTS

At no time in geologic history have lakes occupied more than a minute fraction of the earth's surface—but because lakes form in basins that lie at lower elevations than most soils, lake deposits are much more likely than soils to survive in the face of erosion. Even so, ancient lake deposits are much less common than the many marine deposits that will be described later in Chapter 4.

Lake deposits share a number of typical characteristics by which they can be recognized. For example, sediments around the margins of lakes tend to be coarser than those that lie toward the lakes' centers. This is partially attributable to the fact that the currents of streams and rivers slow down when they meet the waters of a lake, dropping their coarse sediment load near the shore as they do so. Furthermore, the wind-driven waves on the surface of most lakes touch bottom only when they reach the shores, winnowing the sediment there and driving clay-sized particles into suspension. These particles later settle to the bottom toward the lake's center.

Fossils are valuable tools for distinguishing lake sediments from marine sediments. Although fish fossils are found in both kinds of deposits, the presence of fossils representing only marine animals provides strong evidence that ancient sediments did not originate in a lake. Because burrowing animals are not as abundant in lakes as they are in many marine environments (and because waves and currents in lakes are small), the fine-grained sediments that accumulate in the centers of lakes are likely to remain well layered, just as they were when they were laid down (Figure 3-4).

Another clue to the recognition of lake deposits is close association with other nonmarine deposits, such as river sediments. It would be highly unusual to find lake deposits directly above or below deep-sea deposits, or even above or below sediments deposited on the continental shelf, unless the two types of deposits were separated by an unconformity.

GLACIAL ENVIRONMENTS

Even more useful than soils as clues to ancient climatic conditions are certain complex suites of depositional features that are known to develop only in particular climates. The sedimentary features associated with some types of glaciers, for example, are excellent indicators of cold climates. Glaciers that form in mountain valleys seldom leave

FIGURE 3-4   Lake deposits of the early Mesozoic age from The Newark Group of Connecticut. The even laminations are typical of sediments that have accumulated in lakes. The scale is 5 centimeters long. (*Photograph by the author.*)

FIGURE 3-5   Igneous rock smoothed and scratched by glaciers overlain by tillites of the Dwyka Group near Kimberley, South Africa. (*Courtesy of R. F. Flint.*)

enduring geologic records because the mountains through which they move, and on which they leave their mark, stand exposed above the surrounding terrane and therefore tend to suffer rapid erosion on a geologic scale of time. Not so continental glaciers, which leave legible records that survive for hundreds of millions of years.

Today, one continental glacier occupies most of Greenland, and an even larger one occupies nearly all of Antarctica (page 42). Each of these glaciers is thickest in the center, tapering toward the continental margin. Continental glaciers leave traces of their activity on the lowland areas over which they move, and these traces have led to the recognition of widespread intervals of continental glaciation not only in late Neogene time (i.e., during the past 3 million years or so) but also much earlier, during the Late Paleozoic, Ordovician, and Precambrian intervals. Let us now examine the telltale traces of continental glaciation.

As glaciers move, they erode rock and sediment, transporting both in the direction of flow. In this process, pieces of rock are embedded in the ice at the base of a glacier,

where they become tools of further erosion, commonly leaving deep scratches in the bedrock that serve to record the direction of glacial movement. When scratches such as these are found in an area that is presently free of glaciers, they represent excellent diagnostic indicators of ancient glacial activity. Figure 3-5 shows a glacial "pavement"—i.e., bedrock that was scratched and sculpted by glaciers—that formed in Africa more than 200 million years ago. Traces of glacial activity representing the same (late Paleozoic) ice age are also visible in rocks in South America, southern India, Australia, and Antarctica. As will be discussed more fully in Chapter 7, these traces provide evidence that these continental areas, which are now widely separated, once formed an enormous continent that lay above the South Pole and accumulated a large continental glacier.

Glaciers leave records not only of erosion but also of deposition. The mixture of boulders, pebbles, sand, and mud that is scraped by a moving glacier is deposited partly as the glacier plows along and partly when it melts. This heterogeneous material is called **till**. At the position of greatest glacial advance, till ploughed up in front of the glacier is left standing in ridges known as **moraines**. A considerable amount of till, however, after being transported for some distance, is deposited beneath glaciers; thus, till may be deposited over a broad area even as glaciers continue to melt back. In front of a moraine, sediment from a melting glacier is often deposited by streams of meltwater issuing from the retreating mass of ice. Here, the sediment tends to be sorted by size into layers of gravel, cross-bedded sand, and mud, forming well-stratified glacial material known as **outwash**. Figure 3-6 shows till resting on top of outwash in an area of Illinois that was glaciated during the recent (Pleistocene) Ice Age. The lithified till, known as **tillite**, of Figure 3-6*B* rests on the Late Paleozoic glaciated surfaces shown in Figure 3-5. The scratches that can be seen in Figure 3-5 have more than one orientation, indicating that the surface was scoured by more than one glacial advance. Tills that were formed during the earlier glacial retreats must have been removed by subsequent advances. The last glacial retreat must therefore have produced the layer of tillite that blankets parts of the glaciated surface today. This tillite has been protected for millions of years by younger sediments that cover it. Till is not a common type of sediment; thus, when till-like sediment is found to contain scratched cobbles or boulders or to rest on scratched bedrock, its identity is obvious.

In many instances, the meltwaters that issue from a glacier converge to form a lake that sits in front of the

FIGURE 3-6   Glacial till. *A.* Gravelly till of northeastern Illinois that is still unconsolidated, having been deposited during the Pleistocene Epoch; this till rests on cross-bedded outwash deposits. *B.* Similar till of South Africa that has been consolidated into hard rock, having been deposited about 250 million years ago, near the end of the Paleozoic Era. (*A. J. M. Masters, Illinois Geological Survey. B. W. Hamilton, U.S. Geological Survey.*)

FIGURE 3-7   Varved clays. The varves are layers of clay and silt deposited in quiet lake waters. They are present in pairs, one representing deposition during the winter and the other deposition during warm weather. The winter layer is finer grained, consisting of dark clay that settled from the water while ice covered the lake. By counting varves that formed in glacial lakes of Europe, geologists in the early part of this century estimated that Pleistocene glaciers retreated from southern Europe about 8700 years ago. Subsequent calculations based on the rate of decay of radioactive carbon have confirmed this estimate. *A.* Varved clays produced during the Pleistocene Epoch at Toronto, Canada. *B.* Precambrian varves that are about 2 billion years old in southern Ontario, Canada. (*A. Courtesy of P. F. Kerrow, Ontario Department of Mines. B. Courtesy of F. J. Pettijohn.*)

glacier. The sediment that accumulates in such lakes generally consists of alternating layers of coarse and fine sediment called **varves** (Figure 3-7). Each coarse layer forms during summer months, when streams of meltwater carry sand into the lake, whereas each fine layer is formed during the winter time, when the surface of the lake is sealed by ice and all that accumulates on the lake bottom is clay and organic matter that settle slowly from suspension. Each pair of layers in varved sediment thus represents a single year's deposition—a feature that allows geologists to count the number of years represented by a series of layers. In some areas, thousands of years of deposition have been tallied in this manner.

When a glacier encroaches on a lake or ocean, pieces of it break loose and float away as icebergs, some of which are immense (Figure 3-8). As chunks of glacial ice melt in a lake or an ocean, their sediment sinks to the lake or

FIGURE 3-8   Icebergs floating to the open sea after breaking loose from the end of a glacier flowing into a fjord (drowned valley) in eastern Greenland. Icebergs such as these carry coarse sedimentary debris that settles to the sea floor when the icebergs melt. (*Geodetic Institute, Copenhagen.*)

FIGURE 3-9   A large isolated boulder (right of hammer) dropped into fine-grained sediments of Late Precambrian age in Utah, apparently from melting glacial ice. The boulder exhibits scratches that were probably incurred when it was frozen into the base of a glacier. (*From M. D. Crittenden et al., Geol. Soc. Amer. Bull. 94:437–450, 1983.*)

ocean bottom, creating a highly unusual deposit in which pebbles or even boulders rest in a matrix of finer sediments (Figure 3-9). Unlike the tightly packed, coarse material that characterizes glacial till, these **dropstones** occur either singly or scattered throughout the matrix. Very few natural mechanisms other than this so-called ice rafting bring large stones to the middle of a lake or to a sea floor that lies far from land.

When sand grains are trapped beneath a large glacier, they are subjected to enormous pressure from the weight of the moving glacier, resulting in breakage that follows characteristic patterns. In comparison to sand grains found in deserts or along beaches, glacial sand grains tend to exhibit high surface relief (Figure 3-10). Most of these surfaces also display parallel steps and ridges as well as depressions with smoothly curved margins. Although these features can help geologists identify a variety of ancient glacial deposits, they must be used with caution, since glacially transported sand that has moved quickly to a nonglacial environment and has been rapidly buried may undergo little modification and may thus retain its glacial surface texture: the glacier did not necessarily pass over the site where the sand was ultimately deposited.

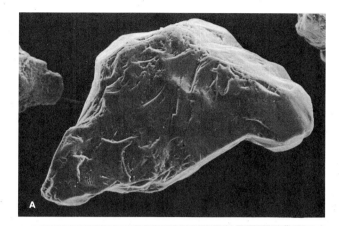

FIGURE 3-10   Scanning electron micrographs of quartz sand grains exhibiting surface textures characteristic of fracturing beneath a glacier. *A.* A grain from glacial deposits of the most recent (Pleistocene) interval of continental glaciation ($\times 105$); note the curved and steplike fracture surfaces. *B.* A closer view of the surface of a similar grain of late Paleozoic age from South Africa ($\times 3600$). (*Courtesy of D. H. Krinsley.*)

that has been ground from larger grains and does *not* contain any of the clay minerals that are commonly found in nonglacial rocks Appendix I. Another important feature is the presence of flat facets on pebbles and boulders; these flat surfaces form when a pebble or a boulder that is frozen into the base of a glacier suffers abrasion along its exposed surface.

DESERTS AND ARID BASINS

Like frigid glacial environments, arid and semiarid basins are characterized by a unique suite of sedimentary deposits. Today, deserts are found not only in tropical regions located between about 20 and 30° of the equator, where dry air descends and absorbs moisture from the earth, but also at slightly higher latitudes in areas far from oceans. Deserts also occur in regions that lie in the rain shadows of mountains.

Deserts contain little or no humic soil, since the dry conditions that characterize deserts support little vegetation, the source of humus. The rain that occasionally falls in deserts leads to erosion and the deposition of sediment, often carrying the chemical products of weathering to desert basins. The subsequent precipitation of evaporite minerals in these basins sets arid regions apart from those with moist climates. In the latter, permanent streams flow great distances without being absorbed into the soil or evaporating into the air; thus, they are frequently capable of reaching the ocean by passing into large rivers. As a consequence, we speak of most humid regions as having **exterior drainage**, or drainage to areas beyond their borders. The runoff from arid regions, in contrast, is too sparse and intermittent to form permanent streams and rivers, and the result is **interior drainage**—a situation in which streams die out through evaporation and through seepage of water into the dry terrane.

Lakes in areas with interior drainage also tend to be temporary and are known as **playa lakes**. Temporary streams bring dissolved salts to these lakes along with suspended sediment, and as the lakes shrink by evaporation, the salts precipitate as evaporite deposits.

Death Valley: A modern example

Death Valley, which is located in Nevada and California, is an excellent example of an arid basin. Death Valley is,

There are additional products of physical abrasion that offer clues to the identity of glacial deposits. One such clue is the presence of **rock flour** between the larger grains of a conglomerate. Rock flour consists of mud-sized material

FIGURE 3-11   Death Valley, California. Alluvial fans spread toward the center of the valley in the upper part of the picture. Braided-stream channels occupy the lower slopes of the fans, and larger braided-stream channels intertwine along the valley floor. White areas in the middle part of the picture represent evaporite deposits on the valley floor. (*C. B. Hunt, U.S. Geological Survey.*)

in fact, the hottest and driest basin in the United States; temperatures of 50°C (~120°F) are not uncommon here, and the average annual rainfall is only about 4 centimeters (~1.6 inches) per year. Because part of the basin floor lies more than 80 meters (~265 feet) below sea level, exterior drainage from Death Valley to the nearby Pacific Ocean is impossible.

Figure 3-11 illustrates a number of ways in which Death Valley typifies arid basins in general. Temporary streams carry sediment down valleys incised into the naked rocks of nearby highlands to form low cone-shaped structures called **alluvial fans**, which spread out onto the floor of Death Valley. These structures form where a mountain slope meets a valley floor, causing streams to slow down and drop much of their sediment. Alluvial fans consist of poorly sorted sedimentary particles that range from boul-

ders to sand near the source area and from sand to mud on the lower, gentler slopes. Most alluvial fans include broad **braided-stream deposits**, which consist of coarse, cross-stratified sediments laid down in a complex network of channels (or **braided streams**), that carry water and sediment during the infrequent rainy intervals. (Alluvial fans and braided stream deposits will be discussed more fully under "River Systems," later in this chapter.) The braided streams of these channels lead to the center of the basin, where their occasional flow of water forms temporary lakes. As the lake waters dry up, evaporite minerals accumulate. The same minerals also accumulate on those parts of the basin floor where groundwater seeps to the surface and evaporates. The evaporites of Death Valley are composed primarily of halite, gypsum, and anhydrite. The alternate wetting and drying in this basin produces large polygonal

FIGURE 3-12   Mud cracks in salt-encrusted mud on the floor of Death Valley. (*J. R. Stacy, U.S. Geological Survey.*)

mud cracks in many areas (Figure 3-12). Around the margins of these "saltpans," calcium carbonate deposits form caliche.

The Green River Formation: Dry-basin deposits of the past

The Green River Formation of Wyoming is renowned for its fine-grained sediments known as **oil shales**, which are rich in organic matter that can be artificially extracted in the form of petroleum (Figure 3-13). These and related deposits date back about 50 million years to the Eocene Epoch, when they formed within basins to the east of the Rocky Mountains and related uplifts. The oil shale itself is a well-laminated sedimentary unit whose dark layers, the ones containing organic matter, alternate with thicker, light-colored layers. This distinct layering of the shale, together with the fossil fishes that are sometimes found within it, suggests that a lake was the site of its deposition. The layers that are rich in organic matter appear to have formed

from blue-green algae that flourished periodically as mats on the lake bottom. Associated with the oil shale are mud-cracked deposits that represent periods in which the lake shrank in size or dried up. Also present in the mud-flat deposits are beds of evaporite minerals that represent dry spells. The fact that lake deposits were more common than evaporites at the basin's center suggests that the annual rainfall was greater in the depositional setting of the Green River Formation than it is in the present century in the vicinity of Death Valley, where lakes are currently small and short-lived.

The Wilkins Peak Member of the Green River Formation, which lies in the Bridger Basin of Wyoming, includes siliciclastics in the form of cross-bedded bodies of rock that truncate each other. These rocks appear to be channel fillings formed by braided streams that frequently changed course during rainy periods, when they deposited particularly large volumes of sediment. Finally, alluvial fans have been detected near the basin margin, where gravel and sand were shed from the Rocky Mountains to the west. (Figure 3-13).

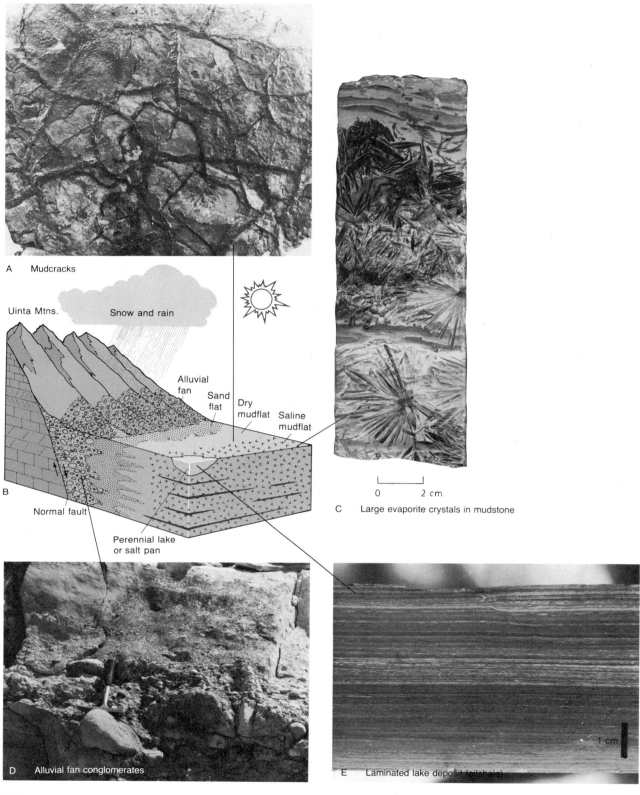

A   Mudcracks

Uinta Mtns.

Snow and rain

Alluvial fan

Sand flat

Dry mudflat

Saline mudflat

B

Normal fault

Perennial lake or salt pan

C   Large evaporite crystals in mudstone

0   2 cm

D   Alluvial fan conglomerates

E   Laminated lake deposit (oil shale)

1 cm

FIGURE 3-13   Rock types and inferred depositional environments of the Wilkins Peak Member of the Eocene Green River Formation in Wyoming. The sediments appear to have accumulated in an arid depositional basin, as shown in the reconstruction. The basin was bordered by a normal fault; alluvial fans of conglomeratic sediment extended into the basin from the highlands west of the fault and passed into evaporite-bearing mud flats. Lakes intermittently occupied the center of the basin, accumulating well-laminated sediments, rich in organic matter, that eventually turned into oil shales. (*Diagram after H. P. Eugster and L. A. Hardie, Geol. Soc. Amer. Bull. 86:319–334, 1975; evaporite photograph courtesy of W. H. Bradley; other photographs courtesy of J. Smoot.*)

Thus, it can be seen that the depositional environments of the Bridger Basin about 50 million years ago resembled those present today in Death Valley. Coarse alluvial-fan sediments near the basin margin graded into braided-stream deposits. The latter gave way to mud flats bordering a lake setting where, during moist periods, sediments were carried in by streams and accumulated in even layers in quiet water. During dry periods, the lakes shrank, leaving evaporites. The main difference between the modern and ancient examples is that rainfall in Death Valley is so meager that lakes are temporary, producing larger volumes of evaporites and mud-cracked sediments than well-laminated lake deposits. Braided-stream deposits are also less well developed in the arid setting of Death Valley than in the ancient environment.

Sand dunes

Sand dunes are hills of sand that have been piled up by the wind (page 60). In some desert areas, including Death Valley, dunes occupy less than 1 percent of the total area— but in areas where dunes are well developed, they form magnificent landscapes. Dunes are familiar sights not only in desert terranes but also landward of the sandy beaches that border oceans and large lakes. Dunes form in the presence of wind and an ample supply of loose sand; when wind velocity and sand supply are adequate, a dune can begin to form over any obstacle that creates a wind shadow in which sand can accumulate. Figure 3-14 illustrates how a dune tends to "crawl" downwind as sand from the upwind side moves over the top of the dune and accumulates on the downwind side. As the prevailing wind direction shifts back and forth, the direction of dune migration shifts as well. This usually leads to the truncation of preexisting deposits, which often causes a new set of beds to accumulate on a curved surface cut through older sets. Thus, dunes are characterized by **trough cross-bedding**. Figure 3-14 shows an idealized cross section through a dune as well as a real cross section through an enormous lithified dune that is more than 200 million years old.

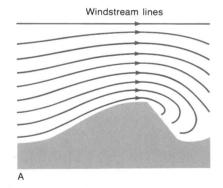

Windstream lines

A

B

C

FIGURE 3-14   Internal structure of dunes. *A.* The wind stream, shown by streamlines, becomes compressed above a dune and, as a consequence, increases in velocity. The dune ceases to grow taller when its height becomes so great that it causes the wind stream to move rapidly enough to transport sand. As sand passing over the dune accumulates on the steep leeward slope, the dune begins to "crawl" in that direction. *B.* Diagrammatic cross section of a dune, showing the crossbedding that results from shifting wind. As the wind direction changes, the shape of the dune is altered by removal of sand, and a new leeward slope forms. *C.* Cross-bedding in Permian dune deposits of the New Red Sandstone of southern Scotland. These giant cross beds are exposed in a quarry wall. (*Institute of Geological Sciences.*)

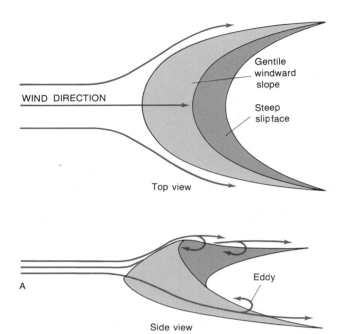

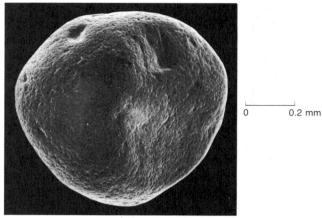

FIGURE 3-16   A rounded, frosted grain of desert sand from a Libyan sand dune. This grain has developed its external character by abrasion from other grains in a dune environment. (*Courtesy of D. H. Krinsley.*)

B

FIGURE 3-15   The formation of a barchan dune, a crescent-shaped dune with "horns" that face downwind. *A.* Drawings show the geometry of an idealized barchan dune. Sand moves up the gentle windward slope and is deposited on the steep slip face. *B.* A Permian dune in southwest Scotland, which shows the bedding of slip-face deposits. Measurements of the orientation of slip-face deposits and of horns of dunes like these in southwest Scotland reveal that the ancient wind direction was from the northeast. (*From K. W. Glennie, Desert Sedimentary Environments, Elsevier, Amsterdam, 1970.*)

Measurements of the orientation of cross-bedding in ancient dunes often serve to indicate the prevailing direction of ancient winds. Figure 3-15, for example, shows an ancient barchan dune—that is to say, a dune that is crescent-shaped in top view, with the "horns" of the crescent pointing downwind. Although individual measurements for this dune differ to some extent because the wind that formed the dune shifted somewhat through time, the prevailing direction of the wind in this example was clearly northeasterly.

Most sand dunes are composed primarily of well-sorted quartz sand, and grains of quartz sand that have spent considerable time in a dune environment tend to become well rounded and frosted (Figure 3-16). This frosted appearance is a product of the microscopic pitting that results when sand grains strike each other at high velocity. A frosted surface can therefore aid in the identification of ancient dune sands. It must be recognized, however, that grains frosted in a dune environment can find their way into aquatic environments. Thus, the presence of frosted grains does not necessarily imply the presence of a dune where the grains are found.

RIVER SYSTEMS

In the previous section, we considered depositional systems in areas of low rainfall. Higher rainfall is more likely to create the features of exterior drainage—small streams

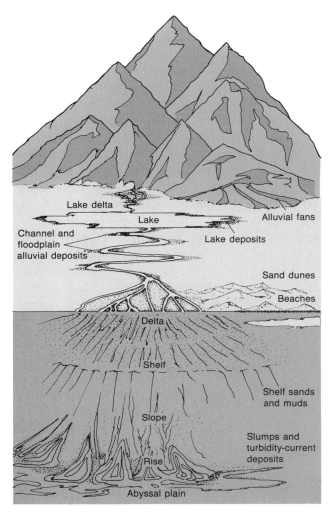

FIGURE 3-17   An idealized representation of the downhill course of sediment transport from mountains to the sea. Along the way, sediment is trapped in many different depositional environments. (*After F. Press and R. Siever, Earth, W. H. Freeman and Company, New York, 1982.*)

that meet to form larger streams, which in turn flow into still larger rivers. Ultimately, the waters of most large rivers reach the sea. Figure 3-17 summarizes the sequence of aquatic environments through which water typically passes as it moves from the headwaters of river systems in hills or mountains to the deep sea, transporting and depositing sediment in the process. Each of these depositional environments is represented by many different sedimentary units in the geologic record. In the remainder of this chapter and in the next one we will describe the diagnostic features of ancient deposits formed in each of these environments— features that enable us to recognize each type of deposit as far back in the geologic record as a few hundred million or even 2 or 3 billion years.

Alluvial fans in moist climates

Sediment in the form of alluvial fans accumulates at the feet of mountains and steep hills in both moist and arid regions. In moist climates, however, water often spreads these coarse sediments farther from their source, producing fans that slope more gently than those of dry basins.

Alluvial fans in ancient rocks are distinguished by their geometry, their sedimentary structures, and the nature of their component sediments. Figure 3-18 illustrates a typical

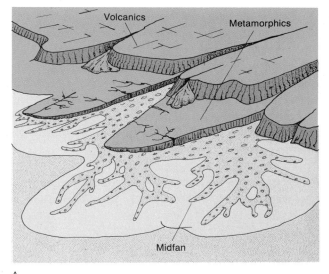

A

B

FIGURE 3-18   Evidence of an alluvial-fan system that existed about 600 million years ago in western Texas. *A.* Reconstruction of the depositional setting. Canyons were cut through volcanic and metamorphic rocks, and coarse, conglomeratic sediments accumulated in them. In front of the highlands, in the midfan region, finer sediments were deposited in and around braided-stream channels. *B.* Well-sorted channel-filling gravels deposited in the upper part of the alluvial fan. (*J. H. McGowan and C. G. Groat, University of Texas Bureau of Econ. Geol. Rept. of Investig. 72, 1971.*)

alluvial fan—one of a group of late Precambrian or early Paleozoic fans that formed in western Texas, New Mexico, and Mexico. These ancient fans are thickest to the north, where they lie in contact with the older rocks from which their sediments were shed, and thin away from their source area. In addition, the grain size of their sediments decreases away from the source area, grading southward within each fan from gravels that dropped from waters near the source to gravels and coarse sands carried by less swiftly flowing waters in the midfan region to coarse sands near the thin edge of the fan.

Channel deposits are found throughout the ancient-alluvial-fan deposits of Figure 3-18. These deposits contain coarse sediments that formed both in the midfan region and, further down the slope, in braided streams that extended beyond the alluvial fans to lower ground. Because braided streams are typical features of such environments, they merit consideration in greater detail.

Braided-stream deposits

Runoff water moves rapidly over the relatively steep, unvegetated terrane that is commonly found on the lower reaches of an alluvial fan and in the region just beyond it. As a result, this water transports large volumes of coarse sediment. In such an environment, a river does not flow as a simple winding waterway like the Mississippi or the Thames. Instead, it follows a complex network of interconnected channels, which give it the name *braided stream* (Figures 3-19 and 3-20). The areas between the channels are elevated "bars" of coarse sediment.

Although braided streams associated with alluvial fans run rapidly because of their steep gradients, their flows tend to vary greatly. Further downstream, a river is the product of so many tributaries (or **feeder streams**) that its rate tends to vary less; when one part of its drainage receives little rain, the chances are that some other part will be well supplied. In contrast, the rate at which a braided river flows close to the headwaters depends on what happens close at hand; strong rains here produce a rapid, strong flow, whereas drought leaves nothing but a trickle of water. At times of strong water flow, the coarse sediment of a braided stream is carried rapidly. The pattern of braiding then changes markedly, since some bars enlarge to the point of choking channels, forcing the flowing water to cut new ones.

FIGURE 3-19   The Muddy River of Alaska, a braided stream that has formed in front of a melting glacier. The stream is choked with coarse glacial sediment released by the meltwater. (*Courtesy of B. Washburn.*)

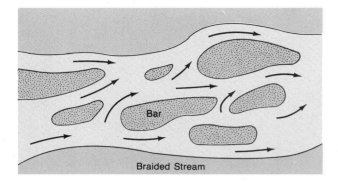

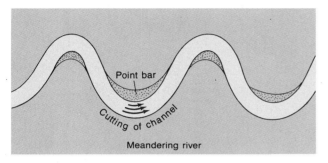

FIGURE 3-20   Map views contrasting the path of a braided stream with that of a meandering river. The arrows show current directions. In a braided stream, the current is divided into many interconnected channels by shifting bars, which are "islands" of coarse sediment. In a meandering stream, the current flows most rapidly near the outside of a bend, as indicated by the long arrow. The fast current on the outside of a bend cuts into the river bank, causing the river to migrate laterally; on the inside of the bend, where the current moves slowly, sediment accumulates to form a point-bar deposit (see Figure 3-22).

The sediments deposited in braided streams consist almost entirely of sand and gravel. Little mud is present because mud tends to wash downstream in the fast currents. Braided-stream deposits and alluvial-fan deposits, which frequently intergrade with one another, are unusual in this regard; few other environments accumulate coarse-grained sediment in such great concentrations.

The coarse sediments of braided-stream deposits are often sorted according to grain size, a feature that helps geologists distinguish such deposits from the glacial till of a moraine. Braided streams form not only in regions adjacent to rapidly eroding highland areas but also in areas where melting glaciers yield large quantities of sediment (Figure 3-19). The gravels positioned below the till in Figure 3-6 were probably deposited in a braided stream that lay

in front of an advancing glacier. It would appear that this glacier moved over the river deposit and then halted, dropping its load of sediment on the stream deposits as poorly sorted, poorly bedded till.

Meandering rivers

Many rivers occupy solitary channels that wind back and forth like ribbons (Figure 3-21). Unlike braided streams, these **meandering rivers** are not choked with sediment—that is to say, the rate at which sediment is supplied to them is low in comparison to the rate at which the water flows. Rivers usually meander most actively in regions far from uplands and in areas characterized by gently sloping terrane. If there is any irregularity in this terrane, the river's path will curve, and centrifugal force will then cause the water to flow most rapidly near the outside of the bend and least rapidly near the inside (Figure 3-20). If this happens, the river will tend to cut into the outer bank of the bend, where the current is swift. On the inside of the bend, the current is so weak that deposition rather than erosion prevails. Sediment in this area accumulates to form what is known as a **point bar.**

The sediment that forms point bars is usually made of sand. Most of this sand is cross-bedded, since large ripples along the river bed where it accumulates migrate in the course of time. In deeper water, where the current is stronger, the sediment in the river channel is coarser (Figure 3-22). Gravel is often found along with coarse sand in the deepest part of the channel, and this gravel moves only at times when the river is flowing strongly.

Because mud tends to move downstream, very little of it accumulates within a meandering-river channel. At flood stage, however, the river overflows its banks, carrying fine sediment laterally to the lowlands adjacent to the river channel. In these areas, which are known as **backswamps**, the spreading floodwaters flow slowly, allowing mud to accumulate before the waters recede. In keeping with the normal pattern of sediment deposition, these floodwaters become progressively slower as they flow away from the channel, and thus they tend to drop the coarser portion of their suspended sediment before they spread far from the channel. Sand is therefore dropped first, followed by silt, and together they form gentle ridges called **natural levees**, which border the channel. Because natural levees and backswamps are inundated only periodically, they tend to dry out and to display mud cracks, many of which have been preserved in the stratigraphic record. Levees and backswamps also become populated by moisture-loving

FIGURE 3-21   The meandering Brazos River of Texas. As the river swings outward in rounding a bend, it cuts its channel on the outside of the curve. As shown in Figures 3-20 and 3-22, deposition occurs in the inside of the channel. (*Courtesy of R. J. Leblanc.*)

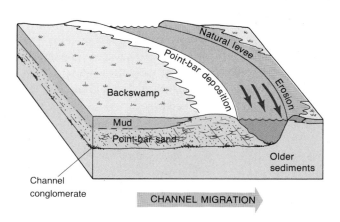

FIGURE 3-22   Deposition of sediment by a meandering river. At each bend, the river migrates outward. The current flows most rapidly on the outside of a bend (*long arrow*); thus, the water cuts into the outer bank. On the inner side of the bend, where the current moves slowly, sand accumulates to form a point bar. As the channel migrates outward, the point-bar sands advance over the coarser, cross-bedded sands deposited within the original channel. Pebbles often accumulate at the base of the channel. Muddy backswamp deposits, which form when the river floods its banks, migrate in their turn over the point-bar sands. The result of this shifting of environments is a sequence in which coarse sediments at the base grade upward into fine sediments at the top.

FIGURE 3-23 Depositional cycle at Ludlow, England, produced by a meandering river. On the right is a diagrammatic section of the vertical sequence of sediments within the cycle showing the trend from coarse sediments at the base to fine sediments at the top. Compare to Figure 3-22. (*Modified from J. R. L. Allen, Sedimentology 3:163–198, 1964.*)

plants, which may leave traces of their roots in the rock record. If after death these plants accumulate in large quantities, they may even leave deposits that eventually turn into coal.

In addition to illustrating the features just described, Figure 3-22 shows how a particular vertical sequence of sediments characterizes meandering-river deposition. This sequence progresses from coarse channel deposits at the base to cross-bedded point-bar sands in the middle to muddy backswamp deposits at the top. Levee sediments sometimes lie between the point-bar and backswamp deposits. In summary, the sequence passes from coarse-grained sediments at the bottom to fine-grained sediments at the top. The reason this coarse-to-fine sequence forms is that, as the channel migrates laterally, the point bar builds out over deeper-channel gravels and the backswamp shifts over older point-bar deposits. As the channel meanders across a broad area, this sequence of deposits forms a broad composite depositional unit. The meandering-river sequence illustrated in Figure 3-22 illustrates **Walther's law**, which states that when depositional environments migrate laterally, sediments of one environment come to lie on top of sediments of an adjacent environment.

In a broad basin that happens to be subsiding, a river may migrate back and forth over a large area many times, piling one coarse-to-fine composite depositional unit on top of another as it does so. Each of these composite units, or **cycles**, lies unconformably on the one beneath it, since the migrating channel in which the basal deposits accumulate removes the uppermost sediments of the preceding cycle. Sometimes, however, a channel cuts deep into the sediments of the preceding cycle, removing not only the uppermost deposits but some lower deposits as well. Thus, many of the cycles that have been preserved in the geologic record of a migrating channel are really only partial cycles.

Figure 3-23 shows a meandering-river cycle in the Devonian System of England. The migrating channel of this cycle removed only the upper part of the backswamp deposits of the preceding cycle. Figure 3-24 represents two of a large sequence of meandering-river cycles exposed in eastern Pennsylvania. Some of these Carboniferous cycles include backswamp deposits that consist of coal, representing a time in the earth's history when the swampy lowlands of many regions were covered with treelike plants whose logs accumulated in large numbers and, after burial, turned to coal.

No river deposits its entire load of sediment in the river valley. In the chapter that follows, we will examine the depositional pattern that results when a river finally discharges its water and remaining sediment into a standing body of water such as the sea.

Shale and coal (overbank)	Sands (point bar)	Cross-bedded coarse sands (channel)
Top ←		Bottom

FIGURE 3-24   Cyclical deposits at Pottsville, Pennsylvania, produced by a meandering river during the Pennsylvanian Period. The beds were tilted to a nearly vertical orientation during Appalachian mountain building. The backswamp deposits forming the upper parts of the cycles here include not only shale but also coal, which formed from the fossil remains of swamp-dwelling trees. A complete cycle (2) is visible in the center of this picture, along with the upper part of the preceding cycle (1) on the right, and the lower part of the succeeding cycle (3) on the left. (*Photograph by the author.*)

CHAPTER SUMMARY

1.   Modern environments where sediment is deposited offer valuable examples of the ways in which ancient sedimentary rocks formed. Sometimes one kind of rock alone serves to identify an ancient environment. Often, however, suites of closely associated rock types are required for this purpose, and these are commonly organized into cycles in which one kind of sediment tends to lie above another.

2.   Ancient soils are sometimes found beneath unconformities, although they may be hard to identify.

3.   Lake deposits, which are much less common than marine deposits, are characterized by horizontal bedding and by an absence of marine fossils.

4.   Glaciers, which plough over the surface of the land, often leave a diagnostic suite of features, including scoured and scratched rock surfaces, poorly sorted gravelly sediment, and associated lake deposits that exhibit annual layers.

5.   In hot, arid basins on the land, erosion of the surrounding highlands creates gravelly alluvial fans. Braided streams flowing from the fans toward the basin center deposit cross-bedded gravels and sands. Beyond these, there may be shallow lake and salt flats where evaporites accumulate. Some arid basins also contain dunes of clean, cross-bedded, wind-blown sand.

6.   In moist climates, meandering rivers leave characteristic deposits in which channel sands and gravels grade upward through point-bar sands to muddy flood-plain sediments.

EXERCISES

1. What kinds of nonmarine sedimentary deposits reflect arid environmental conditions?

2. What kinds of nonmarine sedimentary deposits reflect cold environmental conditions?

3. What kinds of deposits indicate the presence of strong topographic relief in the vicinity of a nonmarine depositional basin?

4. In what nonmarine settings do gravelly sediments often accumulate?

5. Make a drawing to illustrate how a meandering river produces a sedimentary cycle in accordance with Walther's law.

ADDITIONAL READING

Reading, H. G., *Sedimentary Environments and Facies*, Elsevier Publishing Company, New York, 1978.

Reineck, H. E., and I. B. Singh, *Depositional Sedimentary Environments*, Springer-Verlag New York Inc., New York, 1975.

Selley, R. C., *An Introduction to Sedimentology*, Academic Press, Inc., London, 1976.

CHAPTER 4

Marine Sedimentary Environments

In the preceding chapter, our review of nonmarine depositional settings took us on a downhill journey starting with the alluvial fan that forms at the foot of a mountain and ending in a lowland river valley. In effect, we moved down the slope of the hypsometric curve, the graph which depicts the percentages of the earth's surface lying at various positions above and below sea level (Figure 2-1). We will now continue this downhill trip, first examining sedimentary environments along the edge of the sea and then moving toward the deep ocean floor.

DELTAS

If a river does not flow into a larger one, it usually empties either into a lake or into the sea. In doing so, its current dissipates, and it often drops its load of sediment in a fanlike pattern. The depositional body of sand, silt, and clay that is formed in this way is called a **delta** because of its resemblance to the Greek character Δ. Most of the large deltas that have been well preserved in the geologic record formed in areas where sizable rivers emptied into ancient seas.

In Chapter 3, we saw that as moving water slows down, it first drops sand, followed by silt and then by clay. As river water mixes with standing water and begins to slow down, it, too, loses sand first. Silt, which is finer, spreads farther from the mouth of the river, and clay is carried even farther. The typical result is a delta structure that includes **delta plain**, **delta-front**, and **prodelta** deposits (Figure 4-1).

Delta-plain beds, which consist largely of sand and silt, are nearly horizontal except where they are locally cross-bedded. Some delta-plain deposits accumulate within channels. As a river slows down on the surface of the delta, sand builds up on the bottom, causing the river channel to branch repeatedly into smaller channels that radiate out from the mainland. These **distributary channels** are floored by cross-bedded sands. Sand also spills out from the mouths of the channels, forming shoals and sheetlike sand bodies along the delta front. Between the distributaries and separated from them by levees are swamps, which are sometimes dotted by lakes. Here, as in the overbank areas of meandering rivers, muds accumulate and marsh plants often grow, contributing to future coal deposits.

Delta-front (or foreset) beds slope seaward from the delta plain, usually lying in waters that are deeper than those agitated by wind-driven surface waves. These beds consist largely of silt and clay, which can settle under these quiet conditions. Since they lie fully within the marine system, delta-front muds harbor marine faunas that often leave fossil records, but these muds usually contain fragments of waterlogged wood as well. In fact, the presence of abundant fossil wood in ancient marine muds testifies to the presence of both land and a river system near the site of preservation. In short, most ancient subtidal muds that harbor abundant fossil wood debris represent deltaic deposits.

Aerial view of vast ooid shoals beneath the clear waters of the Great Bahama Bank near Cat Cay. Ooids are nearly spherical, sand-sized grains that form on tropical sea floors agitated by waves. The ooids grow while rolling around on the sea floor and are often swept into submarine shoals or dunes by the waves. (*Shell Development Company*.)

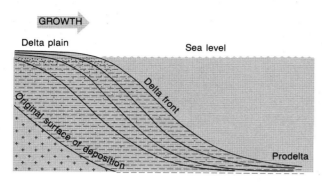

FIGURE 4-1   Diagrammatic cross section of a delta. As river water flows into the sea, it slows down. First sand drops from suspension in the delta plain region of the delta; then silt and clay settle on the delta front; and finally clay settles in the prodelta. As the delta grows seaward, the sandy shallow-water sediments build out over the finer-grained sediments deposited in deeper water.

Spreading seaward at a low angle from the lowermost delta-front deposits are the prodelta (or bottomset) beds, which consist of clay. Even during floods, the freshwater that spreads from distributary channels slows down so rapidly that it loses its silt in the region of delta-front accumulation. It is partly because freshwater is less dense than saline marine water that clay is carried far from distributary mouths. Because the freshwater floats on top of the more dense marine water, it does not mix in quickly; instead, it spreads seaward for some distance, carrying much of its clay.

As a delta **progrades**, or grows seaward, relatively coarse delta-plain deposits build out over finer-grained foreset beds in accordance with Walther's law. These delta-front beds build out over still finer-grained bottomset beds, resulting in a sequence of deposits that coarsens toward the top. As we will now see, the stratigraphic expression of this upward-coarsening sequence varies in accordance with the nature of the delta. We will first consider the nature of a delta that grows into a relatively quiet sea, and we will then examine one that grows seaward in the presence of strong waves.

The famous Mississippi River delta is currently building out into the Gulf of Mexico in an area that is protected from strong wave action. As a result, this delta projects far out into the sea. Because construction of the delta from river-borne sediment prevails decisively over the destructive forces of the sea, this type of delta is sometimes called a **river-dominated** delta.

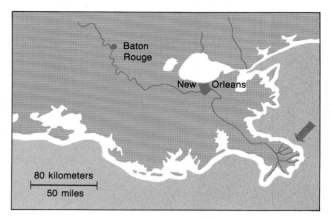

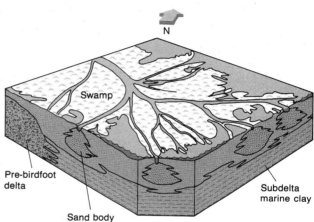

FIGURE 4-2   Structure of the "bird-foot" lobe of the modern Mississippi delta. This lobe, which is identified by the arrow on the map, is the location of active shallow-water deposition where the river meets the sea. Here, as shown in the block diagram below, the river channel divides into many distributary channels. Swamp deposits accumulate between the distributaries. Bodies of sand accumulate in front of the distributaries where the river water meets the ocean and slows down. As the bird-foot delta builds seaward, the distributaries extend over these sand bodies. (*Modified from H. N. Fisk et al., Jour. Sedim. Petrol. 24:76–99, 1954.*)

That portion of the Mississippi delta which is growing at any given time is much smaller than the delta as a whole. This growing portion, or **active lobe**, is the site of the currently functioning distributary channels (Figure 4-2). Many previously active lobes can be identified in the delta-plain portion of the Mississippi delta, and these lobes, which have been dated by the carbon 14 method (see Chapter 5), provide a history of deltaic development during the past few thousand years (Figure 4-3). It is known, for example, that the site of depositional activity (or lobe growth) periodically shifted in this delta when floods caused the river

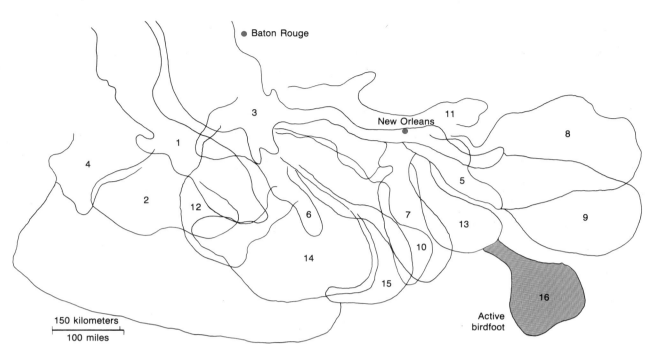

FIGURE 4-3   Lobes of the Mississippi delta. The active bird-foot lobe is numbered 16. Older lobes that are now inactive are numbered 1 through 15, with number 15 being the youngest. The older lobes are now settling, and some are already partially or entirely submerged beneath the sea. (*After T. Elliott, in H. G. Reading [ed.], Sedimentary Environments and Facies, Blackwell Scientific Publications, Ltd., Oxford, 1978.*)

to cut a new channel and to abandon the previously active channel and its distributaries.

The fate of an abandoned delta lobe is, in fact, the key to understanding the stratigraphic sequence produced by a river-dominated delta. In short, an abandoned lobe gradually sinks. It does so for two reasons: first, the sediments from which it is formed compact under their own weight; and second, the lobe is part of the entire delta structure, which is also constantly sinking as a result of the isostatic response of the underlying crust to the weight of the continually growing mass of sediment. After an abandoned lobe settles, a younger lobe will eventually grow on top of it. Each lobe will then consist of the typical upward-coarsening sequence. The result is an accumulation of cycles that differ markedly from those of meandering rivers, which, it will be recalled, become finer-grained toward the top.

Some deltaic cycles in the rock record lack tops or bottoms as a result of erosion that occurs before another cycle is superimposed (Figure 4-4). Figure 4-5 shows the contact between two cycles of Devonian age that have been interpreted as deltaic deposits. Because the delta-building activity of a river-dominated bed is limited to the active lobe at any given time, beds representing such a delta can seldom be traced very far laterally in the rock record.

Two African rivers—the Niger, which flows into the Atlantic, and the Nile, which flows into the Mediterranean—are currently forming deltas that have been less successful than the Mississippi in overcoming the destructive forces of the marine realm (Figure 4-6). These **wave-dominated** deltas are currently forming in areas where wave activity is so strong that it prevents a small active lobe from growing out into the sea. Instead, much of the delta as a whole is traversed by active distributary channels, and the sand that these channels transport to the sea is quickly spread to form beaches and transverse bars. Seaward of these is a zone of silt that grades into finer-grained mud. In effect, the entire delta consists of a single active lobe. This lobe progrades as a unit, representing one large, upward-coarsening stratigraphic sequence.

Ancient deltaic deposits that consist of many cycles are important sources of petroleum, which tends to become trapped in the topset sands of one cycle beneath the impermeable muds at the base of the succeeding cycle. Many buried Cenozoic deltas bordering the Gulf Coast, for example, have yielded large quantities of petroleum.

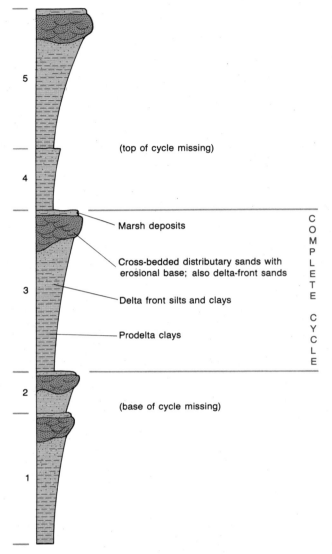

FIGURE 4-4   An idealized stratigraphic section representing five deltaic cycles. Cycle number 3 is a complete cycle; within it, bottomset clays pass upward to topset sands. Each cycle represents an accumulation of sediments resulting from the seaward growth of a deltaic lobe.

THE BARRIER ISLAND–LAGOON COMPLEX

Because deltas form exclusively at river mouths, they occupy only a small percentage of the total shoreline of the world's oceans. Much longer stretches of shoreline are fringed by **barrier islands**, which are linear structures composed largely of clean sand that has been piled up by

waves. Although there are barrier islands that extend laterally from some wave-dominated deltas, most barrier islands derive their sand not from neighboring rivers but from the marine realm itself. They are built by shallow currents called **longshore currents**, which winnow sediments and sweep sand parallel to the shoreline.

Shallow **lagoons** lie behind long barrier islands such as those that border the Texas coast of the Gulf of Mexico (Figure 4-7A). Because they are protected from strong waves, these lagoons act as traps for fine-grained sediment and are usually floored by muds and muddy sands. Rivers that empty into lagoons often build small deltas.

Freshwater from rivers and streams tends to remain trapped in coastal lagoons for some time. Thus, lagoonal salinities in moist climates are often brackish, changing during the year with the changing rate of freshwater runoff from the land. Laguna Madre of Texas is typical of the lagoons found in more arid climates, where temperatures are high (Figure 4-7A). Because they receive little freshwater from rivers and suffer a high rate of evaporation, the ponded waters of this long lagoon are hypersaline. Whether they are brackish or hypersaline, however, the abnormal and fluctuating salinities of lagoons exclude many forms of life that require normal marine salinities. As a result, ancient lagoonal sediments are characterized by fossil faunas of low diversity. Those species that are present in lagoons, however, often occur in large numbers, and some of them are usually burrowers that disturb the muddy sediments of the lagoonal floor, leaving those sediments either mottled or homogeneous and largely devoid of bedding structures.

The Texas barrier islands are unusually long. Padre Island, for example, is more than 70 kilometers (~40 miles) in length (Figure 4-7A). Another way of viewing this is to say that there are long distances between the tidal inlets, or passes through which the tide moves sea water to and from the lagoons. The rarity of tidal inlets is associated with the low tidal range in the Texas region—because the tide rises and falls less than a meter, it moves relatively little sea water to and from the lagoons each day. Along other coastlines fringed by barrier islands, including the North Sea coast of the Netherlands, the tidal range is greater, and tidal inlets are more closely spaced (Figure 4-7B). The Dutch Waddenzee, for example, is largely emptied of water when the tide goes out. Large, branching tidal channels occupy much of the Waddenzee, and between them are intertidal sand flats. Tidal currents here are so strong that little mud accumulates on the broad tidal flats.

The depositional systems just described can be called **barrier island–lagoon complexes**. Whether character-

and changes from time to time. Wind-blown sand often accumulates behind the beach as sand dunes, but these dunes are often eroded before burial and are thus missing from many depositional sequences that represent barrier island–lagoon complexes.

A number of sediments are found along the shores of lagoons. Among these are **tidal flats** formed of sand or muddy sand whose surfaces are alternately exposed and

FIGURE 4-5   Deposits of the Devonian Mahantango Formation west of Harrisburg, Pennsylvania. The beds were tilted during Appalachian mountain building. The thickly bedded, light-colored rocks on the left have been interpreted as delta plain sandstones; the dark shales and siltstones above these are thought to be delta front and prodelta of the next cycle. (*Photograph by the author.*)

ized by a high or a low tidal range, an ancient barrier island–lagoon complex often displays certain diagnostic features, some of which are associated with tidal inlets. When a tidal current flows through an inlet into a lagoon, for example, it slows down as it meets the ponded lagoonal waters and usually drops some of its sediment in the process, thus forming a structure known as a **tidal delta**. The current then divides into two or more channels (Figure 4-8). Because a tidal inlet and its associated channels carries water to and from a lagoon, coarse sediments deposited within the inlet and channels display cross-bedding in which some sets of beds dip landward and others dip seaward (Figure 4-9). The channels also tend to migrate and to produce units that become finer-grained toward the top (Figure 3-22).

The sediments that make up the barrier island itself are usually well-sorted sands. In the beach zone that is washed by breaking waves, deposits tend to have nearly horizontal bedding but often dip gently seaward. Cross-bedding develops in areas where the beach surface is gently irregular

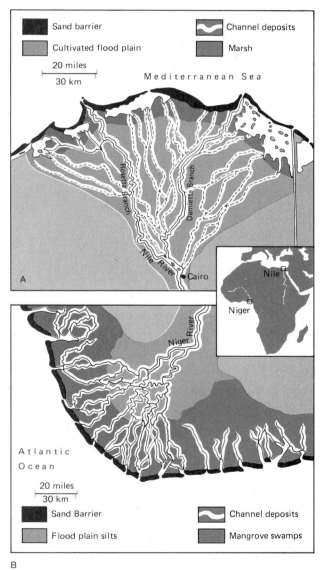

FIGURE 4-6   Two wave-dominated deltas. *A.* The Nile of northern Africa. *B.* The Niger of western Africa. Each delta is molded by marine forces, which smooth its front and throw up eroded sand as marginal barriers. (*After L. D. Wright and J. M. Coleman, Amer. Assoc. Petrol. Geol. Bull. 57:370–398, 1973.*)

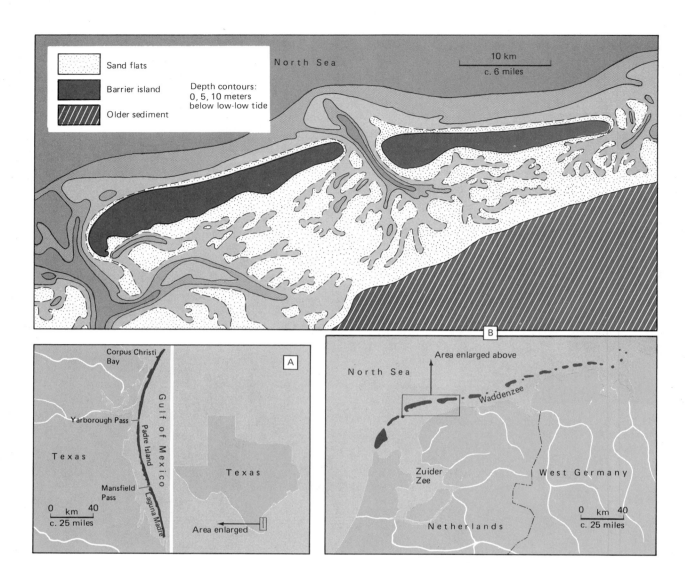

FIGURE 4-7   Barrier islands along the Texas coast (*A*) and the North Sea (*B*). Tides along the Texas coast are weak, and there are few tidal channels or passes. Tides along the European North Sea coast are strong, causing a vertical tidal range of 2.4 to 4.0 meters (~8 to 13 feet), and the currents that these tides produce must flow through many inlets or channels between barrier islands. Thus, the strength of tides determines the length of barrier islands. Large permanent lagoons lie behind the Texas barrier islands, but along the North Sea the lagoons lose nearly all of their water when the tide goes out, leaving sand flats exposed to the air until the tide comes in again. (*B. Modified from L. M. J. U. Van Straaten, Jour. Alberta Soc. Petrol. Geol. 9:203–226, 1961.*)

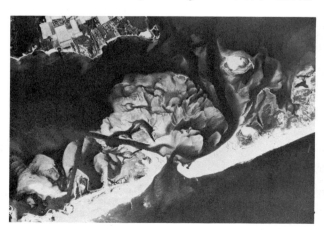

FIGURE 4-8   A tidal delta (*center of picture*) on the lagoonal side of a tidal inlet. The inlet was formed by a hurricane in September 1947. (*Courtesy of M. M. Nichols, U.S. Department of Agriculture.*)

FIGURE 4-9   Cross-bedding in coarse-grained limestone of the Pennsylvanian System in eastern Kansas. (*Courtesy of W. K. Hamblin.*)

FIGURE 4-10   The stratigraphic sequence produced when a barrier island–lagoon complex prograde. As shown in the diagram below, sediments of the lagoon and of the adjacent marshes and tidal flats are superimposed on beach sands of the barrier island. The photographs display an actual record of progradation in the Cretaceous System of northwestern Colorado. The diagram below the photographs identifies the facies present in the Cretaceous outcrop. (*C. D. Masters, Amer. Assoc. Petrol. Geol. Bull. 51:2033–2043, 1967.*)

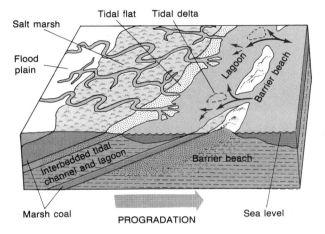

flooded as the tide ebbs and flows. High in the intertidal zone, above barren tidal flats, marshes often fringe one or both margins of a lagoon (Figure 2-29). Here the high rate of production of plant debris leads to the accumulation of peat, which can become coal after a long period of burial.

When a barrier island–lagoon complex receives sediment at a sufficiently high rate, it prograde—that is, it migrates seaward—like the active lobe of a delta. Unlike the migration of a delta, however, this progradation takes place along a broad belt of shoreline. Figure 4-10 illustrates a depositional sequence that now forms part of the Cretaceous System of Colorado. Here, as the shoreline migrated eastward, marsh and tidal flat deposits prograded over sediments of the lagoon and associated tidal channels. These, in turn, built out over deposits of the barrier beaches and over the tidal deltas and marshes behind them. Thus, the horizontal sequence of environments (barrier beach, marsh or tidal delta, lagoon, tidal flat, and marsh) came to be represented by a corresponding vertical sequence of sedimentary deposits in accordance with Walther's law.

If erosion occurs during environmental migration, some of the deposits of a particular environment may be missing from a vertical sequence. However, despite the frequent imperfection of the record it leaves behind, the barrier island–lagoon system illustrates particularly well how the

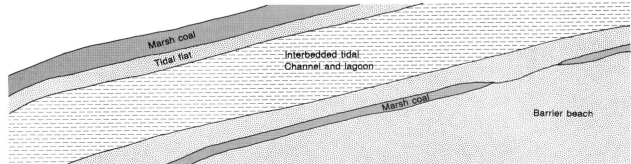

close vertical association of sediments that represent adjacent depositional environments can be used to identify ancient depositional systems. We have seen that vertical associations of rock types are also very useful in the identification of ancient meandering-river deposits (Figure 3-22) and deltas (Figure 4-4).

Offshore shelves and inland seas

Of course, many depositional environments intergrade rather than remaining sharply separated from one another in the manner of marshes and lagoons. As a result, the sedimentary deposits thus produced often intergrade vertically in the geologic record in accordance with Walther's law. We have already seen, for example, that the prodelta clays of a deltaic sequence usually grade upward into silty sediments of the delta front (Figure 4-4). There is often a similar upward-coarsening gradation in the record left by a migrating barrier island because barrier-island sands usually grade into finer sediments in a seaward direction. Because these finer sediments represent progressively deeper environments of a continental shelf or a shallow inland sea, they are deposited on continental crust rather than in the deep sea (Figure 2-28). When a barrier island progrades, it builds out over the finer-grained deposits that represent the deeper water in front, resulting in an upward-coarsening sequence below the package of sediments that represents the beach and lagoon. Such a sequence is well displayed below the beach deposits of Figure 4-11.

More details of this type of sequence are shown in Figure 4-12, which illustrates the vertical profile that would result if progradation were to occur in the modern Gulf of Gaeta, in Italy. Although the upward-coarsening sequence is evident, other typical biological and physical structures in the sediment also display a characteristic vertical order. Very few animals, for example, are able to live in the shifting sands at the edge of wave-ridden beaches, where physically produced bedding structures are much more common than biologically produced burrows. But farther offshore, below the lower limit of wave activity, burrowers are more common, and physical processes have little effect except during major storms; consequently, burrow structures are more abundant in this sediment than are structures produced by physical processes. Thus, in a profile representing progradation of beach sands over offshore muds (Figure 4-12), biologically produced sedimentary structures predominate near the bottom, while physically produced structures prevail toward the top.

Strong tidal currents that sweep over continental shelves can form sand bars well below the limit of wave motion. It

FIGURE 4-11   Offshore shales grading upward into shallow-water and beach sands in the Cretaceous System of Colorado. (Shallow-water deposits of this formation are illustrated in Figure 4-10.) This is a progradational sequence in which the shallow-water deposits built out over the deep-water deposits. (*From H. E. Reineck, Natur. Mus. 101:45–60, 1971.*)

is presumed, however, that more confined inland seas, which are largely cut off from the immense tidal movements of the open ocean, usually experience weak tidal currents. Unfortunately, we have few inland seas to examine in order to verify this assumption, since the level of the ocean is lower today than it has been at most times in the past. The study of sedimentary rocks formed in ancient inland seas does indicate, however, that the sandy deposits that usually fringed the seas generally gave way to muddy sediments in deeper water.

ORGANIC REEFS

In tropical shallow marine settings, if siliciclastic sediments are in short supply, carbonate sedimentation usually prevails. Here, coral reefs are often prominent features. In Chapter 2, we saw that modern coral reefs are rigid structures that rise above the sea floor in shallow waters of high clarity and of normal marine salinity. As we shall see in later chapters, however, large reefs of the geologic past have been formed by organisms other than corals. Because they are produced largely by organisms that secrete calcium carbonate, organic reefs form their own distinct depositional records—as bodies of limestone.

A modern reef has several structural components that form a heterogeneous limestone deposit. The basic framework of a reef consists of the calcareous skeletons of or-

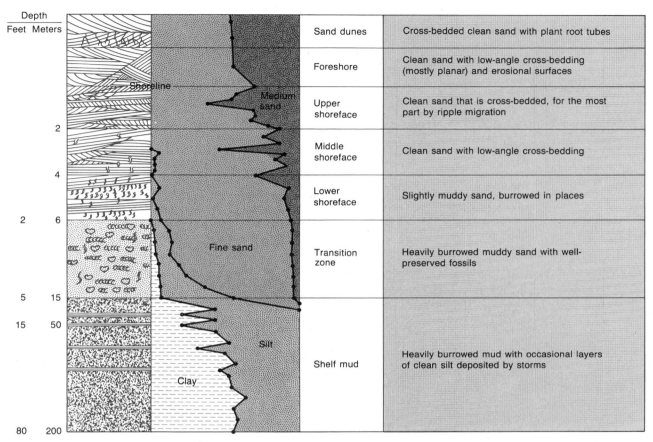

Sand dunes	Cross-bedded clean sand with plant root tubes	
Foreshore	Clean sand with low-angle cross-bedding (mostly planar) and erosional surfaces	
Upper shoreface	Clean sand that is cross-bedded, for the most part by ripple migration	
Middle shoreface	Clean sand with low-angle cross-bedding	
Lower shoreface	Slightly muddy sand, burrowed in places	
Transition zone	Heavily burrowed muddy sand with well-preserved fossils	
Shelf mud	Heavily burrowed mud with occasional layers of clean silt deposited by storms	

FIGURE 4-12   A hypothetical stratigraphic sequence that would result from progradation at the Gulf of Gaeta, Italy. Sand dunes and beach deposits would build out over finer-grained deposits of the offshore area. The column on the left shows how structures within the sediment change with depth; in shallow water, strong wave movements produce cross-bedding and exclude most bottom-dwelling animals. At greater water depths, waves have no effect, and burrowing animals disrupt most sediment layers. (*After H. E. Reineck and I. B. Singh, Depositional Sedimentary Environments, Springer-Verlag New York Inc., New York, 1975.*)

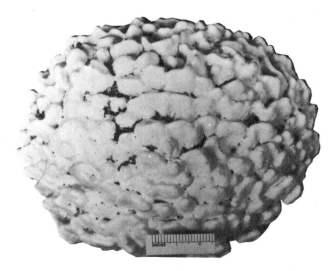

FIGURE 4-13   *Porolithon*, one of several types of coralline algae that grow on reef surfaces, protecting the reef against destruction. (*J. W. Wells, U.S. Geological Survey.*)

ganisms, primarily corals. This framework is strengthened by cementing organisms, including coralline algae, that encrust the surface of the reef (Figure 4-13). Carbonate sediment, which is composed of fragments of the skeletons of reef-dwelling organisms, is trapped within the porous framework, filling many of the voids. With their complex internal structure, reef limestones are for the most part either unbedded or only poorly bedded (Figure 4-14). Even with the presence of infilling debris, reef limestone is highly porous. This accounts for the fact that many ancient buried reefs serve as traps for petroleum, which migrates into them from sediments rich in organic matter.

Because living reefs stand above the neighboring sea floor, they alter patterns of sedimentation nearby. On the leeward side of an elongate reef, there is often a lagoon where waters tend to be relatively calm. This is especially true if the reef has a typical **reef flat**, or horizontal upper surface, that stands very close to sea level (Figure 4-15).

A

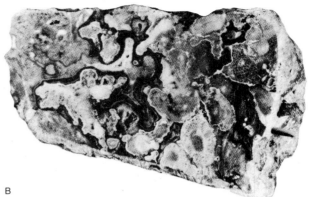

B

FIGURE 4-14   Reef rocks. *A.* A piece of reef limestone obtained by drilling into Bikini Atoll in the western Pacific. This specimen came from a depth of about 25 meters (~80 feet). *B.* A polished slab of reef rock that is more than 200 million years old from the Permian Capitan reef in New Mexico. Both

the old and the young rock include calcareous skeletons of animals and algae as well as particles of sediment trapped between the larger skeletons. (*A. J. W. Wells, U.S. Geological Survey.B. Courtesy of J. A. Babcock.*)

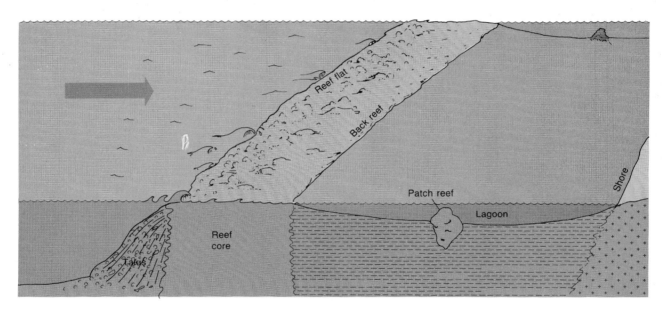

FIGURE 4-15   Diagram of a typical organic reef. The reef grows up to sea level; thus, the reef flat is exposed at low tide. The reef core below consists of reef limestone left beneath the growing surface; talus accumulates at the front of the reef as rubble falls down from shallow water. Waves break across the reef flat, losing energy in the process and leaving a quiet la-

goon behind. Sediment accumulates in the back-reef area and in the lagoon, and here and there patch (or pinnacle) reefs rise up from the lagoon floor. The main reef is of the type known as a barrier reef because its size and position cause it to block waves and thus to protect the shoreline.

Below the living surface of the reef is a limestone core consisting of dead skeletal framework, dead skeletons of cementing agents, and trapped sediment. Extending seaward from the living surface, there is often a pile of rubble called **talus**, which has fallen from the steep, wave-ridden reef front.

Most reefs have the capacity to grow rapidly and to keep pace with rising sea level or a subsiding sea floor. Reefs build upward rapidly enough to remain near sea level even when the sea floor around them is becoming deeper.

Many reefs, in fact, grow so rapidly and are so durable that they build seaward in the manner of a prograding delta (Figure 4-1). Figure 4-16 shows a spectacularly exposed cross section of a Devonian reef in Australia. Although it was built by organisms that have been extinct for hundreds of millions of years, this reef closely resembles many modern reefs in its basic structure—it displays both a seaward talus slope and a leeward reef flat.

As water flows over a reef flat, it is slowed by friction. This gradual dissipation of the energy of motion results in

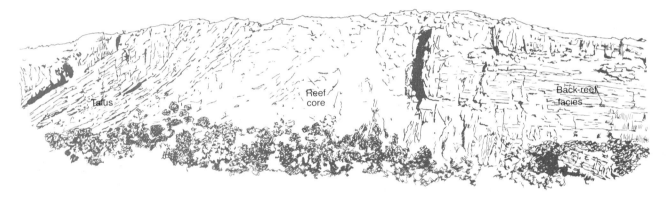

FIGURE 4-16   Outcrop along Windjana Gorge, northwestern Australia, revealing the internal structure of a Devonian reef. The reef core consists of unbedded limestone. The talus is crudely bedded, and the beds slope away from the reef core. The back-reef facies is also crudely bedded, but the beds are approximately horizontal. (*From P. E. Playford, Amer. Assoc. Petrol. Geol. 64:814–840, 1980.*)

what is known as an **energy gradient**. Because different species of framework builders and cementing organisms thrive under different energy conditions, a zonation of species occurs across a typical reef flat from the coralline algae that form a seaward ridge to species that flourish best under the very quiet conditions in the rear. The living reef adjacent to Bikini Island in the South Pacific illustrates these biological zones, which run parallel to the reef front (Figure 4-17). Isolated **patch reefs** or **pinnacle reefs** are often found in lagoons behind elongate reefs (Figure 4-15). Surrounded by quiet water, these small structures tend not to display zonation. Elongate reefs that face the open sea and have lagoons behind them are known as **barrier reefs** (Figure 4-18). Other reefs grow right along the coastline without a lagoon behind, and these are known as **fringing reefs**. Some fringing reefs grow seaward and eventually become barrier reefs.

Perhaps the most curious reefs in the modern world, however, are the circular or horseshoe-shaped structures known as **atolls** (Figure 4-19). Atolls frequently form on volcanic islands and are thus quite common in the tropical Pacific, where such islands are common. Charles Darwin offered an explanation for the origin of the Pacific atolls that is still accepted today (Figure 4-20). According to Darwin, each atoll was formed when a cone-shaped volcanic island was colonized by a fringing reef. The island eventually sank beneath the sea, leaving a circular reef standing alone with a lagoon in the center, where limestone now accumulates in quiet water. Often the reef does not quite form a full circle but is instead broken by a channel on the leeward side, where reef-building organisms do not thrive. Horseshoe-shaped atolls range up to about 65 kilometers (~40 miles) in diameter; during World War II, they served as natural harbors for ships.

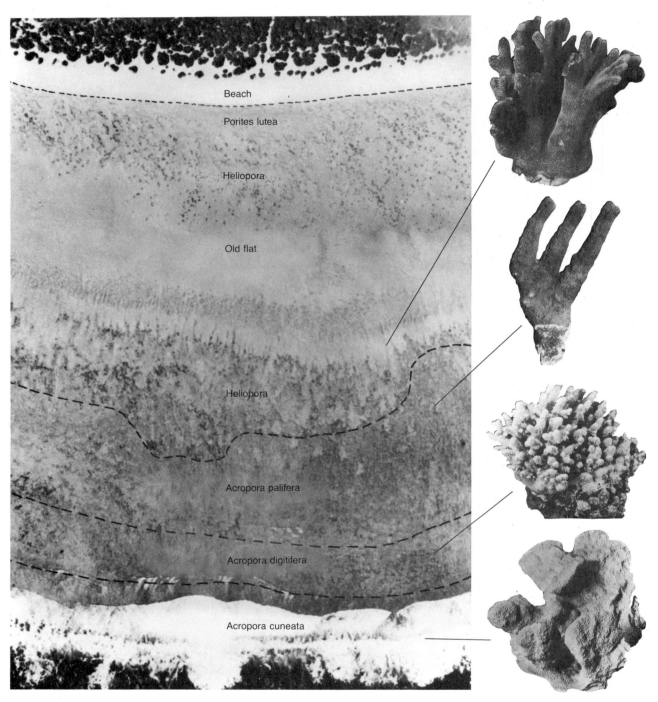

FIGURE 4-17   Aerial photograph showing zonation of coral species across the front of the reef flat on the windward side of Bikini Island in the western Pacific Ocean. The dominant coral species of each zone is shown on the right. (Each of the figured specimens would fit in the palm of your hand.) Waves are breaking across the *Acropora cuneata* zone. *Acropora cuneata* is a stubby coral that is capable of standing up against strong wave action. The strength of the waves weakens toward the rear of the reef flat, and the dominant corals are progressively more fragile. The front of the reef is buttressed by an algal ridge formed of coralline algae; this ridge is not labeled in the figure. (*U.S. Geological Survey.*)

FIGURE 4-18   The barrier reef adjacent to Belize in the Caribbean. A channel in the reef is visible in the foreground, leading to the lagoon on the right. (*Courtesy of R. N. Ginsburg.*)

FIGURE 4-19   Coral-reef atolls of the Tuamotu Archipelago near Tahiti in the South Pacific. The widest segments of the circular atolls face the prevailing winds from the upper right. (*NASA.*)

Ancient atolls that lie buried beneath younger sediments can be identified from the study of cores of sediment brought up from drilling operations—and because porous reef rocks often serve as traps for petroleum, drilling in the vicinity of these atolls is often a profitable venture. Figure 4-21 shows the outline of a subsurface atoll of Late Paleozoic age that has yielded considerable quantities of petroleum in the state of Texas.

CARBONATE PLATFORMS

A carbonate platform is a structure that is formed by the accretion of carbonate sediment and that stands above the neighboring sea floor on at least one of its sides. Like organic reefs, carbonate platforms consist largely of calcium carbonate that originated in shallow tropical waters at or near the site of accumulation. In fact, organic reefs form parts of many carbonate platforms, often growing along the windward margins of these platforms.

In times past, carbonate platforms have stretched along most or all of the eastern margin of the United States—

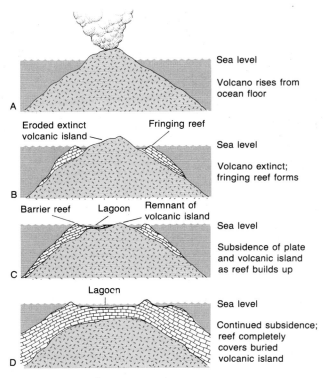

FIGURE 4-20   The development of a typical coral atoll in the Pacific as proposed by Charles Darwin. First, reefs colonize a volcano that stands above sea level. The fringing reef then becomes a barrier reef as the volcano subsides. When the volcano disappears below sea level, all that remains at the surface is a circular reef and central lagoon. (*After F. Press and R. Siever, Earth, W. H. Freeman and Company, New York, 1982.*)

but because climates are cooler today than they have been during most of the earth's history, these structures are not well represented in the modern world. The tropical areas of the modern world do offer several examples, however. In the western Atlantic, for example, a large carbonate platform extends seaward from the Yucatán Peninsula of Mexico. There are also smaller platforms bordering the islands of the Antilles, and the platforms known as the Little Bahama Bank and Great Bahama Bank lie to the east and southeast of Florida (Figure 4-22). Because the varied sediments that are now accumulating on the Bahama banks resemble those of many ancient carbonate platforms, a review of these sediments will serve to introduce carbonate sediments in general.

One general property of carbonate platforms that the Bahama Banks illustrate is rapid deposition of sediments (Figure 4-22). Since mid-Jurassic time, about 170 million years ago, some 10 kilometers (~6 miles) of shallow-water carbonates have accumulated both on the Bahama banks and in southern Florida, which was part of the same car-

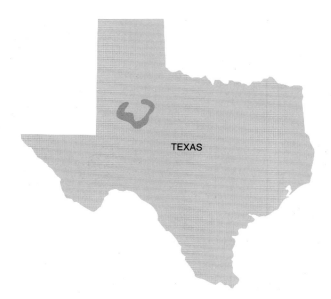

FIGURE 4-21   A late Paleozoic horseshoe-shaped atoll that lies almost a kilometer (~0.7 miles) below the surface of the land. The atoll was discovered when rocks of the region were drilled for petroleum. It would appear that the reef faced prevailing winds toward the south. (*After P. T. Stafford, U.S. Geol. Surv. Prof. Paper No. 315-A, 1959.*)

bonate platform during Cretaceous and earlier times. The heavy buildup of carbonate sediments in this region has caused the oceanic crust to bend downward, and shallow-water Jurassic deposits now lie up to 10 kilometers below sea level. During the Cenozoic Era, the termination of carbonate deposition in some areas led to the development of the deep channels known as the Straits of Florida, Tongue of the Ocean, and Exuma Sound. These channels now separate the Florida peninsula, the Little Bahama Bank, and the Great Bahama Bank.

FIGURE 4-22   Structure of the Bahama banks. The position of the cross section below is indicated by line A–B on the map. Although the Bahamas are now separated from Florida by the Florida Straits, the Bahamas and Florida were part of the same broad carbonate platform during the latter part of the Mesozoic Era. The platform subsided several kilometers beneath the weight of carbonate sediments that accumulated near sea level. During the Tertiary Period, several large channels, including the Florida Straits, Tongue of the Ocean, and Exuma Sound, developed where portions of the subsiding platform failed to maintain a position near sea level. A variety of shallow-water carbonate sediments are currently accumulating on the Great Bahama Bank. (*Map after N. D. Newell and J. K. Rigby, Soc. Econ. Paleont. and Mineral. Spec. Paper No. 5, 1957; cross section after H. T. Mullins and G. W. Lynts, Geol. Soc. Amer. Bull. 88:1447–1461, 1977.*)

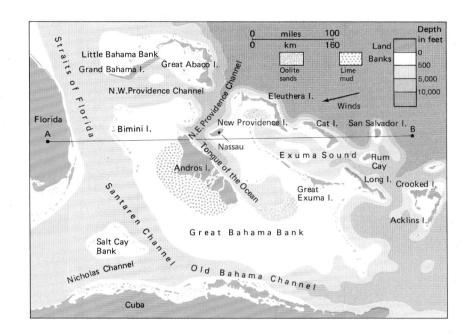

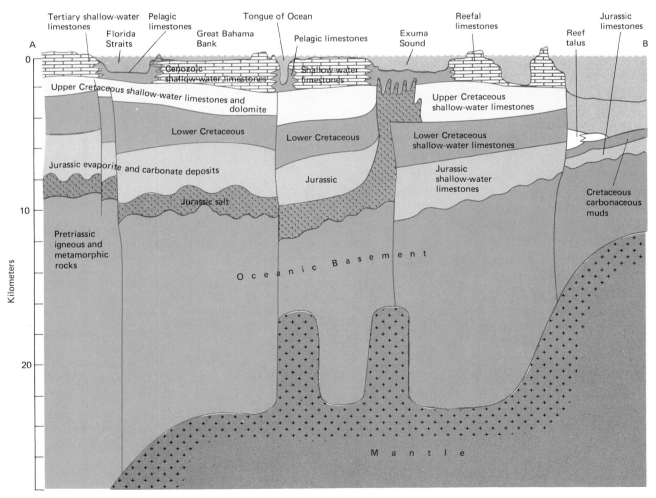

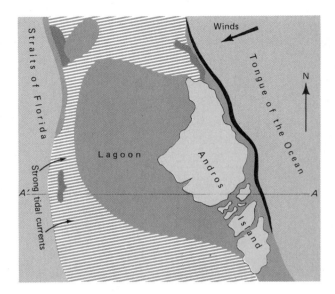

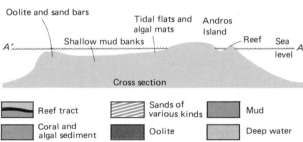

FIGURE 4-23   Map and cross section of the Grand Bahama Bank showing the distribution of sedimentary environments.

A

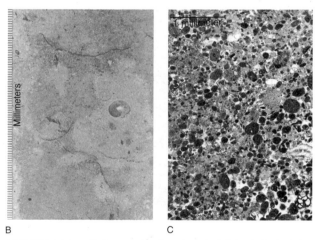

B   C

FIGURE 4-24   Pellets and pelleted mud of the relatively quiet, shallow sea floor west of Andros Island. *A*. Pellets produced by several species of bottom-dwelling marine animals (× 10). *B*. Pelleted mud viewed without magnification. *C*. Magnified view of the pelleted mud; some of the pellets are soft, but others have become cemented by calcium carbonate. (*A*. P. E. Cloud, U.S. Geological Survey. B and C. Courtesy of L. A. Hardie.)

Sediments in the vicinity of Andros Island, the largest of the Bahamas, include a variety of important carbonate types (Figure 4-23). Trade winds from the east-northeast provide water currents and nutrients that are highly favorable to the growth of coral reefs east of Andros Island, which borders Tongue of the Ocean. In the wind shadow of Andros, muddy carbonates accumulate in a quiet, shallow lagoon. These consist primarily of aragonite needles derived from the breakdown of the skeletons of calcareous algae that inhabit the floor of the lagoon. Large quantities of these muddy sediments occur as **fecal pellets**, oval grains composed of mud that has been bound up with mucus after passing through the guts of deposit-feeding worms or other animals. Some of these pellets have become cemented on the sea floor and have thus developed into grains. Many lagoonal deposits that have been studied in the rock record also consist of pelleted and unpelleted lime muds such as those that floor the quiet, shallow seas west of Andros Island (Figure 4-24).

Outside the wind shadow of Andros Island, sediments of coarser grain size predominate. The most spectacular of these sediments are clean sands called **oolites**, which consist of concentrically layered, nearly spherical grains (Figure 4-25*C*). Each individual grain, or **ooid**, forms by accretion—as the grain rolls around on the sea floor, aragonite needles accumulate tangentially on its surface. This rolling action is essential for spherical grain growth, which is why ooids tend to form in areas where there is strong wave activity. In the Bahamas, ooids occur primarily along margins of the Great Bahama Bank, where tidal currents are strong. In the clear, shallow water of these margins, ooids form magnificent submarine dunes (page 82). Only 100,000 years ago, when the seas stood slightly higher than they do today, ooid dunes extended along the southern Florida coastline. Today, these lithified dunes form many of the Florida Keys and also crop out in the Miami area, where their cross-bedding can be seen (Figure 4-25A). Cross-bedded oolites that occur in the rock record represent shallow environments affected by strong tides.

Andros Island is fringed on its leeward (western) side by intertidal carbonate deposits that have numerous counterparts in the rock record. These intertidal sediments consist primarily of pelleted muds that have been washed up by storms from the subtidal sea floor on which they formed (Figure 4-24). The intertidal pellets are hardened to varying degrees by calcium carbonate cement, and they are accompanied by a much smaller volume of skeletal grains.

Still other features of Andros Island characterize carbonate platform environments in general. The intertidal zone west of the island, for example, is bordered by a supratidal beach ridge composed of material that has been thrown up by storm waves (Figure 4-26). Here, tidal channels are bordered by supratidal levees similar to those found along meandering rivers on the land. Beyond the supratidal levees are marshy tidal flats covered by blue-green algae and by sparse mangrove bushes. Shallow ponds also occupy some of the areas between tidal channels. Inland from this intertidal complex is a supratidal marsh, which also consists of blue-green algae and mangrove shrubs.

The seaward-facing slopes of the supratidal beach ridge are wetted periodically and then dried and cemented, causing the surface sediment to form a weak crust. In many places, the crest of the beach ridge is surfaced by partially overlapping, diagonally stacked, flat pebbles (Figure 4-27), which were washed in by storms that ripped them from the seaward-facing slopes. Such diagonally stacked pebbles characterize supratidal settings in areas of carbonate deposition and are frequently found in ancient carbonate rocks.

Bordering the Andros tidal channels in some places are knobby intertidal structures known as **stromatolites** (Figure 4-28), which are produced by threadlike algae that form sticky mats. As shown in Figure 4-28, the algae that form these mats do so by trapping carbonate mud and by growing through it to produce another mat. Repetition of this

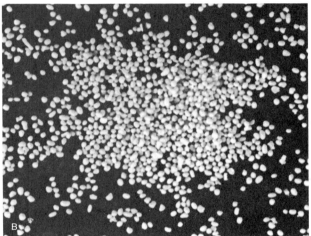

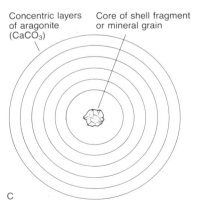

Concentric layers of aragonite ($CaCO_3$)

Core of shell fragment or mineral grain

FIGURE 4-25   A. An outcrop of the Miami Oolite in Miami, Florida. This sediment accumulated about 100,000 years ago as submarine dunes (see page 82). Like terrestrial sand dunes (Figure 3-14), these developed cross-bedding, which is visible here in the foreground. B. Modern ooids from the Great Bahama Bank. Their average diameter is less than ½ millimeter. C. The internal structure of an ooid; as the grain rolls about, concentric layers of aragonite needles accumulate around a central core. (*Photographs by the author.*)

A

FIGURE 4-26   The Three Creeks area along the west coast
of Andros Island, illustrated by an aerial photograph and a block
diagram. Inland of the creeks, which are tidal channels, is an
intertidal marsh composed largely of algae. Ponds occupy some
of the intertidal areas between the channels. (*A. Courtesy of
L. A. Hardie. B. After R. N. Ginsburg and L. A. Hardie, in R. N.
Ginsburg [ed.], Tidal Deposits, Springer-Verlag, Berlin, 1975.*)

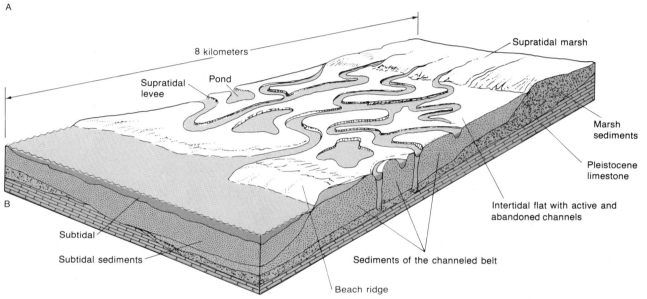

A

B

FIGURE 4-27   The origin of flat-pebble conglomerates. *A.* Flat
pebbles stacked up on the beach ridge illustrated in Figure
4-26. They are formed from fragments of the semihardened
crust that have been torn from the surface of this supratidal
area by occasional storm waters. The pebbles dip toward the
left, the direction from which the disruptive waters flowed. *B.* A

flat-pebble conglomerate of Paleozoic age (scale 5 centimeters
long). (*A. From L. A. Hardie, Sedimentation on the Modern
Carbonate Tidal Flats of Northwest Andros Island, Bahamas.
Johns Hopkins Univ. Press, Baltimore, 1977. B. Photograph by
the author.*)

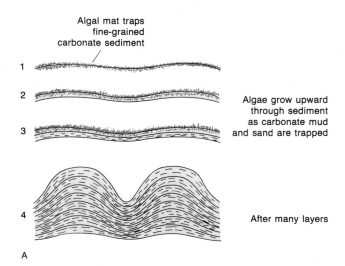

Algal mat traps
fine-grained
carbonate sediment

1

2

3

Algae grow upward
through sediment
as carbonate mud
and sand are trapped

4

After many layers

A

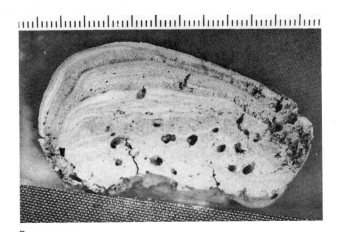

B

C

process on an irregular surface forms a cluster of stromatolites. Stromatolites are produced by blue-green algae, which, as we will see in later chapters, are among the most primitive groups of organisms in the world. The fossil record of stromatolites is unusually ancient, extending back through more than 3 billion years of geologic time.

On the supratidal backslope of the Andros beach ridge are other layered sediments that are also formed by a succession of algal mats. In the supratidal beach ridge setting, each algal mat is formed at a time of flooding by storm waves. In this setting, the surface of accumulation is relatively flat, and the layers are not domed into stromatolites. The horizontal layering of the sediments, together with numerous holes transecting the various layers, gives these sediments the informal name **bird's eye limestone** (Figure 4-29). Although some of the holes in this sediment are tubes produced by burrowing worms, others represent spaces that have been left where strands of algae have decayed, gas bubbles have formed, or sediment has dried and cracked. Currently, well-layered bird's eye limestones occur primarily in supratidal and high intertidal settings; when found in the rock record, they usually represent such settings.

There is a simple reason that stromatolites and bird's eye limestones are found almost exclusively in supratidal and high intertidal settings. Because these environments are exposed above sea level much of the time, they become hot and dry, and relatively few marine animals can survive in them. Thus, there is little to interfere with the tendency of algal mats to form layered structures. In subtidal settings, algal mats grow here and there, but they are eaten by the many grazing animals and damaged by burrowers; as a result, they do not survive well below water and seldom accumulate to form stromatolites or well-layered limestones. As we will see in later chapters, some very ancient stromatolites may have formed subtidally before the origin of animals that feed on filamentous algae.

The supratidal backslopes of tidal channel levees adjacent to Andros Island dry out after occasional flooding by storm-driven seas. As a result, the surface of the sed-

FIGURE 4-28   The growth of stromatolites. *A.* An algal mat traps mud and grows through it to form another layer. The accumulation of several layers leads to the formation of a stromatolite. *B.* Section through a stromatolite that grew intertidally along one of the tidal channels illustrated in Figure 4-26 (scale in millimeters). *C.* A hummocky surface formed by many stromatolites growing along the tidal channel. (*B and C. From L. A. Hardie, Sedimentation on the Modern Carbonate Tidal Flats of Northwest Andros Island, Bahamas. Johns Hopkins Univ. Press, Baltimore, 1977.*)

FIGURE 4-29 Bird's-eye limestones formed in intertidal and supratidal areas where algal layers produce flat mats. Gas bubbles and worm burrows have disrupted the layers here and there, producing holes (dark spots) that give the limestones their bird's-eye appearance. *A.* A section through the sediment behind the beach ridge illustrated in Figure 4-26 (scale in millimeters). *B.* A similar ancient sediment from the Ordovician of Oklahoma; note that some of the algal mats here dried and broke into platy fragments that now sit at slight angles to the bedding. (*A. From L. A. Hardie, Sedimentation on the Modern Carbonate Tidal Flats of Northwest Andros Island, Bahamas. Johns Hopkins Univ. Press, Baltimore, 1977. B. Courtesy of H. Blatt.*)

iment here, which is also bound by an algal mat, is broken by mud cracks resembling those that often form when a mud puddle on the land dries up (Figure 4-30). Mud cracks associated with ancient marine deposits usually represent intertidal or supratidal environments that were alternately wetted by the sea and dried by the sun.

In general, then, a close association in the rock record of flat-pebble conglomerates, stromatolites, well-layered bird's eye limestones, and mud cracks points to a complex of intertidal and supratidal carbonate platform environments. If such deposits are prominently displayed in a carbonate sequence that is hundreds or thousands of meters thick and that represents millions of years of deposition, it is almost certain that they formed on a slowly subsiding carbonate platform resembling the Bahama banks.

FIGURE 4-30   Mudcracks formed in areas only periodically flooded by the sea. *A.* Mud cracks formed along Andros Island on the supratidal surface beyond a levee like the ones illustrated in Figure 4-26. The polygons formed by the cracks are several millimeters across. *B.* Similar mud cracks in a dolomitic carbonate rock of the Carboniferous of Ireland. (*A. From L. A. Hardie, Sedimentation on the Modern Carbonate Tidal Flats of Northwest Andros Island, Bahamas. Johns Hopkins Univ. Press, Baltimore, 1977. B. Courtesy of I. M. West.*)

B   0   1 cm

MARGINAL MARINE SABKAS

Shallow-water carbonate sediments do not accumulate only on platforms that stand high above the surrounding sea floor. Sometimes they fringe the land along gently sloping shelves, as is the case along the southern margin of the Persian Gulf (Figure 4-31). Here, barriers separate elongate lagoons from open marine water. These barriers consist of three kinds of structures: islands with dunes, submarine ooid shoals, and coral reefs. Muddy carbonate sediment accumulates in the quiet lagoons, and inland from these are broad intertidal and supratidal flats.

The intertidal sediments of the southern Persian Gulf are well-layered algal-mat deposits that grade into a sabka. A **sabka** is a flat supratidal area heavily laden with evaporites in the form of nodules, crusts, and crystalline aggregates; halite, anhydrite, gypsum, and mineral grains resembling dolomite are all present. Some of these evaporites form because the arid climate causes marine waters that soak through from the tidal flat and lagoon to evaporate rapidly; however, at least some of the evaporating waters seep into the flat from highlands to the south.

Because of rapid deposition, the supratidal and intertidal sediments of the southern Persian Gulf encroach on the lagoon. As a result, cores obtained by drilling in the supratidal sabka pass down through the sand and evap-

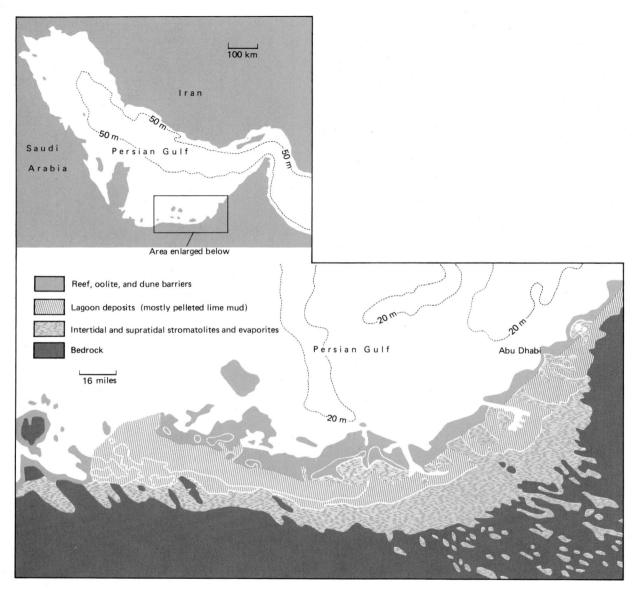

FIGURE 4-31   Sedimentary features of a segment of the Persian Gulf. The slope of the bottom here is gentle. Reefs, ooid shoals, and islands with dunes form a barrier. Behind the barrier lie a lagoon, a broad intertidal area with stromatolites, and a supratidal salt flat. (*After C. G. St. C. Kendall and P. A. d'E. Skipwith, Geol. Soc. Amer. Bull. 80:865–892, 1969.*)

orite layer that is forming here to algal laminations of the tidal flat, which previously extended inland to this position (Figure 4-32*A*). Below these are lagoonal sediments, whose presence reflects a time when the shoreline was still further inland. In other words, as the sediments push the sea back (or prograde), the marginal environments shift along with the shoreline and are superimposed on one another in accordance with Walther's law. In the rock record, remarkably similar sequences can be found serving as a record of the progradation of salt flats in a seaward direction over older intertidal and lagoonal sediments (Figure 4-32*B*).

SUBMARINE SLOPES AND TURBIDITES

One of the greatest advances in the study of sedimentary rocks took place in the middle of the twentieth century with the recognition that certain sedimentary rocks have been produced by **turbidity currents**. A turbidity current is a flow of dense, sediment-charged water moving down a slope under the influence of gravity.

Turbidity currents were first noticed in clear lakes, where flows were observed to form from muddy river water that

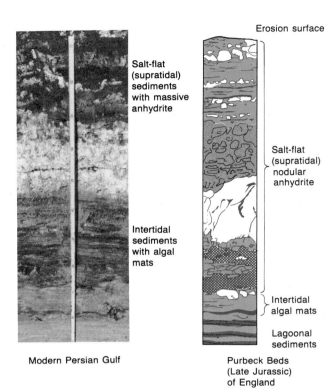

Salt-flat (supratidal) sediments with massive anhydrite

Intertidal sediments with algal mats

Modern Persian Gulf

A

Erosion surface

Salt-flat (supratidal) nodular anhydrite

Intertidal algal mats

Lagoonal sediments

Purbeck Beds (Late Jurassic) of England

B

hugged the lake floor. These currents flowed for a considerable distance toward the center of the lake, slowing down and dropping their sediment only when they spread out and reached gentler slopes. In the 1930s, Dutch geologist Philip Kuenen demonstrated in the laboratory that turbidity currents can attain great speed, especially when they are heavily laden with sediment and are moving down steep slopes. Sediment suspended in a turbidity current behaves as part of the moving fluid, and its presence increases the density of this fluid by as much as a factor of two. Figure 4-33 shows an artificially produced turbidity current flowing down the submarine slope of the island of Jamaica.

When a turbidity current reaches a plane whose slope is more gentle than that from which it originated, it slows down and spreads out, dropping its sediment in the general sequence that we have now seen exemplified again and

FIGURE 4-32   Modern and ancient sabka deposits. *A.* A core taken from the salt flat at Abu Dhabi along the Persian Gulf; the salt-flat deposits have prograded over intertidal sediments. Tape length is 1.8 meters (~6 feet). *B.* A similar progradational sequence in the Jurassic Purbeck Stage of England. The illustrated section is about half a meter (~15 inches) from bottom to top. (*A. Courtesy of K. J. Hsu. B. After D. J. Shearman, Trans. Inst. Min. Metall. B. 75:208–215, 1966.*)

FIGURE 4-33   Side view of a turbidity current flowing down a submerged slope of the island of Jamaica. The propeller of a submarine caused the turbidity current by disturbing sediment along the slope. (*From C. H. Moore et al., Jour. Sedim. Petrol. 46:174–187, 1976.*)

again: First the coarse sediment falls from suspension, and then, much later, the fine material follows. The result is a graded bed that is often characterized by poorly sorted sand and granules at its base and by mud at the top. Such a graded bed is known as a **turbidite**.

A series of turbidity currents in the laboratory produced a succession of graded beds, leading Kuenen and others to recognize that similar turbidite successions are common in nature. This discovery prompted researchers to ask where in the oceans there might be slopes on which turbidity currents could be generated. Continental slopes were the obvious candidates. Sediments moving along these vast slopes would be expected to be deposited in the deep sea, partly along continental rises and partly on abyssal plains. Thus, it was correctly concluded that turbidite successions such as those shown in Figure 4-34 were deposited in deep water.

As we will see in Chapter 8, some ancient marine turbidites were deposited not along continental margins but within continents, in deep interior seaways adjacent to actively forming mountain belts. Thus, they formed in deep-water settings similar to those that border continents.

The hypothesis that most large turbidity currents develop on continental slopes gained strong support from an

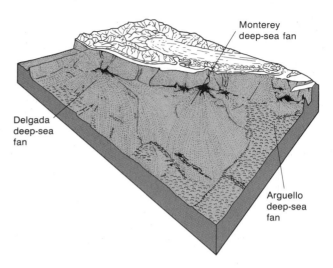

FIGURE 4-35   Deep-sea fans spreading from submarine canyons along the coast of California. The fans consist primarily of turbidite deposits. (*After H. W. Menard, Geol. Soc. Amer. Bull. 71:1271–1278, 1960.*)

FIGURE 4-34   A thick sequence of Miocene turbidite beds separated by thin shale beds along the Santerno River in northern Italy. (*From F. J. Pettijohn et al., Sand and Sandstone, Springer-Verlag, New York, 1973.*)

event that was not fully analyzed until more than 20 years after it took place. In 1929, several transatlantic telegraph and telephone cables on the continental slope and rise off the Grand Banks of Newfoundland were mysteriously broken. An earthquake had taken place just before the cables were severed, and so it was assumed that a slumping of sediments resulting from the earthquake accounted for any breakage that had occurred close to the site of earth movements. Some cables, however, were snapped many minutes later and many kilometers downslope from the site of earth movement. When these events were evaluated more than 20 years after the fact, it was concluded that a turbidity current issuing from slumping sediment in the vicinity of the earthquake was the only agent capable of severing the deeper cables. Cores of sediment removed from the deep-sea floor near these cables provided tangible evidence of this by revealing the presence of a turbidite at the sediment surface. The times at which the various cables broke revealed that the destructive turbidity current reached velocities up to 40 to 55 kilometers (~25 to 34 miles) per hour! The current traveled for at least 700 kilometers (~430 miles), spreading sediment over a huge area of the sea floor.

Turbidity currents that originate near the edge of the continental shelf not only carry sediment to the deep sea but also erode both the continental slope and part of the continental rise. Such currents are also largely responsible for carving the great submarine canyons that incise many

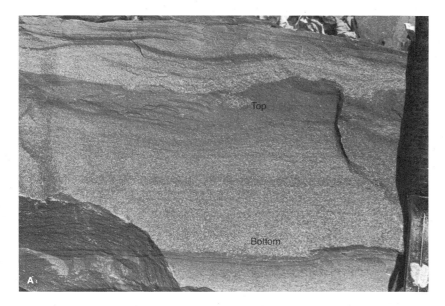

FIGURE 4-36   Ordovician turbidite beds from the eastern United States. *A.* A turbidite bed that grades from coarse at the bottom to fine at the top; the upper surface is irregular because it was disturbed by the succeeding turbid flow and distorted by subsequent compaction. *B.* The bottom surface of a turbidite bed, showing sole marks produced when the sediment forming the bed filled depressions that the turbid flow depositing it scoured the preexisting sediment surface; the current flowed toward the upper left. (*Courtesy of Earle F. McBride.*)

parts of the slope. The turbidity currents slowed down in front of these canyons, dropping part of their sedimentary load to form deep-sea fans that superficially resemble alluvial fans (Figure 4-35). In fact, much of the continental rise actually consists of coalescing submarine fans.

A closer look at turbidite deposits shows that their sands are typically graywackes (Appendix I) that are quite unlike the clean sands of meandering-river deposits. This is only one of several ways in which a cyclical sequence of turbidites differs from the meandering-river cycle, although both show a grading of sediment from coarse to fine within each complete cycle. Another difference is that a turbidite sequence is usually only a few centimeters and seldom as much as a meter thick (Figure 4-34), whereas most com-

plete meandering-stream cycles measure several meters from bottom to top. In addition, turbidites usually lack the cross-bedding that characterizes meandering-river channels.

Turbidites, in fact, exhibit a highly characteristic sequence of sedimentary structures. The base of a turbidite is often irregular because the earliest, most rapidly moving waters of the turbid flow have scoured depressions in the previous sedimentary surface. These scours are subsequently filled in by the first sediments that settle as the current slows. When the base of a lithified turbidite is turned over for inspection, its irregularities, or "sole marks," often reveal the direction of water flow (Figure 4-36).

The characteristic sequence of sedimentary structures above the base of a turbidite is called a **Bouma sequence**,

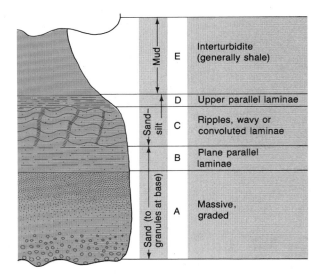

		E	Interturbidite (generally shale)
		D	Upper parallel laminae
		C	Ripples, wavy or convoluted laminae
		B	Plane parallel laminae
		A	Massive, graded

FIGURE 4-37   Diagrammatic representation of a typical turbidite sequence, known as a Bouma sequence. As a turbidity current slows down, it first drops coarse granules, followed by sand and then by silt. Above the silt is mud that accumulates slowly over a relatively long period of time before the next turbidite deposits more coarse sediment. (*After G. V. Middleton and M. A. Hampton, in D. J. Stanley and D. J. P. Swift [eds.], Marine Sediment Transport and Environmental Management, John Wiley & Sons, Inc., New York, 1976.*)

after the man who first recognized it. This sequence includes an upward decrease in grain size (Figure 4-37). Most turbidite beds, however, lack some features of the ideal cycle. Often, for example, the upper part of a cycle has been truncated by scour resulting from the arrival of the next turbidity current. The very top of a complete cycle usually consists of clays that accumulated slowly over a long period of time before the arrival of the subsequent turbidity current. This fine material might be described as representing "background" sedimentation that is periodically interrupted by violent turbid flows.

PELAGIC SEDIMENTS

Turbidity currents and other bottom flows carry mud to the abyssal plain of the ocean well beyond the continental rise. Neither this source nor any other, however, contributes sediment to the deep sea at a high rate; indeed, sediment in most areas of the abyssal plain is accumulated at a rate of about 1 millimeter per 1000 years! In the deep sea,

sparse sediments from turbidity currents are joined by clays from two other sources. One source is the weathering of rocks produced by deep-sea volcanoes. Such clays are less abundant in the Atlantic than in the central Pacific, where volcanoes are common. Clays also reach the deep sea by settling from the pelagic realm, which they reach after traveling through the air as wind-blown dust or moving seaward from the land at very low concentrations in surface waters of the ocean.

Clays settling from the upper part of the water column are one type of pelagic sediment. Other types are contributed by small pelagic organisms whose skeletons become sedimentary particles when the organisms die. Whether clays predominate in any area of the deep sea depends on the extent to which they are diluted by the more rapid accumulation of biologically produced sediments. Some of the latter consist of calcium carbonate, while others consist of silica.

Calcium carbonate predominates in sea-floor sediment at low latitudes (Figure 4-38), where its fine grain size has led oceanographers to refer to it as "ooze." In fact, this **calcareous sediment** has often been termed "globigerina ooze," after the single-celled, amoebalike creature whose skeletons are common constituents. This label is usually not appropriate, however, because other genera of the same group (planktonic foraminifers) are often equally common or even dominant components of deep-sea calcareous ooze (Figure 4-39). Furthermore, skeletons of altogether different kinds of organisms often predominate. The most important are **coccoliths**, armorlike plates that surround **coccolithophores**, the single-celled floating algae that are major components of tropical phytoplankton (Figure 2-31).

Calcareous oozes are abundant only at depths of less than about 4000 meters (~13,000 feet). The reason for this limitation is that calcium carbonate tends to dissolve as it settles through the water column. As pressure increases and temperature declines with depth, the concentration of carbon dioxide increases; as a result, the water becomes undersaturated with respect to calcium carbonate and thus tends to dissolve it. Actually, calcium carbonate begins to dissolve at depths of just a few hundred meters, but the rate of this dissolution is usually too slow to destroy the particles before they reach bottom waters deeper than about 4000 meters, with the exact depth depending on local conditions. Thus, even in the tropics, where planktonic organisms produce calcium carbonate at a high rate near the surface, dissolution of falling particles prevents calcareous

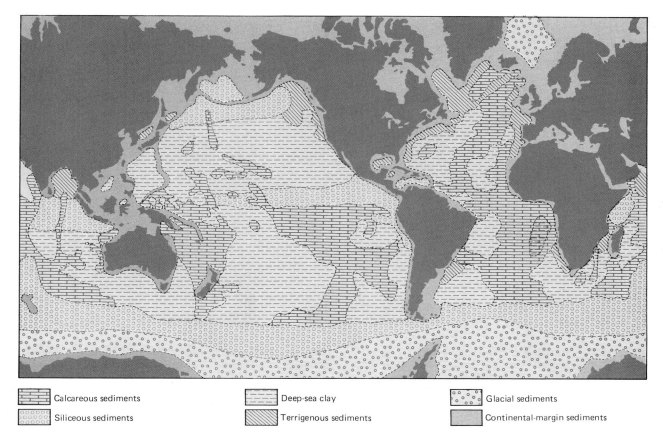

FIGURE 4-38   The global pattern of deep-sea sediments. Calcareous oozes are restricted to low latitudes. Most areas of siliceous ooze lie closer to the poles, although some occur close to the equator in areas of the Pacific and Indian oceans.

(After T. A. Davies and D. S. Gorsline, in J. P. Riley and R. Chester [eds.], Chemical Oceanography, Academic Press, Inc., London, 1976.)

oozes from accumulating on the deeper parts of the deep sea floor. In these areas, deep-sea clays predominate.

In many regions at high latitudes as well as in tropical Pacific regions characterized by strong upwelling, biologically produced siliceous sediments carpet the deep-sea floor. These consist of the skeletons of two groups of organisms that thrive where upwelling supplies nutrients in abundance: the diatoms, a highly productive phytoplankton group found in nontropical waters (Figure 2-31); and radiolarians, which, like planktonic foraminifers are single-celled amoebalike creatures that float in the water and feed on other organisms. The skeletons of both diatoms and radiolarians consist of a soft form of silica similar to that of the semiprecious stone opal. When concentrated as siliceous sediment, these opaline skeletons tend to recrystallize and hence to lose their identity (Appendix I). In the

process, the rock that they form becomes a dense, hard chert composed of finely crystalline quartz. Before recrystallization, the soft sediment is known as diatomaceous earth, which is used as the abrasive component of many of the scouring powders used in kitchens.

The Monterey Formation of California, which is a relatively young (Miocene) unit formed largely of diatom skeletons, illustrates the manner in which a soft, diatomaceous earth is transformed into chert. For reasons that are not understood, different parts of the Monterey Formation are recrystallized to different degrees; in some places, soft, relatively unaltered beds alternate with beds that have been largely recrystallized into chert (Appendix I).

Thick diatomaceous bodies of sediment like the Monterey Formation have formed in marine areas of strong upwelling, where diatom productivity has been unusually

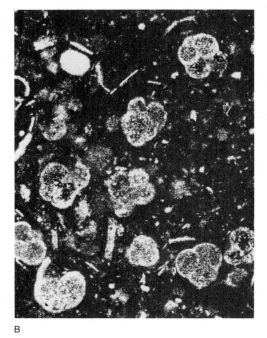

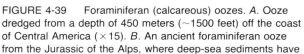

A

B

FIGURE 4-39   Foraminiferan (calcareous) oozes. *A.* Ooze dredged from a depth of 450 meters (~1500 feet) off the coast of Central America (×15). *B.* An ancient foraminiferan ooze from the Jurassic of the Alps, where deep-sea sediments have been elevated high above sea level (scale bar is 0.5 millimeters long). (*A. Courtesy of P. J. Smith, U.S. Geological Survey. B. Courtesy of R. E. Garrison.*)

high. Diatoms did not exist until late in Mesozoic time, however, and this illustrates an important point: The composition of pelagic sediments has changed markedly in the course of geologic time as groups of sediment-contributing organisms have waxed and waned within the pelagic realm of the ocean. We will examine the highlights of these changes in some of the chapters that follow.

CHAPTER SUMMARY

1.   Where a river meets a lake or an ocean, it drops its sedimentary load to form a delta; deltaic deposition typically produces an upward-coarsening sequence as shallow-water sands build out over deeper-water muds.

2.   More widespread than deltas along the margin of the ocean are muddy lagoons bounded by barrier islands formed of clean sand. Seaward of barrier islands, in deeper water, are muddy sediments in which burrowing animals have disrupted most of the original layering.

3.   Coral reefs border shorelines in many tropical areas. A typical reef stands above the surrounding sea floor, growing close to sea level and leaving a quiet lagoon on its leeward side. Most reef limestones are largely unbedded sedimentary units supported by rigid internal organic frameworks. Coral reefs form parts of many carbonate platforms, although these platforms contain a number of other deposits as well (including submarine dunes of cross-bedded ooids and a variety of intertidal and supratidal sediments with characteristic features).

4.   Beyond the edge of the continental shelf, turbidity currents periodically sweep down continental slopes to the continental rise and deep-sea floor, where they spread out, slow down, and deposit graded beds of sediment.

5.   Still farther from continental shelves, only fine-grained sediments accumulate. Clay reaches these deep-sea areas at a very slow rate. In some areas, clay deposition is far surpassed by the accumulation of minute skeletons of planktonic marine life that settle to the sea floor to form calcareous or siliceous oozes.

EXERCISES

1.   Contrast the following three kinds of depositional cycles: the kind produced by meandering rivers, the kind produced by deltas, and the kind produced by turbidity currents.

2.   Draw a profile of a barrier island–lagoon complex, and label the various depositional environments.

3.   Which features of carbonate rocks suggest intertidal or supratidal deposition? Which features suggest subtidal deposition?

4.   What types of sediments and sedimentary structures usually reflect deposition in a deep-sea setting?

ADDITIONAL READING

Reading, H. G., *Sedimentary Environments and Facies*, Elsevier Publishing Company, New York, 1978.

Reineck, H. E., and I. B. Singh, *Depositional Sedimentary Environments*, Springer-Verlag New York Inc., New York, 1975.

Selley, R. C., *An Introduction to Sedimentology*, Academic Press, Inc., London, 1976.

PART TWO
The Dimension of Time

Rocks that represent particular intervals of the geologic time scale can often be recognized throughout the world by characteristic features, the foremost of which—especially for Phanerozoic rocks—are the fossils they contain. The actual ages of these intervals (in years) are estimated by measuring the degree to which radioactive minerals within the rocks and fossils have decayed. The kinds of organisms whose fossils are most useful for dating rocks are ones that existed for short intervals of time, inhabited many different environments, and left behind fossils that are distinctive and abundant.

Of course, fossils would not be so helpful in dating rocks if life on earth were not constantly changing. Organic evolution is the process by which the sweeping changes in organisms have occurred. The fossil record represents a unique repository of information about this process, revealing patterns and rates of evolution and also patterns and rates of extinction.

CHAPTER 5

Correlation and Dating of the Rock Record

A s we saw in Chapter 1, geologists of the nineteenth century divided the rock record into systems such as the Cambrian, Ordovician, and Silurian largely on the basis of fossil occurrences. In fact, fossils continue to be of primary importance in determining the relative ages of rock strata. In this chapter, we will discuss how fossils are used to subdivide the rock record into intervals that represent still smaller units of time. We will also analyze deposits and erosional surfaces that formed over large areas during very brief intervals and that can therefore be used along with fossils to determine whether certain rocks at different sites are approximately the same age. Another important approach we will consider involves the use of naturally occurring radioactive materials, which as we learned in Chapter 1, allow us to estimate the actual ages of rocks and fossils in years.

The study of the age relationships of layered rocks usually begins with the examination of a **stratigraphic section**—a local outcrop or series of adjacent outcrops that

displays the rocks' vertical sequence (Figure 5-1). In the process known as correlation, the fossils of two spatially separate stratigraphic sections—one in England and one in France, for example—are determined to be of the same geologic age. Since William "Strata" Smith first showed that fossils were useful for correlation early in the last century (page 18), the distribution of fossils has become much better known, and the precision of correlation has been greatly improved.

INDEX FOSSILS AND ZONES

Certain fossil species and genera are especially well suited to correlation procedures, and these are called **index fossils** or **guide fossils**. Such fossils have some or all of the following desirable characteristics: (1) they are easily distinguished from other taxa; (2) they are geographically widespread and can thus be used to correlate rocks over a large area; (3) they occur in many kinds of sedimentary rocks and can therefore be found in many places; and (4) they are restricted to narrow stratigraphic intervals, allowing for precise correlation.

Unfortunately, few index or guide fossils exhibit all of these traits to a high degree, as illustrated by the planktonic foraminifer fossils found in late Mesozoic and Cenozoic sediments (page 114). Like living species of planktonic foraminifers, these single-celled creatures resembled amoebas, but, unlike amoebas, they possessed skeletons that settled to the sea floor following death. Planktonic foraminifers meet important index-fossil criteria in two respects: first, they are easily identified under the microscope; and second, they tended to float across large areas

---

An x-ray photograph of the planktonic foraminifer *Hastigerinella digitata* (× 36). Species of planktonic foraminifers, which tend to be distributed throughout the world, are useful for identifying the ages of ancient sediments. Foraminifers are single-celled members of the marine zooplankton that resemble amoebas with skeletons. Filaments of protoplasm project from these skeletons, which are known as tests. You will recall that in places, these tests rain down on the floor of the deep sea to form calcareous oozes (Figure 4-40). Many species of foraminifers are not planktonic but instead live on the sea floor; fossil species of this type tend to be less widely distributed than planktonic species and are therefore less valuable for dating sediments. (*Courtesy of K. Karlin and A. W. H. Bé.*)

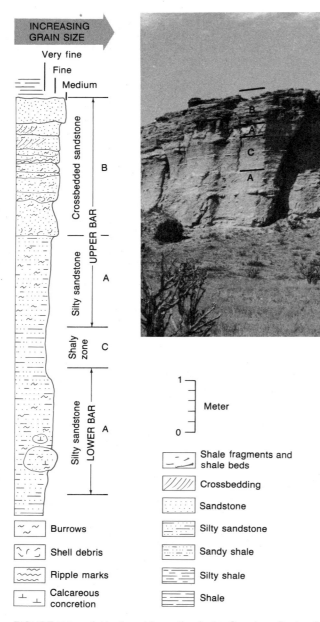

FIGURE 5-1   A stratigraphic section in the San Juan Basin of New Mexico. The sandstone exposed here (unit B) represents an offshore sand bar of Cretaceous age. On the left is a graphic representation of the section, where, in keeping with convention, sandstones are depicted in profile as projecting beyond shales. Sandstones usually stand out farther than other sedimentary rocks in actual outcrops because they are relatively resistant to erosion. (*After N. A. La Fon, Amer. Assoc. Petrol. Geol. Bull. 65:706–721, 1980.*)

of the sea and to settle in many different sedimentary environments. The primary limitation of these organisms is that they lived in offshore areas and are seldom found in nearshore sediments. Moreover, extinct species of planktonic foraminifers exhibit a wide range of geologic longevities. Those that survived for as long as 15 or 20 percent of the Cenozoic Era make poor index fossils, although the earliest and latest appearances of such species do provide information useful for correlation. Other species, however, lived for much shorter intervals and thus provide far more precise correlation.

A segment of the stratigraphic record that is characterized by particular species of index fossils may be formally recognized as a **zone**. Box 5-1 illustrates zones that are based on the occurrence of extinct fossil animals known as **graptolites**, many of which serve as excellent index fossils. Some zones are defined by the presence of single species, while others are distinguished by the presence of two or more species. In general, the **lower boundary** of a zone is defined as the level at which one or more species first appears in the stratigraphic record. Correspondingly, the **upper boundary** is defined as the level at which the same species—or another species or group of species—disappears. Thus, it can be said that a zone is a segment of the stratigraphic record whose upper and lower boundaries are approximately the same age everywhere. A zone is commonly named for a species that characterizes it, but if a zone includes the stratigraphic ranges of several species that belong to a single genus, it may instead be named for this genus. Some zones are further divided into **subzones**.

Although some zones can be found in many parts of the world, others are restricted to a single continent. Moreover, no zone represents exactly the same interval of time throughout the region that it occupies, since no extinct species appeared or disappeared simultaneously in all of the areas that it inhabited. As we will see in the following chapter, it is not unusual for a species to originate in a small area and later greatly expand its geographic range. Often, too, a species or genus has persisted in a restricted region after dying out in most of the areas it previously inhabited (Figure 2-6). Furthermore, many species have experienced complex histories of migration that have in part reflected changing environmental conditions.

Imperfect as the process may be, the recognition of zones often results in correlation with good temporal resolution. If a species existed for only a million years, for example, its discovery in two areas cannot lead to a correlation error that exceeds a million years—even if the species appeared and disappeared at different times in

Box 5-1

Graptolites:

Index fossils

of the Paleozoic

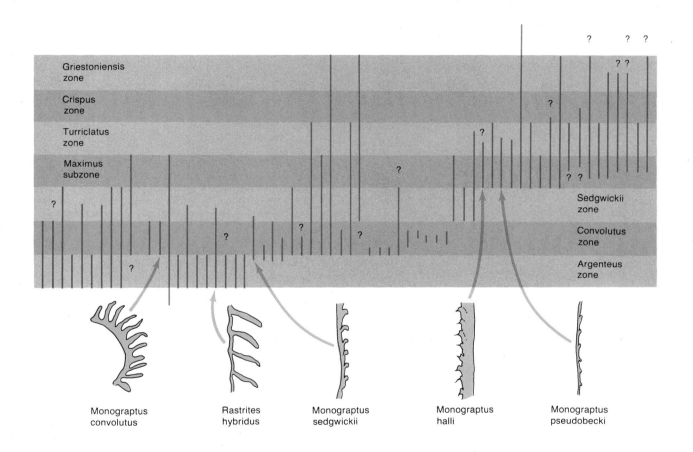

| Monograptus convolutus | Rastrites hybridus | Monograptus sedgwickii | Monograptus halli | Monograptus pseudobecki |

The graptolites are an extinct group of animals whose fossils are extremely useful for correlation. Graptolites included benthic species, but the planktonic species of the group represent better index fossils because many of these species were widespread and existed for only brief intervals of geologic time, especially during the Ordovician and Silurian periods.

The bar graph above depicts zones of the lower part of the Silurian System in the British Isles. Vertical bars represent the known ranges of species, several of which are illustrated. A question mark indicates that a range may be longer than that shown here. All ranges shown are for species that first appear in the *argenteus*, *convolutus*, or *sedgwickii* zone or the *maximus* subzone. Note that many species first appear at the base of a zone and thus help define the zone boundary. The *convolutus* zone is especially valuable for worldwide correlation, since it has been identified not only in many parts of Europe and Asia but also in Nevada and Alaska.

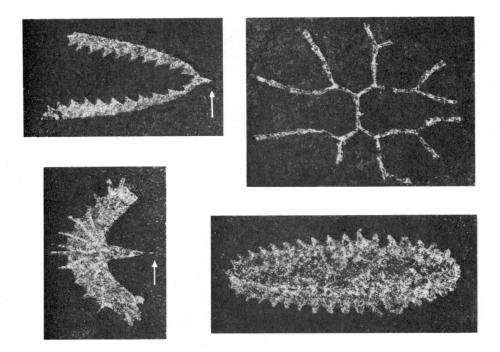

The photographs on this page illustrate four species of graptolites of Ordovician age that, in typical fashion for this group of animals, are fragile forms preserved as carbonized impressions in dark shale. Graptolites became extinct late in the Paleozoic Era. It is not certain what other animals they were related to, but it would appear that they most closely resemble living hemichordates, which are relatives of animals with backbones. Each of the specimens illustrated here represents a planktonic colony that included many individuals. Each individual was housed in a cup, one of the toothlike projections along the saw-toothed margin of the colony. The two photographs on the left show colonies that were attached—probably to floating seaweed—by a threadlike structure whose position is indicated by an arrow.

Studies of the composition and structure of the walls of graptolite colonies suggest that graptolites were closely related to living animals called pterobranchs. Pterobranchs are rare animals today, living on the sea floor as small colonies that consist of branching tubes. (*Photographs courtesy of the U.S. Geological Survey; graph adapted from R.B. Rickards, Geological Journal, 11:153–188, 1976.*) □

different places. Any uncertainties resulting from correlation with such a species would represent less than 2 percent of the Cenozoic Era, less than 1 percent of the Mesozoic Era, or less than 0.5 percent of the Paleozoic Era.

RADIOACTIVITY AND ABSOLUTE AGES

Early geologists such as William Smith were able to establish order for geologic events, but they could not assign actual time spans to the intervals that they recognized and named. Although they could determine, for example, that Cambrian rocks preceded Ordovician rocks and that Ordovician rocks preceded Silurian rocks, they did not know how long ago each of these groups of rocks came into being. The time scale developed by these geologists, in other words, was a **relative** one that had no reference to absolute time measured in years. It was not until the twentieth century that scientists were even able to begin to construct an absolute time scale that could depict whether a particular species existed between 62 and 63 million years ago or between 11.2 and 12.3 million years ago. This time scale rests almost entirely on dates derived from the study of naturally occurring radioactive elements.

Before radioactivity was discovered in the earth's crust, scientists were at odds concerning the age of the earth itself and had very little basis on which to determine the ages of various fossils. Late in the nineteenth century, as we have learned, British physicist Lord Kelvin calculated that the earth could not be older than about 40 million years and might even be as young as 20 million years. Lord Kelvin based his estimate on the assumption that the earth was cooling from a high temperature of formation. Kelvin knew that temperature rises with depth within a mine shaft, and he consequently believed that the high temperatures thus indicated for the earth's interior represented "fossil" heat remaining from the time when the earth came into being. His calculation of the earth's age was based on estimates of the earth's original temperature and its rate of cooling. Geologists, however, were confused by Kelvin's calculation; to them the thickness of sedimentary rocks on earth seemed to indicate that these rocks had been accumulating for well over 40 million years. Only after the discovery of radioactivity was this dispute resolved in favor of the geologists.

In 1895, Antoine Henri Becquerel discovered that the element uranium undergoes spontaneous **radioactive decay**, which means that its atoms change to those of another element by releasing subatomic particles and energy. Geologists soon recognized that this release of energy, which was found to occur not only during the decay of uranium but also during the decay of other naturally occurring radioactive elements, contributes heat to the earth's interior. It thus became evident that the earth had not cooled as rapidly as Lord Kelvin and his followers had believed. Even though the earth is very old, its interior has remained hot because what amounts to a radioactive furnace operates within it.

The second contribution of naturally occurring radioactive elements to our knowledge of the earth's age is quantitatively more precise. Specifically, radioactive elements and the products of their decay act as geologic clocks that enable us to measure the ages of rocks that contain these materials. These clocks are the primary source of the absolute scale of geologic time that now allows us to calibrate the relative time scale, which is based largely on superposition and on fossil occurrences. Before the use of radioactive isotopes, only unusual circumstances, such as the presence of annual varves in lake deposits (Figure 3-7), permitted geologists to estimate intervals of absolute time.

The utility of radioactive elements and their products lies in the fact that each radioactive element decays at its own nearly constant rate. Once this rate is known, geologists can estimate the length of time over which decay in a natural system has been occurring simply by measuring the amount of the radioactive parent element that is left in a rock as well as the amount that remains of the daughter element, or the element produced by the decay process. This procedure is known as **radiometric dating**. It has been found that radioactive decay occurs at a constant exponential or geometric rate, not at a constant arithmetic rate. This means, as Figure 5-2 illustrates, that no matter how much of the parent element is present at the start of decay, half of that amount will survive after a certain time, and then half of this surviving amount will remain after another interval of the same duration, and so on. This characteristic interval is known as the **half-life** of a radioactive element. Thus, in the course of four successive half-lives, the number of atoms of a radioactive element will decrease to one-half, one-fourth, one-eighth, and one-sixteenth of the original number of atoms.

Isotopes are forms of an element that differ in the number of neutrons in their nuclei and hence in their atomic weight. Some elements occur naturally as both radioactive and nonradioactive (or stable) isotopes. As shown in Table 5-1, several radioactive isotopes are abundant in rocks and are therefore useful for geologic dating. Most of these

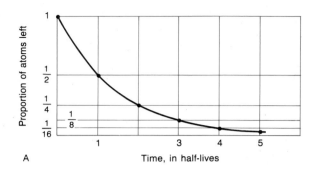

A

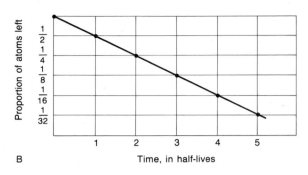

B

FIGURE 5-2   Graphs illustrating the pattern by which atoms are lost through radioactive decay. When plotted on a standard arithmetic scale (A), the number of atoms can be seen to decrease more slowly with each successive interval of time. When the number of atoms is scaled as a geometric progression (B), the plot forms a straight line; half of the atoms present at the beginning of each interval (or half-life) survive to the beginning of the next interval.

isotopes occur in igneous rocks; thus, knowledge of the amounts of parent and daughter elements currently present in an igneous rock allows us to calculate the time that has elapsed since the parent element was trapped when the rock formed by the cooling of a magma. A few minerals that form on the sea floor incorporate radioactive elements as well, and radiometric dates obtained for these minerals represent the interval of time that has elapsed since they formed. Radiometric clocks are often reset by metamorphism, which separates radioactive isotopes from their decay products.

As Table 5-1 illustrates, the half-lives of naturally occurring radioactive isotopes differ greatly, and these differences have a bearing on the ultimate use of various isotopes. More specifically, isotopes with short half-lives are useful only for dating very young rocks, while those with long half-lives are best used for very old rocks. These limitations are related to measurement problems associated with each isotope. In essence, isotopes such as carbon 14 that have short half-lives decay so quickly that their quantities in old rocks are too small to be measured. By the same token, isotopes such as rubidium 87 that have long half-lives decay so slowly that the quantities of their daughter elements in very young rocks are too small to be measured accurately.

Other limitations of radioactive isotopes are imposed by their patterns of occurrence. Potassium, for example, is much more abundant in the earth's crust than is uranium, and thus it is used much more frequently for radiometric dating even though the half-life of uranium 238 is of the same order of magnitude as that of potassium 40. Another problem with uranium dating is that some uranium-bearing minerals incorporate lead into their crystal lattices at the time of their formation, and since this lead is isotopically identical to the product of uranium decay, these minerals often appear older than they actually are. Fortunately, there is one uranium-bearing mineral—zircon—that does not incorporate lead into its crystal lattice when it forms. Radiometric dating of rocks containing zircon therefore tends to be highly accurate.

Table 5-1   Properties of some radiometric isotopes that are commonly used to date rocks

The number after each element name signifies the atomic weight of that element and serves to identify the isotope. Carbon 14, which has a very short half-life (that is, a high rate of decay), is used for dating materials younger than about 70,000 years. The other radioactive isotopes are employed for dating much older rocks.

Radioactive isotope	Approximate half-life, years	Product of decay
Rubidium 87	48.6 billion	Strontium 87
Thorium 232	14.0 billion	Lead 208
Potassium 40	8.4 billion	Argon 40
Uranium 238	4.5 billion	Lead 206
Uranium 235	0.7 billion	Lead 207
Carbon 14	5730	Nitrogen 14

Another uranium–lead dating technique, **fission-track dating**, also circumvents the problem of contaminating lead. When uranium 238 decays, it emits subatomic particles that fly apart with so much energy that they penetrate the surrounding crystal lattice, producing fission tracks. These tracks can be enlarged in the laboratory by acid etching, and can then be counted under a microscope. After these tracks have been counted, the remainder of the uranium can be subjected to a neutron field in the laboratory, causing it to decay completely. The number of tracks thus produced can be compared to the number that formed naturally, and the resulting numerical ratio reveals the age of the mineral.

Argon, the decay product of potassium 40, is an inert gas—that is to say, it is a gas that does not combine chemically with other elements. Argon does, however, become trapped in the crystal lattice of some minerals that form in igneous and metamorphic rocks. The potassium-argon system is best used for dating micas and certain other silicate minerals that incorporate no argon when they form, since all the argon found in these minerals can be assumed to have resulted from decay. A deficiency of the potassium–argon method is that argon leaks from the lattice of a crystal, making it appear that less than the actual amount of decay has occurred. This is especially true under temperatures and pressures slightly higher than those at the earth's surface.

The rubidium–strontium system is not susceptible to the type of leakage that often diminishes the accuracy of the potassium–argon system. This is because strontium, the decay product of rubidium, is a solid rather than a gas. Another useful aspect of rubidium–strontium dating is that rubidium occurs in micas and in feldspars, which are important minerals in igneous and metamorphic rocks. It also occurs in the clay mineral glauconite, which forms on the sea floor and can thus be used for direct dating of some sedimentary rocks.

Carbon 14 dating, or **radiocarbon dating**, is the best-known of all radiometric techniques—but because the half-life of carbon 14 is only 5730 years, this technique can be used only on materials that are less than about 70,000 years old. Most of the materials dated by this method are of biological origin, and many of these consist of wood. Despite its limitations, radiocarbon dating is of great value for dating materials from the latter part of the Pleistocene Epoch—an interval so recent that most other radioactive materials found in its sediments have not decayed sufficiently for their products to be measured accurately. Fortunately, the useful range of carbon 14 extends back far

enough to encompass the entire time interval during which modern humans have existed—as well as the interval during which glaciers most recently withdrew from North America and Europe at the close of the recent Ice Age (Pleistocene Epoch). Thus, radiocarbon dating plays a valuable role in the study of human culture, sometimes being used to date materials that are no more than a few hundred years old.

Carbon 14 is a rare isotope of carbon that forms in the upper atmosphere, about 16 kilometers (~10 miles) above the earth's surface, as a result of the bombardment of nitrogen by cosmic rays. Both carbon 14 and the stable isotope carbon 12 are assimilated by plants, which turn them into tissue. Once a plant dies, however, carbon is no longer incorporated in its tissues, and the carbon 14 that was present at the time of death decays back into nitrogen 14. Thus, the percentage of carbon 14 in the tissues of plants declines in relation to the percentage of carbon 12, and the ratio of the two can be used to determine when the tissue died. A basic assumption in radiometric dating is that the rate of carbon 14 production has been constant during the past 70,000 years and that the ratio of carbon 14 to carbon 12 in the atmosphere has also remained constant. Although no major errors have resulted from this assumption, some minor errors have been noted. As an example, wood used by the ancient Egyptians at times that are well documented in historical records have yielded radiometric dates that are slightly too early. Such errors, which exceed 10 percent for material 5000 years old, apparently result from minor changes in the rate of carbon 14 production in the upper atmosphere.

FOSSILS VERSUS RADIOACTIVITY: THE ACCURACY OF CORRELATION

The fact that the half-lives of radioactive isotopes are well established should not be taken to indicate that radiometric dating offers more accurate correlation of sedimentary rocks than does the use of fossils. For one thing, most minerals that can be dated radiometrically are of igneous origin, and, as shown in Figure 5-3, the dating of igneous rocks often yields only a maximum or a minimum age for associated sedimentary rocks. Clasts of igneous rocks or minerals found in sedimentary rocks can also be dated radiometrically, but only maximum estimates for the ages of sedimentary rocks can be derived in this way.

There are even more fundamental uncertainties inherent to radiometric dating. Most published radiometric dates,

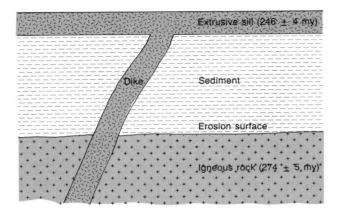

FIGURE 5-3   A diagrammatic cross section of a group of rocks showing how the age of a sedimentary unit can be bracketed by the radioactive dating of associated igneous rocks. In this example, the sediments lie on top of a body of igneous rock dated at 274 ± 5 million years and are therefore younger than this. The sediments are also cut by a dike and covered by a sill of igneous rock dated at 246 ± 4 million years; they are therefore older than this. We can conclude that this sedimentary unit may be as old as 279 million years or as young as 242 million years.

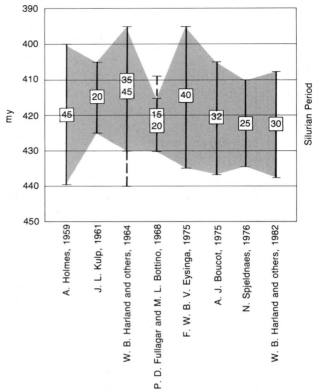

FIGURE 5-4   Estimates of the interval of time represented by the Silurian Period. The date each estimate was made and the author or authors responsible for each estimate are listed. (*Adapted from N. Spjeldnaes, Amer. Assoc. Petrol. Geol., Studies in Geol. 6:341–345, 1978.*)

for example, are followed by the symbol ±, which precedes a number representing a smaller interval of time (e.g., 472 ± 7 million years). This plus-or-minus sign indicates uncertainty that is attributable to possible errors in the measurement of the quantities of parent and daughter elements. Thus, the expression 274 ± 5 million years in Figure 5-3 indicates that errors of measurement leave open a strong possibility that the rock depicted may be as old as 279 million years or as young as 269 million years.

Not indicated by these plus-or-minus figures is the potential error that results from the fact that only parent and daughter atoms actually present in a rock can be detected. In accepting a date, even with a plus-or-minus figure, we are assuming that a dated rock has remained a closed system—i.e., that it has neither lost nor received parent or daughter atoms from some other source. Unfortunately, this is not always the case. Rocks can, in fact, both gain and lose atoms, although loss is a more frequent problem than gain. This is especially true of the potassium–argon decay system, whose daughter element, as we have seen, frequently seeps from rock, leading to underestimation of the amount of decay and hence to underestimation of rock age. Even solid parent or daughter elements representing other decay systems can be leached from rocks by water, leading to similarly inaccurate estimates of age.

These types of errors, which can beset even the most

meticulous radiometric analysis, often add up to sizable total errors, especially when very old rocks are being dated. This point is well illustrated by past estimates made for the beginning and end of the Silurian Period (Figure 5-4). In evaluations made between 1959 and 1968 alone, the duration of the Silurian was halved and then doubled and then halved again. It seems likely that the Silurian began between 440 and 430 million years ago and that it ended between 410 and 400 million years ago—but the exact dates remain uncertain.

Fossil graptolites (Box 5-1) of Silurian age allow for more accurate correlation of sedimentary rocks than do radiometric dates. Many kinds of graptolites floated in the sea, thus spreading quickly over large areas. In addition, graptolites evolved rapidly, and these two factors together have made graptolites highly useful index fossils. Most individual species of graptolites existed for about a million years; thus, correlations based on the occurrences of such species cannot be inaccurate by a larger interval than this. Even widely separated regions that have no fossil graptolite species in common may share certain graptolite genera, and these genera provide for correlations within the Silurian System that are more accurate than correlations based on radiometric dates. Because radiometric dates are not accurate enough to establish the exact times of appearance and disappearance of individual graptolite species, geologists can only estimate very approximately the lengths of time that individual species survived. It can be noted, for example, that there are about 30 successive graptolite zones in the Silurian System and that the Silurian lasted about 30 million years. Therefore, because the geologic ranges of many species coincide or nearly coincide with single zones (Box 5-1), it can be concluded that many graptolite species lived for about a million years.

Although radiometric dating provides a special kind of geologic time scale—specifically an absolute scale, or one based on years—most geologic correlations are still based on fossil occurrences. This is true not only because fossils are more common in sedimentary rocks than are radioactive elements, but also because the analysis of fossils usually allows for greater accuracy. The geologic intervals on either end of the geologic scale of time, however, represent striking exceptions to this general rule. First of all, most rocks that are older than about 1.4 or 1.5 billion years contain few fossils that are well enough preserved and easily enough identified to serve as index fossils. Thus, for these early rocks—most of which are Precambrian, or Cryptozoic—radiometric dates serve as a primary basis for correlation, especially from continent to continent. The second exception is the interval extending from the present back to about 70,000 years ago. For this interval, **radiocarbon dating** offers estimates of age that commonly have plus-or-minus values of only a few percent. Because relatively few species of animals or plants have appeared or disappeared during this brief interval, fossils from the corresponding rocks, have much less value in correlation.

In the following sections, we will examine several other stratigraphic features that in many cases provide more precise correlation than either fossils or radioactive isotopes.

TIME-PARALLEL SURFACES IN ROCKS

Imagine that someone with supernatural powers were, in an instant, to spread a thin layer of durable plastic over many areas of the earth—ocean floors, lake bottoms, and terrestrial lowlands. In keeping with the law of superposi-

tion, this layer would become a **time-parallel surface**—that is to say, it would separate deposits that had formed before it was laid down from accumulations that were laid down afterward. Similarly, if a powerful force were suddenly to lower sea level throughout the world by 200 meters, shoreline deposition would shift seaward for great distances, and the first of the new shoreline deposits thus created would be of nearly the same age on all continents, allowing for accurate global correlation. These two hypothetical occurrences—the formation of a time-parallel surface and a sudden relocation of shorelines—are not far removed from actual events whose geologic records have enabled us to correlate rocks of separate regions. We will consider these two types of events and others in the sections that follow.

Key beds

Key beds or **marker beds** are real sedimentary beds that resemble the hypothetical plastic layer described above. These beds may not be perfectly time-parallel (i.e., of identical age) throughout, but on a geologic scale of time, they are nearly synchronous. Widespread layers of volcanic ash, for example, often function as key beds. Sometimes an ancient ashfall, which may represent either a single volcanic eruption or a series of nearly simultaneous eruptions, can be traced for thousands of square kilometers, providing a time-parallel surface in the stratigraphic record. Figure 5-5 shows an ash bed that formed from a volcanic eruption in the region of the Appalachian Mountains more than 350

FIGURE 5-5   The Center Hill volcanic ash bed, a widespread key bed or time marker within the Devonian Chattanooga Shale in DeKalb County, Tennessee. The ash bed lies near the center of the hammer handle. (*L. C. Conant, U.S. Geological Survey.*)

FIGURE 5-6   A tillite of late Precambrian age representing a key bed for correlation. This particular tillite is from western Scotland. Similar beds produced by an interval of widespread glacial activity are found in many parts of the world. (*British Geological Survey*.)

million years ago, when these mountains were first beginning to form.

Glacial tills such as those described in Chapter 3 also serve as useful marker beds; even tills formed by different glaciers in different parts of the world can be of use if they represent only a brief interval of global cooling. In many regions of the world, layers representing the upper part of the Precambrian interval, whose rocks are otherwise not easily dated, contain glacial tillites (Figure 5-6). The glacial interval that these tillites represent may have lasted a million or even several million years, but such an interval is brief in comparison to the span of time—more than 600 million years—that has elapsed since the glaciation took place.

Certain events produce key beds that allow for correlation within particular basins of deposition. The character of deep-water evaporite deposits in a basin, for example, can reflect aquatic conditions throughout the basin. Because deep-water evaporites (Appendix I) typically occur in widespread horizontal beds (Figure 5-7), an individual bed that differs in chemical composition from beds above and below can often be traced over thousands of square kilometers.

Shifting of depositional boundaries

No environment stretches infinitely far in any direction. Instead, one environment of sediment deposition inevitably gives way to another, and when it does so, there is either

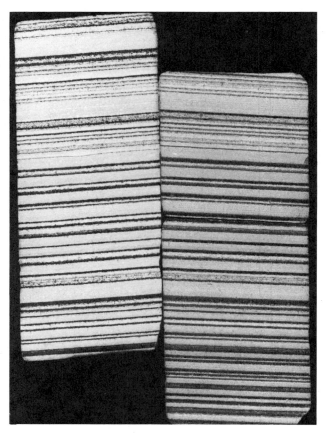

FIGURE 5-7   Two cores taken from the Castile evaporites, which were precipitated in western Texas near the end of the Permian Period. The cores are from localities 14.5 kilometers (~9 miles) apart, yet their lamination matches almost perfectly, allowing for precise correlation. The alternating dark and light bands, which range up to a few millimeters in thickness, probably represent seasonal organic-rich (winter) and organic-poor (summer) layers. If this is true, each pair of bands represents one year, in which case the 200,000 or so paired bands of the Castile Formation represent about 200,000 years of deposition. (*R. Y. Anderson et al., Geol. Soc. Amer. Bull. 83:59–86, 1972.*)

an abrupt or a gradual transition from the first mode of deposition to the second. The set of characteristics of a body of rock representing a particular local environment is called a **facies**, and, accordingly, lateral changes in the characteristics of an ancient body of rock, which reflect lateral changes in the depositional environment, are known as **facies changes** (Figure 5-8). Thus, one facies in a body of rock might consist of a reef of porous limestone containing fossil reef-building organisms in their living positions. This facies might pass laterally into a reef-flank facies characterized by steeply dipping talus deposits of poorly sorted rubble from the reef (Figures 4-15 and 4-16), and this second facies might, in turn, grade into an interreef-basin facies composed of fine-grained clayey limestone.

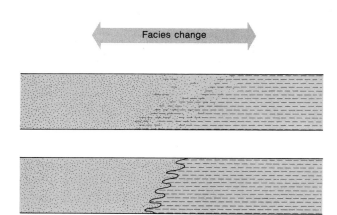

FIGURE 5-8   Diagrams illustrating a facies change, from sandstone on the left to shale on the right. The upper and lower diagrams illustrate two ways in which facies changes are depicted in the stratigraphic cross sections that geologists publish. The depiction in the upper diagram emphasizes the gradational nature of the facies change.

In the previous chapter, we observed that when a sandy beach progrades over muddy offshore deposits, the process of progradation pushes the shoreline seaward. This creates a shoreline migration known as **regression** (Figure 5-9A). In other instances, the shoreline and the facies seaward of it shift in a landward direction in a process known as **transgression** (Figure 5-9B). In still other cases, a transgression is followed by a regression (Figure 5-9C), or vice versa, resulting in a more complex pattern that is very useful for correlation because it contains a time-parallel surface (or a time-parallel line if the facies are depicted in a two-dimensional cross section). This surface (or line) connects the facies boundaries where they were positioned at the time of maximum transgression. Most time-parallel surfaces of this type are useful only for correlating stratigraphic sections that represent single depositional basins—for example, sections that represent different parts of a lake or inland sea. The unconformities discussed in the next section of this chapter can provide evidence for correlation over longer distances.

Unconformities, bedding surfaces, and seismic stratigraphy

Unconformities that represent the same interval of tectonic uplift or nondeposition sometimes constitute nearly time-parallel surfaces. Such unconformities may truncate rocks of many different ages, but the sediments resting directly on top of the erosional surface are often nearly the same age in all parts of a depositional basin. The deposits resting on the surface of an unconformity will rarely be precisely

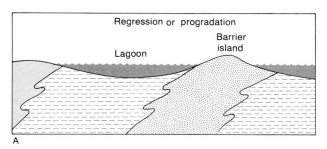

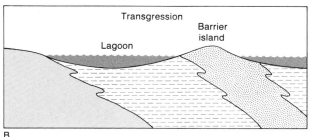

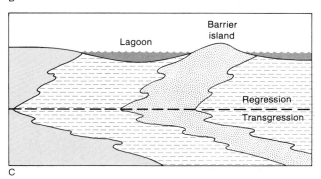

FIGURE 5-9   Correlation based on a stratigraphic pattern in which a regressive depositional sequence follows a transgressive sequence. A. The origin of a regressive (progradational) sequence. B. The origin of a transgressive sequence. C. When a regression follows a transgression, the points of maximum transgression of various facies can be connected to form a line of correlation, as indicated by the broken line in the diagram. A similar line can be constructed for a stratigraphic pattern in which a transgression follows a regression.

the same age, however, and in some cases, the ages of these deposits will vary greatly from place to place. Marine waters that have deserted an area, for example, may later invade it again slowly, after a period of erosion, reaching different parts of the area at different times. On the other hand, if sea level drops suddenly throughout the world and then rises again rapidly, the resulting global unconformities may represent fairly accurate time markers.

It is not known why sea level has dropped periodically in the course of the earth's history. As we will see in Chapter 18, sea level fell repeatedly within intervals of just a few thousand years during the recent Ice Age, when glaciers

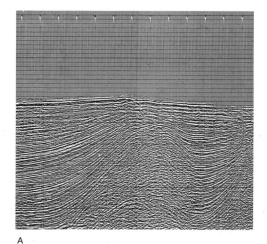

A

Satellite navigation antenna

Underwater phones detect seismic echoes from rock layers

Sounder source

Sea floor

Bottom mud

Bottom mud

Rock layers

Rock layers

B

FIGURE 5-10   *A. A section of the Gulf of Mexico, 30 km long and 10 km deep, in which folded sedimentary layers are revealed by reflected seismic waves. The use of seismic reflections to study sediments and rocks buried beneath the sea floor. Sound waves are generated by a sounder that makes a pneumatic explosion like that of a bursting balloon. The sound waves bounce off surfaces of discontinuity, which include bedding surfaces and unconformities. Underwater phones then pick up the reflections, allowing marine geologists to determine the configuration of buried features. Fossils recovered from sediment ores reveal the ages of individual sedimentary beds. (A. Petty Geophysical Engineering Co. B. After F. Press and R. Siever, Earth, W. H. Freeman and Company, New York, 1982.)*

expanded over the land, "locking up" water and thus removing it from the global hydrological cycle. Nonetheless, there is no evidence that similar continental glaciation took place during earlier intervals in which sea level fell suddenly. It has been suggested that swelling and subsiding of the deep sea floor has moved sea level up and down, but geophysicists have yet to discover any mechanism within the earth that could do this rapidly enough to lower sea level by 50 or 100 meters during intervals of much less than a million and perhaps just a few thousand years.

In general, the global unconformities that have been used most successfully as time markers are those that occur within Mesozoic and Cenozoic sediments lying along continental margins, which are distinguished by the fact that, for the most part, they have remained below sea level since their formation and thus have not been destroyed. Such sediments and unconformities have been studied by means of the seismic reflections that are generated when artificially produced seismic waves bounce off physical discontinuities within buried sediments (Figure 5-10). Some of these discontinuities have been found to be time-parallel bedding surfaces, and others have been identified as unconformities.

A seismic study of discontinuities within the deposits of a continental shelf produces a profile like that shown in Figure 5-11. The interpretation of such a profile is based not only on the positions of stratigraphic discontinuities, but also on lithologic and paleontologic data obtained from drilling operations. The profile in Figure 5-11 shows Cenozoic continental-shelf sediments resting on Mesozoic deposits, which include a prograding siliciclastic package of continental-shelf and slope deposits. Beneath these are a carbonate bank and deposits that accumulated seaward of the bank in deep water.

Figure 5-12 illustrates how seismic profiles can reveal changes in sea level. These changes are recorded in the upward and downward migration of both the shoreline and the continental margin.

When compilations of many seismic profiles are used together with fossil evidence to date the local changes in sea level, global patterns of sea-level change may result. Evidence that sea level rose or fell in many different areas at the same time indicates that the change occurred on a global scale. Estimates of times and amounts of change have yielded a global curve of sea-level changes for the Cenozoic Era (Figure 5-13). Less precise information for earlier intervals, based in part on rocks, fossils, and unconformities visible on the continents, has yielded a less detailed curve for the entire Phanerozoic interval (Figure 5-14).

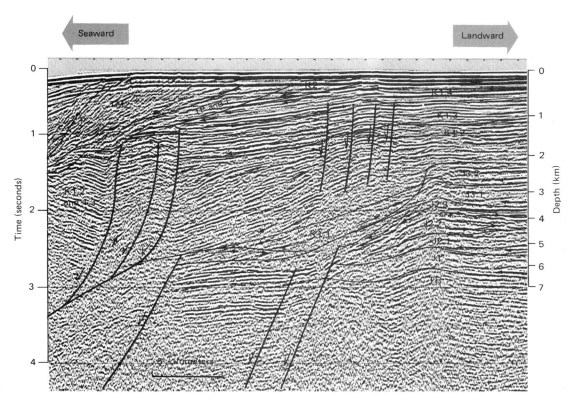

FIGURE 5-11 A seismic section off the shore of northwestern Africa. The fuzzy dark lines indicate surfaces of seismic reflection. The interpretation of these surfaces has been aided through study of sediments and fossils brought up by drilling. The oldest sediments are Triassic in age (TR). Above these are numerous distinct packages of sediment separated by unconformities (*thin lines*) and broken by faults (*thick lines*). The arrows with solid heads indicate landward or seaward building of sediments. A Lower Jurassic package of sediments (J1) is followed by Middle Jurassic packages (J2.1–J2.3) that represent shelf and slope environments. Upper Jurassic and earliest Cretaceous packages (J3.1, J3.2) record the growth of a carbonate bank with a steep seaward slope. This bank is buried beneath three thick sequences of Lower Cretaceous sediments (K1.1–K1.4) separated by unconformities. The oldest of these (K1.1) is a deep-water deposit representing a continental rise. Thin Upper Cretaceous (K2) and Paleocene and Eocene (TP and E) units are followed by a thick prograding sequence of siliciclastic deposits (TM2). (*R. M. Mitchum et al., Amer. Assoc. Petrol. Geol. Mem. 26:53–62, 1977.*)

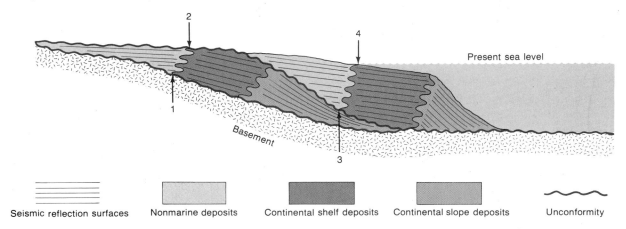

FIGURE 5-12 Diagrammatic cross section showing how relative changes in sea level are recorded in seismic profiles. Two packages of sediment are separated by unconformities. The boundary between nonmarine and marine deposits represents the shoreline (*arrows*), and the movement of this boundary reveals relative rises and falls of sea level. First, sea level rose from position 1 to position 2 or higher during an interval of progradation. Sea level then dropped to position 3 or lower, and an unconformity developed. Progradation followed, and sea level rose to its present position (4). Often, the rising and falling of sea level are also recorded in changes in the position of the continental margin, which separates shelf and slope deposits.

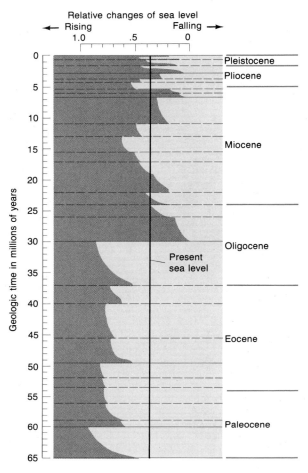

FIGURE 5-13   Estimates, based primarily on seismic stratigraphy, of relative changes in sea level during the Cenozoic Era.
Horizontal segments of the zigzag sea-level curve represent
sudden drops in sea level. The horizontal scale is in arbitrary
units. (*After P. R. Vail et al., Amer. Assoc. Petrol. Geol. Mem.
16:83–97, 1977.*)

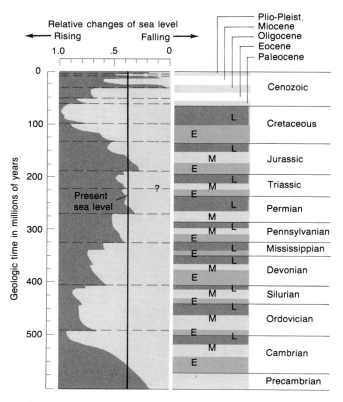

FIGURE 5-14   Estimates of relative changes in sea level during Phanerozoic time (E, M, and L stand for Early, Middle, and
Late). As in Figure 5-13, the horizontal scale is in arbitrary
units, and the sea-level curve is based on seismic stratigraphy.
The Cenozoic portion of this curve differs from Figure 5-13 in
that only large-scale changes are depicted. (*After P. R. Vail et
al., Amer. Assoc. Petrol. Geol. Mem. 26:83–97, 1977.*)

The changes in sea level shown in Figures 5-13 and
5-14 are based on seismic evidence of simultaneous occurrences in many different parts of the world and are thus
assumed to reflect global, or **eustatic**, changes. However,
global changes in sea level are not reflected in seismic
profiles where there is an offsetting tectonic change in the
position of the land (Figures 1-15 and 5-15). For example,
if the land in a local area rises in pace with a eustatic rise
in sea level, the eustatic rise will not be expressed as it
would be if the land were to remain stationary (Figure
5-16*A*). Moreover, if the tectonic uplift exceeds the eustatic
rise in sea level, there will actually be a local lowering of
sea level (Figure 5-16*B*) despite what is happening in the
rest of the world.

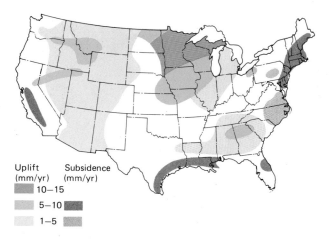

FIGURE 5-15   Rates of uplift and subsidence of the crust today in the United States. (*After S. P. Hand, National Oceanic
and Atmospheric Administration.*)

Time I   Time II

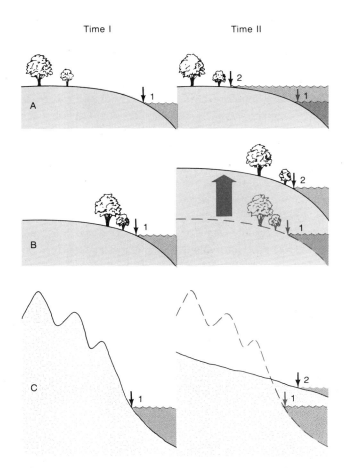

FIGURE 5-16   Diagrammatic cross sections of shorelines showing that a rise in sea level does not necessarily result in a transgression. In each of the three pairs of diagrams, the initial and final positions of sea level are numbered 1 and 2. *A.* The land remains unchanged, and a rise in sea level causes a transgression. *B.* A rise in sea level is accompanied by regression rather than by transgression because the land rises tectonically (*heavy arrow*) more than does sea level. *C.* A rise in sea level is accompanied by regression (progradation) rather than by transgression because sediment eroded from nearby highlands pushes the shoreline seaward.

In addition, vertical sea-level changes have often been mistaken for transgressions and regressions, which are lateral shifts in the position of the shoreline. Of course, a strong correlation exists both between transgressions and eustatic rises and between regressions and eustatic falls— but there is a complicating factor that weakens this correlation, making transgressions and regressions unreliable indicators of eustatic changes. As Figure 5-16C illustrates, the rate of influx of sediment strongly influences whether transgression or regression will occur in a particular area. Thus, regressions have often occurred locally during global

intervals of rising sea level simply because sediment that has been supplied from the land at a very high rate has pushed the sea back from the land. Certainly patterns of transgression and regression offer some evidence of the history of eustatic sea-level changes during Phanerozoic time, but the picture that they supply is much less accurate or detailed than that provided by the seismic-profiling technique illustrated in Figure 5-12.

Magnetic stratigraphy

There is one kind of global event that provides for accurate correlation throughout the world—a reversal in the polarity of the earth's magnetic field. As we have seen, the earth's core consists of dense material made up of iron and other heavy substances. In the outer part of the core, this material is in a liquid state, and its motion generates a magnetic field. As a result, the earth behaves like a giant bar magnet (Figure 5-17). When iron-containing minerals form sedimentary or igneous rocks at or near the earth's surface, they often become aligned with the earth's magnetic field just as a compass needle does when allowed to rotate freely. Small grains of iron minerals that settle from water to become parts of sedimentary rocks often rotate in such a way that their magnetism aligns with that of the earth; and iron minerals that crystallize from lava or magma

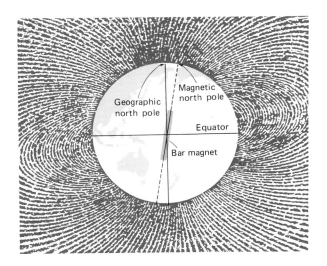

FIGURE 5-17   The earth's magnetic field, illustrated by the lines of force surrounding the earth. This field resembles the one that would be produced by a bar magnet located within the earth with its long axis inclined slightly (11°) from the earth's axis of rotation. (*After F. Press and R. Siever, Earth, W. H. Freeman and Company, New York, 1982.*)

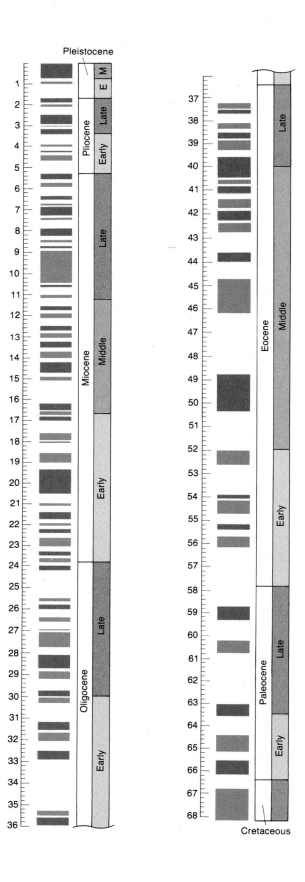

automatically become magnetized by the earth's magnetic field as they grow from the liquid.

It is a startling fact that, for unknown reasons, the north and south magnetic poles of the earth switch positions periodically. Intervals between such **magnetic reversals** vary considerably, but during the Cenozoic Era they have averaged about a half-million years in length. Sequences of magnetized rocks that can be dated radiometrically have revealed the history of magnetic reversals for the Cenozoic Era (Figure 5-18) and for much of the Mesozoic Era as well. Intervals during which the polarity was the same as today are known as **normal intervals**, and intervals during which the polarity was the opposite of what it is today are called **reversed intervals**. Among the rock sequences most useful for magnetic correlation are multiple lava flows, each member of which was magnetized as it cooled. The piles of volcanic rocks that result often provide a detailed history of magnetic reversals.

The primary difficulty encountered in making use of the magnetic record for correlation lies in determining which worldwide event is reflected by a locally observed magnetic reversal. Particularly helpful in this respect is a "signature," that is, a distinctive sequence of reversals. In rocks that are known to be of Eocene age, for example, a pattern that is found to involve a long normal interval flanked by two long reversed intervals can only be Early Middle Eocene in age (Figure 5-18). Used in conjunction with other dating methods, magnetic stratigraphy has greatly improved the geologist's ability to date Cenozoic sediments throughout the world.

THE UNITS OF STRATIGRAPHY

Having examined the ways in which rocks are correlated and dated, we can now appreciate how they are formally classified. As we saw in Chapter 1, geologists divide the stratigraphic record into local three-dimensional bodies of rock that are known as **formations**. Recall that formations are sometimes united into larger divisions known as groups and sometimes include smaller units called members.

FIGURE 5-18   Magnetic polarity scale for the Cenozoic Era. About half of the time the magnetic field has had normal polarity (polarity like that of the present), and about half of the time the polarity has been reversed. (*After W. B. Berggren et al., Geology, in press, 1984.*)

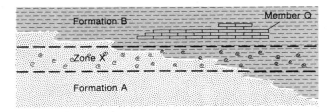

FIGURE 5-19   Diagrammatic cross section showing two formations (A and B) that are separated by a surface that is not time-parallel. The boundaries of the biostratigraphic zone X are approximately time-parallel, and they pass across the boundary between the two formations at an angle. A boundary like the one between formations A and B can result from either transgression or regression (see Figure 5-9). Within formation B, which consists primarily of shale, is a member (Q), a thin layer of limestone.

Groups, in turn, may be united into supergroups. All these entities are known as **rock-stratigraphic units**, and they are given formal names that are often based on local geographic features such as rivers and towns. Formations are delineated on the basis of lithology: they are relatively homogeneous bodies, many of which consist of a single rock type. Others, as we have seen, include two or more rock types in alternating layers; examples of these are the sedimentary rock cycles produced by deltas (Figure 4-4) and by meandering rivers (Figure 3-23). Some formations display modest lateral facies transitions, but a major facies change, even if gradual, usually leads geologists to divide a body of rock into two adjacent formations. A formation is assigned a **type section** at a particular locality, and it is here that its upper and lower boundaries are defined.

Rock-stratigraphic units are designated without regard to time relationships—that is to say, their upper and lower boundaries may or may not be time-parallel. Therefore, geologists have superimposed on the rock-stratigraphic classification another classification system that is based on time relationships revealed by fossil occurrences (Figure 5-19). The basic unit of this **biostratigraphic** system is the zone, which we have already discussed (Box 5-1). As we have seen, the upper and lower boundaries of zones are not perfectly time-parallel surfaces. In most cases, however, they are nearly time-parallel, and thus the biostratigraphic classification of rocks approximates a chronological classification.

It is primarily on the basis of biostratigraphic zones that the earth's stratigraphic record is formally divided into worldwide systems, corresponding to the geologic periods: the Cambrian System, the Ordovician System, the Silurian

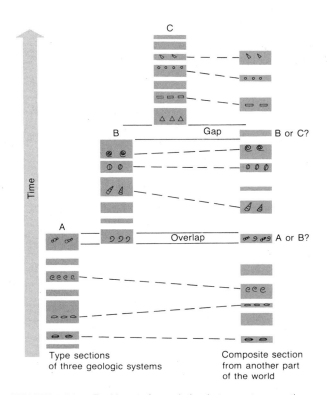

FIGURE 5-20   Problems of correlation between type sections. A, B, and C represent type sections of three geologic systems in different areas. The vertical scale is based on time rather than on sediment thickness. On the right is a composite section showing the systems in yet another, and more fully exposed, area. Gaps resulting from deposition or erosion are found in all these sections. Lines of correlation by means of fossil correspondence (dashed lines) show varying degrees of accuracy, with time lines horizontal. Systems A and B overlap in time, and there is a gap in time between systems B and C. Even if perfect correlation were possible, part of the composite section on the right could not be assigned to a system.

System, and so on. Each system is represented by a single **type section**, which is a local stratigraphic section that serves to define the system. As Figure 5-20 shows diagrammatically, correlation of a type section with sections in other regions is inevitably imperfect. In fact, because of this imperfection, some systems as now recognized may overlap slightly with others. Systems are grouped into **erathems** and divided into **series**, and series are further divided into **stages**. These correspond to eras, epochs, and ages and, like systems, are given the same formal names (Table 5-2). Important stages of the geologic column are listed in Appendix IV.

Are systems, series, and stages biostratigraphic units? It is logical to answer yes because the boundaries of these

Table 5-2   Geologic time units and
time-stratigraphic units

Time-stratigraphic units are bodies of rock that represent time units bearing the same formal name. When an epoch (the time unit) is designated by the term *Early* or *Late*, the corresponding series (the time-stratigraphic unit) is identified by the adjective *Lower* or *Upper*. For example, the Lower Devonian series of rocks represents Early Devonian time.

Time unit	Era	Period	Epoch	Age
Example	Paleozoic	Devonian	Late Devonian	Fammenian
Time-stratigraphic unit	Erathem	System	Series	Stage
Example	Paleozoic	Devonian	Upper Devonian	Fammenian

divisions are based largely on zones, which are biostratigraphic units; but because the imperfections of zonal correlations represent only small errors when they are considered as percentages of large units such as stages, series, and systems, these units have traditionally been treated as if they had time-parallel boundaries. Thus they have been designated as **time-stratigraphic units** (Table 5-2). As we have seen, the absolute ages of the boundaries between these units have been known in an approximate way only after radiometric dating was developed in the twentieth century, and even today most individual boundaries remain imperfectly dated.

CHAPTER SUMMARY

1.   In accordance with the law of superposition, stratified rocks accumulate in such a way that the oldest beds are at the bottom and the youngest are on top.

2.   Because species of plants and animals have appeared and disappeared over large areas of the earth in the course of geologic time, fossils provide a useful

means of correlating widely separated bodies of sedimentary rock.

3.   The most useful species for correlation are called index or guide fossils. Ideally these species are easily identified, widely distributed, abundant in many kinds of rock, and restricted to narrow vertical stratigraphic intervals.

4.   Index fossils are used singly or in groups to define biostratigraphic zones in rocks. The boundaries of most zones are nearly time-parallel.

5.   The scale of geologic time based on fossil occurrences is a relative one in which bodies of rock are simply given an order from oldest to youngest. In contrast, the scale based on radiometric dating is an absolute scale in which events are measured in years.

6.   Errors in radiometric dating are so large that correlations based on radiometric dates are often less accurate than those based on index fossils.

7.   Key beds such as ash falls, glacial tills, and evaporite beds are sedimentary layers that form almost simultaneously over large areas, thus serving as useful time markers.

8.   Unconformities are useful for identifying times in the earth's history when sea level has suddenly dropped throughout the world. The analysis of seismic reflections often allows geologists to recognize these unconformities and their associated bedding surfaces deep within the earth.

9.   Periodic reversals in the polarity of the earth's magnetic field provide excellent markers for correlation of magnetized rocks.

10.   A body of rock that is characterized by a particular lithology or group of lithologies is often recognized as a formal rock unit (a formation, group, or member). Rock units do not necessarily have time-parallel upper or lower boundaries; thus, they are often transected obliquely by biostratigraphic units.

11.   Time-stratigraphic units are composite bodies of rock that have been recognized on a global scale. To the extent that correlations are accurate, time-stratigraphic units (erathems, systems, series, and stages) correspond to units of geologic time such as eras, periods, epochs, and ages.

EXERCISES

1. What factors prevent biostratigraphic zones from having boundaries that are perfectly time-parallel?

2. What factors lead to imperfections in radiometric dating?

3. Construct a diagram resembling Figure 5-9C but depicting the depositional history of a barrier island–lagoon complex that has experienced a regression followed by a transgression.

4. Construct a diagram resembling Figure 5-16B but illustrating that a lowering of sea level need not always lead to regression.

ADDITIONAL READING

Ager, D. V., *The Nature of the Stratigraphic Record*, John Wiley & Sons, New York, 1973.

Eicher, D. L., *Geologic Time*, Prentice-Hall, Inc., Englewood Cliffs, New Jersey, 1976.

Faul, H., *Ages of Rocks, Planets, and Stars*, McGraw-Hill Publishing Company, Inc., New York, 1966.

Vail, P. R., and J. Hardenbol, "Sea-Level Changes during the Tertiary," *Oceanus* 22:71–79, 1979.

CHAPTER 6
Evolution
and the
Fossil Record

The central concept underlying modern biology is that living species have come into being as a result of the evolutionary transformation of quite different forms of life that lived long ago. Indeed, it is often maintained that very little of what is now known about life would make sense in any context other than that of organic evolution. It is important to remember, however, that while the broad definition of *evolution* is "change," *organic evolution* does not encompass every kind of biological change; instead, the term refers only to changes in populations, which are groups of individuals that live together and belong to the same species.

When we examine the broad spectrum of living forms that inhabit our planet, we cannot help but be impressed by the success with which each form functions in its own particular circumstances. Members of the cat family, for example, have razor-edged cheek teeth that allow them to cut meat with great efficiency. Horses, on the other hand, are equipped with broad cheek teeth covered with low ridges, which are well suited to the task of grinding up harsh grasses

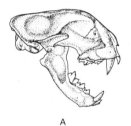

A   B

FIGURE 6-1   Comparison of the teeth of a cat and a horse, both of Pliocene age. *A.* The cat (*Pseudaelurus*) has sharp fangs in front for puncturing the flesh of prey, together with bladelike molars in the rear for slicing meat; upper and lower molars slide past each other like scissor blades. *B.* The horse (*Equus*) has chisel-like front teeth for nipping grass and broad molars for grinding it up; the upper and lower molars meet along broad grinding surfaces.

that contain tiny fragments of silica (Figure 6-1). Plants, too, exhibit a variety of forms and features that differ according to the plants' ways of life. Most tree species that are native to tropical rain forests, for example, have leaves that are, for the most part, waxier than those of plants found in cool regions, and a typical tropical leaf terminates in an elongate tip called a drip point. The drip point and the waxy surfaces help the leaves shed the rainwater that falls on them daily in a rain forest; the waxy surfaces also keep them from drying out in the tropical heat. In contrast, the leaves of another rain-forest plant, the bromeliad, form a cup that acts as a private reservoir for rain (Figure 6-2). Without this feature, the bromeliad would dry up and die, since it lives high above the moist forest floor, attached to trees. Yet another rain-forest plant, the Venus's-flytrap, secretes a sweet nectar that lures insects into the hollow

Evidence from both fossil and living animals convinced Charles Darwin, the great evolutionist, that life evolves. Armadillos provided part of this evidence. Both their fossil and living representatives are restricted to the Americas. The huge animal in the background of this picture is *Glyptodon*, an extinct armadillo from the Pleistocene Series of South America. This animal, which was about 3 meters (~9 feet) in length, almost seems to be looking disgruntled at the success of its living cousin, the smaller nine-banded armadillo in the foreground. Charles Darwin excavated the fossil remains of *Glyptodon* during his expedition to South America. (*The Smithsonian Institution*.)

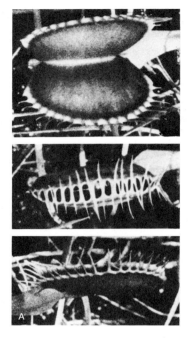

FIGURE 6-2   Leaves that serve unusual functions. *A*. The Venus's-flytrap produces a sweet substance that lures an ant onto a gaping pair of leaves that then snap shut, imprisoning the insect behind the spines on the leaf margins. *B*. Bromeliad plants growing on a branch that has fallen into a tributary of the Amazon River. Bromeliads trap rainwater in cuplike depressions at the bases of their leaves. (*A. Courtesy of S. E. Williams. B. Courtesy of M. J. Balick.*)

space between its leaves. On the margins of these leaves are rows of spines that mesh when the leaves snap shut around an unsuspecting insect. The Venus's-flytrap then devours the insect in a reversal of the normal roles of plant and animal.

These specialized features, which allow animals and plants to perform one or more functions that are useful to them, are known as **adaptations**. Each individual organism possesses many adaptations that function together to equip it for its particular way of life. Before the middle of the nineteenth century, however, the nature of these adaptations was not well understood. At that time, it was assumed that adaptations represented perfect mechanisms that had been specially designed to allow each species to function optimally within its own ecologic niche. Since that time, it has become widely acknowledged that adaptations are fraught with imperfections, many of which stem from evolutionary heritage. An animal or plant, in other words, may develop a useful new feature with which to perform a function, but the evolution of this feature will sometimes be constrained by the structure of the ancestral organism. Evolution can operate only by changing what is already

present; it cannot work with the freedom of an engineer who is designing a new device from raw materials. The business of evolution, in other words, is and always has been remodeling—not new construction.

The fact that evolution works by remodeling is evidenced by the observation that certain organs, such as the cheek teeth of mammals and the leaves of higher plants, serve different functions in different species but nonetheless share a common "ground plan," or fundamental biological architecture. All mammals, for example, possess teeth that are rooted in bone and consist of both dentin and enamel. Similarly, certain types of cells and tissues form the leaves of nearly all flowering plants. Common ground plans suggest common origins, and this is one of the many pieces of evidence that favor the idea that groups of species of the modern world have a common evolutionary heritage: No matter how different the species of a given order or class may be, they share certain basic features that reflect their common ancestry.

We will begin this chapter by reviewing the ideas of Charles Darwin, the man who popularized the idea of evolution. Next we will examine natural selection, which Darwin recognized to be the dominant process of evolution, and discuss current knowledge and theories of how new species come into being. Finally, we will learn more of what the fossil record has revealed about extinction and about rates, trends, and patterns of evolution.

CHARLES DARWIN'S CONTRIBUTION

Few biologists gave serious consideration to the idea of organic evolution until 1859, when Charles Darwin (Figure 6-3) published his great work, *On the Origin of Species by Natural Selection*. One of the most effective ways to appreciate the power of even the simplest kinds of evidence that evolution has occurred is to put yourself in the position of Darwin when in 1831, at the age of 23, he set sail as an unpaid naturalist aboard the *Beagle*, on an ocean voyage that took him around the world. In the course of this trip, Darwin became convinced of evolution and also accumulated much of the evidence that subsequently enabled him to convince others of its validity.

FIGURE 6-3   Charles Darwin in 1849, at the age of 40. At this time, he had already conceived of the idea of natural selection, but it was not until 1859 that he published his classic work, *On the Origin of Species by Natural Selection*. (*Courtesy of G. Hardin.*)

The voyage of the *Beagle*

Some of the observations that Darwin made during his historic voyage were of a purely geologic nature. Darwin had been well tutored in geology by Adam Sedgwick, an expert on the early Paleozoic rocks and fossils of Britain. Moreover, just before he set sail, Darwin was given the first volume of Charles Lyell's *Principles of Geology* by one of his teachers, botanist J. S. Henslow. Although Henslow advised Darwin to read the book but not believe it, Darwin, like most others who studied geology at the time, became convinced that Lyell's uniformitarian approach represented a valid interpretation of the earth's history (page 3). On the voyage of the *Beagle*, Darwin witnessed some remarkable geologic processes—including earthquakes and volcanic activity—that corroborated this notion, allowing him to appreciate more fully than many of his contemporaries how the earth could be transformed continuously by everyday processes. Darwin's conversion to uniformitarian geology paved the way for his subsequent conclusion that natural processes transform life in the same way that they transform the physical features of the earth.

While Darwin's growing uniformitarian view of the earth's history provided a framework for his acceptance of evolution, it was his observation of the geographic distributions of living things that ultimately led him to theorize that many different forms of life possessed a common biological heritage. Darwin was surprised to find South America inhabited by animals that differed substantially from those of Europe, Asia, or Africa. The large flightless birds of South America, for example, were species of rheas (Figure 6-4), which belonged to a different family from the superficially similar birds of other continents—the ostriches of Africa or the emus of Australia. Other unique South American creatures included the sloths and the armadillos. Not only was South America the home of living representatives of these groups, but it was here that Darwin dug up the fossil remains of extinct giant relatives of the living forms (see page 134 and Figure 6-5). Why, he asked, were the rhea birds as well as all living and extinct members of the sloth and armadillo families found nowhere else but in South America?

Darwin was also intrigued to find that species of marine life on the Atlantic side of the Isthmus of Panama differed from those on the Pacific side. In places, the isthmus is only a few miles wide, and it struck Darwin as strange that the marine creatures on opposite sides of this narrow neck of land should differ from each other—unless the various species had somehow come into being where they now lived. If the species had instead been scattered over the

FIGURE 6-4   The rare flightless bird *Rhea darwinii*, which Darwin learned about from the gauchos in Argentina and which was later named after him. Darwin was struck by the fact that the flightless birds of South America bore no close relationship to those of other continents. (*Painting by John Gould from C. Darwin, Zoology of the Voyage of H.M.S. Beagle, Smith and Elder, London, 1838–1843.*)

earth by an external agent, he reasoned, many should have landed on both sides of the isthmus.

Perhaps the most striking of Darwin's observations, however, concerned life forms on oceanic islands. Darwin noted that no small island situated more than 5000 kilometers (~3000 miles) from a continent or from a larger island was inhabited by frogs, toads, or land mammals that had not been introduced by human visitors. The only mammals native to such islands were bats, which could originally

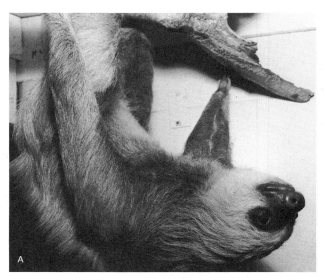

FIGURE 6-5   Unusual South American mammals. *A.* A living two-toed sloth, hanging upside-down in its normal mode of life. This animal is the size of a small dog. *B.* A reconstruction of two of the Pleistocene mammals of South America that Darwin unearthed as fossils. *Megatherium*, the giant ground sloth, was more than 6 meters (~20 feet) in length—larger than an elephant. It ranged northward into the United States. The skeleton of *Glyptodon*, the giant armadillo, is shown on page 134. It interested Darwin that fossil sloths and armadillos had close living relatives in South America but no living or fossil relatives in Europe, Asia, Africa, or Australia. (*A. Courtesy of the Los Angeles Zoo. B. Courtesy of the Field Museum of Natural History, painting by C. R. Knight.*)

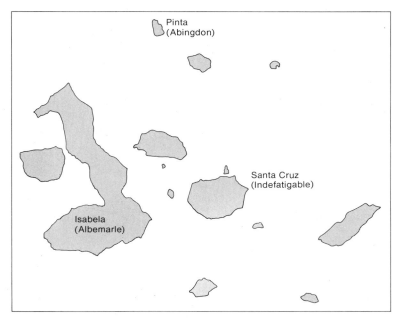

Testudo abingdonii, Pinta I.

Testudo microphyes, Isabela I.

Testudo ephippium, Santa Cruz I.

FIGURE 6-6   Three tortoises, each of which inhabits a different island of the Galápagos, as shown on the map. *Testudo abingdonii (upper right)* which inhabits Pinta Island, has a long neck and a shell that is elevated in the neck region; these features represent adaptations for reaching tall vegetation. (*T. Dobzhansky et al., Evolution, W. H. Freeman and Company, New York, 1977.*)

have flown there. This led Darwin to suspect that species could originate only from other species. Otherwise, why would isolated areas of land be left without important forms of life?

The Galápagos Islands, which lie astride the equator about 1100 kilometers (~700 miles) from South America, played an especially large part in the development of Darwin's new ideas. Darwin found the Galàpagos to be inhabited by huge tortoises, and he thought it curious that the people who lived on the islands could look at a tortoise shell and immediately identify the island from which it had come. The fact that different races of giant tortoises occupied different islands (Figure 6-6) led Darwin to suspect that these distinctive populations of tortoises shared a common ancestry but had somehow become differentiated in form as a result of living separately in different environments. Even more striking were the various kinds of finches that Darwin found in the Galápagos; some types had slender beaks, others had somewhat sturdier ones, and still others had very heavy beaks (similar to those of grosbeaks), which served the function of breaking nuts (Figure 6-7). One kind of Galàpagos finch behaved like a woodpecker but made use of a cactus spine rather than a wood-

pecker's long beak to probe for insects in wood. Furthermore, all of the finches in the Galápagos more closely resembled a species of finch on the South American mainland (the closest large landmass) than they resembled the finches found in other regions of the world.

Darwin subsequently began to ponder whether a population of finches from the South American mainland might have reached the islands and become altered in some way to assume a wide variety of forms. It seemed that the finches had somehow differentiated so as to pursue ways of life that on the mainland were divided among several different families of birds. As Darwin put it in the journal in which he described his scientific work on the voyage:

Seeing this gradation and diversity of structure in one small, intimately related group of birds, one might really fancy that from an original paucity of birds in this Archipelago, one species had been taken and modified for different ends.

FIGURE 6-7   The 14 species of finches that inhabit the Galápagos Islands and nearby Cocos Island. Species A feeds like a woodpecker but employs a cactus spine rather than a long beak to grub for insects in the bark of trees; C, D, and E eat insects; F and G are vegetarians; H is the Cocos Island finch; and I through N feed primarily on seeds, with I specializing on hard seeds that it cracks with its powerful beak. (*From D. Lack, "Darwin's Finches," Scientific American. Copyright ©1953 by Scientific American, Inc. All rights reserved.*)

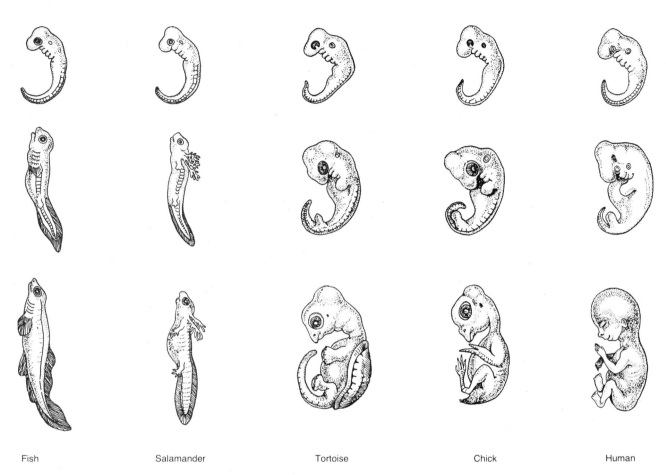

| Fish | Salamander | Tortoise | Chick | Human |

FIGURE 6-8   Early stages in the embryology of various groups of vertebrate animals. Note that in the earliest stage shown here, the embryos of all the groups are nearly identical.

(*After G. Hardin, Biology: Its Human Implications, W. H. Freeman and Company, New York, 1949.*)

Anatomical evidence

When Darwin returned to England and weighed other evidence favoring the evolution of one type of organism from another, he found that certain anatomical relationships seemed to build an especially compelling case. One such piece of evidence was the remarkable similarity of the embryos of all vertebrate animals (Figure 6-8). Darwin was intrigued that Louis Agassiz, a noted American scientist, admitted that he could not distinguish an early embryo of a mammal from that of a bird or a reptile. This, Darwin reasoned, was exactly what could be expected if all vertebrate animals shared a common ancestry: Although adult animals might become modified in shape as they adapted to different ways of life, early embryos were sheltered from the outer world and would thus undergo less change.

Equally convincing to Darwin was the evidence of **homology**—the presence, in two different groups of animals or plants, of organs that have the same ancestral origin but serve different functions. The principle of homology is illustrated by the variations in teeth and leaves discussed earlier in this chapter. Another example is the common origin of the toes of land-dwelling mammals and the wings of bats (Figure 6-9). Bats' wings are actually formed of four toes whose external appearance and bone configuration resemble those of walking mammals. If bats' wings did not share a common biological origin with the feet of walking mammals, why do the two types of organs have similar bone configurations? Examples such as this abound in both the animal world and the plant world.

The existence of **vestigial organs**—organs that serve no apparent purpose but resemble organs that do perform

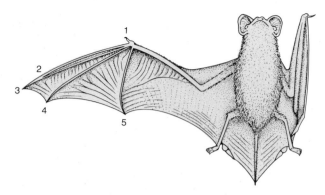

FIGURE 6-9   The extended wing of a bat showing the five digits that are equivalent to the toes of a land-dwelling mammal or the fingers of the human hand.

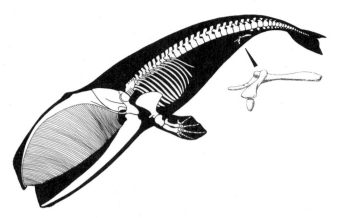

FIGURE 6-10   The skeleton of a baleen whale, the largest living mammal. The enlargement shows the pelvic bones, which resemble those of other mammals but in the whale are only weakly developed and serve no apparent function. (*Drawing by Gregory S. Paul.*)

functions in other creatures—further supported Darwin's argument in favor of evolution. One of the most striking aspects of vestigial organs is that they are usually smaller and less complex than the functional organs that they resemble. Examples of these are the apparently useless bones of modern whales which resemble the functional pelvic bones of other mammals (Figure 6-10). Why would whales possess such bones if they were not remnants of once functional organs that are now in the process of disappearing? Thus, the vestigial bones of whales reflect a biological past in which legs were present.

Natural selection

Darwin also recognized a different type of evidence that pointed to the validity of biological transformation in nature, and this was the fact that animal breeders were known to have produced major changes in domesticated animals by means of selective breeding. If wild dogs could be modified into greyhounds, Saint Bernards, and chihuahuas under domestic conditions, Darwin saw no reason why animals should be anatomically straitjacketed in nature. The question was, what could bring about biological changes under natural conditions?

This question led to Darwin's second great contribution. The first, of course, had been his amassing of an enormous amount of evidence indicating that evolution had occurred in nature. The second was his conception of a mechanism through which this evolution was likely to have taken place. What Darwin proposed was the process of **natural selection**—a process that operates in nature but parallels the artificial selection by which breeders develop new varieties of domestic animals and plants for human use.

Essentially, **artificial selection** in domestic breeding involves the preservation of certain biological features and the elimination of others. In this process, a breeder simply chooses certain individuals of one generation to be the parents of members of the succeeding generation. Darwin recognized that in nature, many more individuals of a species are born than can survive. Accordingly, he reasoned that success or failure here, as in artificial breeding, would not be determined by accident. In nature it would be determined by advantages that certain individuals had over others. Darwin reasoned, for example, that certain individuals would survive longer than others if they were better equipped to find food, avoid enemies, resist disease, or deal with any of a number of environmental conditions. These individuals would then, by virtue of their longevity, produce more offspring than others.

Darwin also recognized, however, that survival was not the only factor influencing success in natural selection, since some individuals with only average life-spans were capable of producing more total offspring than others simply because they bore large litters or shed large numbers of seeds. Thus, as long as there was substantial variation between the individuals of a breeding population in longevity and in rate of reproduction, certain individuals would pass on their traits to an unusually large number of individuals of the next generation. The kinds of individuals that came to predominate as generation followed generation could then be said to be favored by natural selection.

Natural selection is not easy to study in nature because the environment often changes in a variety of complex and unpredictable ways. In England, however, human activity

FIGURE 6-11   Color variants of the peppered-moth species in Britain. The light form is almost invisible on the tree trunk at the left, which is in an unpolluted area, whereas the dark form is quite conspicuous. The reverse is true on the sooty tree trunk at the right. (*Courtesy of H. B. Kettlewell.*)

has accidentally produced a selection process that has been termed **industrial melanism**. Since the middle of the nineteenth century, dark varieties of moths have become more abundant than lighter varieties within many species living in regions whose environment has been blackened by soot from smokestacks (Figure 6-11). Before the industrial revolution, for example, a species known as the peppered moth was represented primarily by speckled individuals—but a dark form of the moth now predominates. In this case, birds have served as the primary agent of natural selection; on industrially blackened tree trunks, they have fed preferentially on conspicuous speckled moths. As a result, it is only in unpolluted areas that the speckled forms prevail, and they do so by virtue of their original camouflage.

GENES, DNA, AND CHROMOSOMES

Darwin faced a major obstacle in convincing others that natural selection could operate effectively to produce evolution. Because he lived before the birth of modern genetics, Darwin was not familiar with the mechanisms of inheritance and thus could not explain how an organism could pass along a favorable genetic trait to its offspring. Although the Austrian monk Gregor Mendel had outlined the basic elements of modern genetics only a few years after Darwin published *On the Origin of Species*, Mendel's work was not acknowledged until the turn of the century, two decades after Darwin's death.

Mendel's most significant contribution to modern genetics was the concept of **particulate inheritance**, which explained how certain hereditary factors, which we now call genes, retain their identities while being passed on from parents to offspring. Mendel's experiments with pea plants demonstrated that individuals possess genes in pairs, with one gene of each pair coming from each parent. In one of his experiments, Mendel employed a true-breeding white-flowered strain of pea plants and a true-breeding red-flowered strain. (**True breeding** signifies that within the

Box 6-1

Mendel's

telltale ratios

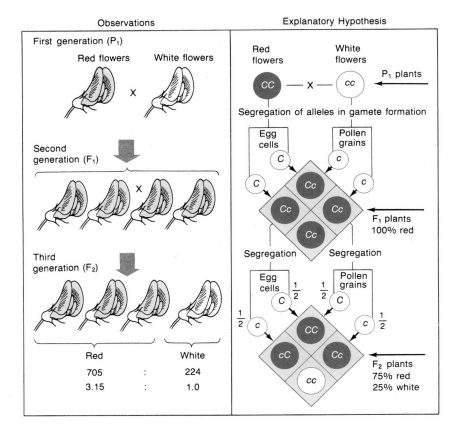

The fact that Gregor Mendel's crosses of pea plants produced both red-flowered and white-flowered varieties in the third generation suggested to him that each parent plant passed a genetic factor of some kind to its offspring. Thus, the white-flowered plants of the first generation had passed a genetic factor on to the third generation by way of the second, even though this factor had produced no white flowers in the second generation. All the plants of the second generation had red flowers, even though each had received a genetic factor for red color from one parent and a genetic factor for white color from the other parent. (Today we would say that the **allele**, or type of gene, responsible for the red-flowered condition is dominant in relation to the allele responsible for the white-flowered condition and that the allele for the white-flowered condition is recessive.)

The reason the third generation was characterized by red- and white-flowered plants in a ratio approximating 3:1 is illustrated in the accompanying diagram. In the pea plant, as in most other higher organisms, reproduction takes place by way of gametes representing two sexes. The male gametes come from pollen grains and each gamete has a single allele for flower color. These grains fertilize the female gametes (or eggs), each of which also has a single allele for flower color. The combination of the two gametes yields a new plant with two alleles for flower color.

Because all pea plants in the second generation of Mendel's experiment possessed one allele for red color (C) and one for white color (c),

these plants produced equal numbers of gametes possessing the "red" allele and the "white" allele. About one-quarter of their progeny had the CC combination and were red; about half had the Cc combination and were red, although they carried a recessive white allele; and about one-quarter had the cc combination and were white. This was how white-flowered plants reappeared in the third generation, forming about one-quarter of the population. (The actual ratio of red to white was not exactly 3:1 but 3.15:1; apparently by chance, slightly fewer cc combinations formed than would have been expected.) (*Illustration after G. G. Simpson and W. S. Beck, Life, Harcourt, Brace & World, Inc., New York, 1965.*)   □

strain, descendants resemble parents throughout a long series of generations.) Mendel's first step was to cross plants of the white-flowered strain with those of the red-flowered strain. The surprising result was that all of the daughter plants had red flowers. When these red plants were crossed with each other, however, they produced both red-flowered and white-flowered descendants. In this third generation, red-flowered plants represented about three-quarters of the plants and white-flowered plants about one-quarter. Box 6-1 explains the reason for this 3:1 ratio.

The most significant aspect of Mendel's work was his discovery that the effect of the gene for white color could surface in the third generation even after its presence had been masked in the second generation. The fact that genes can be preserved in this manner—i.e., as discrete entities that maintain their identity from generation to generation—forms the basis of particulate inheritance. This principle operates even when the effects of a gene are diluted by the presence of one or more other genes—as when, in certain species of flowering plants unlike the pea, the presence of a gene for red color and a gene for white color results in color blending by yielding a pink flower. Here, a cross between pink-flowered plants, which have genes for red flowers and genes for white flowers, will yield some red-, some white-, and some pink-flowered offspring.

Darwin's problem was that he and his contemporaries lived in an era dominated by the erroneous concept of blending inheritance, which held that genetic material is permanently diluted when it is combined with other genetic material through mating. This theory maintained that, barring the influence of nutrition during life, an individual of medium height should result from the mating of a tall individual with a short one. In keeping with this assumption, some of Darwin's opponents argued that any useful new feature that might appear in a species would subsequently be diminished upon being blended with other features as generation followed generation—until the feature eventually disappeared. Since he was unaware of the concept of particulate inheritance, Darwin was unable to counter this argument.

Another discovery that was made during the emergence of modern genetics was that genes can be altered. It is now understood that genes are, in fact, chemical structures that can undergo chemical changes, and these changes, or **mutations**, provide much of the variability upon which natural selection operates. Genes are now known to be segments of long molecules of deoxyribonucleic acid, or **DNA**—a compound that carries chemically coded infor-

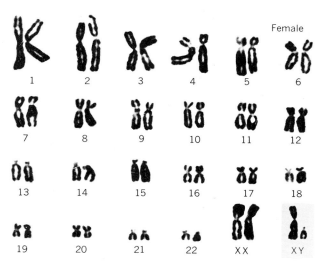

FIGURE 6-12   The complete set of human chromosomes. One member of each of the 23 pairs comes from each parent. The fact that two X chromosomes are present means that this set of chromosomes represents a female; the male condition is determined by the presence of one X and one Y chromosome instead of two X's. (*Courtesy of M. Grumbach and A. Morishima.*)

mation from generation to generation, providing instructions for growth, development, and functioning of organisms. DNA includes four **nucleotide bases**: adenine, thymine, guanine, and cytosine. Its structure and mode of replication are illustrated in Box 6-2.

Changes in one or more of the hundreds of nucleotides that form genes along a strand of DNA are known as **point mutations**. These can result from imperfect replication of the DNA strand during cell division. When this occurs, one of the new cells formed from the preexisting cell inherits a slightly altered copy of the first cell's genetic material. In such a case, imperfect replication of DNA may have resulted from the attachment of a base pair to the wrong partner—adenine to cytosine, for example (Box 6-2). Point mutations can also occur when an already-existing DNA strand is chemically altered by an external agent, which may be a chemical substance or a dose of radiation such as cosmic radiation or ultraviolet light. In any case, point mutations usually produce changes in the structure of the proteins that are coded by the mutated segments of DNA.

In organisms other than bacteria and their close relatives (Figure 1-5), DNA is concentrated within **chromosomes**, which are elongate bodies found in the nucleus of the cell. Most organisms have two of each kind of chromosome, having inherited one of each type from each parent (Figure 6-12). **Chromosomal mutations** also occur,

Box 6-2

DNA,

the genetic code

Within the chromosomes of a cell, deoxyribonucleic acid, or DNA, carries a chemically coded message that contains the information needed for the growth, development, and functioning of the organism. The primary role of DNA is to supply information for the production of enzymes, which are catalysts that function in the production of proteins and mediate most chemical reactions within a cell.

The structure of DNA molecules is that of the famous "double helix," which resembles a twisted rope ladder with rigid steps. The sides of the ladder structure, which scientists speak of as the DNA "backbone," are made up of alternating sugar and phosphate components. Attached to the sugars are components called nucleotides, which form the steps of the ladder structure. Two nucleotide bases, one from each side of the ladder, form each step of the ladder. The two bases are connected by weak hydrogen bonds, and thus the ladder is easily "unzipped" to form two separate strands.

A key feature of DNA is the specific way in which the nucleotide bases pair to form the steps of the ladder. There are four kinds of nucleotide bases: adenine (A), thymine (T), guanine (G), and cytosine (C). *Specific base pairing* is the phrase used to describe the fact that adenine pairs only with thymine and guanine only with cytosine. This phenomenon allows DNA to duplicate itself quite

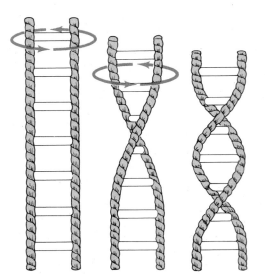

A rope ladder with rigid steps can be twisted to form a structure like that of DNA.

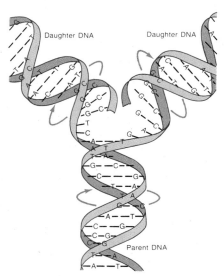

The "unzipping" and duplication of DNA. Separation occurs between the nucleotide bases, *C* and *G* or *A* and *T*.

precisely before cell division so that each daughter cell can be provided with a full quotient of DNA. The duplication of DNA involves an unzipping of the double helix and the building of a new half on each half of the ladder exposed by the division. Perfect duplication is usually ensured by the automatic connection of adenine to thymine and of guanine to cytosine.

The sides of the DNA double helix are formed of sugars (S) and phosphates (P). The steps are formed of either cytosine (C) and guanine (G) or by adenine (A) and thymine (T).

DNA is unzipped not only for duplication but also at other times so that its message can direct the formation of enzymes. This involves participation of the compound ribonucleic acid (RNA), which is similar to DNA and consists of nucleotide bases that are coded by portions of unzipped DNA. RNA carries the DNA message to a site in the cell where it specifies the formation of a particular building block for a protein.

What, then, is a gene on the DNA strand? A **structural gene** is a segment of the strand that codes the formation of a building block of a protein molecule. It appears, however, that some genes are DNA sequences that do not form proteins. Many of these nonstructural genes may function to control structural genes, turning them on and off during the life of the organism and thus causing them to manufacture proteins at the time when they are needed. (*Illustrations after G. Hardin and C. Bajema, Biology: Its Principles and Applications, W. H. Freeman, New York, 1978.*) ☐

and these take the form of changes in the number of chromosomes or changes in the positions of segments of individual chromosomes. Such changes move segments of DNA with respect to each other. The exact importance of these changes to the ways in which DNA controls the development and operation of an organism remains uncertain.

POPULATIONS, SPECIES, AND SPECIATION

When two individuals breed, each normally contributes half of its chromosomes to each offspring by way of a **gamete**, a special reproductive cell that contains only one (rather than a pair) of each type of chromosome. The female transmits this set of chromosomes by way of an egg cell, which is the female gamete, and the male by way of a sperm cell, the male gamete. Similarly, the offspring, if it mates, combines half of its chromosomes with half of those of a member of the opposite sex in order to produce still another generation. The mixing that takes place in this manner, which is known as **sexual recombination**, continually yields new combinations of chromosomes and hence of genes. The process of sexual recombination, in conjunction with the periodic occurrence of point and chromosomal mutations, is responsible for the variability among organisms that provides the raw material for natural selection. Unfortunately, Darwin and his contemporaries had no knowledge of these sources of variability.

In the study of evolution today, a group of interbreeding individuals is referred to as a **population**, and the sum total of genetic components of a population is known as a **gene pool**. Moreover, populations whose members can interbreed if they come into contact form a **species** (page 8). Reproductive barriers between different species keep gene pools separate and thus prevent interbreeding. These barriers include differences in mating behavior, incompatibility of egg and sperm, and failure of offspring to develop into adults successfully.

Not only can a species as a whole evolve in the course of time, but it can also give rise to one or more additional species. The origin of a new species from two or more individuals of a preexisting one is called **speciation**. Because species are kept separate from each other by reproductive barriers, speciation by its very definition entails evolutionary change that produces such barriers. It is widely believed that most speciation events involve the geographic isolation of one population from the remaining populations of the parent species. This isolated population then follows an evolutionary course that causes it to diverge from the parent species in form and way of life—a divergence that may result from such phenomena as the occurrence of unique mutations and the guidance of natural selection by unusual environmental conditions. The development of distinct species of finches on the various Galápagos islands (Figure 6-7) illustrates this principle.

EXTINCTION

Fossils provide the only direct evidence that life has changed substantially over long spans of geologic time. They also offer the only concrete evidence that millions of species have disappeared from the earth, or suffered **extinction**.

The validity of extinction was not widely accepted until late in the eighteenth century. Before this time, fossil forms that no longer seemed to inhabit the earth were thought to live in unexplored regions. In 1786, however, Georges Cuvier, a French naturalist, pointed out that fossil mammoths were so large that living representatives of this group could not possibly have been overlooked. Cuvier thus concluded that mammoths were extinct. His argument was well received, and soon the extinction of many species was accepted as fact.

In general, extinction results from particularly extreme forms of the factors that normally hold populations in check. These limiting factors are predation on a population, competitive interaction with one or more other species, restrictive conditions of the physical environment, or chance fluctuations in the number of individuals (page 30). Changes resulting from one or more of these factors have led to the extinction of most of the species of animals and plants that have inhabited the earth; in fact, of all the species that have existed in the course of the earth's history, only a tiny fraction remain alive today.

There have, in addition, been episodes on a global scale in which large numbers of species have suffered extinction during intervals of just a few million years. These phenomena are called **mass extinctions**. In the chapters that follow, we will review most conspicuous examples of mass extinction in the history of the earth, including the one in which the dinosaurs disappeared. Although the fossil record is imperfect, it nonetheless reveals that at certain times, between 50 and 90 percent of all the species on earth died out during very brief geologic intervals. All mass extinctions, however, have been followed by at least a partial evolutionary recovery in which the number of species on earth has increased again. The causes of mass extinction are by no means fully understood, and different mass extinctions may well have had different causes—but all seem to have affected some groups of living things more severely than others. The kinds of groups that were affected differed from episode to episode.

Species have disappeared not only by dying out, but also by evolving to the point where they have been formally recognized as different species. In this process, which is known as **pseudoextinction**, a species' evolutionary line of descent continues, but its members are given a new name. This often occurs at an arbitrary point, since there is no way of determining precisely when members of the evolving group were no longer able to interbreed with the original members.

Rates of origination and extinction of taxa

One unique contribution of fossils to biological science is that they allow us to assess rates of evolution and extinction. It is only through data derived from the fossil record, for example, that we have been able to measure the rates at which new species, genera, or families have appeared within large groups of animals or plants. We may use the number of species, genera, or families living today as a final number in some calculations, but the geologic record nonetheless provides essential information about the events that have produced the numbers of these living taxa.

Adaptive radiation   At many times in the history of the earth, groups of animals or plants have undergone remarkably rapid evolutionary expansion—that is to say, various phyla, classes, orders, or families have produced many new genera or species during brief intervals of time. Rapid expansions of this kind are known as **adaptive radiations**. In this context, the word *radiation* refers to the pattern of expansion from some ancestral adaptive condition to the many new adaptive conditions represented by descendant taxa. The word *adaptive* expresses the idea that the new shapes that develop in the radiation are functionally related to new ways of life. Figure 6-13 shows how the fossil record permits us to measure the rate at which adaptive radiation has taken place. Here we can see, for example, that the number of families of corals increased rapidly early in Mesozoic time. The coral families depicted in this figure belong to the group known as the hexacorals, which first appeared in mid-Triassic time, about 215 million years ago, and whose living species form the beautiful coral reefs of the modern world (Box 2-1 and Figure 4-17).

Adaptive radiation often occurs in groups of plants or animals within just a few million years of their time of origin. This pattern is typical because such groups often have modes of life that differ from those of the other contemporary animals or plants. This situation tends to prevent ecologic competition from restraining the diversification of the new group. In addition, when a group first evolves, predatory animals may not yet have developed efficient methods for attacking the new group's members, and this protected status permits the new group to form many new species in a short period of time.

Sometimes the extinction of one biological group has allowed for the adaptive radiation of another even though the radiating group was not a new one on earth. Mammals, for example, inhabited the earth during almost all of the Mesozoic Era, but they remained small and relatively inconspicuous until the close of the Mesozoic Era, when the dinosaurs suffered extinction. Unrestrained by competition or predation from dinosaurs, mammals then underwent a spectacular adaptive radiation. Their rise to dominance on the land has led paleontologists to label the Cenozoic Era the Age of Mammals. Most of the living orders of mammals, including the order that comprises bats and the one that comprises whales, came into existence within only about 12 million years of the start of the Cenozoic Era. This represents only about 2 percent of all Phanerozoic time.

Many episodes of adaptive radiation have been preceded by **adaptive breakthroughs**, the appearance of key features that, along with ecologic opportunities, have allowed the radiation to take place. In hexacorals, for example, the rapid growth of the skeletons has been very important, often allowing these creatures to crowd out other animals that inhabit hard surfaces in shallow marine environments (Figure 6-13). Because their skeletons are porous, hexacorals can quickly assume large proportions without having to secrete large volumes of calcium carbonate. Thus, they have had an edge over slowly growing organisms with which they compete for space on the sea floor. Another adaptive breakthrough for the hexacorals was the development of a symbiotic relationship with algae that live in the tissues of the reef-building species and help the hexacorals secrete their skeletons (page 56). A variety of adaptive breakthroughs played major roles in the early Cenozoic adaptive radiation of mammals. The key feature for rodents, for example, may have been gnawing front teeth, which these animals use for eating nuts and other hard seeds. For grazing horses, the origin of grinding cheek teeth, mentioned earlier in the chapter, represented a major adaptive breakthrough.

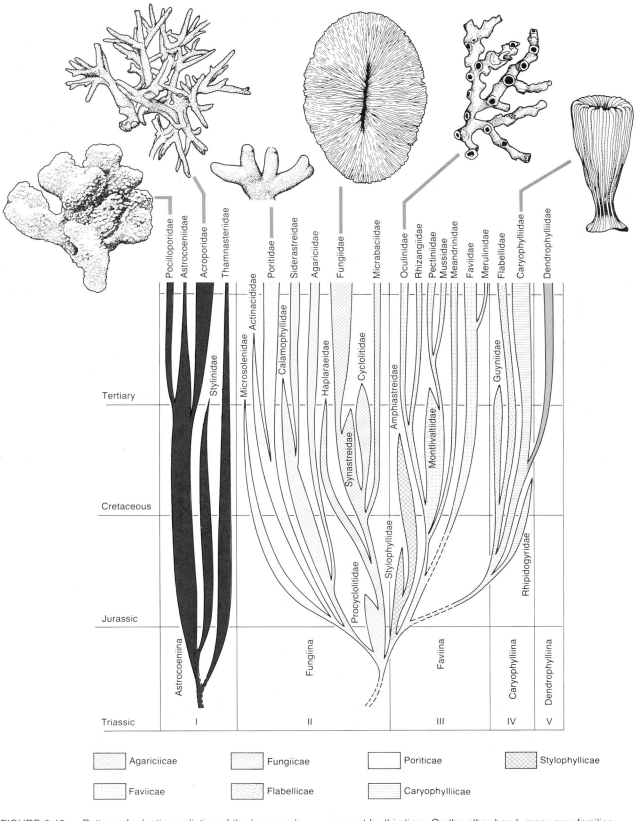

FIGURE 6-13   Pattern of adaptive radiation of the hexacorals since the group originated in mid-Triassic time. This group builds modern coral reefs. Typical representatives of six families are illustrated above. There are five orders of hexacorals (roman numerals), four of which were already present by mid-Jurassic time. Similarly, nearly all superfamilies (*patterns*) were present by this time. On the other hand, many new families (*names printed in color*) formed during the Cretaceous Period. This pattern indicates that the largest evolutionary steps—the origins of orders and superfamilies—occurred early; later, change took place only by smaller steps.

Figure 6-13 illustrates a pattern that typifies adaptive radiation. Note that four of the five modern orders of hexacorals were already present halfway through the Jurassic Period, shortly after the adaptive radiation of hexacorals began. If we lower our sights one taxonomic level, however, we find that families, unlike orders, continued to develop at a high rate well into the Cretaceous Period. Hexacoral genera, which are not depicted in this figure, continued to proliferate rapidly even longer, into the Cenozoic Era. What happened in this adaptive radiation, as in many others, is that early expansion produced large-scale evolutionary divergence at a very early stage. However, after the new orders of hexacorals became established, evolutionary change was more restricted. New families continued to develop at a high rate, but new orders did not. Later, during the Cenozoic Era, few new families evolved, but divergence continued at the genus and species level. This pattern seems to indicate that as a group of animals or plants begins to expand, it quickly exploits any adaptations that its body plan allows it to develop with ease. Later, however, evolution is restricted to the development of variations on the basic adaptive themes that evolved early on. In time, few new families evolve, and eventually only new species.

Local adaptive radiations of the recent past offer special insights into larger adaptive radiations that took place earlier. Many of these local adaptive radiations occurred in isolated environments such as islands or lakes, whose well-defined boundaries prevent the escape of the species within them. When these sites of evolution are of recent origin, they often provide evidence of the remarkably rapid diversification of life. This diversification, in turn, usually reflects the fact that the site of adaptive radiation was uninhabited territory and thus lacked the predators and competitors that might have inhibited the evolutionary diversification.

The Galápagos Islands, where Darwin studied the unique groups of tortoises and finches (Figures 6-6 and 6-7), represents one of the most famous island groups on which adaptive radiation has recently taken place. The Galápagos originated as a result of volcanic eruptions a few million years ago. Many large lakes have also been sites of adaptive radiation in the recent past. Lake Victoria in Uganda, which came into being no more than a few hundred thousand years ago, is an excellent example. About 170 species of fishes of the cichlid group have been identified in this lake, and all but three of these species cannot be found anywhere else in the world. In addition, many of these fishes have highly distinctive adaptations; some are specialized for eating insects, others for attacking other fishes,

Haplochromis chilotes, a specialized insectivore (× 62)

Haplochromis estor, a piscivore (× 21)

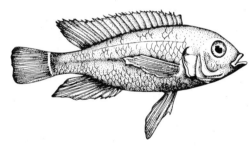

Haplochromis sauvagei, a mollusk eater (× 60)

FIGURE 6-14   Three of the more than 170 species of cichlid fishes that evolved in Lake Victoria, Uganda, within the last few hundred thousand years. The fishes represent many different shapes and ways of life. (*After P. H. Greenwood, Brit. Mus. [Nat. Hist.] Bull. Suppl. 6, 1974.*)

and still others for crushing shelled mollusks (Figure 6-14). Several species of cichlids that still inhabit Lake Victoria look very much like the fishes that seem to have given rise to the great adaptive radiation that has occurred within the lake. In other words, the original species, or descendants very much like them, remain along with much more distinctive products of adaptive radiation.

Given the evidence that dramatic adaptive radiations have taken place in lakes and on islands, it is interesting to consider what may have happened in the aftermath of major extinctions of the past. When the dinosaurs disappeared from the earth at the end of the Mesozoic Era, for example, the great continents of the world must have represented a series of large vacant islands that mammals could colonize on a grander scale than had been possible for them when dinosaurs ruled. For the mammals, which had remained small and inconspicuous in the shadow of the dinosaurs, the opportunity to undergo adaptive radiation must have resembled the evolutionary opportunity that was available to the first cichlids that arrived in Lake Victoria or the first finches that landed on the Galápagos Islands. The opportunity for mammals was greater simply because the entire world, not just a small area, had become available for their occupation.

Thus, small-scale adaptive radiations of the recent past seem to represent useful models for understanding larger adaptive radiations of the more distant past. The most basic lesson here is that when preexisting species do not interfere, a small number of founder species can rapidly produce many new kinds of species, some of which differ substantially from the original forms.

Rates of extinction   The extinction rates of most large groups of animals and plants have varied greatly in the course of geologic time, as indicated earlier. Average rates have also varied greatly from taxon to taxon. An average mammalian species, for example, has survived for just 1 to 2 million years, which means that the extinction rate for mammals has exceeded 50 percent per million years. In contrast, within many groups of marine life, including bivalve mollusks (clams, scallops, oysters, and their relatives), species have on the average existed for 10 million years or more. This means that under ordinary circumstances, only a small fraction of the species within these groups have disappeared every million years.

Groups of animals and plants that have left good fossil records and have also experienced high extinction rates tend to represent useful index or guide fossils (page 115). The ammonoids, which are an order of swimming mollusks with coiled shells, serve as an example of this. As Figure 6-15 illustrates, few ammonoid genera found in any Mesozoic stage are also found in to the next stage—and yet these genera are usually succeeded by a large number of new ones. Thus, the ammonoids experienced both a high rate of extinction of genera and a high rate of formation of new genera.

CONVERGENCE AND ITERATIVE EVOLUTION

One of the most convincing kinds of evidence favoring the adaptive nature of biological form is the occurrence of evolutionary **convergence**—that is to say, the evolution of similar forms in two or more different biological groups.

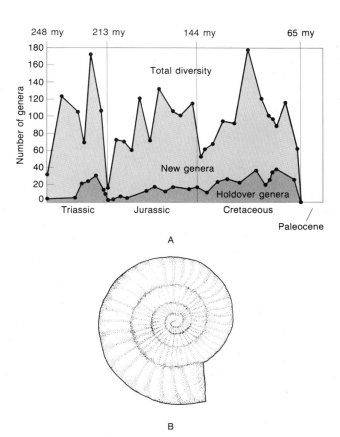

FIGURE 6-15   The appearance and disappearance of ammonoid genera through time. Ammonoids were coiled mollusks related to squids and octopuses but possessing coiled external shells (*B*). They became extinct with the dinosaurs at the end of the Cretaceous Period. As this diagram shows, the ammonoids experienced a high rate of turnover of genera throughout their history. Data plotted for each stage of the Mesozoic Era show that there are few "holdover genera" from one stage to the next. Thus, most genera of each stage are genera that formed during the corresponding age. (*After W. J. Kennedy, in A. Hallam [ed.], Patterns of Evolution. Elsevier, Amsterdam, 1977, pp. 251–304.*)

This principle is strikingly illustrated by the similarity that exists between many of the **marsupial mammals** of Australia and the other kinds of mammals that live in similar ways on other continents (Figure 6-16). Marsupial mammals, which carry their immature offspring in pouches, are the product of a radiation that took place on this isolated island-continent during the Cenozoic Era. That this radiation has been adaptive is indicated by the fact that these marsupials have *di*verged from each other but have simultaneously *con*verged, both in way of life and in body form, with one or more groups of nonmarsupial mammals living elsewhere. The strong similarities between many Australian marsupial mammals and mammals of other re-

MARSUPIALS   PLACENTALS

Tasmanian wolf
(Thylacinus)

Wolf
(Canis)

Native cat
(Dasyurus)

Ocelot
(Felis)

Flying phalanger
(Petaurus)

Flying squirrel
(Glaucomys)

Wombat
(Phascolomys)

Ground hog
(Mormota)

Anteater
(Myrecobius)

Anteater
(Myrmecophaga)

Mole
(Notorcytces)

Mole
(Talpa)

Mouse
(Dasyercus)

Mouse
(Mus)

gions must partially reflect the basic evolutionary limitations of the mammalian body plan and mode of development. It would appear, in other words, that certain adaptations are likely to develop under a variety of circumstances, while others are highly unlikely to evolve. The species illustrated in Figure 6-16 have body forms and modes of life that were likely to evolve when primitive mammals were permitted to undergo an extensive adaptive radiation, and accordingly, all developed at least twice.

Almost all adaptive radiations, however, have produced some surprises as well. Judging from what we see elsewhere in the world, for example, we might have predicted that hoofed, four-footed herbivores resembling deer, cattle, and antelopes would populate the continent of Australia. As it turns out, the Australian equivalents of these large galloping herbivores are kangaroos—animals that hop around on two legs. Apparently, it just happened that the breakthrough represented by the kangaroos' hopping adaptations evolved in Australian herbivorous marsupials before there was an adaptive breakthrough that produced an efficient running apparatus instead. Once the kangaroos had occupied many regions of Australia, there was simply little opportunity for animals resembling deer, cattle, or antelopes to evolve.

Sometimes, too, a taxonomic group gives rise to the same kind of descendant group or groups on more than one occasion. This pattern, which is known as **iterative evolution**, seems to result from the fact that evolution moves more easily in some directions than in others—possibly as a consequence of limitations imposed by the pattern of growth of individuals forming the ancestral group. Fossil planktonic foraminifers (page 108) display a striking example of iterative evolution; both early and late Cenozoic adaptive radiations from one group, the globigerines, yielded similar sets of test shapes (Figure 6-17).

EVOLUTIONARY TRENDS

By examining the evolutionary history of any higher taxon that has left an extensive fossil record, we can observe long-term **evolutionary trends**—general changes that developed over the course of millions of years. Some of these

FIGURE 6-16   Evolutionary convergence between marsupial mammals of Australia and nonmarsupial mammals of other continents. Each of the marsupials is more closely related to a kangaroo than to its counterpart in the other column. (*After G. G. Simpson and W. S. Beck, Life, Harcourt, Brace & World, Inc., New York, 1965.*)

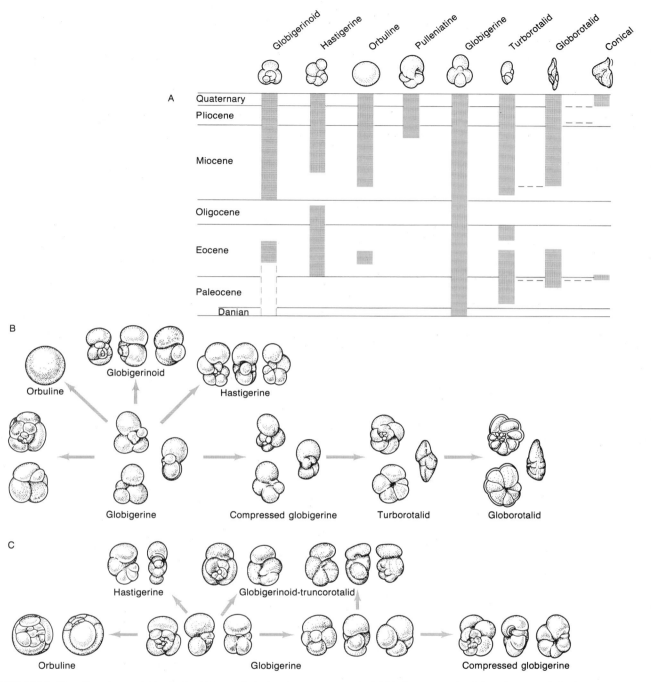

FIGURE 6-17   Iterative evolution in the history of planktonic foraminifers. *A*. Stratigraphic ranges of eight morphologic types. Note that only the globigerine type survived into the Miocene Epoch. *B* and *C*. The sequence of evolution that led from an- cestral globigerines to each of the other morphologic types twice, once in early Cenozoic time (*C*) and once in late Ceno- zoic time (*B*). (*After R. Cifelli, Systematic Zoology 18:159–168, 1969.*)

are changes in form but others simply represent changes in size.

Change of body size

Among the most common evolutionary trends in animals are trends toward increased body size. The tendency for body size to increase during the evolution of a group of animals is known as **Cope's rule** after Edward Drinker Cope, a nineteenth century American paleontologist who observed this phenomenon in his studies of ancient ver- tebrate animals.

There are numerous factors that may cause a group of animals to evolve toward larger body size, but all of these factors have to do with the tendency on the part of large individuals to produce more offspring than smaller ones.

Within species in which males fight for females, for example, larger males often tend to win battles and hence to produce a disproportionately large percentage of offspring. Within other species, larger animals may be especially successful at surviving and reproducing because they are better equipped to obtain food or to avoid predators.

It is important to understand, however, that evolution toward large body size cannot continue indefinitely, since, at some point, further increases of this kind will inevitably cease to be advantageous. For example, a four-legged animal the size of a large building could not run or even stand, since its weight would greatly exceed the strength of its limbs. Indeed, many groups of animals could not gather sufficient food or move efficiently if they were appreciably larger than they are. Few animals that eat only seeds or nuts, for example, are very big; an animal the size of a horse or a cow would find it difficult to gather small, scattered items of food rapidly enough to sustain itself.

Given the fact that increases in body size are advantageous only within limits, the high incidence of size increase in animal evolution would seem to indicate that most animal orders and families evolve from ancestors of less-than-optimal body size. Perhaps the most important reason for this tendency is that large animals within a family, order, or class tend to be complex in form and specialized in way of life. In other words, large size tends to impose many adaptive problems on animals, and the specialized adaptations associated with these problems are not easily altered to produce entirely new adaptations. Thus, large, highly specialized animals tend to represent evolutionary dead ends.

The physical structure of an elephant illustrates some of the problems of being big. An elephant has such a huge body to feed that it must spend most of its time grinding up coarse food with its molar teeth. These requirements dictate that an elephant's teeth and jaws be quite large in relation to the animal's overall size. The elephant's head is therefore large as well—so much so that the neck must be quite short to support it. To compensate for their consequent short reach, members of the elephant family evolved an enormous trunk from an originally short nose, together with long tusks from originally short teeth. These and other unusual features make it unlikely that modern elephants would ever evolve into other very different types of animals.

In contrast, the group of smaller animals from which elephants evolved some 50 million years ago exhibited great evolutionary potential, as evidenced by the fact that it also gave rise to manatees (or sea cows), which are blubbery ocean swimmers. Both elephants and manatees

FIGURE 6-18   The elephant, the manatee, and the hyrax—three animal groups that differ in form but share a common ancestry within the class Mammalia. Outlines below show the relative sizes of the animals.

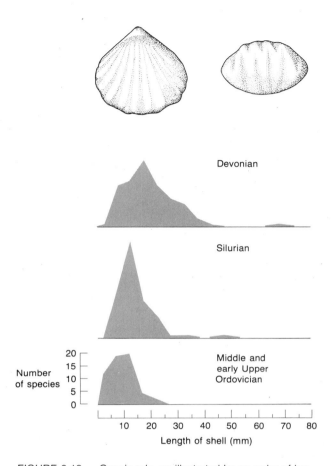

FIGURE 6-19   Cope's rule, as illustrated by an order of brachiopods (the Rhynchonellida). In mid-Ordovician time, this order—of which a typical member is shown in two views above—included few species with shells longer than 25 millimeters (~1 inch). During the Devonian Period, many small species were present, but the maximum size had increased markedly, as had the average size. (*After S. M. Stanley, Evolution, 27:1–26, 1973.*)

belong to a group known as the **subungulates** (Figure 6-18). The evolutionary development of elephants and manatees was made possible by the fact that the ancestral subungulates were small, relatively unspecialized animals that could easily evolve in a variety of directions. The living hyraxes are small, primitive subungulates that did not diverge greatly in form from the animals that long ago gave rise to elephants and sea cows.

When a higher taxon of animals originates with a small body size and then expands to include many species, only some of the new species will be larger than the original ones. Thus, as Figure 6-19 shows for a group of brachiopods, while the maximum body size typically increases, along with average body size, the group continues to include many species of small size.

Ontogeny and phylogeny

The "tree of life" grows as new species develop from older ones by speciation. The formal name for any segment of this branching structure is **phylogeny**. On a smaller biological scale, the sequence of development of an individual, from its origin by the fertilization of an egg to its death, is known as **ontogeny**. Many evolutionary trends that are far more complex than changes in body size entail ontogenic changes.

Almost all species undergo major changes as they develop from embryonic to adult stages, and late in the nineteenth century it became apparent that many of the early ontogenic stages of a species mimic stages of evolution in the ancestry of that species. Unfortunately, this similarity was taken too seriously. With much wishful thinking, some biologists claimed that the evolutionary history of a species could be "read" precisely from its sequence of development. This simplistic notion led to the statement "Ontogeny recapitulates phylogeny." While it is true that the early developmental stages of some species resemble evolutionary stages in the development of their ancestors (Figure 6-20), advanced animals such as mammals do not pass through stages in which they resemble *adult* fishes, amphibians, or reptiles. Thus, the notion that ontogeny usually recapitulates phylogeny in detail is incorrect. Hints of ancestry may be present, but the situation is much more complicated. Ontogeny recapitulates phylogeny in part (see Figure 6-8), but seldom does it do so as faithfully as in the example depicted in Figure 6-20.

One common evolutionary trend that has resulted in many striking adaptive developments involves nothing more than a change in the sequence of development of an organism. In one of the simplest evolutionary changes of this kind, the late stages of development are eliminated so that what once was the juvenile form becomes the adult form. This involves two alterations: first, the transfer of sexual maturation to an early growth stage and, second, the arresting of development at that early stage. This kind of evolutionary change is particularly striking in species that normally undergo a major change in mode of life as they mature. Many members of the vertebrate class Amphibia, for example, experience a metamorphosis that transforms

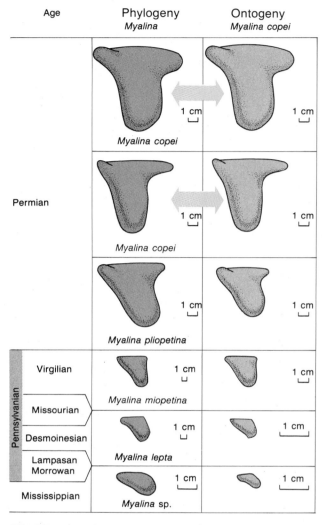

FIGURE 6-20   Strong resemblance between ontogeny and phylogeny for one Late Paleozoic species of bivalve mollusks that superficially resembles a modern marine mussel. The general ancestry of this species, *Myalina copei*, is represented by the sequence of species shown below it on the left-hand side of the diagram. The similar stages of ontogeny of *Myalina copei*, which have been preserved in the fossil record as shells of animals that died young, are shown on the right-hand side of the diagram. Thus, it appears that this single species passed through stages that closely resembled the species that formed part of its evolutionary ancestry. (*After N. D. Newell, State Geol. Surv. Kansas Publ. 10[2], 1942.*)

them from aquatic swimmers early in life to four-legged terrestrial adults that return to the water only to lay their eggs. Many times in the evolutionary history of the amphibians, however, species evolved that remained in the water throughout life (Figure 6-21). In such cases, breeding occurs in the aquatic stage, and even after breeding, the individual fails to metamorphose into a land animal.

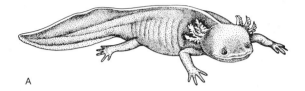

FIGURE 6-21   Aquatic and terrestrial amphibians. *A.* A drawing of the axolotl, which lives its entire life in freshwater. *B* and *C.* Two stages in the ontogeny of a typical salamander. The aquatic larval stage (*upper photo*) closely resembles the adult axolotl, which, in effect, never grows up. (*A. After J. Z. Young, Life of Vertebrates, Oxford University Press, London, 1962. B and C. Courtesy of H. Spencer.*)

Some of the trends that have characterized human evolution have also involved the shifting of juvenile characters to the adult stage. Of course, unlike the amphibians of the previous example, modern humans are not almost identical to the infants of their ancestors. Nonetheless, some juvenile or embryonic features of our ancestors are indeed retained in slightly modified form into our adulthood. For one thing, our hands and feet, with their soft, fleshy surfaces, resemble those of unborn apes and monkeys. For another, our relatively naked skin is apparently a feature that, by means of evolution, has been retained to adulthood from what was previously an immature stage of development. Even the shape of our head, with its flat face and tall forehead, is much more akin to that of a juvenile chimpanzee than to that of an adult chimpanzee (Figure 6-22). Similarly, the large size of our head in relation to our body mimics that of a juvenile chimpanzee or gorilla. Thus, it would appear that our brain became as big as it is by the evolutionary retention into our adulthood of the large relative brain size that characterized the infants of our ancestors. These ancestors were not the species of apes we are familiar with today but were instead apelike forms that are now extinct.

The structure of evolutionary trends

Trends in evolution occur on both small and large scales. A transition from one species to another, for example, represents a simple trend on a small scale. An example of a large-scale trend is the change from the oldest known kind of horse to the modern kind of horse (Figures 6-23 and 6-24). The oldest known horse had four toes on each forefoot and three on each hind foot, had relatively simple molar teeth, and was the size of a small dog. In contrast, the modern horse has a single hoofed toe on each foot, has complex molar teeth, and is a relatively large mammal. Many speciation events separated the ancestral kind of horse from the modern kind. Both the ancestral genus and the modern one have included several species, as have most intermediate genera, and so the transition from one kind to the other represents a complex, large-scale evolutionary trend.

In considering the structure of evolutionary trends, we will focus first on simple trends that cause one species to be transformed into another by a single thread of evolution, and we will then examine complex trends involving many species—trends that consist of many threads of evolution and can thus be viewed as having a fabric.

Some changes from one species to another have occurred in rapid steps of speciation, with the descendant species evolving from the parent species in a relatively short span of time. A probable example is the axolotl, a species whose members remain aquatic throughout life (Figure 6-21). The axolotl evolved from a normal species of salamander—one that is still extant—which underwent metamorphosis from an aquatic juvenile to a terrestrial adult form. The evolutionary transition that produced the axolotl was genetically simple. An axolotl that would normally remain aquatic throughout life can be artificially forced to metamorphose into a terrestrial animal by injecting it with thyroxine, a substance normally produced by the thyroid gland but missing in the axolotl. It thus appears that the axolotl was produced by a speciation event consisting of

FIGURE 6-22   Two stages in the ontogeny of a chimpanzee, showing how the juvenile (*left*) resembles a human far more than does the adult chimp. (*After A. Naef, Naturwiss. 14:445–452, 1926.*)

a simple genetic change that impeded the normal development of the thyroid gland. This change probably occurred quite rapidly on a geologic scale of time. The simple genetic change that produced the axolotl probably spread throughout a single population of the ancestral species. In this respect, however, the axolotl appears to be unusual, since most species that form rapidly from a small ancestral population do not evolve by means of just one genetic change; instead, it appears that several mutations are often entailed.

Other species have evolved slowly by the gradual transformation of an entire species—that is to say, an entire species has changed sufficiently in the course of many generations to be regarded as a new species. Figure 6-25 illustrates an example of this type of evolutionary change in a group of coiled oysters during the Jurassic Period.

Like many biologists and paleontologists of the present century, Darwin believed that gradual trends such as that of the coiled Jurassic oysters produced most large-scale evolutionary trends, including those involved in the origin of the modern horse (Figures 6-23 and 6-24). Recently, however, it has been suggested that gradual trends such as those of the Jurassic coiled oysters are relatively rare. Oysters, for example, are bivalve mollusks, and more than 300 species of bivalve mollusks have been identified from Jurassic rocks of Europe—yet very few of these species exhibit gradual trends like the one illustrated in Figure 6-25. In fact, it is estimated that an average bivalve species living in Europe during the Jurassic Period existed without appreciable change for about 15 million years, or for about one-quarter of the entire Jurassic Period.

Many other animals and plant species have also survived for long geologic intervals. Estimates for various

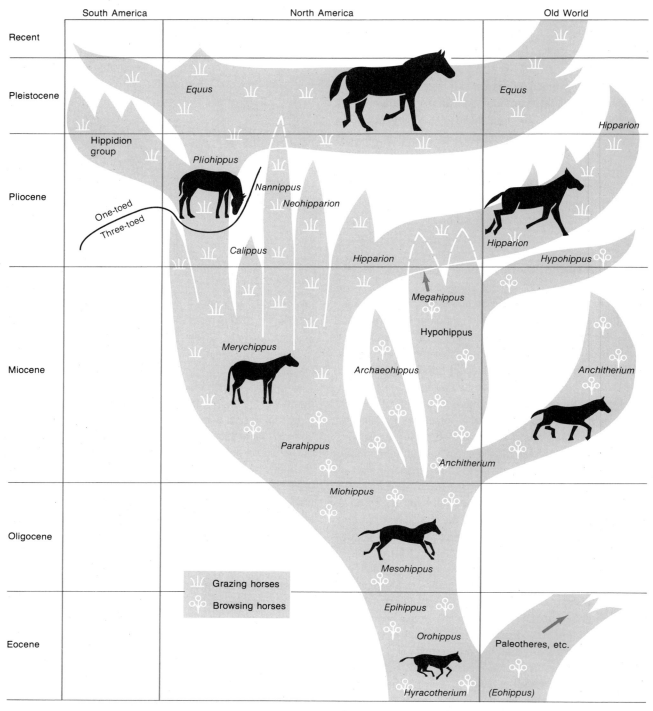

FIGURE 6-23   General pattern of phylogeny of the horse family. The most obvious general trend was toward larger body size. Other important net changes are illustrated in greater detail in Figure 6-24. (*After G. G. Simpson, Horses, Oxford University Press, New York, 1951.*)

biological groups are as follows: Seed-bearing plant species endure an average of 6 million years; freshwater fishes, 6 million years; benthic foraminifers (see page 50), 30 million years; diatoms (page 51), 25 million years; mosses and their relatives, more than 15 million years; beetles, more than 2 million years; and mammals, about 2 million years. These estimates, which represent a wide variety of living things, suggest that most species evolve very slowly; they indicate that an average animal or plant species is not likely to evolve sufficiently to be regarded as a new species even after it has passed through about a million generations. For comparison, note that the oyster species shown in Figure 6-25 changed enough to be regarded as new species within 2 to 4 million years; if an individual

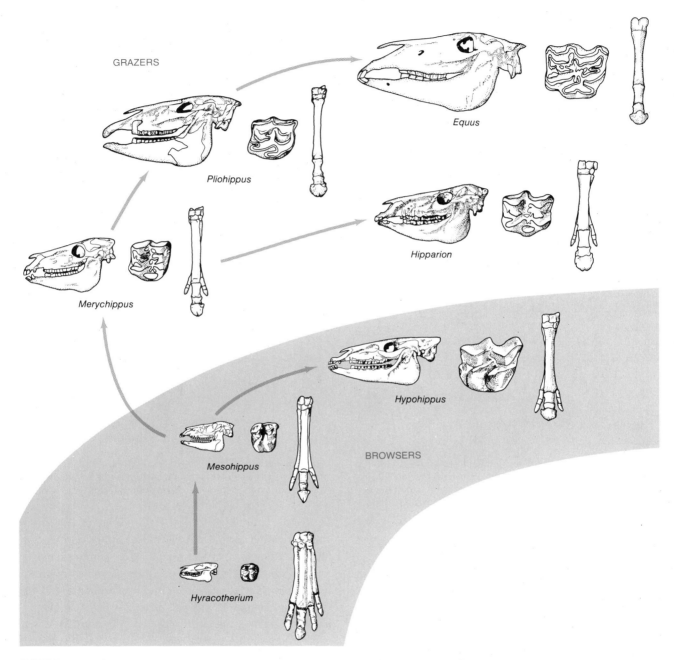

FIGURE 6-24   Large-scale trends in the evolution of the horse family. Heads, molar teeth, and front feet are illustrated here for seven genera. Heads are drawn to the same scale and so are teeth, showing the general increase in size. Feet are not drawn to the same scale. During the history of the horse family, the surfaces of the molar teeth of some members developed complex cusps that were associated with a transition from browsing on soft leaves to grazing on harsh grasses. The number of toes on the front foot was also reduced from four to one. (*After G. G. Simpson, Horses, Oxford University Press, New York, 1951 .*)

oyster required 2 or 3 years to become reproductively mature, then even here evolution required a million generations or so to produce a new species. All of these longevities must be judged in light of how long it has taken for new higher taxa of the same group to develop. Recall, for example, that early in the Cenozoic Era, whales evolved from vastly different small, rodentlike mammals in no more than 12 million years. In comparison to this interval, a

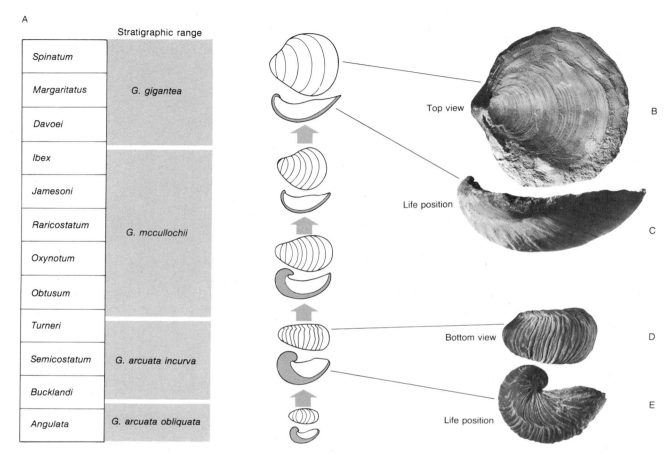

FIGURE 6-25   An apparently gradual trend in a lineage of coiled oysters during an Early Jurassic interval of about 12 million years. During this interval, the shell became larger, but it also became thinner and flatter. These animals rested on the sea floor in the orientation labeled "life position." It may be that the flatter shell was more stable against potentially disruptive water movements. (*After A. Hallam, Philos. Trans. Roy. Soc. London 254B:91–128, 1968.*)

typical survival time of 2 million years for a single mammal species seems quite sizable.

The slow rate of evolution that characterizes many well-established species has led some paleontologists to conclude that most evolution must be associated with speciation—that is, with the rapid evolution of new species from others. Another line of evidence cited in favor of this idea is the evolutionary history that typifies long, narrow segments of phylogeny, segments that undergo little branching but span long intervals of geologic time. If speciation were indeed the site of most evolution, such segments of phylogeny would be expected to exhibit little evolution for the simple reason that they have experienced very little speciation. This is exactly the pattern exhibited by the bowfin fishes (Figure 6-26), a group that has experienced very little speciation and very little evolution during the last 60 million years. The single living bowfin species so closely resembles those of early Cenozoic time that it has been

labeled a living fossil. As it turns out, all of the living species that we know to be at the end of long, narrow segments of phylogeny are living fossils. Other examples of living-fossil animals are the alligator, the snapping turtle, and the aardvark (Figure 6-27). A well-publicized example of a living-fossil plant is the dawn redwood (Figure 6-28), which was thought to be extinct until, in the 1940s, it was discovered living in a small area of China.

FIGURE 6-26   Origin of the bowfin fish (*Amia calva*), a "living fossil" species that is the product of a narrow segment of phylogeny. Bowfin fishes are known from a large number of stratigraphic units ranging back to the Late Cretaceous Series (more than 65 million years old). Since Late Cretaceous time, there have been very few speciation events within the group, and there has been very little evolution. (*After J. R. Boreske, Museum of Compar. Zool. Bull. 146:1–87, 1974.*)

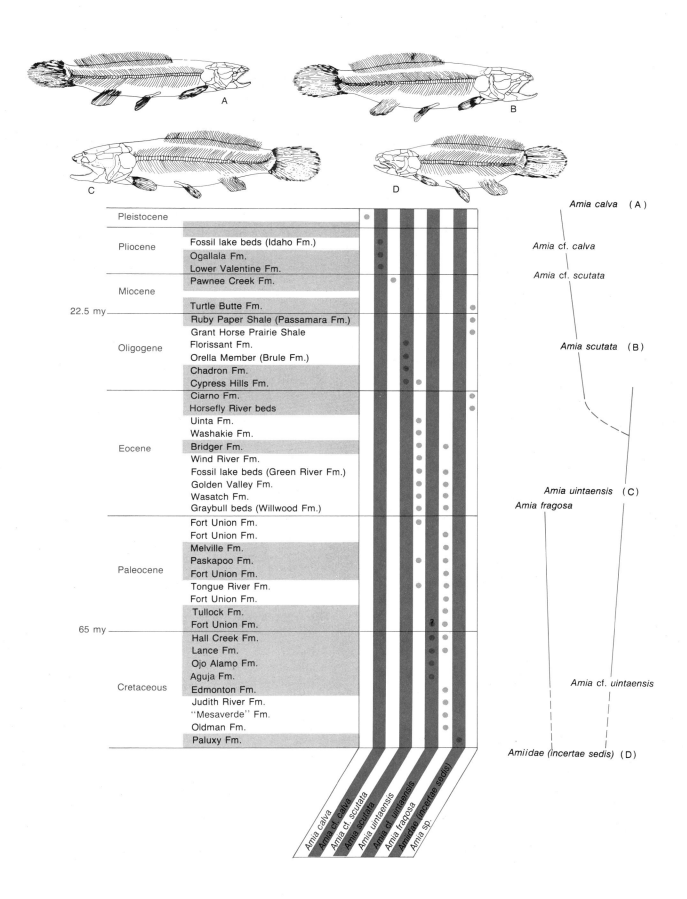

FIGURE 6-27   The living aardvark, shown here, is the only existing species of the order of mammals to which it belongs. Since the beginning of the Miocene Epoch more than 20 million years ago, there have been very few aardvark species living at any time. During this interval, the group has experienced little evolutionary change, and the modern species is thus regarded as a living fossil.

A   B

FIGURE 6-28   The dawn redwood, *Metasequoia*, a living-fossil tree. In the 1940s, at a time when it had been thought extinct for millions of years, this genus was discovered living in the Szechwan province of China. It is now widely grown in the United States as an ornamental tree. *Metasequoia* has a good fossil record in North American deposits ranging in age from Late Cretaceous (the age of the fossils shown here on the right) to Miocene. Recently, it has also been found in the Pliocene Series of Japan. Few species of *Metasequoia* have existed at any time, and the dawn redwood has undergone little evolution. (*A. P. L. Carpenter et al., Plants in the Landscape, W. H. Freeman and Company, New York, 1975. B. Courtesy of L. Hickey and the Smithsonian Institution.*)

The idea that most evolutionary change has been associated with rapid speciation does not necessarily negate the idea that natural selection is the dominant source of evolution, but it has produced considerable controversy. If it is a valid idea, paleontologists must reconsider the mechanisms by which large-scale evolutionary trends are produced. If established species tend to evolve very slowly, surviving for millions of years with little evolutionary change, then large-scale trends such as the one in which small, primitive horses evolved into modern horses (Figures 6-23 and 6-24) cannot result from the gradual modification of species. Instead, such trends would result from two conditions operating singly or together. First, a trend could result from a tendency for speciation events to produce change in a particular evolutionary direction repeatedly. Second, a trend could result from a kind of sorting-out process in which various kinds of new species originated but only certain ones tended to accumulate in the course of time because they produced many descendant species, while others tended to dwindle because they left few descendants. Box 6-3 provides a closer look at these possibilities.

It is important to bear in mind that although the fossil record is not complete enough to reveal the patterns of many evolutionary trends, it does demonstrate that such trends have occurred: It shows that certain ancient groups of animals or plants were the ancestors of certain modern forms. If we had no fossil record, we would know almost nothing about large-scale evolutionary trends. In fact, in the absence of a fossil record, the idea that large-scale

trends occurred at all would perhaps remain controversial among scientists. The argument that modern forms of life evolved from quite different ancient forms could then be based only on the weaker evidence provided by species of the present world (evidence such as that described on pages 137–142 of this book).

The irreversibility of evolution

An evolutionary trend that has resulted from at least several genetic changes is highly unlikely to be reversed by subsequent evolution. This principle is named **Dollo's law** for Louis Dollo, the Belgian paleontologist who proposed it early in the twentieth century. Dollo's law reflects the fact that it is almost impossible for a long sequence of genetic changes in a population to be reversed in an order exactly opposite to the one in which they originally developed. Thus, evolution occasionally produces a species of animals or plants that crudely resembles an ancestor that lived before, but it never perfectly duplicates a species that has disappeared. In other words, once changed by evolution or eliminated by extinction, a species is gone forever.

Box 6-3

The generation of

large-scale trends

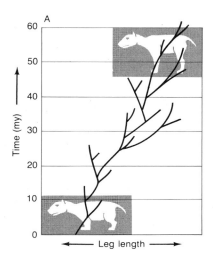

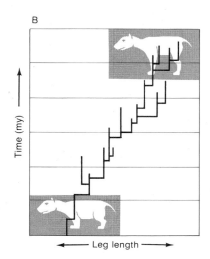

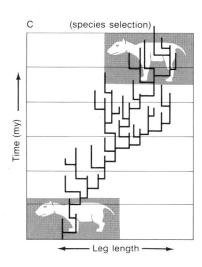

What is the pattern of evolution that produces complex trends in phylogeny such as the one that leads from the earliest horses to modern horses (Figure 6-23)? The three phylogenies shown here illustrate three possible patterns. Any morphologic feature might be represented in diagrams A, B, and C, but for purposes of illustration, what is plotted is the leg length of an imaginary group of mammals. If the gradual transformation of species accounts for most evolutionary change, a general trend can result from a pattern such as that shown in diagram A. On the other hand, if established species undergo little evolution before becoming extinct and if most change is instead associated with speciation, large-scale trends must result from con-

ditions similar to those shown in diagram B or diagram C.

Diagram A is based on the assumption that species (branches of phylogeny) change gradually at significant rates. Rates of evolution are represented by the slopes of the branches. These rates vary, but most branches grow toward the right (that is, they evolve toward longer leg length), thus accounting for the overall trend toward the right.

Diagrams B and C are based on the assumption that species form during very brief geologic intervals and undergo little subsequent evolution. (For simplicity, species are shown here as undergoing absolutely no evolution after they form; the branches of the phylogenies are vertical.)

In diagram B, the trend toward the right results from the simple fact that 15 of the 16 rapid speciation events depicted move toward the right—that is, 15 of the 16 new species have longer legs than their parent species.

In diagram C, the pattern is more complex. Here, the trend moves toward the right even though the number of speciation events producing species with shorter legs equals the number producing species with longer legs. The fact that there is an accumulation of species farther and farther toward the right results from the subtle effects of two factors that are illustrated by the use of an arbitrary time scale on the vertical axis. The first of these factors, as graph I shows, is that species with long legs

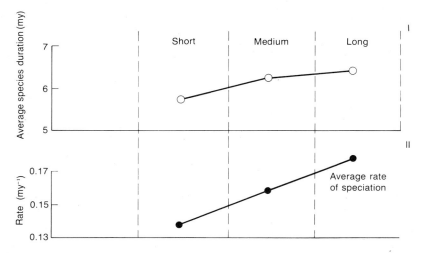

tend to survive longer than those on the left side and thus they tend to give rise to more descendant species. The second of these factors, as graph II illustrates, is that an average species with long legs tends to produce new species at a higher rate *for every million years that it exists*. Actually, either of these traits—a tendency to survive for a long time or a tendency to produce descendant species at a high rate—could result in an accumulation of species with long legs. In complementary fashion, either a tendency to suffer extinction quickly or to produce des-

(or not at all) could by itself cause a dwindling or even total disappearance of species on the left side of the diagram.

The pattern shown in diagram C is labeled **species selection** because it represents a selection process in which species are selected for or against, depending on whether they have long or short durations and produce descendant species at a high or low rate. Species selection is analogous to selection among individuals, but with species instead of individuals as the units of selection. The varying rates of speciation and

tional change in species selection, are analogous to the varying rates of reproduction and death that represent selection at the individual level.

It is important to understand, however, that species selection can operate even if selection at the individual level accounts for the origins of individual species. Species selection simply determines where in phylogeny most individual selection (i.e., most speciation) takes place.

Certainly, all of the mechanisms illustrated in diagrams A, B, and C must contribute to large-scale evolutionary trends. The question is, what is the relative importance of each mechanism? Those who believe that most evolution occurs through the gradual transformation of established species believe that large-scale trends develop primarily as depicted in diagram A. Those who believe that most evolution is associated with speciation events believe that patterns like those depicted in diagrams B and C must prevail. (*Diagram C adapted from S. M. Stanley, Macroevolution: Pattern and Process, W. H. Freeman and Company, New York, 1979.*)   □

CHAPTER SUMMARY

1.   Several lines of evidence convinced Charles Darwin that organic evolution produced the multitudinous species that inhabit the modern world. Among these pieces of evidence were:

The restriction of many closely related groups of species to discrete geographic regions separated by barriers.

Embryological similarities between groups of animals that are dissimilar as adults.

The presence of similar anatomical "ground plans" in animals that live in quite different ways.

The existence in certain animals of vestigial organs that serve no apparent function but resemble functioning organs in other species.

The ability of humans to alter domestic animals and cultivated plants by artificial selection in breeding.

2.   Natural selection is a process in which certain kinds of individuals become more numerous in a population because they produce an unusually large proportion of the total offspring.

3.   A particular kind of individual can be favored by natural selection either because it tends to live for a long time or because it tends to produce offspring at a high rate during its lifetime.

4.   The variability that is the basis of natural selection is generated by two mechanisms: genetic mutation and the generation of new gene combinations by sexual reproduction.

5.   Speciation is the process by which an existing species gives rise to an additional species; it is believed that in most cases, the population that becomes the new species is geographically isolated from the remainder of the parent species.

6.   Extinction is the disappearance of a species from the earth. The agents of extinction are the ecological factors that normally govern the sizes of populations in nature.

7.   Adaptive radiation is the proliferation of many kinds of species from a small ancestral group. Adaptive radiation has usually followed new access to ecological opportunities.

8.   Mass extinction is the disappearance of many species during a geologically brief interval of time.

9.   Under normal circumstances, different rates of speciation and extinction characterize different groups of animals and plants.

10.   One of the most convincing kinds of evidence for the adaptive nature of evolution is evolutionary convergence, or the evolution within two or more higher taxa of species that resemble one another in form and also live in the same way.

11.   Long-term evolutionary trends are evolutionary changes that have developed in the course of millions of years. There are many ways in which such trends can develop, and the relative importance of these depends on what percentage of evolutionary change is associated with rapid speciation events and on what percentage is represented by the gradual transformation of well-established species.

EXERCISES

1.   What geographic patterns suggested to Charles Darwin that certain kinds of species descended from others?
2.   What characteristics make a particular kind of individual successful in the process of natural selection?
3.   What conditions make it likely that a small group of closely related species will increase to a large number of species by means of rapid speciation?
4.   How can evolution proceed by a change in the ontogeny (i.e., the growth and development) of a species?
5.   What are the ways in which an evolutionary trend can develop during the history of a genus or a family?

ADDITIONAL READING

Evolution (Readings from *Scientific American*), W. H. Freeman and Company, New York, 1978.

The Fossil Record and Evolution (Readings from *Scientific American*), W. H. Freeman and Company, New York, 1982.

Gould, S. J., *The Panda's Thumb*, W. W. Norton & Company, Inc., New York, 1980.

Stanley, S. M., *The New Evolutionary Timetable: Fossils, Genes, and the Origin of Species*, Basic Books, Inc., New York, 1981.

Stebbins, G. L., *Darwin to DNA, Molecules to Humanity*, W. H. Freeman and Company, New York, 1982.

PART THREE

Movements of the Earth

The most exciting development in the science of geology during the twentieth century has been the origin of the theory of plate tectonics. According to this theory, plates of lithosphere ascend from the earth's mantle along midocean ridges, slide laterally over the surface of the planet, and descend into the mantle once again. Many plates carry continents with them; the materials that compose continents are of low density, however, and do not descend into the mantle. The movements and interactions of plates deform the earth's crust on a vast scale. They also cause magma to rise through the upper mantle to form igneous rocks on and within the overlying crust.

On continents, mountain ranges are the colossal products of plate motions. Most mountains have formed where two continents on separate plates have collided or where one plate has descended beneath a continent that is situated on the edge of another plate. Continents grow laterally when mountains rise up along their margins, and they break apart when spreading ridges transect them, producing oceans like the modern Atlantic.

CHAPTER 7

Plate Tectonics

The emergence of the theory of plate tectonics during the 1960s fostered a revolution in the science of geology. *Tectonics* is a term that has long been used to describe movements of the earth's crust. Accordingly, the phrase *plate tectonics* refers to the ways in which discrete segments of the earth's crust have been found to move in relation to one another. Whereas continents were once thought to be locked in place by the oceanic crust that surrounds them, the theory of plate tectonics holds that continents move over the earth's surface because they form parts of moving plates (Figure 7-1). Moreover, continents occasionally break apart or, alternatively, fuse together to form larger continents. The theory of plate tectonics explains why most volcanoes and earthquakes occur along curved belts of sea floor, why mountain belts tend to develop along the edges of continents, and why the present ocean basins are very young from a geologic perspective. Many of the kinds of rock deformation discussed in Appendix II also result from plate-tectonic movements.

THE HISTORY OF OPINION ABOUT CONTINENTAL DRIFT

Although the concept of plate tectonics emerged quite suddenly in the 1960s, it resolved many long-standing dis-

The opening of the Red Sea. During the past few million years, Africa and the Arabian Peninsula, which were once united as a single continent, have been moving apart as portions of separate plates of lithosphere, leaving the Red Sea in the gulf thus formed. According to the theory of plate tectonics, currents within the earth's mantle propel the diverging plates.

putes. For many years, the idea that continents move horizontally over the surface of the earth, an idea labeled **continental drift**, failed to receive general support in Europe or North America. In 1944, one prominent geologist went so far as to assert that the idea of continental drift should be abandoned outright because "further discussion of it merely encumbers the literature and befogs the minds of students." Although many geologists may not have read this comment, most agreed with its spirit, and during the 1950s, little attention was given to the possibility that continental drift was a real phenomenon.

When the idea of continental drift first emerged earlier in the twentieth century, it attracted considerable attention, primarily as a result of the arguments of two scientists—Alfred Wegener of Germany and Alexander du Toit of South Africa. We will briefly examine the case that these two men and their followers made and the reasons their arguments were rejected by most of their contemporaries.

Early evidence

The observation that the coasts on the two sides of the Atlantic Ocean fit together like separated parts of a jigsaw puzzle (Figure 7-2), constituted the first evidence that continents might once have broken apart and moved across the surface of the earth. Centuries ago, map readers noted with curiosity that the outline of the west coast of Africa seemed to match that of the east coast of South America. It was not until 1858, however, that Antonio Snider-Pellegrini, a Frenchman, published a book suggesting that a great continent had once broken apart and that the Atlantic Ocean had formed by powerful forces applied from within the earth.

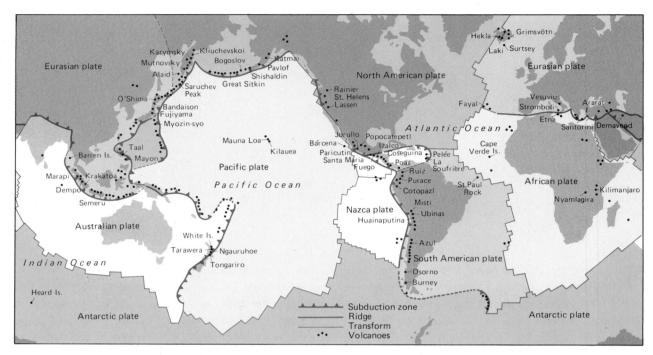

FIGURE 7-1   The distribution of lithospheric plates over the surface of the earth. There are eight large plates and several small ones on earth today. The three kinds of plate boundaries—subduction zones, ridge axes, and transform faults— are shown here; they will be discussed later in the chapter. The locations of volcanoes are also indicated; most lie near subduction zones or ridges. (*F. Press and R. Siever, Earth, W. H. Freeman and Company, New York, 1982.*)

More traditional geologists clung to the idea that large blocks of continental crust could not move over the surface of the earth. Thus, when the distribution of certain living and extinct animals and plants began to suggest former connections between presently separated landmasses, most geologists tended to assume that great corridors of felsic

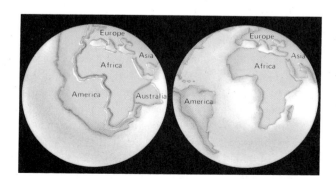

FIGURE 7-2   The previous position of continents (*left*) compared with their present position, according to Antonio Snider-Pellegrini. Snider-Pellegrini was the first to publish the theory that the Atlantic Ocean was formed by the breaking apart of a large continent.

rock (the most abundant material in continental crust) had once formed land bridges that connected continents but later subsided to form portions of the modern sea floor. Today, it is recognized that this could not have occurred, since felsic crust is of such low density that it cannot possibly sink into the mafic rocks that underlie the oceans. Nonetheless, many prominent geologists of the late nineteenth century, including Melchior Neumayr and Eduard Suess, presented schemes of earth history that included the concept of felsic corridors (Figure 7-3). Suess, for example, believed that Atlantis, the sunken continent of Greek mythology, now lay in the North Atlantic sea floor near Greenland.

One phenomenon that led scientists to speculate about ancient land bridges was the similarity between the present fauna of the island of Madagascar and that of India, a land that is now separated from Madagascar by an expanse of ocean nearly 4000 kilometers (~2500 miles) across. Madagascar is characterized by primitive mammals; altogether missing are the zebras, lions, leopards, gazelles, apes, rhinoceroses, giraffes, and elephants that characterize nearby Africa. In contrast, some of the native animals of

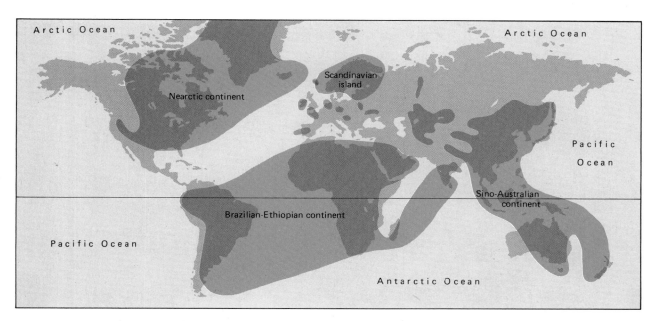

FIGURE 7-3   The former distribution of continents according to Melchior Neumayr, who published his ideas late in the last century. According to Neumayr, continental corridors once connected continents that are now separated by deep oceans. Neumayr thought that these corridors sank to form the floors of the new oceans. We now know that this is impossible, since continental crust is of such low density that it cannot sink into the dense mantle. Furthermore, mafic oceanic crust throughout the world differs from felsic continental crust; continents cannot have turned into the floor of the deep sea.

India closely resemble those of Madagascar. Neumayr and Suess consequently believed that a now-sunken land bridge had once spanned the western part of the Indian Ocean, connecting Madagascar to India.

A second line of evidence for ancient land connections was found in the fossil record. During the nineteenth century, late Paleozoic coal deposits of India, South Africa, Australia, and South America were found to contain a group of fossil plants that were collectively designated the *Glossopteris* flora after their most conspicuous genus, a variety of seed fern (Figure 7-4). After the turn of the century, the *Glossopteris* flora was discovered in Antarctica as well. The occurrence of this fossil flora on widely separated landmasses was one of the facts that led Eduard Suess to suggest that land bridges had once connected all of these continents. Suess introduced the term **Gondwanaland** (Gondwana-Land, in his spelling) to connote the hypo-

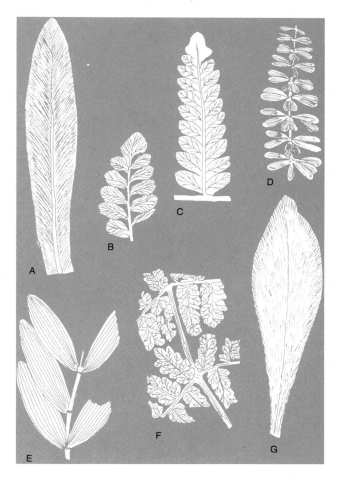

FIGURE 7-4   Representatives of the *Glossopteris* flora, which occupies many southern continents. *A. Glossopteris. B. Merianopteris. C. Sphenopteris. D. Sphenophyllum. E. Schizoneura. F. Sphenopteris. G. Gangamopteris.* (After E. A. N. Arber, British Museum Catalogue of the Fossil Plants of the Glossopteris Flora in the Department of Geology, British Museum, London, 1905.)

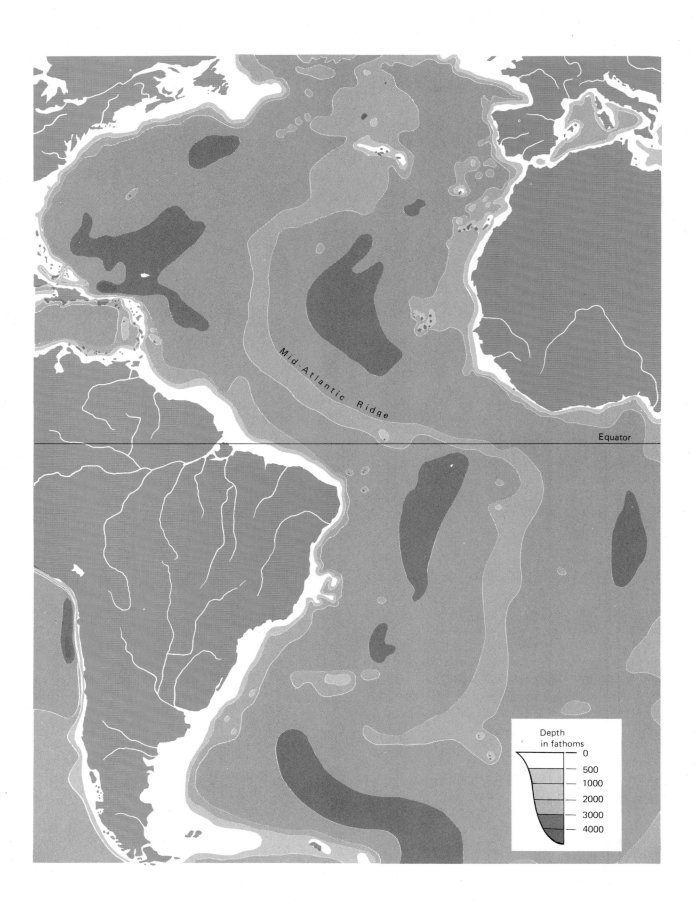

Mid-Atlantic Ridge

Equator

Depth
in fathoms
0
500
1000
2000
3000
4000

thetical continent that consisted of these landmasses and the land bridges that he believed to have connected them. (Gondwana is a locality in India where seams of coal yield fossils of the *Glossopteris* flora.)

Then, in 1908, the American geologist Frank B. Taylor proposed a new explanation for the connection of ancient landmasses. In essence, Taylor hypothesized that the continents had once lain side by side as components of very large landmasses that eventually broke apart and moved across the earth's surface to their present positions. This general idea is central to modern plate-tectonic theory, although the particulars of Taylor's scheme are no longer accepted.

One of Taylor's most significant contributions was his suggestion that the immense submarine mountain chain now called the Mid-Atlantic Ridge, which is one of several **midocean ridges** of the present-day sea floor, marks the line along which one ancient landmass ruptured to form the Atlantic Ocean. The extent of this ridge became evident after soundings were made of the sea floor on the British H.M.S. *Challenger* expedition, a marine exploration undertaken between 1872 and 1876 (Figure 7-5). As its name suggests, the Mid-Atlantic Ridge runs along the center of the Atlantic Ocean floor, generally paralleling the two coastlines that border the Atlantic. These observations led Taylor to propose that the ridge might represent the site at which an ancient landmass had split apart. The stage was thus set for Alfred Wegener's first detailed argument for continental drift.

A twentieth century pioneer: Alfred Wegener

Alfred Wegener was a German meteorologist—a student of weather and climate (Figure 7-6). Perhaps the fact that he was not primarily a geologist gave him the fresh perspective that led to his unconventional interpretation of the earth's history. In 1915, Wegener published the first edition of his book *On the Origin of Continents and Oceans*—but it was not until 1924, when the third edition of this work was translated into many languages, including English, that his ideas became the subject of extensive debate.

FIGURE 7-6   Alfred Wegener, who did not become a scientific hero until more than three decades after his death, when plate-tectonic theory supported his view that the continents have moved laterally for enormous distances.

FIGURE 7-5   A simplified version of the depth chart for the Atlantic Ocean produced by John Murray in 1895 from data gathered on the voyage of the H.M.S. *Challenger*. These data revealed the structure that is now known as the Mid-Atlantic Ridge.

Wegener presented evidence that virtually all of the large continental areas of the modern world were united late in the Paleozoic Era as a single supercontinent, which he labeled **Pangaea**. The idea that Pangaea existed has been revived with the rise of modern plate-tectonic theory, and today nearly all professional geologists accept it as fact—although, as we will see below, they recognize major errors in Wegener's proposed chronology for the fragmentation of this supercontinent (Figure 7-7).

What led Wegener to his belief in Pangaea? The power of his arguments came from his use of several kinds of evidence, which he enlarged upon during his lifetime. Like Taylor, Wegener emphasized the jigsaw-puzzle fit of continents that are now separated by ocean basins. Wegener went much further, however, by positing that virtually all of the major continents on earth today had once been united as a single landmass. In essence, Wegener viewed Eduard Suess's Gondwanaland as part of the larger landmass of Pangaea. Unlike Suess, however, Wegener correctly as-

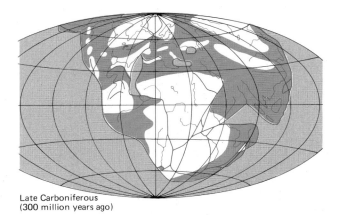

Late Carboniferous
(300 million years ago)

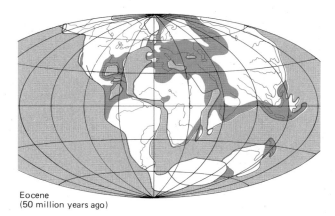

Eocene
(50 million years ago)

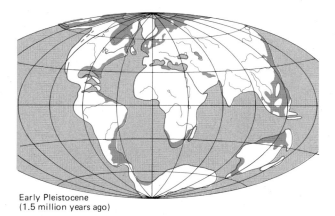

Early Pleistocene
(1.5 million years ago)

FIGURE 7-7   Alfred Wegener's reconstruction of the map of the world for three past times. Africa is placed in its present position as a point of reference. Heavy shading represents shallow seas. Wegener erred in suggesting that Pangaea, the supercontinent shown in the upper map, did not break apart until the Cenozoic Era (*lower two maps*). (*After A. Wegener, Die Eutstehung der Kontinents und Ozeane, Friedrich Vieweg und Sohn, Brunswick, Germany, 1915.*)

sumed that felsic land bridges could not have descended into the more dense oceanic crust and thus could not possibly account for ancient land connections. Instead, like Taylor, Wegener reasoned that continents had broken apart and had drifted about. Wegener also cited the great rift valleys of Africa as possible evidence of newly forming or failed rifts. Wegener's insight was again correct. As we will see later in this chapter, the African rift valleys are now regarded as representing an early stage of continental rifting.

Wegener supported his theory with several additional arguments. He noted, for example, that there were numerous geologic similarities between eastern South America and western Africa, and he also called attention to the many similarities between the fossil biotas of these two widely separated continents. Several extinct groups of animals and plants had been found to occur in the fossil records of two or more Gondwanaland continents, and this led Wegener to argue that these continents must once have lain close together. Even *living* animal and plant groups were shown to exhibit a "Gondwanaland" pattern: A number of species were found to be widely distributed among the southern continents. To Wegener, this fact also seemed to support the thesis that these continents had once been united.

Wegener's arguments were more fully developed by South African geologist Alexander du Toit. Du Toit and others introduced a wealth of circumstantial evidence favoring the idea of continental drift—evidence that was publicized both before Wegener's untimely death in 1930 and during the three decades of controversy that followed.

Among the groups of living animals that seemed to support the concept of Gondwanaland were the earthworms. One genus of earthworm, for example, was found to be restricted to the southern tips of South America and Africa, which lay close together in Wegener's Gondwanaland reconstruction (Figure 7-8). Another genus was encountered only in southern India and southern Australia. Wegener and du Toit wondered how soil-dwelling animals could possibly have such strange distributions in the modern world unless the land areas that these genera inhabited had once been connected. Traditional geologists nonetheless remained skeptical.

The evidence of the fossil record seemed even more compelling. Wegener plotted a map representing the earlier suggestion that the southern continents once formed a large landmass, most of which had sunk to become the floors of the South Atlantic, Indian, and South Pacific oceans. Wegener's map of this mythical landmass is quite striking. Especially when he added to this map the locations of the

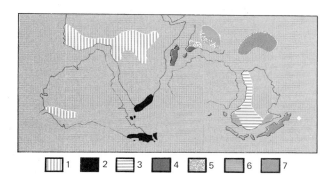

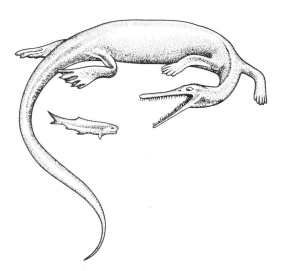

FIGURE 7-8   Geographic distributions of seven genera of living earthworms of the family Megascolecina: (1) *Dichogaster*, (2) *Chilota*, (3) *Megascolex*, (4) *Howascolex*, (5) *Octochaetus*, (6) *Perionyx*, (7) *Pheretima*. Note that each genus occurs on landmasses that are now widely separated. When the continents are brought together in Wegener's reconstruction of Gondwanaland, however, it is easy to see how the earthworms attained their distribution. (*After A. L. du Toit, Our Wandering Continents, Oliver & Boyd Ltd., Edinburgh, 1937.*)

FIGURE 7-10   *Mesosaurus*, a small Early Permian reptile found in freshwater deposits in both South Africa and southern Brazil. The animal measured about 0.6 meters (~2 feet) in length. In the era when continental drift was denied, it was difficult to explain how fossils of *Mesosaurus* could be found on both sides of the Atlantic. (*After H. F. Osborn, in A. S. Romer, Proc. Amer. Philos. Soc., 1968, pp. 335–343.*)

Glossopteris flora (Figure 7-4), the Gondwanaland continents seemed naturally to belong not where they are today but in a tight cluster (Figure 7-9.) It can be seen, for example that 20 of the 27 species of land plants recognized within the *Glossopteris* flora of Antarctica have been found as far away as India. It might be suggested that winds could have spread the plants this far by carrying their seeds, but in fact the seeds of the genus *Glossopteris* itself are several

millimeters in diameter—much too large to have been blown across wide oceans.

Animal fossils also played a role in the debate that resulted from Wegener's work. DuToit noted, for example, that fossils of the reptile *Mesosaurus* (Figure 7-10) occurred at or near the position of the Carboniferous-Permian boundary in both Brazil and South Africa. On both continents, fossils of *Mesosaurus* occur in dark shales along with fossil insects and crustaceans. *Mesosaurus* occupied freshwater and perhaps brackish habitats, and most paleontologists found it difficult to imagine that the animal had somehow made its way across an ocean as broad as the present Atlantic and had then found freshwater depositional settings that were nearly identical to its former habitat.

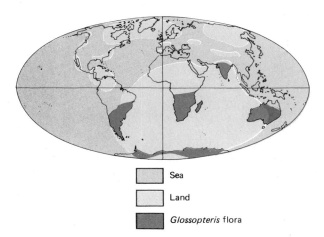

Sea

Land

Glossopteris flora

FIGURE 7-9   Distribution of land and seas during the Carboniferous Period as perceived by the predecessors of Alfred Wegener, who believed that large areas of the present-day ocean floor then stood above sea level. Wegener published this map to show how South America, Africa, India, Australia, and Antarctica seemed to belong together rather than in their present positions. The distribution of the *Glossopteris* flora (not plotted by Wegener in this map) strengthens the impression that the Gondwanaland continents were once united.

The Gondwana sequence

The general stratigraphic context in which the *Glossopteris* flora and *Mesosaurus* are encountered offered further evidence supporting the existence of Gondwanaland. Specifically, Carboniferous and Permian rock units that yield the *Glossopteris* flora form what is known as the **Gondwana sequence**, which occurs with remarkable similarity in South America, South Africa, India, and Antarctica. The Gondwana sequence in du Toit's native South Africa, known as the **Karroo sequence**, serves as a useful example for

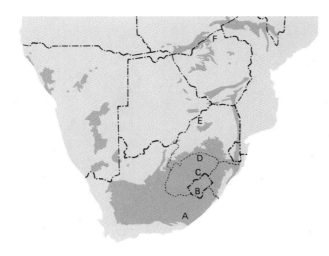

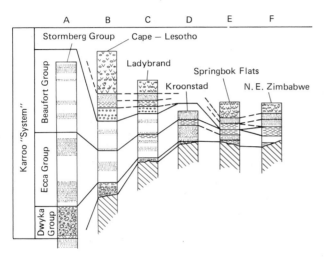

comparison (Figure 7-11). Toward the south, the Karroo sequence begins with the glacial deposits of the Upper Carboniferous Dwyka Tillite. In places, this tillite rests on crystalline rocks that were smoothed and scratched by glacial activity (Figure 3-5). To the south, where it is thickest, the Dwyka group was deposited partly in a marine environment. The upper part of the Dwyka contains the nonmarine carbonaceous shale in which *Mesosaurus* is found. Other shales in the upper part of the Dwyka yield the *Glossopteris* flora as does the Permian Ecca Group, which overlies the Dwyka and consists largely of clastic alluvial-fan deposits. The latest Permian and Triassic Beaufort Group, which follows, contains hundreds of species of fossil reptiles. This is succeeded by a sequence of sediments that in places include dune deposits, which must have formed in a hot, arid environment (Figure 7-12A). Capping the Karroo are extensive basalts of Jurassic age, which are so resistant to weathering that they form prominent highlands (Figure 7-12B).

FIGURE 7-11   Gondwana deposits of South Africa. *Above:* Map of the outcrop area of the Karroo system. *Below:* Stratigraphic sections at six locations. Glacial tillites (Dwyka Group) are present in most sections; they thicken toward the south. Basalts cap many of the sections. (*After A. L. du Toit, The Geology of South Africa, Oliver & Boyd Ltd., Edinburgh, 1937.*)

A   B

FIGURE 7-12   Gondwana sequences of South Africa. *A.* Cliff of the Cave Sandstone in the Orange Free State. The upper part of the cliff consists of dune deposits. *B.* The Great Escarpment of South Africa, where lavas more than a kilometer (nearly a mile) thick can be found. These lavas rest on the Karroo System. (*Courtesy of L. C. King.*)

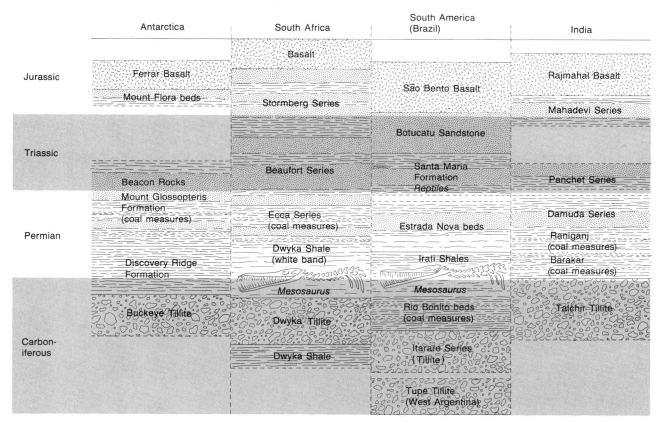

	Antarctica	South Africa	South America (Brazil)	India
Jurassic	Ferrar Basalt / Mount Flora beds	Basalt / Stormberg Series	São Bento Basalt	Rajmahal Basalt / Mahadevi Series
Triassic	Beacon Rocks	Beaufort Series	Botucatu Sandstone / Santa Maria Formation *Reptiles*	Panchet Series
Permian	Mount Glossopteris Formation (coal measures) / Discovery Ridge Formation	Ecca Series (coal measures) / Dwyka Shale (white band)	Estrada Nova beds / Irati Shales *Mesosaurus*	Damuda Series / Raniganj (coal measures) / Barakar (coal measures)
Carbon-iferous	Buckeye Tillite	*Mesosaurus* / Dwyka Tillite / Dwyka Shale	Rio Bonito beds (coal measures) / Itararé Series (Tillite) / Tupe Tillite (West Argentina)	Talchir Tillite

FIGURE 7-13   Correlation of the Gondwana sequences of four continents. In each sequence, glacial tillites are followed by shales and coal beds containing the *Glossopteris* flora. (Modified from G. A. Duamani and W. E. Long, Scientific American, September 1962.)

The Gondwana sequence of Brazil bears an uncanny resemblance to the Karroo of South Africa. At the base of the Brazilian sequence (Figure 7-13) are glacial tillites. These, too, are coarsest at the base and alternate with interglacial sediments, including coals, that yield members of the *Glossopteris* flora. As in South Africa, *Mesosaurus* is found near the base of the Permian in dark shales. Tills occur only as high as the lowermost Permian, but some members of the *Glossopteris* flora persist into the Upper Permian. Much of the Triassic record consists of dune deposits known as the Botacatú Sandstone, which, like the similar dune deposits of South Africa, are succeeded by Jurassic lava flows.

Antarctica and India display Gondwana sequences very similar to those of South Africa and South America (Figure 7-13). Although *Mesosaurus* fossils have not been found in Antarctica or in India, the Gondwana sequences of these two areas do include tillites (again, best developed near the base of each succession), together with coal beds and, of course, the *Glossopteris* flora. The extrusion of Jurassic basalts also terminated deposition of the Gondwana sequences in both Antarctica and India.

Du Toit and other followers of Wegener sought further evidence of ancient land connections in the Gondwana glacial deposits. They measured the orientations of features scoured into underlying bedrock by glaciers and found very interesting patterns (Figure 7-14). They discovered, for instance, that glacial movement in eastern South America was primarily from the southeast, where today no land-

FIGURE 7-14   Map of the modern world showing the locations of late Paleozoic glaciation and the directions in which glaciers flowed. Note that when the continents are not united as Gondwanaland, the glaciers of South America, India, and Australia appear to have flowed from the ocean. (After A. Holmes, Principles of Physical Geology, The Ronald Press Company, New York, 1965.)

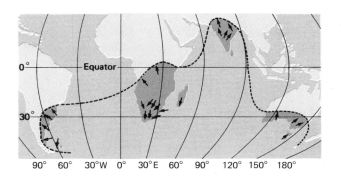

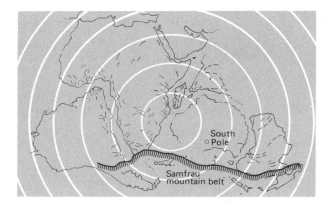

FIGURE 7-15   Alexander du Toit's reconstruction of Gond-wanaland. The short lines indicate regional strikes of folds and faults that align well when the continents are assembled to form Gondwanaland. The Andes mountain chain of South America aligns with mountain systems of South Africa, Antarctica, and Australia to form what du Toit labeled the Samfrau mountain belt of Gondwanaland. (*After A. L. du Toit, The Geology of South Africa, Oliver & Boyd Ltd., Edinburgh, 1937.*)

mass exists that might support large glaciers. In southern Australia, too, there was evidence of glacial flow from the south, where there is now only ocean. Obviously, it would not be difficult to account for such movement if the continents had been united as Gondwanaland at the time the glaciers were flowing. Ice flow would then have radiated from the center of a large continent that might be expected to support large glaciers under cold climatic conditions.

Du Toit correctly deduced from geologic evidence that Pangaea did not form until late in the Paleozoic Era. Before Pangaea was formed, Gondwanaland existed as a distinct supercontinent, and the northern continents were united as Laurasia. (These names, together with the idea of a northern and a southern supercontinent, had actually been proposed in the nineteenth century.)

Du Toit also recognized that if South America, Antarctica, and Australia were assembled as Gondwanaland, the mountain belts along their margins would line up. Even the small Cape Fold Belt at the tip of South Africa would form a link in the chain (Figure 7-15). To this rugged feature of Gondwanaland du Toit gave the name *Samfrau*, a term that was derived from letters of the names *S*outh *Am*erica, *Afr*ica, and *Au*stralia. At the same time, du Toit expanded on Wegener's observation that the ancient geologic features of these now-separate continents would match when the continents were placed in their Gondwanaland positions. Du Toit's plot of these structural trends—including regional strikes of folds and faults—is shown in Figure 7-15.

The rejection of continental drift

Despite mounting evidence favoring Wegener's and du Toit's ideas, geologists of the Northern Hemisphere continued to view the theory of continental drift with considerable skepticism. The primary source of their dissatisfaction lay in the seeming absence of a demonstrated mechanism by which continents could move over long distances. In this regard, Wegener favored a phenomenon that he called the **pole-fleeing force**, which pulls objects near the north and south poles toward the equator. This force operates because the earth's rotation deflects the pull of gravity slightly away from the center of the earth. Wegener's problem in convincing other scientists was that the pole-fleeing force is very weak. Geophysicists of this time knew that both continental crust and oceanic crust were continuous above the Mohorovičić discontinuity (or Moho) and hence they could not imagine how continents could be made to move laterally—to plough through oceanic crust (Figure 7-16) as they assumed was necessary—if they were driven only by the weak pole-fleeing force. Other potential mechanisms of continental motion were also rejected for the same reason. Interestingly, one of the mechanisms considered was convective movement of the earth's mantle—the process that is now widely regarded as the cause of continental movement.

With the geophysicists arguing that continents were locked in place by oceanic crust, geologists who supported Wegener's theory were confronted with a serious challenge. If they wished to continue arguing in favor of continental movement, they would have to advance very strong supporting evidence. The problem, however, was that some fossil evidence seemed to contradict the notion of continental drift. Specifically, Wegener had suggested a brief timetable for drift in which he proposed that Pangaea, which

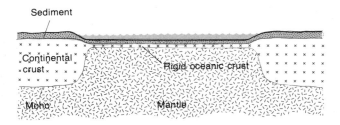

FIGURE 7-16   Diagrammatic cross section of the earth's crust showing the rigid oceanic crust that separates the continents. The Moho is the discontinuity between the crust and the mantle. In 1909 it was discovered that the Moho existed beneath the oceans and hence that there was crust in oceanic areas. The assumption was then made that this rigid crust locked the continents in place.

FIGURE 7-17 Broad distribution of animals on Mesozoic continents. During the Mesozoic Era, several of the present continents formed Gondwanaland and thus were all colonized by certain groups of dinosaurs (*colored*) and by smaller reptiles. To the north, North America, Greenland, and Eurasia formed the supercontinent Laurasia. Laurasia was also populated throughout by a particular set of dinosaurs and smaller reptiles. (*From B. Kurten, "Continental Drift and Evolution." Copyright © 1969 by Scientific American, Inc. All rights reserved.*)

incorporated virtually all the modern continents, had survived into the Cenozoic Era (Figure 7-7). Thus, paleontologists looked for evidence that the world's biotas had evolved into increasingly distinctive geographic groupings since the start of the Cenozoic Era. No such evidence was found.

We now know that Wegener made a dating error that misled paleontologists of his time. The rifting of Pangaea had actually begun near the start of the Mesozoic Era—much earlier than Wegener believed. Continents could not have moved far enough during the brief Cenozoic Era to have allowed biotas to diverge greatly; instead, continents have been relatively widely dispersed since the very beginning of this era. It was recently pointed out that during the Mesozoic Era, when many continents *were* in contact with one another, various kinds of dinosaurs and other animals spread over large areas of the earth (Figure 7-17).

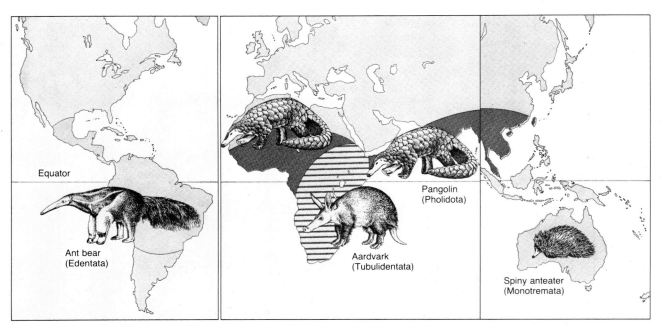

FIGURE 7-18   Narrow distribution of ant-eating mammals on Cenozoic continents. Continents of the Cenozoic Era are smaller and more widely dispersed than those of the Mesozoic Era. Adaptive radiations of mammals have taken place largely independently on the separate continents. Thus, four different kinds of ant-eating animals have evolved, each representing a particular order of mammals. The order to which each figured animal belongs is indicated in parentheses below the animal's common name. (*From B. Kurten, "Continental Drift and Evolution." Copyright © 1969 by Scientific American, Inc. All rights reserved.*)

In contrast, modern mammal faunas form more distinctive regional groupings because they have evolved during the Cenozoic Era, when the continents have been more widely separated (Figure 7-18).

Questions about the mechanism and timing of continental drift were not all that troubled geologists, however. They were reluctant to adopt the hypothesis of continental drift simply because its chief proponent, Wegener, was a meteorologist—an outsider to their own ranks. Two eminent geologists of Wegener's time—Reginald Daly, an American, and Arthur Holmes of England—supported the drift hypothesis, but their views had little bearing on the opinions of their contemporaries. Consequently, continental drift retained a degree of popularity only in South America, South Africa, and Australia, where du Toit and his colleagues saw firsthand the remarkable resemblance between the various Gondwana sequences.

Ironically, after new data supporting plate tectonics had finally brought continental drift into favor, an exciting fossil find was made at Coalsack Bluff in Antarctica. This was the discovery in 1969 of the genus *Lystrosaurus*, an animal classified as a member of the mammal-like reptile group, which will be described in Chapter 15. *Lystrosaurus* was a heavy-set herbivorous animal with beaklike jaws (Figure 7-19). This genus, which is now represented by numerous

FIGURE 7-19   Reconstruction of *Lystrosaurus*, the mammal-like reptile now known from Antarctica as well as from Africa and southeast Asia. *Lystrosaurus* was a herbivorous animal with short legs, beaklike jaws, and a pair of short tusks. (*From E. H. Colbert, Wandering Lands and Continents, E. P. Dutton & Company, Inc., New York, 1973.*)

Antarctic specimens, is a characteristic element of Lower Triassic faunas of Africa and India. *Lystrosaurus* may have been semiaquatic, like a modern hippopotamus, but it was obviously no oceangoing swimmer. One has to wonder whether an earlier discovery of *Lystrosaurus* fossils in Antarctica might have revived enthusiasm for continental drift before the advent of plate-tectonic theory.

The puzzle of paleomagnetism

In the late 1950s, interest in continental movements was renewed as a result of new evidence derived from **paleomagnetism**—or the magnetization of ancient rocks at the time of their formation. We have already seen how the earth's magnetic field has reversed its polarity on many occasions. During the 1950s, geophysicists attempted to ascertain whether the north and south magnetic poles not only reversed their positions but also wandered about periodically. To explore this possibility, these researchers attempted to determine the previous positions of the magnetic poles by using magnetized rocks as compasses for the past.

As we learned in Chapter 5, a magnetic field frozen into a rock is similar to that which is "read" by a compass. The angle that a compass needle makes with the line running

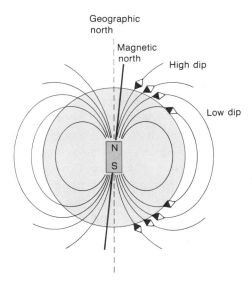

Geographic north

Magnetic north

High dip

Low dip

N
S

FIGURE 7-20   Diagrammatic illustration of the structure of the earth's magnetic field. The core has north and south poles and thus behaves like a bar magnet. The north-south axis has a declination of 15° from the earth's north-south geographic axis. Curved lines represent magnetic lines of force. These lines of force have high dips near the poles and low dips near the equator.

to the geographic north pole is called the **declination**. Today, each magnetic pole lies about 15° from the geographic pole, creating only a small declination (Figure 7-20). A compass needle not only points at the north magnetic pole but, if allowed to tilt in a vertical plane, also dips at a particular angle. As Figure 7-20 shows, the dip of a compass needle varies with the distance of the compass from the magnetic pole. The dip is lowest near the equator, where the lines of force of the earth's magnetic field intersect the earth's surface at a low angle.

Paleomagnetism in a rock also has both an inclination and a dip, which serve to indicate the apparent direction of the north magnetic pole at the time when the rock was first magnetized, as well as the distance between the rock and the pole. It is important to understand, however, that neither a compass needle nor the magnetism of a rock reveals anything about longitude (position in an east-west direction).

When geologists first began to measure rock magnetism, they found that recently magnetized rocks exhibited a magnetism that was consistent with the earth's current magnetic field. The magnetism in older rocks, however, had different orientations. As data accumulated, it began to appear that the earth's magnetic pole had wandered. A plot of the pole's apparent positions for rocks of various ages in North America and in Europe showed that the pole seemed to have moved to its present position from much farther south, in the Pacific Ocean. However, another striking fact emerged to cast doubt on this hypothesis. The path obtained from European rocks differed in detail from that obtained from North American rocks (Figure 7-21*A*). It was recognized that this pattern might actually reflect a history in which the north pole did not wander at all; instead, as Wegener had suggested, the continents of Europe and North America might have moved in relation to the pole and to one another, carrying with them rocks that had been magnetized when the continents were in different positions. This possibility led to the use of the cautious phrase *apparent polar wander* to describe the pathways that geophysicists plotted.

Tests were conducted to examine the possibility that continents rather than poles had moved. It was hypothesized, for example, that if North America and Europe had once been united and had drifted over the surface of the earth together, they should have developed identical paths of apparent polar wander during their joint voyages. The test, then, was to fit the outlines of North America and Europe together along the Mid-Atlantic Ridge to determine whether, with the continents in this position, their apparent-

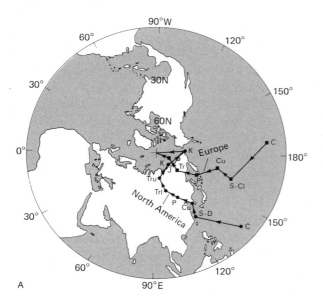

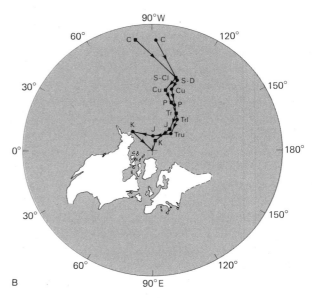

FIGURE 7-21   Apparent-polar-wander paths for North America (*circles*) and Europe (*squares*). *A.* Plot of polar-wander paths based on the assumption that the continents have remained in their present positions. *B.* Plot for North America and Europe juxtaposed, as postulated for Paleozoic time by Wegener and his followers. Here the Paleozoic and Mesozoic apparent-polar-wander paths for the two continents nearly coincide, suggesting that the continents were united during the Paleozoic Era. The time-rock units represented are Cretaceous (*K*), Triassic (*Tr*), Upper Triassic (*Tru*), Lower Triassic (*Trl*), Permian (*P*), Upper Carboniferous (*Cu*), Siluro-Devonian (*S-D*), Silurian to Lower Carboniferous (*S-Cl*), and Cambrian (*C*). (*After N. W. McElhinny, Paleomagnetism and Plate Tectonics, Cambridge University Press, London, 1973.*)

polar-wander paths coincided. As Figure 7-21*B* shows, the apparent-polar-wander paths of North America and Europe did coincide almost exactly for both Paleozoic and early Mesozoic time. This evidence strongly suggested that the continents had indeed drifted apart, carrying their magnetized rocks with them.

THE RISE OF PLATE TECTONICS

During the late 1950s, these new paleomagnetic data generated widespread discussion of continental drift—especially in Great Britain, where many of the data had been assembled. Even so, most geologists continued to doubt the validity of continental movement. There were two reasons for this continuing skepticism: First, many paleomagnetic measurements were known to be imprecise; and second, the belief persisted that no natural mechanism could move continents against oceanic crust. Then, in 1962, American geologist Harry H. Hess published a landmark paper entitled "History of Ocean Basins," proposing a novel solution to this problem.

Sea-floor spreading

In essence, Hess suggested that the felsic continents had

not ploughed through the dense mafic crust of the ocean at all but that instead, the *entire* crust had moved. Hess's ideas were highly unconventional (he labeled his contribution "geopoetry"), but the manner in which he compiled his facts exemplifies the way geologists assemble circumstantial evidence to construct theories. In the following summary of Hess's paper, the critical facts and inferences appear in italics.

During World War II, Hess commanded an American naval vessel in the Pacific, and he seized upon this opportunity to pursue his geologic interests. In order to study the configuration of the ocean floor, for example, Hess kept his ship's echo-sounding equipment operating for long stretches of time. While profiling the bottom in this way, he discovered curious flat-topped seamounts rising from the floor of the deep sea (Figure 7-22), and he named them **guyots** after the nineteenth century geographer Arnold Guyot. On the basis of their size and shape, Hess concluded that guyots were volcanic islands that had been truncated by wave erosion near sea level. Two decades later, shallow-water fossils of Cretaceous age were recovered from the tops of some guyots, proving that the *guyots had indeed once stood near sea level.* How the ocean floor on which they sat had subsided to such great depths, however, remained a mystery.

Another piece of evidence that Hess pondered was the

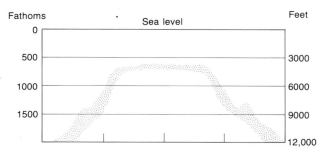

FIGURE 7-22   One of Harry Hess's echo-sounding records showing the shape of a guyot near Eniwetock Atoll. Hess obtained the record by bouncing sound waves off the sea floor and recording their travel time. This guyot, or flat-topped volcanic seamount, is about 15 kilometers (~9 miles) across. (*From H. H. Hess, Amer. Jour. Sci. 244:772–791, 1946.*)

apparent youth of the ocean basins. At the time, it was estimated that sediment was being deposited in the deep sea at a rate of about 1 centimeter ($\sim\frac{1}{2}$ inch) per thousand years—-but at this rate, 4 billion years of earth history would theoretically produce a layer of deep-sea sediment 40 kilometers (~25 miles) thick. In fact, the average thickness of sediment in the deep sea today is only 1.3 kilometers (less than 1 mile). Thus, allowing for some compaction, Hess estimated that *the layer of sediment existing in the deep sea represented only about 260 million years of accumulation*—a figure that might therefore approximate the average age of the sea floor. (Hess's calculation was of the right order of magnitude, but we now know that the average age of the sea floor is even younger than 260 million years; in fact, little or none of the sea floor is as old as 200 million years.)

Hess found support for the idea of a youthful sea floor in his observation that *there are only about 10,000 volcanic seamounts (volcanic cones and guyots) in all the world's oceans*. Hess knew that when a volcano has been eroded below wave base, it withstands further erosion very effectively—and so he assumed that if the oceans were permanent features, their oldest volcanic seamounts should still be extant. Given the fact that there were only 10,000 volcanic seamounts in modern oceans, Hess further reasoned that if the oceans were nearly as old as the earth— say 4 billion years old—an average of only one volcano would have formed every 400,000 years or so. The existence of so many obviously young volcanoes indicated to Hess that new volcanoes appear much more frequently— perhaps at a rate of one every 10,000 years. Thus, the relatively small number of volcanic seamounts in modern oceans suggested to Hess that current ocean basins are much younger than the earth.

Like many earlier workers, Hess noted the central lo-

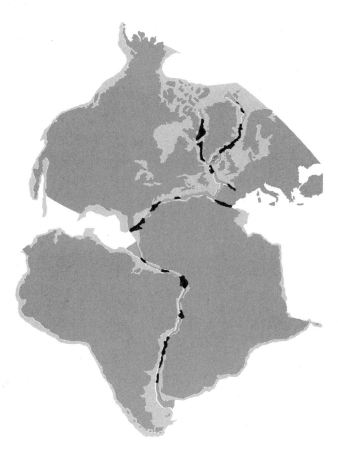

FIGURE 7-23   Computer-generated "best-fit" union of continents that now lie on opposite sides of the Atlantic. This fit was calculated by Sir Edward Bullard and coworkers at the University of Cambridge. The fit was made along the 500-fathom line of each continental slope. Overlaps of continental margins are shown in black. Gaps between margins are shown in color. (*After P. M. Hurley, Scientific American, April 1968.*)

cation of the Mid-Atlantic Ridge. Making use of oceanographic information that had not been available earlier, he also noted that *other midocean ridges tend to be centrally located within ocean basins*. (A "best fit" restoration of continents along the Mid-Atlantic Ridge, calculated after Hess's paper was written, is shown in Figure 7-23.) Four other curious facts about these ridges seemed significant to Hess:

1.   They are characterized by a high rate of upward heat flow from the mantle to neighboring segments of sea floor.
2.   Seismic waves from earthquakes move through the ridges at unusually low velocities.
3.   Along the crest of each ridge there is a deep furrow.
4.   Volcanoes frequently rise up from midocean ridges.

Hess developed a hypothesis that seemed consistent with all of these observations. Essentially, he suggested that midocean ridges represent narrow zones where oceanic crust forms as material from the mantle moves upward and undergoes chemical changes. Hess further maintained that as this material rises, it carries heat from the mantle to the surface of the sea floor. The expanded condition of the warm, newly forming crust thus accounts for the swollen condition of the crust here—that is, for the presence of a ridge.

Hess then revived a geophysical concept that had been discussed by a number of earlier researchers—namely, the idea that material within the earth's mantle rotates by means of large-scale thermal convection. The material of the mantle, existing at high temperatures and pressures, must flow in the manner of a very thick liquid. As we have seen, the mantle is heated by the decay of radioactive isotopes within it and is cooled from above by loss of heat through the crust. Consequently, the upper part of the mantle, being cool, is more dense than the lower part and thus tends to sink while the lower part tends to rise. In a deep body of liquid, the result is **convection**, the same circular overturning movement that heat from the sun creates in the earth's atmosphere. Hess proposed that the earth's liquid-like mantle is divided into **convective cells** (Figure 7-24) whose low-density material forms crust as it rises and cools. This crust then bends laterally to become one flank of a ridge (Figure 7-25). The furrow down the center of many ridges could then be explained as the site at which newly formed crust separates and flows laterally in two directions. Similarly, volcanoes along midocean ridges would represent the rapid escape of mantle material at certain sites, while the low velocity of earthquake waves passing through a ridge would result from the fact that the rocks of

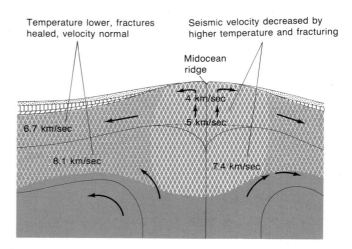

FIGURE 7-25   Hess's model of the structure of a midocean ridge. Arrows show the flow of new crust derived from the convecting mantle. The newly formed crust carries heat from the mantle. This factor, along with fracturing of the rock as it bends laterally, results in low velocities (given in kilometers per second) for seismic waves passing through the ridge. The elevation of the ridge results from the hot, swollen condition of the newly formed crust. (*After H. H. Hess, in Petrologic Studies: A Volume in Honor of A. F. Buddington, Geol. Soc. Amer., 1962.*)

the ridge exist at a high temperature and are extensively fractured where they bend laterally to form the basaltic sea floor.

How do the guyots that Hess discovered fit into this theory? According to the scheme, the sea floor adjacent to a midocean ridge (together with anything attached to this sea floor) moves laterally, away from the spreading center. The volcanoes that frequently form along midocean ridges sometimes grow upward to sea level, as is the case with Ascension Island in the Atlantic. As a volcano moves laterally from the ridge along with the crust on which it stands, it moves away from the source of its lava. It then becomes an inactive seamount, and its tip is quickly planed off by erosion. The sea floor gradually deepens away from midoceanic ridges, since newly formed material of the crust cools and therefore shrinks as it moves laterally away from the ridge. Thus, a truncated seamount is gradually transported out into deep water as if it were on a conveyor belt, and it then becomes a guyot (Figure 7-26). Assuming that the Atlantic Ocean has developed by sea-floor spreading since the end of the Paleozoic, Hess calculated a spreading rate of about 1 centimeter per year.

Continents can be viewed as enormous bodies that float in oceanic crust by virtue of their low density. They would be expected to ride passively along like guyots. Here, then, was Hess's explanation for the fragmentation of continents:

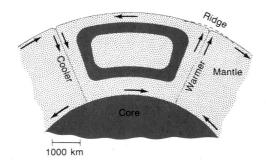

FIGURE 7-24   Convective motion within the mantle as envisioned by Harry H. Hess. Midocean ridges form where the upward-flowing limbs of two adjacent convection cells approach the surface. (*After H. H. Hess, in Petrologic Studies: A Volume in Honor of A. F. Buddington, Geol. Soc. Amer., 1962.*)

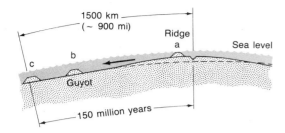

FIGURE 7-26   Hess's interpretation of the manner in which a typical guyot is formed. First, a volcano builds a cone along a midocean ridge. The cone initially stands partly above sea level, and its tip is later planed off by wave erosion (a). The resulting flat-topped structure moves laterally with the spreading crust and is carried gradually downward (b and c). It is carried downward because the newly formed crust beneath it cools while moving away from the ridge. (*After H. H. Hess, in Petrologic Studies: A Volume in Honor of A. F. Buddington, Geol. Soc. Amer., 1962.*)

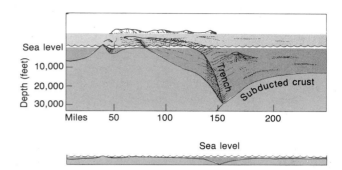

FIGURE 7-27   Section through the Tonga Trench in the Pacific Ocean. The view is northward. In the upper picture, vertical distances are exaggerated by a factor of 10. The lower diagram is drawn without vertical exaggeration. The island toward the left is Kao, which represents a dormant volcano. The oceanic crust to the right (or east) of the trench is moving down into the mantle beneath the oceanic crust to the left. (*Modified from R. L. Fisher and R. Revelle, Scientific American, November 1955.*)

He reasoned that when convective cells in the mantle change their locations, the upwelling limbs of two adjacent cells must sometimes come to be positioned beneath a continent. Convective spreading should then **rift** the continent into two fragments and move them apart from the newly formed spreading center. New ocean floor should subsequently form at the same rate on each side of the spreading center. Hess further maintained that the spreading center would continue to operate along the midline of the new ocean basin—and thus persist as a midocean ridge—as long as the convective cells remained in their new location.

If oceanic crust forms and flows laterally without an enormous change in thickness, however, it must disappear somewhere. Hess postulated that it must be swallowed up again by the mantle along the great **deep-sea trenches** that exist at certain places in the ocean floor (Figure 7-27). Movement of crust into the mantle along one side of a trench provided Hess with a ready explanation for the fact that *the earth's gravitational field here is unusually weak*; the presence here of low-density crustal rock in place of dense mantle rock would be expected to weaken the gravitational force exerted by the earth on objects at or above its surface. Hess estimated that the formation of new crust along midocean ridges and the simultaneous disappearance of crust into the deep-sea trenches would produce an entirely new body of crust for the world's oceans every 300 or 400 million years.

Hess's hypothesis of **sea-floor spreading** had two great strengths. First, by asserting that continents move along *with* oceanic crust, it overcame the objection that continents could not move *through* such crust. Second, the hypothesis was consistent with a variety of facts, the most

important of which are italicized above. Most of these facts had not previously made sense.

The triumph of paleomagnetism

Despite the strong circumstantial evidence that favored Hess's hypothesis, his publication in 1962 created no great stir within the geologic profession. What was needed was a really convincing test of the basic idea. Such a test was soon found. It was based on the well-known fact that the earth's magnetic field has periodically reversed its polarity (page 129). In 1963, the British geophysicists Fred Vine and Drummond Matthews reported that newly formed rocks lying along the axis of the central ridge of the Indian Ocean were magnetized while the earth's magnetic field was in its present polarity. This came as no surprise, since it was known that such "normal" magnetization also characterized other midocean ridges. It turned out, however, that seamounts on the flanks of the Indian Ocean ridge were magnetized in the reverse way. Vine and Matthews concluded that this pattern might confirm Hess's sea-floor-spreading model. They reasoned that if crust is now forming along the axis of any midocean ridge, as this crust crystallizes from the molten mantle, it must become magnetized with the magnetic field's present polarity. In older crust lying at some distance from the ridge, however, reversed polarity should be encountered, and beyond this, in even older crust, the polarity should be normal again.

Magnetic "striping" had, in fact, recently been observed on many parts of the sea floor. The stripes are called **anomalies** because their magnetism, if normal, adds to the earth's present magnetic field and, if reversed, weakens the magnetic field. Thus, the presence of magnetic stripes causes measurements of regional magnetism to be abnormal or anomalous.

The striping patterns were soon put to a more rigorous test. During the 1960s, a time scale was developed for late Cenozoic magnetic reversals. This scale was based on measurements of the magnetic polarity of terrestrial rocks of known age. It was assumed that the spreading rate for each midocean ridge had remained reasonably constant over the past 4 or 5 million years. It was then found that the relative widths of sea-floor stripes were proportional to the time intervals that these stripes were thought to represent—that is to say, long intervals of normal polarity were represented by broad stripes, while narrow intervals were represented by narrow stripes. Thus, the detailed patterns of striping were found to match the known timing of magnetic reversals (Figure 7-28).

Because of inhomogeneities in the oceanic crust, magnetic anomalies do not form perfectly smooth striping pat-

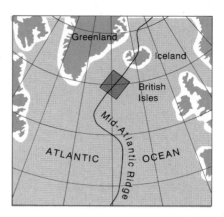

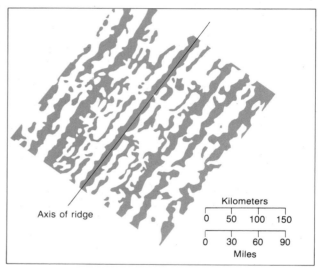

FIGURE 7-29   Magnetic-anomaly pattern of the North Atlantic sea floor. Symmetrical striping is revealed by measurement of the strength of the magnetic field at many locations from a ship. The position of the area represented in the lower diagram is shown in the map above. (*From A. Cox et al., "Reversals of the Earth's Magnetic Field." Copyright © 1967 by Scientific American, Inc. All rights reserved.*)

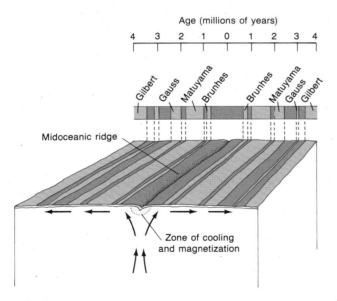

FIGURE 7-28   Magnetic anomaly patterns of the sea floor fit the prediction that they represent magnetic reversals. Above is the time scale for known magnetic reversals of the past 4 million years. The labels (Gilbert through Brunhes) represent intervals of normal and reversed polarity, which have been dated by identifying the polarity of terrestrial rocks whose ages are known. The relative widths of these intervals are remarkably similar to those of the magnetic-anomaly stripes on either side of a midocean ridge. Assuming that the rate of sea-floor spreading has not varied greatly during the past 4 million years, this correspondence is exactly what we would expect if the striping resulted from magnetic reversals. (*After A. Cox et al., Scientific American, February 1967.*)

terns on the sea floor. Nonetheless, the patterns can be striking, as Figure 7-29 illustrates. This figure also shows that the Mid-Atlantic Ridge continues northward through Iceland, where it stands above sea level. Thus fortunate condition enables us to study the ridge easily.

At places in Iceland, the furrow down the center of the ridge can be seen to be a structural **graben** (Figure 7-30)—that is to say, a valley bounded by normal faults that have allowed a central "fault block" to slip downward (Figure 7-31). Grabens form where the crust is extending— as is the case along midocean ridges, where crust is forming and moving laterally. As the crust periodically breaks apart along midocean ridges, lava moves upward to fill the

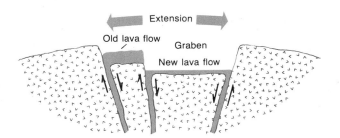

FIGURE 7-31   Diagram illustrating a graben of the sort that a midocean rift represents. As tension breaks the crust and spreads it laterally, some blocks of crust sink back into the zone of extension along faults. Lavas (*black pattern*) moving upward along the faults fill in the space produced by rifting and also flow laterally and solidify on the floor of the graben.

FIGURE 7-30   Thingvellir Graben in Iceland. This is a portion of the Mid-Atlantic Rift that is dramatically exposed above sea level. As the rift separates, lava squeezes upward to form new basaltic crust. (*Icelandic Airlines.*)

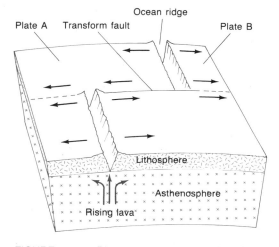

FIGURE 7-32   Diagram showing a transform fault. The central part of this fault is a plate boundary along which two plates slide past one another. In this example, Plate A and Plate B are separating at a rift, and the transform fault offsets this ridge. Arrows indicate the opposite directions of plate motion between the two segments of the ridge. (*After F. Press and R. Siever, Earth, W. H. Freeman and Company, New York, 1982.*)

space thus formed, producing new oceanic crust. Extruded lavas have also been photographed along submarine mid-ocean ridges. These lavas sometimes form pillow structures, the hummocky shapes characteristic of lava cooling under water (Appendix I).

Midocean ridges and neighboring features of the sea floor have now been mapped in considerable detail. It turns out that these ridges are frequently offset along enormous strike-slip faults called **transform faults** (Figures 7-32 and 7-33). The famous San Andreas Fault of California is a transform fault that happens to cut across the edge of the North American continent. Earthquakes caused by movement along the San Andreas have caused considerable damage in California (Figure 7-34).

It is now recognized that the boundary between the crust and the mantle—the Moho (Figure 7-16)—is not the surface along which the "skin" of the earth moves. This surface, which lies well below the Moho, is the boundary between the asthenosphere (the partially molten part of the mantle) and the lithosphere (the upper mantle and the crust). The asthenosphere-lithosphere boundary is situated closer to the surface beneath midocean ridges, where high temperatures keep mantle material in a molten state even at

very shallow depths (Figure 7-35). Figure 7-36 shows the current configuration of the crust and upper mantle in the vicinity of the Atlantic Ocean, which is still growing by sea-floor spreading.

Few major scientific ideas are entirely new, and so it was with Hess's idea of sea-floor spreading. In 1931, the British geologist Arthur Holmes had proposed that basaltic crust operated as if it were a conveyor belt that moved continents by stretching and carrying them laterally. Holmes further proposed that the stretched basaltic crust descends into the mantle along deep-ocean trenches. In these im-

FIGURE 7-33   Artist's representation of the North Atlantic sea floor based on the studies of Bruce C. Heezen and Marie Tharp of the Lamont-Doherty Geological Observatory. Depths shown are in feet below sea level. Transform faults are conspicuous features. (Painting by Heinrich C. Berann courtesy of the National Geographic Society.)

FIGURE 7-34   View of the Stanford University campus immediately after the California earthquake of 1906, which resulted from movement along the San Andreas Fault. Ironically, the overturned statue represents a geologist, Louis Agassiz. Agassiz established the fact that the earth recently experienced an ice age. (*Stanford University Archives.*)

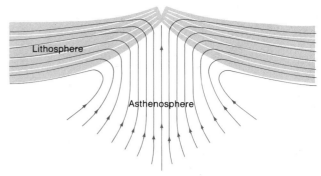

FIGURE 7-35   Schematic diagram illustrating how the boundary between the asthenosphere and the lithosphere is elevated by high heat flow toward the surface along a midoceanic ridge. Arrows indicate movement of material from the asthenosphere to the lithosphere. Solid lines represent lines of equal temperature. They are deflected toward the surface beneath the ridge, where hot material from the asthenosphere rises toward the surface. (*After S. Uyeda, The New View of the Earth, W. H. Freeman and Company, New York, 1978.*)

portant respects, Holmes's scheme of continental movement resembled that of Hess. The major difference was that Holmes considered midocean ridges to be the remains of continental masses that had rifted apart. Apparently, he did not consider the possibility that new crust might form along these ridges.

Another interesting story concerns the misfortunes of a Canadian geologist named L. W. Morley. Morley developed the same model for magnetic anomalies that Vine and Matthews published, but the manuscript in which he outlined his model was rejected by the two journals to which it was submitted in 1963 for publication. One reviewer of the manuscript cynically commented that "such speculation makes interesting talk at cocktail parties." Because Vine and Matthews were fortunate enough to have had their paper accepted for publication, they were the ones who ultimately received recognition in the scientific world. Radically new ideas are not easily established in science.

Perhaps the most important lesson offered by the early debate about continental drift is that circumstantial evidence must never be taken lightly in geology. There is a parallel in the history of geologists' assessment of the age of the earth (page 21). Recall that late in the last century, the British physicist Lord Kelvin, assuming that the earth had no internal heat source, claimed that the planet was much younger than geologists believed. At the turn of the century, the discovery of radioactivity proved him and his followers wrong, but until then their calculations perplexed the geologists, who had compiled considerable evidence of the earth's extreme antiquity. It was also geophysicists who generated the most powerful arguments against continental drift. When the Moho was discovered, physicists fixed their attention on the idea that continents could not plough through strong oceanic crust, and the circumstantial evidence favoring the idea of continental drift was therefore given less attention than it deserved.

Subduction at deep-sea trenches

Following the publication of Hess's seminal paper, deep-sea trenches have attracted much interest. These trenches are the sites of the lithosphere's descent into the asthenosphere—a process now called **subduction**. It was soon discovered that most of the trenches of the modern world encircle the Pacific Ocean (Figure 7-37). Trenches were also found to be associated with two other geologic features: volcanoes and **deep-focus earthquakes**. The latter are earthquakes that originate more than 300 kilometers (~190 miles) below the surface of the earth (page 14). In areas far from deep-sea trenches, deep-focus earthquakes are rare. Near trenches, however, both shallow- and deep-focus earthquakes occur frequently.

The typical spatial relationship between trenches, volcanoes, and earthquake foci is shown in Figure 7-38. As this figure illustrates, the earthquake foci fall along a narrow, nearly planar zone that angles down from the trench. A little thought will reveal that the earthquake foci in Figure 7-38 must lie along or within the segment of a plate that

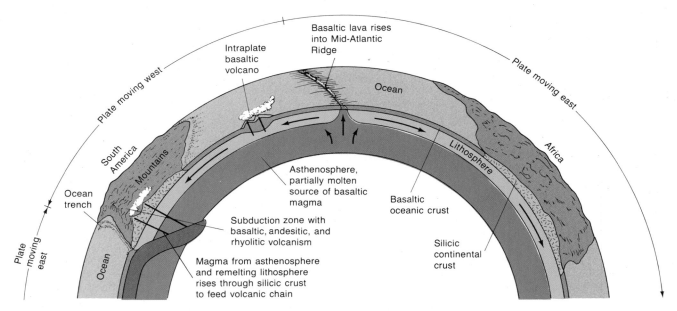

FIGURE 7-36   Schematic cross section of the lithosphere and the asthenosphere in the vicinity of the Atlantic Ocean. Note that the lithospheric plate that includes South America is moving westward from the Mid-Atlantic Ridge. At the same time, the lithospheric plate beneath the eastern Pacific Ocean is moving eastward; where it meets South America along a deep-sea trench, this oceanic plate moves downward into the astheno-sphere. (*After F. Press and R. Siever, Earth, W. H. Freeman and Company, New York, 1982.*)

is descending into the asthenosphere. The descent of a plate causes earthquakes to originate deeper than they do elsewhere within the lithosphere, since a descending plate is a portion of lithosphere that is moving down into the asthenosphere. It is not known exactly how movements of the descending plate cause earthquakes or even how the plate moves downward—whether it simply slides downward or is forced downward by the opposing plate.

Chains of volcanic islands often run parallel to deep-sea trenches (Figure 7-37). They are positioned in this way because the descending slab of lithosphere undergoes partial melting (Figure 7-39), and any molten material that is less dense than the asthenosphere rises toward the surface as magma. Some of this magma solidifies within the crust to form intrusive igneous bodies that are several kilometers in diameter. The rest reaches the surface to emerge in volcanic eruptions.

A band of subducted lithosphere is called a **subduction zone**. Subduction zones border much of the Pacific Ocean (Figure 7-37). The volcanoes associated with these zones form what is known as the "Ring of Fire" around the Pacific.

Deep-sea trenches and the chains of volcanoes associated with them tend to exhibit arcuate shapes when viewed from above. The volcanoes that rise above sea level form **island arcs** (Figure 7-40). Lithosphere tends to descend along a curved path for a reason that appears to be quite simple: Because the earth is nearly spherical, lithosphere that is depressed into the less rigid asthenosphere tends to bend downward along a curved line in the way that a circular dent forms in a Ping-Pong ball when you press it with your thumb (Figure 7-41).

Relative and absolute plate movements

Today the earth's lithosphere is divided into eight large plates and several small ones (Figure 7-1). We have already discussed three kinds of plate boundaries. To recapitulate, an actively spreading ridge represents one kind of boundary, and a subduction zone (or trench) represents a second. Plates move away from ridges, along which they form, toward subduction zones, where they are destroyed. As two plates move, they scrape past each other along the third kind of plate boundary, a **transform fault** (Figure 7-32). It is important to note, however, that not all transform faults form plate boundaries; some represent short "tears" in plate margins along spreading ridges (Figure 7-33).

The configuration of lithospheric plates has changed throughout earth history. From time to time, patterns of convection have changed within the mantle, causing new ridges and subduction zones to form and old ones to disappear. Although we cannot reconstruct the entire history of plate motions from geologic evidence, much of this history is known for Cenozoic time. The history of spreading along the Mid-Atlantic Ridge is particularly well understood, as are the histories of spreading along the East Pacific Rise west of South America and its continuation south of Australia (Figure 7-42). Thus, for example, we know how far plates have moved apart since Pliocene, Eocene, and Jurassic times, and this information reveals the rates at which these plates have moved away from midocean ridges.

The complex history of continental movement described by plate tectonics differs in an important way from the pattern that Wegener envisioned. Wegener thought that the enormous landmass of Pangaea existed as a stable crustal feature for hundreds of millions of years and that its fragmentation was a single event. Plate tectonics, on the other hand, entails continuous movement of most landmasses in relation to one another. As a plate moves over the surface of the earth, some of its neighboring plates may move toward it while others move away from it, and these neighboring plates seldom move in the same direction. Thus, the rate and direction of movement of a plate in relation to one neighboring plate usually differs from its rate and direction of movement relative to another neighboring plate. Inasmuch as all plates are moving, then, no piece of lithosphere represents a perfectly immobile block against which the movement of all others can be assessed.

Analysis of the movement of one plate in relation to another is illustrated in Figure 7-43. Such movement can be evaluated as a rate of rotation about an axis that passes through the center of the earth. The rate of rotation might be measured in degrees per million years, for example. The position of the pole can be determined through measurement of the orientation of transform faults along which the two plates move against each other. The transform faults will form segments of great circles that are oriented at right angles to the axis of rotation.

Today, the enormous Pacific plate is moving to the northwest *relative* to the plates that neighbor it on the north and northeast (Figure 7-44). How can we measure the *absolute* movement of a plate? Absolute movement is movement relative to a fixed feature, such as an immobile point at the surface of the earth's mantle. Such immobile points appear to have been discovered in the form of **hot spots**. A hot spot is a location where heating and igneous activity occur within the crust as a result of the arrival of a **thermal plume**, or a column of material that rises from the mantle. Often a volcano forms at the surface above the plume. Some hot spots have existed for millions of years. One is responsible for the geysers and recent volcanic

activity at Yellowstone National Park in Wyoming. Since most of the mantle is in a state of convective motion, it is remarkable that the plumes that form hot spots are nearly stationary. As a plate moves over a plume, its successive positions are commonly recorded as a chain of volcanoes such as the one constituting the Hawaiian Islands (Figure 7-45). More than a century ago, American geologist James Dwight Dana argued that the Hawaiian Islands increase in age toward the northeast because erosion has had an increasingly pronounced effect in this direction (the islands are progressively smaller). From radiometric dating, we now know that Hawaii, the largest and easternmost of the islands, is less than 1 million years old, while the small northwestern island of Kauai is about 5.6 million years old, and a long train of even older submarine seamounts extends northwestward beyond Kauai. This age pattern would seem to indicate that the Pacific plate is moving in a west-northwest direction over a stationary hot spot. Note that this also approximates the direction of movement of the Pacific plate in relation to the plates that border it on the north and northeast (Figure 7-44).

A search of the entire globe has turned up more than 100 hot spots that have been active within the last 10 million years (Figure 7-46). Several of these have been used to estimate the absolute directions and rates of plate motion. Many lithospheric plates are moving at a rate of only 5 centimeters per year or more. You can appreciate how slow this is by recognizing that it is about how rapidly your toenails grow!

The rifting of continents

Eastern Africa provides us with a model of continental rifting in progress. Here a system of rift valleys formed during the Cenozoic Era extends southward from the Red Sea and the Gulf of Aden (Figure 7-47). The rift valleys and the basins harboring the Red Sea and Gulf of Aden are grabens formed by extension and breaking of the crust (page 186).

A close look at the place where the Red Sea meets the Gulf of Aden and the north end of the African rift system reveals an interesting area of Ethiopia known as the Afar Triangle. The three rift systems here form what is known as a **triple junction**—a junction between three plates or incipient plates. Such junctions are common features of the earth's crust (Figure 7-1). The plate boundaries at a triple junction may differ from one another. Each boundary may consist of a spreading zone, a subduction zone, or a

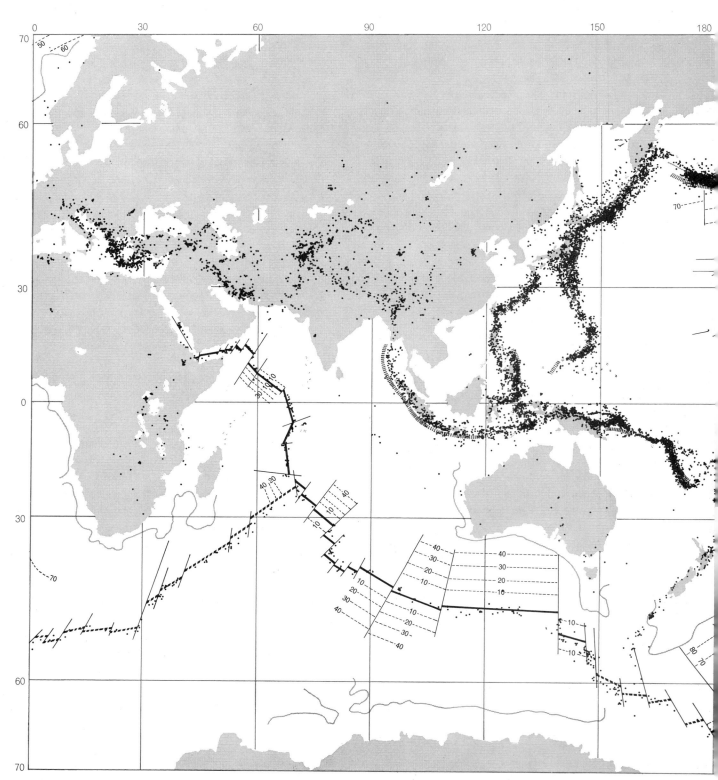

FIGURE 7-37   Distribution of deep-sea trenches and of centers of earthquake activity between 1961 and 1967. Note that many trenches have curved shapes when viewed from above, and that earthquake centers are concentrated along trenches. (*After M. Barazani and J. Dorman, Seismol. Soc. Amer. Bull. 59:369–380, 1969.*)

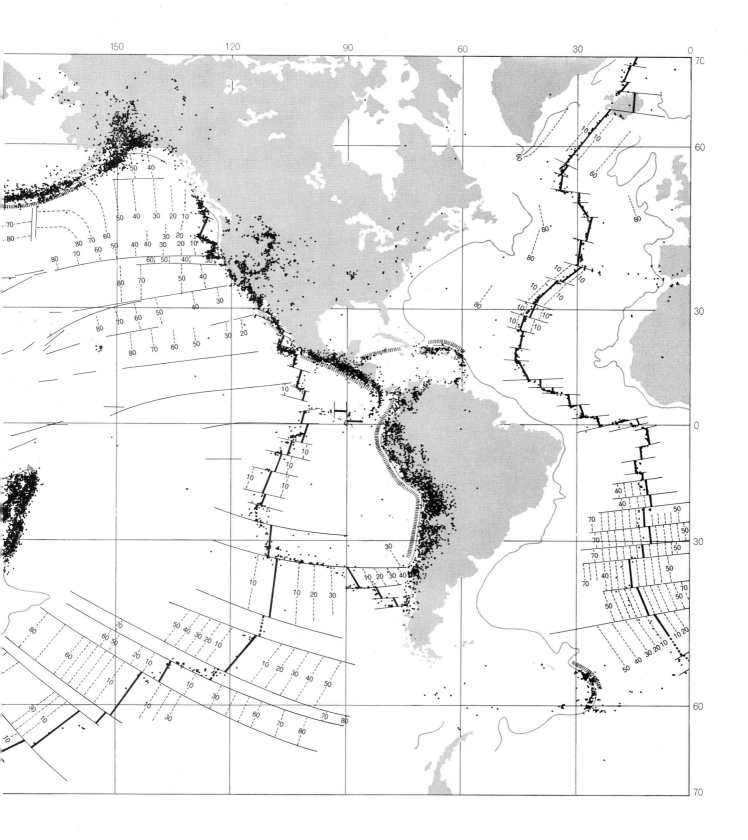

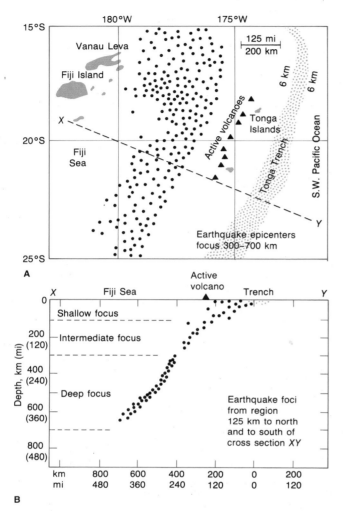

A

B

FIGURE 7-38   Geologic features associated with the Tonga Trench. *A.* Schematic map showing that active volcanoes and epicenters for deep-focus earthquakes are positioned in zones parallel to the trench. *B.* Earthquake foci are seen to lie along a sloping plane, which is actually the plate that is moving down into the asthenosphere. Periodic movement of the plate causes the earthquakes. (*After P. J. Wyllie, The Way the Earth Works, John Wiley & Sons, Inc., New York, 1976.*)

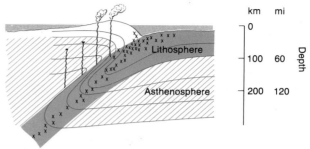

FIGURE 7-39   Volcanic activity produced by subduction. The subducted plate partially melts after reaching a critical depth. Magma of low density rises and reaches the surface to form volcanoes. Crosses represent earthquake foci. Curved lines are lines of equal temperature. Note that subduction of the lithosphere, which is initially cold, causes these lines to bend downward. (*After S. Uyeda, The New View of the Earth, W. H. Freeman and Company, New York, 1978.*)

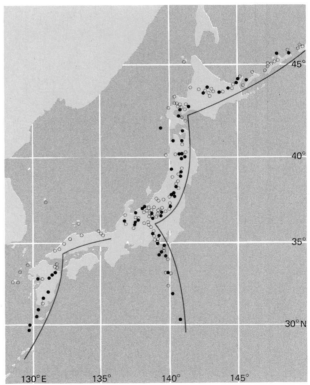

FIGURE 7-40   Island arcs that lie partly along the islands of Japan. Black circles represent active volcanoes; hollow circles represent young but inactive volcanoes.

transform fault. At the Afar Triangle, the junction happens to involve three spreading zones.

When rifts develop, they often begin as three-armed lesions that resemble the one centered near the Afar Triangle. Long before the advent of plate tectonics, geologists noticed that at the locations of three-armed grabens, continental crust is frequently elevated into a dome. In the context of plate tectonics, it seems evident that doming preceded three-armed rifting and thus represents the first result of the development of a hot spot.

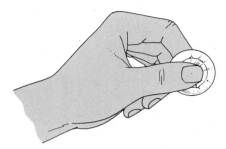

FIGURE 7-41 A thumb pressed against a Ping-Pong ball causes a circular dent to form. Because the earth's crust is also a nearly spherical shell, it usually bends downward along curved lines. This appears to be why deep-sea trenches tend to be curved in map view and why the volcanoes associated with them form island arcs.

It appears that when a large continent is rifted apart, the jagged line along which it divides often represents a composite structure formed from arms of several three-armed rifts. The rifting that formed the Atlantic Ocean illustrates this phenomenon (Figure 7-48). A three-armed rift usually contributes two of its arms to the composite rift, while the third arm becomes a failed rift—a plate-tectonic dead-end. Before it ceases to be active, this third arm forms a graben or a system of grabens that projects inland from the new continental margin formed by the other two arms. Some of the world's largest rivers, including the Mississippi and the Amazon, flow through valleys located in failed rifts that border the Atlantic basin.

It is not uncommon, however, for all three arms of a three-armed rift to develop into segments of plate boundaries. Note, for example, that the Mid-Atlantic Ridge terminates at a triple junction of ridges in the South Atlantic. Similarly, although the rift arm that projects into Africa has not yet divided the continent, such division may take place in the future. The failure of this rift to divide Africa thus far is probably a result of its orientation, which presents it with the task of dividing a much larger continental segment than the small segments that have already been severed to form the Red Sea and Gulf of Aden (Figure 7-47).

It is not surprising that many hot spots are situated on or very near midocean ridges (Figure 7-46). These may be the only surviving hot spots of a larger number that formed three-armed rifts that subsequently became active spreading ridges. The southern portion of the Mid-Atlantic Ridge has shifted to a position slightly to the west of three surviving hot spots that appear to have played a role in its origin.

It has been suggested that doming and three-armed rifts tend to develop when continental crust remains nearly stationary for some time over a hot spot. Conversely, when a continent is moving rapidly over the asthenosphere, it is not greatly affected by a plume rising from the mantle. The idea here is that the effect of the plume is smeared out, and the plume thus fails to "burn through" the continent. A nearly stationary continent, however, may be more severely affected by a plume rising from the mantle because heat beneath it becomes intense, blistering the crust and rupturing it into a local three-armed rift.

The modern continent of Africa may provide an illustration of this concept. Africa has certainly become increasingly distant from South America while the Mid-Atlantic Ridge has been in operation, but this does not mean that it is moving in relation to the underlying mantle. It has been proposed that the rifting of Africa during the past 25 million years has resulted from the continent's near immobility above the mantle; perhaps the African plate has remained almost motionless while the plates to the east, south, and west have moved rapidly away from it (see Figure 7-1). Not only is there a concentration of hot spots in Africa, but there are also many regions of high elevation, and the continent as a whole stands unusually high above sea level. These features and the rifts of eastern Africa may indicate that the African continent is about to be torn asunder by newly forming rift systems.

Just how the mantle plumes that form hot spots relate to mantle convection remains uncertain. The problem is that hot spots appear to be stationary while most of the mantle is in convective motion. Perhaps plumes form where upward-flowing segments of convection cells meet before they produce a complete rift system (Figure 7-24).

CLUES TO ANCIENT PLATE MOVEMENTS

A great deal has been learned about plate motions and interactions during Cenozoic time through the study of magnetic striping, directions and relative rates of plate movement, and hot spots. As we look far back in geologic time, however, these methods cease to be useful, since the evidence on which they are based has been obscured or destroyed by processes like subduction, metamorphism, and erosion. We must therefore rely on other geologic clues to tell us how plates moved and interacted hundreds of millions of years ago. Even for the Cenozoic Era, complex

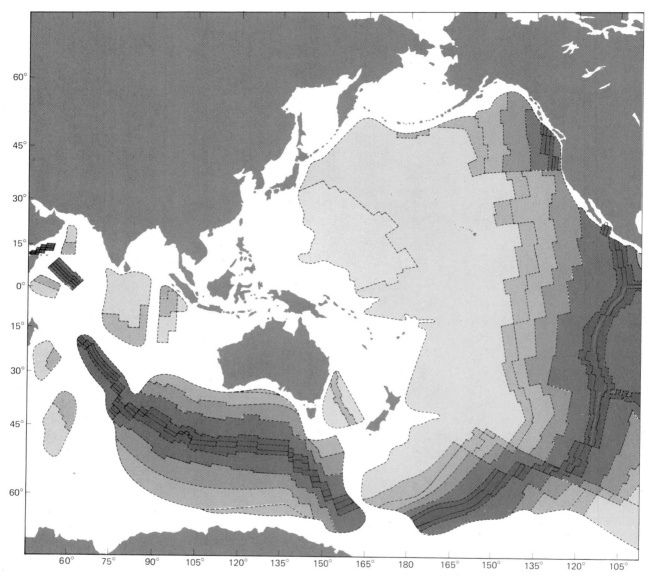

FIGURE 7-42   The ages of the world's ocean basins. No part of the sea floor is older than Mesozoic. (*After W. C. Pitman et al., Geol. Soc. Amer. Map and Chart MC-6, 1974.*)

motions of plates in certain regions (for example, the Mediterranean) can be deciphered only by the study of special clues. Let us examine some of these clues now.

Signs of ancient rifting

As we have seen, a midocean ridge is often associated with block faulting that is expressed by a graben running along the ridge's midline (Figures 7-30 and 7-33). When a spreading zone first passes beneath continental crust,

however, it seldom produces faults that cut cleanly through. Instead, the extension tends to break the thick continental crust into a complex band of fault blocks. Today this is happening in Africa, where a system of rift valleys passes southward from the Red Sea (Figure 7-47). Each valley is a long, narrow, downfaulted block of crust associated with mafic volcanoes that have welled up from the mantle. (Rifting on the land often produces mafic dikes and flood basalts.) Some of the rift valleys cradle great lakes such as Lake Tanganyika. These rift valleys have been in existence

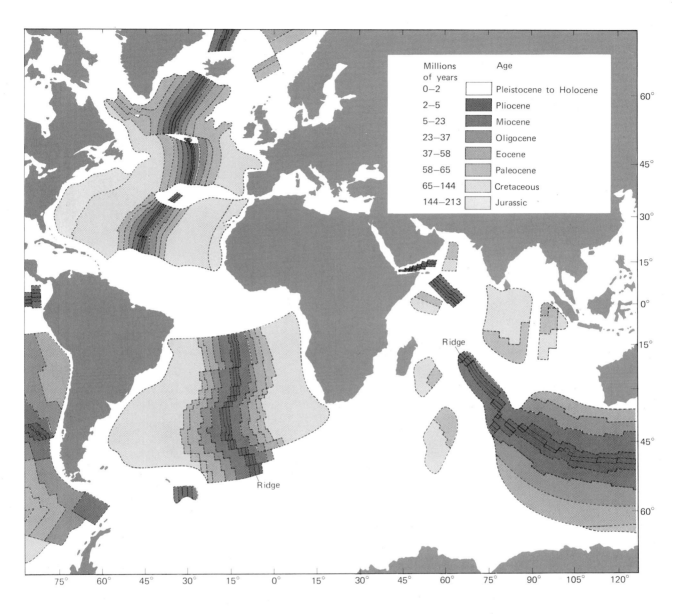

Millions of years	Age
0–2	Pleistocene to Holocene
2–5	Pliocene
5–23	Miocene
23–37	Oligocene
37–58	Eocene
58–65	Paleocene
65–144	Cretaceous
144–213	Jurassic

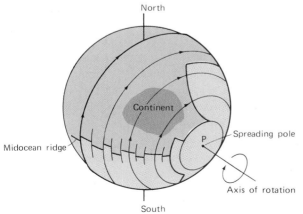

FIGURE 7-43   Movement of a plate over the earth's surface. The plate that includes the continent is moving away from another plate along the midocean ridge. The transform faults along the ridge form segments of great circles that indicate the rotational movement of the plate. All of the great circles lie in planes that are perpendicular to the axis of rotation of the plate.

FIGURE 7-44   Movement of the Pacific plate. Small arrows show the directions of movement of the plate in relation to its neighboring plates. The large arrow shows the absolute direction of movement of the Pacific plate. (*After P. J. Wyllie, The Way the Earth Works, John Wiley & Sons, Inc., New York, 1976.*)

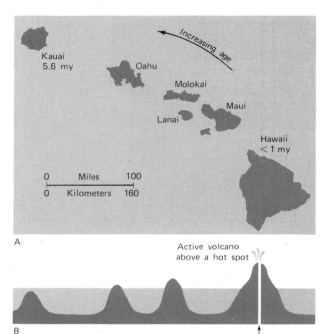

FIGURE 7-45   Formation of a chain of islands as an oceanic plate moves over a hot spot. *A*. The major Hawaiian islands. The islands increase in age toward the northwest. They have formed, one after the other, as the Pacific plate has moved northwestward over a hot spot in the mantle. *B*. Diagram showing how volcanic islands form, one after the other, as an oceanic plate moves over a hot spot.

only since early in Miocene time (for less than 20 million years).

Before the sea enters a continental rift valley, nonmarine clastic sediments often accumulate rapidly to great thicknesses. The reason for this is that rapid subsidence of downfaulted basins creates a landscape of high topographic relief in which subsequent erosion is rapid as well. The sediments typically include conglomerates derived from the steep valley walls and also red beds and alluvial-plain deposits. Lakes that form within the elongate valleys also leave a sedimentary record. In arid climates, temporary lakes leave accumulations of nonmarine evaporites. If rifting continues long enough, a rift valley becomes so wide and so extended that it opens up to the sea. Because inflow from the sea tends at first to be restricted or sporadic, saline waters within the rift valley evaporate more rapidly than they are renewed. Under such circumstances, marine evaporites form. Waters from the Indian Ocean, for example, only recently gained full access to the currently widening Red Sea. Beneath a thin veneer of marine sediment in the Red Sea, geologists have found evaporites that formed during an earlier time, when there was only a weak connection to the larger ocean, or perhaps even earlier, when arid nonmarine basins were present.

The margins of the Red Sea also exhibit geologic features that typify the early stages of continental rifting. The Afar Triangle, for example, is a forbidding region, where temperatures remain very hot throughout the year and fierce nomadic tribes threaten intruders. What geologic evidence has been gathered from the region indicates that not long ago, the Afar Triangle was part of the Red Sea floor. Most of the rocks in this area are basalts similar to those that form oceanic crust. Flat-topped volcanoes that resemble wave-truncated guyots are common, and fossil coral reefs less than 2 million years old indicate that uplift of the region has been quite recent. Much of the topography is the product of block faulting, and in places there are great thicknesses of evaporite deposits (Figure 7-49).

In summary, then, regions of continental rifting are characterized by normal faults, mafic dikes and sills, and thick sedimentary sequences within fault block basins that often include lake deposits, coarse terrestrial deposits, and evaporites followed by oceanic sediments. An ancient example of this suite of features can be found in fault basins of the

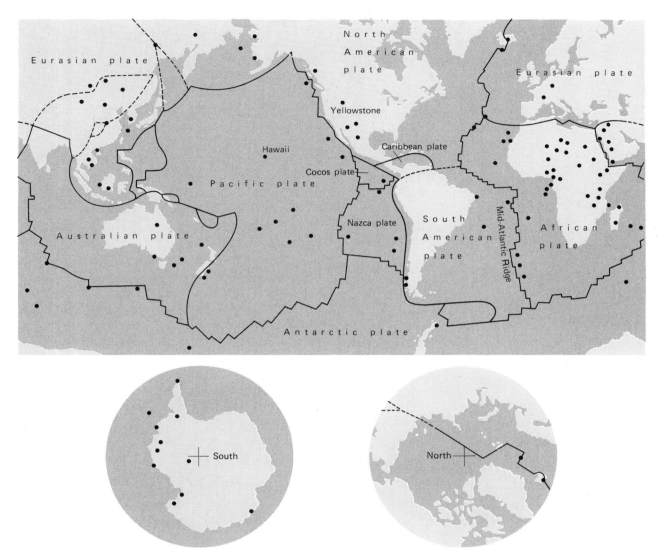

FIGURE 7-46   All of the world's identified hot spots. Note the concentration of hot spots in Africa. (*After K. C. Burke and J. T.* *Wilson, Scientific American, August 1976.*)

eastern United States. The basins are labeled in Figure 7-48 as the Newark and Connecticut rifts. These are actually failed rift arms—rifts that never became part of the composite rift that formed the Atlantic Ocean. As we will see in Chapter 16, these rifts passed far inland and never opened widely enough to allow the sea to invade, although they contain other sedimentary sequences that typify incipient continental rifts.

Evidence of ancient subduction zones

Subduction zones also leave telltale signs in the rock record. One clue is provided by igneous rocks. Some of the magma rising from a subducted plate reaches the surface of the crust and erupts volcanically, while some of it cools within the crust as intrusions. (The volcanic rocks tend to be andesites, which are more dense than the intrusives, which are often granites; see Appendix I.) Another, very specific clue is found in the belt between the site of igneous activity and the deep-sea trench, where plate convergence typically creates a zone of intensely deformed rocks (Figure 7-50). Most of these are deep-ocean sediments such as dark muds and graywackes, with bits of ocean crust mixed in. Some have been scraped from the descending plate. Others previously descended with the plate to depths as great as 30 kilometers (~20 miles) and rose again,

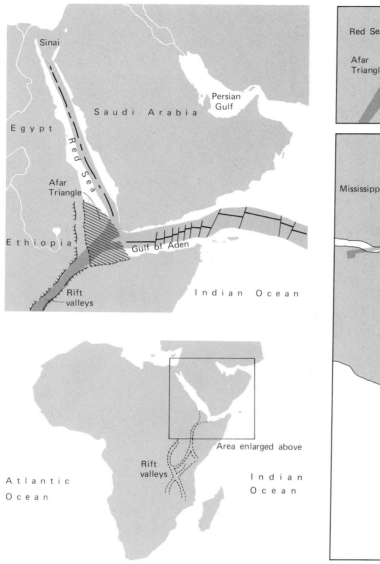

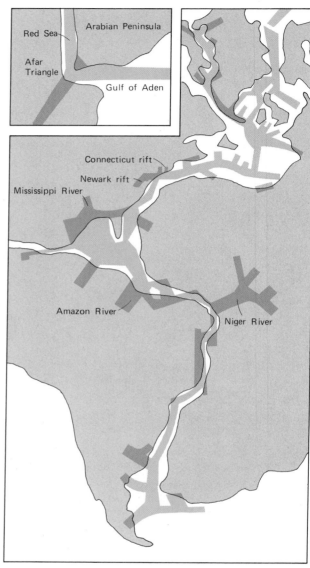

FIGURE 7-47   Three-armed rift along the northeast margin of Africa. Two of the arms represent new oceans: the Red Sea and the Gulf of Aden. The third is beginning to break the continent of Africa apart along Africa's famous rift valleys. The Afar Triangle is a small region of oceanic crust that has been elevated to become land. (See also the photograph on page 168.)

FIGURE 7-48   Ancient three-armed rifts that become apparent when we reassemble continents now bordering the Atlantic Ocean. Many of these rifts contributed two arms to the fracture zone that became the Mid-Atlantic Ridge, and some of them are now occupied by large rivers. (*After K. C. Burke and J. T. Wilson, Scientific American, August 1976.*)

apparently because of their low density. The whole is characteristically metamorphosed at low temperatures (because of the shallow depth of deformation) and at high pressures (because of the force of plate convergence). This chaotic, deformed mixture of rocks is called a **mélange** (the French word for mixture). The mélange shown in Fig-

ure 7-51 formed when a subduction zone passed beneath the western margin of North America.

Sometimes two continental masses converge along a subduction zone. Because they are felsic in composition, neither can descend into the dense mantle. As we will see in the next chapter, this is one cause of mountain building;

FIGURE 7-49   Tilted fault blocks in the Afar Triangle. The white areas are crusts of evaporite deposits. (*H. Tazieff, Scientific American, February 1970. Photograph by H. Tazieff.*)

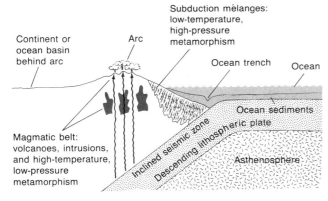

FIGURE 7-50   Diagrammatic cross section of a subduction zone showing the zone of intense deformation between the trench and the volcanic arc. Here sediments are deformed into mélanges and are metamorphosed. (*After F. Press and R. Siever, Earth, W. H. Freeman and Company, New York, 1982.*)

often it leads to the formation of enormous faults and extensive folding. Some slices of sea floor usually end up on the surface of the combined continent that is formed by the plate convergence. These segments of sea floor, known

as **ophiolites**, consist of deep-sea sediments such as turbidites, black shales, and cherts, which are often underlain by pillow lavas that formed the deeper oceanic crust. Other important components of ophiolites are ultramafic crystalline rocks that originated in the mantle and once lay beneath the lavas and deep-sea sediments. In effect, ophiolites are samples of ancient ocean basins that have been conveniently elevated for our study. It appears that many ophiolites now lying between sutured continents became attached to the margin of one continent before the actual collision, when the continent moved against oceanic crust along a subduction zone. The continent, being of low density and therefore highly resistant to subduction, was presumably overridden by the margin of the opposing plate, which then became an ophiolite. Because ophiolites often mark the positions of vanished oceans that once lay between continents, they are key features in our recognition of plate convergence along subduction zones.

Transform faults in the rock record

We have considered the geologic characteristics of *di-*

FIGURE 7-51   A mélange of the Franciscan sequence in California. Large blocks of exotic material are visible in the dark, metamorphosed deep-sea sediment. This mélange formed during the Mesozoic Era, when deep-sea sediments were pushed against the margin of the continent along a subduction zone. (*J. Schlocker, U.S. Geological Survey.*)

verging plate margins and of *con*verging plate margins. Unfortunately, boundaries along which plates moved past one another are much more difficult to recognize. Transform faults leave no diagnostic record except for the strike-slip movement of the rocks through which they pass. The offset along these faults is commonly very large—tens or hundreds of kilometers—and this kind of strike-slip displacement in ancient rocks can sometimes be linked to transform-fault systems.

CHAPTER SUMMARY

1.   Large plates of lithosphere cover the earth. Plate movement over the asthenosphere accounts for much deformation and breakage of lithospheric rocks.

2.   Similarities of rocks and fossils on opposite sides of ocean basins provide strong circumstantial evidence that continents have moved horizontally over the surface of the earth. The idea that this continental drift actually occurs was strongly opposed for several decades—partly because it was assumed that continents would have to move over or through the oceanic crust that lies between them.

3.   It is now recognized that oceanic crust moves right along with the continents. Circular, convective motion within the mantle causes the plates to move.

4.   Mantle material is extruded along midocean ridges, spreading laterally in both directions to form the oceanic crust. Oceanic crust is recycled back to the mantle by means of subduction along deep-sea trenches. Geologists can measure the rate of sea-floor spreading by observing how rapidly the sea floor increases in age away from a midocean ridge. Spreading rates are high enough that all segments of the modern sea floor are of Mesozoic and Cenozoic age. All of the sea floor that formed during the Paleozoic has been consumed along deep-sea trenches.

5.   Fracturing of continents often begins with the doming of continental crust in several places. Each dome then fractures to form a three-armed rift system. The joining of some of the rift arms then produces a fracture that cuts across the entire continent.

6.   Block faulting and deposition of thick siliciclastic sequences and evaporites often mark the beginning of continental fracturing.

7.   Continental material is of such low density that it cannot be consumed by the mantle. As a result, continents sometimes converge along deep-sea trenches.

8.   When a continent encroaches on a deep-sea trench, a slice of sea floor is frequently pushed up onto the continental surface; when such a seafloor remnant, called an ophiolite, is found within a modern continent, it marks the position of an ancient ocean that disappeared when two continents were united.

EXERCISES

1.   List as many pieces of evidence as you can in support of the idea that continental drift has taken place.

2.   What is apparent polar wander? Draw pictures of the earth showing how the movement of a continent can produce apparent wander of the north magnetic pole.

3.   Why are most volcanoes that have been active in the last few million years positioned in or near the Pacific Ocean?

4.   How can a hot spot indicate the absolute direction of movement of a plate?

5.   What are failed rifts, and how are they important to our understanding of the breaking apart of continents? (Hint: Refer to Figure 7-48.)

6.   What geologic features enable us to recognize ancient continental rifting, and what features enable us to recognize ancient subduction zones?

ADDITIONAL READING

Burke, K. C., and J. T. Wilson, "Hot Spots on the Earth's Surfaces," *Scientific American*, August 1976.

Continents Adrift (Readings from Scientific American), W. H. Freeman and Company, New York, 1972.

Glen, W., *Continental Drift and Plate Tectonics*, Charles E. Merrill Books, Inc., Columbus, Ohio, 1975.

Hallam, A., "Continental Drift and the Fossil Record," *Scientific American*, November 1972.

Hallam, A., *A Revolution in the Earth Sciences: From Continental Drift to Plate Tectonics*, Oxford University Press, New York, 1973.

Uyeda, S., *The New View of the Earth: Moving Continents and Moving Oceans,* W. H. Freeman and Company, New York, 1978.

Wyllie, P. J., *The Way the Earth Works*, John Wiley & Sons, Inc., New York, 1976.

CHAPTER 8
Mountain Building

Untll recently, the process of mountain building was a mystery to geologists. Compounding the puzzle was the fact that, as we saw in the previous chapter, few geologists before the 1960s believed that continents move laterally over the earth's surface. What processes cause mountains to grow, and why do mountains occur in so many different positions on continents? Before the origin of plate-tectonic theory, contradictory ideas emerged. North American geologists, for example, have long been struck by the fact that one chain of mountains, the Cordilleran system, parallels the west coast of their continent and another, older chain, the Appalachian system, parallels the east coast. Before the 1960s, some North American geologists concluded from this symmetrical pattern that mountains tend to form along the margins of relatively stable continental masses known as **cratons**. Most European geologists, however, held a different view. Seeing the Ural Mountains standing between Europe and Asia within the largest landmass on earth, they argued that long mountain chains could indeed rise up in the center of a continent. Plate-tectonic theory explains both situations: It is true that long mountain chains form only along continental margins; continents can unite, however, resulting in the formation of a mountain chain where the margins are welded together within a newly formed, large landmass.

The unification, or **suturing**, of two continents occurs when they meet along a subduction zone. In fact, according to the theory of plate tectonics, subduction zones are the key to the origin of mountain chains on continents. As we will see, however, not all mountain-building events result from continental suturing; a mountain chain can also form along the margin of a solitary continent that has been rafted up against a subduction zone.

PLATE TECTONICS AND OROGENESIS

The process of mountain building is known as **orogenesis**, and a particular orogenic episode is termed an **orogeny**. Orogenesis is a complex process that is only partially understood; nonetheless, the simple idea that plates move, carrying continents with them, points to one mechanism of orogenesis: the collision of two continents. Ultimately, two properties of continental crust—its great thickness and its low density relative to the density of the asthenosphere—lead to mountain building by collision. Continents are simply too thick and too buoyant to be subducted. How this causes continents to collide is shown in Figure 8-1. When a continent rides up against a deep-sea trench, its resistance to subduction forces a reversal in the direction of subduction: The oceanic plate opposite the continent is forced to descend into the mantle. If a second continent is riding on the newly subducting plate, it will eventually collide with the first continent along the subduction zone, and the two continents will be welded together. The juncture formed in this way is called a **suture**.

Continental suturing appears to produce mountain chains because subduction along a trench wedges the margin of one continent beneath the margin of the other. The thickness of the continental crust is thus doubled along the suture (Figure 8-1D). Often, some oceanic crust is pinched up along the suture, forming an ophiolite (page 201).

Of course, continental collision is not the only process by which mountains are built. It was crustal rifting rather than continental collision that created the great midocean ridges; these ridges are, in effect, submarine mountain chains that form along rift zones rather than subduction

---

The tip of Mount Everest is the highest location on earth. Mount Everest forms part of the great Himalayan mountain system in southern Asia. (*The Royal Geographic Society.*)

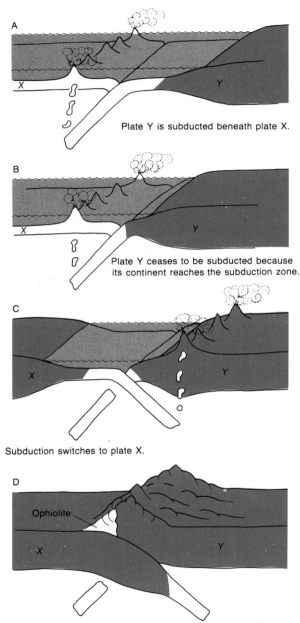

A

Plate Y is subducted beneath plate X.

B

Plate Y ceases to be subducted because its continent reaches the subduction zone.

C

Subduction switches to plate X.

D

Ophiolite

The continents are sutured when the continent of plate X joins the continent of plate Y at the subduction zone. Subduction then ceases.

FIGURE 8-1   Diagrammatic representation of suturing and mountain building where two continents (*shown in color*) meet along a subduction zone.

zones (Figure 7-33). Even along a subduction zone, however, a mountain chain can form at the margin of an isolated continent. The central Andes, for example, which fringe the east coast of South America, have risen to great elevations

while the Nazca oceanic plate has been subducted beneath them.

The Andes display many features that typify mountain chains in general. Figure 8-2 illustrates the most important of these features for an idealized mountain chain rising along the margin of a continent. The next two sections will discuss these features and the way in which they can develop, even in the absence of continental collision.

THE ANATOMY OF A MOUNTAIN CHAIN

When a continent rides up against a subduction zone, it overrides the igneous arc associated with the zone (page 190), causing magma to rise into the continental crust (Figure 8-2). Some of the magma reaches the surface, forming a chain of volcanoes that elevate the crust, often forming mountain peaks. Some magma also cools within the crust, forming plutons of igneous rock (see Appendix I). In keeping with the principle of isostasy (page 15), the addition to the base of the crust of large volumes of low-density igneous rock causes the crust along the igneous arc to bob upward. This vertical movement contributes to the elevation of the mountain chain. At the same time, a root of crustal rock produced by the igneous arc extends downward beneath the mountain chain, balancing the weight of the mountains by displacing dense rocks in the asthenosphere.

An igneous arc like that just described forms the core of a typical mountain chain. Extending along either side of this core is a belt of regional metamorphism (see Appendix I). This belt consists of crustal rocks that have been metamorphosed by heat from the core and locally intruded by igneous activity issuing from the core. Rocks of this **metamorphic belt** (Figure 8-2) are also deformed by other processes that will be described shortly.

Metamorphism dies out in either direction from the igneous core, and the metamorphic belt gives way to a **fold-and-thrust belt** (Figure 8-2). The folds of the fold-and-thrust belt are typically overturned away from the core of the mountain, reflecting the fact that the prevailing forces of deformation come from the direction of the core. Because of their distance from the igneous core, the preexisting sedimentary rocks in the fold-and-thrust area are largely unaffected by metamorphism and are folded less severely than the rocks in the metamorphic belt. Nevertheless, during the process of deformation, these sedimentary rocks behave in a more brittle, less plastic manner than do the rocks in the metamorphic belt. Thus, the folds are broken by enormous thrust faults along which large slices of crust have moved away from the core.

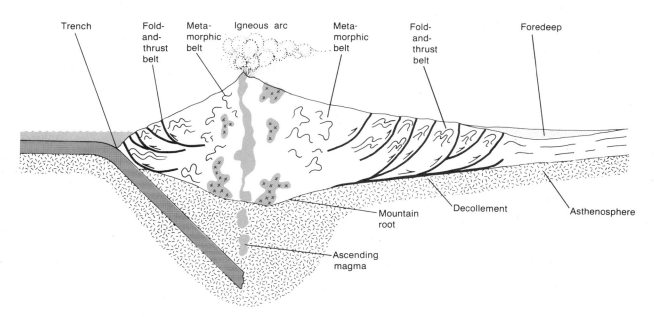

FIGURE 8-2   The configuration of an idealized mountain chain forming where an oceanic plate is being subducted beneath the edge of a continent. This cross section illustrates the general symmetry of the mountain chain. Metamorphism dies out both toward the sea and toward the land from the central igneous arc, and beyond each metamorphic belt is a fold-and-thrust belt. Beyond the inland fold-and-thrust belt, the crust is warped downward to form a foredeep, where sediments from the mountain system accumulate.

When a mountain belt forms along a continental margin, the fold-and-thrust belt on the oceanic side is typically narrower than the equivalent belt on the continental side. This is so because the igneous arc usually lies quite close to the continental margin, and there is less material on that side to be deformed. In the larger, interior fold-and-thrust belt, the thrust sheets usually slide along a basal thrust surface known as a **decollement**. The rocks beneath such a surface are commonly referred to as **basement rocks**. A look at Figure 8-2 will reveal how the telescoping action of folding and thrusting above a decollement shortens and thickens the crust.

FOREDEEP DEPOSITION

Inland from an actively forming mountain chain, the crust typically buckles gently downward, producing an elongate basin, or **foredeep**, whose long axis lies parallel to the mountain chain (Figure 8-2). The foredeep forms rapidly and is usually so deep initially that the sea floods it either through a gap in the mountain chain or through a passage around one end of the chain.

The foredeep typically deepens so quickly that the first sediments to accumulate within it are black shales and turbidite deposits, collectively known as **flysch** (Figure 8-3A). As the mountain system evolves, folding and thrust faulting move progressively farther inland, and mountain building proceeds toward the continental interior. As this happens, flysch deposits, too, become folded and faulted. At the same time, the mountain core rises progressively higher, shedding sediments more and more rapidly. Eventually, sediment chokes the foredeep basin, pushing marine waters out and leaving only nonmarine depositional settings in their place. These comprise alluvial fans along the mountain chain and also rivers, floodplains, and related environments farther inland. Collectively, the sediments of all these settings are termed **molasse**. Molasse deposits can accumulate to great thicknesses; as deformation continues, they too can sometimes be folded and faulted. By the time molasse begins forming, the foredeep may no longer be a topographic basin but may instead appear as a broad depositional surface sloping away from the mountain front (Figure 8-3B). As the foredeep subsides beneath the accumulating sediments, however, it remains a *structural* basin (see Appendix II).

The depositional transition from the deep-water flysch to nonmarine molasse occurs during the evolution of most mountain systems. Later, even molasse deposition comes to an end. This happens when igneous activity ceases in the core of the mountain chain, so that the core becomes leveled by erosion and the source of the molasse sediment

Fold-and-thrust belt

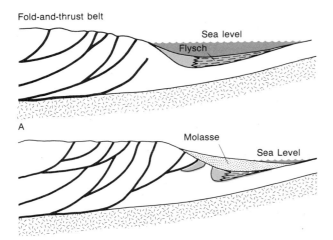

A

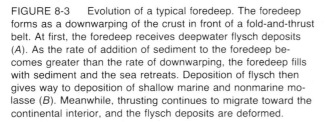

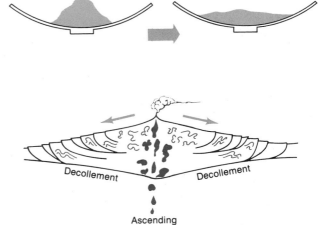

FIGURE 8-3   Evolution of a typical foredeep. The foredeep forms as a downwarping of the crust in front of a fold-and-thrust belt. At first, the foredeep receives deepwater flysch deposits (A). As the rate of addition of sediment to the foredeep becomes greater than the rate of downwarping, the foredeep fills with sediment and the sea retreats. Deposition of flysch then gives way to deposition of shallow marine and nonmarine molasse (B). Meanwhile, thrusting continues to migrate toward the continental interior, and the flysch deposits are deformed.

FIGURE 8-4   Formation of a fold-and-thrust belt by gravitational spreading. Above: Spreading of pudding heaped high on a plate. Below: Igneous activity thickens the felsic crust of an idealized mountain chain (shown in cross section). The thickened crust spreads to produce folding and thrusting in each direction. In both examples, spreading moves material uphill.

disappears. Folding and thrusting also cease because, as will be described in the next section, these activities result from the presence of an elevated mountain core.

THE MECHANISM OF DEFORMATION

Why have folding and thrust faulting occurred in the central Andes and in other mountain ranges not associated with continental collision? It is noteworthy that in these mountain systems, deformation processes typically proceed in both directions from the axis of uplift, just as they do when continents collide (Figure 8-2). This phenomenon implies that horizontal pushing of one continent against another is not a prerequisite for folding and thrust faulting. It may be that the only process necessary to initiate folding and thrusting is regional elevation of the mountain core. Mechanical evidence supports this hypothesis: It can be shown that rocks standing at a high elevation along an axis tend to spread laterally away from the central axis, in the directions in which the surfaces slope (Figure 8-2).

The central axis of the mountain chain is the igneous core, which becomes elevated for two reasons. First igneous volcanic rocks pile on top of the continental surface parallel to its margin. Second, intrusive igneous rocks form along the arc at the base of the crust; because these intrusive rocks are of lower density than mantle rocks, they tend to rise isostatically (Figure 1-19).

Gravity, acting upon the elevated rocks, is the force that causes the spreading. When plutonic and volcanic activity have built the central core of a mountain chain to a critical thickness, this core and the sedimentary units that flank it will spread laterally. As they move, the sedimentary units deform by folding and by breaking along thrust faults (see Appendix II). Enormous slices of rock move over one another in the process. (This gravitational explanation for the motion of thrust sheets also applies to the movement of many glaciers.) The deeper thrust faults of a mountain belt normally join the decollement, the boundary that separates the lowest thrust sheets from the nearly immobile rocks below. In nearly every case, the decollement of a thrust belt slopes *upward* in the direction of thrusting. This means that the thrust sheets are not really sliding downhill but instead are flowing laterally in the way that a mound of pudding will spread, if piled high on a surface (Figure 8-4). It does not matter whether the surface beneath the pudding is flat or upwardly concave. If the pudding has the proper consistency, it will flow "uphill."

When two continents collide, there is often another process, in addition to igneous activity, that contributes to the formation of an elevated central axis from which folding and thrusting can proceed. This process is the doubling of the crust that occurs when the edge of one continent is wedged beneath the edge of another (Figure 8-1D).

Obviously, the process of mountain building can be fully understood only in terms of plate tectonics. Before the emergence of plate tectonics, theories of mountain building centered on the concept of the geosyncline. Because of its historical importance, this earlier attempt at defining the mechanisms of mountain building is described in Box 8-1.

The preceding discussions dealt with ideal mountain systems. Actual mountain systems, however, frequently lack one or more of the features described above. The following sections describe the histories of four real mountain systems: the Alps, the Himalayas, the Andes, and the Appalachians. The Rocky Mountains and other mountain ranges of western North America, which form the so-called Cordilleran system, have a history so long and complex that it will be treated in installments in subsequent chapters.

THE YOUTHFUL ALPS

The Alps of southern Europe belong to a great series of mountain chains of Cenozoic origin that stretch from Spain and North Africa to Indochina (Figure 8-5). The highest of these mountains are the Himalayas. All of these Eurasian chains formed as a consequence of the northward movement of fragments of Gondwanaland, the immense southern continent that broke apart during the Mesozoic Era (page 178). The Alps and other Cenozoic mountains of the Mediterranean region (Figure 8-6) formed as the African plate moved northward against the Eurasian. In most of this zone of orogenic activity, the African continent has not actually been sutured to the Eurasian continent; however, many small plates of lithosphere have been caught between the African and Eurasian cratons, and these **microplates** have been pushed against one another and against the Eurasian craton.

The Alps are a relatively young mountain system that formed largely during the Eocene and Oligocene epochs. They are positioned along what was until early Cenozoic time a part of the southern margin of the Eurasian landmass. This history is confirmed by the presence within the Alps of ophiolites that represent Mesozoic and early Cenozoic oceanic crust. The ophiolites occur mainly within the Pennine Alps—at the base of the famous Matterhorn, for example—and in equivalent zones to the west (Figures 8-7 and 8-8).

Structure and composition of the Alps

The Alps, which are crescent-shaped in map view, can be divided into zones that roughly parallel the regional strike (Figure 8-7). Each zone represents a well-defined segment

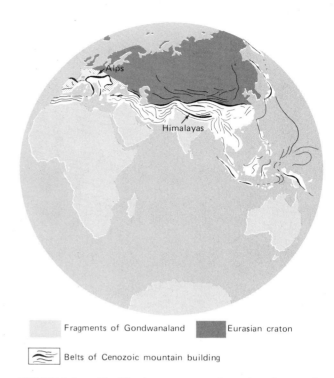

Fragments of Gondwanaland   Eurasian craton

Belts of Cenozoic mountain building

FIGURE 8-5   The Himalayas are one of a series of mountain chains, including the Alps, that formed along the southern margin of Eurasia when fragments of Gondwanaland moved northward against the large northern continent during the Cenozoic Era. (*After H. Cloos, Einführung in die Geologie, Verlag von Gebruder Borntraeger, Berlin, 1936.*)

along the gradient of deformation that typifies mountain chains (Figure 8-2). For example, the igneous core of the system lies in the zone known as the Southern Alps, where plutonic rocks have been exposed by erosion. North of the igneous core, in the Pennine zone, rocks were metamorphosed and flowed in a plastic manner while being deformed under intense heat and pressure. These rocks constitute the metamorphic belt of the mountain system.

Still farther north, in the zones known as the Helvetic Alps, the Molasse Basin, and the Jura, rocks behaved in a less brittle fashion during Alpine orogenic activity. These zones constitute a fold-and-thrust belt resembling that of the idealized mountain range depicted in Figure 8-2. The sedimentary rocks within this belt have moved northwestward along large thrust faults. The lowest of these faults join a decollement at the base of the mountain system. At the northwest extremity, in the Jura Mountains, folding, rather than faulting, prevails above the decollement because most of the thrust faults of the Alps terminate at the surface farther to the southeast. Where erosion has cut deeply into the Jura, the decollement can be seen to rest

Box 8-1

Geosynclines

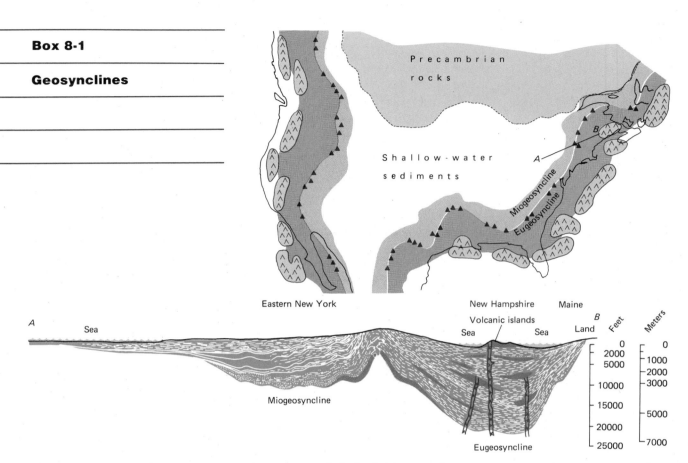

Eastern New York

New Hampshire   Maine

The miogeosyncline and eugeosyncline that were supposed to exist along eastern North America early in the Ordovician Period. (*After M. Kay, Geol. Soc. Amer. Memo. 48, 1951.*)

Before the emergence of plate-tectonic theory in the 1960s, ideas about mountain building were dominated by the concept of the geosyncline. This concept originated in 1859, the year Charles Darwin published *On the Origin of Species*, when James Hall of the New York Geological Survey pointed out that the sedimentary formations that had been deformed to produce the Appalachian Mountains were much thicker than those of the lowlands of the continental interior to the west. He noted, however, that the Appalachian sediments gave evidence of having been deposited in very shallow water and even in nonmarine environments. These facts implied that the Appalachian sediments had accumulated

in an elongate belt of subsidence and that the deposition was unusually rapid, generally keeping pace with the subsidence. On the more stable continental interior, sediments were also laid down in shallow water, but here the crust had been more stable: There had been less subsidence because there was less accumulation. Hall believed that the great weight of sediment deposited along the continental margin had caused the large amount of subsidence there. Eventually, as a result of processes that Hall did not attempt to analyze, the subsidence was reversed and the enormous sedimentary accumulations were uplifted to form the Appalachian Mountains.

James Dwight Dana, a contem-

porary of Hall, introduced the term *geosynclinal* (later shortened to *geosyncline*) to describe the long depositional belt from which a mountain chain like the Appalachians was considered to have formed. Dana also argued that it was not simply the loading of sediments that accounted for their great thickness in a geosyncline, but that shrinkage of the earth's interior warped the geosyncline downward, resembling a wrinkle in the crust.

For more than a century, the concept of the geosyncline was debated and modified. Some adherents broadened the definition of the term to include large troughs of deposition of many different kinds. The two kinds of geosyncline that seemed to correspond most closely to Hall's original concept were the *miogeosyncline* and the *eugeosyncline*. The miogeosyncline was defined as a shallow marginal seaway from which the sea frequently spilled over onto the continental interior. Because the miogeosyncline was marginal and therefore was supported by neighboring continental crust on only one side, it would subside readily and hence could accumulate great thicknesses of sediment. Carbonate sediments and clean sands are characteristic "miogeosynclinal" deposits. A eugeosyncline was believed to exist on the oceanward side of and parallel to the miogeosyncline; it was thought to be a depositional belt in deeper water, where black shales, turbidites, cherts, and submarine lavas accumulated. In some depictions, the eugeosyncline was separated from the miogeosyncline by one or more islands, which shed a great deal of sediment. Much of the eugeosynclinal sediment, however, was thought to have been derived from volcanoes rising up within the eugeosyncline itself.

Geosynclinal concepts developed in an era when most geologists believed that continents did not move laterally. Even then, however, it was recognized that these concepts left some important questions unanswered. One puzzling aspect was that geosynclines could not be found accumulating in most areas of the modern world. The Gulf of Mexico seemed to represent one example of a geosyncline, but it was not a good one. Here, the shelf area bordering the southern United States could be viewed as a developing miogeosyncline, but the deep-water setting beyond the shelf was not typical of a eugeosyncline because volcanoes were not common in the offshore area. A second problem was the absence of an acceptable explanation for how thick sedimentary deposits in a region like the Gulf of Mexico could be uplifted and deformed to produce a mountain chain.

The emergence of plate-tectonic theory provided an entirely new framework for the study of mountain building. What had been called miogeosyncline was revealed to be simply a continental shelf where thick piles of sediments accumulate. Here, limestone or clean sands are likely to form a considerable part of the sedimentary record. What had been called a eugeosyncline was, it turned out, nothing more than the deep-sea environment at the edge of the continental shelf—particularly the continental rise, where turbidites are deposited by water carrying sediment down the continental slope. Volcanoes, which had been thought to be an integral part of eugeosynclines, are not usually found here. Rarely except during mountain building, when a continent rests against a subduction zone, does the deep-sea floor bordering a continent take on the full character attributed to the eugeosyncline, including the presence of volcanoes and sediments derived from them. □

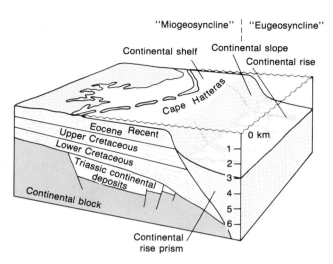

Modern interpretation of "miogeosynclinal" and "eugeosynclinal" environments as settings along any normal depositional slope. (*After R. S. Dietz, Scientific American, March 1972.*)

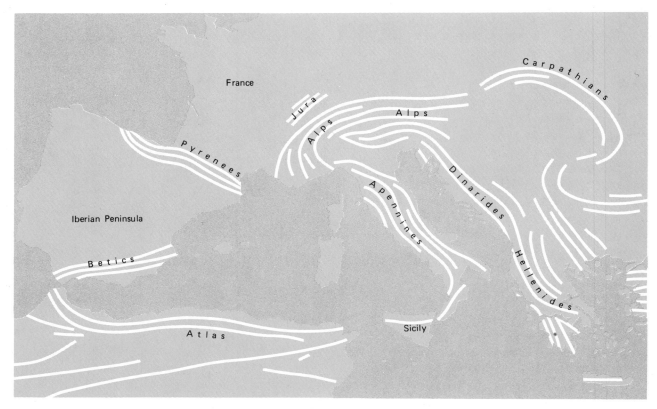

FIGURE 8-6   Regional trends of mountain belts of the Medi-   *Europe, Elsevier Publishing Company, Amsterdam, 1969.)*
terranean region. (*After M. G. Rutten, The Geology of Western*

upon Triassic evaporites, which provided a smooth surface over which the rocks slid northwestward during their deformation (Figure 8-9).

It might be expected that the gradient from the igneous rocks of the Southern Alps to the northern fold-and-thrust belt of the Jura would be mirrored by a similar gradient extending southward from the igneous belt. Unfortunately, the Po River Valley in northern Italy, where this southern segment of the mountain belt may exist, provides no revealing outcrops. Erosion in this region has subdued the mountainous terrane, and younger sediments have buried the rocks that were affected by the Alpine orogeny.

Flysch and molasse

In the fold-and-thrust belt of the Alps (that is, in the Helvetic Alps and the Jura) another feature characteristic of mountain chains is evident. This is the transition from deep-water marine flysch deposits (mainly black shales and turbidites) to nonmarine molasse deposits. In fact, the generic names of these rock types derive from the formal names of particular stratigraphic units in the Alps. Flysch (mainly black shales and turbidites) is found primarily within the Helvetic

Alps, while molasse is found primarily within the Molasse Plateau farther to the northwest (Figure 8-7).

Origin of the Alps

Having reviewed the general structure of the central Alps, and some aspects of their development, we can now step back and examine their origins in a broader context, particularly in relation to plate-tectonic movements. The histories of the Alps and related chains of the Mediterranean region are complex and by no means fully understood. In the case of the Alps, however, we do have a broad understanding of the pattern of orogenesis.

Figure 8-10 depicts events leading to the formation of the Alps from a time early in the Mesozoic Era. At the start of the Mesozoic Era, Africa and Eurasia were both part of the supercontinent Pangaea (Figure 7-7). Then, during the Triassic Period, large-scale rifting began in what is now the Mediterranean Sea. A sea named the **Tethys** had inundated this region before rifting began, but its waters had remained shallow because the region rested on continental crust. As block faulting and volcanism disrupted carbonate banks here during Triassic and Jurassic times, shallow-

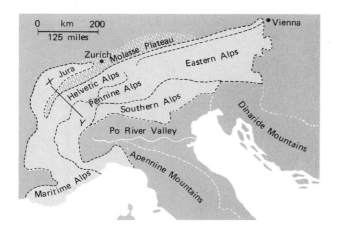

FIGURE 8-7 Map view of and cross section through provinces of the Swiss Alps. The line of cross section (X–Y) is shown on the map. Thrusting from the southwest deformed the Eurasian plate as well as the flysch and earlier molasse that were deposited in front of the advancing mountain belt. The margin of the Adriatic plate, which collided with the Eurasian plate, was also thrust up over the Eurasian plate. Today, fragments of the Adriatic plate remain high in the Alps; one such fragment includes the famous Matterhorn. (*Partly after S. E. Boyer and D. Elliott, Amer. Assoc. Petrol. Geol. Bull. 66:1196–1230, 1982.*)

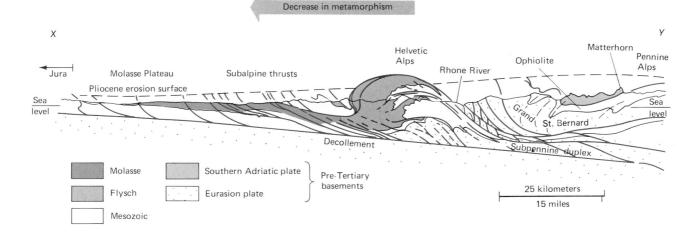

FIGURE 8-8 The Matterhorn. This beautiful Alpine peak, which has ophiolites at its base (Figure 8-7), has been shaped by glacial activity. (*Swiss National Tourist Office.*)

water carbonate deposition gave way to accumulations of fine-grained pelagic sediments on quieter, deeper sea floors within the Tethys (Figure 8-10).

By the middle of the Mesozoic Era, rifting was concentrated along a ridge within a deep seaway, the Penninic Ocean, which had only recently begun to form. The Eurasian plate remained to the north of this narrow ocean; to the south were the African plate and a few microplates, also fragments of Pangaea. As shown in Figure 8-11, two of these microplates moved northward against Eurasia during Late Cretaceous time. Today, one represents the Iberian Peninsula (Spain and Portugal), while another, the Adriatic plate, includes the peninsula of Italy. The details of microplate movements in Mesozoic and Cenozoic time are not well understood, but apparently, during the Mesozoic Era, the Adriatic plate became sutured to Eurasia in the area of the Balkans. Then, early in the Cenozoic Era, the Adriatic plate moved northward, attaching what is now peninsular Italy to Eurasia and thereby leading to the origin of the Alps.

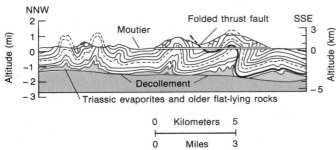

FIGURE 8-9   Cross section through the Jura Mountains, which represent a fold-and-thrust belt. Thrusting has caused younger rocks to slide along a decollement over Triassic evaporites.

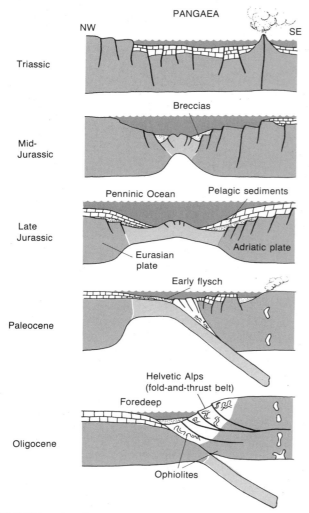

FIGURE 8-10   Plate movements thought to have led to the origin of the Alps. Rifting in the Mediterranean region during the Triassic Period produced the Penninic Ocean during the Jurassic Period. The Adriatic plate was produced by this rifting; it was a fragment of the Eurasian crust. During the first half of the Cenozoic Era, the direction of movement of the plates changed, and the Adriatic plate rode up over the Eurasian plate along a subduction zone. (*After V. J. Dietrich, Geology 4:147–152, 1976.*)

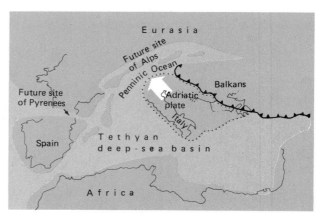

FIGURE 8-11   Inferred geography of the Mediterranean region near the end of the Cretaceous Period. By this time, Africa had split away from Eurasia, and the Adriatic plate, which included the modern Italian peninsula, was moving northward. As it moved, it closed the Penninic Ocean and approached the coast of Eurasia, where its arrival early in the Cenozoic Era was associated with the origin of the Alps. Microplates that formed islands near Eurasia toward the end of the Cretaceous Period later moved to their modern positions as the islands of Corsica, Sardinia, and Sicily.

Ophiolites, flysch, molasse, and radiometrically dated igneous rocks all offer evidence about the timing of events in the Alpine orogeny. First, it is evident that the Adriatic plate, having moved northward from its initial position of attachment to the Balkan region, made contact with Eurasia in the vicinity of Switzerland; this occurred during the Eocene Epoch, about 45 million years ago. In the collision, the edge of the Adriatic plate rode up over the edge of the Eurasian craton. The crust was thickened by this doubling and by the accumulation of igneous rocks above the subducted plate. From the elevated crust, folding and thrusting proceeded toward the northwest.

The oldest flysch deposits now found in the Alps were deposited within the Penninic Ocean (Figure 8-10). The collision of the Adriatic and Eurasian plates during the Eocene Epoch destroyed this ocean, but it left deep marine waters standing in a foredeep atop the continental crust. Here, younger flysch deposits accumulated, and as igneous activity and folding and thrusting pushed progressively farther northwestward, these flysch deposits were deformed. Finally, during the Oligocene Epoch, sediments shed from the mountains pushed marine waters from the foredeep, and deposition of nonmarine molasse prevailed. There is little doubt that molasse was shed while mountain building was still in progress, because some molasse was folded and overridden by thrust sheets. The Alpine orogeny did not die out until Late Miocene time, between about 10 and

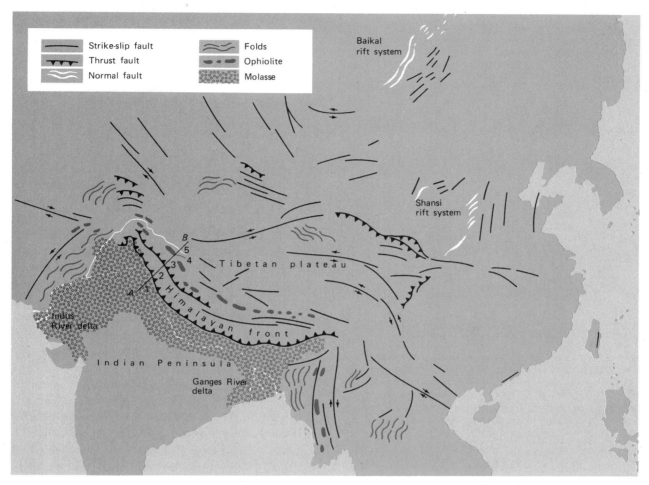

FIGURE 8-12   Geologic features of the Himalayan region. The high-standing Tibetan plateau is bounded by thrust faults, especially in the south, and molasse is being shed southward from the plateau. Numerous strike-slip faults throughout the region seem to have permitted the Asian crust to squeeze eastward, as the Indian peninsula has pushed northward. At the same time, extension of the crust toward the east may account for the development of the Baikal and Shansi rift systems. An inferred geologic cross section along line A–B is shown in Figure 8-14. (*Modified from P. Molnar and P. Tapponier, Scientific American, April 1977.*)

5 million years ago, when the youngest molasse deposits accumulated.

The events discussed in the preceding section refer only to the central Alps. The Maritime Alps (Figure 8-7) and the many other Cenozoic mountain ranges of the Mediterranean region (Figure 8-6) have different histories. In every case, however, mountain building resulted from movement of the African plate in relation to the Eurasian plate or from associated microplate movements.

THE LOFTY HIMALAYAS

The Himalayas are even younger than the Alps, having formed primarily during the Neogene Period. They also cover a much larger area than the Alps (Figure 8-12). Partly because of their youth, the Himalayas are the tallest mountain range on earth.

Structure of the Himalayas

The Himalayan front rises abruptly from the flat Ganges Plain; not far from the front, Mount Everest, the tallest mountain on earth, towers to 8848 meters (~5.5 miles) above sea level (page 204). Even the broad Tibetan plateau, which lies to the north of the Himalayan front (Figure 8-12) stands at an average elevation of about 5 kilometers (~3 miles) above sea level—higher than any mountain peak in the 48 contiguous United States.

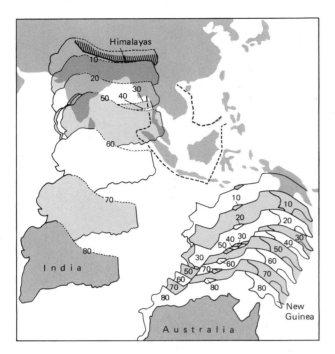

FIGURE 8-13   Northward movement of the Indian craton between 80 and 10 million years ago. Numbers show the times (in millions of years before the present) when geographic boundaries reached various positions. The small Indian craton was a part of the Australian plate, as were Australia and New Guinea, both of which also moved northward. (*After C. McA. Powell and B. D. Johnson, Tectonophysics 63:91–109, 1980.*)

Like the Alps, the Himalayas owe their origin indirectly to the breakup of Gondwanaland. Recall that the Indian peninsula, which projects southward from the Himalayas, was originally a fragment of Gondwanaland. By late in the Mesozoic Era, this fragment was moving northward as an island continent within the large Australian plate (Figure 8-13). The collision of this Indian craton with Eurasia resulted in the uplifting of the Himalayas.

The high-standing Himalayan mountain system is maintained in approximate isostatic balance by a deep root. Thus, the Himalayan crust is extremely thick. This thickness could have developed in any of several ways: One possible mechanism is shortening of the crust as a result of thrust faulting; another is a buildup of igneous plutons within the crust. Perhaps the most likely mechanism for crustal thickening in the Himalayas, however, is a doubling of the crust caused by the wedging of a large segment of the former Indian craton beneath Eurasia. More specifically, it appears that the asthenosphere and felsic lithosphere of the Indian craton became wedged beneath the felsic lithosphere of Eurasia (Figure 8-14). A large, curved fold-and-thrust belt parallels the northeastern margin of India (zones 2 and 3 of Figures 8-12 and 8-14), forming

the main Himalayan mountain chain. Stretching to the south of this belt is a broad zone of molasse that is still accumulating today as debris is shed from the mountains. Geologic features of the fold-and-thrust belt show that it is apparently formed from a segment of the Indian craton that has been sutured to the Eurasian landmass.

Origin of the Himalayas

If the Indian craton did push northward beneath the Eurasian landmass, then the fold-and-thrust belt parallel to the margin was probably formed from a sliver of crust that broke from the craton before it began to pass beneath Eurasia. The suture between the Indian craton and Eurasia (zone 4 of Figures 8-12 and 8-14) contains ophiolites. Some of these ophiolites, along with remains of what appear to be island arc volcanics, were attached to the Indian craton shortly after the start of the Cenozoic Era close to 60 million years ago, when the Indian craton was still far from Eurasia (Figure 8-13). Thus, the small craton must have collided with an island arc before it encountered Eurasia.

When did the Indian craton arrive? Animal fossils provide some clues. When the Age of Mammals began, about 65 million years ago, India was an isolated island continent. New groups of mammals that evolved in Eurasia could not have reached India until the two continents neared each other or actually made contact. The fossil record of the Indian peninsula indicates that the first Cenozoic mammals arrived in Middle Eocene time, about 45 million years ago. There is no evidence that the large-scale mountain building associated with collision began at this time, however. During Eocene time, shallow seas covered much of the Indian craton, and limestones were laid down over large areas. Molasse, which blankets the limestones, is of Late Miocene age and must have formed soon after folding and thrust faulting began. Flysch is not present beneath this molasse, apparently because the rapid uplift of the mountains quickly produced large volumes of sediment. Sediments in the Indian Ocean provide additional evidence of the timing of orogenic activity. The oldest deep-sea turbidites deposited offshore from the Indus and Ganges rivers (Figure 8-12) date to the Middle Miocene. The rivers themselves cannot be much older, and they came into being when the Himalayas began to form.

What happened during the nearly 40 million years between the arrival of mammals on the Indian craton (Middle Eocene) and the onset of mountain building (Late Miocene)? One likely explanation is that mammals passed from Asia to the Indian craton when the northeastern corner of that craton collided with Indochina about 45 million years

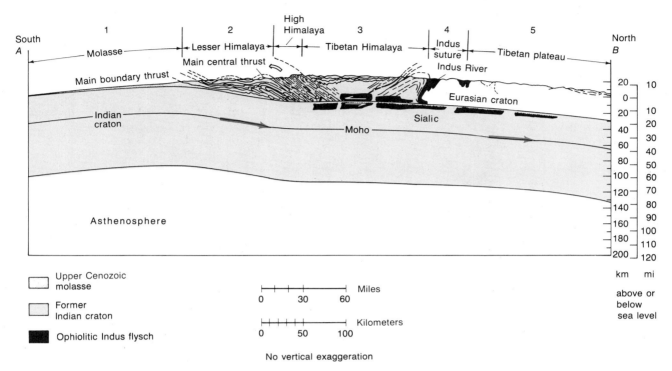

FIGURE 8-14   Diagrammatic cross section illustrating how the Himalayas may have formed when the Indian craton wedged beneath the Eurasian craton. The location of this cross section is shown in Figure 8-12. At the surface, the terrane south of the Indus suture (2–3) represents a segment of Indian crust that was apparently attached to Eurasia before the remainder of the Indian craton broke from this segment and began to slide northward beneath it. A band of ophiolites lies along the suture. (*After C. Powell and B. D. Johnson, Tectonophysics 63:91–109, 1980.*)

ago. At about the same time, the northward progress of the Indian craton slowed (Figure 8-13), perhaps because of its collision with Indochina. The craton did not make contact with Eurasia near its present position until about 15 million years ago, when the oldest molasse suggests that mountain building began. Indeed, much of the Himalayan uplift has occurred during the last 5 million years.

Figure 8-12 shows that thrust sheets have moved outward in several directions from the broad area of thickened crust—that is, from the Tibetan plateau and neighboring areas. This pattern is exactly what we would expect from the principle of gravitational spreading illustrated in Figure 8-4. The region of the Tibetan plateau, including the area to the east, is also characterized by numerous large, left-lateral strike-slip faults (Figure 8-12). It has been suggested that, as the Indian craton pushes northward into the Eurasian landmass, it is squeezing the larger landmass toward the east. According to this notion, Asia can be compared to a tube of toothpaste being squeezed by a thumb (India) toward the tube's open end (the Pacific Ocean). Slices of crust move past one another along strike-slip faults. Two large rift systems, the Baikal and the Shansi, may represent the extension of certain blocks of crust as others drag past

them. Earthquakes still rumble through the Himalayan region as a result of movement along faults, and there is every reason to believe that mountain building here is far from over.

THE ANDES: MOUNTAIN BUILDING WITHOUT CONTINENTAL COLLISION

The Ring of Fire that encircles much of the Pacific Ocean is in most areas a product of the subduction zones that lie between segments of oceanic crust (Figure 7-37). In certain segments of the ring, however, subduction is instead occurring along blocks of continental crust. In Japan, for example, volcanic activity is centered on a slice of continental crust that has become separated from the Asian landmass, and in the eastern Pacific Ocean, subduction zones have encroached extensively on continental borders during much of Phanerozoic time. Until recently, subduction and associated igneous activity affected the entire west coast of North America, and a subduction zone still operates all along the South American coast, where the Andes mountain system continues to form (Figure 8-15).

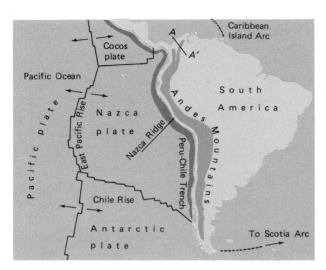

FIGURE 8-15   Map of the Andean mountain chain, showing its position adjacent to a subduction zone, the Peru-Chile Trench. To the north, the Andean chain is connected to the Caribbean Island Arc; to the south, it is separated from the Scotia Arc by a transverse fault. (*Modified from D. E. James, Scientific American, August 1973.*)

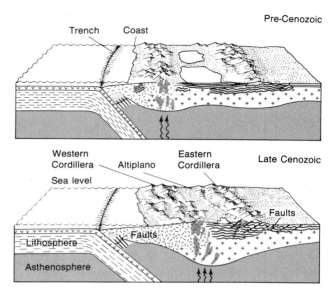

FIGURE 8-16   Formation of the Andes, illustrated diagrammatically. During Cretaceous and Paleocene time (*above*), igneous material was added to the crust from the oceanic plate descending along the marginal trench. During Neogene time, igneous activity shifted farther east. Thrust faulting has occurred both east and west of the area of igneous activity. (*From D. E. James, "The Evolution of the Andes." Copyright © 1973 by Scientific American, Inc. All rights reserved.*)

The Andes system is the longest continuous mountain chain in the world. Where the southern Andes meet the sea, a transverse fault connects them structurally along the sea floor to the Scotia Arc, an island arc that circles back to meet a related active mountain chain in Antarctica. Similarly, where the northern Andes disappear into the ocean, the igneous arc responsible for their formation loops out into the Caribbean Sea.

Mountain building along the west coast of South America has a long history—one that extends well back into the Paleozoic Era. During Paleozoic time, considerable thicknesses of marine sediment were laid down along an ancient continental margin where the Andes now stand, extending the continental margin westward. Widespread unconformities in the area attest to mountain building during the Devonian Period, and the continental margin remained elevated throughout late Paleozoic time—particularly in the eastern Andes, where many late Paleozoic rocks are of nonmarine origin.

Post-Paleozoic mountain building in the Andean region has not caused the continental margin to extend appreciably seaward. Instead, it has involved igneous activity and uplift within the Paleozoic boundary of the continent. The present pattern of mountain building began early in the Mesozoic Era, when a subduction zone came to lie beneath the margin of South America (Figure 8-16). Enormous volumes of igneous rock have since risen from the subducted

oceanic plate and have been added to the Andean crust, thickening it in some places to more than 70 kilometers (~45 miles). When Charles Darwin sailed around the world on the *Beagle*, he noted the presence of Cenozoic marine fossils at high altitudes in the Andes. These fossils offered proof that the Andes had been elevated greatly during recent geologic time. Darwin also saw, firsthand, that the movements occurred in pulses: He witnessed earthquakes during which land along the seacoast was suddenly raised several feet. From a distance, Darwin also observed the eruption of Andean volcanoes. We now know that, for about the last 200 million years, the Andean crust has been thickened by the addition of igneous material below; throughout that period, it has also been bobbing up isostatically. At the same time, volcanic rocks have been piled on top.

The details of Andean mountain building are complex and only partially understood, but certain general patterns are of interest. For example, igneous activity has steadily shifted toward the continental interior during Mesozoic and Cenozoic time; in other words, magma has ascended at positions farther and farther inland (Figure 8-16). Probably because today the subducted plate descends at a low angle, the zone of plutonism and volcanism is now centered about 200 kilometers (~125 miles) inland from the coast. Earlier, the subducted plate probably descended at a steeper angle, thus reaching the depth of partial melting nearer the coast.

In cross section, the Andes display a symmetrical pattern, with fold-and-thrust belts flanking a central zone of uplift where igneous and metamorphic rocks predominate (Figure 8-16). In fact, the cross section shown in Figure 8-2, with its eastward-dipping subduction zone, might be viewed as an idealized representation of the Andes. Island arcs have at times collided with the northern and southern Andes; however, the central Andes, at least during the latter half of the Cenozoic Era, have escaped collision. In the central Andes, the buildup of a core of igneous rocks has apparently been sufficient to cause folding and thrusting toward both the east and the west.

THE APPALACHIANS: AN ANCIENT MOUNTAIN SYSTEM

The Appalachians are a much older system than any of the three youthful, rugged mountain belts we have examined so far. Their internal structures were formed mainly during the Paleozoic Era more than 200 million years ago. The heavily eroded surface of the Appalachians gives them a look of antiquity. In fact, even the modest topographic relief that the mountain chain displays is largely of Cenozoic origin; by late Mesozoic time, the original Appalachians had been worn down. The hills and mountains that we now call the Appalachians were formed by subsequent rejuvenation of the region—a process consisting of gentle uplift followed by slow erosion of resistant bodies of rock and more rapid erosion of weak ones. Because the original Appalachians formed so long ago, little is known about the geologic framework in which they developed; we know more of the settings that gave rise to younger mountain belts. When geologists study the Alps, for example, paleomagnetic data from the sea floor help them to estimate past locations and orientations of important lithospheric plates. Similarly, excellent outcrops facilitate study of the Alps and the Andes. In contrast, much of the heavily eroded Appalachian system now lies beneath thick Mesozoic and Cenozoic deposits of the Atlantic Coastal Plain and the continental shelf. This buried portion of the Appalachians can be studied only through the use of complex and costly techniques such as seismic analysis or drilling.

In our examination of the Appalachian mountain system, we will review the system's geologic provinces, which resemble the zones of the Alps; the two large-scale sedimentary cycles of the Appalachian region, each of which includes flysch and molasse and reflects a separate interval of mountain building; and, finally, the plate movements that caused the intervals of mountain building.

Provinces of the Appalachians

Eastern North America is divisible into numerous discrete geologic provinces (Figures 8-17 and 8-18). Along the east coast, sediments of the Coastal Plain overlap the heavily eroded Appalachian mountain belt. In the Chesapeake Bay region, these sediments overlap the Piedmont Province. The Piedmont rocks typify the central and southern Appalachians in that they are metamorphosed, intensely deformed, and intruded by scattered igneous plutons. A less deformed, relatively unmetamorphosed zone, the Valley and Ridge Province, lies closer to the continental interior than does the Piedmont. Still farther inland are the slightly uplifted and only mildly deformed Appalachian plateaus.

The gradient from intense deformation in the Piedmont to weak deformation in the plateaus resembles the gradients in the Andes (Figure 8-16) and the Alps (Figure 8-7) on the inland side of the axis of igneous activity; thus, it would appear that the exposed Appalachians represent the inland flank of a mountain system. We know little about the structure of the central axis and the other flank, however, since for the most part they lie buried beneath sediments of the Coastal Plain and the submerged continental shelf. Before delving into the history of the Appalachians, let us take a closer look at the present structure of the Valley and Ridge and Piedmont provinces as well as the narrow Blue Ridge Province, which lies between them.

The Valley and Ridge Province is so named because of the topographic expression of its many structural folds (Figure 8-19); the Piedmont, on the other hand, exhibits little topographic relief (Figure 8-18). The Valley and Ridge Province, which has been described as the most elegant folded mountain belt on earth, is what many nongeologists think of as the Appalachian Mountains. In fact, this province constitutes the western fold-and-thrust belt of the Appalachian system. Here, sedimentary rocks have been folded and transported toward the continental interior as immense thrust sheets. In the southern Appalachians, many thrust faults are exposed at the surface. Restoration of these thrust sheets to their approximate original positions shows that the strata above the decollement in many places have been "telescoped" to half their original horizontal span or less (Figure 8-20). In the north, fewer thrust faults reach the present erosion surface, and folding is a more conspicuous feature of the landscape. As in the Alps, many of the thrust faults descend to a basal decollement that slopes uphill in the direction of thrusting. It is clear from this evidence that the lowest Appalachian thrust sheets, like those of the Alps, moved uphill toward the interior of the continent.

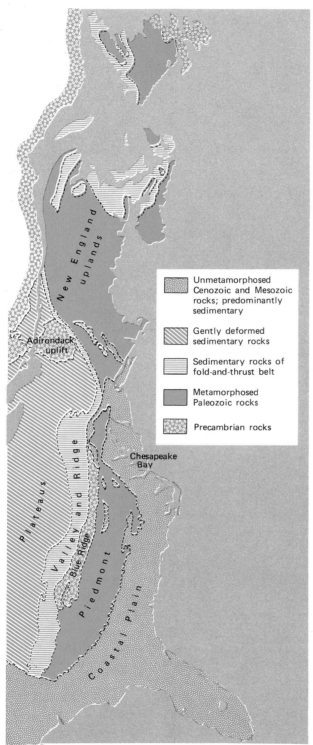

FIGURE 8-17   Geologic provinces of the Appalachian region. A portion of the original Appalachian mountain system has been leveled by erosion and buried beneath the Coastal Plain. (*Modified from G. W. Colton, in G. W. Fisher et al. [eds.], Studies of Appalachian Geology: Central and Southern, John Wiley & Sons, Inc., New York, 1970.*)

The Piedmont Province includes the zone of intense deformation to the southeast of the fold-and-thrust belt. As we will see later in this chapter, many segments of Piedmont terrane were formed from blocks of crustal rock that were welded to the North American craton during the orogenic events that formed the Appalachians. Because of their proximity to the eastern margin of the craton during orogenesis, these blocks suffered intrusion from igneous arc activity and also experienced metamorphism and intense folding. Rocks of the Piedmont, like similarly positioned rocks in the Alps, behaved in a less brittle fashion than those of the nearby fold-and-thrust belt; they flowed extensively in a plastic manner and were contorted to form complex patterns (Figure 8-21).

The Piedmont is separated from the Valley and Ridge by the narrow Blue Ridge Province. Geographers have applied the name *Blue Ridge* to a conspicuous chain of hills in Virginia (Figure 8-22); however, the Blue Ridge Province also includes the Great Smoky Mountains farther south and the Green Mountains of New England. This province consists primarily of a band of crystalline rocks of the Precambrian basement that moved upward along faults during the formation of the Appalachian system.

Tectonic cycles of deposition in the Appalachians

The stratigraphic record in the Valley and Ridge Province reflects a history of sediment deposition that reveals much information about the sequence of mountain building in the Appalachians. This orogenic sequence includes three major orogenies, each of which contributed to the formation of the mountain system.

The most obvious feature of the block of sedimentary rocks within the Valley and Ridge Province is its extreme thickness along the southeastern margin of the province; the block thins toward the continental interior (Figure 8-23). Within this block of sediments are two large packages, one resting on top of the other. Each of these packages represents a **tectonic cycle** of deposition consisting of three units much like those in the Alps. The first of these units represents stable, preorogenic conditions, and the second and third, orogenic conditions.

More specifically, the basal unit of the tectonic cycle contains shallow-water sedimentary deposits consisting primarily of carbonates laid down on a shallow platform. This sequence formed during a period of tectonic quiescence in the region, when there was little mountainous terrane to supply siliciclastic sediments.

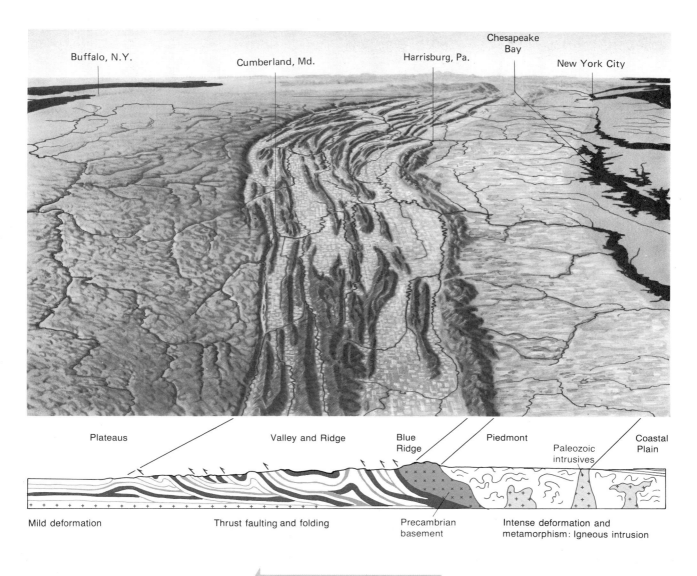

Buffalo, N.Y.   Cumberland, Md.   Harrisburg, Pa.   Chesapeake Bay   New York City

Plateaus   Valley and Ridge   Blue Ridge   Piedmont   Paleozoic intrusives   Coastal Plain

Mild deformation   Thrust faulting and folding   Precambrian basement   Intense deformation and metamorphism: Igneous intrusion

Decreasing intensity of deformation

FIGURE 8-18   Aerial view and idealized cross section of the Appalachian region. The aerial view is toward the northeast. Sediments of the Coastal Plain lap up on the worn-down, eastern portion of the Appalachian system. To the west of the Coastal Plain, the low-lying Piedmont Province is separated from the Valley and Ridge Province by the conspicuous Blue Ridge Province. (*Modified from J. S. Shelton, Geology Illustrated, W. H. Freeman and Company, New York, 1966.*)

The second unit consists of deep-water flysch deposits that record the existence of a foredeep where the carbonate bank once stood. This dramatic change in depositional setting reflects the onset of an episode of mountain building to the southeast, since it is in this direction that the flysch deposits become thicker and more coarsely grained.

The third unit of the tectonic cycle of deposition contains shallow marine and nonmarine siliciclastics, including some red beds. These molasse deposits began to form when the influx of sediments from mountain building increased, pushing back the waters of the foredeep (Figure 8-3).

Let us now look in greater detail at the events recorded by the two tectonic cycles in the central Appalachians.

The first tectonic cycle of deposition

The first tectonic cycle began during latest Precambrian time with the deposition of siliciclastic sediments derived from the continental interior along the stable margin of the North American craton. During the Early Cambrian Period, seas transgressed progressively farther over the craton, and by the end of the period, the siliciclastics along the

FIGURE 8-19   Oblique aerial photographs of structures in the Valley and Ridge Province in Pennsylvania. *A.* Eastward view of folds in the vicinity of the Juniata River (*foreground*) and the Susquehanna River (*background*). *B.* Syncline crossed by the Susquehanna River near Harrisburg. The upturned flanks of the syncline include resistant beds that form high ridges. (*J. S. Shelton, Geology Illustrated, W. H. Freeman and Company, New York, 1966.*)

eastern continental margin had given way to carbonate deposits (Figure 8-23), producing an immense carbonate platform that stretched from Newfoundland to Alabama (Figure 8-24). We know that in many places this platform formed a steep edge on the continental shelf because deepwater deposits to the southeast contain blocks that tumbled down from the shelf (Figure 8-25). Stromatolites, mud cracks, and other indications of intertidal or shallow subtidal conditions indicate that the platform itself maintained a position near sea level; in this respect, it resembles the modern Bahama Banks, which are, however, much smaller structures (Figure 4-22). As the Cambrian Period progressed, the seas transgressed farther and farther over the continental interior; still, the carbonate platform formed the shelf edge throughout Cambrian and part of Ordovician time.

During mid-Ordovician time, the depositional framework along the east coast changed drastically. Carbonate deposition ceased, and flysch deposits such as the Martinsburg Formation and other deep-water units were laid down. Black shale deposition predominated at the outset of this activity; later, turbidites were prevalent (Figure 8-23). That all these deposits derived from an eastern source is indicated by sole markings in the Martinsburg turbidites (Figure 4-36).

The orientation of the foredeep and the arrival of siliciclastic sediments from the southeast suggest that by this time mountain building had begun to the southeast. Radiometric dating of igneous rocks in the Piedmont Province confirms the existence of a Middle and Late Ordovician

FIGURE 8-20   Cross section of the Valley and Ridge Province in Tennessee and Virginia. *A.* Actual cross section without vertical exaggeration. The strata shown in black belong to the Cambro-Ordovician Knox Group. *B.* The strata restored to their original breadth—that is, the distance they spanned before they were folded and thrust toward the west. Gaps represent segments of strata removed by erosion. (*After D. Roeder et al., Univ. Tennessee, Studies in Geol. Sci. 2, 1978.*)

A

```
     6 mi
   |—————|
     10 km
```

(horizontal and vertical scale)

B

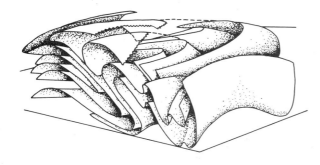

FIGURE 8-21   Typical structural pattern of rocks in the Piedmont Province. The folds are so complex that it is difficult to represent them even diagrammatically, as in this figure. (*After O. T. Tobisch and L. Glover, Geol. Soc. Amer. Bull. 82:2209–2230, 1971.*)

episode of mountain building now known as the **Taconic orogeny**.

Eventually, the rate of sediment supply from the eastern area of uplift exceeded the rate of subsidence. Near the end of the Ordovician Period, flysch gave way to molasse in the form of shallow marine and nonmarine clastics, some of which were red beds. (*Molasse* is, in fact, a term most commonly used in Europe; Americans more often use the phrase **clastic wedge**, which describes the geometry of a body of clastic sediments that thins away from an upland source area.) Thus, in the central Appalachians, the Juniata and Tuscarora formations consist of coarse shallow marine and nonmarine deposits in a clastic wedge or molasse sequence that tapers out toward the northwest. Cross-bed-

FIGURE 8-22   The Blue Ridge Mountains of Virginia. This view is to the southeast from the Shenandoah Valley.

(*Virginia State Travel Service.*)

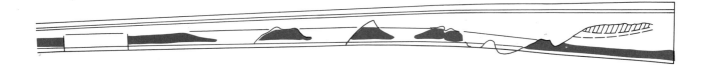

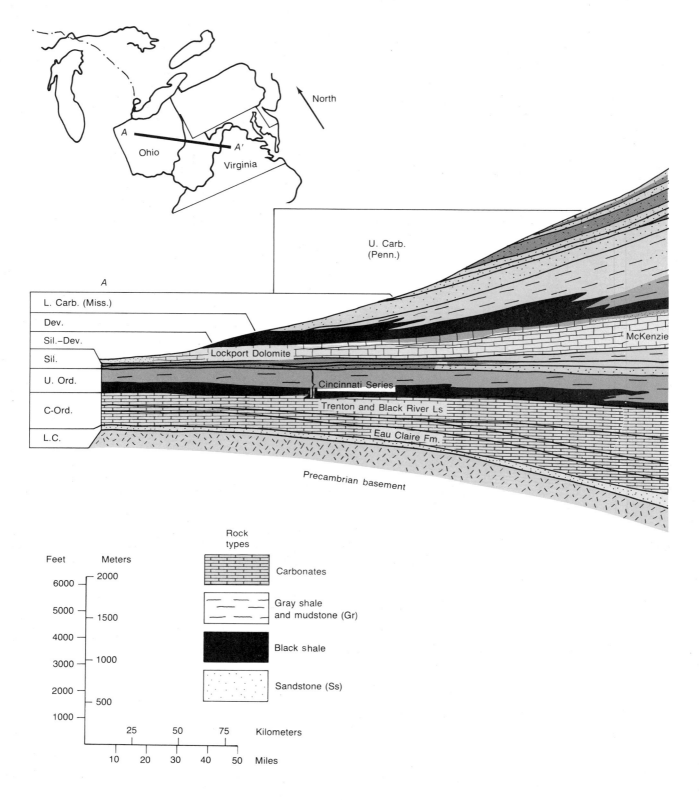

North

Ohio

Virginia

A

U. Carb.
(Penn.)

L. Carb. (Miss.)

Dev.

Sil.–Dev.

Sil.

U. Ord.

C-Ord.

L.C.

Lockport Dolomite

McKenzie

Cincinnati Series

Trenton and Black River Ls

Eau Claire Fm.

Precambrian basement

Rock
types

Carbonates

Gray shale
and mudstone (Gr)

Black shale

Sandstone (Ss)

Feet	Meters
6000	2000
5000	1500
4000	1000
3000	
2000	500
1000	

25   50   75   Kilometers

10   20   30   40   50   Miles

FIGURE 8-23   Stratigraphic cross section through the central Appalachians west of the Blue Ridge Province, with vertical exaggeration. Folds and faults are not shown. The thickest deposits lie to the southeast, in the Valley and Ridge Province. The thinnest deposits lie to the northwest, in the plateau region west of the Appalachians. This package of Paleozoic strata represents two tectonic cycles of deposition. Each cycle contains a

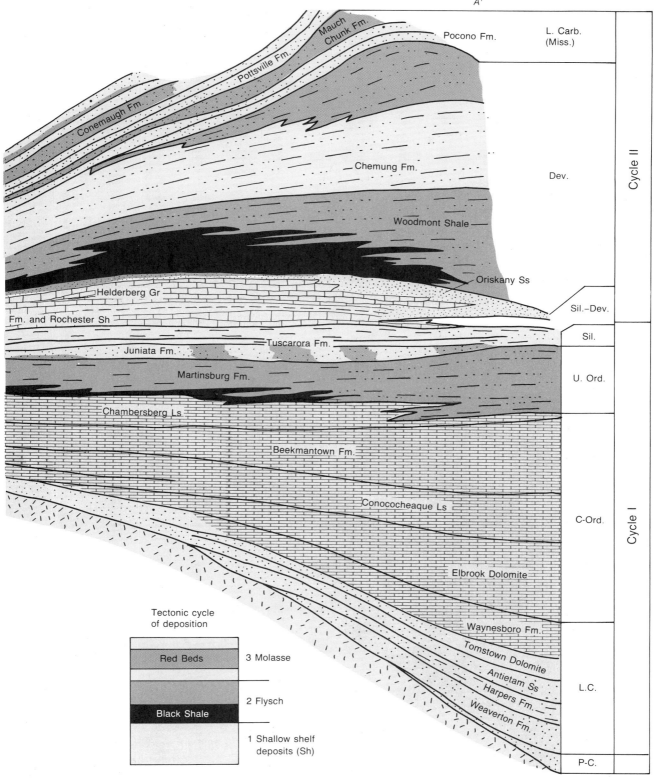

A'

L. Carb. (Miss.)

Pocono Fm.

Mauch Chunk Fm.

Pottsville Fm.

Conemaugh Fm.

Chemung Fm.

Dev.

Cycle II

Woodmont Shale

Oriskany Ss

Helderberg Gr

Sil.–Dev.

Fm. and Rochester Sh

Sil.

Tuscarora Fm.

Juniata Fm.

U. Ord.

Martinsburg Fm.

Chambersberg Ls

Beekmantown Fm.

Conococheaque Ls

C-Ord.

Cycle I

Elbrook Dolomite

Waynesboro Fm.

Tomstown Dolomite

L.C.

Antietam Ss

Harpers Fm.

Weaverton Fm.

P-C.

Tectonic cycle of deposition

Red Beds	3 Molasse
	2 Flysch
Black Shale	
	1 Shallow shelf deposits (Sh)

sedimentary record of mountain building to the east. In the origin of each cycle, flysch and then molasse was deposited in a foredeep that formed where a shallow shelf previously existed.

(*Modified from G. W. Colton, in G. W. Fisher et al. [eds.], Studies of Appalachian Geology: Central and Southern, John Wiley & Sons, Inc., New York, 1970.*)

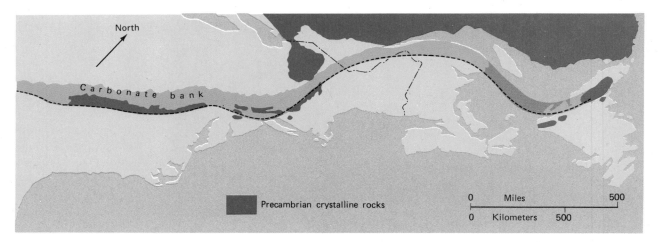

FIGURE 8-24   The dashed line shows the easternmost outcrops of shallow-water Cambro-Ordovician deposits that represent an ancient carbonate bank along the eastern margin of North America. (*After H. Williams and R. K. Stevens, in C. A. Burk and C. L. Drake [eds.],The Geology of Continental Margins, Springer Publishing Company, Inc., New York, 1974.*)

FIGURE 8-25   A bed of breccia within the Frederick Limestone of central Maryland. This bed is interpreted as a deposit of the continental slope below the huge carbonate platform that bordered the continent of North America during much of the Cambrian Period. The clasts of the breccia are pieces of shallow-water limestone that slid over the edge of the platform into deep water. (*Courtesy of R. DeMicco.*)

ding and other features reveal that this was indeed the primary direction of transport (Figure 8-26). This clastic wedge represents the final unit of the first tectonic cycle of deposition in the Appalachians. During the Silurian Period, coarse sediments ceased to arrive from the east, signaling that the Taconic orogeny had ended and that the Taconic Mountains had been effectively leveled by erosion.

Figure 8-27 summarizes the events that produced the first tectonic cycle, as recorded in the New England area, well to the north of the cross section shown in Figure 8-23. The sequence of events in the northern Appalachians resembles the sequence in the southern Appalachians with one major difference: In the north, from New York State to Quebec, blocks of lower Paleozoic sedimentary rocks many kilometers in width and length slid from the eastern highlands into the foredeep.

The second tectonic cycle

After the leveling of the Taconic Mountains during the Silurian Period, the eastern margin of North America was once again an area of low relief. During the remainder of the Silurian and part of the Devonian, shallow-water carbonates of the Helderberg Group accumulated over a broad shelf along this margin. These sediments are succeeded in places by a thin body of clean, well-sorted quartz sand. This unit, known as the Oriskany Formation (Figure 8-23), was deposited along the beach that fringed the Devonian sea.

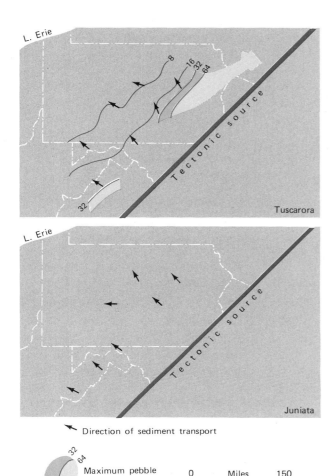

FIGURE 8-26   Directions of river currents that deposited molasse units of Upper Ordovician and Silurian age. Pebbles in the Tuscarora Formation decrease in size away from the source area. (*After L. D. Meckel, in G. W. Fisher et al. [eds.], Studies of Appalachian Geology: Central and Southern, John Wiley & Sons, Inc., New York, 1970.*)

The shallow-water deposits at the base of the second tectonic cycle of deposition, like those at the base of the first cycle, were succeeded by flysch deposits, including the Woodmont Shale of Figure 8-23, during Early Devonian time. This activity signaled the origin of a new foredeep as well as a new episode of mountain building, the **Acadian orogeny**. As in the first cycle, flysch deposition here was characterized by the initial accumulation of black mud throughout most of the foredeep and later by the predominance of turbidite deposition.

Near the beginning of Late Devonian time, as the Acadian orogeny intensified, clastic wedges built westward, pushing back the waters of the foredeep. Thus began the molasse phase of the second tectonic cycle of deposition.

The great thickness of the Late Devonian clastic wedge in the northern and central Appalachians (Figure 8-23) suggests that the Acadian orogeny elevated very large mountains to the east. During this Late Devonian phase of molasse deposition, sands accumulated in nonmarine environments to the east and spread to marine environments of the foredeep farther west. In Late Devonian time, these sediments formed what is often called the Catskill delta, the thickest accumulation of which is in New York State (Figure 13-39). In fact, the Catskill was not a single delta but rather a complex of alluvial, deltaic, and submarine environments. Some of the meandering-river deposits are red beds resembling some of the molasse units of the first Appalachian tectonic cycle. In places, meandering-river sequences are piled one on top of another to great thicknesses (Figure 8-28).

Remarkably, the molasse phase of the second tectonic cycle of deposition continued more than 50 million years, from Late Devonian time through most of the Carboniferous Period. Early in Carboniferous time, the rate at which sediment was shed from the highlands to the east seems to have been reduced for a time, suggesting that the Acadian highlands may have been subdued by erosion. However, the rate of influx increased again, leading to the accumulation of thick siliciclastic units like the Mauch Chunk and Pottsville formations (Figure 8-23; rocks of the Pottsville, a meandering-river deposit, are also illustrated in Figure 3-24). This evidence suggests that there was a lull in mountain-building activity followed by another large pulse of orogeny, during the Carboniferous Period. Additional support for this inference is supplied by dating of igneous and metamorphic rocks in the Piedmont, which reveals a pulse of igneous activity centered in Late Carboniferous time and perhaps extending into the Permian Period. This pulse is called the **Alleghenian orogeny** (or, less commonly, the **Appalachian orogeny**).

Thus, the thick molasse sequence of the second tectonic cycle represents sediment shed from mountains that formed during two successive orogenic episodes, the Acadian and the Alleghenian. The second followed so closely upon the first that, although the supply of siliciclastic sediment slowed in the interim, it never fully stopped. It seems evident, therefore, that some highlands remained in the region of orogenic activity and that the continued accumulation of molasse right up to the start of the Alleghenian orogeny prevented the foredeep from deepening sufficiently to permit deposition of flysch when the Alleghenian orogeny began.

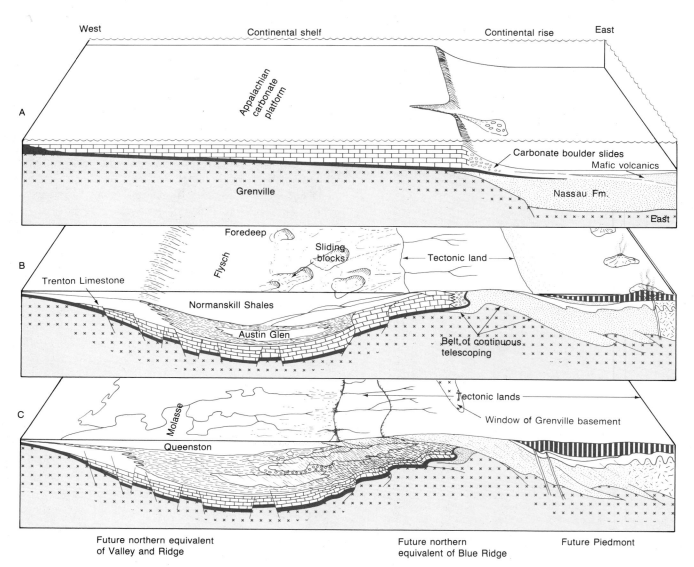

West   Continental shelf   Continental rise   East

A

Appalachian carbonate platform

Carbonate boulder slides
Mafic volcanics

Grenville   Nassau Fm.

East

B

Foredeep
Flysch   Sliding blocks   Tectonic land

Trenton Limestone

Normanskill Shales

Austin Glen   Belt of continuous telescoping

C

Tectonic lands

Window of Grenville basement

Molasse

Queenston

Future northern equivalent   Future northern   Future Piedmont
of Valley and Ridge   equivalent of Blue Ridge

FIGURE 8-27   Schematic block diagrams showing the Cambro-Ordovician history of the continental margin of North America now represented by western New England. *A.* Initially, a carbonate bank formed the continental margin, as it did to the south. *B.* Later in Ordovician time, volcanism and uplifts produced land areas east of the earlier continental margin. At the same time, the carbonate platform was warped downward to form a foredeep, where flysch was deposited. Large blocks of sediments periodically slid westward from the newly formed land area into the foredeep. *C.* Near the end of Ordovician time, deposition of flysch was succeeded by deposition of molasse. (*After J. M. Bird and J. F. Dewey, Geol. Soc. Amer. Bull. 81:1031–1060, 1970.*)

Plate movements and Appalachian orogenesis

Having examined the structure of the Appalachians and the history of the sedimentary deposits associated with them, we can now consider the plate movements responsible for the growth of the Appalachian mountain system.

As we have seen, early in the Mesozoic Era the large landmasses of the world were united as the supercontinent Pangaea (Figure 7-7). This union of Paleozoic continents led to the building of the Appalachians—that is, to the Taconic, Acadian, and Alleghenian orogenies. How these earlier continents became assembled is the subject of the following sections.

One difficulty in reconstructing the plate movements that ultimately produced Pangaea is that a key form of evidence, paleomagnetism, gives information only about the latitudinal position and orientation of ancient continents; it tells us nothing about continental movements in an east-

FIGURE 8-28   Cyclical meandering-river deposits of the Catskill Formation at Peters Mountain, Pennsylvania. Each cycle contains deposits that grade upward from coarse sandstone (C) to fine mudstone (F). (*F. J. Pettijohn and P. E. Potter, Atlas and Glossary of Primary Sedimentary Structures, Springer Publishing Company, Inc., New York, 1964.*)

west direction (page 181). This limitation is evident when we consider plate positions during Late Ordovician time, when the Taconic orogeny took place.

The Taconic orogeny   Early in the Paleozoic Era, North America was separated from Europe and Africa by what has become known as the Iapetus Ocean. North America was at that time part of a large continent known as Laurentia, which lay to the northwest of northern Europe, which in turn formed a separate continent that geologists call Baltica (Figure 8-29). Greenland was also part of Laurentia and was attached to North America, while England, which was not yet attached to Scotland, lay far to the south. The border of eastern North America consisted of the long carbonate bank on which were deposited the sediments of the first Appalachian cycle. As we have seen, during the Ordovician Period, the carbonate bank collapsed and a foredeep was formed. At the same time, the continental

margin must have encountered a subduction zone, because mountain building had begun to the east adjacent to an igneous arc that left a record of volcanic rocks (Figure 8-27).

It is uncertain whether the eastern margin of Laurentia encountered another continent, perhaps Baltica, along the subduction zone during the Ordovician Period. In Norway, which lay along the margin of Baltica, there are in fact igneous and metamorphic rocks of Ordovician age, along with an associated fold-and-thrust belt. Nonetheless, it remains possible that Laurentia and Baltica merely approached one another without making contact, with each experiencing mountain building along a marginal subduction zone. Considerable narrowing of the Iapetus Ocean between Laurentia and Baltica is suggested by the composition of shallow-water marine faunas of Ordovician age in North America and Europe; while Lower Ordovician faunas of the two continents are quite distinct, Upper Ordovician

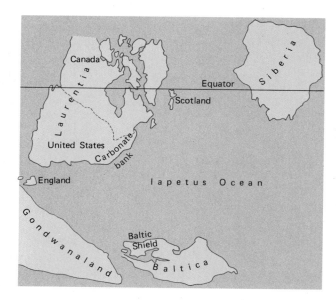

FIGURE 8-29 Positions of landmasses and the Iapetus Ocean in Middle Ordovician time. (*After C. R. Scotese et al., Jour. Geol. 87:217–268, 1979.*)

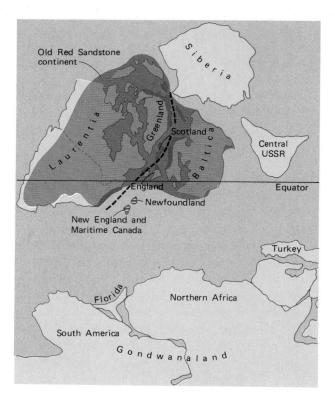

FIGURE 8-30 World geography during Caledonian mountain building. The suture (*broken line*) marked the disappearance of the Iapetus Ocean by Early Devonian time. Newly formed mountains (the Appalachians and Caledonides) are shown in color. A broad sea lay between Gondwanaland and the northern continents. (*Modified from C. R. Scotese et al., Jour. Geol. 87:217–268, 1979.*)

faunas are similar. Apparently, the Iapetus Ocean was narrow enough by Late Ordovician time that many species were able to cross from one side to the other.

Plate movements and the Acadian orogeny   We do not know whether the Taconic orogeny was associated with continental collision between Laurentia and Baltica; however, there is clear evidence of such a collision during the middle Paleozoic Acadian orogeny. The Acadian event is known as the **Caledonian orogeny** in Europe, where it affected Scandinavia and united Scotland and England into what we now know as Great Britain (Figure 8-30). The Devonian molasse deposits in eastern North America and Britain are strikingly similar; apparently, these deposits spread in opposite directions from a single mountain chain that stood along a suture between Laurentia and Baltica.

Many nonmarine deposits of the Old Red Sandstone Formation of northern Britain resemble sediments of the Upper Devonian Catskill clastic wedge of New York and Pennsylvania (Figures 8-28 and 13-39); likewise, similar fossil freshwater fishes of Devonian age are found in Britain and eastern North America. The Acadian/Caledonian suturing of Baltica to North America formed what is called the Old Red Sandstone continent, which we will discuss in more detail in Chapter 13.

Dates of igneous activity, folding, and foredeep deposition all reveal that Acadian/Caledonian orogenic activity began in the north and spread southward. This activity was under way in Greenland and Scandinavia by mid-Silurian time, but it did not begin in the central Appalachian region until mid-Devonian time.

The Acadian/Caledonian suturing may have been the first contact between Laurentia and Baltica during Phanerozoic time. It is important to note, however, that this would have been the second meeting of these continents if the Taconic orogeny involved similar suturing. Furthermore, if suturing did occur during the Taconic orogeny, then there must have been a subsequent rifting apart of the continents prior to the resuturing during the Acadian/Caledonian orogeny.

Plate movements and the Alleghenian orogeny   The Acadian/Caledonian suturing left the Old Red Sandstone continent separated from Gondwanaland, which still lay far to the south, by a vast tropical seaway. The closure of this seaway and the attachment of Gondwanaland to the Old Red Sandstone continent in Late Carboniferous time were associated with the Alleghenian orogeny in the Appalachian region. These were two of three primary suturing

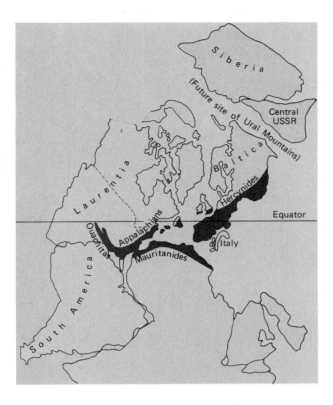

FIGURE 8-31 Alleghenian mountain building (*colored area*) during Late Carboniferous time. Northward movement of Gondwanaland late in Paleozoic time closed off the western end of the Tethys Sea. Mountain building occurred not only here, where Gondwanaland was sutured to North America, but also to the east, along the southern margins of Great Britain and Eurasia. For other possible plate positions at this time, see Box 8-2. (*Modified from C. R. Scotese et al., Jour. Geol. 87:217–268, 1979.*)

events in the creation of Pangaea; the third, which took place slightly later, was the attachment of Europe to Asia, accompanied by the elevation of the Ural Mountains (Figure 8-31).

What Americans call the Alleghenian orogeny extended across southern Europe, where it is known as the **Hercynian orogeny** or the **Variscan orogeny**. The Mauritanide mountains of northwestern Africa formed on the southern side of the suture, opposite the Appalachian and Hercynian uplifts (Figure 8-31). Adjacent to what is now the Gulf of Mexico, the Ouachita Mountains formed as an extension of the Appalachians. Figure 8-32 reviews the major events of Appalachian mountain building, placing the Alleghenian orogeny in the perspective of preceding events.

One question remains the subject of lively debate today: Exactly where did Gondwanaland become attached to the Old Red Sandstone continent? This question, which is addressed in Box 8-2, may soon be answered. However,

many smaller questions about plate movements in relation to the origin of the Appalachians may never be fully resolved. As we will see in the following section, one such mystery concerns the manner in which fragments of crust were accreted to the margin of North America as part of the Appalachian mountain belt.

Exotic crustal blocks in the Appalachian system   We have observed that microplates played important roles in Cenozoic orogenic activity of the Mediterranean region. Examples include the Adriatic plate, which took part in the formation of the Alps, and the microplate that includes Spain (Figure 8-11). Similarly, microplates played important roles in the development of the Appalachian mountain system.

In recent years, it has been recognized that the seaward portion of the Appalachian mountain belt includes large and small blocks of terrane that were welded onto the North American craton during the orogenies that produced the Appalachians. Some of these exotic blocks may have existed previously as discrete microplates; others may have been parts of the larger plates with which North America came into contact. Each of these microplates must have been positioned in the Iapetus Ocean at some time before its closure. Some were probably slivers of crust that broke away from North America as a result of rifting or transform faulting and were later reattached at different locations by suturing; others may have been pieces of Baltica that remained attached to North America when Pangaea broke apart during the Mesozoic Era; others may have been slivers of the Iapetus Ocean floor; and still others may have come from volcanic-island arc systems that were accreted to North America when it was rafted against the subduction zones above which these arc systems formed.

Individual exotic blocks in the Appalachians can be identified by unique internal features that distinguish them from neighboring terranes. Many blocks are separated from one another by ophiolites and mélanges, signs of ancient subduction. Stratigraphic relationships enable us to estimate when some of these blocks became attached to North America (Figure 8-33). It is important to remember, however, that in some cases one exotic block may have joined another before the combined blocks became attached to North America.

Figure 8-34 shows the locations of some of the exotic blocks of crust within the Appalachian system. Most of these blocks are themselves composite blocks consisting of several smaller pieces of crust that were united before or during attachment to North America. The recognition and interpretation of these blocks is a controversial topic, but one general pattern is clear: Just as we would expect,

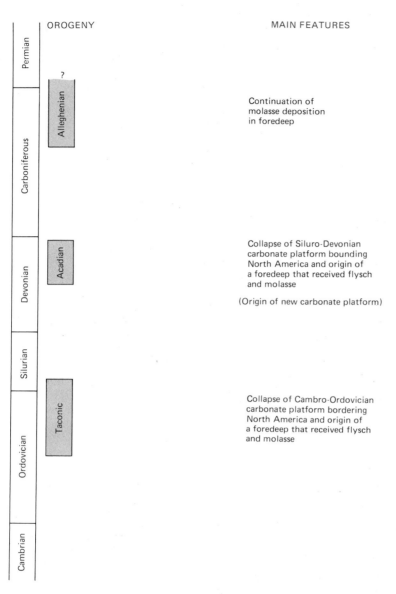

 OROGENY   MAIN FEATURES   PLATE TECTONIC EVENTS

Continuation of
molasse deposition
in foredeep

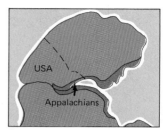

Suturing of Old Red Sandstone
continent to Gondwanaland

Collapse of Siluro-Devonian
carbonate platform bounding
North America and origin of
a foredeep that received flysch
and molasse

(Origin of new carbonate platform)

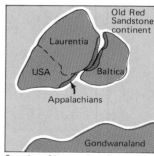

Suturing of Laurentia to Baltica to
form Old Red Sandstone continent

Collapse of Cambro-Ordovician
carbonate platform bordering
North America and origin of
a foredeep that received flysch
and molasse

Convergence, but not necessarily
contact, of Laurentia and Baltica

FIGURE 8-32   Summary of Appalachian orogenic events.

the exotic blocks that now lie closest to the interior of North America were added to the larger craton before or at the same time as blocks that now lie nearer the continental margin.

In the area extending from eastern New York State to New Brunswick, Nova Scotia, are several discrete crustal blocks affected by the Taconic orogeny. The Taconic ranges (Figure 8-34), for example, are formed from blocks of deep-water sediments that were elevated by the Taconic orogeny and that slid westward over deposits of the continental shelf (Figure 8-27).

Larger blocks of exotic crust were affected by the Acadian orogeny, which means that these blocks must have been attached to North America before or during mid-Paleozoic time. For example, the rocks of the Avalon Peninsula of Newfoundland and related terranes were part of composite blocks added to the craton after the Taconic orogeny and before or during the Acadian orogeny. On the Avalon Peninsula of Newfoundland, the exotic terrane consists for the most part of late Precambrian volcanics and sedimentary rocks overlain by Early Paleozoic sediments. Long ago it was noted that Cambrian trilobite faunas of the

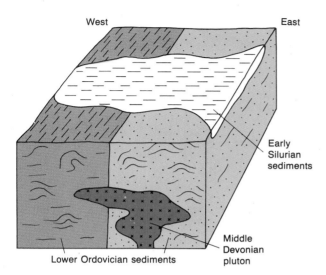

FIGURE 8-33   Recognition of exotic blocks of crust in the Appalachians. In this hypothetical illustration, two adjacent blocks can be distinguished from one another because they are characterized by different kinds of Lower Ordovician sediments that do not intergrade (shale is in the western block and sandstone is in the eastern block). The Early Silurian sediments that blanket the two blocks of crust and the Middle Devonian pluton that cuts through both of them are not deformed in the manner of the Lower Ordovician sediments. This pattern indicates that the two blocks were deformed and welded together during the Late Ordovician Taconic orogeny.

FIGURE 8-34   Exotic crustal blocks in the Appalachians. As would be expected, the older blocks (those accreted to the craton before the Late Ordovician Taconic orogeny) lie inland from blocks accreted during the mid-Paleozoic Acadian orogeny. Most of the large, exotic crustal blocks are actually composite terranes consisting of two or more smaller slivers of crust. The Cambrian trilobite *Paradoxides*, which is found in some of the exotic blocks, is a genus that is also found in Europe but not in rocks of the original North American craton. The Taconic ranges and other small exotic blocks that rest on rocks of the original craton seem to have slid to their positions from the east during the Taconic orogeny (Figure 8-27). (*Modified from H. Williams and R. D. Hatcher, Geology 10:530–536, 1982.*)

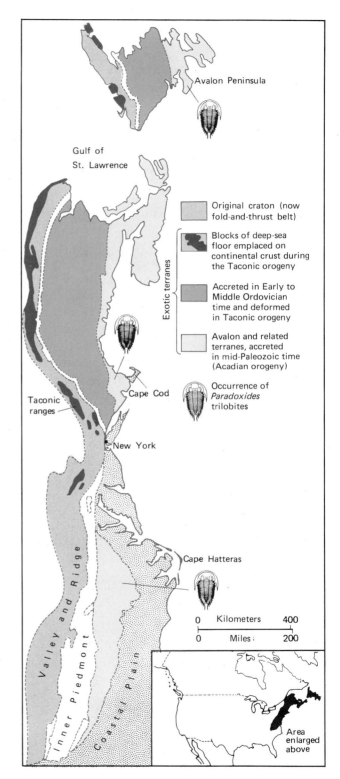

Avalon and related terranes contain genera like *Paradoxides* that are also found in European faunas (Figure 8-34). It seemed strange that typical American faunas should be found in western Newfoundland just a short distance inland from these exotic faunas. Plate tectonics explains the puzzle with the revelation that the Avalon terrane became attached to North America during the Acadian orogeny.

Paleomagnetic evidence has revealed a startling fact about the Avalon-like terranes of New Brunswick and New England. During mid-Carboniferous time, these terranes moved northeastward along the margin of the continent from about the latitude of Florida 1500 kilometers (~900

Box 8-2

Where was Gondwanaland

in Late Carboniferous

time?

It is widely agreed that early in the Mesozoic Era the configuration of Pangaea was such that Africa lay to the south of where the Mediterranean now lies and South America lay to the south of what is now the Gulf of Mexico (A). Ever since the Atlantic Ocean began to open at a slightly later time, these north-south alignments have undergone only minor modifications.

In the past, it has generally been assumed that when Pangaea originally began to form late in the Carboniferous Period, it took the configuration shown in Figure A. This idea is now being challenged. For example, paleomagnetic data for Gondwanaland suggest that in mid-Carboniferous time, South America lay well to the north of the Gulf of Mexico. Paleomagnetic data, of course, offer information only about latitudinal position, not about longi-

tude, but the only way to locate South America as far north as the paleomagnetic data seem to demand is to shift it laterally; otherwise it would overlap with North America. Two geographic solutions to the problem are illustrated in Figures B and C. In the more extreme solution, the "hump" of northern South America is located in the Mediterranean region (C). In the more moderate solution, which does not place Gondwanaland quite as far north as the paleomagnetic data demand, the "hump" of northern South America lies between the eastern United States and southern Europe (B).

If the configuration shown in Figure A represents Pangaea after the beginning of the Triassic Period, the earlier configurations postulated in Figures B and C both require major movements of continental crust during the Triassic Period. Specifically,

the land mass that was formerly Gondwanaland would have to have moved southwestward relative to the northern portion of Pangaea, from its position in Figure B or Figure C to its position in Figure A. This scenario implies that one or more enormous right-lateral strike-slip faults operated early in the Triassic Period. Unfortunately, any such faults would probably have been located in areas now positioned near the margins of continents—that is, in areas later deformed by mountain building in southern Eurasia or presently covered by the Mediterranean Sea and the Atlantic Ocean. As a result, there is little direct geologic evidence with which to evaluate the alternatives illustrated in Figures A, B, and C. Whether or not we should accept the dramatic implications of the paleomagnetic data is currently under debate. □

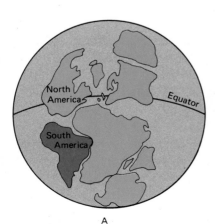

A

B

C

miles) away! It is not certain whether all the Avalon terranes share this history; it is evident, however, that all of the Avalon terranes were sutured to North America when Pangaea formed or shortly before and remained attached to North America when Pangaea broke apart.

In the southern Appalachians, the segment of the Piedmont lying inland of the Avalon-like block is of uncertain origin, but seismic studies suggest that this inner Piedmont is underlain by sedimentary rocks like those of the Valley and Ridge Province further inland. Thus, it appears that the highly metamorphosed and deformed rocks of the southern inner Piedmont have been thrust hundreds of kilometers inland over rocks of the fold-and-thrust belt. This movement seems to have occurred during the Alleghenian orogeny, when igneous plutons were also emplaced in the Piedmont. At the same time, farther inland, rocks of the Valley and Ridge Province, which range in age from latest Precambrian to Carboniferous, underwent the folding and thrusting that gives them their present configuration.

As we will see in later chapters, the Appalachians were largely leveled by erosion during the Mesozoic Era. The moderate elevation of the Valley and Ridge Province that we see today has resulted from secondary uplift probably caused by isostatic adjustment during the Cenozoic Era.

CHAPTER SUMMARY

1.   A mountain chain tends to form where the edge of a moving continent encounters a subduction zone. This situation is causing the elevation of the Andes today.

2.   Mountain chains can also form when two continental blocks converge along a subduction zone. The Himalayas and the Alps are two chains that have formed in this way.

3.   Oceanic crust is being subducted along the west coast of South America and is melting within the mantle to release magma of low density. As this magma rises and cools within the continental crust, the thickened crust rises isostatically to form the axis of the Andes. This uplift and the accumulation of volcanic rocks leads to a spreading of the crust, which in turn produces folding and thrusting in both directions from the axis of uplift. Metamorphism declines in both directions from the igneous axis of the mountain chain.

4.   The Himalayas have been elevated because the crustal block that now forms peninsular India has been forced beneath the margin of Asia, doubling the thickness of the crust.

5.   The Alps and the Appalachian mountain system resemble each other in many ways. Each displays a sedimentary sequence characteristic of mountain building. The sequence begins with shallow-water sediments that have been deposited along a continental border. When mountain building began along the border, a foredeep was created by a downwarping of the crust inland from the orogenic activity. This foredeep became the site of deposition for flysch in the form of black shales and graywackes of deep-water origin. In both the Alps and the Appalachians, as the foredeep became choked with sediment from the young mountains, deep-water flysch deposits gave way to shallow marine and nonmarine clastic sediments, or molasse.

6.   In the Appalachians, three episodes of mountain building are evident in the sedimentary record. The first destroyed an enormous early Paleozoic carbonate platform that had extended along most of the eastern margin of North America. The second and third were associated with the coalescing of continents to form the supercontinent Pangaea. As the Appalachians formed, small blocks of crust were added to the margin of North America.

EXERCISES

1.   How does the old idea of the geosyncline relate to modern ideas about plate tectonics and mountain building? (Refer to Box 8-1.)
2.   How can mountain chains form without continental collision?
3.   Are the ophiolites found within a mountain chain episode normally older or younger than the foredeep deposits?
4.   Examine a world map or globe to locate mountain chains that are not discussed in this chapter. Then locate these chains on the plate-tectonic map of the world (Figure 7-1). See if you can figure out how the presence of each mountain system might relate to plate-tectonic processes. (Some of the answers will appear in the chapters that follow.)

ADDITIONAL READING

James, D. E., "The Evolution of the Andes," *Scientific American*, August 1973.

Molnar, P., and P. Tapponier, "The Collision between India and Eurasia," *Scientific American*, April 1977.

Sullivan, W., *Continents in Motion: The New Earth Debate*, McGraw-Hill Publishing Company, Inc., New York, 1974.

PART FOUR

The Precambrian World

The earth came into being about 4.6 billion years ago, forming—with the other planets of our solar system—from a whirling cloud of dust. During the first billion years or so of earth history, continents were small, volcanism was widespread, and life consisted of bacteria and their lowly relatives. During this Archean Eon, radioactive elements were abundant in the lithosphere and released heat at a high rate. By the beginning of the Proterozoic Eon, however, the earth had cooled, larger continents were forming, and modern plate tectonic processes were in operation. In the course of Proterozoic time, cells more advanced than bacteria arose within aquatic environments, and from them multicellular plants and animals evolved.

CHAPTER 9

The Archean Eon of Precambrian Time

Since the last century, the interval of earth history that preceded the Phanerozoic Eon has been known as the Precambrian. Although the term *Precambrian* has no formal status in the geologic time scale, it has traditionally been employed as if it did. The Precambrian includes nearly 90 percent of geologic time, ranging from 4.6 billion years ago, when the earth formed, to the start of the Cambrian Period about 4 billion years later.

Two eons are formally recognized within the Precambrian: the Archean and the Proterozoic. The Archean Eon, which is the subject of this chapter, includes about 45 percent of the earth's history—the interval from about 4.6 to 2.5 billion years ago. The Archean Eon was an important span of time during which, as we will learn in this chapter, the earth underwent enormous physical change and life developed on its surface. Many details of Archean history, however, remain unknown or poorly understood.

In fact, much less is known of both the Archean and Proterozoic eons than of the succeeding Phanerozoic Eon. One reason for this discrepancy is that, although the Precambrian constitutes such a large span of geologic time, Precambrian rocks form less than 20 percent of the total area of rocks exposed at the earth's surface (Figures 9-1 and 9-2). Erosion has destroyed many Precambrian rocks, and metamorphism has altered others to the degree that

they can no longer be dated and therefore recognized as Precambrian. Still other Precambrian rocks lie buried beneath younger sedimentary and volcanic rocks that were laid down upon the surface of the earth. Furthermore, index fossils are seldom found in Precambrian rocks because primitive organisms without durable skeletons predominated until the end of Precambrian time. As a result, stratigraphic correlation of Precambrian rocks has been based largely on radiometric dating.

Most geologic information about the Precambrian has been derived from **cratons**, which are large portions of continents that have not experienced tectonic deformation since Precambrian or early Paleozoic time. All the continents of the present world include cratons that consist primarily of Precambrian rocks (Figure 9-2). A **Precambrian shield** is a largely Precambrian portion of a craton that is exposed at the earth's surface. The largest of these is the vast Canadian Shield, which has recently (geographically speaking) become more fully exposed by the action of Pleistocene glaciers (page 42). Although shields contain some sedimentary rocks, they consist primarily of crystalline (igneous and metamorphic) rocks. As we will see, mountain belts that formed during Precambrian time left traces that can be recognized within Precambrian shields, but their elevated topography was destroyed long ago by

---

Precambrian rocks, which represent the first 90 percent or so of earth history, are best exposed in broad, low-lying shield areas of continents. This photo shows an oblique aerial view of the Canadian Shield, revealing the effects of Pleistocene glaciation. Note the smooth topography and the many lakes that occupy areas scoured by glaciers. The Archean Eon, which represents the early portion of the Precambrian Era, spanned more than 2 billion years. (*Royal Canadian Air Force.*)

PRECAMBRIAN TIME

Billion years ago

4.6		2.5		1.6	0.9	0.59
			Early		Middle	Late
	Archean Eon			Proterozoic Eon		

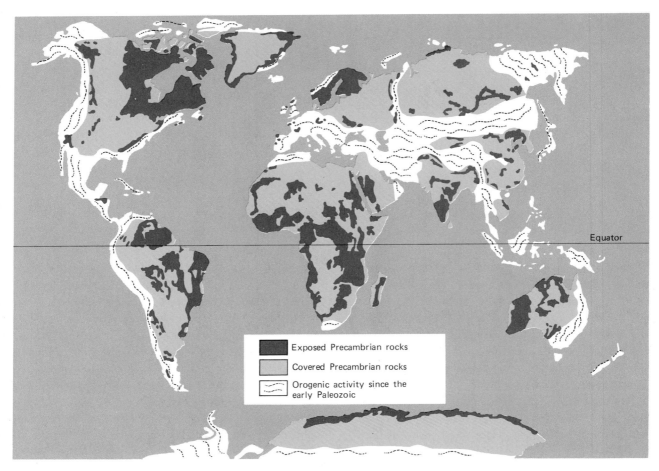

FIGURE 9-1   Distribution of Precambrian rocks in the modern world. Note that these form or underlie most of the cratonic area of the modern world. (*After A. M. Goodwin, in B. F. Wind-ley [ed.], The Early History of the Earth, John Wiley & Sons, Inc., New York, 1976.*)

erosion. Today, the only Precambrian rocks that stand at high elevations within mountain ranges are those that have been uplifted by Phanerozoic orogenies.

The Archean Eon, as previously stated, was a time of major change in the structure of the earth itself, beginning with the events that gave the planet its basic configuration, in which the mantle and crust surround the core (page 14). Early in Archean time, the earth's crust seems to have

FIGURE 9-2   Diagram illustrating how Precambrian shields are often flanked by younger (Phanerozoic) mountain belts and sediments.

differed from what it is today. The earth was much hotter than it is now, and there were apparently no large felsic cratons. Moreover, plate-tectonic processes may not have operated precisely as they do in the modern world. By the end of Archean time, however, large cratons had begun to form, and plate-tectonic processes were modifying these cratons in the same way that they do now. The transition to this modern style of tectonism ushered in the Proterozoic interval of Precambrian time, which will be discussed in the two chapters that follow. In the absence of useful biostratigraphic data, the boundary between the Archean and Proterozoic intervals is defined by its absolute age of 2.5 billion years before the present.

We will begin our review of Archean events by considering when and how the earth and other planets of the solar system may have formed. We will then review evidence that suggests how the earth changed during the remainder of Archean time.

AGES OF THE UNIVERSE AND THE PLANET EARTH

All seriously considered theories of the origin of our solar system are based on the premise that its planets formed almost simultaneously. Part of the evidence for simultaneous origin is the observation that the planets' orbits around the sun lie in nearly the same plane. Had the planets formed independently, we would expect their orbits to occupy different planes.

Precisely when the planets came into being has been a more difficult issue for scientists to resolve. Astronauts report that the most beautiful object they see from space is the earth, whose surface is partially blanketed by swirling white clouds set against the blue background of extensive oceans (Figure 9-3). But while the earth's water is aesthetically pleasing and necessary for life, its abundance near the earth's surface makes rapid erosion inevitable. Continuous alteration of the crust by erosion and also by igneous and metamorphic processes make unlikely any discovery of rocks nearly as old as the earth. Thus, geol-

FIGURE 9-3   The earth viewed from the moon. The swirling white masses are clouds. (*From W.J. Kaufmann, Planets and Moons, W.H. Freeman and Company, New York, 1979.*)

ogists have had to look beyond this planet in their efforts to date the origin of the earth. Fortunately, we do have samples of rock that is generally believed to represent the primitive material of the solar system. These samples are **meteorites**—extraterrestrial objects that have been captured in the earth's gravitational field and have subsequently crashed into our planet (Figure 9-4; see also page 2).

Some meteorites consist of rocky material and, accordingly, are called **stony meteorites** (Figure 9-5A and B). Others are metallic and have been designated **iron meteorites** even though they contain subordinate amounts of elements other than iron (Figure 9-5C and D). Still others consist of mixtures of rocky and metallic material and are thus called **stony-iron meteorites**. Meteorites come in all sizes, from small particles to the small planets known as asteroids; no asteroid, however, has struck the earth during recorded human history. Many meteorites appear to be fragments of larger bodies that have undergone collisions and broken into pieces.

Meteorites have been radiometrically dated by means of several decay systems, including rubidium-strontium, potassium-argon, and uranium-thorium (page 120). The fact that the dates thus derived tend to cluster around 4.6 billion years suggests that this is the approximate age of the solar system. After many meteorites had been dated, it was gratifying to find that the oldest ages obtained for rocks gathered on the surface of the moon also approximated 4.6 billion years. Ancient rocks can be found on the moon because the lunar surface, unlike that of the earth, has no water and is characterized by only weak tectonic activity.

FIGURE 9-4   Meteor Crater, near Flagstaff, Arizona. This huge, fresh crater was created by the impact of a large meteorite only a few thousand years ago. Numerous fragments have been collected nearby. One of these is shown in Figure 9-5. (*Smithsonian Institution.*)

FIGURE 9-5   Meteorites. *A* and *B*. An entire stony meteorite and the polished surface of a stony meteorite. These fell in a meteorite shower near Parral, Mexico, in 1969. *C* and *D*. An entire iron meteorite from Pottstown, Pennsylvania, and the sliced surface of an iron meteorite from Canyon Diablo, Arizona. (*Smithsonian Institution.*)

The universe, however, seems to be much older than the solar system. Stars tend to be clustered into enormous galaxies within the universe, and many of these galaxies have disclike spiral shapes (Figure 9-6). We see our own galaxy, the Milky Way, edge on as a cloudy band of stars, but if it were viewed from a distant perspective along a line of sight perpendicular to the plane of the spiral arms, it, too, would have a spiral shape.

It is now widely believed that the universe is expanding. Evidence of this movement comes from the so-called red shift, a shift toward longer wavelengths in the spectrum of colors that reach us from stars moving away from the earth (Figure 9-7). This phenomenon is an example of the **Doppler effect**, as is the progressive lowering of a train whistle's pitch, as the locomotive recedes from a person standing near the tracks. Studies of red shifts indicate that nearly all galaxies are moving away from each other at extraordinary rates; indeed, calculations have shown that between 10 and 20 billion years ago, all of the galaxies occupied the same spot. This has been taken to indicate the approximate date of the so-called big bang—the primordial explosion of matter from which stars have formed. As we will see, however, the earth and other planets of our solar system formed long after the big bang took place. In order to understand the origin of the earth, it is important that we know something of our solar system's other planets, which formed at the same time as the earth.

THE PLANETS

The planets of our solar system differ from one another in many respects. The earth, for example, has unique features that allow it to support living creatures; no other planet can now sustain life as we know it. Like Mercury, Venus, and Mars, the earth is relatively small and close to the sun (Figure 9-8). All four of these planets are composed of rocky material, but despite their high density, they are so small that they possess little angular momentum (which is the product of mass, velocity, and distance from the center of the sun). The next four planets, which are known as the Jovian planets, are Jupiter, Saturn, Uranus, and Neptune. These planets consist largely of frozen elements and compounds that exist as gases on the surface of the earth—primarily hydrogen and helium for Jupiter and heavier compounds such as ammonia (NH_3) and methane (CH_4) for Uranus and Neptune. Because they have larger masses and lie farther from the sun, Jupiter, Saturn, Uranus, and Neptune have more angular momentum than the earth,

FIGURE 9-6   The configuration of galaxies. *A.* Spiral galaxy M51, which is sometimes called the Whirlpool Galaxy because of its distinctive appearance. A small satellite galaxy is attached. *B.* A spiral galaxy viewed edge on. The arms spiral outward from the central bulge. (*A. Hall Observatories. B. Kitt Peak National Observatory.*)

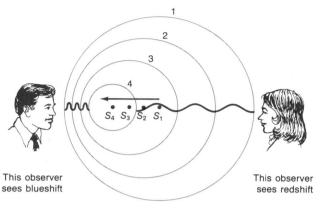

This observer sees blueshift

This observer sees redshift

FIGURE 9-7   The Doppler effect. The points labelled S_1 to S_4 mark successive positions of the light source. A corresponding circle shows the distance light has radiated from each position of the light source. The wavelength of the light is compressed in the direction the light source is moving and stretched in the opposite direction. (*After W. J. Kaufmann, Galaxies and Quasars, W. H. Freeman and Company, New York, 1979.*)

Mercury, Venus, or Mars. Neptune was not discovered until 1846, shortly after its existence was predicted on the basis of its measurable perturbation, by gravitational attraction, of the motion of Uranus. A belt of asteroids in orbit around the sun lies between the inner and outer planets. Pluto, the outermost planet, was not discovered until 1930, and even today much less is known about it than other planets.

It is interesting to compare the earth to the other three inner planets. Although all four planets are rocky, their densities differ. The earth is the second most dense of these planets, with an overall specific gravity of 5.5 (that is to say, it is 5.5 times as dense as water). It is remarkable that Mercury, whose mass is only slightly more than one-twentieth that of earth, has a specific gravity of 5.7, making it even denser than the earth. Mercury's small mass indicates that its gravitational force is much weaker, and thus its inner regions are under less pressure. The fact that Mercury is nonetheless relatively dense suggests that it contains a high concentration of heavy elements such as iron. Externally, Mercury differs greatly from the earth; it has the harshest climate of all planets, with daytime temperatures reaching 425°C and nighttime temperatures dropping to −175°C. In addition, Mercury has virtually no atmosphere. The topography of the planet resembles the surface of the moon, with myriads of craters (Figure 9-9).

Venus is much larger than Mercury. It is often one of the brightest heavenly bodies in our evening sky, orbiting close enough to the sun to reflect intense solar radiation and close enough to the earth to allow us to experience its brightness. Venus is only slightly smaller and less dense

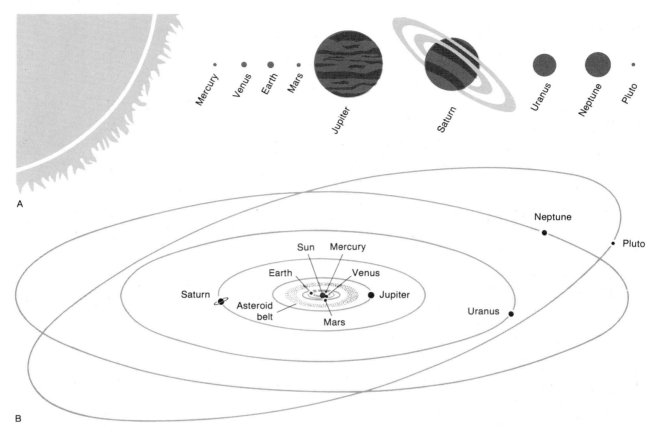

FIGURE 9-8   The solar system. *A*. The relative sizes of the sun and planets; the earth is one of four rocky planets closest to the sun. *B*. A diagram of the planetary orbits. The fact that the orbits lie in nearly the same plane represents evidence that the planets formed simultaneously from a rotating dust cloud. (*After F. Press and R. Siever, Earth, W. H. Freeman and Company, New York, 1982.*)

than the nearby earth and, in fact, is sometimes called the earth's twin. For this reason, it was once believed that Venus's surface might resemble that of the earth. We now recognize, however, that Venus is much hotter than the earth, often reaching temperatures of 450°C. Carbon dioxide predominates in the Venusian atmosphere, while water vapor and free oxygen are rare, and there are also swirling clouds of strongly acidic compounds (Figure 9-10). The carbon dioxide of Venus's atmosphere is present at pressures 90 times greater than those at the surface of the earth. Apparently, this gas has been driven from Venusian rocks by the planet's exceedingly high temperatures. The surface of Venus, like that of Mercury, is heavily cratered.

Mars has a specific gravity of 3.9, which is relatively low. Like Venus, its atmosphere consists mostly of carbon dioxide, but this atmosphere is under low pressure and is much less dense than that of the earth. Under such a low atmospheric pressure, liquid water cannot survive on the surface of Mars, since if introduced it would quickly evaporate; thus, what little water exists on Mars is primarily locked up on ice caps at the poles, where temperatures are about −125°C (Figure 9-11). These polar ice caps, however, differ from those of the earth in that they grow by addition of carbon dioxide "snow" during the Martian winter and shrink by loss of carbon dioxide during the Martian summer. Mars rotates at almost the same rate as the earth; thus, a day on Mars is only 41 minutes longer than a day on earth.

The surface of Mars displays not only numerous craters of varying sizes, but enormous volcanic cones as well. It also exhibits channels that look very much like water-formed features on earth, displaying meanders and what appear to be river mouths enlarged in a downstream direction (Figure 9-12). Although temperatures at the equator approximate what we regard as "room temperature" on earth, the dry conditions that predominate on Mars would appear to be hostile to life as we know it. Still, the presence of abundant frozen water and evidence that it once flowed in a

FIGURE 9-9   Mercury as photographed by the *Mariner 10* spacecraft. Craters are the most conspicuous features. This image is a mosaic of several pictures. (*NASA.*)

FIGURE 9-10   Venus as photographed by the *Mariner 10*. The top of the cloud cover lies 65 kilometers (~40 miles) above the surface of the planet. (*NASA.*)

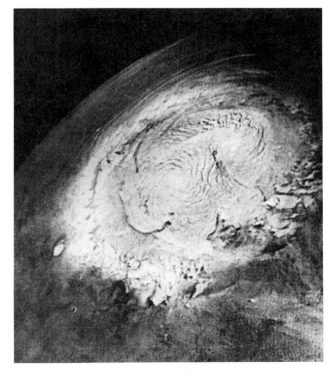

FIGURE 9-11   The north polar cap of Mars, photographed by *Mariner 9* during Martian springtime. This cap shrinks during the Martian summer by evaporation of carbon dioxide. (*NASA.*)

liquid state suggests the possibility that at some time in the past, life may have existed on the so-called red planet. Perhaps the life of Mars will have to be studied by paleontologists rather than by biologists!

ORIGIN OF THE SOLAR SYSTEM AND EARTH

It is interesting to note that we know more about the origin of distant stars than we do about the origin of our own solar system. This is because our solar system formed long ago, and we have no other examples of young solar systems to observe at close range. Scientists can train their telescopes on multitudes of large stars in various stages of development—but unfortunately, these stars are too far

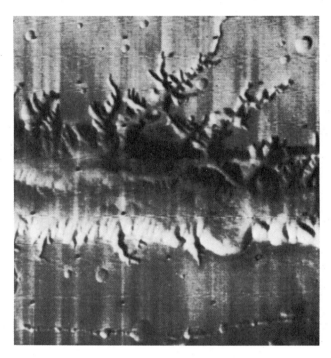

FIGURE 9-12   A segment of Mars's Grand Canyon, which runs about 4000 kilometers (~2500 miles) nearly parallel to the planet's equator. The branching pattern and downstream enlargement of the canyon's tributaries are clearly visible. The main canyon may represent some kind of tectonic rift. (*NASA.*)

away to allow us to observe how planets may be forming in association with them.

At one time, it was widely believed that planets formed when a star passed close to the sun, tearing away matter that subsequently condensed into planetary bodies. Subsequent calculations have shown that matter removed in this way would either fall back to the sun or fly into outer space. Today, most experts favor some variation on the **solar nebula theory**, which holds that the solar system formed from a cloud of particulate matter called cosmic dust. The details of this theory, however, are widely debated.

Galaxies, which are enormous clusters of stars, form by the gravitational collapse of dense clouds of gas (mainly hydrogen) into stars. Our galaxy, which is believed to have originated less than 10 billion years ago, is made up of approximately 250 billion sunlike stars. Even after a galaxy forms by the establishment of some stars, secondary stars are continually born within the spiral arms, where galactic matter is concentrated. Pressure varies in time and space within a spiral arm as matter contracts and expands.

Although all stars form in essentially the same manner, their histories vary according to their size. The sun, for example, is apparently the remnant of a star that imploded, or collapsed violently, to form heavy elements. After this

collapse, a **supernova**—or an exploding star that casts off matter of low density—was formed (Figures 9-13 and 9-14). What remained was a dense cloud that condensed as it cooled after the explosion. This dense cloud, or **nebula** (Figure 9-15), is assumed to have had some rotational motion when it formed and must then have rotated more and more rapidly as it contracted, just as an ice skater automatically spins more rapidly (conserving angular momentum) as he pulls in his arms.

It is possible that the sun formed alone from a nebula and then captured a cloud of cosmic dust that formed the planets. Alternatively, the sun and the planets may have formed from the same dust cloud. What does seem certain, however, is that the planets formed either during or soon after the birth of the sun. In fact, the sun and the planets appear to have originated during an interval of time no longer than about 100 million years. This has been deduced from the excessive concentration in some meteorites of the isotopes xenon 129 and plutonium 244. ("Excessive" in this case means present in an amount greater than is normal for meteorites.) Excessive amounts of these isotopes in a meteorite indicate that some of the short-lived parent isotopes were originally present in the meteorite material and then decayed to xenon 129 and plutonium 224. Thus, the planet or planetlike body of which such a meteorite is a fragment must have formed soon after the elements of the solar system came into being; otherwise, the short-lived parent isotopes of xenon 129 and plutonium 244 could not have been incorporated into the material of meteorites in a sufficiently high concentration to leave excessive amounts of their daughter products. The conclusion is that the sun and the heavy elements that were synthesized in its formation cannot be very much older than the meteorites, the moon, and the planets, which have ages of about 4.6 billion years.

Whether it was captured or formed with the sun by the gravitational collapse of galactic matter, however, the dust cloud that became the planets must have been rotating. Today, the angular momentum of the solar system resides less in the sun than in the planets, as evidenced by the fact that the planets are positioned far from the sun and rotate rapidly even though they contain little mass. It is not known precisely why this distribution of angular momentum developed, but it is known why the outer planets formed from volatile (i.e., easily vaporized) elements. It would appear that these elements were expelled from the hot inner region of the nebula to solidify in colder regions far from the sun. More dense materials were left behind, and these formed the inner planets, including the earth. When the rotating dust cloud reached a certain density and rate of

A

B

FIGURE 9-13 The spiral galaxy NGC 4725 as it appeared in 1931 (*A*) and on May 10, 1940 (*B*), when the star to which the arrow points erupted into a supernova. (*Hale Observatories*.)

FIGURE 9-14 The Crab nebula, the beautiful remnant of a supernova explosion that was observed by Chinese astronomers 900 years ago. (*Lick Observatory*.)

FIGURE 9-15   A nebula in which stars are being born within the constellation Orion. Nebulas such as this are the visible features that outline the arms of spiral galaxies. (*Lick Observatory*.)

rotation, it automatically flattened into a disc, and the material of the disc then segregated into rings, which later condensed into the planets (Figure 9-16). Each planet began to form by the aggregation of material within one of these rings. The aggregates eventually reached the proportions of asteroids, which subsequently coalesced to form planets.

It would appear that the sun passed through the so-called T Tauri stage that some other stars can be seen to represent today. This stage is an early interval in the history of a star during which matter is emitted at an exceedingly rapid rate (Figure 9-17). The powerful "solar wind" that characterizes this stage would have driven a large percentage of preexisting hydrogen and helium (light elements) out of the solar system. Obviously this did not occur until the outer planets had already trapped the hydrogen and helium that forms so much of their bulk.

After the planets formed, asteroids also remained in orbit around the sun. While some of these asteroids—including those that form the asteroid belt (Figure 9-8)—survive to this day, most have been swept up by larger planets. Others have undoubtedly had their motions so severely disturbed by near collisions that they have passed out of the solar system.

The origin of the moon, which circles the earth, is still debated. Was the moon an independent body that was captured by the earth's gravitational field? Was it produced by the separation of a large chunk of the earth? Or was it accreted from matter in orbit around the primitive earth? Lunar rocks obtained in the Apollo space program exhibit an isotopic composition of oxygen that is remarkably similar to that of terrestrial rock. This suggests that the moon formed near the earth from the same portion of the solar nebula. (Meteorites, in contrast, which are presumed to form in a variety of regions, show widely varying isotopic compositions of oxygen.) Thus, we have some evidence that the moon is not a totally foreign body. If it was captured, it was previously a near neighbor.

HOW THE EARTH AND ITS FLUIDS BECAME CONCENTRICALLY LAYERED

The ideas described earlier in this chapter explain the manner in which our planet may have come into being, but they do not account for many of the earth's particular features. How, for example, did the concentric structure of the solid earth and the fluids above it develop? How did metal and rock become segregated into core, mantle, and crust? Have there always been continents and ocean basins? How did the oceans and the atmosphere develop? These are questions that we will consider in this section.

The core, mantle, and crust

The concentric layers of the earth, from the core to the atmosphere, are arranged according to density, with the least dense layer, the atmosphere, situated on the outside. To understand how this layering may have developed, let us begin with the earth's core, which is thought to be composed primarily of iron (page 14). The traditional view has been that the earth formed by **homogeneous accretion**—that is, that the condensed material that formed the planet initially aggregated in a haphazard fashion. When we begin with this assumption, however, our problem is to explain

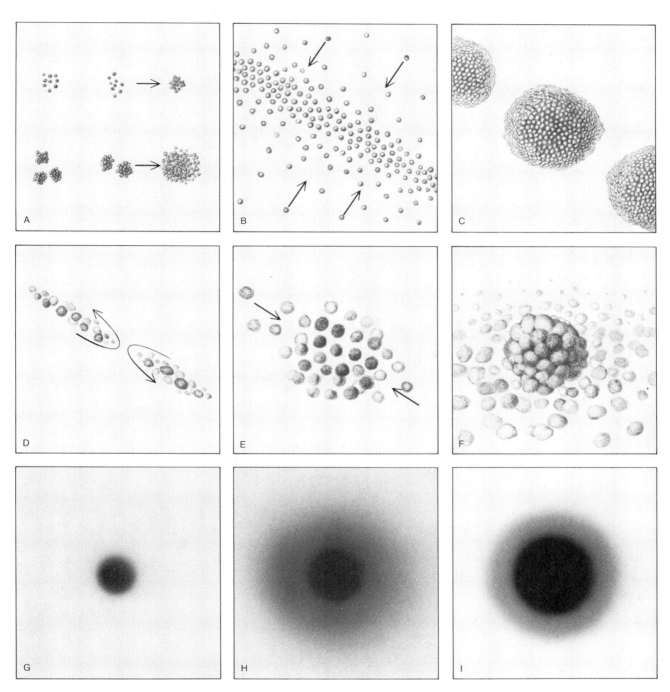

FIGURE 9-16   How a planet forms. The process begins when dust grains collide and stick together, forming larger and larger clumps (A). The clumps move toward the central plane of the nebula (B) and collect into bodies the size of asteroids (C). These bodies gravitate into clusters (D), collide (E), and form the nucleus of a planet (F). The planet grows (G), and as its mass increases, it may attract gas from the nebula (H). If the planet is large enough, it may attract so much gas and draw it in so closely that the gas forms a dense shell representing most of the planetary mass (I). (From A. G. W. Cameron, "The Origin and Evolution of the Solar System." Copyright © 1975 by Scientific American, Inc. All rights reserved.)

FIGURE 9-17   A cluster of stars and gas within the constellation Monoceros. Many of these stars are in the T Tauri stage, in which powerful stellar winds issuing from them are blowing away nearby nebular material. (*Hale Observatories.*)

how iron and nickel, which are dense, abundant metals sank to the earth's center to form the primary constituents of the core. The explanation would be simple if the primordial earth were known to have been a molten mass, in which case the high-density materials would automatically have sunk toward the center while less dense components would have floated toward the surface and formed the crust. The mantle, which is of intermediate density, would then have remained between crust and core.

Was the earth initially molten? The earth's early thermal history is difficult to evaluate because its temperature of formation is assumed to have been controlled by the rate of accretion of earth materials, about which little is known. Rapid accretion would concentrate a great deal of heat, but slow accretion would allow components to cool while falling into place. Many experts currently believe that the earth accreted components in a solid state—but even if this was the case, the earth must have developed a liquid interior when it heated to a critical temperature upon reaching a certain size. At this point, gravitational collapse would inevitably have moved heavy elements (primarily nickel

and iron) to the center. Such contraction would then have released an enormous amount of energy—enough, some would claim, to raise the temperature of the earth by as much as 1200°C! Under such conditions, the mantle would also have segregated in a liquid state, and radioactive elements (especially isotopes of uranium, thorium, potassium, and rubidium) would then have moved to positions near the surface as a consequence of their tendency to combine chemically with other elements in low-density silicate minerals.

It is not universally accepted that the earth's core formed secondarily by liquefaction and gravitational collapse. Some have suggested that the primordial earth formed instead by **inhomogeneous accretion**, with a naked core developing first as a dense nebular condensate upon which the less dense silicates of the mantle and crust later collected. What is certain, however, is that the earth's core formed early in the planet's history. Remnant magnetism found in the oldest rocks now recognized, which are about 3.8 billion years old, has established a minimum age for the earth's magnetic field—which, as we have seen, is generated in the outer, liquid portion of the core.

The atmosphere   The asteroids that coalesced to form the earth were too small for their gravitational fields to have held gases around them as atmospheres. We can thus conclude that the earth did not inherit its atmosphere from these ancestral bodies; instead, the gases that the primitive earth retained as an atmosphere must have been emitted from within the planet after it formed, many of them while it was in a liquid state so that they could easily escape to the surface. If the core, mantle, and crust became differentiated when the earth first became liquefied, extensive **degassing**—or loss of gases to the earth's surface—would have accompanied this differentiation.

Degassing by way of volcanic emissions has continued to the present, albeit at a much lower rate than early in the earth's history. The chemical composition of gases released from modern volcanoes provides a general picture of what the early atmosphere contained: primarily water vapor, hydrogen, hydrogen chloride, carbon monoxide, carbon dioxide, and nitrogen. In the modern world, plant photosynthesis is responsible for most of the oxygen in the atmosphere today. In the absence of plants, little oxygen entered or formed within the early Archean atmosphere, and these small quantities were quickly removed by oxidation of iron minerals and other materials at the earth's surface. The early atmosphere would thus have been inhospitable to animal life.

Hydrogen and helium are the only elements of low enough density to have escaped from the earth's gravitational field during the period of rapid degassing. It is therefore strange that more dense volatile elements and compounds, such as carbon, nitrogen, water, and noble gases (neon, argon, and their chemical relatives), are also very rare in the earth and its atmosphere today in comparison to their abundance elsewhere in the solar system. Clearly, these more dense volatiles must have escaped from planetary condensates before the latter coalesced to form the earth. Although they are rare, carbon, nitrogen, and water are more abundant in the earth than the other volatiles, and it seems evident that they were preferentially retained as a result of being partially locked up in solid compounds. Noble gases rarely combine with other elements.

The oceans   The rapid degassing of the liquid earth released hot clouds of water vapor. Initially, the great heat of the earth would have kept the water in a gaseous state. Only when the planet cooled to the point at which its surface temperatures fell below 100°C (boiling point of water) would water have fallen as rain and remained on the earth as lakes, rivers, and oceans.

Like modern rain, the rains that formed the earliest oceans are assumed to have contained few salts. Salts accumulated in early sea water by the reaction of the water and carbon dioxide dissolved within it with natural minerals such as clays and carbonates. Calculations show that sea water should have attained a salinity comparable to that of the modern world early in Archean time. Since that time, salts have been precipitated on the sea floor as evaporite sediments about as rapidly as they have been added to the oceans, and thus the salinity of sea water has not varied greatly. At the same time, the global hydrological cycle has moved water but not salts from the oceans to the atmosphere and back again (Figure 2-8).

Continental crust   As we learned in Chapter 1, continents consist primarily of thick felsic crust and are surrounded by the thinner mafic crust of ocean basins. An important question is how the two kinds of crust came into being. To understand the answer, it is important to recognize that the earth's interior cooled from its fully molten state primarily by means of convection, which carried hot material to the surface. The primitive crust probably formed rapidly during this brief interval of cooling as magma flowed to the surface at a high rate. Most of the early crust was composed of mafic material that rose from the denser ultramafic mantle. Mafic materials at the earth's surface weather more rapidly

than do felsic materials (see Appendix I). The result in early Archean time was that erosion and weathering of mafic rocks standing above sea level left a residue of felsic minerals, especially clay. These minerals, having accumulated along the margins of mafic islands, were then turned into felsic crystalline rocks by the metamorphism and melting that accompanied continuing igneous activity. It is assumed that quartz was an abundant mineral in many of these felsic rocks.

Inasmuch as mafic minerals are preferentially destroyed and are not replaced by other minerals, the sequence of weathering, erosion, metamorphism, and remelting serves, in effect, as a natural "machine" for the production of granite and other felsic crystalline rocks of the sort that form most continental crust. This machine began to operate as soon as the crust had cooled enough for felsic igneous and metamorphic rocks to form within it. (Recall that felsic rocks form at lower temperatures than do mafic rocks; see Appendix I.) Since that time, the machine has continually increased the volume of continental crust by extracting felsic material from more mafic material—primarily mafic material that has moved up from the mantle to form oceanic crust.

Continental crust was probably present earlier than 4 billion years ago, but doubts have been expressed about the validity of the few radiometric ages that have been found to be this great. Of the Archean rocks whose ages are widely accepted, the oldest are metamorphic rocks of various kinds found in a small area near the southern tip of Greenland. Among these are rocks that were metamorphosed about 3.8 billion years ago after they had accumulated as sediments sometime earlier (Figure 9-18). In fact, metamorphism has undoubtedly altered many bodies of rock that are older than 3.8 billion years and reset their radiometric clocks to give younger ages. Other early Archean rocks have been remelted or destroyed by erosion.

THE GREAT METEORITE SHOWER

Before moving to a general discussion of surviving Archean rocks, we will consider what must have been one of the most remarkable episodes in earth history: the pelting of the earth by large numbers of meteorites and asteroids over a period of several hundred million years. Given the rarity of early Archean rocks, our only evidence that this extraterrestrial shower took place is provided by other planets and, particularly, by the moon. These other bodies of the

FIGURE 9-18   Some of the oldest known rocks on earth. These banded iron formations from the Isua area of southern Greenland were deposited as sediments at least 3.8 billion years ago. The individual beds, which are deformed and weakly metamorphosed, average a centimeter or so in thickness. Beds of chert (fine-grained quartz) alternate with darker beds composed of iron minerals. (*P. E. Cloud, Scientific American, September 1983.*)

solar system have not undergone the weathering and erosion that constantly alter the surface of the earth.

Earthbound observers have long commented on the moon's pockmarked appearance. The moon's craters, which are known as **maria**, (singular, mare), the Latin word for "seas," were first sketched by Galileo. It is only from manned lunar exploration and from photographs provided by artificial satellites, however, that we have gained detailed knowledge of these enormous craters. The maria that face the earth have an average diameter of approximately 200 kilometers (~125 miles) and are distributed in a crescent-shaped pattern (Figure 9-19). Initially, it was not known whether the maria were craters produced by volcanoes or were indeed the impact scars of huge meteorites, but meteoritic origin has now been established by detailed study of the maria and neighboring lunar terrane. The entire surface of the moon is, in fact, scarred with craters, most of which are much smaller than the enormous maria. The lunar highlands surrounding the maria consist of rock fragments that testify to the pulverization of the lunar crust by the impact of falling meteorites. The maria facing the earth

FIGURE 9-19   Crescent-shaped arrangement of dark maria on the side of the moon that faces the earth. These maria are craters produced by the impact of huge meteorites between 3.9 and 4.6 billion years ago, when the earth must have been similarly bombarded. (*Lick Observatory.*)

are floored by immense flows of dark volcanic basalts, which give the maria their dusky appearance. These basalts are themselves scarred by smaller craters, and the relative ages of successive lava flows can be deduced from the density of cratering (Figure 9-20).

The dating of associated rocks has revealed that most large lunar craters are quite old, ranging in age from slightly less than 4.0 billion years to about 4.6 billion years. Many rocks of the lunar highlands exhibit ages of approximately 4.0 billion years, and it is believed that their radiometric clocks were reset some 600 million years after the moon formed and that the highlands formed with the rest of the moon about 4.6 billion years ago. The clocks may have been reset by the intense meteorite bombardment whose effects are so evident. (During collisions, the kinetic energy, or energy of motion, of a meteorite would have been turned into heat that melted rocks.)

It has been estimated that during the early cataclysmic interval of lunar history, the frequency of meteorite impacts was more than 1000 times the present rate. Radiometric dating shows that basalts flooring the lunar maria tend to be slightly younger than the maria themselves; most of these basalts range in age from 3.2 to 3.9 billion years. The relatively modest cratering of the basalts shows that shortly after 4 billion years ago the intensity of meteorite showering subsided to something like its present low level.

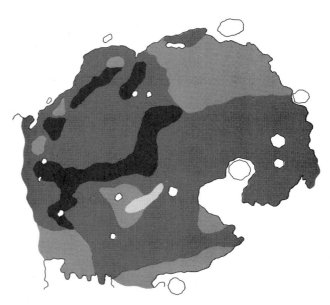

FIGURE 9-20   Recognition of the relative ages of lava flows in Mare Imbrium. The lava flows that have the highest density of cratering are the oldest. These are indicated by the darkest pattern in the diagram on the right. The lightest pattern repre-

sents the least densely cratered (youngest) lava flows. (*From J. A. Wood, "The Moon." Copyright © by Scientific American, Inc. All rights reserved. Photograph by Lick Observatory.*)

Not only the large maria but most of the lunar craters have been shown to have formed early in the history of the solar system. We can see that craters of all sizes also formed on planets of the solar system whose surfaces have not since been as heavily altered as that of the earth (Figure 9-9). These craters formed during a time when the solar system was, in effect, being cleared of many small remaining pieces of solid material that had consolidated from the solar nebula but had not been assimilated into planets or captured as planetary satellites. It seems an inescapable conclusion that the primitive earth was subjected to the same kind of meteorite bombardment as was its neighboring moon. The composition and structure of the earth's crust must have been altered by the material contributed by this enormous shower of meteorites. Unfortunately, however, the rock record that might have provided us with details has been all but destroyed by igneous and metamorphic activity as well as by weathering and erosion.

ARCHEAN ROCKS

Before the advent of radiometric dating, scientists found it difficult to establish even relative ages for Precambrian rocks. Early on, however, geologists concluded on the basis of the principle of superposition (page 17) that a particularly distinctive association of rocks seemed to repre-

sent the oldest Precambrian interval, which became known as the Archean.

Most Archean rocks can be divided into two groups: the even-grained, high-grade metamorphic rocks known as **granulites** and the bodies of rock known as **greenstone belts**. Greenstone belts, which are unique to the Archean Eon, are dominated by volcanic rocks and by sedimentary rocks that are derived primarily from volcanics. The rocks of greenstone belts are frequently metamorphosed, and the metamorphic mineral chlorite gives them the greenish color from which they derive their name. Granulites, many of which have formed since Archean time as well as during it, result from such severe metamorphism that they teach us little about environments of the Archean. Greenstones are more revealing, and it is upon them that we will focus in this section.

Greenstones and ancient sediments

Archean greenstones include or are associated with a limited variety of rocks characterized by a particular structural pattern. The composite features of greenstone belts are without counterpart in the modern world and thus reveal that the earth's crust in Archean time differed from what it is today.

The Canadian Shield and areas to the south include several small areas of Archean terrane and three large areas: the Superior, Slave, and Wyoming provinces (Figure 9-21). The largest uninterrupted greenstone belt here is the Abitibi Belt of southern Ontario and Quebec (Figure 9-22). The Abitibi greenstones resemble those found in other regions of the world in that they occur as podlike bodies of volcanic rock associated with sedimentary deposits. In cross section, a typical greenstone pod is seen to have the configuration of a syncline (Figure 9-23). Granitic rocks often surround a body of greenstone, and sometimes they also intrude it, proving that they are of younger age.

Typical greenstone basins such as those depicted in Figure 9-24 reveal an interesting pattern: Volcanics at the base of the sequence are more mafic than those at the top. In the Canadian Shield, the upward trend is normally from mafic to felsic, but older greenstone belts (those exceeding 3 billion years in age) tend to display a fuller sequence, ranging upward from ultramafic through mafic to felsic. Ancient greenstones that include basal ultramafics are especially well exposed in Western Australia and southern Africa (Figure 9-24). Figure 9-25 illustrates the sequence of development of a major greenstone belt of southern Africa. In all cases, sedimentary rocks are most common at or near the top of the volcanic sequence.

The typical ultramafic-to-felsic sequence may reflect the operation of the "machine" for the production of sialic crust described earlier in this chapter. In this machine, weathering rapidly destroys ultramafic and mafic minerals, leaving a residue of felsic products such as clays. Some of these felsic minerals, when buried deep within the earth, become incorporated by magma. As greenstone belts evolved, their magmas may have incorporated more and more sialic materials as these materials continued to form.

The sediments associated with the Archean volcanics represent a strikingly consistent suite of rock types. Like the greenstones, they have been metamorphosed and have often been extensively invaded by granite as well. Even so, primary sedimentary features can be distinguished, and they reveal that the original deposits consisted primarily of graywackes, sediments of volcanic origin, and dark mudstones now altered to slate. Recall that graywackes and dark mudstones of post-Archean age often represent deep marine environments adjacent to continents. The Archean examples occur both in large belts adjacent to the greenstone synclines and within the synclines themselves (Figure 9-22). The volcanics characteristically display relict pillow structures, indicating submarine cooling of the original lava (Figure 9-26). The volcanic belts may have extended along the margins of early landmasses, whereas mud-

FIGURE 9-21   Distribution of identifiable Archean rocks in North America. Their largest areas of exposure are in the Canadian Shield. The three largest Archean provinces of North America are labeled the Slave, Superior, and Wyoming provinces. (*Modified from W. R. Muehlberger et al., Amer. Assoc. Petrol. Geol. Bull. 51:2351–2380, 1967.*)

stones and graywackes formed seaward in basins. Certainly the mudstones represent the "background" sedimentation of sedimentary basins, while the graywackes represent turbidite deposits containing material brought in from shallower water.

Other characteristic but less abundant Archean sediments are coarse conglomerates containing large, rounded cobbles that suffered considerable stream or beach abrasion (Figure 9-27). These conglomerates are not crossbedded like stream gravels, however; they seem instead to have been dumped into place, perhaps by enormous turbidity flows traveling down steep slopes. It is therefore not surprising that the conglomerates tend to occur adjacent to or folded into greenstone belts; they seem to represent nearshore facies. The conglomerates contain pieces of greenstone, but also pieces of granites and other plutonic rocks from unknown source areas. While very few granitic rocks now found in the Archean terrane of the Canadian Shield are older than the sediments, the granitic cobbles in the conglomerates seem to point to the presence of some ancient granitic crust located either between sedimentary basins or on islands within them. The graywackes, like the

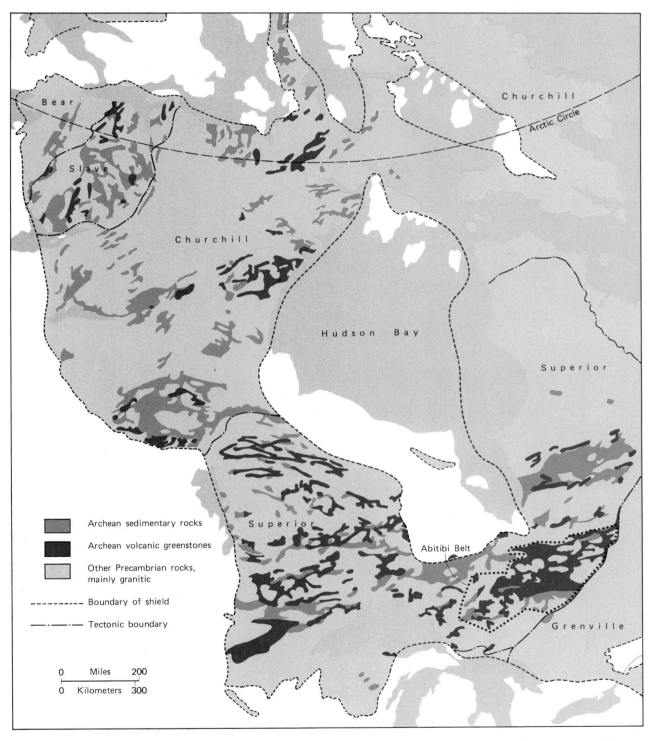

FIGURE 9-22   Archean volcanic and sedimentary rocks of the Canadian Shield. The volcanic rocks underwent metamorphism and are preserved as greenstones. The Abitibi Belt contains the largest concentration of greenstone units in the shield. (*After A. M. Goodwin, in B. F. Windley [ed.], The Early History of the Earth, John Wiley & Sons, Inc., New York, 1976.*)

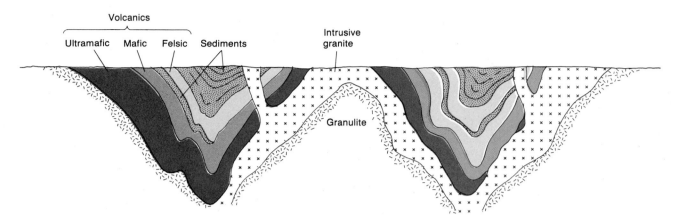

FIGURE 9-23   Diagrammatic cross section of two adjacent Archean greenstone "pods" within granulitic terrane. Each pod has a synclinal configuration. The oldest greenstone volcanics are ultramafic in composition, and these are overlain by mafic and then felsic volcanics; sediments are most abundant above the volcanics. Granites have invaded the volcanic rocks.

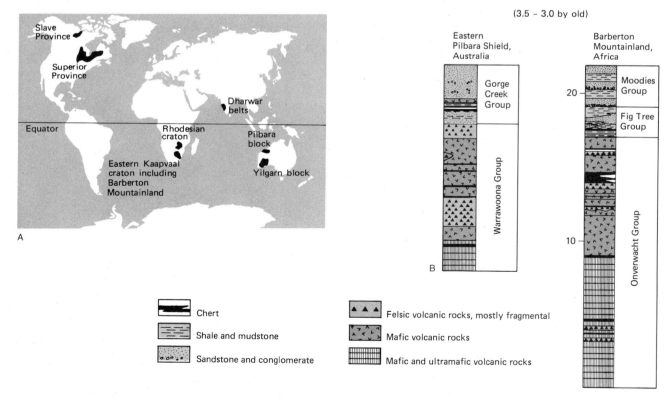

FIGURE 9-24   Shield areas containing well-preserved greenstone belts (A) and representative stratigraphic sections through two greenstone sequences (B). In both of these sequences, there is a transition from ultramafic and mafic volcanics to felsic volcanics. (After D. R. Lowe, Ann. Rev. Earth and Planet. Sci. 8:145–167, 1980.)

North   Sea level   South

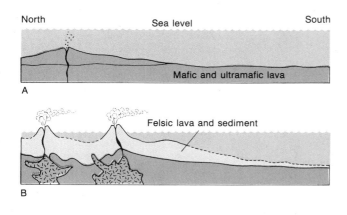

Mafic and ultramafic lava

A

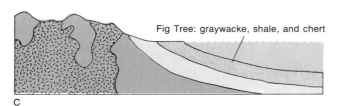

Felsic lava and sediment

B

Fig Tree: graywacke, shale, and chert

C

FIGURE 9-25   Formation of a greenstone sequence: The geologic history of Barberton Mountainland during deposition of the Onverwacht and Fig Tree Groups (see Figure 9-24). *A.* The initial phase of ultramafic and mafic volcanism, which produced the lower Onverwacht. *B.* Building of a platform by the felsic extrusions and sediment accumulation that formed the upper Onverwacht. *C.* Deposition of the Fig Tree sediments as a result of uplift and erosion in the south and a deepening of the basin. (*After D. R. Lowe and L. P. Knauth, Jour. Geol. 85:699–723, 1977.*)

FIGURE 9-26   Archean pillow lava, Sioux Lookout area, northwestern Ontario. The "pillows" have been planed off by erosion. Compare to Figure AI-10, which shows much younger pillow lavas. (*Courtesy of F. J. Pettijohn.*)

granitic cobbles, contain abundant quartz, indicating derivation from felsic crystalline rocks.

In summary, the Archean rocks seem to indicate the presence of basins of moderate depth flanked by volcanoes that spewed out lava and by small, steep-sided blocks of felsic crust. Under ordinary circumstances, muds were deposited offshore, but turbidity flows periodically dumped conglomerates near the shore and carried finer erosional material (future graywacke) out toward the centers of basins.

Cherts and iron-rich sedimentary rocks known as **banded iron formations** are also found in the Archean sedimentary belts. These structures are occasionally widespread but are seldom of great thickness. Banded iron formations are rare in Phanerozoic rocks but became far more prevalent shortly after the end of Archean time, and thus they will be discussed at some length in the following chapter. It is notable, however, that the banded iron formation at Isua, West Greenland, represents one of the oldest known bodies of rocks on earth. Like other banded ironstones of the Precambrian, it consists of iron-rich layers alternating with

quartz layers (Figure 9-18). The Isua rocks are believed to have originated by chemical precipitation in marine basins, and the quartz within them is thought to have existed initially as chert precipitated from sea water.

The rarity of nonmarine and continental-shelf deposits

It is a striking fact that most Archean sediments are of deep-water origin. These include graywackes, mudstones, iron formations, and sediments derived from volcanic activity. In contrast, carbonates and quartz sandstones, which are abundant in younger sedimentary sequences, are rarely found in Archean sediments (Figure 9-28). Today, of course, carbonates and clean sands accumulate extensively in shallow marine settings along the continental shelves of large cratons.

While sediments of shallow seas are rare in Archean terranes, terrestrial and freshwater deposits are virtually unknown. Two hypotheses can be proposed to explain this

FIGURE 9-27   Coarse conglomerate of the Archean Manitou Series, Mosher Bay, northwestern Ontario. The pebbles, cobbles, and boulders in these rocks testify to the presence of continental crust nearby. (*Courtesy of F. J. Pettijohn.*)

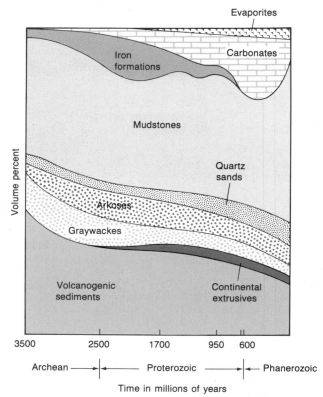

FIGURE 9-28   Changes in the relative proportions of rock laid down on cratonic surfaces in the course of geologic time. Notice the increase in evaporites, carbonates, and quartz sands after the Archean. (*Modified from F. T. Mackenzie, in J. P. Riley and G. Skirrow [eds.], Chemical Oceanography, Academic Press, London, 1975. Based on data from A. B. Ronov.*)

phenomenon: (1) such sediments failed to form extensively during Archean time, suggesting that terrestrial and freshwater environments were rare; or (2) such deposits actually formed in many areas but were later almost entirely destroyed. How can we evaluate these two possibilities? The answer can be found in the relative vulnerability of various kinds of sediments to erosion.

All sedimentary deposits tend to be cycled through the earth's crust; whether they become lithified or not, they are likely to be destroyed following deposition by erosion, igneous activity, or metamorphism. This is especially true of sedimentary evaporites, which are vulnerable to chemical solution and thus tend to be destroyed even more rapidly than sedimentary silicates. Certainly this may partially account for the rarity of evaporites in the Archean record.

Quartz sandstones and carbonate rocks, on the other hand, are chemically and physically resistant rocks, and thus a lack of inherent durability cannot explain their scar-

city in Archean terranes. One possible reason for this scarcity is that Archean cratonic deposits, including quartz sands, formed at higher elevations than basin deposits and were thus particularly vulnerable to erosion. We have seen, however, that in more recent times, crustal subsidence has caused shallow marine sediments such as those of the Bahamas (page 96) and nonmarine sediments such as those of rift valleys (page 198) to be buried quite deeply in very short spans of time. Why, then, were no similar sequences preserved in Archean time? Preservational bias may partially account for the scarcity of Archean cratonic sedimentary deposits (especially evaporites), but other factors must also have been operating. These appear to have been related to the tectonic framework of the Archean world: There are no widespread continental or continental-shelf deposits in the Archean record simply because there were no large continents during most of Archean time. This condition will be the topic of the following section.

ARCHEAN TECTONICS

The lack of evidence supporting the existence of large continents, taken together with the floodlike character of many Archean pillow basalts and the relative abundance of graywackes, has led many workers to postulate that cratonic bodies existed in the Archean Eon only as small, steep-sided felsic protocontinents along which conglomerates accumulated. If such protocontinents were present, they must have been separated by numerous marine basins that accumulated lava and volcanic sediments to become greenstone belts. It is not clear whether the sedimentary basins were semi-isolated structures scattered over an ancient crust or bodies whose waters covered most of the globe and were interrupted only by protocontinents of felsic crust or volcanic material scattered here and there. The second configuration would account for the unusual sedimentary record of the Archean Eon; if basins covered most of the earth's surface, conglomerates and graywackes would have been derived from the erosion of scattered felsic protocontinents, and the absence of extensive shelf deposits (carbonates and quartz sandstones) would reflect the absence of large continental shelves.

The history of radioactive decay supports the notion of a fragile, mobile Archean crust with many areas of weakness forming basins between or around small protocontinents. Recall that the earth's heat source is constantly diminishing through time as radioactive isotopes decay without renewal (page 21). Because these decay rates are constant, geologists can calculate the approximate difference between the earth's rate of heat production today and those of times past. Such estimates show that during the Archean, the rate of heat flow to the earth's surface was much higher than it is today. Near the end of Archean time, the total rate of heat production was perhaps twice as high as it is now (Figure 9-29), and earlier it was even higher. This substantial rate of heat flow must have resulted in a less stable crust. Under such conditions, lavas may have burst quickly to the surface, flooding basinal areas that we now see as greenstone belts.

LARGE CRATONS APPEAR

Radiometric dating shows that during early Proterozoic time, large bodies of magma were intruded into cratonic rocks. It can therefore be concluded, in accordance with the principle of intrusive relationships, that cratons of moderate

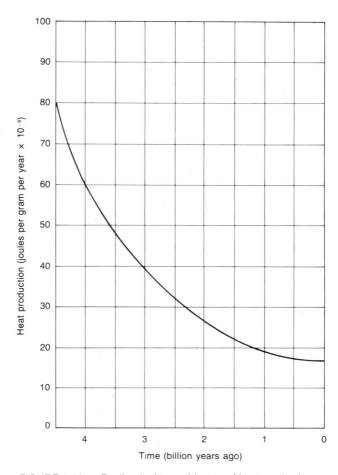

FIGURE 9-29   Decline in the earth's rate of heat production through time. (*After W. H. K. Lee, PhD thesis, University of California at Los Angeles, 1967.*)

proportions had come into being very late in Archean time. The hardened magma often forms upright tabular structures, or dikes (see Appendix I), that range in age from about 2.1 to 2.5 billion years. Large mafic dikes of this age, often occurring in swarms associated with other mafic intrusives, have been identified in North American, Greenland, the Baltic Shield, South America, southern Africa, India, and Australia. In many parts of the world, evidence also indicates that major metamorphic episodes occurred slightly earlier, about 2.7 to 2.3 billion years ago. Geologists do not yet understand the origins of this metamorphic interval, but they do know that it resulted in the resetting of many radioactive clocks and in the consolidation of many crustal elements into sizable cratons.

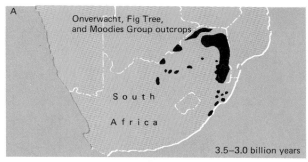

FIGURE 9-30   The distribution of some of the world's oldest known extensive cratonic rocks, in the Archean of southern Africa. Siliciclastic deposits of the Pongola Basin (B) and Witwatersrand Basin (C) accumulated on a broad crust composed partly of greenstone sequences of the Onverwacht, Fig Tree, and Moodies groups (A). (Adapted from C. R. Anhaeusser, Philos. Trans. Roy. Soc. London A273:359–388, 1973.)

There is also evidence, however, that "cratonization" did not occur simultaneously throughout the world. In most areas, typical Archean greenstone associations formed until approximately 2.5 billion years ago, but in southern Africa a large craton was already present about a half million years earlier (Figure 9-30). Here, the Barberton Mountainland greenstone sequence (Figures 9-24 and 9-25), whose age exceeds 3 billion years, is immediately overlain by a substantial body of clastic sediments known as the Pongola Supergroup, and this is followed by the Witwatersrand sequence, which is famous for the gold deposits that accumulated as detrital material within it (Figure 9-30). The Pongola Supergroup consists of deposits that are 3 billion years old but are nonetheless strikingly similar to intertidal

sequences of younger portions of the stratigraphic record (Figure 9-31). The presence of a broad intertidal belt during Pongola deposition indicates that a sizable land mass existed 3 billion years ago in southern Africa.

The Witwatersrand sequence offers further evidence that cratons did not appear simultaneously. This sequence, whose sediments range in age from about 2.5 to 2.8 billion years, consists of alternating coarse-grained and fine-grained sediments that cover about 40,000 square kilometers and have a maximum thickness of nearly 8 kilometers (~5 miles). These deposits accumulated in nonmarine environments on the surface of a craton. The coarsest sediments, which are widespread conglomerates, are most common in the upper division of the Witwatersrand, and it is from these sediments that gold is mined (Figure 9-32). Gold is a very dense metal and thus tends to settle from moving water with much larger particles. During Witwatersrand deposition, bits of gold accumulated with larger silicate pebbles. The gold-bearing gravels apparently formed bars within braided streams (page 76) that seem to have washed enormous quantities of sedimentary debris from highlands across alluvial fans into one or more lakes (Figure 9-33). Lake environments were the sites of accumulations of mud and, occasionally, of sand as well. The alternation of coarse and fine sediment throughout the Witwatersrand seems to reflect periodic uplifts in the source area followed by rapid erosion and deposition of gravels.

It is remarkable that the Witwatersrand sequence, like the Pongola Supergroup, has largely escaped metamorphism. Both in thickness and in areal extent, the Witwatersrand is rivaled by no other recognized Archean sequence of shallow-water sedimentary deposits. It could only have formed along the margin of a sizable landmass. The idea that a substantial craton existed is supported by the occurrence below the Witwatersrand of large granitic bodies of rock of the type that characterize younger cratons. The ages of these granites range back to perhaps 3.4 billion years. The Barberton Mountainland sequence (Figure 9-24) offers additional evidence that cratonization began in southern Africa over 3 billion years ago. Here, the transition from the Onverwacht Group to the Moodies Group records a shift from deep-water deposition to essentially continental, clastic deposition over a fairly broad area. Thus, it appears that the region we now call southern Africa may have been one of the first modern cratons to form. As we will learn in the next chapter, many large continents were in existence a few hundred million years after the formation of the South African craton and the familiar tectonic mechanisms of the Phanerozoic Eon were widespread.

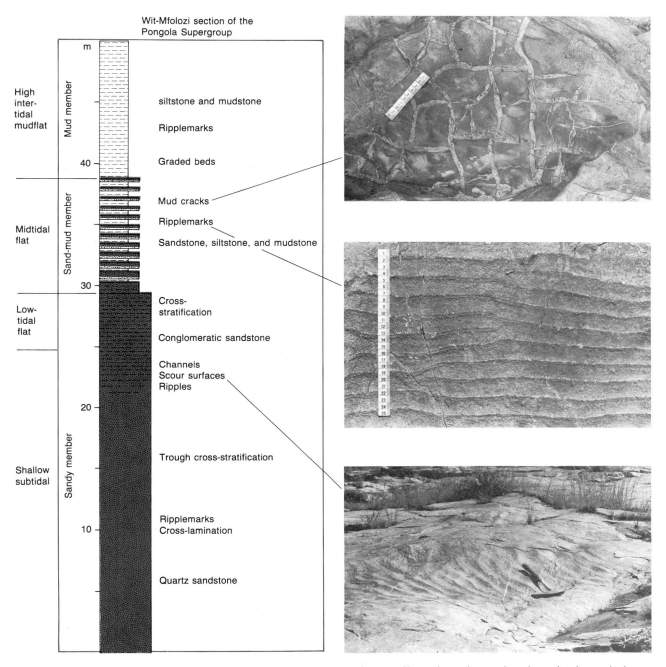

FIGURE 9-31 Depositional environments represented within a stratigraphic section of the Pongola Supergroup of southern Africa. The section represents a 3-billion-year-old regressive sequence in which a tidal flat prograded seaward over subtidal environments. In the lower, sandy member, cross-stratification and symmetrical ripples are common. Some of these ripples are double-crested, reflecting the ebb and flow of tides. Tidal channel deposits floored by pebbles are present, particularly in the upper part. Above them, the sand-mud member bears ripples and mud cracks that suggest a shallower, midtidal environment. The uppermost mud member bears smaller mud cracks as well as mud chips and appears to represent a high intertidal mud flat that lay landward of the zones where sand settled from tidal waters. (*After V. von Brunn and D. K. Hobday, Jour. Sedim. Petrol. 46:670–679, 1976.*)

FIGURE 9-32   Casts of ripples on the ceiling of a gold mine within the Witwatersrand sequence, South Africa. (*Courtesy of R. B. Hargraves.*)

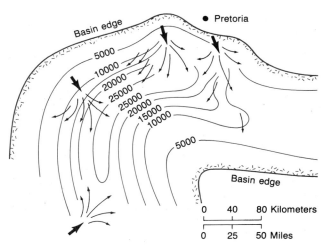

FIGURE 9-33   Thickness (in feet) of siliciclastic sediments of the Witwatersrand Basin of South Africa (see Figure 9-30). The positions of alluvial fans near the source area are shown by arrows. (*After F. J. Pettijohn et al., Sand and Sandstone, Springer-Verlag, Berlin, 1972. Based on B. B. Brock and D. A. Pretorius.*)

Even at the close of Precambrian time, the large continents remained unusual in one important respect: They were barren of advanced forms of life. Long before the first large cratons existed, however, living cells had begun to populate the marine realm, where they remained at a primitive stage of development for perhaps a billion years or more. These early forms of life are the subject of the following section.

ARCHEAN LIFE

Of all the planets in our solar system, only the earth is well suited to life as we know it. One of the reasons is that its size is right. On a much larger planet, the gravitational pull on the atmosphere would be so great that the resulting atmospheric density would exclude sunlight, which is the fundamental source of energy for life. A much smaller planet, on the other hand, would lack sufficient gravitational attraction to retain an atmosphere with life-giving oxygen. In addition, the earth's temperatures are such that most of its free water is liquid, the form that is essential to life. Even Venus, our nearest neighbor closer to the sun, is much too hot to allow water to survive in a liquid state, whereas Mars, our nearest neighbor farther from the Sun, has an atmosphere so thin that liquid water would evaporate from the planet's surface almost immediately.

In this section, we will first examine the fossil evidence of early life on earth, and we will then review some of the fundamental chemical steps in the origin and early evolution of life. Finally, we will consider possible scenarios for the evolution of the earliest single-celled organisms.

Fossil evidence

We can never hope to possess clear fossil evidence of the origin and early evolution of life on earth, since cells break down easily, and even chemical components that might be indicative of past events can quickly deteriorate beyond recognition. Further complicating this issue is the fact that seemingly dense rocks are in fact riddled with minute cracks and pores into which contaminants can intrude. Thus, while molecules found in Archean cherts might represent primary organic compounds, researchers cannot rule out the possibility that these substances actually seeped into the rock at much later dates. Nevertheless, certain facts do suggest that life is at least as old as the oldest rocks now known. For one thing, graphite, a mineral consisting of pure carbon, is present even in the oldest known sedimentary rocks, the banded iron formations of Isua, Greenland—perhaps reflecting concentration of carbon by organisms (Figure 9-18). In fact, carbon found in sedimentary rocks of Archean age tends to have nearly the same ratio of isotopes (^{13}C to ^{12}C) that characterizes biological systems today.

A

B

FIGURE 9-34   Some of the oldest known geologic structures thought to be stromatolites (*A*. Top view; *B*. Cross section). These layered structures, which are from the Warawoona Group of the Pilbara Shield of Western Australia, are between 3.4 and 3.5 billion years old. (*Courtesy of D. R. Lowe.*)

More direct evidence of very early cellular life is provided by stromatolites, the internally layered structures shaped like domes, mounds, or pillars that were described in Chapter 4. Today, stromatolites form along the margins of warm seas but are scarce in comparison to their former abundance. Recall that within a well-preserved stromatolite, layers of carbonate sediment rich in organic matter alternate with layers of purer carbonate sediment (Figure 9-34). Stromatolites usually grow side by side in large numbers; some stromatolites, especially those of the geologic past, have attained heights of several meters.

Because structures resembling stromatolites sometimes form by deposition of layered sediment in the absence of algae, we cannot always identify fossil stromatolites with certainty. The oldest structures that are thought to be stromatolites occur in the Pilbara Shield of Australia in rocks 3.4 to 3.5 billion years old (Figure 9-34). Stromatolites are known to occur in slightly younger rock units

such as the Bulawayan Group of Rhodesia, for which indirect dating gives an age of approximately 2.8 billion years, and the 3-billion-year-old Pongola Supergroup (Figure 9-31). They are also found in a few other Archean units, but in general they are poorly represented in comparison to their abundance in Proterozoic sediments. The rarity of Archean stromatolites probably reflects the rarity of Archean shelf deposits. We cannot be certain, however, that the organisms that formed Archean stromatolites were photosynthetic, and therefore we cannot draw definite conclusions concerning the environments in which these stromatolites formed. Certain threadlike bacteria that are closely related to blue-green algae form stromatolite-like structures today in hot springs such as those of Yellowstone National Park, but most of these bacteria are not photosynthetic; that is to say, they do not synthesize food using light as an energy source. Thus, we cannot exclude the possibility that similar bacteria formed the stromatolite-like structures found in Archean rocks.

Actual fossils of blue-green algae and bacteria have also been tentatively identified in Archean rocks. Among the oldest cells yet known are filaments from a rock at North Pole, Western Australia, where the apparent stromatolites of Figure 9-34 are also found. These filaments, which are thought to be approximately 3.5 billion years old (Figure 9-35), are about the same size as those of modern blue-green algae (Figure 9-36A) and display similar transverse partitions. Somewhat younger spheroidal (nearly spherical) structures resembling other types of living bacteria or blue-green algae occur in the Fig Tree Group of southern Africa. These structures, which are about 3 billion years old, are preserved in what appear to be various stages of cell division (Figure 9-37).

Bacteria and blue-green algae seem to be the only life forms represented in Archean rocks, and they resemble each other so closely that many experts now advocate changing the name of blue-green algae to blue-green bacteria. Because bacteria and blue-green algae (Figure 9-36) differ markedly from all other groups of cellular life in the modern world, they are placed by themselves in the kingdom **Monera** (Figure 1-5). Recall that Monera are single-celled organisms characterized by a primitive kind of cell that does not have a nucleus and whose DNA is not clustered into discrete chromosomes. This type of cell, which is known as a **prokaryotic** cell, also lacks certain other internal organlike structures (organelles) that are present in more advanced forms of life. Life forms that do possess chromosomes as well as nuclei and other organelles are

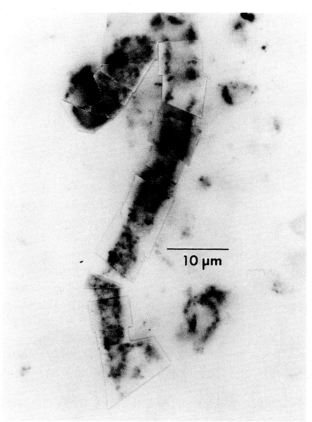

FIGURE 9-35   A fossil filament from North Pole, Western Australia, where apparent stromatolites are also found (see Figure 9-34). This and associated filaments display distinctive transverse partitions and may be between 3.4 and 3.5 billion years old, in which case they may be the oldest known fossil cells of blue-green algae. (*Courtesy of S. M. Awramik.*)

known as **eukaryotes**. Of all the cellular forms of life in the world today, only bacteria and blue-green algae are prokaryotes. All other cellular taxa are eukaryotes, and fossils of these are unknown from the Archean and early portion of the Proterozoic record. Thus, with respect to the history of life on earth, the Archean might well be labeled the Age of Prokaryotes.

Chemical evidence

Although the fossil record will never tell us a great deal about the earliest stages of organic evolution, such information has been derived from other sources. In the early 1950s, for example, researchers found that they could readily produce amino acids in the laboratory by sending electrical sparks through sealed vessels containing ammonia, methane, hydrogen, and steam (Figure 9-38). These sparks may

duplicate what lightning did in the Archean world. Similar laboratory results have since been obtained with simple starting mixtures such as carbon monoxide, nitrogen, and hydrogen. In some cases, ultraviolet light has been substituted for electrical sparks as a source of energy. **Amino acids** are the building blocks of proteins, and proteins are important compounds in living systems. Thus, experiments demonstrating the generation of amino acids under conditions that might well have characterized the primitive earth are highly significant. It has also been found that the heating of amino acids drives off water and links them into chains called **polypeptides**, which resemble proteins but are less complex. This, too, could have occurred in many parts of the primitive earth.

Recent discoveries suggest that the formation of amino acids need not necessarily have begun on earth. Radiospectroscopy has detected the presence in outer space of many compounds that represent intermediate steps in the formation of amino acids—compounds such as formic acid and methylamine, which can combine to give the amino acid glycine.

Meteorites offer additional evidence of extraterrestrial synthesis of molecules that are precursors of important biological compounds. Stony meteorites containing compounds of carbon are termed **carbonaceous chondrites**. In 1969, fragments of the famous Murchison meteorite of Australia were collected on the same day the meteorite landed, and they yielded numerous organic compounds, including amino acids. A remarkable feature of the Murchison and other carbonaceous chondrites is that each of their amino acids tends to be represented by more or less equal proportions of left-handed and right-handed structural configurations. (An amino acid molecule is three-dimensional and can exist in either of two mirror-image forms, just as a screw or bolt can have left-handed or right-handed threads.) Amino acids constructed by living systems on earth happen to have a left-handed configuration, which suggests that those of the Murchison meteorite are indeed extraterrestrial. Supporting evidence comes from the presence in these meteorites of organic compounds that are not found on earth and from the occurrence of carbon in a higher isotopic ratio of ^{13}C to ^{12}C than is known in any terrestrial biological system.

While amino acids and proteins are essential features of terrestrial life, they do not account for a basic aspect of living systems: the capacity for self-replication. On earth, this capacity resides within the DNA and RNA of cells. Recall that the genetic messages of chromosomes are

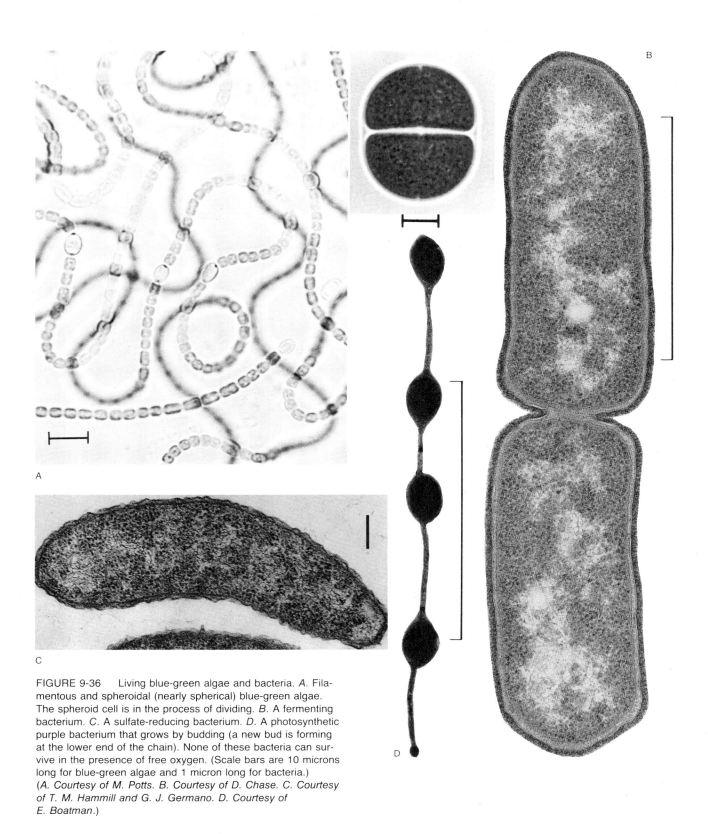

FIGURE 9-36   Living blue-green algae and bacteria. *A*. Filamentous and spheroidal (nearly spherical) blue-green algae.
The spheroid cell is in the process of dividing. *B*. A fermenting
bacterium. *C*. A sulfate-reducing bacterium. *D*. A photosynthetic
purple bacterium that grows by budding (a new bud is forming
at the lower end of the chain). None of these bacteria can survive in the presence of free oxygen. (Scale bars are 10 microns
long for blue-green algae and 1 micron long for bacteria.)
(*A. Courtesy of M. Potts. B. Courtesy of D. Chase. C. Courtesy
of T. M. Hammill and G. J. Germano. D. Courtesy of
E. Boatman*.)

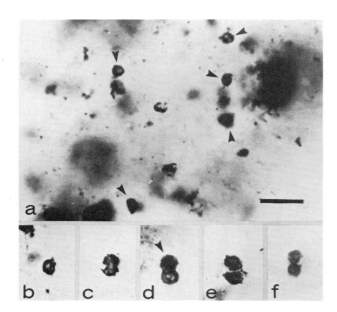

FIGURE 9-37   Microfossils older than 3 billion years from the lowest part of the Fig Tree Group of South Africa (scale bar is 10 microns long). Many of the cells are in some stage of division (*b* through *f*). They may represent blue-green algae or bacteria. (*A. H. Knoll and E. S. Barghorn, Science 198:396–398, 1977.*)

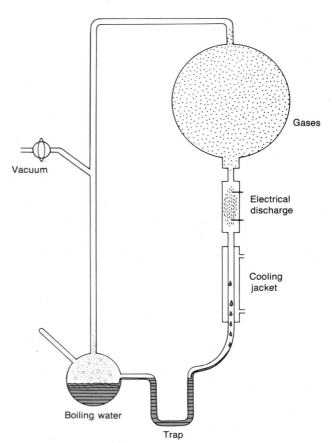

FIGURE 9-38   Experimental apparatus of S. L. Miller in which amino acids were produced by circulating ammonia (NH_3), methane (CH_4), water vapor (H_2O), and hydrogen past an electrical discharge. Amino acids accumulated in the trap. (*From G. Wald, "The Origin of Life." Copyright © 1954 by Scientific American, Inc. All rights reserved.*)

chemically encoded in DNA; these messages are transcribed by RNA, which is similar in structure to DNA, and are then carried by RNA to regions of the cell where, according to the chemical prescription, proteins are constructed from amino acids. Both RNA and DNA consist of structures called bases, which lie along a chain formed of alternating sugars and phosphate structures (Box 6-2). Nucleic acids are not complex chemical structures, and it does not seem at all preposterous to imagine that the first nucleic acid formed in nature by the assembly of sugars, phosphates, and nucleotide bases; but we do not know exactly how this happened or precisely how the codes for protein synthesis came to reside in DNA.

Another problematic step was the origin of cell-like bodies having the ability to build true proteins. Although such bodies have not been created in the laboratory, simpler spherical bodies have been produced. These are structures that have formed from proteinlike compounds in water or salt solutions, especially when the liquid has been cooled. Some of these spherical structures aggregate into chains that superficially resemble those of the living bacteria *Streptococcus*, forms of which cause infections in humans (Figure 9-39). A major question that remains, however, is

how nucleic acids might have been incorporated into spherical structures composed of proteinlike compounds.

The earliest forms of cellular life that employed nucleic acids to build proteins and to reproduce may have obtained their nutrition in either of two ways: They may have been animal-like (consumers) or plantlike (producers). Consumers obtain food directly from the environment, whereas producers manufacture their own food from simple raw materials that they obtain from the environment (page 31). We will consider two scenarios for early Archean evolution—one in which the earliest single-celled organisms are animal-like and the other in which they are plantlike. Modern bacteria live in many different ways, and as very prim-

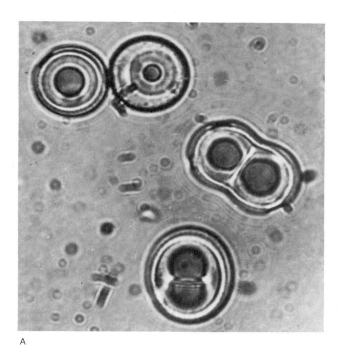

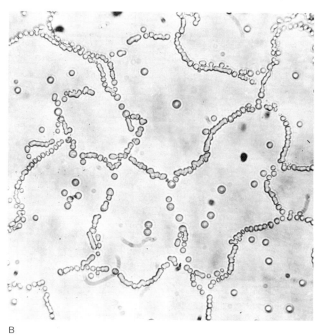

A

B

FIGURE 9-39   Spherical bodies formed in the laboratory by cooling of watery liquids containing proteinlike chemicals. Although they are simpler in chemical composition, some of these bodies show concentric layering and undergo division much like living cells (A). Others spontaneously assemble into chains that superficially resemble modern *Streptococcus* bacteria (B). (*Courtesy of S. W. Fox.*)

itive organisms, some of which may have survived from Archean time with little modification, they offer clues about evolution during Archean time.

Animal-like beginnings?

It has been argued that because the earliest true cells were very simple forms of life they must have been consumers—that is to say, they must have required food and energy from their environment in order to maintain themselves and to reproduce. Like all modern cells, early cells probably employed a compound known as ATP (adenosine triphosphate) as a source of energy. Unlike modern cells, however, early cells may have obtained the ATP that they needed from their environment—by eating it, in effect. In the laboratory, ATP is easily produced from simple gases, and it may well have formed inorganically in the Archean world. Even very primitive animal-like cells must have had at least a limited ability to synthesize other essential compounds, although they fed on amino acids, sugars, and other molecules necessary for their existence. It has often been assumed that these organisms lived in a lake or an ocean filled with a sort of natural "soup" of essential molecules.

In time, according to this scenario of animal-like beginnings (Figure 9-40), certain cells developed the ability to ferment organic compounds—that is, to break down such compounds into simpler ones and to use the energy liberated in this way to build some of the compounds that they needed. In other words, they became plantlike. Many bacteria employ fermentation today (Figure 9-36B). Most of them ferment sugar or cellulose (the material that forms the cell walls of plants) into products such as ethanol (ethyl alcohol). Energy liberated by fermentation is stored as ATP until it is used to build compounds essential to the operation of the bacterial cell.

Some bacteria that are fermenters also obtain energy by converting compounds that contain sulfate into others that contain sulfide—that is to say, they remove oxygen atoms that have been attached to sulfur atoms. These bacteria (Figure 9-36C) cannot tolerate oxygen, and most therefore live in the muds of swamps, ponds, or lagoons. The hydrogen sulfide that these bacteria liberate into their muddy environments generates the characteristic "rotten

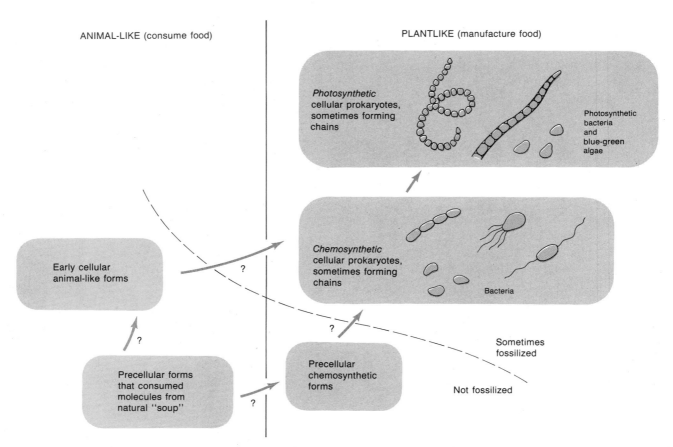

FIGURE 9-40   Sequence of nutritional changes in the early evolution of life according to scientists who believe that the earliest forms of life fed on a natural "soup" of complex molecules.

According to this scenario, the earliest chemosynthetic forms might have been either cellular or precellular.

egg" smell of those settings. If, as is widely believed, the earth's atmosphere lacked oxygen during Archean time, such **sulfate-reducing bacteria** may have occupied a much wider range of environments than they do today.

Bacteria that conduct fermentation or reduce sulfate are said to engage in **chemosynthesis**. The origin of the energetically more efficient process of **photosynthesis** in living organisms represented an important breakthrough that altered the ecosystem profoundly. Photosynthesis is the process by which single-celled algae and multicellular plants employ the green pigment chlorophyll to transform the energy of sunlight into chemical energy. This energy is then used to transform carbon dioxide and water into energy-rich sugar. When it is later released from the sugar, the energy also fuels essential chemical reactions.

Presumably, blue-green algae were not the only Archean Monera that conducted photosynthesis. Forms called **photosynthetic bacteria** probably existed as well. Today these organisms are represented by the purple and green

bacteria that inhabit moist areas lacking free oxygen, a substance that is poisonous to them (Figure 9-36D). Like blue-green algae, photosynthetic bacteria employ sunlight to produce sugar. Unlike blue-green algae, however, they use hydrogen sulfide rather than water as a raw material, and sulfur rather than free oxygen, is a by-product of their photosynthesis (Table 9-1).

The evolution of the blue-green algae, which do release free oxygen, had a profound effect on the global ecosystem. The oxygen that was liberated by these organisms accumulated in the atmosphere to form the reservoir from which humans and other animals were later able to breathe. In the following chapter, we will examine the current controversy over the time of accumulation of the earth's atmosphere.

Plantlike beginnings: Another alternative

Recently, some scientists have proposed a new set of ideas

Table 9-1   Comparison of photosynthesis in blue-green algae (and advanced plants) and in photosynthetic bacteria

Photosynthetic bacteria employ hydrogen sulfide rather than water as raw material.

	Different compounds		Similar compounds				Different compounds
Blue-green algae and advanced plants	Water (H_2O)	+	carbon dioxide (CO_2)	energy of sunlight $\longrightarrow$	sugar $(C_6H_{12}O_6)$	+	oxygen (O_2)
Photosynthetic bacteria (green and purple)	Hydrogen sulfide (H_2S)	+	carbon dioxide (CO_2)	energy of sunlight $\longrightarrow$	sugar $(C_6H_{12}O_6)$	+	sulfur (S)

about the evolution of primitive cells. This set of ideas conflicts with some of the ideas outlined in the previous section—particularly the notion that the first cells were consumers that fed on molecular soup. The new ideas are based on the discovery during the 1970s of hot springs located along midocean ridges in the deep sea. According to the new scenario, the full series of steps from the origin of complex molecules necessary for life to the origin of cellular organisms took place within similar hot springs early in Archean time.

Especially well studied are modern hot springs along the ridge adjacent to the Galápagos Islands, where Charles Darwin gained insight into the process of organic evolution (Figure 6-6). Because of the many fractures in the newly formed oceanic crust adjacent to a midocean rift (Figure 7-33), sea water seeps far down into the crust, where it eventually reaches magma rising up from the mantle along the rift. Under great pressure resulting from the weight of water above, the water is then heated to temperatures that may sometimes reach 1000°C. The hot water subsequently rises along the rift and reaches the surface through a narrow conduit, a roughly cylindrical fracture system called a vent. The water cools as it rises so that it reaches the overlying sea water, which is nearly freezing, at temperatures ranging from 10 to several hundred degrees centigrade.

The water that rises from a midocean vent is rich in various dissolved compounds, some of which are derived from the magma below. These include carbon dioxide (CO_2), ammonia, methane, and hydrogen sulfide (H_2S). Interestingly these are some of the compounds from which many chemosynthetic bacteria derive their energy in the modern oceans. Accordingly, it has been suggested that such bacteria resemble the earliest forms of cellular life and that these forms evolved in the hot waters of deep-sea vents in Archean oceans. By this argument, the earliest cellular organisms would have been plantlike, but they would have been chemosynthetic rather than photosynthetic.

Figure 9-41 shows how such vents might have been settings for all the major steps in the early evolution of life. High temperatures deep in a vent near the magma could have led to the formation of amino acids from simple compounds such as hydrogen, methane, and ammonia. Higher in a vent, under moderate-to-high temperatures, more complex building blocks might have formed, just as they do in the laboratory (Figure 9-38). And still higher in the vent, spherical cell-like bodies might have originated by cooling of the vent waters, just as they do in the laboratory when similar fluids are cooled (Figure 9-39). Chemosynthetic cellular forms might also have evolved in the vicinity of the vents, deriving their energy from carbon dioxide, ammonia, methane, or hydrogen sulfide supplied by the vent waters.

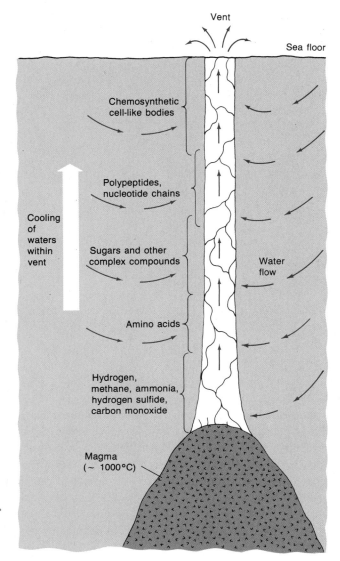

FIGURE 9-41 Possible sequence of evolution of early life within a deep-sea vent. Sea water seeps into the vent, which is a roughly cylindrical fracture system above a body of magma. In the vent, the water is heated to high temperatures and then rises. Increasingly complex structures would have originated toward the top of the vent as the rising water cooled. Ultimately, simple chemosynthetic organisms might have evolved near the sea floor, deriving their energy from hydrogen, methane, or other simple compounds formed deep within the earth. (*After J. B. Corliss et al., Oceanologica Acta, pp. 59–69, 1981.*)

By invoking chemosynthetic nutrition for early cellular life, the deep-sea–vent scenario eliminates the need for some type of molecular soup as food for the earliest cells. (We have no evidence that such a soup ever existed.) The deep-sea–vent scenario also derives appeal from the fact

that the Archean sea floor must have been studded with numerous hot-water vents; as we have seen, igneous activity, which centered in greenstone belts, was more widespread than it is today.

Whatever the early history of life may have been, one thing seems certain: Missing from the Archean world of prokaryotic life were animals and advanced animal-like cells that fed upon bacteria and blue-green algae. The origin and early evolution of these advanced consumers will be discussed in the chapter that follows.

CHAPTER SUMMARY

1. Radiometric dating has revealed that stony meteorites, which represent the primitive material of the solar systems, are 4.6 billion years old, as are the most ancient moon rocks. This, then, is the apparent age of the earth and other planets of the solar system.

2. The earth originated by condensation of material that had been part of a rotating dust cloud. The stratification of the earth into crust, mantle, and core may have developed either because material of high density accreted before material of low density or because dense material sank to the center when the earth liquefied from its heat of formation.

3. The Archean interval of earth history extended from the time of the earth's origin to approximately 2.5 billion years ago.

4. Between its time of formation and slightly later than 4 billion years ago, the earth was pelted by large numbers of meteorites. During the same interval, meteorites produced most of the large craters that are still visible on the moon, which has a less active surface than the earth.

5. The most readily studied Archean rocks occur in greenstone belts. Greenstones are metamorphosed volcanic rocks that have synclinal configurations. They seem to have formed along the margins of basins adjacent to small continents, from which were derived dark mudstones and graywackes associated with the volcanics.

6. Large continental landmasses apparently did not form until late in Archean time. The oldest of those now recognized are in southern Africa, where shallow marine and nonmarine siliciclastic sediments were

spread over sizable landmasses between 3 and 2 billion years ago.

7. The formation of large continents was inhibited in early and middle Archean time by the abundance of active elements whose decay produced heat at a high rate; under conditions of high heat flow, the earth's crust was easily broken by faults that allowed lava to reach the surface in areas that now represent greenstone belts.

8. Blue-green algae formed stromatolites during Archean time. Blue-green algae and bacteria are also represented in Archean rocks by fossil cells. These two groups represent the most primitive forms of cellular life on earth today, lacking cell nuclei and chromosomes.

9. There is a difference of opinion among scientists as to whether the earliest forms of life were animal-like in their nutrition, feeding on complex molecules, or plantlike, manufacturing their own food from simple chemical components.

EXERCISES

1. Why might we expect the earth to be nearly the same age as its moon and the material that forms meteorites?
2. What geologic features characterize greenstone belts?
3. What types of sedimentary rocks were rare in the Archean Eon? What does this suggest about the nature of cratons during Archean time?

4. Why did magma rise from the mantle to the surface of the earth at a higher rate during Archean time than it does today?
5. What features make the earth a more hospitable place than other planets for life as we know it?
6. What are stromatolites? From what we know of their formation today, why might we expect them to have been present early in the history of the earth?
7. What are some of the ways in which bacteria obtain their nutrition? How may the modes of life of certain living bacteria shed light on early evolution?

ADDITIONAL READING

Kaufmann, W. J., *Planets and Moons*, W. H. Freeman and Company, New York, 1979.

Lowe, D. R., "Archean Sedimentation," *Annual Review of Earth and Planetary Sciences*, 8:145–167, 1980.

Margulis, L., *Early Life*, Science Books International, Boston, 1982.

Moorbath, S., "The Oldest Rocks and the Growth of Continents," *Scientific American*, March 1977.

Silk, J., *The Big Bang: The Creation and Evolution of the Universe*, W. H. Freeman and Company, New York, 1980.

CHAPTER 10

Global Events of
the Proterozoic

The Proterozoic Eon, which succeeded the Archean 2.5 billion years ago, was in many ways more like the subsequent Phanerozoic Eon, in which we live. We have already seen a foreshadowing of this difference between the Proterozoic and Archean eons in the origin of large cratons late in Archean time. The persistence of large cratons throughout the Proterozoic Eon produced an extensive record of deposition in broad, shallow seas—a pattern that differed substantially from the Archean record of deep-water deposition, which is now largely confined to greenstone belts and adjacent areas. In addition, a larger number of Proterozoic than Archean sedimentary rocks remain unmetamorphosed and therefore accessible for study.

In this chapter, we will learn about the major events that have been recorded in the extensive deposits of Proterozoic age. These deposits not only document ancient mountain-building events strikingly similar to those of the Appalachians and other younger orogenic belts but also reveal records of major intervals of glaciation, at least one of which seems to have affected most of the world. Also present in Proterozoic rocks is a fossil record of organic evolution that reveals a transition from the simplest kinds of single-celled life at the start of the Proterozoic Eon to more advanced single-celled forms and, finally, to multicellular plants and animals, some of which belonged to modern phyla.

This array of global events is the subject of the present chapter. Next, in Chapter 11, we will view the Proterozoic world on a regional scale and learn how each of the modern continents began to take shape.

A MODERN STYLE OF OROGENY

As we discussed in Chapter 9, cratons of modern proportions first began to form about 3 billion years ago, late in Archean time. This was also the time when the oldest known sedimentary deposits were laid down over substantial continental areas in southern Africa (page 259). Although mountain-building processes resembling those of the Phanerozoic world were undoubtedly in operation by this time, it is in rocks about 1 billion years younger that geologists have found the oldest well-displayed remains of a mountain system that is thoroughly modern in character. This is the Wopmay orogen of Canada, which formed along the margin of an early continent that developed between about 1.8 and 2.1 billion years ago, over a large area now approximately 100 kilometers (~60 miles) to the west of Hudson Bay. Today, remarkably well preserved sedimentary rocks of this orogen are exposed along the low-lying surface of the Canadian Shield as a result of continental glaciation that has repeatedly scoured the orogenic belt over the past 2 million years.

The present Epworth Basin, which is the site of the Wopmay orogen, lies along the western margin of the

Stromatolites are present in Archean rocks but are much more abundant in Proterozoic rocks, where they are the most conspicuous fossils. Large stromatolites are rare today, but examples in Shark Bay, Western Australia (inset), closely resemble 2-billion-year-old examples. The long axes of the living forms parallel the prevailing wave direction. The Proterozoic stromatolites are from the Great Slave Lake area of Canada. Their long axes must have lain parallel to the prevailing wave direction and perpendicular to the ancient shoreline. (Courtesy of P. H. Hoffman.)

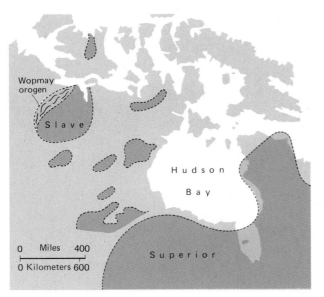

FIGURE 10-1 The Wopmay orogen, which formed about 2 billion years ago along the margin of the Slave Province of northwestern Canada. The Slave Province and other regions of Archean terrane are shown in brown. (*After P. Hoffman, Philos. Trans. Roy. Soc. London A233:547–581,1973.*)

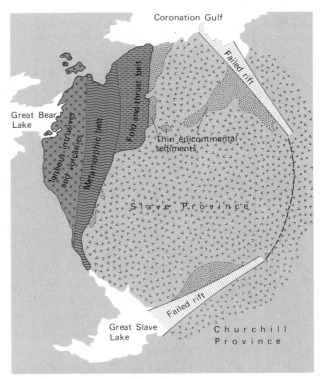

FIGURE 10-2 The three principal belts of the Wopmay orogen (*shaded*) and associated features. While sediments accumulated along the margin of the Slave craton where the orogen later developed, thinner sequences accumulated inland on the craton, and thick sequences in failed rifts that cut into the craton. (*After P. Hoffman, Philos. Trans. Roy. Soc. London A233:547–581, 1973.*)

geologic region known as the Slave Province (Figures 10-1 and 10-2) and is floored by an ancient fold-and-thrust belt (Figure 10-3). Although it has long been planed off by erosion, this ancient zone of deformed rocks bears a striking resemblance to the younger belts described in Chapter 8. In the Wopmay orogen, thrusting was toward the east, and igneous intrusions associated with the deformation now lie primarily within the Bear Province to the west. A belt of metamorphism lies between the igneous belt and the fold-and-thrust-belt. To the east, epicontinental sedimentary rocks continuous with those of the fold-and-thrust belt are flat-lying except where they lie adjacent to local upwarps of basement rock. Like sedimentary deposits of younger fold-and-thrust belts, those of the Wopmay belt

show a clear relationship to tectonic history. Near the end of Archean time, before the orogen was formed, most of what we now call the Slave Province existed as a discrete craton. Then, early in the Proterozoic Eon, thin sedimentary sequences developed on the interior of this craton, and thick shelf deposits accumulated along its western margin. As in younger mountain belts, these were succeeded by flysch and then by molasse deposits. In fact, the thick sequence of deposits in the Wopmay fold-and-thrust belt closely

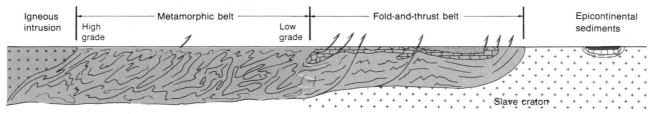

FIGURE 10-3 A cross section of the northern part of the Wopmay orogen, depicted with no vertical exaggeration. Compare the three belts as shown here in cross section with their

appearance in map view (Figure 10-2). (*After P. Hoffman, Philos. Trans. Roy. Soc. London A233:547–581, 1973.*)

FIGURE 10-4 Diagram showing the sequence of development of sediments in the fold-and-thrust belt of the Wopmay orogen; numbers refer to depositional units described in the text. Units 1 and 2 represent marine deposition along a shallow continental shelf. Units 3 and 4 are deep-water deposits, including flysch, that accumulated when the shelf foundered as mountain building began to the west. Unit 5 consists of shallow-water deposits transitional between flysch below and molasse above. Unit 6, the molasse phase of deposition, followed the exclusion of marine waters by a heavy influx of sediment from the west. (*After P. H. Hoffman, in M. R. Walter [ed.], Stromatolites, Elsevier Publishing Company, Amsterdam, 1976.*)

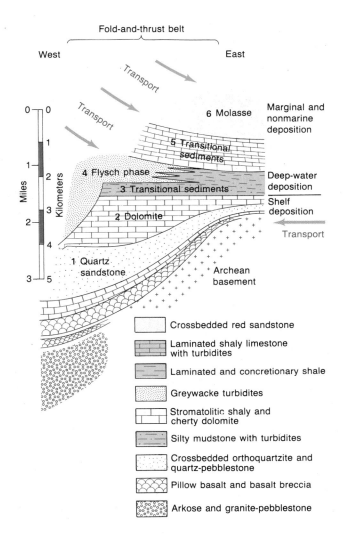

resembles that found in each of the tectonic cycles of the Appalachian orogen (Figure 8-23). The Wopmay sequence has the following characteristics:

1. The first thick deposit, which formed along the edge of the continental shelf, is a quartz sandstone that prograded toward the basin (Figure 10-4). This unit resembles the sequence of Lower Cambrian shelf sands at the base of the first Appalachian tectonic cycle. The quartz sandstones of the Wopmay orogen grade westward into deep-water mudstones and turbidites that now lie within the metamorphic belt.

2. Next come rocks in which stromatolites and dolomite predominate. These represent a Proterozoic carbonate platform that resembled the Cambro-Ordovician platform of the first Appalachian tectonic cycle. In the Proterozoic platform, sedimentary cycles record repeated progradation of tidal flats across a shallow lagoon (Figure 10-5A). Laminated dolomite representing the lagoonal environment forms

A

B

FIGURE 10-5 Shelf deposits of the Wopmay orogen.
A. Cycles of the Rocknest Formation, displayed at a location in the fold-and-thrust belt where deformation has rotated the bedding so that it is nearly vertical. Each cycle is progradational (regressive), with the massive upper beds representing stromat-

olites that fringed the shoreline. These prograded over lagoonal deposits, which form the lower part of each cycle. B. Close-up view of the stromatolites that form the upper parts of cycles. (*Courtesy of P. H. Hoffman, Geological Survey of Canada.*)

the base of each cycle, while at the top of each cycle are oolitic or stromatolitic deposits that must have formed in environments fringing the lagoon on its landward side (Figure 10-5*B*). It would appear that the shoreline seldom if ever migrated westward as far as the shelf margin, since cycles are less evident here, while at the same time enormous stromatolite mounds are present. These mounds formed a persistent barrier behind which the fine-grained deposits of the lagoon were trapped; thus, stromatolites bounded the lagoon on both its landward and seaward margins. The present metamorphic zone consists of a thinner sequence of mudstones that represent deeper environments beyond the shelf edge, together with beds of dolomite breccia that contain blocks as large as 50 meters (~165 feet) in length. It is obvious that these beds were transported down the steep slope in front of the shelf edge by catastrophic submarine debris flows (compare Figure 8-25).

3. The carbonate-platform deposits give way to transitional mudstones, which reflect a downwarping of the platform as a foredeep formed.

4. The mudstones are followed by flysch deposits (turbidites) similar to Upper Ordovician flysch of the first Appalachian tectonic cycle. A westward thickening of the Wopmay flysch and the inclusion within it of particles derived from plutonic rocks to the west indicate that the source area of siliciclastics now lay seaward of the orogenic belt, just as it did in the development of each Appalachian tectonic cycle when the foredeep formed. It is clear that the Wopmay foredeep formed when plutonism and orogenic uplift began in the offshore area to the west.

5. The deep-water turbidites grade upward into beds containing mud cracks and stromatolites, both of which represent shallow-water environments and thus point to a shallowing of the foredeep.

6. The influx of sediments eventually pushed marine waters from the Wopmay foredeep, and the Wopmay cycle, like younger tectonic cycles, ended with an interval of molasse deposition. The Wopmay molasse consists largely of river deposits in which cross-bedding is conspicuous.

Two kinds of evidence suggest that the Proterozoic Wopmay orogen had the same pattern of formation as a modern orogenic system. First, the parallel igneous, metamorphic, and fold-and-thrust belts resemble similarly arranged belts of younger mountain ranges (Figure 10-3). Second, within the fold-and-thrust belts, shallow-water shelf deposits are succeeded by flysch deposits that give way to molasse deposits.

Another indication that normal plate-tectonic processes were operating in northern Canada about 2 billion years ago is evidence of continental rifting. East of the Wopmay orogen, slicing into the platform from the northwest and southwest, are deep, narrow troughs containing thick sedimentary sequences (Figure 10-2). These troughs represent failed rifts—that is, rift systems that ceased to be active before they cut across the entire Slave craton.

It seems likely that the westward-facing edge of the Slave craton along which the Wopmay orogen formed came into being more than 2 billion years ago, when a rift system broke apart a slightly larger continental mass. More or less simultaneously, the two rifts that later failed began to form at right angles to the newly forming shelf edge. Shortly thereafter, about 2 billion years ago, the shelf edge must have foundered when it was rafted up against a subduction zone. Then, as we have seen for younger orogens, igneous activity elevated the crust seaward of the shelf edge, and mountain building began.

PROTEROZOIC GLACIAL DEPOSITS

The fact that the Slave terrane along the Wopmay orogen behaved like rigid continental crust when it was rifted and deformed indicates that by 2 billion years ago the earth was markedly cooler than it had been 3 billion years ago, when magmas were pushing up from the mantle to the surface throughout much of what is now the Canadian Shield. That climates in this region were also quite cool approximately 2 billion years ago is shown by evidence that glaciers spread over the land, as we will now see.

Early Proterozoic glaciation

Just to the north of Lake Huron in southern Canada are some of the most spectacularly exposed of all ancient glacial deposits: those of the Gowganda Formation, which forms part of the Huronian Supergroup. Well-laminated mudstones in this formation almost certainly represent varves that formed in standing water in front of glaciers. In Chapter 3, these deposits were compared to the strikingly similar glacial varves that formed nearby, where Toronto is now located, just a few thousand years ago (Figure 3-7). Some of the laminated Gowganda mudstones contain dropstones—pebbles and cobbles that appear to have fallen to the bottom of a sea or lake from ice that melted as it floated out from a glacial front (Figure 10-6). Whether the mudstones accumulated in an ocean or in a lake is not clear, but it is evident that they alternate with tillites, which seem to represent the periodic encroachment of glaciers

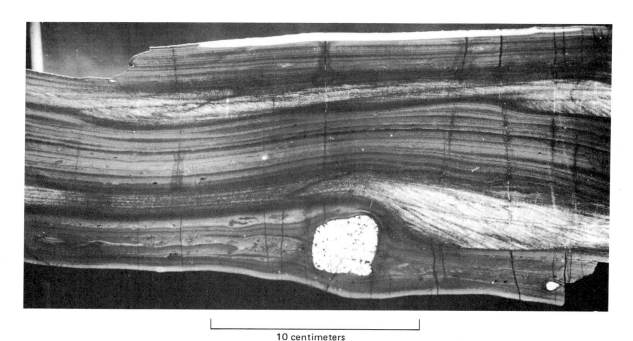

10 centimeters

FIGURE 10-6   Fine-grained sedimentary rock of the Gowganda Formation containing an igneous pebble that apparently dropped from floating ice. The layers in the sedimentary rock appear to represent varves. (*Courtesy of D.A. Lindsey, U.S. Geological Survey.*)

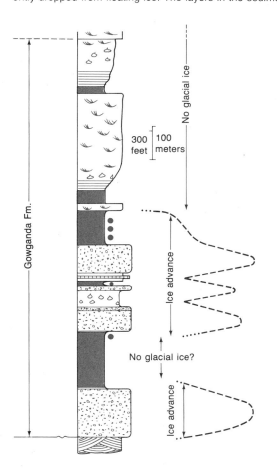

into the body of water (Figure 10-7). Some of the pebbles and cobbles of these tillites are faceted or striated from having slid along at the bases of moving glaciers.

Although the exact age of the Gowganda deposits has not been determined, it is known that the unit is slightly more than 2 billion years old, since it rests on 2.6-billion-year-old crystalline rocks and is intruded by rocks 2.1 billion years in age. Tillites of similar age are found elsewhere in Canada as well as in southern Africa and India. Thus, it would appear that there was an interval of extensive continental glaciation not long after the transition from Archean to Proterozoic time.

The late Proterozoic glacial interval

Rocks ranging in age from less than 1 billion years to about 2.3 or 2.4 billion years show little evidence of glacial activity,

FIGURE 10-7   Stratigraphic section of the Gowganda Formation of southern Canada in which the position of tillites is shown. These tillites appear to represent two major glacial advances, the second of which was divided into three pulses. Between these glacial advances, fine-grained, varved sediments with dropstones accumulated (dots). (*After G. M. Young, in M. J. Hambrey and W. B. Harland [eds], Earth's Pre-Pleistocene Glacial Record, Cambridge University Press, 1981.*)

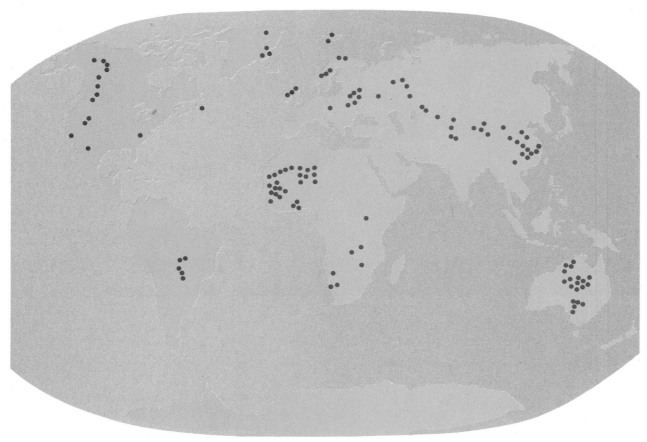

FIGURE 10-8   Distribution of glacial deposits of late Proterozoic age. These deposits, which formed between about 1 billion
and 600 million years ago, are found in most parts of the world.

*(After M. J. Hambrey and W. B. Harland, Earth's Pre-Pleisto-
cene Glacial Record, Cambridge University Press, Cambridge,
1981.)*

indicating that the early Proterozoic period of glaciation was
followed by an interval of about 1.5 billion years in which
continental glaciers were rare or absent. Then, 800 or 850
million years ago, in late Proterozoic time, there occurred
what may have been the greatest glacial episode in the
entire history of the earth. This episode was so extensive
that tillites and other glacial deposits of late Proterozoic
age can be found on all major continents of the world except
Antarctica (Figure 10-8). In North America, for example,
late Proterozoic tillites are widely distributed along both the
Cordilleran and the Appalachian orogens, while in Scotland
they are well exposed at several localities (Figure 5-6).
Many of these tillites contain striated and faceted pebbles
and boulders, and some are associated with varve-like
mudstones.

   In general, late Proterozoic glacial units range in age
from 1 billion years to about 600 million years. Although
precise dating of individual tillites has been difficult to accomplish, it is strongly suspected that the earth experi-

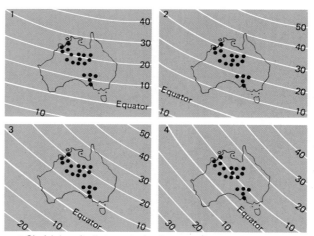

• Glacial deposits that accumulated at some time
   during the total interval represented

FIGURE 10-9   Latitudinal positions of Australia at four successive times during the interval in which it was subjected to
late Proterozoic glaciation. The glaciated regions lay within 30°
of the ancient equator at all times. *(After M. C. McWilliams and
M. W. McElhinny, Jour. Geology 88:1–26, 1980.)*

enced several discrete glacial episodes during the long interval that these units represent. At least two and possibly three or more of these episodes are known to have occurred in Africa, for example, and the last took place about 600 million years ago, when glacial deposits accumulated in many parts of the world.

The global distribution of late Proterozoic tillites is puzzling in that it is difficult to comprehend how glaciers could have spread over so many continents. Indeed, the limited paleomagnetic data now available suggest that even regions lying close to the equator experienced some degree of continental glaciation during this time. Almost all of Australia, for example, seems to have lain within 30° of the equator throughout the latter part of Proterozoic time (Figure 10-9), and yet glaciation here was extensive (Figure 10-8). It is possible that most of the earth was at times quite cold late in Proterozoic time.

ATMOSPHERIC OXYGEN

In the last chapter, we learned that the earth's primitive atmosphere, which formed by the degassing of the planet, contained little or no free oxygen. It is not known precisely when atmospheric oxygen reached its present level, but it is now widely agreed that a moderate level was reached about 2 billion years ago, early in Proterozoic time. Thus, the presence of very low levels of atmospheric oxygen seems to be another characteristic that distinguishes the Archean world from all subsequent times except the earliest part of the Proterozoic Eon. In this section, we will review the evidence supporting this conclusion, but first we will explain why the atmosphere serves as a reservoir for free oxygen today.

It would appear that the concentration of oxygen in the earth's atmosphere has not varied substantially over the past few hundred million years despite the fact that oxygen is continuously removed from the atmosphere by processes such as oxidation of minerals at the earth's surface and respiration. (Respiration is the process by which organisms employ oxygen to obtain energy from their food; in effect, organisms use oxygen to burn up their food in the same manner that fire oxidizes a flammable material by causing it to combine with oxygen and thus to release energy.) The atmospheric reservoir of oxygen persists because oxygen is continuously returned to it by photosynthesis and, to a lesser extent, by the breakdown of water in the upper atmosphere by the sun's rays (Figure 2-7).

A number of factors prevent the concentration of atmospheric oxygen from increasing to a significant degree. Were oxygen to build up much beyond its present level, for example, weathering would become more intense and would consequently "soak up" excess oxygen. Furthermore, if plants markedly increased their biomass on earth and thus began to liberate more oxygen, animals and simpler respiring organisms, including bacteria, would also increase in biomass because more plants and plant debris would be available for them to feed on. The resulting increase in the rate of oxygen consumption by respiration would then offset the increase in oxygen production by plants.

With these factors in mind, let us now examine the evidence supporting the assumption that low levels of atmospheric oxygen existed until early Proterozoic time.

The case for early anaerobic evolution

One piece of evidence suggesting that the earth's early atmosphere was anoxic (oxygen-free) lies in the fact that the chemical building blocks of life could not have formed in the presence of atmospheric oxygen. Indeed, chemical reactions that yield amino acids from simpler compounds in the laboratory (page 266) are inhibited by even smaller amounts of oxygen than those found in the earth's atmosphere.

Furthermore, oxygen has been found to be toxic to the most primitive living bacteria, including purple photosynthetic bacteria and bacteria that obtain energy from fermentation (page 267). As we have seen, photosynthetic bacteria as well as bacteria that derive energy from fermentation rather than from respiration are currently restricted to anoxic habitats such as swamps, ponds, and lagoons (page 267) and are consequently termed *anaerobic*. The fact that the simplest living cellular organisms are anaerobic suggests that the earth's first forms of cellular life had similar metabolisms, and it also suggests that these organisms may have evolved in this way because there was virtually no atmospheric oxygen when they came into being. If this was the case, the subsequent buildup of atmospheric oxygen limited these bacteria to the environments that they now occupy.

The oxidation state of minerals

Also widely cited as evidence of an anoxic Archean atmosphere is the distribution of uranium and iron minerals in Precambrian rocks. Under the atmospheric conditions of

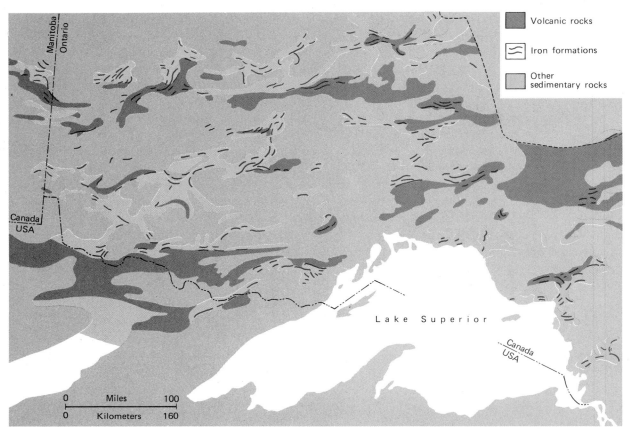

FIGURE 10-10   Distribution of banded iron formations of Archean age in the southwestern Canadian Shield. Many of these are mined for their iron. The volcanic and sedimentary rocks shown here form greenstone belts. (*After A. M. Goodwin, Econ. Geol. 68:915–933, 1973.*)

the modern world, the uranium oxide mineral uraninite (UO_2) is readily oxidized further and quickly dissolves from rocks. The iron sulfide mineral pyrite (FeS_2) also disintegrates readily when it is exposed to the modern atmosphere. Today, uraninite and pyrite are seldom found in the sediments of rivers and beaches, and they are similarly rare in ancient siliciclastic sediments younger than about 2 billion years. In contrast, both minerals are relatively abundant in buried nonmarine and shallow marine siliciclastic deposits older than 2 billion years; uraninite, for example, has been found in such rocks on five continents.

Significantly, red beds (see Appendix I) display the opposite pattern: They are never found in terranes older than 2.2 or 2.3 billion years. Hematite, a highly oxidized iron mineral, gives red beds their color (Appendix I). Often, the hematite found in red beds has formed secondarily by oxidation of other iron minerals that accumulated with the sediments. During Phanerozoic time, which has been characterized by the presence of a well-oxygenated atmosphere, this secondary oxidation has often occurred within a few millions or tens of millions of years after deposition. It would appear that oxidation of this type did not occur early in the earth's history.

Some researchers have argued that red beds cannot be found in Archean sedimentary sequences because the latter formed mainly in deep marine basins, whereas most red beds are of nonmarine origin. The Huronian and Witwatersrand sequences are especially significant in this regard because they formed more than 2 billion years ago on sizable cratons (page 260). Although both of these sequences comprise large volumes of nonmarine sediments and would therefore be expected to include some red beds under modern atmospheric conditions, both lack red beds but contain substantial quantities of detrital pyrite and uraninite. This condition has been widely cited as evidence that oxygen concentrations remained low until 2 billion years ago or later.

Banded iron formations are other sedimentary rocks whose distribution may reflect the history of the earth's atmosphere. Rocks of this type are rare except in Precam-

FIGURE 10-11   A tectonically deformed banded iron formation that is approximately 2 billion years old, forming Jasper Knob in northern Michigan. (*Courtesy of F. J. Pettijohn.*)

brian terranes. Although banded iron formations are present in Archean terranes (Figure 10-10) and, in fact, are among the oldest known rocks on earth (Figure 9-18), most accumulated early in the Proterozoic Era between about 2.5 and 1.8 billion years ago. The term *banded iron formation* refers to a bedding configuration in which laminae of chert that is sometimes contaminated by iron alternate with laminae of other minerals that are richer or poorer in iron than the chert (Figure 10-11). The iron in these formations may take any of a variety of forms, including iron oxide, iron carbonate, iron silicate, and iron sulfide—but the present chemical form of the iron is not necessarily the form in which it was originally deposited, and in many cases there is evidence that the mineralogy of the iron has altered over time. Banded iron formations account for most of the iron ore that is mined in the world today. One reason for their economic value is that many contain iron in the form of magnetite (Fe_3O_4), whose iron-to-oxygen ratio is higher than that of hematite (Fe_2O_3).

The origin of banded iron formations is problematic. In some of them, iron minerals accumulated as chemical muds, while in others they were deposited as sand-sized grains, and cross-lamination is sometimes present in the coarser

accumulations. Some of the constituent particles of these formations, such as those that became cross-laminated, were loose grains. It is known that most banded iron deposits are marine, since they often occur in close conjunction with marine deposits such as offshore turbidites, mudstones, and pillow lavas. Some researchers believe that nearby volcanism provided the silica and iron of banded iron formations, while others maintain that most of the original iron minerals in these formations were washed into marine basins from the land.

The fact that the iron in banded iron formations was at least weakly oxidized when these structures first formed implies that some oxygen was present in the environment at this time. In fact, some iron formations are underlain by red beds that are even more highly oxidized. This suggests that atmospheric oxygen existed at a relatively high level between 2.2 and 1.8 billion years ago, attaining greater concentrations in some depositional environments than in others but remaining sufficiently abundant to produce at least weakly oxidized iron minerals in many environments. In any case, the abundance of red beds in rock sequences younger than banded iron formations testifies strongly in favor of a relatively high level of oxygen during the past

1.8 billion years, when banded iron formations have no longer been forming.

Stromatolites and oxygen

It is widely agreed that photosynthesis caused atmospheric oxygen to build up during Precambrian time. Before such a buildup could occur, however, natural reservoirs known as **oxygen sinks** had to be filled. Essentially, oxygen sinks are chemical elements and compounds that combine readily with oxygen and that are believed to have been present in the crust or atmosphere immediately after the earth formed. Sulfur and iron are two of the most important oxygen sinks. (Note how, on a much smaller scale in the modern world, iron that we extract from naturally occurring compounds rusts when exposed to atmospheric oxygen.)

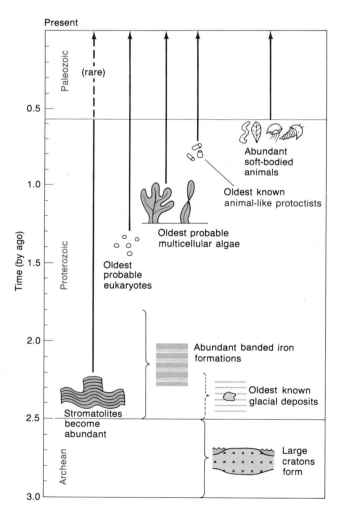

Today photosynthesis liberates oxygen at such a high rate that it could refill the atmosphere within just a few thousand years. Given the probable presence of photosynthetic bacteria and blue-green algae by at least 3.5 billion years ago, how could it have taken nearly 2 billion years for atmospheric oxygen to fill major sinks and then approach a high level? The answer may well be that there were simply not enough stromatolites to liberate oxygen at a high rate until 2.3 or 2.2 billion years ago, when the geologic record reveals that stromatolites became very abundant (Figures 10-12 and 10-13). The reason for their earlier rarity may be attributable to certain geologic factors. Perhaps the absence of large continents and continental shelves during Archean and early Proterozoic time restricted the growth of stromatolites to the steep margins of small Archean landmasses offering little space for colonization.

It is also likely that during Archean time nutrients were not easily elevated from the deep sea to shallow water for use by stromatolites. Today, upwelling occurs along the margins of continents where currents drag surface waters away from the shore (page 54), and nutrient-laden waters from great depths replace these surface waters. During Archean time, however, the absence of large continents may have caused upwelling to be weak and nutrient supplies sparse. Then, as continents grew, more extensive upwelling may have accelerated the rate of photosynthesis and oxygen production.

Volcanism and oxygen

The widespread volcanism that seems to have characterized the hot Archean earth may also have impeded the accumulation of atmospheric oxygen during this time, since some of the gases emitted by volcanoes are quickly oxidized, removing oxygen from the atmosphere. Carbon monoxide (CO), for example, is oxidized to carbon dioxide (CO_2). Today there is less volcanic activity, and so this type of oxidation removes oxygen from the atmosphere at a slow rate—but during the Archean Eon, gasses liberated

FIGURE 10-12   Evolutionary developments during Proterozoic time, with arrows signifying survival of biological forms to the present time. Also shown are developments in the physical environment. All of the occurrences shown in this diagram are based on what we know from the rock record.

FIGURE 10-13   A large reeflike mound built by stromatolites in Proterozoic rocks of Victoria Island, Arctic Archipelago, Canada. Such structures are common in Proterozoic rocks. This mound, which is 6 meters (~10 feet) in diameter, has been exposed by erosion. (*G. M. Young, Precambrian Res. 1:13–41, 1974.*)

by the widespread volcanic activity may have rapidly soaked up oxygen produced by blue-green algae, thereby suppressing the buildup of atmospheric oxygen.

LIFE OF THE PROTEROZOIC EON

Even as late as the early Proterozoic interval, after an estimated 1.5 billion years of evolution, it would appear that the earth was populated exclusively by single-celled prokaryotic forms of life. In the course of Proterozoic time, however, a great expansion of life issued from these simple forms. Figure 10-12 outlines what we know of this expansion and depicts other Proterozoic events described earlier in the chapter. We will soon examine what is known about advanced forms of life that evolved before the start of the Cambrian Period, but first let us review the fossil evidence supporting the continued dominance of prokaryotes early in Proterozoic time. This evidence takes the form of stromatolites and microscopic cell outlines.

Stromatolites

As emphasized in the preceding discussion of the Proterozoic atmosphere, stromatolites first became abundant in the fossil record about 2.3 or 2.2 billion years ago, perhaps as a result of an increase in the areal extent of continental shelves. Stromatolites remained abundant throughout the rest of Precambrian time and did not decline in number until early in the Paleozoic Era, when animals diversified on a large scale and probably inhibited stromatolite growth. Proterozoic rocks contain stromatolites of many different shapes, some of which form reeflike structures (Figure 10-13). Attempts to use these fossils for stratigraphic correlation, however, have been hindered by the fact that filamentous blue-green algae of a single type often produce stromatolites of different shapes in different environments (page 272 and Figure 10-14). Similar ecologic influences on form in Precambrian time have made it difficult to identify evolutionary trends that might make fossil stromatolites useful for dating purposes.

A

FIGURE 10-14   Living and Proterozoic stromatolites. The living forms (A) are from an area of Shark Bay, Western Australia,
where there is strong wave action—a factor that accounts for
the pillarlike shape of these stromatolites. Compare these to the
stromatolites shown on page 272. These pillarlike stromatolites
resemble many fossil forms, including Proterozoic examples
from the Coppermine River area of Canada (B). (*Courtesy of
P. H. Hoffman.*)

B

Fossil prokaryotic cells

One of the most interesting fossil discoveries since the
middle of this century has been the identification of cell
remains preserved in Precambrian rocks. Such remains
were first identified in a formation of Ontario and northern
Minnesota known as the Gunflint Chert. The Gunflint fossils, which are about 1.9 billion years old, display a wide
variety of shapes (Figure 10-15) and appear to represent
bacteria or blue-green algae, as do the members of other
microscopic fossil assemblages of early Proterozoic age.
Analysis of these fossils has revealed not only that prokaryotic forms of life persisted from Archean time into the
Proterozoic Eon, but also that many of the forms found in
Precambrian rocks closely resemble forms that are alive
today.

The earliest eukaryotes

All forms of life except blue-green algae and bacteria are
eukaryotes—that is, forms whose cells contain chromosomes, nuclei, and other advanced internal structures.
Unfortunately, we do not know exactly when the first eu-

karyotic cell evolved. One promising approach to solving
this problem involves size comparisons of fossilized cells
of sphere-shaped algae. While very few living prokaryotic
algae with a spherical shape are larger than 20 microns in
diameter, spherical algae with diameters exceeding 60
microns are invariably eukaryotic. In the fossil record, mi-
crofossil assemblages older than 1.4 billion years are dominated by small cells whose diameters seldom exceed 10
microns, and it is only in rocks younger than 1.4 billion
years that cells with diameters larger than 60 microns are
abundant. These findings suggest that the eukaryotic condition evolved about 1.4 billion years ago or slightly earlier.

By far the most convincing evidence supporting the
presence of eukaryotes during Proterozoic time is provided
by fossil cells called **acritarchs**. These near-spherical or
many-pointed forms are the dominant group of algal plankton found in the Paleozoic fossil record (Figure 10-16).
Some acritarchs are believed to have represented the resting stages (or cysts) of dinoflagellate cells. Although dinoflagellates constitute one of the most important groups
of planktonic algae today (page 50) and are known to be
eukaryotic, we do not know for certain that all acritarchs
were dinoflagellates. In fact, some experts believe that

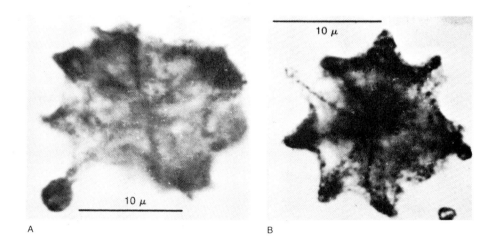

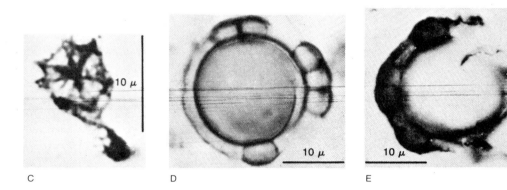

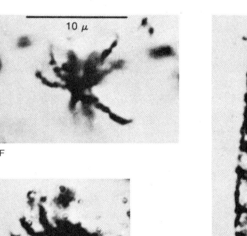

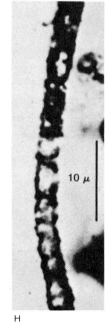

FIGURE 10-15   Precambrian flora of prokaryotic cells from the early Proterozoic Gunflint Formation of the Lake Superior region. *A* through *C. Kakebekia umbellata*, an umbrellalike form that closely resembles a living bacterium. *D* and *E. Eosphaera tyleri*, a spheroidal form that seems to have budded off small daughter cells. *F* and *G. Eoastrion simplex*, a star-shaped form. *H. Gunflintia grandis* (threadlike blue-green algae). (*Courtesy of E. S. Barghoorn.*)

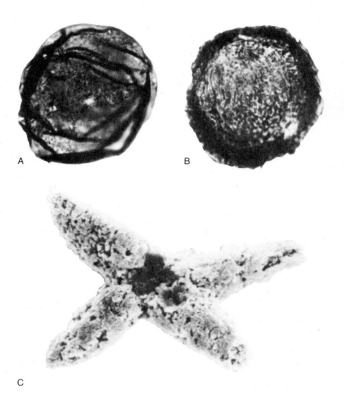

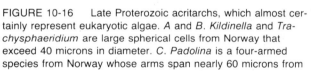

FIGURE 10-16   Late Proterozoic acritarchs, which almost certainly represent eukaryotic algae. *A* and *B. Kildinella* and *Trachysphaeridium* are large spherical cells from Norway that exceed 40 microns in diameter. *C. Padolina* is a four-armed species from Norway whose arms span nearly 60 microns from tip to tip. *D. Chuaria* is an exceptionally large form from the Grand Canyon in which the cells, flattened on a bedding surface here, exceed 1 millimeter in diameter. (*A through C. Courtesy of G. Vidal. D. Courtesy of T. D. Ford.*)

Proterozoic acritarchs were not dinoflagellates but instead belonged to the group known as green algae. In any event, the large size of many Proterozoic acritarchs, together with the chemical composition of their cell walls, suggests that acritarchs were eukaryotic (Figure 10-16). It is also significant to note that some Proterozoic acritarchs had cell walls whose complex patterns resembled those that today are restricted to eukaryotes. Interestingly, the fossil record of acritarchs, like that of large cells in general, begins in rocks about 1.4 billion years old.

Despite the antiquity of the acritarch fossil record, acritarchs are rarely found in rocks older than about 850 or 900 million years. In contrast, they are both abundant and diverse in younger rocks. In fact, the fossil record reveals that acritarchs underwent a rapid adaptive radiation between 900 or 850 and 700 million years ago. For this reason, acritarchs are very useful for dating rocks of latest Precambrian age.

It is interesting to note that acritarchs suffered a mass extinction at the time of worldwide glaciation about 600 million years ago (page 279). Only a few simple spheroidal types survived this crisis, which may have resulted from worldwide cooling.

Early evolution of eukaryotes

Because the fossil record of single-celled eukaryotes lacking skeletons is very poor, most of our ideas about the early evolution of this group are based on evidence derived from living organisms. The most remarkable of these ideas, which was not widely accepted until the 1970s, is that the eukaryotic cell arose from the union of two or more prokaryotic cells, at least one of which came to reside within another. According to this theory, the prokaryotic cell that first came to live inside another was altered in minor ways to form a structure called a **mitochondrion**, one or more

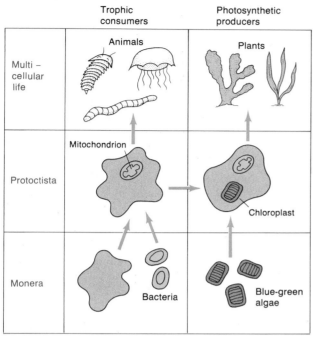

FIGURE 10-17   The probable sequence of major events leading from Monera to multicellular animals and plants. The first protoctist apparently evolved when one moneran engulfed but failed to digest another, which then became a mitochondrion. A plantlike protoctist evolved when an animal-like protoctist engulfed but failed to ingest a blue-green algal cell, which then became a chloroplast.

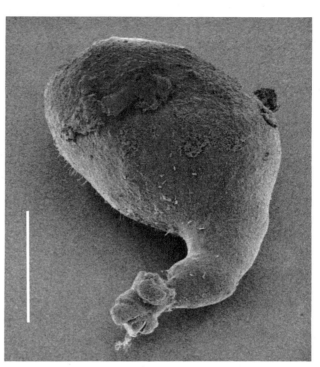

FIGURE 10-18   *Pelomyxa palustris*, which is considered to be the most primitive known eukaryote. This single-celled creature lacks mitochondria but is colonized by bacteria that serve the same respiratory function that mitochondria fulfill in more advanced eukaryotes. (*Courtesy of E. W. Daniels.*)

of which is present in nearly all eukaryotic cells (Figure 10-17). Mitochondria are, in fact, the structures that allow cells to derive energy from their food by means of respiration. In mitochondria, complex compounds are broken down by oxidation, which yields energy and, as a by-product, carbon dioxide. A mitochondrion is apparently the slightly modified evolutionary descendant of a small bacterium that became trapped within a larger one, as evidenced by the presence in mitochondria of DNA and RNA, both of which are essential to the independent existence of any cell (Box 6-2). It is assumed that the smaller cell that became a mitochondrion was eaten by the larger one but proved resistant to the digestive processes of the predator cell.

There is one very primitive single-celled organism that seems to illustrate an early stage in the evolution of the eukaryotic condition. This is *Pelomyxa palustris*, a species that inhabits the bottom of freshwater ponds (Figure 10-18). Individuals of this species have nuclei, which constitute a eukaryotic trait, but the DNA within these nuclei is not arranged into chromosomes. Furthermore, mitochondria are lacking in *Pelomyxa* cells; in their place are bacteria that perform essentially the same respiratory function. *Pelomyxa* dies if the bacteria living inside it are killed with an antibiotic. Although *Pelomyxa* itself may not have been ancestral to normal eukaryotes, it serves to illustrate the condition that may have preceded the transformation of intracellular bacteria into mitochondria. As the host, *Pelomyxa* benefits from having its food broken down to supply its energy, while its intracellular bacteria benefit both from the abundance of food available and from the protection afforded by the host.

Single-celled eukaryotes and a few simple multicellular organisms constitute the kingdom Protoctista (Figure 1-5), and single-celled animal-like members of this kingdom (that is, consumers rather than producers) are called **protozoans**. It is believed that the earliest protoctists were protozoans, since their immediate ancestors were consumers that had eaten bacteria that became mitochondria. Among

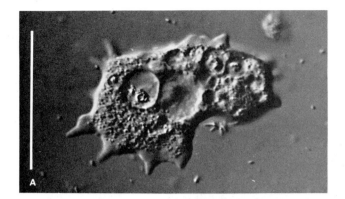

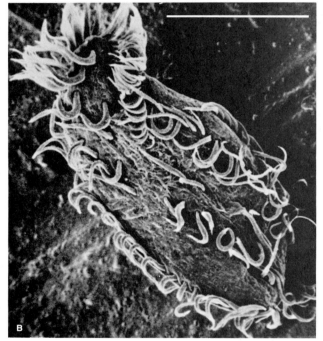

the more familiar living protozoans are amoebas and ciliates (Figure 10-19). A minority of protozoan groups, including foraminifers (page 50), have skeletons that can be fossilized.

When did the protozoans first evolve? The oldest known protozoans in the fossil record, a group of organisms with rigid, vase-shaped skeletons, are no older than about 0.8 billion years (Figure 10-20). Protozoans without these durable skeletons, however, are known to have existed earlier, although they did not leave a fossil record. We can infer this from the fact that the fossil record of acritarchs extends back to about 1.4 billion years. Acritarchs and other plantlike protoctists almost certainly evolved from protozoans, and so it can be assumed that the latter evolved more than 1.4 billion years ago.

Interestingly enough, it is widely agreed that plantlike protoctists, like protozoans, evolved as a result of the union of two kinds of cells. In this major step in the evolution of the Protoctista, a protozoan consumed and retained a spherical or oblong blue-green algal cell. This cell then became an intracellular body known as a **chloroplast**, which served as the site of photosynthesis both in plantlike protoctists and in higher plants, which evolved from them (Figure 10-17). The similarity between blue-green algae and chloroplasts is striking. In both, for example, the pigment

FIGURE 10-19   Living animal-like protoctists. *A.* A naked amoeba (genus *Amoeba*) from the Atlantic Ocean (scale bar is 50 microns). Amoebas engulf particles of food by changing their shape. *B.* A ciliate (genus *Gastrostyla*). The hairlike cilia are used for feeding on bacteria and other ciliates (scale bar is 100 microns). (*Courtesy of F. C. Page and J. Grim.*)

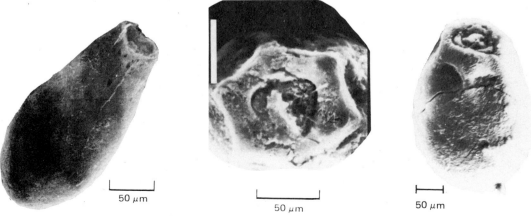

50 μm   50 μm   50 μm

FIGURE 10-20   Vase-shaped fossils that appear to represent   of the Grand Canyon. (*Courtesy of B. Bloeser.*)
animal-like protoctists from the late Proterozoic Chuar Group

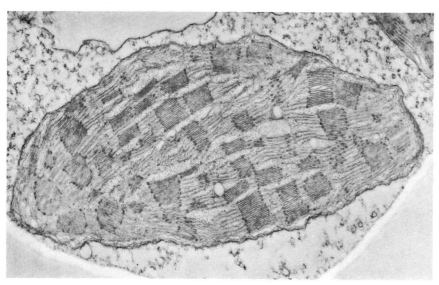

FIGURE 10-21   Section through a chloroplast. The layered structures where chlorophyll resides are similar to those of blue-green algae. (*Courtesy of A. E. Vatter.*)

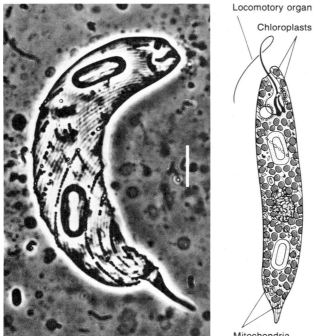

Locomotory organ

Chloroplasts

Mitochondria

FIGURE 10-22   *Euglena*, a plantlike (photosynthetic) protoctist, which, like many animal-like protoctists, is mobile. A *Euglena* cell contains large numbers of chloroplasts. (*Courtesy of G. F. Leedale.*)

chlorophyll, which absorbs sunlight and permits photosynthesis, is located on layered membranes (Figure 10-21).

In many different kinds of protoctists, photosynthesis is conducted within chloroplasts, and it is generally believed that plantlike protoctists evolved many different times when protozoans retained within their cells the blue-green algae that they had eaten. Thus, mobile protozoans may have evolved into mobile photosynthetic forms such as *Euglena* (Figure 10-22) while immobile protozoans evolved into immobile photosynthetic forms.

Multicellular algae

Algae (singular, alga) is a term that encompasses a variety of forms. Photosynthetic protoctists, for example, are called algae, as are the prokaryotic blue-green algae and a large group of multicellular plants. The multicellular plants that are known as algae differ from more advanced land plants in that they lack multicellular reproductive structures to protect their eggs and embryos.

Today, algae that are composed of many connected cells can be found in both freshwater and marine environments. Some groups form lettucelike carpets along rocky seashores, whereas others comprise the great kelps that rise up from some sea floors in twisted ribbons tens of meters long. Even these large forms, however, are fleshy structures that decay easily and are thus unlikely to be fossilized. Some ribbonlike fossils of Proterozoic age are probably segments of blue-green algal mats that only superficially resemble multicellular algae. Carbonaceous structures from 1.3 billion-year-old rocks of the Belt Supergroup of Montana, on the other hand, may well be true multicellular algae, since they are branched. Ribbons and oval remains from the 0.8- or 0.9-million-year-old Little Dal

Group of northwestern Canada almost certainly fall into this category (Figure 10-23).

Multicellular animals

Multicellular animals evolved from animal-like protists rather than from multicellular plants. It is unlikely that any kind of stationary multicellular alga would evolve into an animal that must gather food. On the other hand, it is likely that

FIGURE 10-23   Apparent multicellular algae from Montana (A) and northwestern Canada (B through D). The branched and oblong shapes of these carbonaceous fossils could not have formed by the fragmentation of mats of blue-green algae. The scale bars for B, C, and D are each 10 millimeters long. (Courtesy of M. R. Walter and H. J. Hoffman.)

certain mobile, predatory animal-like protists evolved into higher animals simply by developing multicellular body forms.

Before there were single-celled or multicellular eukaryotic consumers, the earth's ecosystem was relatively simple. Because the primary photosynthetic producers—blue-green algae and eukaryotic algae—did not suffer predation, they multiplied in aquatic settings to an extent that was limited only by the supply of nutrients essential to their growth. Seas and lakes were, in effect, saturated with algae—a situation that may have slowed evolution by leaving very little room for the origin of new species. Although the first organisms to feed on algae must have been animal-like protists, today these forms are feeble in this role in comparison to multicellular animals, which because of their size can consume algae rapidly. Thus, the origin of eukaryotic consumers, especially multicellular animals, added a new trophic level to many aquatic ecosystems (page 53). Exactly when multicellular animals evolved remains uncertain, but we now have evidence that it was not until late in Proterozoic time that these organisms first appeared. For many years, paleontologists have searched well back into the Precambrian interval for skeletons of multicellular animals, but their search has been unsuccessful. The only skeletons that have been found have come from rocks situated close to or above the Precambrian-Cambrian boundary—which, as we will see, cannot be recognized precisely in most areas.

Even during the last century, it was acknowledged that fossil skeletons appear in the stratigraphic record quite suddenly near the base of the Cambrian System. The complexity and variety of fossilized Cambrian life gave rise to speculation that multicellular animals had a long Precambrian history during which they lacked hard parts and therefore left no fossil record. One highly effective means of testing this hypothesis is based on the fact that soft-bodied multicellular animals—those that lack hard parts—have always crawled over the sea floor or burrowed into it and, in doing so, have often left trace fossils in the form of tracks, trails, and burrows in sedimentary rocks (page 11). It was therefore reasoned that if soft-bodied invertebrate animals had existed for a long Precambrian interval, some of them would have left such trace fossils. A search for trace fossils in Precambrian rocks has since turned up a striking pattern—such fossils have been found only in rocks less than about 1 billion years old. For example, 1.3-billion-year-old sedimentary rocks of the Belt Supergroup of Montana, which has yielded some of the multicellular algae of Figure 10-23, exhibit no tracks or trails. As Figure 10-24 illustrates, Belt mudstones are often strikingly well layered in com-

25 mm

FIGURE 10-24   Laminated siltstone from the 1.3-billion-year-old Greyson Shale (Belt Supergroup) in Montana, showing a typical absence of burrowing structures from rocks older than late Proterozoic. (*Courtesy of C. A. Byers.*)

parison to younger deposits, in which laminae are often disrupted or destroyed by burrowing animals.

Although the oldest undisputed Precambrian trace fossils are thought to date back less than 1 billion years, the age of these fossils has never been precisely determined. Such fossils do, however, display a general evolutionary pattern. The oldest trace fossils are geometrically simple structures; often they are tubes made by wormlike animals that burrowed through the sediment (Figure 10-25). In several regions of the world, as stratigraphic sections progress upward toward and into the Cambrian System, simple trace fossils are replaced by more and more complex types whose shapes also vary increasingly (Figure 10-26). This pattern of increase in both complexity and variety seems to represent the initial evolutionary diversification of multicellular animals in the world's oceans. In many regions, the oldest trace fossils are found above the youngest Precambrian

FIGURE 10-25   The undersurface of a bed of sandstone, which displays fillings of primitive, simple, tubelike burrows from Upper Precambrian rocks of Norway. (*Courtesy of N. L. Banks*.)

glacial sediments, which suggests that little multicellular animal life existed before the final glacial episode.

Additional evidence supporting the presence of multicellular life in the latest Precambrian interval has come to light since 1950. In many parts of the world, imprints of soft-bodied animals have been found in sandstone below the oldest rocks containing fossil skeletons (Figure 10-27). Many of these fossils have been thought to represent jellyfishes or sea pens, although this assignation has recently been questioned. Jellyfishes and sea pens are living groups that belong to the phylum **Coelenterata** (the same phylum as corals). Whereas jellyfishes float in the water, sea pens are stalked creatures that stand upright on the sea floor (Figure 10-28). The Ediacara fauna of Australia is the most famous of late Precambrian "soft-bodied" faunas and was also the first to be recognized. In addition to the possible coelenterates, the Ediacara fauna includes several addi-

B

C

FIGURE 10-26   Undersurfaces of sandstones with fillings of relatively complex burrows. These fossils occur in Norway in younger strata than those of Figure 10-25. *A* and *B*. Fillings of feeding burrows. *C*. Filling of a shallow burrow on which can be seen scratch marks left by the legs of the animal that dug it. (*Courtesy of N. L. Banks*.)

FIGURE 10-27   Representatives of the late Precambrian Ediacara fauna of Australia. *A*. A problematic flat, segmented form (×1). *B*. An animal that appears to be intermediate in form between a segmented worm and an arthropod (×1.7). *C*. A strange form of uncertain biological relationships (×1.35). *D*. An animal that may be a jellyfish (×0.7). *E*. An animal that may be a sea pen (×0.6). *F*. A form that may be related to the echinoderms (×1.2). (*Courtesy of M. F. Glaessner.*)

FIGURE 10-28   A reconstruction of the Ediacara fauna (see Figure 10-27), based on the interpretation that some Ediacara fossils represent sea pens and some, jellyfishes. The stalked forms standing upright on the seafloor are sea pens, and the floating, bell-like animals are jellyfishes. Unlike the other forms represented here, these two groups of coelenterates have close living relatives.

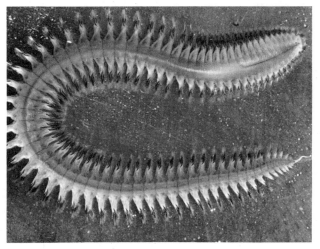

FIGURE 10-29   A living annelid worm. Like most marine annelids, this worm has a pair of small, leglike flaps on each of its many segments, but it lacks true legs. (*L. Margulis and K. V. Schwartz, Five Kingdoms, W. H. Freeman and Company, New York, 1982.*)

tional kinds of animals that cannot be related with certainty to any living group. Fossils of soft-bodied animals such as those of the Ediacara fauna are rare in rocks of Phanerozoic age. It seems likely that the late Precambrian faunas that were preserved escaped destruction only because few predators or scavengers at that early time were capable of devouring their carcasses quickly.

It is puzzling that multicellular animals, including those that form the Ediacara fauna, seem not to have undergone substantial evolutionary diversification until late in Proterozoic time. In fact, the first multicellular animal may not have originated until considerably after 1 billion years ago. We have just seen that the presence of large cells in the fossil record suggests that the eukaryotic cell evolved more than 1.4 billion years ago. Thus, an interval of several hundred million years seems to have separated the origin of the eukaryotic cell from the diversification of multicellular animals. What could account for this evolutionary delay? It has been postulated that protozoans were too primitive to give rise to the multicellular animals before very late in Proterozoic time, perhaps because these organisms lacked sexual reproduction or other advanced eukaryotic features. This idea appears plausible but is unsupported by biological or geologic evidence. An alternative hypothesis is that atmospheric oxygen accumulated slowly, failing to reach the levels necessary to support higher life until late in Proterozoic time. This notion is also feasible, but to date it, too, lacks factual support. Certainly, however, there is geologic evidence—such as the initiation of widespread red-bed formation—supporting the hypothesis that levels of atmospheric oxygen were increasing by about 2 billion years ago.

What is clear is that animals more advanced than coelenterates were well established before the end of Proterozoic time. Among these advanced groups were the segmented worms known as **annelids**, which include modern earthworms as well as many kinds of marine and freshwater species (Figure 10-29). Annelids undoubtedly formed many of the tubelike fossil burrows of late Proterozoic age, perhaps including the simple ones shown in Figure 10-25. Also present were early members of the phylum **Arthropoda**, which includes modern crabs, lobsters, insects, and spiders. Arthropods have external skeletons and jointed legs, a set of which presumably made the scratches visible in the burrow shown in Figure 10-26. Coelenterates, annelids, and arthropods not only continued to flourish into Paleozoic time but have also remained important up to the present time.

CHAPTER SUMMARY

1.   **At least as early as 2 billion years ago, plate-tectonic processes formed mountain belts with characteristics similar to those of Phanerozoic mountain belts.**

2. Continental glaciers spread over parts of Canada and other regions more than 2 billion years ago and also covered many parts of the world between 1 billion years and 600 million years ago.

3. The presence of red beds in sedimentary sequences younger than 2.2 or 2.3 billion years, together with the rarity in such sequences of the easily oxidized minerals pyrite and uraninite, suggests that atmospheric oxygen had reached a moderate level by Proterozoic time. The concentration of atmospheric oxygen earlier than this is the subject of much debate.

4. Banded iron formations, which formed in the presence of oxygen, accumulated in great abundance within marine basins between about 2.5 and 1.8 billion years ago. They may have formed when the concentration of oxygen in the atmosphere was lower than it is today.

5. Stromatolites first became abundant about 2.3 or 2.2 billion years ago. Their success at this time, which may have resulted from the growth of continental shelves, probably led to a buildup of atmospheric oxygen.

6. Large cells that probably represent some of the oldest eukaryotes are found in rocks that are about 1.4 billion years old.

7. Acritarchs are fossils of single-celled planktonic algae that almost certainly were eukaryotic. They underwent an adaptive radiation between 800 and 700 million years ago and then suffered a mass extinction at the time of the last major Precambrian glacial episode, about 600 million years ago.

8. The oldest fossils that appear to represent predatory protists are about 800 million years old. The oldest fossils of multicellular animals are about the same age.

EXERCISES

1. Compare the basic features of the Wopmay orogenic belt to those of the Appalachian orogenic belt described in Chapter 8.

2. What kinds of geologic evidence suggest that glaciers were present on earth more than 2 billion years ago?

3. List as many differences as you can between the Archean world and the world as it existed 1 billion years ago.

4. What arguments favor the idea that little atmospheric oxygen existed on earth until 2 billion years ago?

5. What reasons do we have for believing that multicellular animals did not exist 2 billion years ago?

ADDITIONAL READING

Glaessner, M. F., "Pre-Cambrian Animals," *Scientific American*, March 1961.

Harland, W. B., and M. J. S. Rudwick, "The Great Infra-Cambrian Ice Age," *Scientific American*, August 1964.

Margulis, L., *Early Life*, Science Books International, Boston, 1982.

Schopf, J. W., "The Evolution of the Earliest Cells," *Scientific American*, September 1978.

CHAPTER 11

Proterozoic Cratons: Foundations of the Modern World

A lthough geologists have long attempted to determine how the continents of the modern world originated, the histories of these continents have been traced back only into Proterozoic time. Indeed, the configurations and relative positions of most older, Archean microcontinents will probably never be known. At the same time, uncertainties remain even about the histories of large Proterozoic cratons. The difficulty is that depositional patterns and structural trends of rocks more than half a billion years old are often obscured by erosion, metamorphism, or burial; paleomagnetic data are also sparse and difficult to interpret. Nonetheless, geologists have reconstructed partial histories for most large blocks of Proterozoic crust and have thus gained some knowledge of how they became part of the Phanerozoic world that forms the subject of the final seven chapters of this book.

In the present chapter, we will learn how North America grew episodically during Precambrian time and how this continent also appears to have lost terrane through continental rifting. We will then review what is known of the ancient landmass of Gondwanaland, which, during Proterozoic time, included the oldest block of continental crust that has been identified in the modern world. Finally, we will examine the Precambrian terranes that during Phanerozoic time became assembled into the great northern continent of Eurasia.

Before we discuss the histories of individual cratons, however, let us consider how most cratons undergo changes in size. As we have seen, size increase on a large scale results from the **suturing** of major cratons, which takes place along a subduction zone and is usually accompanied by mountain building in the vicinity of the suture (page 205).

Cratonic growth on a smaller scale, which is known as **continental accretion**, also entails mountain building, but this process occurs at the margin of a single large craton. As noted earlier, marginal accretion can result either from the attachment of a **microplate** to a large craton along a marginal subduction zone (page 209), or from the compression and metamorphism of sediments that have accumulated along a continental shelf. The latter is sometimes referred to as **stabilization**, since it thickens the crust and hardens both unconsolidated sediments and soft sedimentary rocks (Figure 11-1).

Stabilization is a cannibalistic process inasmuch as some of the sediment that is deposited and stabilized along a continental margin is derived from the interior of the continent by erosion. On the other hand, limestone that accumulates along continental margins is precipitated from sea water or secreted by organisms and thus represents an external contribution to the mass of the continent—as do the igneous rocks and oceanic crust that become welded to a continental margin resting along a subduction zone.

Orogenic processes do not simply add material to continents; they also alter preexisting crust. Regional metamorphism, for example—which sometimes operates in conjunction with structural deformation—**remobilizes** continental crust. As we saw in Chapter 5, metamorphism often alters the character of preexisting rocks beyond recognition and resets radiometric clocks so that the age of the crust can no longer be determined. In reviewing the Precambrian

The Black Hills, a blisterlike dome whose core is an isolated uplift of Archean rocks in South Dakota. Here at Mount Rushmore, the busts of four presidents have been carved from the ancient rocks. (*National Park Service.*)

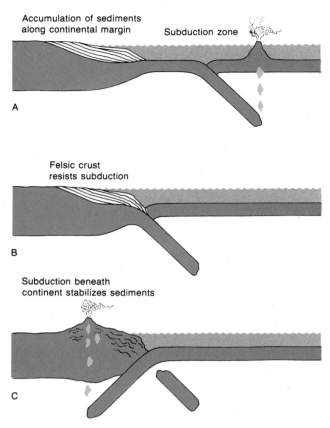

FIGURE 11-1   Stabilization of sediments that have accumulated along the margin of a continent (*A*). The continent comes
to rest along a subduction zone. Because the continent is of low
density and resists subduction, the direction of subduction
switches (*B*). Igneous activity then adds rock to the continental
margin and metamorphoses the sediments that have collected
there (*C*).

FIGURE 11-2   Major geologic features of North America. The
Canadian Shield ends where sediments of the interior lowlands
lap over it on the south and west. The Cordilleran, Ouachita,
and Appalachian orogens flank the North American on the west,
south, and east.

history of individual Proterozoic cratons, we will encounter
many examples of crustal remobilization.

How do continents decrease in size? They can shrink
by erosion, but this process operates so slowly that it is of
little overall significance. Far more important is the process
of continental rifting, which operates on many scales. It can
remove a small sliver of crust, or it can divide a large craton
in half. Although continental rifting that took place more
than half a billion years ago is difficult to document, there
is evidence suggesting that certain major rifting events occurred late in Proterozoic time.

NORTH AMERICA, GREENLAND, AND NORTHERN BRITAIN AS LAURENTIA

Although it lies close to North America, Greenland today

is a continent in its own right. Until late in Phanerozoic time,
however, Greenland was attached to North America, and
these two continents were primary components of a large
landmass known as Laurentia. As we will see, the crust
that now forms northern Great Britain was also part of
Laurentia during Proterozoic time, and it may be that Siberia and Scandinavia were also attached to this great landmass until mid-Proterozoic time. Forming the core of
Laurentia was the crustal block that now represents most
of the North American craton. This ancient block is well
exposed today as the largest Precambrian shield in the
world: the Canadian Shield.

The Canadian Shield constitutes a large portion of the
North American craton, including a small part of the northern United States (Figure 11-2). Precambrian rocks also
underlie the interior of the continent to the south, where
they are overlain by a relatively thin veneer of Phanerozoic

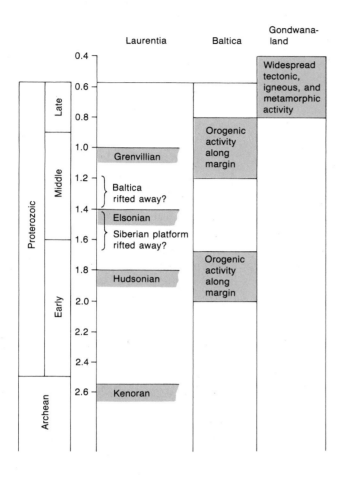

FIGURE 11-3   Major geologic events in Laurentia, Gondwanaland, and Baltica after 2.7 billion years ago.

sedimentary rocks. Rocks obtained from wells that penetrate the Phanerozoic cover have provided a good picture of the general distribution of buried Precambrian rocks. These rocks, together with the exposed rocks of the Canadian Shield, reveal that in the course of Proterozoic time Laurentia gained territory by continental accretion, but also lost territory through continental rifting. Figure 11-3 depicts the sequence of major geologic events on this continent and in other regions during Proterozoic time.

Precambrian provinces and continental accretion

As we learned in Chapter 8, the Appalachian and Ouachita orogenic belts flank the North American craton on the east and south, while the Cordilleran belt flanks it on the west. We have also seen that new crust was added to the margin

of the continent during the development of the Appalachian mountain system in the Paleozoic Era (page 219). As we will see in later chapters, the late Paleozoic events that produced the Ouachita mountain system also increased the size of North America, as did the Paleozoic, Mesozoic, and Cenozoic events that produced the Cordilleran system.

Evidence that North America was growing by accretion during Proterozoic time began to appear decades ago, when the recognition of structural trends and regional lithologic patterns permitted geologists to recognize natural geologic provinces within the Canadian Shield. More recently, reliable radiometric dates for rocks of the Canadian Shield and for subsurface rocks bordering the shield have yielded a much more detailed picture. These dates are obtained from crystalline rocks, and thus they represent episodes of igneous and metamorphic activity. As it turns out, such activity occurred throughout Precambrian time but was concentrated primarily in four major intervals, which are depicted in Figure 11-3 along with comparable intervals on other cratons. The igneous and metamorphic rocks that were produced during each interval are not spread randomly throughout the North American craton, however; instead, they tend to be localized in discrete belts that are formally recognized as geologic provinces (Figure 11-4).

The arrangement of the Precambrian provinces tells the story of continental accretion. To understand this story, however, it is important to recognize that each date shown in Figure 11-4 represents the youngest interval of widespread activity within a province. How might we expect provinces of various ages to be arranged within the North American craton? In general, we would expect older provinces to lie in the center of the craton and younger provinces to lie near the margin. In a general way, this is the pattern that we find within the North American craton. Note, for example, that the ancient Superior Province, which has remained stable since latest Archean time (about 2.6 billion years ago), is flanked by younger provinces. The Central Province, which has remained largely stable for the last 1.4 billion years, borders it on the south. Both the Superior and Central provinces are bordered on the east by the Grenville Province, which represents a long marginal orogen that was active about 1.0 billion years old. The Grenville Province, in turn, is bounded by the younger Appalachian orogenic belt, which, as we saw in Chapter 8, is of Paleozoic origin.

The nature of the Grenville and Appalachian orogens illustrates that while an orogenic episode may add crust to the margin of a craton, it usually also deforms and metamorphoses older cratonic crust that lies alongside the or-

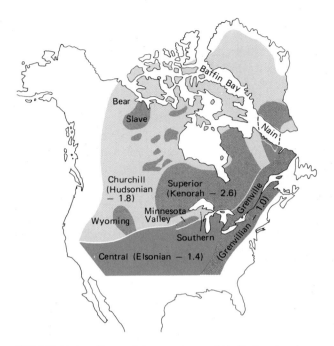

FIGURE 11-4   Precambrian provinces of North America. In parentheses are the names and terminal dates (in billions of years) of the intervals of final stabilization or remobilization of the provinces' rocks. Archean terranes were stabilized during the Kenoran orogenic interval, and younger terranes were stabilized around them in a pattern of continental accretion.

ogenic belt. Thus, while the **Grenville** orogenic episode probably accreted some new crust to the eastern margin of North America, it also metamorphosed a large segment of crust that had previously represented terrane of the Superior and Slave provinces. The Appalachian orogenic activity, in turn, overlapped the Grenville Province. In fact, the Blue Ridge and similar uplifts to the north and south consist of crystalline rocks of Grenville age that were caught up in the Appalachian deformation (Figure 8-18).

In the following sections, we will address in greater detail the Proterozoic distribution of metamorphism in the North American craton as well as other interesting aspects of the accretion and modification of the craton.

Stabilization of Archean crust: The Kenoran interval

Recall that podlike bodies of greenstone characterize many

Archean terranes throughout the world. In Chapter 9, we observed that the largest group of Archean greenstones of the Canadian Shield occurs in the Superior Province (Figure 9-22). Within this province are numerous bodies of crystalline rock whose radiometrically determined ages cluster around 2.6 billion years (Figure 11-4). Widespread igneous activity produced these rocks during the **Kenoran** orogenic interval (Figure 11-3).

The Kenoran orogenic events at the close of Archean time appear to have united many small Archean crustal blocks to form a large, stable craton that now constitutes the Superior Province. The smaller Wyoming Province is characterized by uplifts such as the Beartooth Mountains (Figure 11-5) and the Black Hills, a domelike structure of Precambrian basement rocks that is elevated high above the Great Plains (Figure 11-6); here the giant presidential busts of Mount Rushmore are sculpted from Archean rocks (page 296).

In the Southern Province, which lies primarily in Minnesota, are rocks that yield some of the oldest radiometric ages ever obtained on earth: nearly 3.7 billion years. This small province was continuous with the Superior Province until about 1.8 billion years ago, when the two provinces were separated by a narrow zone of metamorphic activity.

The Archean rocks that form the Nain Province of northeastern Canada and Greenland are separated from the Superior terrane by a narrow band of remobilized rocks (Figure 11-4). Greenland, of course, is now a discrete continent, but its Precambrian terranes align with those of Canada on the opposite side of Baffin Bay. As we will see in Chapter 16, Greenland did not separate from North America until late in the Cretaceous Period, less than 100 million years ago. The southern tip of Greenland is considered to be part of the Nain Province, which means that it was continuous with the North American segment of this province at the end of the Kenoran orogenic interval. The Slave Province to the northwest would appear to be the remnant of another landmass that existed at the end of the Kenoran interval.

The Hudsonian interval

The preceding section explains why geologists believe that sizable landmasses had formed from small Archean crustal blocks by the beginning of Proterozoic time, about 2.5 million years ago. What followed in early Proterozoic time was an interval of relative tectonic quiescence that lasted until

FIGURE 11-5   A U-shaped glaciated valley in the Beartooth Mountains of Wyoming. Archean rocks form the core of the Beartooth uplift. The glacial features formed within the last 2 million years during the recent Ice Age. (*National Park Service*.)

slightly less than 2 billion years ago. Then came another pulse of widespread orogenic activity in Canada that has been labeled the **Hudsonian** interval. Mountain building in the Wopmay orogen (page 273) occurred at this time, accompanied by orogenic activity in a horseshoe-shaped belt that now encircles the peninsular region east of Hudson Bay (Figure 11-7). In the area of this second orogen, which is now represented in the Labrador Trough, banded iron formations are among the most conspicuous sedimentary units. Also deformed and metamorphosed during the Hudsonian interval was the Animikie sequence near Lake Superior. Mining of banded iron formations here has provided the United States with most of its iron ore (Figure 11-8).

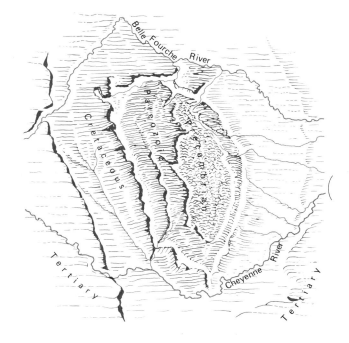

FIGURE 11-6   Sketch of the Black Hills in aerial view, showing the domelike structure.

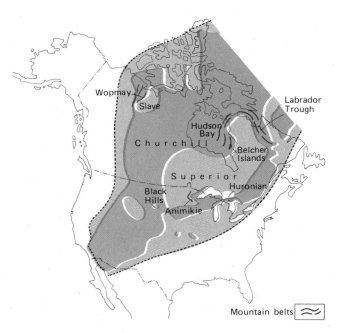

FIGURE 11-7   The North American craton at the end of the Hudsonian orogenic interval, 1.8 billion years ago. Broken lines show the approximate boundaries of the craton. Brown shading indicates regions where there was orogenic activity after Archean time but before 1.8 billion years (the end of the Hudsonian interval). The Huronian sedimentary sequence is identified, as are several mountain belts. (*Modified from W. R. Muehlberger et al., Amer. Assoc. Petrol. Geol. Bull. 51:2351–2380, 1967.*)

Many crystalline rocks within the Churchill Province of western North America yield radiometric dates of about 1.8 billion years, which indicates that Hudsonian orogenic activity occurred here as well (Figure 11-7). Magnetic anomalies and deviations in the earth's gravitational field in this province reveal structural trends in Precambrian rocks lying beneath younger sediments. Distinctive north-south trends within the Wyoming and Superior provinces differ from those of the Churchill Province, suggesting that the latter provinces were united in the intervening zone (which now lies within the Churchill Province) when orogenesis occurred here about 1.8 billion years ago.

Largely spared from the deformation and metamorphism of the Hudsonian interval was the Huronian sequence of sediments east of Lake Superior. Here, preserved with little alteration, are magnificent varved deposits that represent the Proterozoic glacial episode that took place more than 2 billion years ago (Figure 3-7).

The Elsonian interval

Figure 11-7 shows that most of the present Canadian Shield had accreted by the end of the Hudsonian orogeny about 1.8 billion years ago. The stability that resulted is reflected in the fact that igneous and metamorphic rocks with dates appreciably younger than 1.5 billion years are all but unknown in the central provinces of the Canadian Shield (the Superior, Churchill, and Slave provinces). Rocks ranging in age from 1.4 to 1.5 billion years, however, are common in large areas along the eastern and southern flanks of the early Proterozoic craton (Figure 11-9). Drill cores show that much of the midcontinent of the United States is also underlain by Elsonian-age rocks. Thus, although details are not yet clear, it would appear that, soon after the close of the early Proterozoic interval, the craton grew toward the east and south.

Middle Proterozoic rifting in central and eastern North America

The greatest disturbance of the central North American craton during the last 1.4 billion years was an episode of continental rifting that took place between about 1.3 and 1.0 billion years ago. This episode, which ultimately failed,

FIGURE 11-8   Aerial view of mining of iron formations of the Proterozoic Animikie sequence in the Mesabi Range, Minnesota. (*C. F. Park and R. A. MacDiarmid, Ore Deposits, W. H. Freeman and Company, San Francisco, 1975.*)

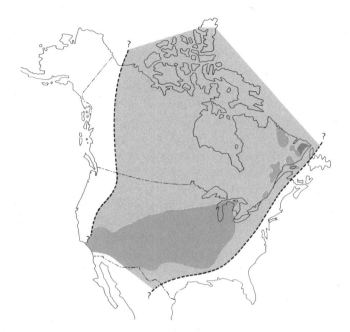

was characterized by faulting and by mafic igneous activity in which lavas poured into downwarped basins along a belt that extended from the Great Lakes region to Kansas (Figure 11-10). Some of these lavas, which are known as Keweenawan basalts, are exposed near the southern border of the Canadian Shield. They contain ore deposits of native copper, a mineral consisting of the element copper uncombined with other elements (Figure 11-11). Similar basin basalts lie to the southwest beneath the sedimentary cover of the Midwest, as has been revealed both by the examination of rock cuttings taken from deep wells and by the

FIGURE 11-9   The North American craton at the end of the Elsonian orogenic interval, 1.4 billion years ago. Broken lines show the approximate boundaries of the craton. Brown shading indicates regions where there was orogenic activity between 1.8 and 1.4 billion years ago. (*After W. R. Muehlberger et al., Amer. Assoc. Petrol. Geol. Bull. 51:2351–2380, 1967.*)

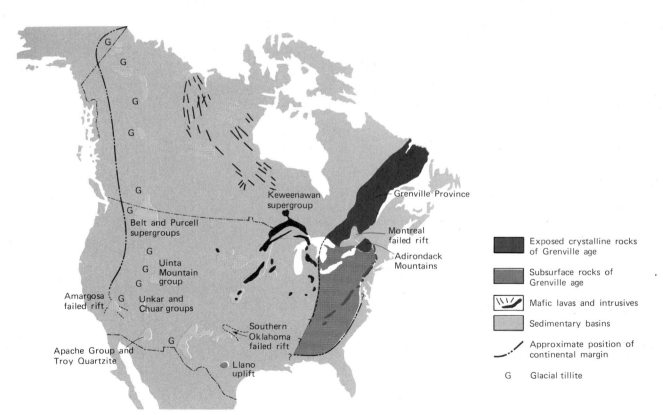

FIGURE 11-10   The North American craton about 850 million years ago, showing major geologic features developed after 1.4 billion years ago. Many of the glacial tillites (G) are actually slightly younger than 850 million years. The Keweenawan Supergroup, which includes basic volcanics, accumulated in a rift that ultimately failed. Some of the sedimentary basins of western North America were failed rifts. The Grenville Province represents an orogenic belt that was active slightly before a billion years ago. (*Modified from J. H. Stewart, Geology 4:11–15, 1976.*)

FIGURE 11-11   Columnlike joints in the Edwards Island flow, a body of volcanic rock within the Keweenawan Supergroup of northern Michigan. (*N. K. Huber, U.S. Geological Survey*.)

detection from the earth's surface of strong magnetism. Because these basalts are rich in iron and magnesium and are therefore of high density, their presence is also associated with a feature known as the Midcontinental Gravity High, which is a local increase in the earth's gravitational field as measured from the surface. While the basalts were forming, numerous basic dikes were also emplaced across the Canadian Shield to the north (Figure 11-10), and a large plutonic igneous body, the Duluth Gabbro, was intruded within the area of the Keweenawan eruptions.

The configuration and composition of Keweenawan rocks and their subsurface counterparts are reminiscent of the Triassic Newark Series, whose configuration indicates the presence of a failed rift in the eastern United States (page 199). The Keweenawan volcanics, for example, are associated with red siliciclastic rocks and alluvial-fan conglomerates in what appear to be downfaulted troughs— configurations that tend to occur in newly forming continental rifts. Thus, it would appear that about 1.3 billion years ago, a spreading center formed beneath the late Precambrian craton and began to rift it apart. It is possible that this spreading center intercepted the eastern margin of the ancient craton, but this remains uncertain. In any event, it is obvious that the Keweenawan rifting was abortive. Rifting ceased before the continent was split but left its mark in the enormous volumes of mantle-derived lavas that were disgorged along a belt more than 1500 kilometers

(~900 miles) long and 100 kilometers (~60 miles) wide!

We have seen that most of the present Canadian Shield had accreted by the end of the Hudsonian orogenic interval about 1.8 billion years ago (Figure 11-7). The emplacement of large dikes within the Churchill Province between 1.3 and 1.0 billion years ago, together with the rifting and the mafic extrusions of the Keweenawan Supergroup, indicates that the craton was rigid at this time. Paleomagnetic data provide further evidence that what we now call the Canadian Shield has behaved like a rigid block since the Hudsonian interval or even earlier: Rocks within the various provinces of the shield show similar paths of apparent polar wander for the last 2 billion years (see page 181).

Let us now examine what happened along the eastern and western margins of the large North American craton which had formed by mid-Proterozoic time.

The Grenville interval in Eastern North America

Recall that the Grenville orogenic belt formed along the east coast of North America (Figure 11-10). The igneous and metamorphic activity of the Grenville orogeny ended about 1 billion years ago (Figure 11-3). The resulting crystalline rocks and others are best exposed in the Canadian portion of the Grenville Province (Figure 11-4). To the south, most crystalline rocks of Grenville age are buried, but some crop out here and there—especially in the Adirondack uplift of New York State and, as noted above, in the Blue Ridge Mountains and other uplifts lying between the Valley and Ridge Province and the Piedmont. In the Gulf Coast region, most of the Grenville Belt lies beneath Phanerozoic cover, but in central Texas crystalline rocks of Grenville age appear at the surface as a small, isolated prominence known as the Llano uplift.

Much evidence indicates that between about 1.4 and 1.0 billion years ago, a major episode of continental rifting took place in the eastern half of North America, creating a new continental margin in the area of the present northern Appalachian Mountains. The enormous failed rift of the midcontinent region, where the Keweenawan rocks accumulated, represents one piece of evidence, and another is a failed rift of similar age in the vicinity of Montreal (Figure 11-10). The rocks of the Montreal failed rift include volcanics situated in northeasterly trending dikes. These failed rifts must have projected from a continental margin that probably came into existence about 1.4 or 1.3 billion years ago, precisely when the failed rifts began to form. The newly formed continental shelf must have lain to the southeast of the present Grenville Province, since sediments laid down along this shelf are now buried beneath the Piedmont Province and the Coastal Plain.

What landmass was rifted away from eastern North America to form the new continental margin? One possibility is that this landmass, or at least part of it, was the Baltic Shield. There is paleomagnetic evidence indicating that Scandinavia may have lain close to eastern North America about 1 billion years ago, and, as we will see in the next chapter, there is also a Precambrian province of Grenville age in eastern Scandinavia.

Figure 11-12 summarizes the events that produced the rocks of the Grenville Province. As this figure illustrates, rifting was followed first by deposition along the newly formed shelf and then by the formation of the Grenville orogen. Deposition and stabilization of the sediments of the Grenville belt added a marginal increment of unknown proportion to eastern North America. It is not known, however, whether the Grenville orogeny, between about 1.1 and 1.0 billion years ago, involved continental collision or simply the rafting of eastern North America against a subduction zone.

In any event, after the orogeny a new depositional margin formed along the eastern margin of North America. This margin seems to have developed earlier in the north, where marine deposition began about 800 million years ago, than in the southern United States, where it did not commence until about 600 million years ago. The body of water that bordered eastern North America during this interval of deposition was the Iapetus Ocean, which was discussed in Chapter 8 (Figure 8-29). Recall that the latest Precambrian and early Paleozoic sediments of the first tectonic cycle of the Appalachian Mountain system accumulated along the margin of the Iapetus Ocean and that in Cambro-Ordovician time, these sediments were carbonates deposited along a vast carbonate platform (Figures 8-24 and 11-12F).

The British Isles

Modern geology was born in Great Britain more than a century ago, and it was there that most of our formal geologic systems were originally defined. Thus, it is a lucky accident that this small group of islands happens to contain rocks representing every Phanerozoic period, together with a variety of rocks of Precambrian age. Precambrian rocks of the British Isles are best exposed in northern Scotland

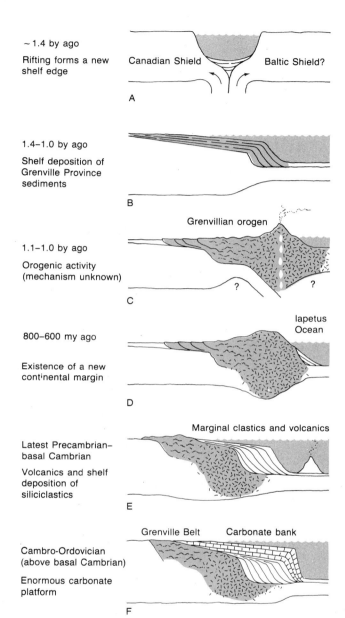

~1.4 by ago
Rifting forms a new shelf edge

Canadian Shield   Baltic Shield?

A

1.4–1.0 by ago
Shelf deposition of Grenville Province sediments

B

Grenvillian orogen

1.1–1.0 by ago
Orogenic activity (mechanism unknown)

?   ?

C

Iapetus Ocean

800–600 my ago

Existence of a new continental margin

D

Marginal clastics and volcanics

Latest Precambrian–basal Cambrian

Volcanics and shelf deposition of siliciclastics

E

Grenville Belt   Carbonate bank

Cambro-Ordovician (above basal Cambrian)

Enormous carbonate platform

F

FIGURE 11-12   Sequence of events along the eastern part of the North American craton from middle Proterozoic time until early Paleozoic time. Sediments of the Grenville Province were deposited before 1.1 billion years ago along a shelf margin that may have formed when a continental block (possibly the Baltic Shield) rifted away from North America. After the Grenville orogeny deformed the continental margin, a clastic sequence and then a carbonate bank formed along the edge of the Iapetus Ocean.

(Figure 11-13), where James Hutton did his pioneering work in uniformitarian geology, as well as in the northern part of Ireland. In these northern areas, Precambrian rocks

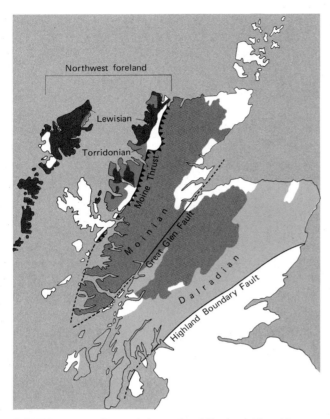

FIGURE 11-13   Precambrian rocks of Scotland. The oldest ones are the crystalline rocks of the Lewisian sequence. The younger rocks are sedimentary. The stratigraphic sequence of these rocks is shown in Figure 11-15.

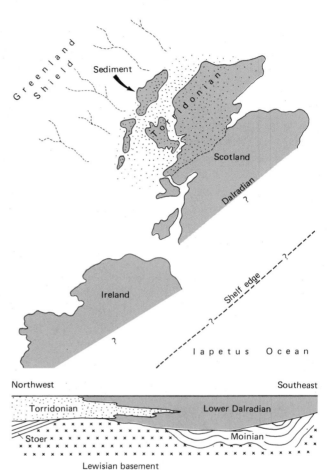

FIGURE 11-15   Stratigraphic cross section of Precambrian rocks in northern Scotland and Ireland (*above*) and the generalized geography of northern Scotland and Ireland when Torridonian and Dalradian sediments were deposited near the end of the Precambrian (*below*); these regions were attached to Greenland along the margin of the Iapetus Ocean.

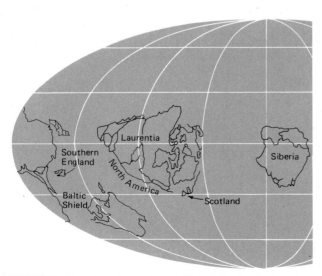

FIGURE 11-14   Late Cambrian position of Scotland, when it was still adjacent to Greenland and North America. Southern England lay at some distance from Scotland. (*After C. Scotese et al., Jour. Geol. 87:217–277, 1979.*)

form part of an ancient shield that during late Precambrian and early Paleozoic intervals was connected to the Canadian and Greenland shields as part of Laurentia (Figure 11-14). At this time, southern Great Britain was a separate landmass. As we learned in Chapter 8, northern and southern Great Britain became attached to each other during the middle Paleozoic Caledonian orogeny (Figure 8-30). As we will learn in later chapters, all of Britain (as well as the rest of Europe) remained attached to Greenland until early in the Cenozoic Era, less than 60 million years ago.

Figure 11-15 illustrates the attachment of northern Britain to Greenland during late Proterozoic time and the general sequence of Precambrian rocks in northern Britain.

FIGURE 11-16   An outcrop of Lewisian gneiss in Scotland. (*Courtesy of J. Watson.*)

metamorphism may have disguised even older Archean rocks. Lewisian rocks are separated from younger Precambrian rocks by an unconformity. Metamorphic rocks (or gneisses), which make up the bulk of the Lewisian complex (Figure 11-16), crop out spectacularly throughout northern Scotland in what is known as the Northwest Foreland (Figure 11-17).

Southeast of the main Lewisian outcrops in Scotland, metamorphosed sediments known as Moinian are exposed over a large area. These rocks, which are about 1.2 billion years old, were originally poorly sorted siliciclastic deposits of the Grenville Belt, deposited when Scotland was attached to Greenland and North America. They were then metamorphosed and deformed during the Grenville orogeny about 1 billion years ago and were ultimately transported northwestward along what is known as the Moinian Thrust Fault during the Caledonian orogeny (Figure 11-13). In the Northwest Foreland, another package of sediments, the Stoer Group, was deposited on top of the Lewisian slightly later than the Moine sequence was laid down, as a consequence of which the Stoer Group is less severely deformed.

The older rocks in this area, which are known collectively as the Lewisian sequence, range in age from late Archean to early Proterozoic (about 2.9 to 1.7 billion years), although

FIGURE 11-17   Landscape of the Northwest Foreland of the Assynt region of Scotland. The mottled rocks in the foreground are of Lewisian gneiss. The mountain silhouetted against the clouds is formed of Torridonian sandstone that unconformably overlies the Lewisian (see text below and Figure 11-18), but its peaks on the right are capped by Cambrian quartzites that rest unconformably on the Torridonian. (*A Scotman photograph.*)

Between about 900 and 800 million years ago, an erosion surface developed on the deformed Moine and Stoer groups, exposing the older Lewisian rocks in some places. On top of this irregular surface were deposited thick sequences of sediment that have traditionally been labeled Torridonian in the northwest and Dalradian in the southeast. The Torridonian sequence, which accumulated between 800 and 650 million years ago, is often separated from underlying rocks by a striking angular unconformity (Figure 11-18) and includes several distinct packages of sedimentary rocks, the most conspicuous of which consist of red arkoses. The mineral composition of some Torridonian sediments suggests that these sediments were shed from the Precambrian basement of Greenland (Figure 11-15).

The lower part of the Dalradian sequence to the southeast correlates with the Torridonian. The Dalradian sequence, however, ranges upward into the Cambrian without significant interruption. Late Proterozoic glaciation is represented within the Dalradian sequence by extensive tillites both in Scotland and in the northern part of Ireland, where as many as 17 separate glacial advances have been recognized. The Dalradian sequence contains a varied group of sediments, including stromatolitic marine limestones. Its upper (Cambrian) portion, however, is dominated by turbidites and by pillow lavas, indicating that part of Scotland was once deep sea floor at the western margin of the Iapetus Ocean.

FIGURE 11-18   Unconformity between the Torridonian sediments of Loch Assynt, northern Scotland, and Lewisian gneiss below. (*Courtesy of J. G. C. Anderson.*)

Western North America

Middle and Late Proterozoic events in western North America differed greatly from those in the east. In several areas from southern Canada to southern Arizona, for example, sequences of relatively fresh Precambrian sedimentary rocks of this age unconformably overlie older and more highly altered rocks. The younger rocks appear to have been deposited in local basins that probably lay near a newly formed continental margin (Figure 11-10). Above these basin

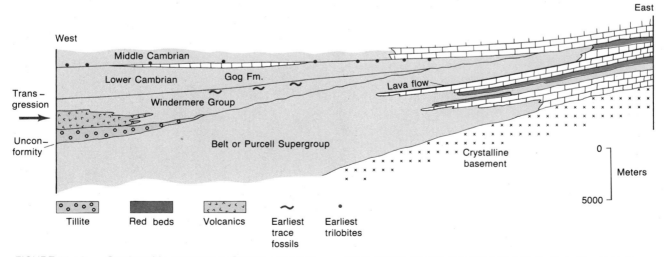

FIGURE 11-19   Stratigraphic sequence in Canada just north of Montana, ranging from the middle Proterozoic through the Middle Cambrian. The Belt Supergroup accumulated in a large basin (Figure 11-10). (*Modified from P. B. King, The Evolution of North America, Princeton University Press, Princeton, New Jersey, 1977.*)

FIGURE 11-20   Stromatolitic limestones of the Belt Group, Helena, Montana (*R. Rezak, U.S. Geological Survey.*)

deposits are still younger Precambrian and Lower Cambrian shelf sediments that extend in a much more continuous belt from northern Canada to northern Mexico.

In the northern United States, the largest of the basin sequences is the famous Belt Supergroup (Figure 11-10), which ranges in age from about 0.9 to 1.5 billion years. The northern extension of this formation is known in Canada as the Purcell Supergroup—but for simplicity we can apply the name *Belt* to the entire sequence. In general, the Belt thickens toward the west (Figure 11-19), sometimes reaching a thickness of 16,000 meters.

The Belt formed in a northwesterly trending basin resembling a failed rift. Sandstones increase in abundance toward the western part of the sequence, while limestones increase toward the east, where deposition occurred in shallower water. In general, however, mudstones predominate. The Belt apparently formed as a result of sediment accumulation in very shallow water during rapid subsidence. Salt crystals and mud cracks, both of which indicate a drying up of shallow bodies of water, are present in Belt sediments, and shallow-water stromatolites are also abundant in the limestones (Figure 11-20).

Above the Belt are the Windermere Group and its equivalents (Figure 11-19). These shelf deposits, which include upper Precambrian and lower Cambrian deposits, unconformably overlie the Belt and other localized basin deposits and have a much more extensive north-south distribution. Toward the west, the deposits begin with rocks that represent late Proterozoic glaciation, including tillite as well as pebbles and cobbles that exhibit what appear to be glacial striations or scratches. Near the top of the Windermere are found the earliest trace fossils of the region. These burrows, which occur well below the first shelly faunas, appear to be very late Precambrian in age.

A remarkably similar sequence of rocks is found as far south as Death Valley, California (Figures 11-21 and 11-22). Here the lower part of the sequence, which resembles the Belt Supergroup, accumulated in a structure known as the Amargosa failed rift (Figure 11-10). A tillite in the upper part of the sequence, like that recorded in the Windermere Group, represents late Proterozoic glaciation. Subsequently, seas inundated a broader area over which the Noonday Dolomite and younger deposits were laid down.

As Figure 11-10 depicts, additional localized sedimentary units were deposited in late Proterozoic time to the south and east. Thus, the interval represented by the Belt Supergroup (approximately 1.5 to 0.9 billion years ago) was a time of localized basin deposition in western North America. In many areas, however—including the Belt basin—sediments deposited during this interval were soon weakly deformed. The more widespread latest Precambrian and Early Cambrian sequences—the Windermere and its equivalents—were then deposited during a marine transgression along the full length of the continental margin.

It seems evident that a rifting episode preceded the interval of Belt deposition, since a new continental margin

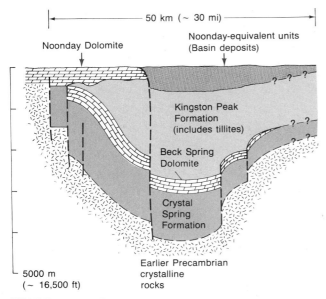

FIGURE 11-21   Diagrammatic cross section of the Amargosa failed rift after deposition within it had ceased. The dashed vertical lines are normal faults. (*After L. A. Wright et al., Calif. Div. Mines and Geol. Spec. Rept. 106:7–18, 1976.*)

FIGURE 11-22   View from the southwest of the western side of the Panamint Range, with Death Valley in the distance. The Crystal Spring and Kingston formations of the Amargosa failed rift (Figure 11-21) form most of the dark outcrops. (*J. H. Maxon, National Park Service, courtesy of B. W. Troxel.*)

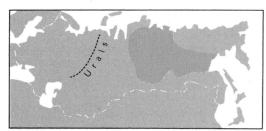

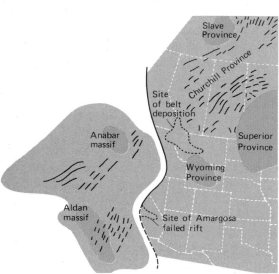

was produced, and another failed rift cut into the continent in the Death Valley region. The other new marginal basins may have represented failed rifts as well. In this regard, two major questions remain to be resolved. First, what produced the modest deformation of these units in many areas before the younger Windermere and its equivalents were spread over them? And second, what happened to the landmass that was rifted away about 1.5 billion years ago, leaving a new continental margin along which the Belt and other marginal basin sequences formed (Figure 11-10)?

The first question remains unanswered; we simply do not know what tectonic mechanism disturbed the Belt and its equivalents. For the second question, however, geologists have a likely answer. The Siberian platform, which now lies far across the Pacific, exhibits a number of fea-

FIGURE 11-23   Possible connection between the North American craton and the Siberian platform before the deposition of the Belt Supergroup about 1.6 billion years ago. The outline of Asia (*upper left*) shows the present position of the Siberian craton. (*Modified from J. W. Sears and R. A. Price, Geology 6:267–270, 1978.*)

tures indicating that it may have been the landmass that was rifted away. Rotation of the Siberian platform through less than 90° orients it so that it fits rather neatly against the margin of the North American continent that existed when the Belt and its equivalents were deposited (Figure 11-23). In this arrangement, the structural trends of the two cratons also align. The Aldan and Anabar massifs, which are Siberian blocks of Archean age, then become partners of the Slave and Superior provinces of North America and of the Wyoming Province. (**Massif** is a name commonly used for durable, crystalline crustal blocks that are smaller than Precambrian areas recognized as shields.) The region between the Anabar and Aldan massifs, which aligns with the Churchill Province, experienced orogenic activity between 1.7 and 2.1 billion years ago—at about the time when Hudsonian orogenesis affected the Churchill Province. The crystalline rocks of the Siberian platform are also overlain by middle and late Proterozoic sedimentary rocks that resemble those of the North American craton. In the interior of the Siberian craton, basal conglomerates and sandstones are followed by stromatolitic carbonates. Thicker deposits of the shelf margin begin with basal sandstones and tillites that pass upward into alternating siliciclastics and stromatolitic carbonates. In places, marginal failed rifts have been recognized. In a general way, this configuration is a mirror image of what we find in western North America, and it thus tends to support the hypothesis that the Siberian craton is the missing continent. Still, the so-called Siberian connection remains unproven.

THE EARLY HISTORY OF GONDWANALAND

As we learned in Chapter 7, Gondwanaland was the great southern continent that fragmented during the Mesozoic Era to form South America, Africa, peninsular India, Australia, and Antarctica. Similarities among the life, strata, and geologic structures of these five landmasses formed the strongest early argument in favor of continental drift (page 174).

We have seen that Laurentia experienced considerable marginal accretion during the Elsonian, Grenville, and Appalachian orogenic episodes (Figure 11-4). Gondwanaland, too, experienced marginal accretion, which was largely confined to belts along what now constitute the margins of South America, Antarctica, and Australia. Gondwanaland remained intact throughout the Paleozoic Era, and Lau-

rentia and Gondwanaland became united as the dominant elements of the supercontinent Pangaea near the beginning of the Mesozoic Era more than 200 million years ago. Subsequently, the fragmentation of Pangaea produced the continents of the modern world, including North America and Greenland, which broke off from the part of Pangaea that had once been Laurentia.

When Pangaea broke apart, the large portion that had once been Gondwanaland suffered much more severe fragmentation, fracturing into five fragments—four of which represent large continents today. The reassembly of these five fragments into Gondwanaland reveals interesting patterns of orogenesis and metamorphism (Figure 11-24). Recall from Chapter 7, for example, that when continents are assembled into the hypothetical landmass of Gondwanaland, those mountain belts in South America, Antarctica, and Australia that formed after early Paleozoic time line up neatly to form what was known as the Samfrau chain.

Inland from the Samfrau mountain chain in Gondwanaland was an older zone of Proterozoic and early Paleozoic age that represented a vast network of igneous, metamorphic, and tectonic activity in Gondwanaland between about 800 and 400 million years ago (Figure 11-24). What is striking about this network of activity is that much of it seems to have taken place as regional metamorphism within

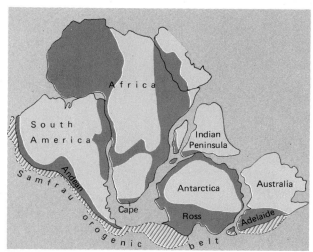

Mobile belts active 800–400 million years ago

Samfrau orogenic belt—active after 400 million years ago (after early Paleozoic time)

FIGURE 11-24   Modern continents reassembled into their relative positions within Gondwanaland. Note that many of the mobile belts that were active in late Proterozoic and early Paleozoic time correspond to the lines along which Gondwanaland later broke apart.

rather than along the margins of Gondwanaland; thus far, no evidence of continental suturing has been found along the interior metamorphic zones. If suturing did not occur, however, the metamorphic zones cannot be explained by conventional plate-tectonic processes, and their origin remains a mystery. In Figure 11-24, the zones representing igneous, metamorphic, or tectonic activity between 800 and 400 million years ago have been collectively labeled mobile belts rather than orogenic belts because those belts that display only metamorphism may not have formed parts of conventional mountain systems.

It is interesting to note that some 400- to 800-million-year-old mobile belts outline continental blocks that coincide with the present-day continents derived from Gondwanaland. In other words, when Gondwanaland finally broke apart during the Mesozoic Era, many of the rift zones tended to follow the mobile belts of the earlier interval of instability. It would appear that zones of crustal weakness from this earlier interval persisted for hundreds of millions of years.

The earlier history of Gondwanaland is difficult to decipher; indeed, it is not even certain that Gondwanaland existed much earlier than 800 million years ago. Thus, for the purposes of our discussion, we will consider the Precambrian rocks of Gondwanaland in terms of the five modern continents where they are found today. The Precambrian record of Africa will be our first subject.

Africa

As described in the previous chapter, the southern part of modern Africa includes the oldest cratonic block of large proportions now recognized in the world. This block, which is more than 3 billion years old, may have been the crustal core around which Gondwanaland grew. Radiometric dating reveals that several regions in modern Africa consist of crustal elements older than 3 billion years; by early Proterozoic time (about 2 billion years ago), nearly all of the modern craton was in existence (Figure 11-25).

During mid-Proterozoic times, deformation continued on a smaller scale. Then, from the end of the Precambrian interval to Paleozoic time, a series of orogenic events rejuvenated not only a large percentage of Africa but much of Gondwanaland as well.

Approximately 400 million years ago, Africa became a remarkably stable landmass, and it remained so until very recently. This stability can be attributed to the fact that much of Africa lay in the interior of Gondwanaland and was thus protected by South America on one side and by India and Antarctica on the other. Africa's northwestern margin and southern tip, which were exposed along the perimeter of Gondwanaland, were deformed during the Paleozoic Era: and, as is often the case, these continental margins were rafted against subduction zones. Today the surface

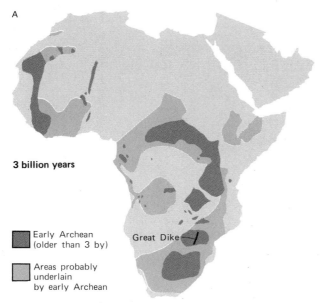

A

3 billion years

■ Early Archean (older than 3 by)

□ Areas probably underlain by early Archean

Great Dike —/

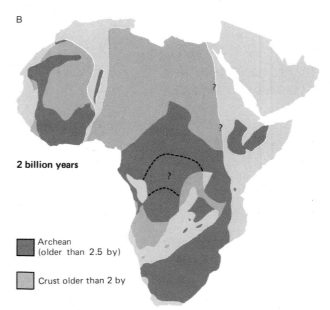

B

2 billion years

■ Archean (older than 2.5 by)

□ Crust older than 2 by

FIGURE 11-25   Precambrian crust of Africa. *A*. Crustal areas present at the end of early Archean time. *B*. Crustal areas present 2 billion years ago; by this time, most of the modern landmass was present. (*After A. Kröner, Precambrian Research, 4:163–213, 1977.*)

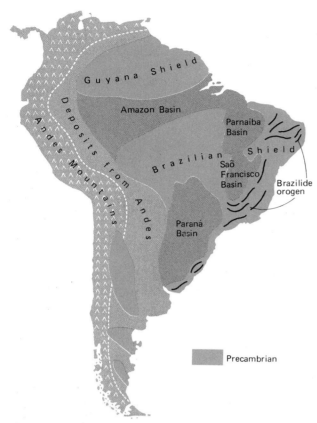

FIGURE 11-26   Major geologic features of South America. Two large Precambrian shields are found in the north and three small Precambrian massifs, in the south. The Andes Mountains and sediments shed from them dominate the western margin of the continent.

rocks are uncommon throughout South America, the relative ages of these shields are difficult to assess. The most widespread Archean units are found in the northern part of the Guyana Shield, and a small Archean region lies near the Atlantic margin of the Brazilian Shield. One of the reasons Archean dates are relatively scarce here is that both shields were extensively metamorphosed about 2 billion years ago. Along the coast of the Brazilian Shield is a younger orogenic zone, the Brazilide orogen. This area has been radiometrically dated at 650 to 450 million years and thus represents one of the many orogenic episodes that affected Gondwanaland between 800 and 400 million years ago (Figure 11-24).

There is a striking difference between the Phanerozoic behavior of Precambrian terranes in North and South America. Even at the beginning of Phanerozoic time, it would appear that the Precambrian rocks of North America had been eroded to low relief (page 221), and certainly the Canadian Shield is a low-lying area today (page 238). In contrast, the South American shields and massifs have remained positive topographic features throughout Phanerozoic time. To the east of the tall Andes Mountains, which have been uplifted intermittently throughout most of Phanerozoic time, depositional basins lie between the Precambrian massifs (Figure 11-26). Although some marine sediments in this area accumulated during Paleozoic time, most of the sediments here are much younger Cenozoic clastics that have been shed from the Andes. The three southern Precambrian massifs lie within this sedimentary belt, but these durable massifs have tended to stand above the other terrane east of the Andes and, as a result, have received relatively little sedimentary cover.

These high-standing massifs provide only one example of the strong influence that the Precambrian shields and massifs have exerted over the Phanerozoic topography of South America. Another example concerns the Amazon Basin, which is one of the most interesting regions of the continent. The course of the Amazon River did not develop by accident. The elongate basin in which the great river now flows has a history tracing back at least to late Precambrian time, when it was the depositional site of extensive sediments derived from the stable Guyana and Brazilian cratons. The basin persisted through much of the Paleozoic interval, although it was periodically invaded by marine waters that separated the Guyana and Brazilian shields. Thus, the passage of the modern Amazon through the less resistant sedimentary terrane between the shields is only the most recent evidence of weakness in this region. To the east and south, three other shallow basins—the

of Africa stands high above sea level, and, as we have seen (Figure 7-47) and will learn more fully in Chapter 18, the continent is fractured by rift systems that developed relatively recently. It is possible that a few million years hence, Africa will be divided into two or more smaller continents. On the other hand, the rifting may fail, as did the rifting that began to divide North America about 1.3 billion years ago (Figure 11-10).

South America

South America, Africa, and Antarctica, which were attached to each other for hundreds of millions of years as part of Gondwanaland, have histories that are interrelated in many ways. There are five general areas of Precambrian terrane in South America (Figure 11-26); the two in the north are the large Guyana and Brazilian shields. Because Archean

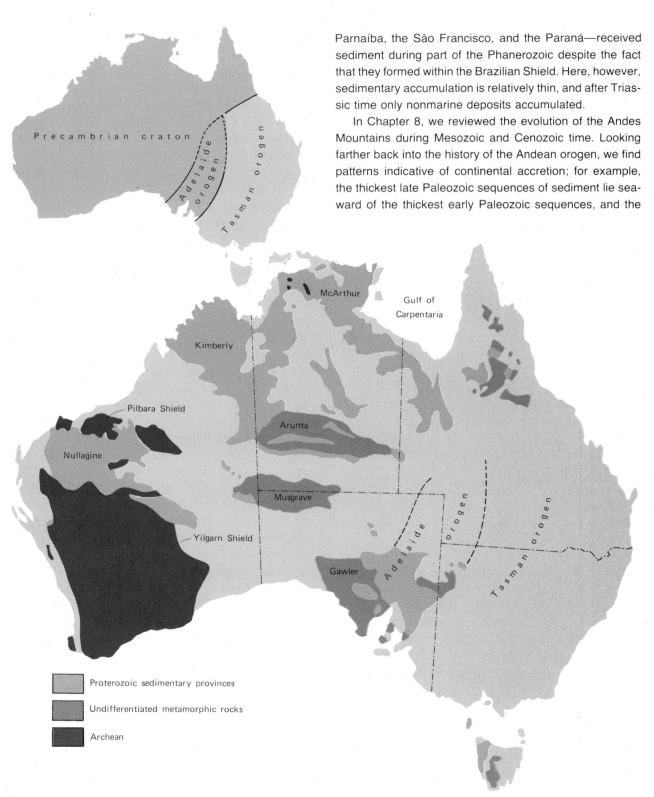

Parnaíba, the São Francisco, and the Paraná—received sediment during part of the Phanerozoic despite the fact that they formed within the Brazilian Shield. Here, however, sedimentary accumulation is relatively thin, and after Triassic time only nonmarine deposits accumulated.

In Chapter 8, we reviewed the evolution of the Andes Mountains during Mesozoic and Cenozoic time. Looking farther back into the history of the Andean orogen, we find patterns indicative of continental accretion; for example, the thickest late Paleozoic sequences of sediment lie seaward of the thickest early Paleozoic sequences, and the

Precambrian craton

Adelaide orogen

Tasman orogen

McArthur

Gulf of Carpentaria

Kimberly

Pilbara Shield

Arunta

Nullagine

Musgrave

Yilgarn Shield

Gawler

Adelaide orogen

Tasman orogen

Proterozoic sedimentary provinces

Undifferentiated metamorphic rocks

Archean

FIGURE 11-27   Major geologic features of Australia. Unaltered Archean rocks are found only in the west. The continent accreted eastward in the course of time. The Adelaide orogen includes Proterozoic sediment deformed during Cambrian time; farther east, the Tasman orogen was stabilized later in Phanerozoic time. (*Modified from D. A. Brown et al., The Geological Evolution of Australia and New Zealand, Pergamon Press, Oxford, 1968.*)

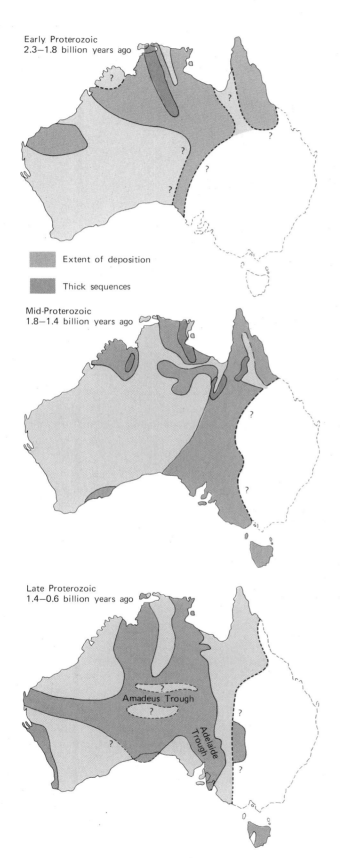

Early Proterozoic
2.3—1.8 billion years ago

Extent of deposition

Thick sequences

Mid-Proterozoic
1.8—1.4 billion years ago

Late Proterozoic
1.4—0.6 billion years ago

Amadeus Trough

Adelaide Trough

thickest Mesozoic sequences lie even closer to the Pacific margin. More generally, it is important to note that when Gondwanaland is reconstructed, the Andes form a segment of the Samfrau mountain belt (Figure 11-24).

Australia

The continent of Australia reveals a striking pattern of accretion along the margin of what was once Gondwanaland. The only large area of exposed Archean rocks in Australia is the Yilgarn Shield, which lies near the western margin of the continent (Figure 11-27). Here and in the neighboring Pilbara Shield are found the familiar greenstones and granites that characterize Archean terranes throughout the world. Metamorphic Precambrian belts to the east—the Arunta, Musgrave, and Gawler belts—probably include altered Archean deposits as well, indicating that the western two-thirds of the present continent existed at the end of Archean time or soon thereafter. Before we consider the pattern of eastward accretion that followed in Proterozoic and Paleozoic time, however, let us examine the vast accumulations of Proterozoic sedimentary deposits that have been found in Australia.

Sedimentary successions of Proterozoic age are more extensively exposed in Australia than anywhere else in the world, and this is why Australia has been and will continue to be an especially rich hunting ground for fossils of Proterozoic age. Figure 11-28 shows the widespread distribution of sedimentary deposits for three intervals of Proterozoic time. Early in the Proterozoic interval, approximately 2.3 to 1.8 million years ago, extensive marine sedimentary deposits accumulated in northern Australia. In the west there developed banded iron formations similar to those that formed in many other parts of the world about 2 billion years ago. Then, about 1.8 billion years ago, many areas were disturbed by orogenic activity. Sedimentation during the remainder of Proterozoic time was most extensive in what is now the central part of Australia. It is interesting to note, however, that Proterozoic deposits younger than about 1.8 billion years include no turbidites or other deposits indicative of deep water. For more than a billion years, deposition of shallow-water marine sediments, including stromatolites, prevailed over large areas of the Aus-

FIGURE 11-28   Deposition of sedimentary deposits in Australia during three Proterozoic intervals. These deposits form the most complete and extensive Proterozoic sedimentary record to be found on any continent in the world. (*After D. A. Brown et al., The Geological Evolution of Australia and New Zealand, Pergamon Press, Oxford, 1968.*)

FIGURE 11-29   The Pound Quartzite forms the plateau in the upper part of this photograph. The site of the Ediacara fossil fauna of soft-bodied animals (Figure 10-27) was deposited in the Adelaide Trough of southeastern Australia near the end of Proterozoic time. (*N.C. Winter, South Australian Government Tourist Bureau.*)

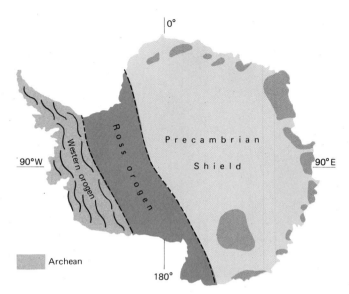

FIGURE 11-30   Major geologic features of Antarctica. Like Australia, this continent accreted asymmetrically so that today a Precambrian shield is positioned at one end.

tralian craton. In addition, tillites, varved deposits, and scoured pavements offer evidence of late Proterozoic glacial activity in many areas.

The stabilization of late Proterozoic sediments in the Adelaide Trough, which lay near the eastern margin of the Proterozoic craton, led to considerable eastward accretion of the craton (Figures 11-27 and 11-28C). Among the sedimentary units found here is the Pound Quartzite (Figure 11-29), which has yielded the Ediacara fauna of soft-bodied marine organisms (Figure 10-27). The orogenic stabilization of the Adelaide sediments occurred in Cambrian time, representing yet another segment in the network of orogenic activity that took place in Gondwanaland between 800 and 400 million years ago.

Eastward continental accretion continued with the late Paleozoic uplift of the Tasman orogenic belt adjacent to the Adelaide Belt. Like the Andes, this zone represents part of the Samfrau mountain system of Gondwanaland. Volcanism here has continued into the Neogene Period.

Antarctica

Because about 99 percent of its surface is covered by ice, Antarctica is less well understood from a geologic stand-

point than any other continent. Small outcrops have nonetheless revealed the general geologic history of this continent, which entails asymmetrical marginal accretion resembling that of Antarctica's Gondwanaland neighbor, Australia.

East Antarctica consists of a large Precambrian shield in which several Archean regions have been identified (Figure 11-30). The Ross orogen to the west of the shield (Figure 11-31) contains some Precambrian rocks of uncertain age together with Cambro-Ordovician sedimentary units. The latter were deformed, intruded, and metamorphosed during Ordovician time, shortly after the deformation of the Adelaide belt of Australia. Like the Adelaide episode, the Ordovician Ross episode was one of the many events that affected Gondwanaland between 800 and 400 million years ago (Figure 11-24).

Later in Phanerozoic time, Antarctica underwent further accretion through stabilization of crust in what is known as the Western orogen. Like the Andes of South America and the Tasman Belt of Australia, this orogen represents part of the Samfrau system that was fragmented by separation of the three continents during the Mesozoic Era.

India

The peninsula of India, though now joined to Asia (Figure 11-32), was once wedged between Antarctica and Africa

FIGURE 11-31   Royal Society Range within the Ross orogen of Antarctica. (*U.S. Navy.*)

as part of Gondwanaland (Figure 11-24). When Gond-wanaland broke apart during the Cretaceous Period, the Indian peninsula began a long northward journey. As we have seen, this journey ended in the Cenozoic Era with collision against the Asian continent and uplift of the Himalayan Mountains (Figure 8-5).

It is possible that almost all of the present craton of peninsular India existed in Archean times. We cannot prove that this is the case, however, because extensive Protero-zoic metamorphism has left only a few areas of recogniz-able Archean terrane. The largest of these is the Dharwar Shield, which forms much of the southern portion of the peninsula; smaller areas of Archean crust lie to the north (Figure 11-33). The Archean terrane of India, like that else-where, is typified by areas of greenstones and granites as well as by areas of high-grade metamorphic rocks.

Most of the Proterozoic rocks of India are crystalline. The unaltered sedimentary rocks of India are primarily non-marine and are therefore difficult to date by means of fossils, but they are known to span an interval of time beginning about 1.4 or 1.2 billion years ago and extending into the Cambrian. These rocks were deposited near the margin of Gondwanaland during a period in which the tip of what is now the modern Indian peninsula experienced tectonism and plutonism as part of the widespread mobi-lization to which we have repeatedly referred (Figure 11-24).

EURASIA: A COMPOSITE LANDMASS

The vast modern continent of Eurasia is a patchwork quilt of ancient continental blocks that have been joined together during the Phanerozoic Eon. What is now Eurasia was an array of small continents at the end of Precambrian time. Thus, while Gondwanaland broke apart during Phanero-zoic time, Eurasia was assembled into a large continent.

We have already seen how two peninsulas became attached to Eurasia late in the assembly of this great land-mass. These are the Arabian and Indian peninsulas, which had been part of Gondwanaland (page 311). Like the Indian peninsula, the Arabian peninsula was attached late in the Cenozoic Era. This western suturing event took place when Africa collided with Eurasia and mountain systems, includ-ing the Alps, were uplifted in the Mediterranean region (page 209).

The greatest suturing event in the assembly of Eurasia, however, occurred much earlier, at the end of the Paleozoic Era, when the Siberian and Russian platforms, which had been separate cratons, were united along the Ural Moun-tains (Figure 11-34). The assembly of eastern Asia was more complex and is poorly understood. It would appear, however, that this region formed from the union of several microplates late in the Paleozoic Era.

Today, the total area of Precambrian rocks exposed in Eurasia is relatively small (Figure 11-34) since a consid-

FIGURE 11-32   The Indian peninsula as photographed from   visible on the horizon. (*NASA.*)
the satellite Gemini XI. The Himalayan Mountains are barely

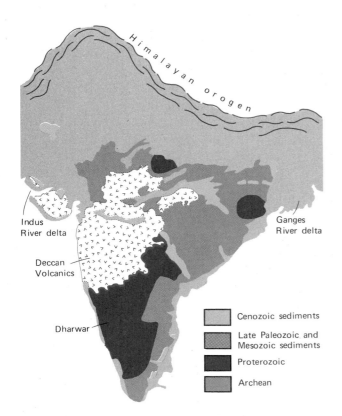

FIGURE 11-33   Major geologic features of India. A Precambrian shield forms most of the Indian peninsula. Cenozoic sediments from the Himalayas lap over the shield from the north. On the west, the Cenozoic Deccan Volcanics cover Precambrian rocks.

erable amount of Precambrian crust has been blanketed by Phanerozoic sedimentary rocks. The exposed Archean terranes of Eurasia are especially meager, and they do not exhibit greenstone belts as well developed as those found in other parts of the world. Nonetheless, there remain four principal shields that include Archean rocks. Two of these, the Anabar and Aldan massifs, lie within the Siberian platform. Although much of the platform is now covered with younger sediments, the Anabar and Aldan massifs belong to a single large block of Precambrian crust. As we have seen, this block may have been attached to western North America during the Proterozoic Eon (Figure 11-23).

The part of Eurasia that lies west of the Ural suture also displays two sizable shield areas that contain Archean rocks. One of these, the Ukrainian massif, lies in the southern part of the Russian platform. More than half of the surface of this massif is Archean. Furthermore, it is continuous with

FIGURE 11-34   Major geologic features of Eurasia, the largest continent of the modern world. Eurasia formed by several events of crustal suturing. Today its Precambrian rocks are exposed only in small Precambrian shields and massifs. The largest of these is the Baltic Shield.

FIGURE 11-35   Glacially scoured Precambrian rocks of the   rocks at Lofoten, Northern Norway. (*Courtesy of C. Schlinger.*)
Baltic Shield. These outcrops are of Archean metamorphic

FIGURE 11-36   Age provinces of the Baltic Shield (ages are
given in billions of years). The Caledonides form a Paleozoic orogen that was once connected to the Appalachians. Note that
the crust forming the Baltic Shield and the Caledonides displays
a general pattern of accretion toward the west.

the Baltic Shield beneath Phanerozoic sediments. As the
largest Precambrian shield area in all of Eurasia, the Baltic
deserves a closer examination.

The Baltic Shield, like the Canadian Shield, has been
scoured by Pleistocene glaciers (Figure 11-35). Its eastern
portion constitutes a sizable province of Archean rocks
(Figure 11-36). The shield and, presumably, the entire Archean crustal block underlying the surface deposits of the
Russian platform grew by continental accretion in what is
now a westerly direction. The eastern (Archean) portion of
the shield is flanked on the west by a zone of rocks that
was stabilized 1.7 to 2.0 billion years ago; to the west of
this lies a zone that was stabilized 0.8 to 1.2 billion years
ago.

As we noted earlier, the Baltic Shield may represent at
least part of a crustal block that was rifted away from eastern North America between about 1.3 and 1.0 billion years
ago (Figure 11-12). The youngest Precambrian province
of the Baltic Shield is about the same age as the Grenville
Province of North America, and, like the Grenville, it may
have formed through an episode of continental accretion
shortly after Baltica and Laurentia split apart.

CHAPTER SUMMARY

1.   Cratons grow by marginal accretion and by suturing of cratonic blocks to one another, and they decrease in size largely by rifting.

2.   During Proterozoic and part of Phanerozoic time, North America, Greenland, and northern Britain were united as Laurentia.

3.   Rocks of the Canadian Shield reveal that during the course of Proterozoic time, Laurentia grew by several episodes of continental accretion. The last of these was the Grenville event, which affected the eastern margin of North America.

4.   All of the fragments of Gondwanaland include Precambrian crust that was united within Gondwanaland during Proterozoic time. Late in Proterozoic time, Gondwanaland experienced a series of widespread orogenic episodes that continued into the Paleozoic Era.

5.   Eurasia is a composite craton that includes many different pieces of Precambrian crust. The largest Precambrian segment is the Baltic Shield, which may have been attached to North America until late in Proterozoic time, as may Siberia. The Arabian Shield and the peninsula of India once formed portions of Gondwanaland.

EXERCISES

1.   When South America, Antarctica, and Australia are positioned relative to one another as they were within Gondwanaland, where are their Precambrian shields and massifs located in the collective landmass? Where are their Phanerozoic mountain belts located? How does this reconstruction reveal a pattern of continental accretion?

2.   How does the Appalachian orogenic belt of North America relate to the Grenville orogenic belt?

3.   What evidence of continental accretion is found in Scandinavia?

4.   What evidence is there of a "Siberian connection" with North America?

5.   What was the largest craton in existence at the end of Proterozoic time? What is the largest craton in existence today? What elements of crust were transferred from the first craton to the second?

ADDITIONAL READING

Kummel, B., *History of the Earth*, W. H. Freeman and Company, New York, 1970.

Read, H. H., and J. Watson, *Introduction to Geology*, vol. 2: *Earth History*, John Wiley & Sons, Inc., New York, 1975.

PART FIVE

The Paleozoic Era

An evolutionary explosion of invertebrate animals with skeletons ushered in the Paleozoic Era, or "interval of ancient life." Before long, fishes also evolved, and midway through the era some of them developed jaws. Multicellular plants invaded the land, soon to be joined by scorpions, insects, and amphibians (descended from fishes), and later by reptiles (descended from amphibians). Mass extinctions punctuated the history of Paleozoic life, the final one striking at the close of the era. One of the early mass extinctions coincided with the first of two major Paleozoic glacial ages. Paleozoic time saw the formation of several mountain ranges, including the Appalachians, together with their counterparts in Europe, and early mountain ranges in the American West.

CHAPTER 12

The Early
Paleozoic World

C hapter 1 described how the Cambrian and Ordovician systems were established more than a century ago in what is now Wales. The rock record of these early Paleozoic systems includes the first tectonic cycle of the Appalachian mountain belt, discussed in Chapter 8, as well as evidence of widespread mountain building in Gondwanaland, mentioned in Chapter 11. Early Paleozoic rocks of marine origin are also well displayed on the broad surfaces of cratons, reflecting the fact that, with brief interruptions, sea level rose to a high level in the course of the Cambrian Period and remained high during most of Ordovician time. Among the revelations of the resulting rock record is that life in the oceans diversified rapidly at the end of Proterozoic time. Despite brief episodes of mass extinction during Cambrian time, the marine ecosystem near the end of the Ordovician Period encompassed about as many families of animals as at any subsequent interval in the Paleozoic Era.

LIFE

The story of early Paleozoic biotas is essentially one of life in the sea. It is presumed that certain simple kinds of pro-

tists and fungi had made their way into freshwater habitats by this time, but no fossil record of early Paleozoic freshwater life is known. The terrestrial realm, too, was barren of all but the simplest living things. Before middle Paleozoic time, neither insects nor vertebrate animals occupied the land.

The organisms that we will discuss in this chapter were also distinguished by the fact that they were the first biotas in the history of the earth to leave a conspicuous fossil record—one that, in many areas, is plainly visible even to a casual observer because it includes a great variety of external shells and other kinds of skeletons composed of durable minerals.

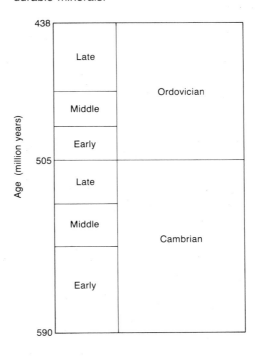

Out with the old and in with the new: This reconstruction of a Cambrian sea floor shows trilobites crawling in channels between stromatolite heads. Stromatolites flourished early in Paleozoic time, but by the end of the Ordovician Period they were rare. It would appear that stromatolites could not thrive in the face of the grazing and burrowing activities of animals that evolved near the start of the Paleozoic Era. This reconstructed Cambrian scene therefore represents a time of overlap in which higher animals had begun to diversify but stromatolites still flourished. (*New York State Museum.*)

In Chapter 10, we learned that the first major adaptive radiation of multicellular marine animals occurred during the last few tens of millions of years of the Proterozoic Eon. Nearly all of the creatures that emerged during this time, however, were soft-bodied; many, for example, were naked jellyfishlike animals or worms. During the earliest segment of Cambrian time, the seas became populated by a different kind of fauna—one consisting of small shelled animals, some of which failed to survive into later Cambrian time. We will begin by describing this fauna, which was the world's oldest diverse group of shelled animals; next we will focus on the more conspicuous and enduring Cambrian biota of shelled animals that followed; and we will conclude the section by describing the different, more highly diversified biota that characterized the Ordovician Period.

The Tommotian Stage

The first diverse biotas of animals with skeletons are found in rocks of the **Tommotian Stage**, an interval of approximately 15 million years. This stage was not formally added

FIGURE 12-1 Fossils that represent the Tommotian fauna, which is the oldest diverse skeletonized fauna on earth. All of the specimens shown here are small; none exceeds a few millimeters in length. A to C. Fossils that appear to represent primitive mollusks with coiled shells. D to K. Of the remaining specimens, only I can be assigned to a familiar group of animals; it is a spicule (skeletal element) from a sponge. Fossil H seems to represent the tube of a wormlike animal. (A to C and H to J. From S. C. Matthews and V. V. Missarzhevsky, Jour. Geol. Soc. London 131:289–304, 1975. D to F, G, and K. Courtesy of S. Bengston.)

to the base of the Cambrian System until the 1970s, when its fossil record was brought to the attention of the general scientific community. Before this time, the exceedingly small Tommotian fossils had been overlooked by all but a few paleontologists—but they have since been found on many continents. Some belong to groups of animals that still live in the ocean, such as sponges and mollusks (Figure 12-1). Also found in Tommotian rocks, however, is a host of strange skeletal elements that cannot be assigned to any living phylum and that seem to be unrelated to any group of fossils found in rocks younger than Cambrian age.

The development of the types of skeletons that characterize Tommotian faunas constituted a major evolutionary event. Although skeletons are known to support soft tissue and to facilitate locomotion, such adaptive functions cannot explain why so many different kinds of skeletons developed suddenly in the early part of Tommotian time. It has been suggested that a chemical change within the oceans triggered the production of these skeletons, but this hypothesis does not explain why some skeletons were composed of calcium carbonate and others of calcium phosphate, two compounds with quite different chemical properties. The rapid evolution of various kinds of external skeletons is probably in part attributable to the fact that animals needed protection from enemies; the first multicellular animals must have fed on single-celled creatures and might also have fed on larger plants.

The effective predation of some animals on others marked a change in the basic structure of ecosystems. There are probably two ways in which carnivorous animals prevented Phanerozoic soft-bodied animals from leaving a conspicuous fossil record comparable to that found in the latest Precambrian Ediacara Sandstone of Australia (Figures 10-27 and 10-28). First, predators must have drastically reduced the populations of soft-bodied species, making preservation less common. Second, during Phanerozoic time, many carnivores have scavenged on dead animals; only in the absence of advanced scavengers could animals of the Ediacara type have lain on sandy sea floors long enough to have been buried intact.

Later Cambrian marine life

The brief Tommotian interval of Cambrian time was followed by the evolution of many larger marine animals with hard parts, the most conspicuous of which were the **trilobites**. These many-legged arthropods (Figure 12-2) are by far the most conspicuous Cambrian fossils found above the Tommotian Stage. Many Cambrian trilobite species survived for only short spans of geologic time— 1 million years or less. Because of this and because most trilobite species have proven easy to identify, trilobites have served as the principal index fossils for Cambrian strata. Most trilobites crawled or swam along the sea floor, frequently leaving conspicuous trace fossils. Some of these traces represent the scratching or digging activity of the trilobites' many appendages, while others are trackways that represent paths of locomotion (Figure 12-3). It would appear that other small trilobites either swam or floated in the water.

Algal stromatolites were also more abundant during the Cambrian and Ordovician intervals than in subsequent times. The reconstruction shown on page 324 depicts trilobites crawling among stromatolite heads, although there is some evidence that trilobites were relatively uncommon in the environments that stromatolites favored.

Double-valved suspension feeders known as **brachiopods** are abundant in Cambrian strata as well, but their representatives are small and of limited variety (Figure 12-4). These organisms, which are also found in modern seas, resemble but are unrelated to bivalve mollusks. **Mollusks** are also common in Cambrian strata, but they, too, are small, and many advanced molluscan groups that became highly significant later in Paleozoic time were still inconspicuous or absent. **Echinoderms** were represented in the Cambrian by a remarkable variety of classes (Figure 12-5), but none of these closely resembled modern echinoderms such as starfishes, sea urchins, and sea cucumbers (Figure 2-32). A few other groups of fossils that are well represented in younger Paleozoic rocks also occur in Cambrian strata, but in small numbers. These include **conodonts**, which were apparently toothlike structures belonging to a group of swimming animals (Figure 12-6), and **ostracods**, a group of bivalved arthropods that has survived to recent times (Figure 12-7).

It can be assumed that many important groups of soft-bodied animals flourished during the Cambrian Period without leaving fossil records. One indication of the variety of such forms is provided by the unique sample of animals preserved in the Middle Cambrian Burgess Shale, which lies in the Rocky Mountains of British Columbia. Later in this chapter, we will examine the environment in which the Burgess Shale formed—but at this point, it will be sufficient to note that this black shale accumulated in a deep-sea environment that was virtually free of oxygen, resulting in the absence of the types of bacteria that normally lead to the decay of soft tissue. As a consequence, soft-bodied

Upper
Cambrian

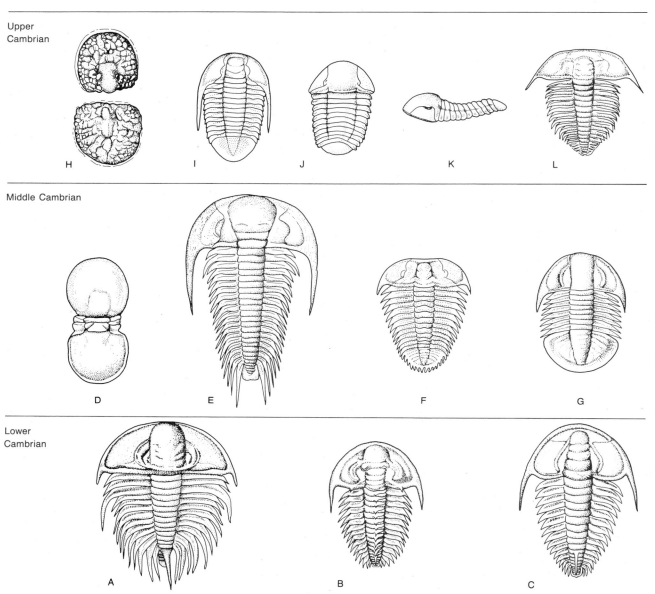

Middle Cambrian

Lower
Cambrian

FIGURE 12-2   Typical Cambrian trilobites, which are highly useful in stratigraphic correlation. Trilobites were arthropods (invertebrate animals with segmented bodies and jointed legs). The external skeleton of the trilobites was calcified to protect the animal. Its soft body and many legs were positioned beneath its flexible, jointed armor. Trilobites had mouth parts for chewing small pieces of food. Most species crawled over the sea floor, but some burrowed in sediment, and a few small species, including those labeled *D* and *J*, were planktonic. *A. Olenellus. B. Holmia. C. Redlichia. D. Lejopyge. E. Paradoxides. F. Blackwelderia. G. Glossopleura. H. Glyptagnostus. I. Saukia. J* and *K. Illaenurus. L. Olenus.* (*After Treatise on Invertebrate Paleontology, Part O, R. C. Moore [ed.], Geological Society of America and the University of Kansas Press, Lawrence, Kansas, 1959.*)

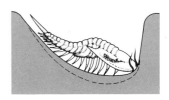

A

B

FIGURE 12-3   Cambrian trilobite tracks. *A*. A track preserved as a cast on the underside of a bed of sediment.
Scratches made by the appendages of the trilobites are clearly
visible. *B*. A diagrammatic cross section of a burrow being dug
by a trilobite (*above*), together with a drawing depicting a cast

of this kind of burrow (*below*); the cast was made by sediment
deposited from above. (*A. Courtesy of T. P. Crimes. B. After
A. Seilacher, in T. P. Crimes and J. C. Harper [eds.], Trace
Fossils, Seel House Press, Liverpool, 1970.*)

animals that were washed into this deep-water setting were
preserved as impressions and carbonizations of their soft
tissues (Figure 12-8). Most impressive among the Burgess
Shale species is the array of non-trilobite **arthropods**, which
are distributed among 30 genera or so—more than twice
the number of trilobite genera in the Burgess fauna. Second

in variety in this unusual fauna are the **polychaete worms**;
six families, all of which survive today, have been found in
the Burgess Shale, illustrating that these worms were already highly diversified in Cambrian time. In fact, this important group probably produced many of the burrows found
in rocks of latest Precambrian age (Figure 10-26). Even

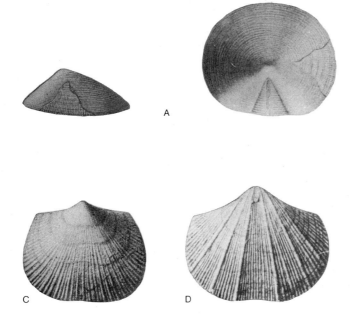

A

B

C

D

FIGURE 12-4   Cambrian brachiopods (suspension-feeding
animals housed in bivalved shells). A modest number of brachiopod species live in modern seas, primarily on the surface of
the substratum. Most fossil forms lived in a similar manner. *A*
and *B* are inarticulate brachiopods, forms in which the two
halves of the bivalved shell are not connected by interlocking
teeth. *A*. Side and top views of a cap-shaped shell. *B*. Exterior
and interior views of a spoon-shaped shell. *C* and *D*. Articulate
brachiopods, forms in which the two valves of the shell are connected by interlocking teeth. Only a few kinds of articulate brachiopods lived during the Cambrian Period. (*From James Hall's
volumes of the New York State Natural History Survey [1862–
1894].*)

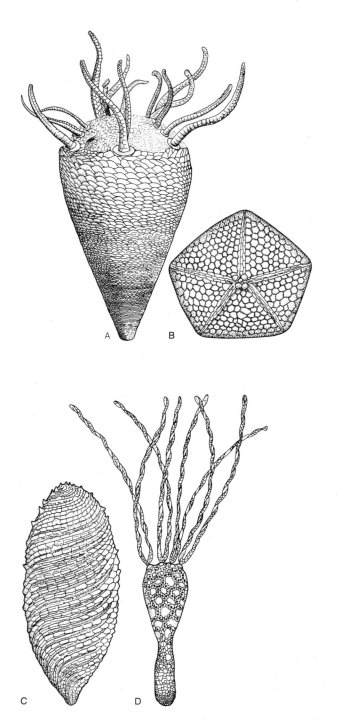

FIGURE 12-5 Strange Cambrian echinoderms that show no close relationship to any younger group. *A, B,* and *D.* Attached forms that were apparently suspension feeders. *C.* A flexible form that probably burrowed in sediment. Most living echinoderm groups, including starfishes and sea urchins, display a fivefold radial symmetry similar to that which can be seen in *B.* (*After Treatise on Invertebrate Paleontology, Part U, R. C. Moore [ed.], the Geological Society of America and the University of Kansas Press, Lawrence, Kansas, 1966.*)

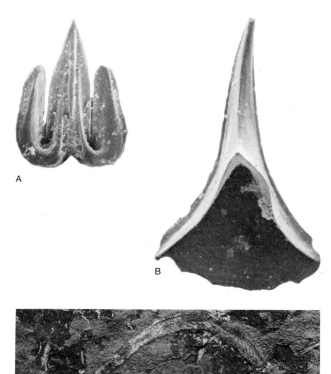

FIGURE 12-6 Cambrian conodonts. *A. Westergaardodina* (×75). *B. Furnishina* (×45). *C.* Only in the 1980s were the remains of an entire conodont animal found in the fossil record. This unique fossil is a late Paleozoic impression of an eel-like swimmer (×3). Conodonts, which are thought to have been teeth, are found in the head region (*arrow*). (*A and B. Courtesy of K. J. Müller. C. From D. E. G. Briggs et al., Lethaia 16:1–14, 1983.*)

so, Burgess Shale worms were of restricted variety in one sense—none had jaws like those found in late Paleozoic and younger rocks or in certain families of living polychaetes.

Most of the Cambrian animals described above were herbivores that fed on algae. The abundance of algal stromatolites on the sea floor was noted earlier, and in the water column above were single-celled planktonic algae that must also have been important producers in Cambrian seas. Modern forms of planktonic algae, however, were not yet present; instead, the phytoplankton known as ac-

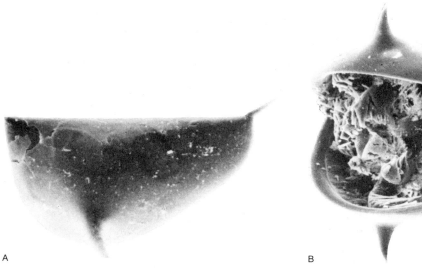

A   B

FIGURE 12-7   The Cambrian ostracod *Vestrigothia*. *A*. Lateral view of the bivalved shell (×110). This external shell is the part of the ostracod that is usually fossilized. *B*. This remarkable specimen is a shell that is gaping open to reveal many appendages that were well preserved because soon after the animal died they were coated with phosphatic material (×100). (*From K. J. Müller, Lethaia 12:1–27, 1979.*)

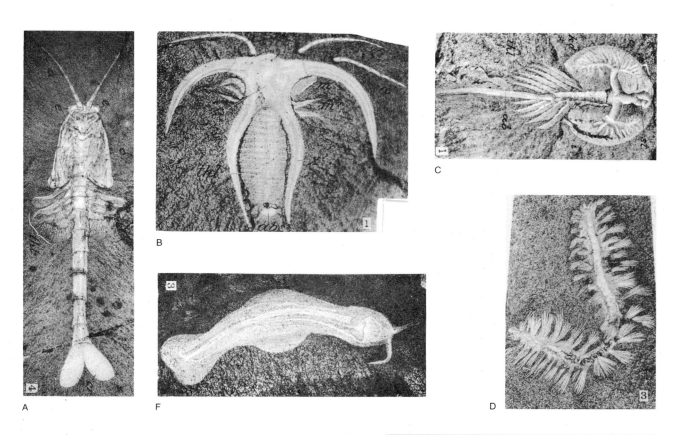

FIGURE 12-8   Animals without durable shells from the Burgess Shale of British Columbia. *A*, *B*, and *C*. Arthropods (specimen *B* is related to the trilobites). *D*. A polychaete worm. *E*. An especially important animal that is intermediate in form between polychaete worms and arthropods; it had a wormlike body but possessed walking legs that resembled those of the arthropods. *F*. A worm of uncertain biological relationships. All pictures ×3 except *A* (×1.5) and *E* (×4). (*Smithsonian Institution.*)

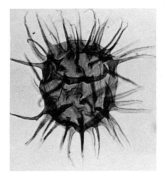

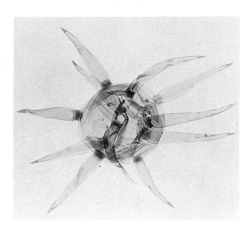

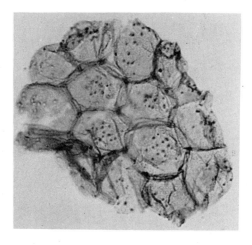

FIGURE 12-9   Acritarchs from Ordovician deposits of Oklahoma. Two single-celled spiny varieties, above, contrast with
the cluster of cells shown below. (*H. Tappan, The Paleobiology
of Plant Protists, W. H. Freeman and Company, New York,
1980.*)

ritarchs flourished during the Cambrian Period and persisted throughout the entire Paleozoic interval (Figure
12-9). The cysts in which acritarchs were encased during
resting stages are all that remains of these organisms in
the fossil record. As we have seen, acritarchs were conspicuous in the late Precambrian record as well (Figure
10-16). It is possible that other types of phytoplankton also
played major roles in early Paleozoic seas without leaving
fossil records.

What carnivores existed in Cambrian time? Some of
the wormlike animals and arthropods of the Burgess Shale
fauna were equipped with small pincers or specialized mouth
parts that were adapted to killing or chewing up small animals (Figure 12-10). Trilobites, too, had small jawlike structures that were capable of crushing tiny species of prey.
Jellyfishes, which were equipped with stinging cells (Box
2-1), must also have captured small victims. One group of
predators that arose very late in the Cambrian Period were
the **nautiloids** (Figure 12-11). Like the modern mollusks
with which they are united in the class **Cephalopoda**,
nautiloids were predators whose tentacles were used to
grasp prey and whose beaks served to tear the prey apart.
Numerous nautiloid species have been found in very late
Cambrian strata in China, but only a few have been found
elsewhere in the world. Cambrian nautiloids were quite
small; most measured between 2 and 6 centimeters (~1.0
and 2.5 inches) in length.

Entirely missing from Cambrian seas, as far as we know,
were large carnivores—animals that could break large shells
or tear apart large prey in the way that crabs and jawed
fishes do in the modern world. Bony plates of very small
fishes have recently been found in Cambrian rocks (Figure
12-12), but these did not belong to biting fishes. They represent the remains of a large group of jawless fishes that,
as we will see in the following chapter, groveled in mud for
small items of food both during the Cambrian and during
middle Paleozoic time.

During Early Cambrian time, the **archaeocyathids**
flourished. These cone-shaped creatures of uncertain biological relationship were attached to the substratum by
the tips of their skeletons (Figure 12-13). The vaselike shape
and sedentary habits of archaeocyathids suggest that these
animals were suspension feeders that at least superficially
resembled sponges. The archaeocyathids were partially
responsible for the origin, in Early Cambrian time, of the
world's first organic reefs (Figure 12-14). The main framework builders were archaeocyathids, but algae and other
organisms of unknown biological relationships also con-

A

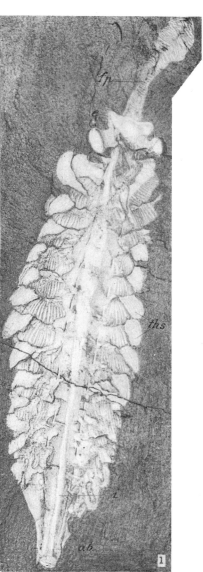

B

FIGURE 12-10   Fossil remains of Cambrian predators from the Burgess Shale. *A*. The clawlike appendage of an arthropod (×2). *B*. An arthropod (*right*) with a single pincerlike appendage (×2). (*The Smithsonian Institution*.)

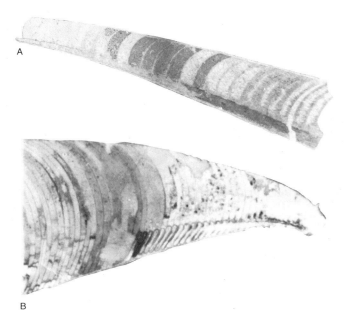

A

B

FIGURE 12-11   Late Cambrian nautiloids. These cephalopod mollusks were quite small (*A*. ×2.7; *B*. ×2.5), and, although many such animals existed near the end of Cambrian time, no species of large body size evolved until Ordovician time. Partitions called septa divide a nautiloid shell into chambers that housed gas for buoyancy. Septa were added as the shell grew. The body of the animal was housed within the open chamber at the end of the cone-shaped shell. Like the living pearly nautilus, Cambrian nautiloids presumably fed on other animals that they caught with tentacles (see Figure 12-16). (*From J. Chen and C. Teichert, Palaeontographica 181(A):1–102, 1983*.)

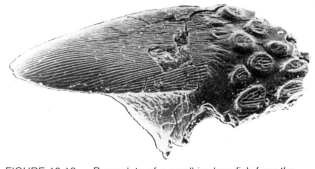

FIGURE 12-12   Bony plate of a small jawless fish from the Cambrian System of Wyoming. (*U.S. Geological Survey*.)

FIGURE 12-13   Reconstruction of three archeaocyathid species. All three were vase- or bowl-shaped. It seems likely that these animals were similar to sponges in that they pumped water through their porous walls, but it is not certain whether they were closely related to any group of modern organisms. (*After I. T. Zhuravleva, Akad. Nauk. U.S.S.R., Geol. Geofiz. Novosibirsk, 2:42–46, 1960.*)

tributed to the solid structures. These early reefs, which stood above the sea floor as low mounds, ceased to form before the end of Early Cambrian time, when archaeocyathids became extinct. All that remained until mid-Ordovician time were very small, inconspicuous reef-like structures formed by the organisms that had been secondary contributors to the archaeocyathid reefs.

The trilobites originated near the end of Tommotian time or shortly thereafter and immediately underwent a remarkable adaptive radiation. Of some 140 families of trilobites recognized in Paleozoic rocks, more than 90 have been found in Cambrian strata.

Many of the types of marine animals that appeared during the Cambrian Period were strange groups that included only a few genera and species; indeed, some are classified as discrete classes or even as phyla. The early Paleozoic history of the phylum Echinodermata illustrates this phenomenon. Today, this phylum includes only a few large groups, such as starfishes and sea urchins—but during Cambrian and Ordovician time, quite a number of bizarre echinoderm classes evolved (Figure 12-5). None was represented by many species or genera, and most survived for only a short time. Many body plans were "tried out" in this manner, with only a few succeeding in the face of predation and competition from other forms of life. This pattern is sometimes referred to as evolutionary "experimentation," with the understanding that what is happening is not a planned event but rather a development produced blindly by nature. Of the many groups of invertebrates that appeared during Tommotian and later Early Cambrian time, only a few—such as sponges, snails, brachiopods, and trilobites—flourished long afterward.

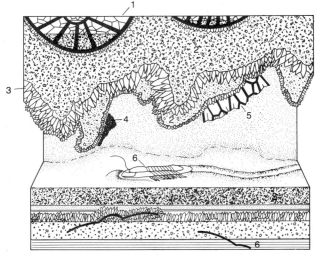

FIGURE 12-14   One of the world's oldest organic reefs, located in the Lower Cambrian System of Labrador. The photograph (*above*) shows sediment in the floor of a cavity of the reef (scale in centimeters). The diagram (*below*) shows the composition of the reef. This reef was constructed by several kinds of organisms, the most important of which were archaeocyathids (1) and calcareous algae (2). Cavities were encrusted with crystals of the mineral calcite (3), which were precipitated from sea water, and by organisms whose biological relationships were uncertain (4, 5). Trilobites (6) left tracks on sediment flooring cavities in the reef and also left fossil remains within the sediment. (*D. R. Kobluck and N. P. James, Lethaia 12:193–218, 1979.*)

The Cambrian adaptive radiation of marine animals with skeletons was not without interruption. During the latter part of Cambrian time, several mass extinctions eliminated most of the trilobite species in North America and in at least some other regions of the world. We will examine these extinctions later in this chapter, in the context of paleogeography.

The Ordovician adaptive radiation

The last of the Cambrian mass extinctions, at the very end of the Cambrian Period, eliminated large numbers of nautiloid and trilobite genera. This was a crisis from which the trilobites never fully recovered. Trilobites are found in many Ordovician strata (Figure 12-15), but not in abundances or diversities comparable to those of many Cambrian limestones. The Ordovician Period was instead characterized by the adaptive radiation of many other groups of animals. Some of these almost certainly originated in Cambrian time but failed to diversify markedly until the Ordovician, while others probably did not appear until Ordovician time. The Ordovician adaptive radiation populated the seas with many classes and orders of animals that continued to flourish in later Paleozoic periods. Indeed, it is possible that all animal phyla from subsequent geologic intervals were present in the seas by the end of Ordovician time.

An interesting aspect of Late Ordovician fauna is that most of the skeletonized members were animals that lived on the surface of the sediment rather than within it (Figure 12-16). It is difficult to move about and obtain oxygen within sediment, and in Ordovician time relatively few kinds of animals had developed methods of coping with these problems. Let us take a closer look at life in Ordovician seas.

Three groups, which were all present in Cambrian times but were not highly diversified, provide especially important Ordovician index fossils: the articulate brachiopods, the graptolites, and the conodonts.

Articulate brachiopods (Figure 12-17) are the most conspicuous group of well-preserved fossils both in Ordovician rocks and in all younger Paleozoic systems as well. These animals were immobile suspension feeders that rested on or were partially buried in sediment or that attached to solid objects, and most lived in subtidal settings.

Graptolites were especially common in Ordovician and Silurian times (Box 5-1). They are most frequently found in black shales—partially because they were too fragile to be easily preserved in sand and partially because many of

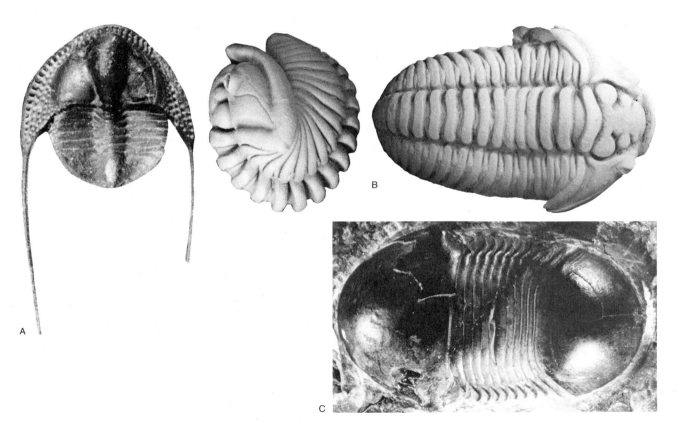

A

B

C

FIGURE 12-15 Typical Ordovician trilobites, demonstrating features not found in the earliest trilobites of the Cambrian Period. *A*. A form whose long spines may have served to discourage potential predators. *B*. A form that could roll up tightly for protection. (The earliest trilobites could not do this.) *C*. A large, smooth form that was probably adapted for burrowing in the sediment (see Figure 12-3). (*A and B. Smithsonian Institution. C. R. J. Ross, U.S. Geological Survey.*)

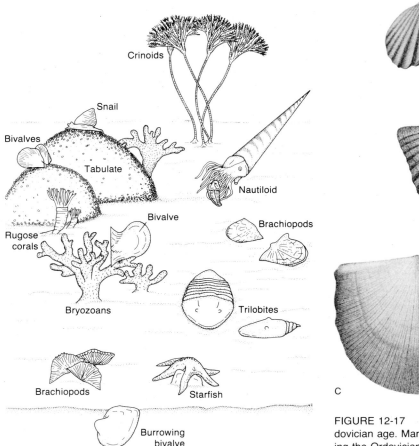

FIGURE 12-16   Life of a Late Ordovician sea floor in the area of Cincinnati, Ohio. Fossils of many of the groups of animals represented here can be seen in Figures 12-15 and 12-17 through 12-25. Note that at this early stage of Phanerozoic evolution, relatively few animals lived within the sediment. On the left, a snail crawls over a large tabulate colony, and two bivalve mollusks are attached to another tabulate colony by threads that give the bivalves stability. Another bivalve is similarly attached to the branch of a bryozoan colony. Two solitary rugose corals, lodged alongside one of the tabulate colonies, have their tentacles outstretched for food. Stalked crinoids are waving about in the center of the picture, suspension feeding with their arms. To their right, a large nautiloid prepares to eat a trilobite that it has trapped in its tentacles; below the nautiloid's eye is a spoutlike siphon that is used to expel water for jet propulsion. Two kinds of brachiopods live as suspension feeders on the sea floor. In the right foreground are trilobites of a type that left trace fossils indicating a burrowing mode of life. In the central foreground, a starfish prepares to devour a bivalve by prying apart the shell halves with its sucker-covered arms; then, by extruding its stomach, the starfish can digest the bivalve within its opened shell.

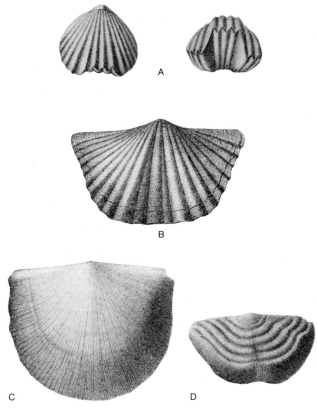

FIGURE 12-17   Representative articulate brachiopods of Ordovician age. Many more varieties of these animals existed during the Ordovician Period than during the Cambrian Period. (*From James Hall's volumes of the New York State Natural History Survey [1892–1894].*)

them were oceanic plankton that sank to muddy deep-sea floors after death. Since most individual species existed for less than a million years, fossil graptolites (like fossil trilobites) are especially useful for correlation.

The wide distribution of conodonts suggests that these toothlike structures also represent elements of creatures that floated or swam. The recent discovery of the first carbonized impression of the conodont animal reveals the presence of fins, which suggests a swimming mode of life (Figure 12-6). The broad distributions and short stratigraphic ranges of individual conodont species makes these fossils ideally suited for correlation. Geologists can also extract conodonts from limestones in large numbers simply by dissolving the rock with acid—a process that does not attack the phosphatic material of which these fossils are composed.

A

B

FIGURE 12-18   Ordovician animals that stood upright on the sea floor. *A.* One of the oldest kinds of rugose corals. Like modern hexacorals (Box 2-1), this primitive coral probably used tentacles to capture its prey. *B.* An early crinoid (sea lily); animals like this were suspension feeders attached to the sea floor by a flexible stalk formed of disclike elements of calcite that interlocked like poker chips and were encased in flesh. A modest variety of similar crinoids inhabit modern seas. Figure 12-16 includes reconstructions of rugose corals and crinoids. (*A. From B. D. Webby, Lethaia 4:153–168, 1971. B. From D. R. Kolata, Paleont. Soc. Mem. 7, 1975.*)

Joining the brachiopods as important sedentary animals of Ordovician sea floors were the **rugose corals**, which are sometimes known as horn corals because of their shape, and the **crinoids**, which are sometimes called sea lilies, even though they were animals rather than plants (Figure 12-18). Primitive corals and crinoids had existed during the Cambrian Period, but at such low diversities that they were insignificant members of the ecosystem.

Three groups of colonial animals with sturdy skeletons attained importance on Ordovician sea floors. Of these, the **bryozoans**, or moss animals (Figure 12-19), were the most conspicuous. The others, the **stromatoporoids** and the **tabulates** (Figure 12-20), attained their greatest importance as reef builders in middle Paleozoic times. The recent discovery of close living relatives of stromatoporoids reveals that these fossil forms were sponges with dense

A

B

FIGURE 12-19   Ordovician bryozoans. Bryozoans, which inhabit oceans and lakes of the modern world, are informally known as moss animals. They are colonial, living as groups of interconnected individuals (polyps) that occupy pores in rigid skeletons. Polyps employ tentacle-bearing organs for suspension feeding. *A.* A massive colony in which some of the individual animals were positioned on star-shaped mounds. *B.* A colony of finger-sized branches in which a number of brachiopods are nestled. (Figure 12-16 shows Ordovician bryozoan colonies as they grew on the sea floor.) (*A. R. S. Boardman, Smithsonian Institution.*)

C

FIGURE 12-20   Paleozoic reef builders. *A*. A stromatoporoid colony of Devonian age from Iowa ($\times 0.6$). *B*. A cross section showing the layered structure of a similar stromatoporoid (scale in centimeters). Colonies similar to these are found in Ordovician rocks. For many years, geologists could not determine what kind of organisms stromatoporoids were. Then, in the 1960s, sponges that secrete similar skeletons were discovered living on tropical reefs, and it was thus learned that stromatoporoids were sponges. (Note that the root of the word *stromatopo-* *roid* is *strom*, which means "layered." We have also encountered it in *stromatolite*, the name for layered structures produced by algae.) *C*. An example of one of these modern sponges, from Jamaica. *D*. A tabulate. Tabulates, which were apparently colonial corals of a primitive type, are named for the horizontal platforms (tabulae) within their skeletons (scale in centimeters). (Figure 12-16 shows a tabulate growing on an Ordovician sea floor.) (*A, B, and D. Photographs by the author. C. Courtesy of W. Hartman.*)

skeletons. Some tabulates may also have been sponges, although most of them were probably coral-like coelenterates. Because tabulates became extinct at the end of the Paleozoic Era, we may never be certain of their biological relationships.

In addition to trilobites, the mobile epifauna of Ordovician time included new varieties of snails (or gastropod mollusks) as well as the first **echinoids** (or sea urchins), which differed from modern echinoids in that they had flexible bodies (Figures 12-21 and 12-22). The ostracods on Ordovician sea floors (Figure 12-7) were also more diverse than those of Cambrian time.

The bivalve mollusks, which were apparently represented by only a few tiny species during the Cambrian

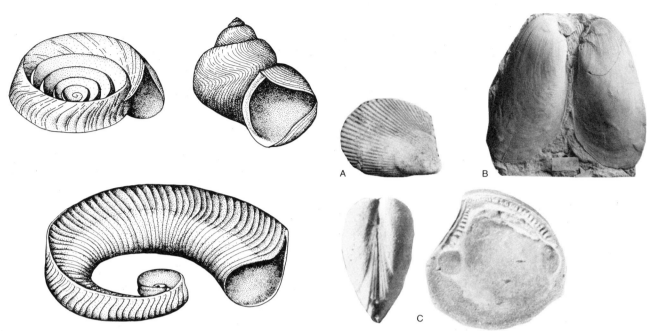

FIGURE 12-21   Snails (gastropods) that inhabited Ordovician seas. The shells at the upper left and bottom probably belonged to animals that were largely stationary and rested on the sea floor in the positions shown here. The shell in the upper right belongs to a group of crawling snails represented in modern seas by a few surviving species. Figure 12-16 shows this type of snail moving over a tabulate colony. (*After Treatise on Invertebrate Paleontology, Part I, R. C. Moore [ed.], the Geological Society of America and the University of Kansas Press, Lawrence, Kansas, 1960.*)

FIGURE 12-23   Bivalve mollusks of Ordovician age. *A.* A species that was attached to the surface of the substratum by threads that it secreted. *B.* A species that lived partly buried in sediment, attached by threads. *C.* A species that burrowed in the sediment. The species labeled *A* is depicted near the left edge of Figure 12-16. (*U.S. Geological Survey.*)

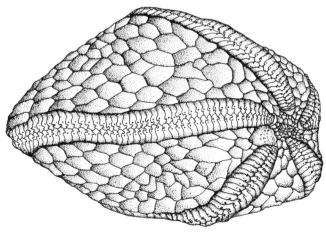

FIGURE 12-22   An Ordovician echinoid (sea urchin; × 2) that differed from living echinoids in its flexibility. In life, the globular skeleton was covered with spines, which served for locomotion. The animal probably fed on algae or plant debris. Its mouth was positioned underneath (at the "south pole") and the anus was positioned near the "north pole." (*After E. W. McBride and W. F. Spencer, Philos. Trans. Roy. Soc. London 229B:91–136, 1939.*)

Period, radiated during Ordovician time to develop a variety of forms adapted to modes of life on or within the substratum (Figure 12-23). During Ordovician time, burrowing bivalves attained the position that they hold today—they became the most successful group of burrowing suspension feeders with shells.

Jawless fishes persisted from Cambrian into Ordovician time but continued to contribute only fragmentary fossil remains to the rock record, so we know little about their shapes. During early Paleozoic time, before fishes acquired jaws, predation on large animals was a role monopolized by invertebrates. As depicted in Figure 12-16, two groups seem to have dominated here—the starfishes (Figure 12-24) and the nautiloids (Figure 12-11). All five of the living orders of starfishes were already present in Ordovician time. The rapidity of adaptive radiation during this second period of the Paleozoic Era is illustrated by the early history of the bryozoans (Figure 12-25). Very few

FIGURE 12-24   An Ordovician starfish—one of the first of this class of animals. In Figure 12-16, a starfish is shown feeding on a bivalve. (*Smithsonian Institution*.)

FIGURE 12-25   Bryozoan reef in the Middle Ordovician System of New York State. The notebook near the center of the photograph is about 20 centimeters (~8 inches) tall. This small reef apparently represents a time in which the only reefs being built were of simple structure. More complex reefs displaying several stages of development (see Figure 12-27) developed later in Middle Ordovician time. (*M. Pitcher, Proceedings of the North American Paleontological Convention, Allen Press, Lawrence, Kansas, 1969.*)

bryozoans are found in Lower Ordovician rocks, but Middle Ordovician rocks, which represent a period of some 20 million years, record the rapid development of diverse bryozoan faunas. Within the Simpson Group of Oklahoma, for example, early Middle Ordovician strata contain just five or six bryozoan species. Shale and limestone units, which alternate with sandstones, yield these and younger bryozoan species, and each successive interval of shale and limestone yields a larger variety of species than the previous one. Finally, in the uppermost parts of the Simpson Group, the variety of species rivals that of diverse Late Ordovician bryozoan faunas.

The adaptive radiation of many groups of marine invertebrates during the Ordovician Period is shown in Figure 12-26, where the number of marine invertebrate families is plotted against time. The darkest portion of the graph represents families belonging either to classes that are restricted to the Cambrian System or to classes that are better represented here than in younger systems. These families can be said to form the Cambrian fauna, although even this fauna is divisible into the strange, early Tommotian fauna and the succeeding Cambrian fauna, dominated by trilobites.

What is most striking about the post-Cambrian fauna is that it consisted of slightly more than 400 known families by the end of the Ordovician Period—approximately the same number that characterized all subsequent intervals of Paleozoic time. In contrast, diversity in the Cambrian

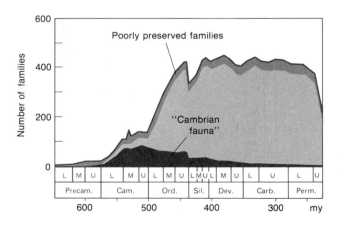

FIGURE 12-26   Changes in the number of families of marine invertebrates through the Phanerozoic Era. The "Cambrian fauna" consists of families that are found only in the Cambrian System or are best represented there. The expansion of life that produced the Cambrian fauna was followed by a new adaptive radiation in the Ordovician Period, and this Ordovician expansion produced a general level of diversity that was maintained until the end of the Paleozoic Era. (*After J. J. Sepkoski, Paleobiology 4:223–251, 1978.*)

Period had leveled off at about 150 families. Two points can be extrapolated from these observations: First, the great adaptive radiation of the Ordovician Period greatly augmented the diversity of the marine ecosystem; and second, while some old families died out and other new ones appeared later in the Paleozoic Era, something seems to have held the number of families in check from Ordovician time until the end of the era. At least three factors may have operated to accomplish this: (1) environments may have become filled with life to the point where new forms could no longer easily evolve; (2) the evolution of increasingly effective predators may have made it more and more difficult for new forms to evolve; and (3) most of the animals that existed may have been too specialized to give rise easily to other totally new types. We cannot at present assess the relative importance of these factors.

Reefs of a new type   By Middle Ordovician time, adaptive radiation had produced a number of new reef-building animals. Thus, some 40 million years after the archaeocyathid reefs of the Cambrian had ceased to grow, a new era of reef building began. Some of the first Middle Ordovician reefs were built by bryozoans in northeastern North America (Figure 12-25). Stromatoporoids and tabulates also contributed to later reef building, subsequently expanding to dominate organic reefs during Silurian and Devonian time. Although many Ordovician reefs were small mounds or patch reefs similar to those of the Cambrian Period (Figure 12-14), others exceeded 100 meters (~300 feet) in length and 6 or 7 meters (~20 feet) in height. Many Ordovician reefs, in addition, display several stages of development. The community that inhabited and added to a reef at each developmental stage was followed by another of different composition. Often in the final stage massive stromatoporoids were dominant (Figure 12-27).

Animal life and the decline of stromatolites   Of all the Phanerozoic periods, only the Cambrian and Ordovician were characterized by an abundance of stromatolites. This abundance was carried over from late Precambrian times, when stromatolites were widespread. During Cambrian time, stromatolites continued to blanket enormous areas of the sea floor (Figure 12-28)—but by the end of the Ordovician interval, large stromatolites were rare.

The location of the few areas of active stromatolite growth in the modern world offers some clues as to what happened to stromatolites during the Ordovician Period. The types of algae that form stromatolites occur widely in modern seas, but only in supratidal areas and in hypersaline lagoons do

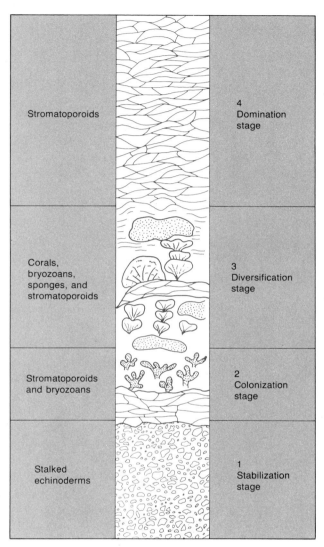

FIGURE 12-27   Stratigraphic section through a reef in the Crown Point Formation of Vermont. Four stages of development are shown. First, stalked echinoderms stabilized the sea floor (stage 1). Then the stabilized sea floor was colonized by stromatoporoids and by branching bryozoans (stage 2). These were later joined by a variety of reef-building forms (stage 3). Eventually, massive stromatoporoids dominated the reef (stage 4). (*After L. P. Alberstadt et al., Geol. Soc. Amer. Bull. 85:1171–1182, 1974.*)

they prosper well enough to form conspicuous stromatolitic structures. Marine animals are largely absent from both of these kinds of habitats. In more normal marine environments, animals burrow through algal mats and also eat them; this destruction is so severe that the mats do not survive long enough to form stromatolites. Experiments have shown that when animals are excluded from small

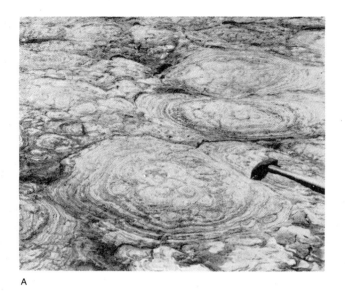

A

areas of sea floor in tropical climates, algal mats flourish, just as they did long ago. It seems evident that the great adaptive radiation of Ordovician life (Figure 12-26) produced a variety of animals that tended to prevent stromatolites from developing in all but unusual habitats. As a result, the character of shallow sea floors was forever altered.

Did plants invade the land?   The following chapter will reveal that in Silurian time, small multicellular plants were well established in moist terrestrial environments. Although similar plants probably invaded the land during the Ordovician Period, the evidence is not yet conclusive. It consists of fossilized sheets of cells similar to those that cover the surfaces of modern land plants as well as structures that closely resemble the spores released by primitive (non-seed-bearing) land plants of the modern world (Figure

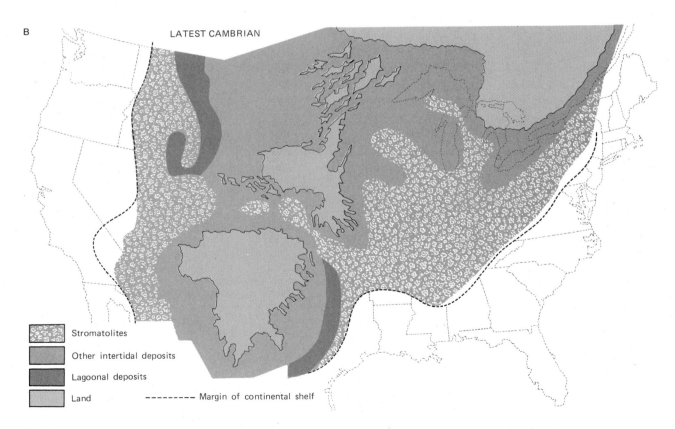

FIGURE 12-28   The great abundance of stromatolites during the Cambrian Period. *A*. An erosional surface displaying Upper Cambrian stromatolites at Saratoga, New York. *B*. The widespread occurrence of stromatolites in North America during the latest part of the Cambrian Period, when shallow seas spread across much of the continent. These stromatolites occupied vast intertidal and supratidal areas. (*A. New York State Geological Survey. B. Modified from C. Lochman-Balk, Geol. Soc. Amer. Bull. 81:3191–3224, 1970.*)

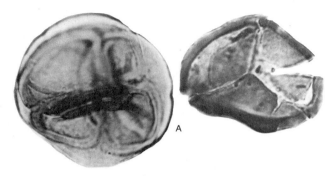

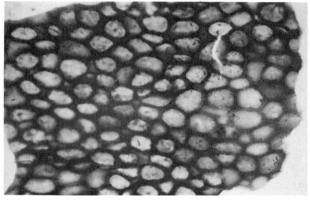

FIGURE 12-29   Late Ordovician fossils that may represent plants that lived on the land. *A.* Spores that resemble those of modern land plants (×670). *B.* A sheet of fossil cells that resemble those covering the surfaces of some land plants (×384). (*From J. Gray et al., Geology 10:197–201, 1982.*)

12-29). If land plants did evolve during the Ordovician Period, it seems likely that they were restricted to moist habitats, as are mosses, ferns, and other spore-bearing plants of modern times.

Terminal Ordovician mass extinction

The Ordovician Period concluded with one of the greatest mass extinctions in all of Phanerozoic time. In North America, considerably more than half of the species of brachiopods and bryozoans died out, and on a broader geographic scale, 100 families of Ordovician marine animals failed to survive into the Silurian Period. A mass extinction is a geographic phenomenon, striking biotas over a large area of the globe. For this reason, we will examine the causes of the terminal Ordovician mass extinction—together with the causes of the Cambrian mass extinctions—in the context of paleogeography.

PALEOGEOGRAPHY OF THE CAMBRIAN WORLD

Because Cambrian rocks are so widespread and are found on so many continents, geologists have a much clearer picture of the Cambrian world than of the late Proterozoic world. This paleogeographic framework gives us considerable insight into the repeated mass extinctions of Cambrian marine life.

Continents of the Cambrian world

Rock magnetism and other geologic evidence strongly suggest that most cratons became fused into one giant supercontinent near the end of Precambrian time. Although the position of Baltica with respect to such a configuration is uncertain, the fact that orogenies affected Baltica about 1 billion years ago suggests that this craton may have been connected to Laurentia along the contemporaneous Grenville belt of deformation, which bordered eastern North America (Figure 11-10). The arrangement of continents late in Cambrian time, 50 or 60 million years later, was strikingly different. By this time, Gondwanaland and smaller landmasses occupied equatorial zones (Figure 12-30). Some of these landmasses actually sat astride the equator, and no continent lay near either pole. This positioning of continents helps explain why Cambrian limestones are common on many cratons—most shallow-water deposits of Cambrian age accumulated in tropical or near-tropical climatic zones.

The Cambrian Period was notable for the progressive flooding of continents. The stage for this trend was set near the end of Precambrian time, when most of the earth's cratons stood largely exposed above sea level. In fact, the oceans probably stood lower in relation to the cratons during latest Precambrian time than at any time during the Paleozoic Era. As a result, only scattered local areas on modern continents yield a continuous record of shallow-water deposition across the Precambrian-Cambrian boundary. In Chapter 11, we discussed several of these areas: the Death Valley region of California and Nevada (see Figure 11-21), the Rocky Mountain region of southern Canada (see Figure 11-19), the Southern Highlands of Scotland (see Figure 11-15), and central Australia (see Figure 11-28). Other areas in which marine deposition continued uninterrupted into Cambrian time are eastern North America, northern Wales, China, and Siberia.

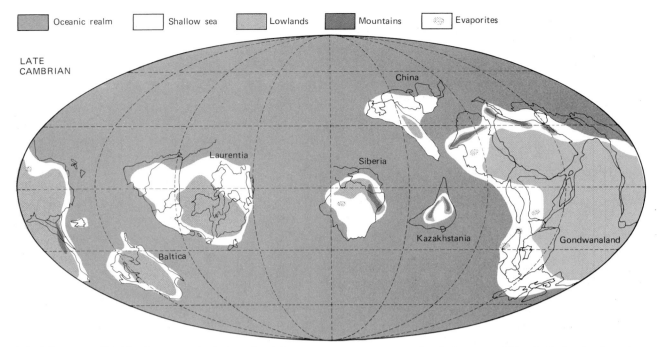

FIGURE 12-30   World paleogeography in Late Cambrian time. Continents were positioned at low latitudes, and many were inundated by shallow seas. (*After R. K. Bambach et al., American Scientist 68:26–38, 1980.*)

As the Cambrian Period progressed, many parts of Gondwanaland remained above sea level—partially as a result of regional uplifts caused by orogenic activity between 800 and 400 million years ago (Figure 11-24). Some smaller cratons, however, show evidence of continued encroachment of Cambrian seas to a point where little of their total area remained exposed late in Cambrian time (Figure 12-31). This flooding represented one of the largest and most persistent sea-level rises of the entire Phanerozoic Eon. It was interrupted in North America only by a modest regression at the end of the Early Cambrian and by another during Late Cambrian time.

A characteristic pattern of deposition

At the beginning of Cambrian time, as the seas began to encroach on broadly exposed continents, siliciclastic sediments were eroded from the continents and accumulated around the continental margins. Examples of this pattern, which were described in earlier chapters, are the Windermere Formation of western North America (Figure 11-19), the sandstones and shales that form the first tectonic cycle in the eastern Appalachians (Figure 8-23), and the Dalradian sequence of Scotland (Figure 11-15).

When the seas encroached further over most continents during Middle and Late Cambrian times, a characteristic sedimentary pattern emerged. To illustrate the nature of this pattern, let us examine the geography of Laurentia,

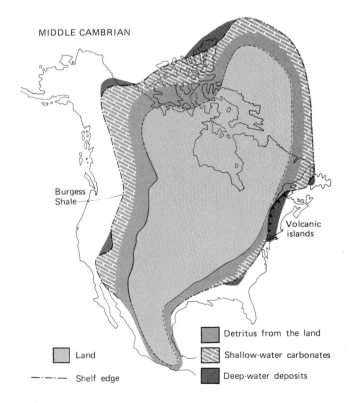

MIDDLE CAMBRIAN

Burgess Shale

Volcanic islands

Land
—·—·— Shelf edge

Detritus from the land

Shallow-water carbonates

Deep-water deposits

FIGURE 12-31   Concentric pattern of sediment deposition around the margin of Laurentia during Middle Cambrian time. Note the location of the Burgess Shale, renowned for its fauna of soft-bodied invertebrates, at the base of the Middle Cambrian continental shelf in western Canada. (*After A. R. Palmer, American Scientist 62:216–224, 1974.*)

the landmass that included North America, Greenland, and Scotland (Figure 12-30).

At all times during the Middle and Late Cambrian, some part of central Laurentia stood above sea level (Figure 12-31). Around the margin of the land, belts of marine deposition were arranged in a concentric fashion. The innermost belt was the site of deposition of siliciclastic sediments derived from the craton. This belt was essentially the same as the marginal siliciclastic belt that surrounded the continent during earliest Cambrian time, but it had shifted inland along with the shoreline. Seaward of this belt were broad carbonate platforms that were often fringed by reefs or stromatolites and that, as we have seen for eastern North America, terminated along a steep slope. Muds and breccias derived from the platform accumulated in deep water near the base of the steep continental slope (Figure 8-25), and subduction zones that lay close to cratons contributed volcanics to the deep-water belt from the opposite direction.

Trilobites, the dominant shell-bearing animals of Middle and Late Cambrian oceans, were distributed around continents in a pattern corresponding to the arrangement of the sedimentary belts. Certain groups of these trilobites are found primarily in deep-water deposits. These include small, blind forms that some experts believe were planktonic and also the genus *Paradoxides*, which characterizes the exotic blocks of oceanic volcanic terrane that became attached to eastern North America early in Paleozoic time (page 233). Other groups were largely restricted to the broad carbonate-floored shallow seas of continents; trilobites of this type are found over large areas of the North American craton, which was extensively flooded by warm tropical seas during Late Cambrian time (Figure 12-30).

Periodic mass extinctions of trilobites

The trilobite species that inhabited warm, tropical seas— including some small species thought to have been planktonic—were the ones that suffered in the repeated mass extinctions of the Cambrian Period. Each mass extinction was followed by an adaptive radiation that restored the diversity of shallow-water trilobites to a high level (Figure 12-32). These events are well documented for North America but thus far have been recorded elsewhere only in Australia.

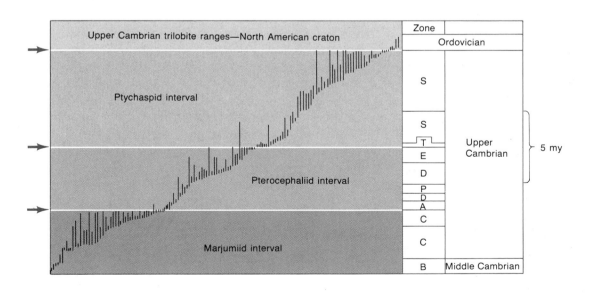

FIGURE 12-32   Repeated adaptive radiations and mass extinctions of Cambrian trilobites. The vertical bars indicate the stratigraphic ranges of important species in Middle and Upper Cambrian deposits of North America. The ranges form clusters and thus delineate three successive adaptive radiations, each of which was terminated by a mass extinction (*see arrows at left*). A stratigraphic interval representing approximately 5 million years is shown on the right. Note that many trilobite species survived for less than 1 million years. (*After J. H. Stitt, Oklahoma Geol. Surv. Bull. 124:1–79, 1977.*)

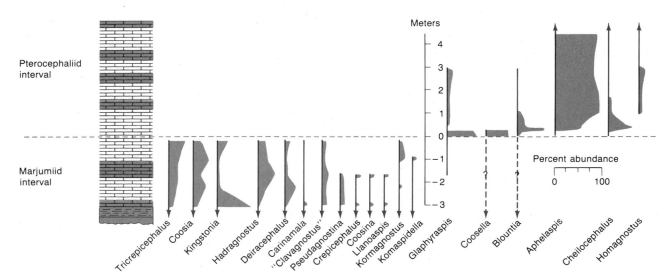

FIGURE 12-33   The stratigraphic pattern of trilobite mass extinction recorded in limestones at the top of the Cambrian Bonanza King Formation of Nevada. Many genera of the Marjumiid interval (see Figure 12-32) disappear at the "0 meter" level. Just above this level, only a few forms are found. Arrows indicate that species are also found below or above the occurrences displayed in the diagram. (*After A. R. Palmer, Alcheringa 3:33–41, 1979.*)

Each adaptive radiation of Cambrian trilobites occupied several million years, but each extinction was quite sudden (Figure 12-32). The fossil record reveals that each extinction took place during the deposition of a thin layer of sediment; thus, it can be assumed that each must have occurred over an interval of no more than a few thousand years. The transition from one adaptive radiation to another followed a characteristic pattern that is illustrated in Figure 12-33. The mass extinction is recorded in a layer of limestone or shale just a few centimeters thick. Above this "disaster" layer, in sedimentary beds just a meter or so thick, a variety of new trilobite genera join species that survived the mass extinction. These beds appear to represent a brief time of biotic adjustment in which opportunistic species flourished for a short time and then dwindled in the face of competition from new, more successful forms. The beds above contain a different fauna that consisted of about half as many species. From this fauna there issued a new adaptive radiation that lasted several million years—until another mass extinction reinitiated the cycle.

What led to the periodic mass extinction of trilobites? Elimination of shallow-water habitats by lowering of sea level can be ruled out, since the extinctions did not coincide with regression of seas from most areas of Laurentia. In fact, the only sizable regression of Late Cambrian time occurred between two mass extinctions (Figure 12-32). Because the adaptive radiation that preceded each mass extinction took place in association with the deposition of tropical limestones, it has been suggested that a sudden, temporary cooling of the seas was the agent of the trilobites' periodic, massive death. This idea gains support from evidence that the adaptive radiation following each mass extinction issued from a group of trilobites that lived offshore in cool, deep waters marginal to the continent. These offshore trilobites did not suffer in the mass extinction. Such evidence is only circumstantial, however, and the case for temperature as an agent of mass extinction remains unproven.

PALEOGEOGRAPHY OF THE ORDOVICIAN WORLD

Figure 12-34 summarizes major events that took place on earth during early Paleozoic time. Notice that after its general rise during the Cambrian Period, sea level remained high during most of Ordovician time, flooding broad cratonic areas. The largest regression of this period occurred at the end of Early Ordovician time but does not appear to have had any major effect on marine life. There is abundant evidence, however, that a cooling of climates contributed to the major marine extinction that took place at the close of the Ordovician Period. Before discussing the probable cause of this extinction, let us examine how the movement of two

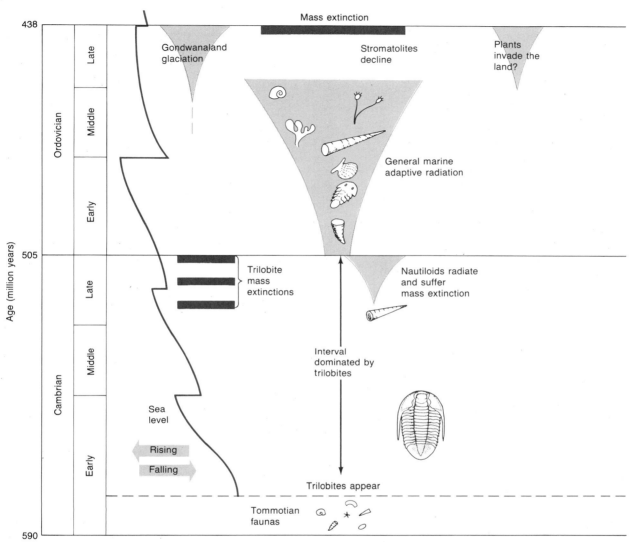

FIGURE 12-34   Major events of early Paleozoic time. During this time, there were three distinctive intervals in the history of life: (1) the Tommotian (pretrilobite) interval of the Cambrian, which was typified by very small animals; (2) the remainder of the Cambrian, which was dominated by trilobites; and (3) the Ordovician Period, when many groups of marine animals ap- peared and stromatolites declined. Note that neither the large drop of sea level at the end of Early Ordovician time nor the substantial drops during Cambrian time coincided with mass ex- tinctions. The terminal Ordovician mass extinction coincided with the climax of glaciation in Gondwanaland.

major continents, Gondwanaland and Baltica, had pro- found local climatic consequences.

Baltica moves northward

As late as Middle Ordovician time, the center of Baltica lay just a few degrees from the South Pole. The Ordovician temperature gradient from equator to pole was nonetheless gentle enough to allow diverse marine faunas to occupy the shallow seas of Baltica. During the latter half of the Ordovician Period, Baltica and the isolated microcontinent of England moved toward the equator (Figure 12-35). It was this movement that brought Baltica close to the eastern margin of Laurentia when the first Paleozoic tectonic cycle of the Appalachian region ended with orogeny and the deposition of flysch and clastic wedges (Figure 8-23). As Baltica and England moved toward Laurentia, the brachio-

Oceanic realm	Shallow sea	Lowlands	Mountains	Evaporites

MIDDLE
ORDIVICIAN

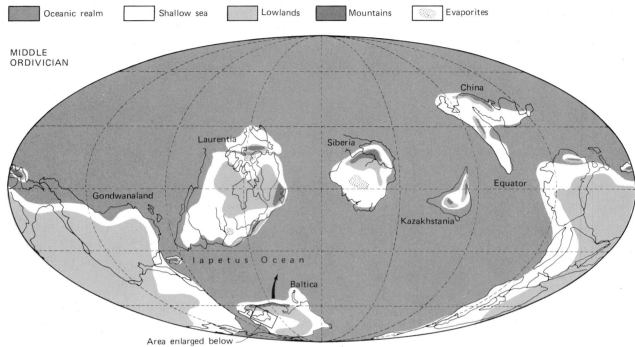

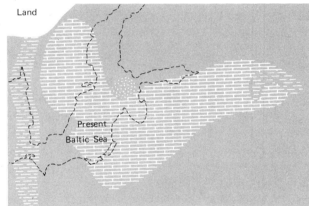

FIGURE 12-35   Movement of Baltica northward during the Ordovician Period. The map shows the paleogeography of the Middle Ordovician, and the large arrow shows the direction in which Baltica moved during Late Ordovician time. The enlargement of the Baltic Sea region shows the widespread deposition of marine limestones in this area during Middle Ordovician time. As Baltica moved northward, these limestones became more tropical in character. (*Plate-tectonic reconstruction after R. K. Bambach et al., American Scientist 60:26–38, 1980.*)

pods, trilobites, and graptolites of these two northward-moving continents became increasingly similar to those of Laurentia. Although this change would appear to reflect the convergence of these continents, the fauna of Baltica and England may have changed simply because these continents moved from high latitudes toward equatorial regions, where conditions were more tropical.

The northward movement of Baltica had other effects as well. In Middle Ordovician time, Baltica seems to have lain within about 30° of the South Pole. Figure 12-35 shows that the continent's northern margin, which is now the Atlantic coast of Scandinavia, was bordered by mountains at this time. Inland from the marginal mountain belt, a broad

shallow sea inundated the area today occupied by the Baltic Sea. Figure 12-35 also shows the present distribution of deposits laid down within this ancient sea. Clastic sediments were shed from highlands to the north, and muds from this region were deposited in a long belt to the west of which limestones accumulated in shallower water. In Middle Ordovician time, however, almost all of the limestones that accumulated in this area were composed of skeletal grains and of carbonate mud that may also have been derived from skeletons of marine life. It was not until latest Ordovician time that numerous grains formed by the precipitation of calcium carbonate from sea-water, grains resembling ooids and cemented pellets similar to grains

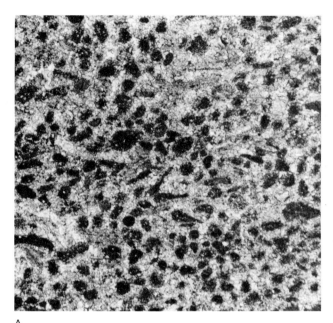

A

B

FIGURE 12-36   Upper Ordovician limestones from south-central Sweden, reflecting tropical conditions. *A*. Pelleted lime- stone (×10). (Compare to Figure 4-24.) *B*. Oolitic limestone (×10). (Compare to Figure 4-25). (*Courtesy of S. Stridsberg.*)

that now accumulate in the Bahamas (page 98). These Bahama-like limestones (Figure 12-36) continued to form through most of the Silurian Period. Even before the plate-tectonic movement of Baltica was recognized, geologists knew that Middle Ordovician limestones were probably deposited under climatic conditions that were temperate or subtropical and, further, that the change to the Bahama-like deposition signaled a transition to tropical conditions. Now that we know that Baltica traveled rapidly northward during Late Ordovician time and ultimately straddled the equator by mid-Silurian time, we can easily see why the climate of this continent underwent such changes. Thus, while Gondwanaland experienced a major glacial episode near the end of the Ordovician Period, the climate of Baltica was warmer than it had been 20 million years earlier. Before the advent of plate-tectonic theory, this unusual local pattern must have confused geologists considerably.

Glaciers in Gondwanaland and marine mass extinction

While Baltica moved toward the equator, Gondwanaland moved poleward. Thus, while Baltica became warmer, parts of Gondwanaland became much cooler. Gondwanaland lay

squarely on the equator late in Cambrian time (Figure 12-30), but by mid-Ordovician time only its northern margin was equatorial (Figure 12-35). Several million years before the end of the Ordovician Period, glaciers grew in and around the south polar region of Gondwanaland, and, as the period came to a close, the glacial episode reached a climax that was accompanied by a mass extinction in the marine realm.

Because the bulk of the evidence for extensive glaciation lies in remote desert regions of northern Africa, it was not discovered until the 1970s. At this time, evidence was uncovered indicating that extensive glaciation had occurred near the close of the Ordovician Period (Figure 12-37). Three or four levels of glacial deposits as well as a remarkable variety of glacial features, including ancient moraines (page 66), were found in the central Sahara Desert together with scratch marks on Upper Ordovician rocks and ice-rafted boulders (Figure 12-38). These features and less extensive ones on other continents point to the occurrence of widespread glaciation late in Ordovician time. Possible tillites in North and South America may be of Silurian age; confirmation of the age and glacial origin of these deposits would indicate a continuation of the glacial episode beyond Ordovician time. In any event, this episode seems to have peaked at the close of the Ordovician Period, causing sea

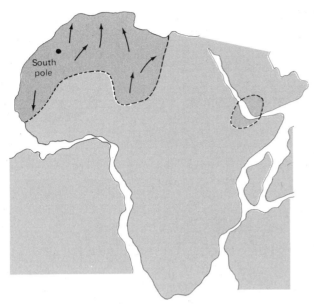

FIGURE 12-37   The known distribution of Late Ordovician glacial activity in northern Africa (*shaded area*). The position of the Late Ordovician South Pole is also shown, as are directions of ice flow (*arrows*).

of cold regions usually mix with waters from warmer regions. Thus, glaciers may accumulate on a large body of land in cold polar regions while an ocean in the same location would remain ice-free.

It has been suggested that the lowering of sea level at the end of the Ordovician Period contributed to the mass extinction of marine life by reducing the area of shallow sea floor. General doubts about the efficacy of such a mechanism are raised by the fact that large numbers of modern species survive on small areas of sea floor. For example, the shallow sea floor around the small, isolated Hawaiian Islands today supports about a thousand species of shelled mollusks. It addition, as noted above, the Early Ordovician drop of sea level failed to cause a mass extinction even though this drop was greater than that which took place at the close of the period.

The fate of the Ordovician graptolites strongly suggests that the cooling of the seas played an important role in the Late Ordovician mass extinction. During Ordovician time, most graptolite species were restricted to particular bands

level to drop suddenly—albeit less profoundly than it had fallen at the end of Early Ordovician time.

Without question, the movement of Gondwanaland over the South Pole contributed to the Ordovician glacial episode. In general, oceans tend to remain warmer than the land in cold regions. This is partially attributable to the fact that the albedo (the reflectance of sunlight) is usually higher for land than for water (page 37). Moreover, ocean waters

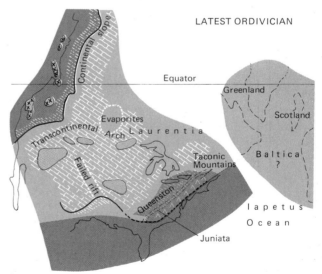

FIGURE 12-39   Pattern of deposition in North America during latest Ordovician time, when glaciers were spreading over parts of Gondwanaland. The Iapetus Ocean narrowed and may have closed altogether as Baltica and Laurentia converged. In eastern North America, the Juniata and Queenston formations formed as clastic wedges of sediment shed inland from newly formed mountains (see Figure 8-23). Carbonate sediments were laid down in the shallow seas that inundated almost all of the craton of North America. These deposits blanketed the early Paleozoic failed rift that had extended inland from the Gulf Coast. Several low islands remained along the Transcontinental Arch. The continental margin bordering western North America remained stable, but volcanic islands lay offshore.

FIGURE 12-38   Late Ordovician glacial sediments on the Arabian Peninsula. The arrow points to a cavity where an ice-transported boulder about 1 meter (~1 yard) in diameter has weathered out of the tillite. (*Courtesy of H. A. McClure.*)

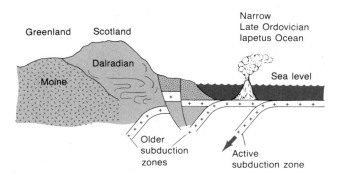

FIGURE 12-40   Subduction along the southwestern border of Scotland at the margin of the Iapetus Ocean (see Figure 12-35). The Precambrian Moine and Dalradian sequences (Figure 11-15) were deformed, and rocks of volcanic origin accumulated offshore. (*After A. H. G. Mitchell and W. S. McKerrow, Geol. Soc. Amer. Bull. 86:305–315, 1975, and T. R. Owen, The Geological Evolution of the British Isles, Pergamon Press, Oxford, 1976.*)

of latitude, as are many modern planktonic organisms (Figure 2-35). At the end of this period, these bands shifted toward the equator as cold temperatures did the same, and tropical groups of graptolites suffered most severely in the mass extinction.

REGIONAL EXAMPLES

Having viewed Early Paleozoic history from a distance, we will now focus on several interesting regional developments. First we will discuss the paleogeography of eastern Great Britain and North America late in Ordovician time, when the Iapetus Ocean narrowed or closed altogether and mountains were uplifted in Scotland and North America. We will then outline the histories of western and central North America during Cambro-Ordovician time. Finally, we will delineate the Cambro-Ordovician evolution of the Adelaide and Tasman mobile belts, which bordered the portion of Gondwanaland that was to become eastern Australia.

The borders of a narrowing Iapetus Ocean

As we have seen, the first Paleozoic tectonic cycle of the Appalachians ended in Late Ordovician and Silurian time with the deposition of clastic wedges that were shed westward from a newly forming mountain belt produced by the Taconic orogeny (Figures 8-23 and 8-26). Figure 12-39 shows the regional geography when the latest Ordovician clastic wedges were forming. As this figure illustrates, the Juniata and Queenston formations spread westward from

FIGURE 12-41   Ordovician pillow lavas of Ayrshire, southwestern Scotland, where subduction led to a narrowing of the Iapetus Ocean. (*Institute of Geological Sciences, British Crown copyright.*)

what might be called the Taconic Mountains. Beyond the clastic belt, the central interior of the United States (southern Laurentia) was flooded by shallow seas in which limestones accumulated.

We have also seen that the Iapetus Ocean had narrowed greatly by this time, as Baltica approached Laurentia. The convergence of these continents may have temporarily sealed off the Iapetus Ocean. At the very least, the continents moved close enough to one another that their shallow-water marine faunas became quite similar. Subduction introduced volcanism to the southern oceanic margin of Scotland, and the Dalradian and Moine sequences were uplifted, metamorphosed, and deformed (Figure 12-40). England, which was not yet attached to Scotland, lay somewhere to the south, between Laurentia and Gondwanaland. As England approached Laurentia, shelf sediments accumulated along its northern margin. Farther offshore, in the oceanic realm and along the border of Scotland, graywackes, volcanic deposits, and black muds were laid down (Figure 12-41); graptolites were fossilized

A

B

FIGURE 12-42   Outcrops of the Cambro-Ordovician Hales Limestone in Nevada, representing the ancient continental slope of western North America (see Figure 12-39). *A*. A graded bed of coarse fragments of shallow-water limestone that were transported down the continental slope. *B*. Well-bedded black limestone representing deposition under quieter conditions on the continental slope. (*H. E. Cook and M. E. Taylor, U.S. Geological Survey.*)

in these muds, and those of the Upper Ordovician deposits are quite similar to graptolites found in eastern North American deposits of the same age.

Stability in western North America

We have seen that a marine carbonate platform rimmed the craton of Laurentia during Cambrian time (Figure 8-24). We have also learned that in the east, the bank persisted into Late Ordovician time, when orogeny caused it to sink and give way to environments of flysch and then molasse deposition (Figure 8-23). In western North America, the marginal carbonate bank survived much longer. Parts of it sank into deep water during Silurian time, but orogeny did not strongly affect the western margin of the craton until the end of the Devonian Period.

Thus, throughout Cambro-Ordovician time, a stable continental shelf bounded western North America, passing

diagonally across what is now southern California (Figure 12-39). Coarse, poorly sorted sediments derived from shallow-water environments accumulated at the base of a steep platform edge, just as they did in the east (Figure 8-25). In many places, there are great thicknesses of these limestones, composed in part of debris from shallow-water stromatolites and invertebrates (Figure 12-42A). Deposits interpreted as representing quieter conditions in deep water beyond the platform include black limey mudstones and limestones; individual beds of these units can be traced for tens of meters (Figure 12-42B).

To the north, in British Columbia, the continental rise at the base of the carbonate bank was the site of preservation of the famous Burgess Shale fauna of soft-bodied animals described earlier in this chapter. It is said that this fauna was discovered by a curious accident. In 1909, Charles Walcott, the secretary of the Smithsonian Institution and an expert on Cambrian fossils, was riding along a narrow mountain trail when his horse stumbled over a block of shale in which Walcott caught a glimpse of a spectacularly well preserved fossil. Careful examination of the strata above the trail turned up the 2-meter-thick (~7-foot-thick) layer from which the fossil-bearing block had fallen. Walcott subsequently initiated a quarrying operation and removed nearly all of the fossil-bearing material (Figure 12-43).

Since Walcott's day, the mode of formation of the Burgess assemblage has been extensively studied. The fact that the fauna is so well preserved indicates that it was entombed in an oxygen-free environment from which destructive bacteria and scavenging animals were excluded. Stratigraphic evidence further indicates that the Burgess Shale was deposited at the foot of the steep carbonate shelf. In fact, the escarpment above the site of deposition is still preserved in cross section in the mountainside above

the site of the fossiliferous beds, and it rises some 200 meters (~650 feet) above them (Figure 12-44). Presumably the carbonate bank stood close to sea level, so this figure of 200 meters approximates the depth at which the Burgess fauna was preserved. The Burgess fauna was collected from a series of turbidite beds. Within each of these beds, calcareous siltstone grades upward into the fine-grained mudstone. The beds apparently formed when turbid flows descended the escarpment from one or more channels in the carbonate bank. The animals found in the Burgess Shale probably lived on the continental slope and were swept farther down the slope by the turbid flows. It is possible that an anaerobic environment of preservation was created by rapid, protective burial, but because several flows produced the same result, it seems more likely that the entire site of preservation was an anaerobic basin near

FIGURE 12-43　A fossil-collecting party from the Smithsonian Institution working at their quarry in the Burgess Shale on Mount Wapta, British Columbia. Figures 12-8 and 12-10 show specimens from the quarry. (*The Smithsonian Institution*.)

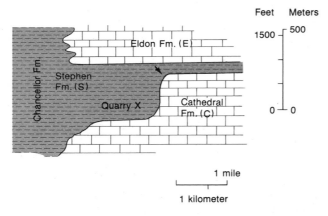

FIGURE 12-44   Location of the Burgess Shale Quarry within the Stephen Formation in British Columbia. The formations labeled in the diagram are identified in the photograph by letters. The arrow in the diagram points to the edge of the ancient continental shelf. The Burgess Shale accumulated below this shelf edge at the foot of the steep slope. (*Photograph from W. H. Fritz, Proc. N. Amer. Paleont. Conv. J:1155–1170, Allen Press, Lawrence, Kansas, 1969.*)

the foot of the continental slope—a depression filled with stagnant water from which oxygen had become depleted. The Santa Barbara Basin off the coast of California may represent a modern analog. In any case, we must be grateful for this spectacular glimpse of the soft-bodied marine life of Middle Cambrian time. Perhaps further investigation will lead to the discovery of even more diverse faunas near Walcott's site.

Movements of the sea in central Laurentia

During Middle and Late Cambrian time, carbonate deposits accumulated in shallow seas and on tidal flats over large areas of the Laurentian craton, and stromatolites were widespread (Figure 12-28). At all times during this period, however, at least a small, central area of the craton remained partly exposed. This area is known as the Transcontinental Arch (Figure 12-39). Meanwhile, the southern Oklahoma failed rift persisted from late Precambrian time (Figure 11-10). Through Middle Cambrian time, it accumulated volcanics, but by Late Cambrian time subsidence had slowed and volcanism had ceased; carbonates then

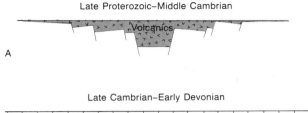

FIGURE 12-45 Diagrammatic cross sections of the southern Oklahoma failed rift (see Figure 12-39 for location). *A.* Cross section representing the rift when it was active, between late Proterozoic and Middle Cambrian time. At this time, lavas were welling up from the mantle along normal faults that formed where the crust was under tension. Beginning in Late Cambrian time, the rift ceased to be active and became blanketed with shallow-water limestones (*B*). (*After P. Hoffman et al., Soc. Econ. Paleont. Mineral. Spec. Publ. 19:38–55, 1974.*)

blanketed the trough (Figure 12-45). This general geographic picture, which is depicted in Figure 12-39, remained essentially unchanged through Early Ordovician time.

Late in Early Ordovician time, a major change took place. The marginal shelves of Laurentia persisted, but the Transcontinental Arch became broadly exposed and eroded after the shallow seas moved out to occupy only a narrow zone near the cratonic margin. This regression reflected the global lowering of sea level depicted in Figure 12-34. Later, in Middle Ordovician time, the seas again encroached well onto the arch, but as a result of the interval of widespread erosion, these advancing seas spread well-sorted sands over many nearshore marine environments east of the Transcontinental Arch. The St. Peter Sandstone of the central United States (Figure 12-46) is one of the nearly pure quartz sandstones formed during this interval. The widespread deposition of such clean sand may seem strange to a modern observer, since there is no region on earth

A

B

FIGURE 12-46 *A.* St. Peter Sandstone in Calhoun County, Illinois. This is the most famous of several Middle Ordovician formations of clean sandstone deposited on the North American craton. *B.* The sandstone as seen microscopically in thin section; all of the grains are quartz, but some appear dark because the section is viewed in polarized light. (*Photograph of outcrop from the Illinois Geological Survey, courtesy of P. E. Potter; photograph of thin section courtesy of F. J. Pettijohn.*)

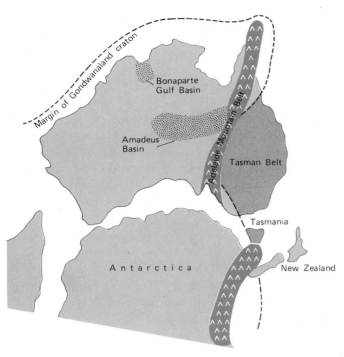

FIGURE 12-47   Paleogeography of the Australian region of Gondwanaland in Late Cambrian time. While mountains formed along the eastern margin of the continent, volcanism occurred offshore in the Tasman Belt.

where a comparable pattern of deposition can now be found. What we must remember, however, is that continents today stand unusually high above sea level. A special sequence of events not found in the modern world led to the Middle Ordovician pattern. First, a craton of very low relief was partly exposed by regression of the seas. Much of the sand that was then washed from the remaining lowland was probably eroded from older sandstone deposits and then further rounded and sorted by deposition and redeposition. Renewed transgression of the seas probably resulted in further reworking and concentration of the sands along the shorelines of the low islands that remained.

Late in Ordovician time, before the final regression resulting from glacial activity, sea level throughout the world may well have stood as high as it ever stood again until the middle of the Cretaceous Period nearly 400 million years later. In North America, the Late Ordovician transgression left only local elevations of the Transcontinental Arch standing as small islands surrounded by broad, shallow seas. At this time, there were no extensive land areas to provide siliciclastic detritus, and carbonate sedimentation prevailed.

Continental accretion in Australia

In contrast to the stable east and west coasts of Laurentia, the eastern margin of Australia during Cambrian time was the site of mountain building—activity that marked the beginning of substantial marginal growth of the Australian portion of Gondwanaland. The initial site of orogeny was the Adelaide mobile belt (Figure 12-47). Here, along the continental margin of Gondwanaland during late Precambrian time, the famous Ediacara fauna of early multicellular life was preserved in shallow-water sandstones. After an interruption, shallow marine deposition continued in Cambrian time, but what is now the eastern margin of Australia was then deep sea.

Late in Cambrian time, after enormous thicknesses of sediment had accumulated, the shelf deposits of the Adelaide Belt were folded, and intense metamorphism occurred along the continental margin—creating a mountain-building episode much like that of Late Ordovician time in the Appalachian region. This episode occurred during an interval of widespread orogeny in Gondwanaland (Figure 11-24), and it affected Antarctica as well. During Ordovician time, mountains were thus left standing along the margin of Australia, and from them sediment was shed into a seaway that extended across what is now the entire continent. Orogenic activity persisted into Ordovician time. Meanwhile, volcanism was widespread in the offshore Tasman Belt, apparently reflecting the subduction activity that caused the Adelaide Belt to deform. Here we see the beginnings of continental accretion. Throughout the Paleozoic Era, there was volcanism in the Tasman Belt, and by the end of the era this belt stood above sea level, having been added to the margin of Gondwanaland alongside the worn-down mountain belt of the Adelaide region. This is one part of the world where there is little question that substantial continental accretion took place during Paleozoic time. Much of the crust of eastern Australia today consists of dark sediments and mafic rocks that were part of the early Paleozoic sea floor.

CHAPTER SUMMARY

1.   **The early Paleozoic world was populated by three successive faunas of marine invertebrates: first the earliest Cambrian Tommotian fauna of small animals, many of which are not known from later intervals; then the main Cambrian fauna, which was dominated by**

trilobites; and finally, the Ordovician fauna. The latter resembled faunas of later Paleozoic time in the diversity of its components, which included many kinds of brachiopods, mollusks, echinoderms, and other marine invertebrates.

2. During the Cambrian Period, trilobites suffered periodic mass extinctions, the last of which occurred at the close of the period.

3. During Cambro-Ordovician time, stromatolites declined, apparently because of the grazing and burrowing activities of new groups of animals.

4. During the Ordovician Period, there developed a highly successful type of reef community that was dominated by tabulates, stromatoporoids, and colonial rugose corals. This community went on to thrive throughout almost all of middle Paleozoic time.

5. Early in the Cambrian Period, many continents stood unusually high above sea level—but as the period progressed, the continents were progressively flooded. Siliciclastic deposits fringed the land, and carbonates were laid down across the expanding continental shelves and along marginal banks.

6. Late in the Ordovician Period, the Iapetus Ocean narrowed as Laurentia and Baltica moved closer together. At the same time, subduction led to mountain-building episodes both in Scotland and in eastern North America.

7. Whereas the carbonate platform bordering eastern North America was destroyed by Late Ordovician mountain building, the carbonate platform that bordered western North America remained intact into middle Paleozoic time.

8. In Cambro-Ordovician time, mountain building occurred along the eastern margin of Australia, representing an early phase of continental accretion that continued throughout the Paleozoic Era.

9. Late in the Ordovician Period, after an interval during which the seas had stood unusually high throughout the world, global sea level dropped and continental glaciers spread over the part of Gondwanaland that had moved to a south polar position. At the same time, a mass extinction eliminated many groups of marine life.

EXERCISES

1. Why do geologists know more about the life that colonized early Paleozoic sea floors than about the life that floated and swam above these sea floors?

2. What fossil evidence suggests that there were more effective predatory animals during the Ordovician Period than in Cambrian time?

3. Which continental regions are likely to have had warmer climates in Late Cambrian time than today? (Compare Figure 12-30 with a map of the modern world.)

4. Give two reasons why deposition of shallow-water limestone was more widespread during the Late Cambrian Period than it is today. (Question 3 supplies one hint, and another comes from Figures 12-30 and 12-28.)

5. How did the Cambro-Ordovician history of the eastern margin of North America differ from that of the western margin? (Your answer will be more complete if you reexamine the discussion of the Appalachian Mountains in Chapter 8.)

6. Why does the early Paleozoic geology of Scotland have little in common with that of Scandinavia, which is its geographic neighbor today?

ADDITIONAL READING

Bassett, M. G., [ed.], *The Ordovician System: Proceedings of a Palaeontological Association Symposium, Birmingham, September, 1974*, The University of Wales Press and National Museum of Wales, Cardiff, 1974.

Levi-Setti, R., *Trilobites: A Photographic Atlas*, The University of Chicago Press, Chicago, 1974.

Morris, S. Conway, and H. B. Whittington, "The Animals of the Burgess Shale," *Scientific American*, July 1979.

Palmer, A. R., "Search for the Cambrian World," *American Scientist* 62:216–224, 1974.

CHAPTER 13
The Middle
Paleozoic World

The oceans of the world stood high during most of Silurian and Devonian time, leaving a widespread sedimentary record on every continent. Marine deposition was interrupted in one region, however, by the most profound plate-tectonic event of middle Paleozoic time: the suturing of Baltica to Laurentia along a zone of mountain building. As we learned in Chapter 8, this orogenic event produced the Caledonide mountain belts of Europe and the Acadian orogen of the Appalachian region.

In the northern British Isles, Silurian rocks were tilted by the Caledonian orogeny, and thus an angular unconformity separates them from the overlying Devonian sediments. It was farther south, in Wales, however, that in 1835 Adam Sedgwick and Roderick Murchison founded the Silurian System, along with the Cambrian. Five years later, these stratigraphic pioneers formally recognized the Devonian System, naming it for the county of Devon, along the southern coast of England. They recognized that the fossils in this system were intermediate in character (we would now say intermediate in evolutionary position) between those of the Silurian System below and those of the Carboniferous System above. (The Carboniferous System, though younger than the Devonian, had been recognized earlier in the century.)

The broad, shallow epicontinental seas of Silurian and Devonian time teemed with life. In the tropical zone, a diverse community of organisms built reefs larger than any that had formed during early Paleozoic time. More advanced predators were also on the scene, including the first jawed fishes—some of which were the size of large modern-day sharks. The Devonian Period was also distinguished by the progressive colonization of land habitats by new forms of life. Plants, for example, were restricted to marshy environments in Silurian time but were forming large forests by Late Devonian time. The oldest known insects are also of Early Devonian age, and near the end of the Devonian Period the first vertebrate animals crawled up onto the land, the fins of their ancestors having been transformed into legs. Shortly before the end of the Devonian Period, however, a wave of mass extinction swept away large numbers of aquatic taxa, leaving an impoverished biota in their place during the final 3 or 4 million years of the Devonian Period and the earliest segment of late Paleozoic time.

Cliffs of the Devonian Old Red Sandstone Formation at St. Ann's Head, Pembroke County, Wales. The Old Red Sandstone is a well-known formation that was studied during the birth of modern geology in Great Britain; it has yielded many of the most important specimens of early jawed fishes. (*Institute of Geological Sciences, British Crown copyright.*)

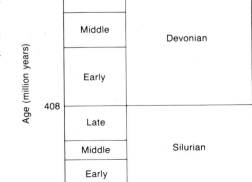

LIFE

The profound mass extinction at the close of the Ordovician Period was followed by the evolutionary recovery of many of the decimated taxa. We will begin our discussion of middle Paleozoic life by examining the nature of this re- covery and the ways in which it surpassed the Ordovician adaptive radiation, yielding superior reef builders and swimming predators. We will then consider how plants ex- panded their ecologic role on the land, and we will conclude the section by discussing how animals invaded the terres- trial realm.

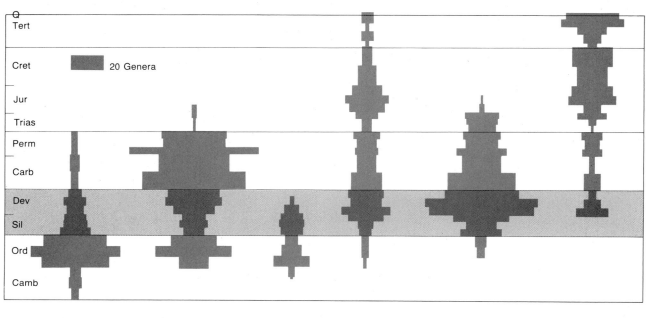

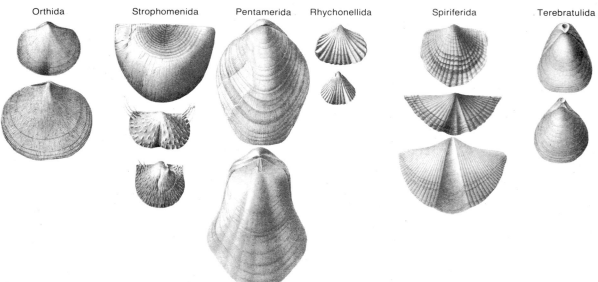

FIGURE 13-1   Middle Paleozoic articulate brachiopods. *Above:* The diversity (numbers of genera) of the six major or- ders of Paleozoic brachiopods, showing that all were well repre- sented in middle Paleozoic (Silurian and Devonian) time. *Be- low:* Typical middle Paleozoic representatives of each brachiopod order. (*Diversity data from Treatise on Invertebrate Paleontology; brachiopod illustrations from James Hall's vol- umes of the New York State Natural History Survey [1862–1894].*)

Aquatic recovery

Most of the marine taxa that had flourished during the Ordovician Period rediversified after the terminal Ordovician mass extinction to become prominent members of the Silurian and Devonian marine biota. One group of marine animals that failed to recover fully, however, was the trilobites, which were less conspicuous in middle Paleozoic than in early Paleozoic seas. For other groups, recovery took the form of renewed adaptive radiation. The orthids, which were the largest early Paleozoic group of articulate brachiopods, failed to rebound fully from the terminal Ordovician crisis, but other groups of articulate brachiopods soon underwent adaptive radiations. In fact, as Figure 13-1 shows, all of the important Paleozoic articulate brachiopods, including the orthids, were well represented in middle Paleozoic seas.

Other marine groups that recovered during middle Paleozoic time were the bivalve and gastropod mollusks (Figure 13-2). The bivalves, in fact, expanded their ecologic role by invading nonmarine habitats; some of the oldest known freshwater bivalves are found in the Upper Devonian strata of New York State. On the surface of the sea floor, bryozoans (Figure 12-19) rediversified after the terminal Ordovician crisis, and crinoids (Figure 12-18) increased in variety. Acritarchs were the dominant group of fossil phytoplankton in middle Paleozoic time, just as they had been in late Precambrian and early Paleozoic time (Figures 10-16 and 12-9). One of the most spectacular Early Silurian adaptive radiations, however, was that of the graptolites (Box 5-1), which had nearly disappeared at the end of the Ordovician Period. The number of species of graptolites known in the British Isles increased from about 12 to nearly 60 during just the first 5 million years or so of Silurian time.

Luxuriant reefs   Most of the Silurian radiations of marine life did not vastly alter marine ecosystems but instead represented a refilling of niches. Builders of organic reefs, however, did diversify in new ways, and they also produced reefs of much greater size than any of Cambro-Ordovician age. In the last chapter we saw that a new kind of organic reef developed in mid-Ordovician time: While the earliest reefs of the Middle Ordovician were formed entirely by bryozoans (Figure 12-25), later Middle Ordovician reefs were more complex, with tabulates and stromatoporoids playing dominant constructional roles. Reef communities of this second general type, which we can call **tabulate-strome reefs**, diversified and persisted for about 120 mil-

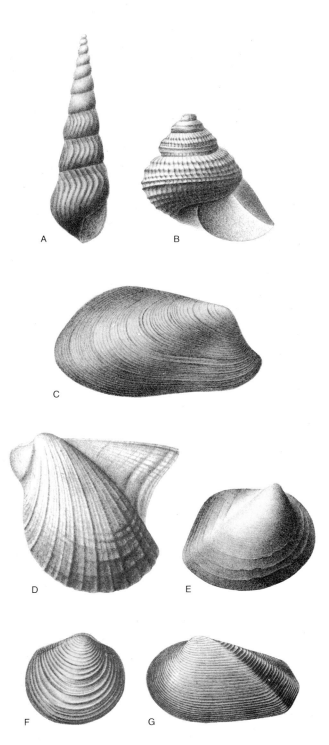

FIGURE 13-2   Typical mollusks that inhabited middle Paleozoic sea floors. *A* and *B*. Two snails. *C* and *D*. Two bivalves that attached to the substratum by threads. *C* closely resembles certain modern species that live partially buried in sediment, and *D* is related to modern wing shells that are attached to objects standing upright on the sea floor. *E* through *G*. Species that burrowed in sediment. (*From James Hall's volumes of the New York State Natural History Survey [1879–1885].*)

lion years, until late in the Devonian Period. The success of these reefs was a result of mid-Paleozoic adaptive radiations of tabulates, colonial rugose corals (Figure 13-3), and stromatoporoids (Figure 12-20).

During the Silurian Period, tabulate-strome reefs occasionally attained heights of 10 meters (~65 feet) above the sea floor, but most stood no taller than 5 meters, and few were longer or wider than 3 kilometers (~2 miles). During Devonian time, however, tabulate-strome reefs assumed enormous proportions. We will examine some of these large reefs later in this chapter. In areas of strong wave action, the growth of tabulate-strome reefs followed a characteristic ecologic succession (Figure 13-4). First, sticklike tabulates and rugose corals colonized an area of subtidal sea floor. A low mound was then formed when these fragile forms were encrusted by platy and hemispherical tabulates and colonial rugose corals. Finally, as

the mound grew up toward sea level, stromatoporoids and algae encrusted the seaward side, forming a durable ridge that resembled the algal ridges of many modern reefs (Figure 4-17). Tabulates and colonial rugose corals occupied a zone of quieter water behind the ridge, and beyond them was a lagoon in which mud-sized sedimentary grains accumulated along with coarser skeletal debris from the reef. Pockets of fossils preserved in reef rock reveal that a wide variety of invertebrate life inhabited tabulate-strome reefs; brachiopods and bivalve mollusks attached themselves to a typical reef, snails grazed over it, and crinoids and lacy bryozoans reached upward from its craggy surface. Although its fauna would look unusual to us today, a living, fully developed tabulate-strome reef (Figure 13-5) would certainly seem as colorful and spectacularly beautiful as the coral reefs that flourish in the modern tropics (page 90; Box 2-1).

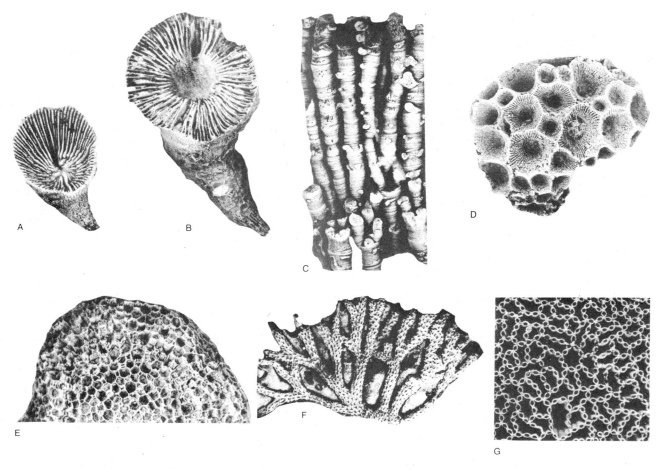

FIGURE 13-3   Rugose corals and tabulates of middle Paleozoic age. A and B. Solitary rugose corals, or "horn corals." C and D. Colonial rugose corals of the kind that contributed to reefs. E through G. Colonial tabulates that were important reef builders. All pictures ×1, except G (×3). (E. C. Stumm, Geol. Soc. Amer. Mem. 93, 1964.)

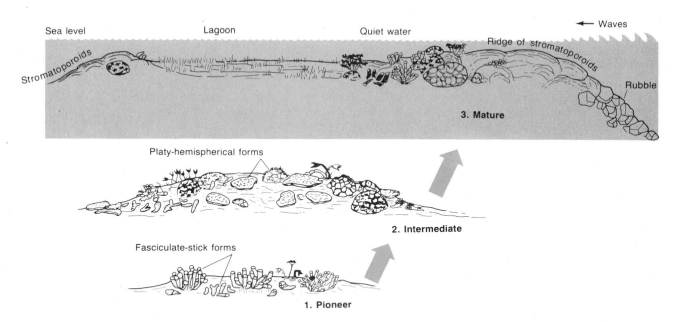

FIGURE 13-4 Ecologic succession of a typical Devonian reef. (*1*) The pioneer community consisted of fragile, twiglike rugose corals and tabulates. (*2*) Broad and moundlike tabulate colonies dominated during the intermediate stage of development. (*3*) In the mature stage, the reef grew close to sea level, and waves broke against a ridge of massive, encrusting stromatoporoids; behind them were species that were adapted to quieter water, and leeward of these was a lagoon populated by fragile, twiglike species. The leeward side of the lagoon was bounded by a small stromatoporoid ridge. Below is a photo of a moundlike reef formed by tabulates and rugose corals. The fossils were collected in Michigan (see Figure 13-31) and were then reassembled at the Smithsonian Institution to re-create the reef. This moundlike reef represents the intermediate stage (stage 2). (*Diagrams after P. Copper, Proc. Second Internat. Coral Reef Symp. 1:365–386, 1975.*)

FIGURE 13-5   Reconstruction of a Middle Devonian reef community. The flowerlike structures are the tentacle-bearing mouth regions of corals. In the central foreground are a coiled ammonoid, a straight-shelled nautiloid, and a large, spiny trilobite. The remains of such reefs are found in the region of the Great Lakes and the Ohio River. (*Field Museum of Natural History.*)

New swimming animals   Perhaps the greatest change in the nature of aquatic ecosystems during middle Paleozoic time resulted from the origin of new kinds of nektonic (swimming) animals, many of which were predators. The most important of these among the invertebrates were the **ammonoids**, which were coiled cephalopod mollusks that evolved from straight-shelled nautiloids during Early Devonian time (Figure 13-6). After giving rise to the ammonoids, the nautiloids persisted at low diversity. The ammonoids, in contrast, diversified rapidly, and because their species were distinctive, widespread, and relatively short-lived, they serve as important guide fossils in rocks ranging in age from Devonian to latest Mesozoic. (Ammonoids died out along with the dinosaurs at the end of the Mesozoic Era.)

The **eurypterid** arthropods were a second important group of invertebrate predators that proliferated during middle Paleozoic time. These distant relatives of scorpions were swimmers, and many had claws. Although the eurypterids appeared in the Ordovician Period and survived until Permian time, their most conspicuous fossil record is in Middle Paleozoic rocks, and it seems evident that their greatest ecologic impact was in the Silurian and Devonian periods (page 322 and Figure 13-7). Unlike ammonoids, eurypterids ranged into brackish and freshwater habitats.

Other swimmers that were adapted to both marine and nonmarine conditions were the fishes, the major groups of which are illustrated in Figure 13-8. Whereas only fragments of fish skeletons have been found in early Paleozoic sediments (page 332), the Silurian and Devonian systems have yielded diverse, fully preserved fish skeletons, many of them from freshwater deposits of lakes and rivers. The Devonian Period has, in fact, been called the Age of Fishes. This label is somewhat misleading, however; although many kinds of marine and freshwater fishes lived during the middle Paleozoic interval, no more varieties were present at

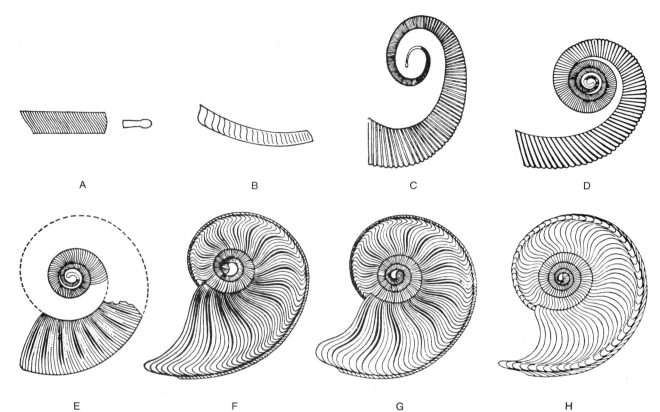

FIGURE 13-6   Shells of Lower Devonian cephalopod mollusks from the Hunsrück Shale of Germany reveal the apparent evolutionary sequence leading from nautiloids to early ammonoids. *A* and *B*. Fragments of nautiloids of the group that evolved into ammonoids. *C* through *H*. Early ammonoid species representing various degrees of coiling. The bulblike shape of the earliest part (tip) of each shell, together with other shared features, suggests that these species are closely related to one another; the coiling sequence displayed here apparently represents the evolutionary sequence. (*After H. K. Erben, Biol. Rev. 41:641–658, 1966.*)

FIGURE 13-7   Eurypterids found in Upper Silurian deposits of New York State. *A. Pterygotus. B. Stylonurus. C. Hughmilleria. D. Carcinosoma.* Note the long pincerlike appendages of *Pterygotus.* (*Field Museum of Natural History.*)

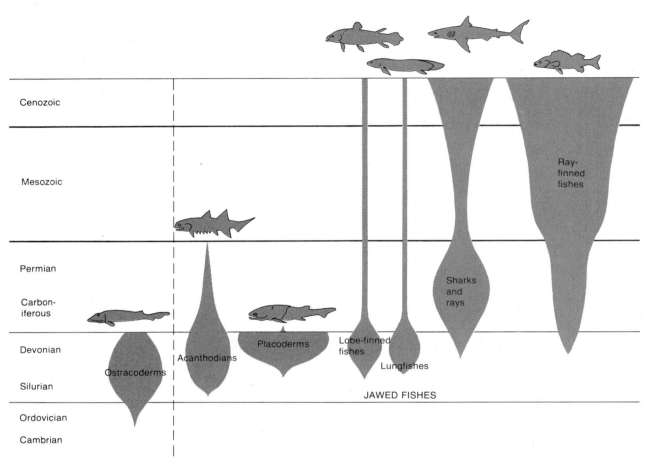

FIGURE 13-8   Geologic occurrence of various kinds of fishes. By Devonian time, all the major groups living today were in existence; no ostracoderms and few placoderms survived beyond the Devonian Period. (*Modified from E. H. Colbert, Evolution of the Vertebrates, John Wiley & Sons, Inc., New York, 1980.*)

this time than during later geologic periods. The label simply reflects the fact that there were no vertebrates on earth other than fishes until the very end of the Devonian Period. From the standpoint of vertebrate evolution, we might more accurately label the Devonian the Age of Almost Nothing but Fishes.

We do not know when fishes first occupied freshwater. All known Cambro-Ordovician fish remains, however, have been found in marine deposits. This does not prove that fishes evolved in the ocean, but it does lend support to the idea. In any event, most Silurian fish remains, unlike most Cambro-Ordovician fish fossils, come from freshwater deposits. One of the most conspicuous groups of these were the **ostracoderms**, whose name means "bony skin." Ostracoderms were small animals with paired eyes like those of higher vertebrates. Lacking jaws and covered by bony armor, ostracoderms did not resemble modern fishes. In

addition, ostracoderms had small mouths and could only have consumed small items of food. Many of these fishes, such as *Hemicyclaspis* (Figure 13-9), also had flattened bellies that were apparently adapted to a life of scurrying along the bottoms of lakes and rivers. The upper fin of the asymmetrical tail of *Hemicyclaspis* was elongate; when wagged back and forth for swimming, this structure would have pushed the animal's head downward rather than upward. In contrast, the ostracoderm know as *Pteraspis* (Figure 13-10) had a curved belly and an elongate lower fin in its tail that would have lifted it upward, suggesting a life of more active swimming well above lake floors and river bottoms. Ostracoderms not only lacked jaws for chewing but also lacked paired fins, which impart stability and control during movement, and bony internal skeletons; it is assumed that they possessed cartilaginous internal skeletons. These animals continued to thrive throughout most

FIGURE 13-11 Restoration of *Euthacanthus,* a member of the most primitive group of jawed fishes, which are known as acanthodians. This animal was about 20 centimeters (~8 inches) in length. (*Field Museum of Natural History.*)

FIGURE 13-9 Reconstruction of *Hemicyclaspis,* a Devonian ostracoderm (jawless fish) that was about 20 centimeters (~8 inches) long. The upper picture shows the asymmetrical tail, and the lower picture shows the flattened belly; both of these features probably relate to a life of scurrying along the floors of lakes and rivers. (*Field Museum of Natural History.*)

FIGURE 13-10 *Pteraspis,* a Devonian ostracoderm (jawless fish) that was about 6 centimeters (~2¼ inches) long. Note the difference between the asymmetry of the tail here and in *Hemicyclaspis* (Figure 13-9). *A.* Upper and lower surfaces of a well-preserved specimen from the Old Red Sandstone of Wales. *B.* Side view of a restoration. (*A. E. I. White, Philos. Trans. Roy. Soc. London B225:381–457, 1935. B. Field Museum of Natural History.*)

of the Devonian Period but disappeared at the end of this interval.

Late in Silurian time, a second, quite different group of small marine and freshwater fishes made their appearance. These were the **acanthodians,** elongate animals with numerous fins supported by sharp spines (Figure 13-11). The acanthodians appear to have been the first fishes to possess several features that they passed on to more advanced, modern fishes: Their fins were paired; scales rather than bony plates covered their bodies; and, most important, they had jaws. With the origin of jaws, a wide variety of new ecologic possibilities opened up for vertebrate life— possibilities that related primarily to the ability to prey on other animals. Unlike ostracoderms, many acanthodians must have been predators that fed on small aquatic animals. As shown in Figure 13-12, it seems evident that the jaws of acanthodians evolved from the gill supports of ancestral fishes; in fact, the first teeth on such jaws were probably modified scales. The modification of one pair of gill arches into a simple jaw is a classic example of evolutionary "opportunism"—the evolutionary alteration of an existing structure to perform an entirely new function. To return to the analogy employed in Chapter 6, an engineer with the capacity to design living systems from inanimate objects might have produced a different jaw apparatus, but the engineer, unlike the process of evolution, would not have been forced to employ structures that were already present in a preexisting animal.

Although acanthodians disappeared near the end of Devonian time, the evolutionary legacy that they left was of great ecologic significance. We do not know precisely how other, more advanced groups of fishes were related to acanthodians, but during the Devonian Period a great adaptive radiation of descendant jawed fishes added new levels to the food webs of both freshwater and marine habitats; soon very large fishes were feeding on smaller fishes, and these in turn fed on still smaller fishes. At the top of this food web were the largest representatives of the

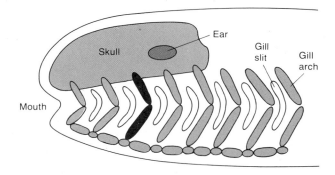

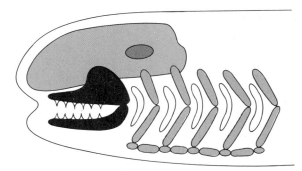

FIGURE 13-12   Origin of vertebrate jaws. Jaws (*black structures in lower picture*) evolved in fishes from a pair of gill supports (*black structures in upper picture*). Note how a pair of gill supports happened to be positioned in such a way that it could easily be converted into a jointed jaw. Teeth were presumably added to the jaw secondarily by the evolutionary modification of scales.

FIGURE 13-13   The massive, armored skull of *Dunkleosteus,* a placoderm fish of Late Devonian age. This formidable species of jawed fish attained a length of more than 10 meters (~30 feet). Note the bony teeth and the armor protecting the eye. (*Field Museum of Natural History.*)

group, the **placoderms**. These heavily armored, jawed fishes made their appearance during the Devonian Period but almost disappeared before the beginning of Carboniferous time. A few placoderms are known from Lower Devonian freshwater deposits, and a wide variety of freshwater species existed by mid-Devonian time. Only secondarily did placoderms make their way into the oceans, and here they were not highly diversified until Late Devonian time. During their brief stay on earth, however, this remarkable group of fishes expanded to include a wide variety of species. *Dunkleosteus*, a Late Devonian marine genus, attained a length of some 10 meters (~30 feet). Like other placoderms, it had armorlike bone protecting the front half of its body (Figure 13-13), but its unarmored tail was exposed to attack, remaining flexible for locomotion. Figure 13-14 depicts *Dunkleosteus* pursuing *Cladoselache*, a shark commonly found with it in black shales of northern Ohio.

Sharks were, in fact, among the most important groups of fishes in Devonian seas. Unknown in rocks older than

mid-Devonian, the sharks may have been the last major group of fishes to evolve. Devonian sharks were primitive forms, and few greatly exceeded 1 meter (~3 feet) in length.

Also making an appearance in Devonian time were the **ray-finned fishes**. These jawed forms, which attained only modest success during the Devonian Period, went on to dominate Mesozoic and Cenozoic seas. They include most of the familiar modern marine and freshwater fishes, such as trout, bass, herring, and tuna. The phrase *ray-finned* refers to the fact that the fins of these fishes are supported by thin bones that radiate from the body—the bones that can be seen through the transparent fins of living fishes. The oldest mid-Devonian ray-finned fishes, like *Cheirolepis* (Figure 13-15), differed from modern representatives in that they had asymmetrical tails and diamond-shaped scales that did not overlap.

Cheirolepis and many other freshwater fishes of Devonian age have been found in deposits that accumulated around the Old Red Sandstone continent, a land area formed, as we will see later in this chapter, by the closing of the Iapetus Ocean. This ancient landmass included parts of modern North America, Greenland, and western Europe. Many of the best fossil-fish specimens come from the segment of this continent that now forms part of Scotland.

The origin of the ray-finned fishes was an event of great significance, but so was the origin of a related group of jawed fishes—one that included the lungfishes and the lobe-finned fishes.

FIGURE 13-14   The giant placoderm *Dunkleosteus* (see Figure 13-13), shown in pursuit of the Late Devonian shark *Cla-* *doselache. (Drawing by Z. Burian under the supervision of Professor J. Augusta.)*

FIGURE 13-15   *Cheirolepis,* a primitive ray-finned fish of mid-Devonian age. The tail of this animal was strongly asymmetrical, and the small, diamond-shaped scales did not overlap each other. The animal was about 4 centimeters (~10 inches) long. (*After A. S. Romer, Vertebrate Paleontology, University of Chicago Press, Chicago, 1966.*)

FIGURE 13-16   The lobe-finned fish *Eusthenopteron,* a large animal that exceeded 50 centimeters (~20 inches) in length. This unusually well preserved specimen is from Upper Devonian deposits at Scaumenac Bay, Canada. (*Swedish Museum of Natural History.*)

The Devonian Period was the time of greatest success of the **lungfishes**, which survive today via only three genera—one in South America, one in Africa, and one in Australia. (Presumably this fragmented distribution reflects the Mesozoic breakup of Gondwanaland.) The Australian genus, *Neoceratodus,* so closely resembles the Triassic genus *Ceratodus* that it is commonly referred to as a "living fossil." The surviving lungfishes are named for the lungs that allow them to gulp air when they are trapped in stagnant pools during the dry season. Such lungs presumably served a similar function in Devonian time.

Similar lungs were present in Devonian **lobe-finned fishes**. These fishes derive their name from their paired fins, whose bones are not radially arranged, as in ray-finned fishes, but are instead attached to their bodies by a single shaft (Figure 13-16). Most lobe-finned fishes inhabited freshwater, but one unusual group, the coelacanths, invaded the oceans. A single coelacanth genus survives in the oceans today; this was discovered in 1952 in deep waters near Madagascar (see Appendix III, Figure AIII-5). Lobe-finned fishes, like lungfishes, declined after the Devonian Period but left a rich evolutionary legacy. As we will learn later in this chapter, lobe-finned fishes are the ancestors of all terrestrial vertebrates, including humans; their lungs were the predecessors of our own.

The great diversification of jawed fishes and, to a lesser extent, the expansion of ammonoids and eurypterids must have had a profound effect on many relatively defenseless aquatic animals. These predators may have contributed to the middle Paleozoic decline in trilobite diversity: About 80 families of trilobites are known from the Ordovician (far fewer than are known from the Cambrian), but only 23 families have been found in Silurian deposits. It seems likely that the weakly calcified external skeletons of trilobites offered little resistance to the jaws of fishes, and certainly trilobites had no mechanism for rapid locomotion. In the last chapter, it was suggested that nautiloids, which evolved at the very end of the Cambrian Period, suppressed the diversification of trilobites during the Ordovician Period. Continuing with this line of reasoning, we must suspect that more advanced cephalopods and jawed fishes were largely responsible for the further decline of trilobites in mid-Paleozoic time. The small, apparently defenseless ostracoderms, which died out late in the Devonian Period, must also have served as easy prey for jawed fishes. Ostracoderms even lacked the ability to burrow in sediment that characterized at least some of the trilobites (Figures 12-3 and 12-16).

Plant life: Invasion of the land

It is difficult to imagine how the landscape looked in Precambrian and early Paleozoic times, before there were conspicuous terrestrial plants. Certain terrestrial environments must have been populated by algae and by other simple plants and plantlike organisms, but there were no forests or meadows, and there must have been large areas of barren rock and soil with little or no humus (decayed organic matter). Thus one of the most important events revealed by the fossil record of Silurian and Devonian life is the invasion of terrestrial habitats by higher plants.

The basic requirements for the terrestrial existence of large multicellular plants are very different from those of plants that live in water. Unlike water, air is a fluid of much lower density than the tissues of a plant; if it is to stand upright in air, a plant must have a rigid stalk or stem. Tall plants must also be anchored by a root system or by a buried horizontal stem, which serves the further indispensable function of collecting water and nutrients from the soil. Furthermore, most large land plants of the modern world

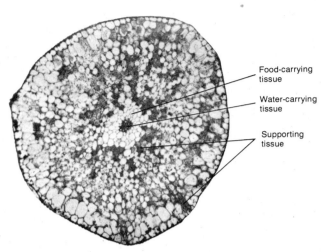

FIGURE 13-17   Section through the stem of *Rhynia*, a primitive Early Devonian genus of vascular plants. The supporting tissue surrounds tissue that was formed into tubes for carrying food downward as well as similar hollow tissue for carrying water and dissolved minerals upward. (*A. C. Seward, Plant Life Through the Ages, Cambridge University Press, New York, 1931.*)

are **vascular**, which means that their stems have one set of special vessels to carry water and nutrients upward from their roots and another to distribute the food that the plants manufacture for themselves. Most large modern plants also bear leaves, which serve to capture the sunlight that assists in the manufacture of food.

The first upright plants to make their way onto the land lacked the roots, vascular systems, and leaves that made their descendants so successful. Essentially, these plants were simple rigid stems, the underground portions of which served as roots and the aboveground portions of which served to transport materials and manufacture food without benefit of hollow vessels or leaves. Fragments of such early plants have been found in Silurian rocks. Silurian plants seem to have been pioneers that lived near bodies of water, and they may actually have been semiaquatic marsh dwellers rather than fully terrestrial plants.

The first major adaptive breakthrough for life on land, before the evolution of roots and leaves, was the origin of vascular tissue. Two kinds of vessels developed—one for the transport of water and nutrients and another for the transport of manufactured food. Figure 13-17 shows the vessels of a stem of the Early Devonian genus *Rhynia*. A few kinds of vascular plants are found in nonmarine deposits of latest Silurian age (Figure 13-18). These had branched stems as well as bulbous organs that shed spores.

As noted in the previous chapter, **spores** are reproductive structures that can grow into new adult plants when

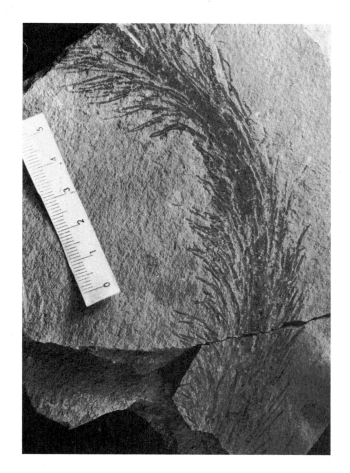

released into the environment. Spore-bearing plants are present in the modern world; ferns are a familiar example. Most spore-bearing plants, including ferns, have a life cycle that involves two very different kinds of adult plants. Of the two, the spore-bearing form is normally the larger (Box 13-1). The Silurian fossils of vascular plants are accompanied by characteristic spores. The fossil record of similar spores extends well back into the Ordovician System (Figure 12-29), but while this suggests that upright land plants existed much earlier than the Late Silurian, the older spores may in fact represent aquatic or semiaquatic species. In some Early Devonian forms, solitary spore-bearing organs stood atop upright stalks, while other species displayed clusters of spore-bearing organs in similar positions; and in still other species, organs were arrayed along the upright stalks (Figure 13-19).

FIGURE 13-18   One of the oldest known vascular plants, a specimen of the genus *Baragwanathia* from the Upper Silurian of Victoria, Australia. The scale, in millimeters, is about 2 inches long. (*Courtesy of M. J. Garratt.*)

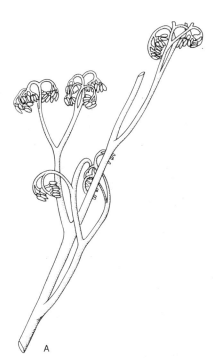

A   B   2 Centimeters   C

FIGURE 13-19   Early Devonian vascular plants displaying spore-bearing organs. *A.* Drawing of *Dawsonites,* in which clusters of organs are suspended at the end of the stem. *B.* Photograph of a fossil of *Bucheria,* in which the organs are arranged in rows near the end of the stem. *C.* Reconstruction of *Rhynia* (see Figure 13-17), in which there is a single organ at the end of the stem. (Scales in centimeters.) (*B. Princeton University Museum of Natural History. C. Field Museum of Natural History.*)

Box 13-1

A life cycle

typical of a middle

Paleozoic land plant

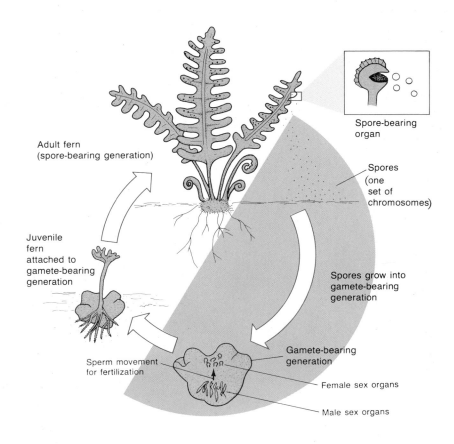

Adult fern
(spore-bearing generation)

Spore-bearing
organ

Spores
(one
set of
chromosomes)

Juvenile
fern
attached to
gamete-bearing
generation

Spores grow into
gamete-bearing
generation

Sperm movement
for fertilization

Gamete-bearing
generation

Female sex organs

Male sex organs

The familiar fern plant (shown here at the top of the cycle) resembles most other kinds of organisms in that it possesses two sets of chromosomes. The plant sheds spores, each of which has only one set of chromosomes. Once on the ground, a spore grows into an inconspicuous gamete-bearing plant that seldom grows larger than $\frac{1}{4}$ inch (~6 millimeters) in diameter. This small plant has two kinds of gamete-bearing organs: one that produces eggs and another that produces sperm. Like the plant that bears them, each gamete has one set of chromosomes. A sperm then unites with an egg to produce a juvenile fern plant with two sets of chromosomes. The fern plant grows right on top of the gamete-bearing plant but attains a much larger size, and eventually the small gamete-bearing plant dies.

Ferns and their relatives are restricted to moist habitats because their sperm must travel through moisture to eggs on the surface of the gamete-bearing plant. ☐

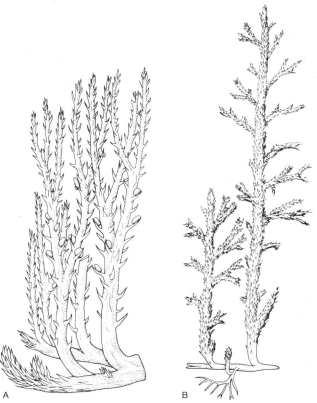

FIGURE 13-20   The reconstruction of *Protolepidodendron* (*A*) and *Asteroxylon* (*B*), both primitive Devonian forms of lycopods. The living lycopod, *Selaginella,* is growing in a flower pot (*C*). In contrast to these oldest and youngest representatives, many lycopods of late Paleozoic time were large trees. (*Photograph C courtesy of Dr. T. E. Weier.*)

Regardless of whether land plants existed much before the latest Silurian interval, it was apparently near the end of the Silurian Period that vascular tissues evolved. As a result of this physiological breakthrough, a great adaptive radiation took place in Early Devonian time. The Devonian plants that resulted were still relatively low, creeping forms that lacked well-developed roots and leaves, but during Early and Middle Devonian time several shapes evolved. The vascular tissues of early vascular plants like *Rhynia* were confined to a narrow zone of the stem (Figure 13-17) and were thus mechanically weak and inefficient at conducting liquid. By Late Devonian time, however, some plants had developed vascular tissues that occupied a larger volume within the stem and were therefore mechanically stronger and more efficient at conduction. Plant groups with these useful traits also evolved roots for support and for effective absorption of nutrients and also leaves for capturing large quantities of sunlight. These plants seem to have competitively displaced plants like *Rhynia*, which were less efficient at obtaining nutrients, synthesizing food, and growing to a large size.

Certain of the small plants that arose during Early and Middle Devonian times are classified as **lycopods**. This group includes the tiny club mosses of the modern world (Figure 13-20), but as we will see in Chapter 14, some late Paleozoic lycopods grew to the proportions of trees, and their petrified remains supply much of modern society's coal. These large forms vanished by the end of the Paleozoic Era, however, and only tiny creeping lycopods resembling the primitive types of the Early Devonian Period have survived to this day.

Because of the limitations imposed by their mode of reproduction, these Early Devonian land plants must have formed low marshes along bodies of water (Figure 13-21). Like living ferns and other present-day spore-bearing plants, their reproductive cycle would have restricted them to damp habitats, since the sperm of such plants can travel to the egg only under moist conditions (Box 13-1). As the Devonian

FIGURE 13-21   Early Devonian landscape showing some of the first land plants, which still bordered the bodies of water. The plants labeled *A* and *B* are similar to those shown in Figure 13-19; those labeled *C* are similar to those shown in Figure 13-20. (*Drawing by Z. Burian under the supervision of Professor J. Augusta.*)

Period progressed, however, the appearance of a second adaptive "innovation," the **seed**, liberated land plants from their dependence on moist conditions, allowing them to invade many habitats. Today, most large land plants grow from seeds.

Box 13-2 shows that a seed is actually a juvenile stage of the spore-bearing generation of a plant, formed after the gamete-bearing organs are fertilized. The seed remains small enough that it can move by air, water, or animal host to favorable ground, where it can sprout into the adult form of the spore-bearing generation. An important aspect of the reproductive cycle of seed plants is that their spores are not released to the environment but are instead retained on the spore-bearing plant. This means that the gamete-bearing generation, which is formed from a spore, does not have to survive alone in a moist environment, as it does in ferns and their relatives (Box 13-1). Instead, both the male and female varieties of the gamete-bearing generation are protected. The female variety, the **embryo sac**, lives like a parasite on the large spore-bearing generation.

The male variety, the **pollen**, also forms there and, when exposed to the environment as it moves to reach the embryo sac of the same or a different plant, has little difficulty surviving. Pollen is supplied with a waxy protective coating and is resistant to many forms of environmental stress, including temperature change and drought.

Before the seed could evolve, a preliminary evolutionary step was necessary. This was the development in a single species of two kinds of spores, one of which gave rise to a female gamete-bearing plant and the other to a male gamete-bearing plant. (Later, in seed plants, the female came to be retained as the embryo sac and the male as the pollen grain.) Before this step was taken, a single kind of gamete-bearing plant produced both male and female gametes, as occurs in the modern fern shown in Box 13-1. The earliest evidence of the separation of sexes comes in the form of spores of two different sizes found in association with a plant of mid-Devonian age (Figure 13-22). Since spores grow into plants of the gamete-bearing generation, we can assume that the smaller spores grew into

Box 13-2

A life cycle of a seed

plant, which does not

require a moist setting

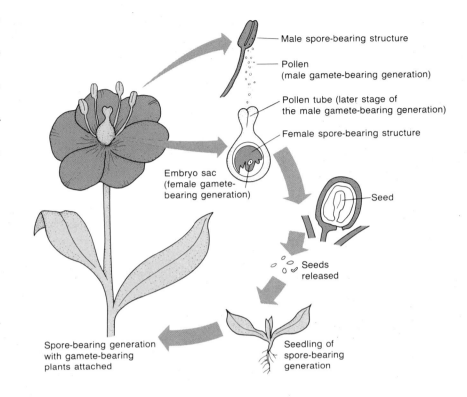

Male spore-bearing structure

Pollen
(male gamete-bearing generation)

Pollen tube (later stage of
the male gamete-bearing generation)

Female spore-bearing structure

Embryo sac
(female gamete-
bearing generation)

Seed

Seeds
released

Spore-bearing generation
with gamete-bearing
plants attached

Seedling of
spore-bearing
generation

This particular example illustrates the cycle of a flowering plant. Flowering plants did not evolve until the Mesozoic Era, but for these, as for other seed plants, the spore-bearing generation is the one familiar to us—the one having roots, stems, branches, and leaves. This generation has both male and female spore-bearing structures. Unlike ferns and their relatives (Box 13-1), seed plants do not release their spores into the environment. Instead, the spores are retained by the spore-bearing generation and grow into the male and female forms of the gamete-bearing generation *within the tissues of the spore-bearing generation*. Thus, the male and female forms of the gamete-bearing generation (the pollen and embryo sac) live like parasites on the spore-bearing generation. The pollen is then released and lands above the embryo sac, where it forms a pollen tube that penetrates the plant tissues in order to reach the embryo sac. In the embryo sac, a female gamete is then fertilized by a male gamete formed within the pollen tube. This union produces a seed. Each gamete has one set of chromosomes, and thus the seed ends up with two.

The seed is, in effect, a juvenile stage of development of the spore-bearing generation—a stage in which the plant remains small and well protected so that it can either await favorable conditions (often the end of winter or of the dry season) or move to a favorable location before growing into a seedling.

Because fertilization in seed plants takes place without the requirement that sperm travel to eggs through moisture in the environment, seed plants, unlike ferns and their relatives, can occupy relatively dry habitats. In seed plants, the two forms of the gamete-bearing generation remain in the protection of the spore-bearing generation. Only the pollen (male variety) is released for a time, but it is extremely durable and able to withstand a wide range of environmental conditions. □

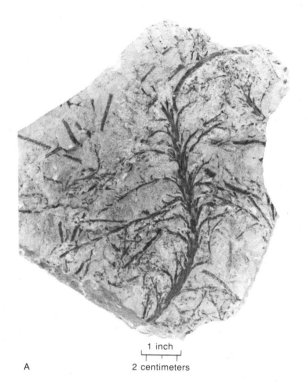

A

1 inch

2 centimeters

B

C

plants of one sex and the larger spores into plants of the other sex. Because females generally need more energy to produce their large eggs than males need to produce their small sperm, it is assumed that the large spores grew into large female plants and the small spores into male plants.

As we will see in Chapter 16, advanced seed plants with flowers did not evolve until Cretaceous time. Flowers attract insects and birds, which aid in fertilization by unknowingly carrying pollen from flower to flower. More primitive, flowerless seed-bearing plants lack this sophisticated mechanism of fertilization, relying instead on less efficient agents—primarily wind—to carry pollen from plant to plant.

Flowerless seed plants originated in Late Devonian time and soon became important elements of Late Paleozoic terrestrial floras (Figure 13-23), opening up a new world to the plant kingdom. During Late Devonian time, for the first time, dry land was invaded on a vast scale. Seed plants soon grew into large trees with strong, woody stems (Figure 13-24). These events changed the face of the earth. Trees formed the world's first forests, where there had been only barren land before.

One of the consequences of the spread of terrestrial vegetation was that, for the first time in earth history, plants carpeted the soil and gripped it with their roots, thereby stabilizing it against erosion. Braided-river deposits, which reflect rapid erosion (page 76), characterize the clastic wedges of Precambrian and early Paleozoic age. Only in middle Paleozoic time, when vegetation first stabilized the land, did rivers begin to meander and to deposit sediment in orderly cycles (page 77) on a large scale.

Animals move onto the land

The Rhynie Chert of Scotland is a Lower Devonian formation that has yielded not only a variety of beautiful fossil plants but also some of the oldest known fossil arthropods, including scorpions and flightless insects.

Arthropods probably invaded dry land in Late Silurian time, before some of their terrestrial representatives were preserved in the Rhynie Chert—but it was not until the Late

FIGURE 13-22   The Middle Devonian plant *Chaleuria* (*A*), which produced spores of two sizes. The larger spores (*B*) are believed to have produced female plants, and the smaller ones (*C*), are thought to have produced males (× 140). (*Courtesy of H. N. Andrews.*)

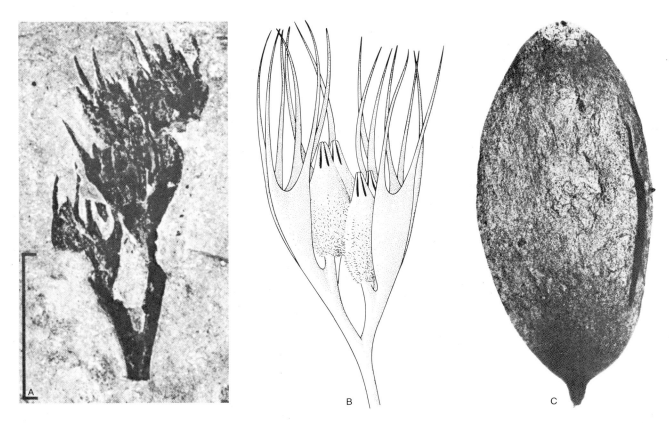

FIGURE 13-23   Upper Devonian fossils that represent some of the oldest known seeds. These seeds were borne at the ends of branches, as seen in the fossil specimen A. The individual seeds, four of which are shown in the drawing (B), were about 1 centimeter ($\sim\frac{1}{4}$ inch) in length. The internal portion of the seed that grew into a new plant is shown in photograph C. (J. M. Pettitt and C. B. Beck, University of Michigan Paleont. Contrib. 22:139–154, 1968.)

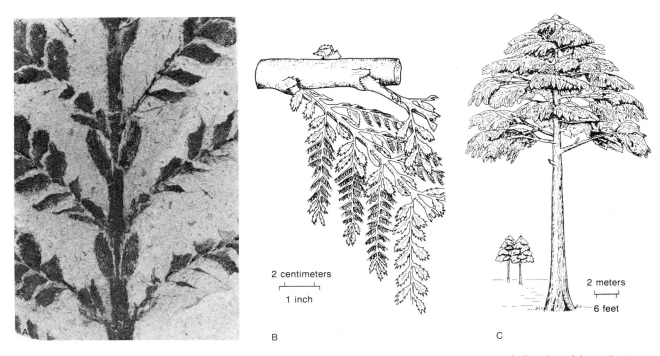

FIGURE 13-24   Archaeopteris, one of the very oldest kinds of large tree. A. A segment of a branch with leaves. B. Recon-struction of a branch with leaves. C. Drawing of the entire tree. (After C. R. Beck, Biol. Rev. 45:379–400, 1970.)

Devonian interval that vertebrate animals made a similar transition. It has long been recognized from anatomical evidence that the four-legged vertebrates most closely related to fishes are the **amphibians**—frogs, toads, salamanders, and their relatives. That amphibians represent the most primitive four-legged vertebrates is also suggested by the fact that these animals are legless and aquatic early in life. They hatch from eggs in water, spend their juvenile existence there, and then usually metamorphose into air-breathing, land-dwelling adults. Living amphibians are small animals that differ substantially from the large fossil amphibians found in Upper Paleozoic rocks. Their evolutionary history began late in the Devonian Period.

In eastern Greenland, the remains of unusual vertebrate animals have been found in uppermost Devonian rocks of

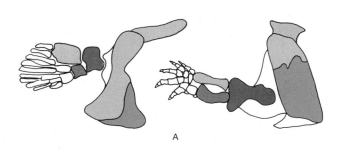

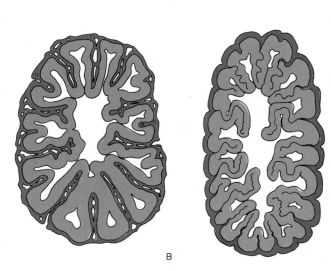

FIGURE 13-25   Resemblance between lobe-finned fishes (*left*) and amphibians (*right*). *A*. Comparison of the shoulder and limb bones; shading identifies particular bones that the two groups have in common. *B*. Comparison of cross sections of teeth, showing the unusual, complex structure found in both groups.

the Old Red Sandstone continent. These fossils, many of which are assigned to the genus *Ichthyostega*, represent creatures that are strikingly intermediate in form between lobe-finned fishes and amphibians: The lobe fin itself is formed of an array of bones resembling that found in amphibians; similarly, the complicated teeth of lobe-finned fishes closely resemble the teeth of early amphibians (Figure 13-25). These features alone strongly suggest that amphibians were derived from lobe-finned fishes, but additional features make the derivation a certainty. *Ichthyostega* had four legs, as do amphibians, but its skull structure was remarkably like that of a lobe-finned fish. The creature also had a fishlike tail—one feature of its ancestors that was probably of no use on land (Figure 13-26). Because of this intriguing combination of features, *Ichthyostega*, which was not discovered until the present century, represents what is commonly termed a "missing link."

The lung, which was occasionally used by early fishes to breathe air, was available for exploitation long before amphibians evolved. The way in which the lung became a full-time supplier of oxygen represents yet another example of the "opportunism" of evolution. Unlike the example of the gill arches that evolved into jaws earlier in vertebrate evolution, the origin of lungs involved the use of a preexisting structure that required very little evolutionary modification to open up an entirely new mode of life.

There was an interval of more than 80 million years between the time when vascular plants appeared on land (Late Silurian or earlier) and the time when the first amphibians evolved (latest Devonian). It is not surprising that vascular plants colonized the land before vertebrate animals, since a food web must be built upward from the base; animals cannot live on land in the absence of an adequate supply of edible vegetation.

It has often been argued that there must have been some reason for fishes to leave the water and evolve a habit of life on the land. The reason often suggested is that the climate became drier, thereby causing bodies of water to shrink and forcing fishes with lungs to spend more and more time breathing air as they left their aquatic habitat to move from pond to pond. While this is a plausible hypothesis, there is no reason to believe that any environmental change necessarily forced fishes out of water. The land was entirely free of vertebrate life before this transition occurred; in the absence of competition, any population of lobe-finned fishes that developed rudimentary adaptations for life on land might have taken a step toward a more terrestrial existence. In other words, one or more adaptive

FIGURE 13-26   The primitive amphibian *Ichthyostega,* whose legs contrast to the fins of the approximately contemporary lobe-finned fish *Eusthenopteron* on the left (see also Figure 13-16) and the lungfish *Rhynchodipterus* on the right. The trunk and branches belong to the tree fern *Eospermatopteris.* (*Drawing by Gregory S. Paul.*)

"breakthroughs" may have triggered the change even in an unchanging environment. Certainly, there is no need to suggest that this all-important ecologic transition depended on a global change of climate.

Amphibians evolved so late in Devonian time that they played no significant role in the ecosystem of that period. It was the Carboniferous and Early Permian that might be called the Age of Amphibians. Nevertheless, the dominant animals of that interval descended from the Late Devonian *Ichthyostega* or from similar taxa not yet discovered.

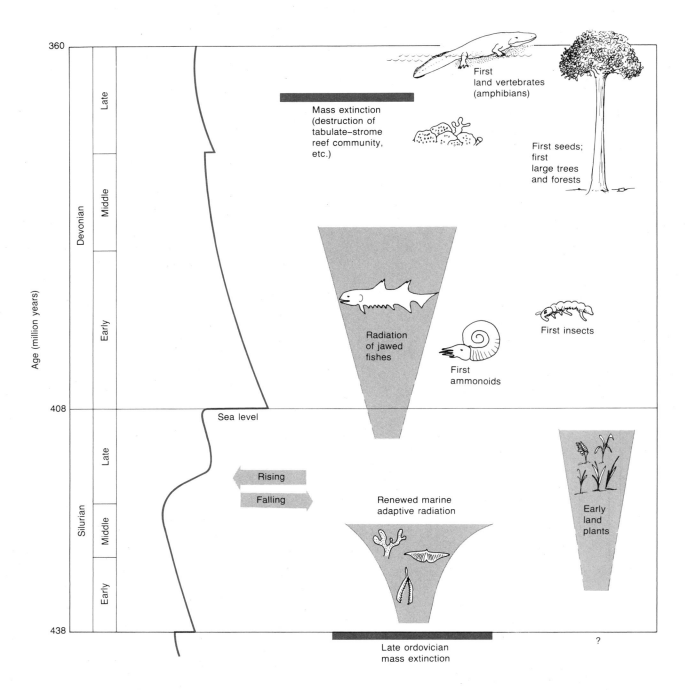

FIGURE 13-27   Major events of middle Paleozoic time. Early in the Silurian Period, adaptive radiation replenished marine life after the terminal Ordovician mass extinction. Although the seas generally stood high during middle Paleozoic time, a major drop in sea level occurred at the end of the Silurian. New groups of swimming predators, including jawed fishes and ammonoids, became conspicuous early in the Devonian Period, when arthropods also invaded the land and the first insects appeared. During Late Devonian time, a mass extinction struck marine faunas, and vertebrates moved onto the land shortly after the first large trees evolved there and the first large forests developed.

PALEOGEOGRAPHY

In general, the Silurian and Devonian were periods when sea level stood high in relation to the surfaces of major cratons (Figure 13-27). Early in Silurian time, sea level rose from its low position at the end of the Ordovician Period— a rise that is thought to have resulted from the melting or partial melting of the extensive polar glaciers that had formed late in Ordovician time. Simultaneously, many marine invertebrate taxa underwent the adaptive radiations discussed earlier. In many parts of the world, regression then occurred toward the end of the Silurian Period. This did not happen, however, in the central Appalachian region, where shallow-water limestones were deposited continuously across the Siluro-Devonian boundary (Figure 8-23).

The widespread occurrence of organic reefs, carbonates, and evaporites strongly suggests that middle Paleozoic climates were relatively warm. Because they were especially warm during the Devonian Period, it has been suggested that shallow-water tropical marine communities were more widespread at this time than at any other time in the Phanerozoic Eon. Climates were not only warm but in many areas were relatively dry; in fact, larger volumes of evaporite deposits accumulated in middle Paleozoic than in early Paleozoic time. Note that evaporites of Silurian age lie within 30° or so of the ancient equator but that some Devonian evaporites formed farther north and south, apparently reflecting the widespread distribution of warm climates during the Devonian Period (Figure 13-28).

Continents and oceans

Figure 13-28 displays the positions of continental areas for Middle Silurian time and for late Early Devonian time, about 65 million years later.

An important new geographic feature to appear during Devonian time was the **Old Red Sandstone** continent, named for a well-known, largely Devonian sandstone unit of the British Isles (page 358). This was a continent of high relief that formed when Laurentia and Baltica were welded together as the Iapetus Ocean disappeared. Earlier, we saw that this collision began in the north during the Silurian Period and progressed southward, ending late in Devonian time (page 230). Paleomagnetic evidence reveals that by the end of the Devonian Period, the gap between the Old Red Sandstone continent and Gondwanaland had narrowed. The proximity of these continents explains the fact that North America, Europe, and North Africa share at least 80 percent of their Late Devonian genera of marine invertebrates. (As we saw in Chapter 8, Gondwanaland moved even closer to this continent during Late Carboniferous time, later colliding with North America to form the giant supercontinent of Pangaea.)

Biogeographic provinces of the Devonian Period

During Silurian time, conspicuous marine groups like the brachiopods and graptolites were remarkably cosmopolitan in that a number of species and genera inhabited the shallow seas of many continents. At this time, the continents formed a relatively tight cluster. By Early Devonian time, however, mountain building associated with the formation of the Old Red Sandstone continent left an embayment that projected from the south far into the interior of what is now North America (Figure 13-28). During the first half of the Devonian Period, this area became the Appalachian province, and it was characterized by a distinctive fauna. Further transgression of the seas connected it with the west coast, however, so that the Appalachian marine faunas were blended again with others.

At the same time, possibly because of a cooling of polar regions, a discrete marine province formed along the margin of southern Gondwanaland. This southern realm, which was called the Malvinokaffric Province, was populated by a fauna that appears to have lived in cool water. Within this province early in Devonian time, the Paraná Basin (now central South America) was partly barred from the open ocean by newly formed mountains in the Andean region (Figure 13-29). The center of the Paraná Basin lay perhaps only 15° from the South Pole, so it is not surprising that, while stromatoporoids, tabulates, and rugose corals formed reefs throughout most of the shallow-water areas in the world, they were not found in this basin. Also missing were bryozoans and ammonites. Burrowing bivalves formed a large percentage of the marine species in the Paraná Basin, just as they do in polar regions today.

Late Devonian mass extinction

One of the most devastating mass extinctions of marine life in all of Phanerozoic time took place near the end of the Devonian Period. Geologists divide the Upper Devonian Series into two stages—the Frasnian Stage and the Famennian Stage. The great extinctions occurred late in Frasnian and early in Famennian time.

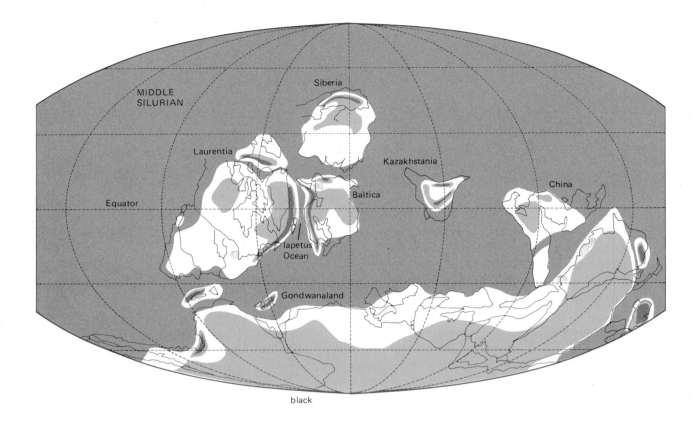

black

Oceanic realm   Shallow sea   Lowland   Mountains   Evaporites

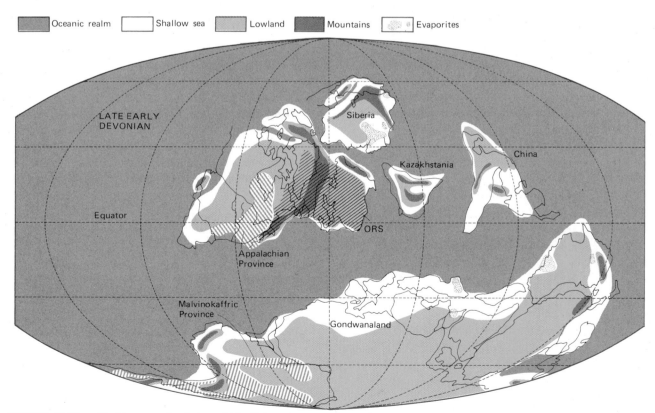

FIGURE 13-28   World geography during middle Paleozoic time. During this interval a broad tropical seaway developed north of Gondwanaland. Note how the Old Red Sandstone continent (ORS) also formed during this interval by the union of Baltica and Laurentia; the Appalachian biogeographic province became isolated to the west. Shallow seas near the South Pole formed the Malvinokaffric Province, where species were apparently adapted to cold temperatures. (*Modified from R. K. Bambach et al., American Scientist 68:26–38, 1980.*)

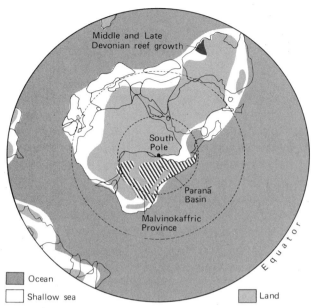

Ocean

Shallow sea

Land

FIGURE 13-29   The position of Gondwanaland in late Early Devonian time. The Paraná Basin, which lay near the South Pole on what is now the continent of South America, supported cold-water marine faunas. Later in Devonian time, extensive reef growth took place near the equator along what is now the western margin of Australia. (*Modified from C. R. Scotese et al., Jour. Geol. 87:217–277, 1979.*)

Within the pelagic system, acritarchs, the only group of phytoplankton with an extensive Devonian fossil record, suffered heavy losses, and placoderms, the dominant pelagic carnivores, almost all disappeared.

The demise of the middle Paleozoic reef community exemplifies an important geographic pattern of the Late Devonian mass extinction: Tropical taxa were most severely affected. In contrast, the polar communities of the Malvinokaffric Province of South America (Figure 13-29) were largely unaffected. This bias against tropical species suggests that an episode of global cooling may have led to the mass extinction. During such an episode, the cold-adapted forms could simply have migrated toward the equator, where the disappearance of tropical conditions may have destroyed reef builders, reef dwellers, and other species that were intolerant of cool temperatures. Fossil evidence in New York State supports the hypothesis that cooling was the agent of mass extinction. Here, glass sponges, which today live in cool waters, experienced an evolutionary expansion while many previously diverse groups of marine animals were on the decline. After the mass extinction, when the groups that had suffered heavy losses began to rediversify, the glass sponges dwindled, perhaps in response to a return of warm climatic conditions.

On the land, vascular plants appear to have been unaffected by the Late Devonian crisis, but it is difficult to interpret the meaning of their persistence, since they differed so much from modern plants. Freshwater biotas suffered heavy losses, however; placoderms nearly disappeared from lakes and rivers, and both lungfishes and lobe-finned fishes were greatly reduced in their diversity.

In the marine realm, brachiopods were hard-hit; only about 15 percent of Frasnian brachiopod genera are found in Famennian strata. Ammonoids experienced a similar decline, and many types of gastropods and trilobites disappeared as well. Two complex communities seem to have suffered almost total collapse: the tabulate-strome reef community and the pelagic community (plankton and nekton). The tabulate-strome community had enjoyed its greatest faunal diversity during Middle Devonian time, 100 million years or so after coming into existence. During much of the Frasnian Age, reef species were already somewhat diminished, but at the end of Frasnian time the reef community suffered truly catastrophic extinctions. Tabulates, stromatoporoids, and rugose corals are rarely seen in Famennian strata, and tabulate-strome reefs are virtually nonexistent both here and in younger Paleozoic strata.

REGIONAL EXAMPLES

The Silurian and Devonian periods were times of widespread reef development and carbonate deposition, but they were also times of orogeny for the continents that bordered the Iapetus Ocean. The regional case studies presented below reflect this variety of conditions. We will look first at the transformation of eastern North America from an Early Silurian highland to a Middle Silurian carbonate shelf and at the formation of reefs and evaporite deposits farther to the west. Next we will take a tour of the Old Red Sandstone continent, which formed later with the Devonian closure of the Iapetus Ocean. We will then review reef growth, mountain building, and other events in western North America. Finally, we will shift our focus to Gondwanaland, giving particular emphasis to an enormous Devonian reef complex on the present continent of Australia.

Eastern North America: Carbonates, reefs, and evaporites

The first tectonic cycle of the central Appalachians ended with the deposition of clastic wedges of sediment shed

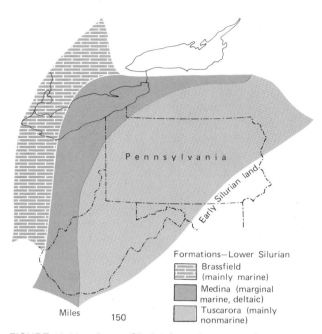

FIGURE 13-30   Lower Silurian formations west of the Appalachians. The coarse, largely nonmarine Tuscarora Formation passes westward to finer-grained deltaic deposits of the Medina Formation, and these give way to marine limestones of the Brassfield Formation. (*Modified from K. J. Mesolella, Amer. Assoc. Petrol. Geol. Bull. 62:1607–1644, 1978.*)

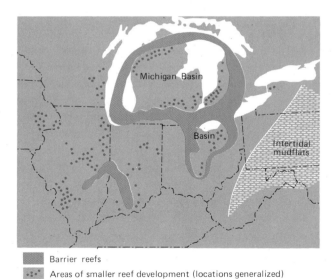

Barrier reefs

Areas of smaller reef development (locations generalized)

FIGURE 13-31   Middle Silurian reefs of the Great Lakes region. Barrier reefs encircled the Michigan Basin and a smaller basin in Ohio. They also flourished in southern Indiana and Illinois. Extensive mud flats now lay to the east in the Pennsylvania region, which contrasts with the environments of coarse clastic deposition that occupied this area in Early Silurian time (Figure 13-30). (*Modified from K. J. Mesolella, Amer. Assoc. Petrol. Geol. Bull. 62:1607–1644, 1978.*)

westward from mountainous areas that formed along eastern Laurentia late in the Ordovician Period (Figure 8-23). The Silurian Period began with a continuation of this pattern, with sediments spreading northwestward into the area that now forms Ohio and southern Ontario (Figure 13-30). As the eastern highlands were subdued by erosion, the site of clastic deposition became more broadly flooded by shallow seas, and late in Silurian time the deposition of shallow-water carbonates initiated the second tectonic cycle of the central Appalachians.

To the west, tabulate-strome reefs dotted shallow epicontinental seas (Figure 13-31). Here, however, the pattern of sedimentation and reef development changed drastically from Middle to Late Silurian time. First, during the Middle Silurian Period, two basins accumulated muddy carbonates. One of these was the Michigan Basin, and the other lay in what is now north-central Ohio. These basins were bounded by large barrier reefs and were populated by scattered pinnacle reefs. At this time, siliciclastic muds were still accumulating to the east on broad tidal flats.

FIGURE 13-32   Mud cracks in intertidal deposits of the Middle Silurian Wills Creek Formation in western Maryland. A broad area of intertidal deposition lay to the east of the reef growth in the Great Lakes region. (*From F. J. Pettijohn and P. E. Potter, Atlas and Glossary of Sedimentary Structures, Springer-Verlag, New York, 1964.*)

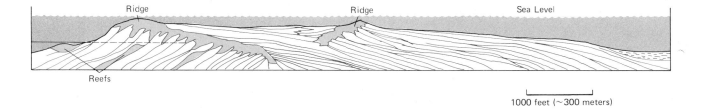

1000 feet (~300 meters)

FIGURE 13-33   Diagram of the Thornton Reef of northern Illinois. This circular reef is exposed in a limestone quarry. The upper diagram represents a cross section of the reef along the axis indicated in the lower diagram, which represents a map view. A durable stromatoporoid ridge faced the prevailing direction of wind-driven waves. Behind the ridge was a zone of tabulates and rugose corals, and behind this were a reef flat and a lagoon. A weaker stromatoporoid ridge bounded the lagoon on the leeward side. (*After J. J. C. Ingels, Amer. Assoc. Petrol. Geol. Bull. 47:405–440, 1963.*)

During Late Silurian time, this pattern changed. To the east, siliciclastic mud deposition gave way to carbonate sedimentation. Central Pennsylvania and neighboring areas were now the sites of accumulation of supratidal, intertidal, and shallow subtidal carbonates (Figure 13-32). At the same time, the Michigan and central Ohio basins came to be only weakly supplied with sea water and thus turned into evaporite pans in which dolomite, anhydrite, and halite were precipitated. The resulting deposits are a major source of rock salt today. In the Michigan Basin, it would appear that marginal reefs grew upward during Middle Silurian time to such an extent that they restricted the flow of water into

the basin. In time, evaporation and, possibly, a slight lowering of sea level led to the exposure and consequent death of the reefs. Although a weak flow of sea water into the basin replenished the water that had been lost by evaporation, the rate of evaporation was so high that evaporite minerals were precipitated around the margins of the basin and even at considerable depths within it. At first, the center of the basin was moderately deep, but the accumulation of evaporites caused shallowing to the point where the sea was eventually excluded altogether.

The evaporite basin in Ohio extended into Pennsylvania and New York State, shifting eastward through time. Because the conditions within the evaporite basins were so inhospitable, Late Silurian reef growth came to be restricted to the southwest, in Indiana and Illinois. Here a number of reefs have been studied in detail. The most famous is the Thornton reef of northern Illinois (Figure 13-33). The structure of this reef indicates the direction of the prevailing winds at the time of growth: The stromatoporoid ridge obviously faced waves advancing from the southwest.

As we have seen in Chapter 8, carbonate deposition of the second tectonic cycle in the central Appalachians was interrupted by the onset of orogenic activity and continental suturing during the Devonian Period. The Old Red Sandstone continent, which formed by this suturing, is the subject of the following section.

The Old Red Sandstone continent

One of the classic angular unconformities in the geologic record occurs in Scotland between Devonian beds of the Old Red Sandstone and the nearly vertical Silurian beds on which they rest. It was from this example, at the locality shown in Figure 13-34, that James Hutton recognized the meaning of stratigraphic unconformity in 1788. Thus, it came to be understood that the Old Red Sandstone was deposited after a Silurian episode of mountain building—as we now know, after the mountain building that occurred during the suturing of Baltica and Laurentia. The Old Red crops

FIGURE 13-34 Angular unconformity between the Old Red Sandstone, above left, and Silurian rocks, below left, at Siccar Point, Berwickshire, Scotland; the Silurian rocks were tilted and folded when Baltica collided with Laurentia to form the Old Red Sandstone continent (see Figure 13-28). The Old Red Sandstone was subsequently deposited near the margin of this continent. (*Institute of Geological Sciences, British Crown copyright.*)

out over large areas of Scotland (Figure 13-35). Chunks of its distinctive red sandstones can be found in parts of Hadrian's Wall, which was built by the Roman emperor Hadrian across northern England in the second century A.D.

The Old Red includes not only rocks of Early, Middle, and Late Devonian age but also rocks representing the Late Silurian and earliest Carboniferous. For a long time, geologists found it puzzling that such a large volume of sediment could have been shed from highlands in the British Isles when most of the isles' area formed a depositional basin. Now the puzzle has been solved within the framework of plate tectonics by the reassembly of the landmass that formed when Laurentia, the British Isles, and Baltica were united during mid-Paleozoic time. Portions of this landmass that stood above sea level formed what has been called Laurasia, or the Old Red Sandstone continent (Figure 13-28). The British Isles, which lay near the southeastern margin of this landmass, were the site of extensive deposition of freshwater molasse deposits derived from mountains in the area that is now Greenland and Labrador. Recall that the orogenic activity associated with the for-

mation of this continent proceeded from north to south (page 230). Molasse deposition resulting from tectonic activity known as Caledonian in Britain was under way in northern Britain early in Devonian time, but in the northeastern United States the oldest molasse (clastic-wedge) deposits, resulting from the interval of deformation known as Acadian, are of mid-Devonian age (Figure 8-23). The proximity of Britain and eastern North America during the latter half of the Devonian Period explains another long-standing problem as well—the similarity of freshwater fishes and early land plants of the two regions.

During much of Devonian time, an arm of land may have extended southwestward across what is now the western interior of North America; here the Transcontinental Arch persisted from early Paleozoic time (Figure 13-36), and there was little or no sediment accumulation. The Devonian equator passed through the southern part of the Old Red Sandstone continent. As is the case today (Figure 2-11), prevailing trade winds must have blown from the east. Minor coal deposits, which formed from early land plants, are found in the east, where moist air must have risen as it passed landward and dropped moisture gained

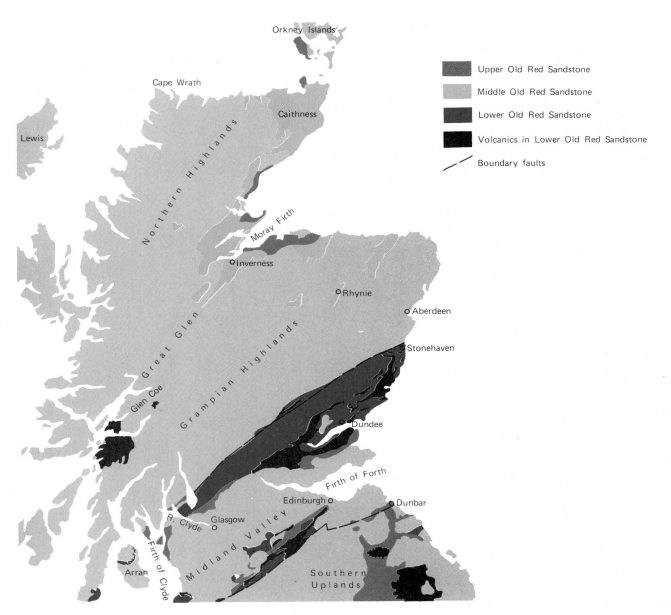

FIGURE 13-35   Geologic map of Scotland, showing the widespread occurrence of the Old Red Sandstone Formation. The
Old Red spans the entire Devonian Period and includes a bit of
the Silurian, below, and the Carboniferous, above. The Old Red
is entirely nonmarine and has yielded numerous beautifully preserved freshwater fishes.

from passage over seas to the east (Figure 13-36). Evaporites are found here and there near the eastern margin
of the continent but are best developed in the rain shadow
to the west, in the area of North America where the Rocky
Mountains now stand. The climate was at least intermittently hot and dry enough along the east and west coasts
of the Old Red Sandstone continent that caliche nodules
formed abundantly in low-lying areas that are now represented by ancient soils. In the east, rainfall was probably
not only heavier but also seasonal, as in the tropics today.

Representing the southern margin of the Old Red Sandstone continent, deposits in the British Isles reflect a spectrum of environments ranging from dunes (Figure 13-37)
to alluvial deposits to marine mud flats, barrier islands, and
subtidal sea floors. Not all of the rock units are red. In the
north, only freshwater units are found. Marine influence

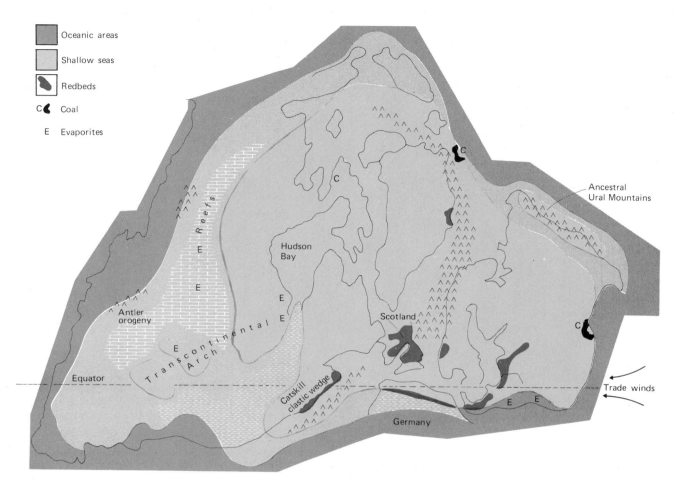

FIGURE 13-36   The Old Red Sandstone continent during Late Devonian time. The central mountain belt running from north to south formed as Baltica converged with Laurentia (see Figure 13-28). Red beds, including those that formed the Old Red Sandstone Formation (Figure 13-35), were concentrated in the south near the equator. Deep-water deposits accumulated in what is now central Germany. Shales accumulated in North America west of the Catskill clastic wedge, and limestones, reefs, and evaporites formed farther west. The Antler orogeny affected the western margin of the continent.

increased southward, and in northern Germany, which had formed a southern segment of Baltica, marine deposition prevailed (Figure 13-38). Alluvial sediments in Britain range from poorly sorted mud flat deposits and gravelly cross-bedded sands that accumulated near source areas to meandering river sequences farther from the area of sed-

FIGURE 13-37   Cross-stratified Devonian sandstones of the Dingle Peninsula of southwestern Ireland. These were dune deposits formed in the southern part of the Old Red Sandstone continent, attesting to arid conditions. (*From R. R. Horne, Geol. Mag. 108:151–158, 1971.*)

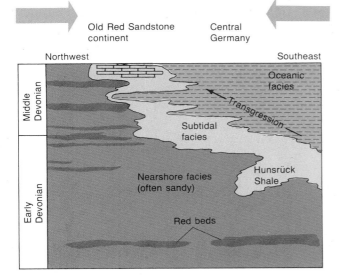

FIGURE 13-38   Stratigraphic cross section through the Lower and Middle Devonian of northern Germany. Nearshore facies accumulated along the southern margin of the Old Red Sandstone continent (Figure 13-36). The region now forming central Germany lay offshore in the oceanic realm. Note that a transgression occurred during the interval represented in the diagram. (*After R. Goldring and F. Langenstrassen, Spec. Papers Paleont. 23:81–97, 1979.*)

iment supply, where more mud was deposited than sand.

In Germany, red beds and sands were confined to the northwestern region (Figure 13-38). Near shore, they accumulated along estuaries and lagoons inhabited by bivalves, ostracods, fishes, eurypterids, and a few kinds of brachiopods; here, plant debris from the land also came to rest. On outer shelves, which were characterized by normal marine salinities, a greater variety of brachiopods lived along with trilobites, crinoids, bryozoans, corals, and mollusks—all of which apparently could not live near shore. To the southeast, black shales accumulated beyond the edge of the continental shelf. Among these was the Hunsrück Shale, which is famous for its fossils of soft-bodied life and for the ammonoids illustrated in Figure 13-6. Apparently this shale, which is black because of its high carbon content, accumulated in a deep, oxygen-free basin resembling that in which the similar Burgess Shale of Cambrian age was preserved (Figure 12-44). As can be seen from the cross section shown in Figure 13-38, a transgression that took place in northern Germany during Early and Middle Devonian time shifted marine environments inland (northward).

The package of Devonian rocks in the northeastern United States reflects the presence of environments similar to those represented in northern Germany. There are two important differences between these two regions, however. First, offshore deposits in the northeastern United States reflect a quiet, relatively shallow interior seaway, as opposed to the deep-sea environments bordering the continental margin in Germany. Second, the rate of sediment supply in the northeastern United States was much greater than that of crustal subsidence. Thus, instead of a transgression, there was a westward regression in this area. This regression is recorded in the enormous Catskill clastic wedge, in which nonmarine red beds in the east give way to finer-grained marginal marine and marine deposits to the west (Figure 13-39).

The Catskill clastic wedge is sometimes referred to as the Catskill delta, but this is an inappropriate designation. The Catskill wedge does include a number of deltaic sequences that characteristically display coarsening-upward deltaic cycles (Figure 4-5), but it is really a composite of these sequences and others, including meandering-river deposits in which fine-grained sediments occur near the top (Figure 8-28). Sandstones and conglomerates were laid down in braided streams near highlands, meandering-river deposits formed downstream, and tidal-flat deposits with marine faunas accumulated along the shoreline. Offshore were barrier islands and sandy shelves, and farther offshore, muddy sea floors.

Late in the Devonian Period, a muddy seaway lay between the Old Red Sandstone continent and the Transcontinental Arch, extending northward to the Hudson Bay area (Figure 13-36). The giant placoderm *Dunkleosteus* flourished in this sea, together with other fishes now preserved in the black shales of northern Ohio (Figures 13-13 and 13-14). Near the end of Devonian time, the deposition of black muds extended westward, leaving a remarkably large area of eastern and central North America blanketed with these sediments (Figure 13-40). One formation consisting of such sediments is the Chattanooga Shale, which is found in Tennessee and neighboring areas (Figure 13-41). It would appear that while algae, fishes, and other forms of life flourished near the surface of the "Chattanooga Sea," lower levels were stagnant and depleted of oxygen. The decay of algal remains and other organic debris must have used up whatever oxygen reached the sea floor, and much organic matter was buried without decay. Apparently this seaway lay in the wind shadow of the mountains to the east, so that its waters were not well stirred by wave action.

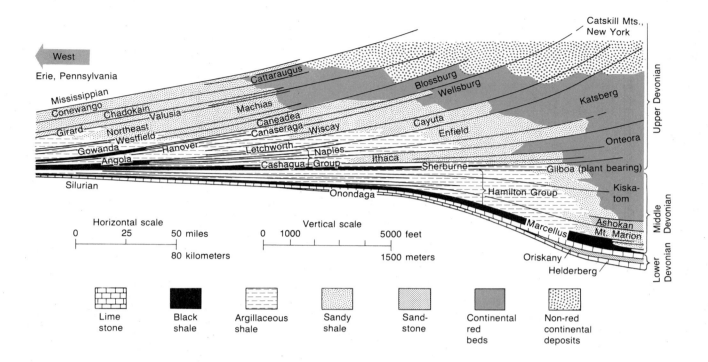

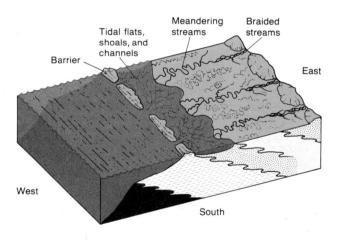

FIGURE 13-39   Devonian sedimentary rocks of New York State. *Above:* Stratigraphic cross section showing the formations. Coarse deposits shed from mountains to the east constitute a clastic wedge that is thinning to the west. *Below:* Environments of deposition of the Catskill clastic wedge and associated deposits of New York State. Braided streams meander seaward from the foot of mountains to the east. These empty into tidal channels behind barrier islands. Muds are deposited offshore. Note that the sequence is regressive (compare to Figure 13-38). (*Upper diagram after P. B. King, The Geological Evolution of North America, Princeton University Press, Princeton, New Jersey, 1977; lower diagram after J. R. L. Allen and P. F. Friend, Geol. Soc. Amer. Spec. Paper 106:21–74, 1968.*)

Reef building and orogeny in western North America

Along the continental shelf west of the Old Red Sandstone continent in what is now western Canada, tabulate-strome reef complexes developed during the latter half of the Devonian Period (Figure 13-42). Reefs here took the form of elongate barriers, atolls with central lagoons, and platforms. Many of the reefs that formed in this area now lie buried beneath the surface, and their porous textures have created important traps for petroleum. The Devonian reefs of western Canada were varied in form and organic composition, but typically they displayed the sequence of development illustrated in Figure 13-4. The Alexandra reef complex, which grew far to the north (Figure 13-43), is an example of this pattern. Here reef growth began with the development of a meadow of sticklike corals on a sea floor of burrowed mud. These corals were then overgrown by colonial tabulates, which in turn gave way to massive stromatoporoids. Behind the reef, carbonate mud accumulated in quiet water. South of the belt of reef growth, carbonates were deposited in shallow seas, just as they had been early in Paleozoic time (Figure 13-44).

During Frasnian time, subsidence of the reef area allowed encroachment of deep-water muds onto the shelf

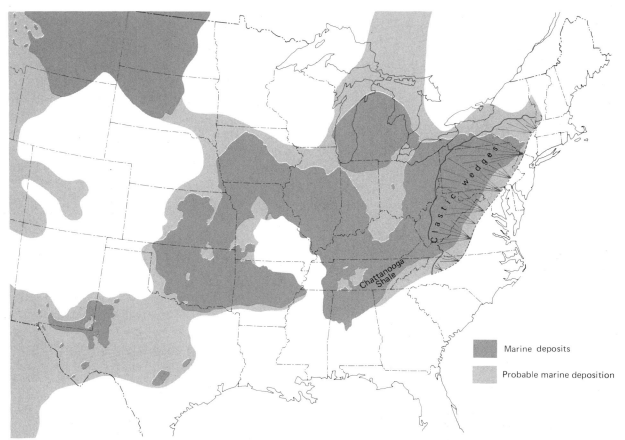

FIGURE 13-40   Broad distribution of the Chattanooga Shale and contemporaneous Upper Devonian deposits in the eastern United States. Most of these marine deposits are black shales that formed in the lee of the mountains in the Appalachian region (see Figure 13-36). (*After L. C. Conant and V. E. Swanson, U.S. Geol. Surv. Prof. Paper 357, 1961.*)

where reefs had flourished late in Middle Devonian time (Figure 13-42). Then, at the end of Frasnian time, the worldwide faunal crisis discussed earlier in the chapter brought an end to tabulate-strome reef building throughout the world.

Through all of Silurian and nearly all of Devonian time, the western margin of North America remained approximately where it had been during early Paleozoic time (Figure 13-36). In middle Paleozoic time, however, an island arc stood offshore. Ophiolite sequences (graywackes, shales, cherts, and volcanics) in the Klamath Mountains and in the Sierra Nevada of present-day northern California record the presence of this **Klamath Arc** (Figure 13-45A). Rocks in Nevada show that this simple geographic picture became more complex between Middle Devonian and Early Mississippian time; they reveal closure of the basin between the Klamath Arc and the craton. In central Nevada, deep-

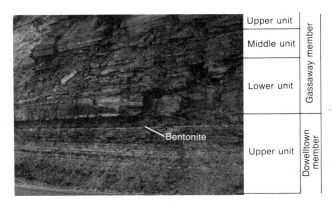

FIGURE 13-41   Roadside outcrop of the Chattanooga Shale in De Kalb County, Tennessee. The volcanic bentonite bed (*arrow*) represents a single ashfall. (*U.S. Geological Survey.*)

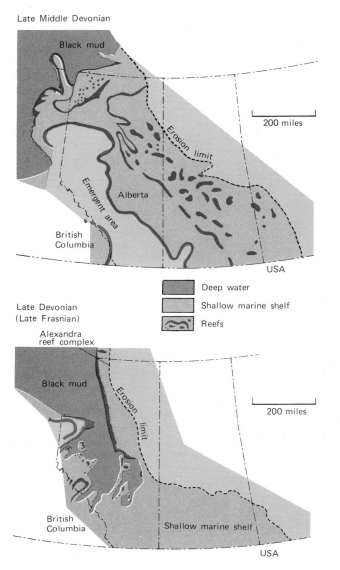

Late Middle Devonian

Late Devonian
(Late Frasnian)

FIGURE 13-42   The distribution of reefs in western Canada during the latter part of Devonian time. Late in Frasnian time, reef growth ceased, and black mud spread onto the continental shelf. Northeast of the erosion limit, Middle and Late Devonian rocks have been eroded away. (*After E. R. Jamieson, Proc. N. Amer. Paleont. Conv. J:1300–1340, 1969.*)

FIGURE 13-43   (*Above*) Outcrops of the Upper Devonian Alexandra reef complex in western Canada. The top photograph shows a section showing ecologic succession. At the base are muddy deposits; above these are deposits formed by a coral meadow. Next come massive tabulates and rugose corals, and, at the top, stromatoporoids represent the climax stage of development (compare to Figure 13-4). The lower picture shows lagoonal deposits that formed behind the Alexandra reef, now exposed along the Hay River Gorge. (E. Magathan and the Geological Survey of Canada.)

sea deposits like those of northern California can be seen to have been thrust as far as 160 kilometers (~100 miles) onto the craton (Figure 13-45*B*). The principal thrust fault along which this movement occurred, the Roberts Mountains Thrust, has since been broken and displaced by subsequent block faulting. The deep-water deposits that moved onto the craton along the Roberts Mountains Thrust range in age from Cambrian to Devonian and have a thickness

that exceeds 10 kilometers (~6 miles). This tectonic episode apparently resulted from the closure of the basin between the Klamath Arc and the craton. One interpretation of the tectonic mechanism is shown in Figure 13-45. According to this view, the ocean basin closed by subduction, and the deep-sea sediments and volcanics were elevated and then slid inland over the continental margin along the Roberts Mountains Thrust.

FIGURE 13-44   Paleozoic formations of the Bighorn Canyon. Note the prevalence of carbonates, which form the light-colored cliffs, in the Montana region during an interval spanning more than 200 million years. The Gallatin Limestone and Gros Ventre formations are Middle and Upper Cambrian deposits; the Big-horn Dolomite is Upper Ordovician; the Jefferson Limestone and Three Forks Shale are Middle and Upper Devonian; the Madison Limestone is Lower Mississippian; and the Amsden Formation is Upper Mississippian and Lower Pennsylvanian. (*U.S. Geological Survey.*)

The event that produced the Roberts Mountains Thrust is known as the **Antler orogeny**. This was the first sizable episode of mountain building in the Cordilleran region of North America during Phanerozoic time. The remainder of the Cordilleran story, which will appear in the chapters that follow, has mountain building as its dominant theme.

The Devonian Great Barrier Reef of Australia

Any discussion of Devonian reefs must include the enormous, spectacularly well exposed tabulate-strome complex of the Canning Basin in western Australia (Figure 13-46). The outcrop of this reef complex ranges some 350 kilometers (~220 miles) along the northern margin of the Canning Basin (Figure 13-47). Like the Canadian reefs described above, these Australian reefs are of Middle and Late Devonian age.

By Late Devonian time, part of the Tasman Belt had been stabilized, and there was nonmarine deposition over a broad area of what is now eastern Australia (Figure 13-47). In other words, continental accretion had continued in this direction (compare Figure 13-47 to Figure 12-47).

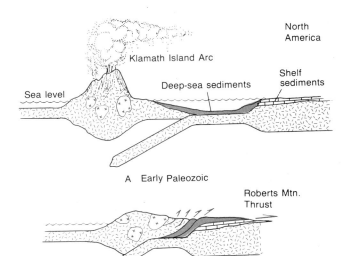

FIGURE 13-45   Diagram showing the likely mechanism by which the Klamath Arc was added to the North American continent by the Antler orogeny during Late Devonian and Mississippian time. The basin between the craton and the Klamath Arc (A) closed. As the continental crust was thrust beneath the volcanic crust of the Klamath Arc, deep-sea sediments slid onto shallow-water carbonates along the Roberts Mountains Thrust (B).

FIGURE 13-46   Aerial view showing the Napier Range of the Canning Basin of Western Australia. Most of the ridges consist of durable carbonates of the enormous reefs that grew along the basin during the latter half of the Devonian Period. (*Department of Lands and Surveys, Perth, Western Australia.*)

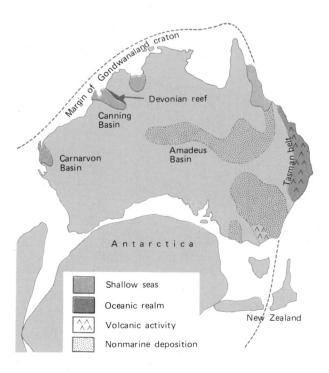

Marine deposition in Australia was restricted to a few marginal basins, including the Canning Basin, where the Devonian Great Barrier Reef developed. This name for the reef alludes to the modern Great Barrier Reef, which, by coincidence, also lies in northern Australia (Box 2-1). The structure of a typical Devonian reef of the Canning Basin is shown in Figure 4-16. Facies representing the reef core and facies representing the reef flat are illustrated in this figure. As one might expect, both the reef margin and the reef flat were constructed by tabulates, stromatoporoids, and rugose corals. An unusual and unexplained feature of

FIGURE 13-47   Paleogeography of the Australian portion of Gondwanaland in Late Devonian time. Except for a small area that still formed part of a marginal subduction zone, eastern Australia had by this time been stabilized; it stood largely above sea level, with nonmarine deposits accumulating in two large basins. In the west, shallow seas were also of limited extent. The Devonian Great Barrier Reef occupied the northern edge of the Canning Basin, one of the three marginal basins. (*Modified from D. A. Brown et al., The Geological Evolution of Australia and New Zealand, Pergamon Press, Oxford, 1968.*)

these reefs, however, is that column-shaped stromatolites also contributed to their growth. As we have seen, few stromatolites have formed since Ordovician time except in environments that are inhospitable to animals. Buried parts of the reef were strengthened through submarine cementation by calcium carbonate, but chunks of its surface occasionally tumbled down the reef's slope to become part of the rubble that accumulated there. The steep forereef slope supported some forms of life, including algae and sponges along with smaller numbers of crinoids and brachiopods. The flat-lying sediments of the basin in front of the reef yield fossils that, for the most part, represent swimming and floating animals such as fishes, crustaceans, radiolarians, nautiloids, and ammonoids. Also present here are conodonts, which apparently belonged to swimming animals.

Reef growth was possible in the Canning Basin because the basin was situated near the equator, far from the colder regions of Gondwanaland (Figure 13-28). In Australian seas, as seen in other low-latitude settings, near the end of Frasnian time there was heavy extinction; nearly all reef-building animals and much of the associated fauna disappeared. During the Fammenian interval that followed, reef building resumed in the Canning Basin, but algae were primarily responsible for this activity. Here, as elsewhere, the era of the tabulate-strome reefs had come to an end.

CHAPTER SUMMARY

1.   Many important middle Paleozoic groups of marine life, depleted during the great mass extinction at the end of the Ordovician Period, reexpanded at the start of the Silurian. Other forms, like the ammonoids and the jawed fishes, were new. Jawed fishes and mollusks occupied freshwater as well as marine habitats.

2.   The Silurian Period witnessed the invasion of the land by vascular plants, followed in Devonian time by the invasion of arthropods (scorpions and insects) and vertebrate animals (amphibians). During the Devonian Period, spore-bearing plants were joined by seed-bearing plants, which did not require moist habitats for reproduction and thus were not restricted to the fringes of aquatic habitats.

3.   During most of middle Paleozoic time, world climates were relatively warm, and the seas stood high in relation to the surfaces of continents.

4.   The Old Red Sandstone continent was formed by the unification of Laurentia and Baltica with the closure of the Iapetus Ocean along the chain of mountains that arose in the Caledonian orogeny. Freshwater sediments of the new landmass yield well-preserved fish fossils.

5.   Tabulate-strome reefs flourished throughout middle Paleozoic time and were especially abundant in the Great Lakes region of North America, in western North America, and in western Australia.

6.   Shortly before the end of the Devonian Period, a great mass extinction eliminated many forms of marine life, including bottom-dwelling animals such as those that form the tabulate-strome reef community and most acritarchs, which were floating algae. Species that occupied cold regions seem to have survived preferentially, suggesting that changing climatic conditions may have been the cause of the extinction.

EXERCISES

1.   In what important ways did invertebrate life change between Ordovician time and Devonian time?

2.   What animals have the oldest extensive fossil record in freshwater sediments?

3.   In what way did terrestrial environments of the Late Devonian Period look different from those of Early Devonian time?

4.   Buried ancient reefs are commonly porous structures that serve as traps for petroleum. If you wanted to drill for oil in Devonian reefs, what geographic regions would seem promising?

5.   Why did large quantities of sediment accumulate in the south-central part of the Old Red Sandstone continent during the Devonian Period?

ADDITIONAL READING

Andrews, H. N., *Studies in Paleobotany*, John Wiley & Sons, Inc., New York, 1961, chaps. 2, 3, 8, 9, 13, 16.

Colbert, E. H., *Evolution of the Vertebrates*, John Wiley & Sons, Inc., New York, 1980, chaps. 2–5.

House, M. R., et al., (eds.), *The Devonian System: Special Papers in Palaeontology* 23:1–353, 1979.

CHAPTER 14
The Late Paleozoic World

The late Paleozoic interval of geologic time included the Carboniferous Period, when new groups of animals and plants exerted influence over the accumulation of sediments, and the subsequent Permian Period, when many of these organisms died out in the greatest mass extinction in all of Phanerozoic time. Before this event, during Carboniferous time, skeletal debris from Carboniferous marine organisms accumulated to form widespread limestones, and spore-bearing trees stood in broad swamps, contributing their wood to the formation of coal when they died.

The late Paleozoic world was marked by major climatic changes that today are reflected in the distribution of rocks and fossils. Glaciers, for example, spread over the south polar region of Gondwanaland during the Carboniferous Period and then disappeared during Permian time. A general drying of climates at low latitudes during the Permian Period led to a contraction of coal swamps and to extinction of spore-bearing plants and amphibians, both of which required moist conditions. At the same time, seed plants and animals called mammal-like reptiles inherited the earth, and evaporites accumulated in many areas.

In addition to the great mass extinction, another major event took place near the end of the Paleozoic Era. This was the attachment of Gondwanaland to the Old Red Sandstone continent, accompanied by mountain building in Europe and in eastern North America. By the time this major

suturing event was completed, almost all of the supercontinent of Pangaea was in place.

The Carboniferous System was formally recognized in Britain in 1822, early in the history of modern geology. Its name was chosen to reflect the system's vast coal deposits, which had long been mined for fuel. Actually, it is only the upper part of the Carboniferous System that harbors enormous volumes of coal; the lower part contains an unusually large percentage of limestone. Recognizing this distinction, American geologists late in the nineteenth century began

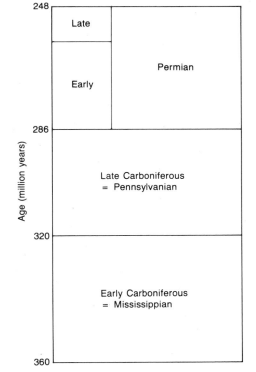

During Late Carboniferous time, logs of spore-bearing trees accumulated in broad swamps to form coal deposits that are now exploited by modern societies. In the scene shown here, ferns and seed ferns form the undergrowth beneath lycopod trees. On the fallen log in the foreground is a cockroach, one of the many kinds of insects that evolved during Late Carboniferous time. (*Field Museum of Natural History.*)

referring to the lower, limestone-rich Carboniferous interval as **Mississippian** because of its excellent exposure along the Mississippi River Valley and to the upper, coal-rich interval as **Pennsylvanian** because of its widespread representation in the state of Pennsylvania. Soon the Mississippian and Pennsylvanian were informally recognized as separate systems in North America, and in 1953 the United States Geological Survey officially granted them this status.

Although this difference between the upper and lower parts of the Carboniferous System is evident elsewhere in the world, most geologists in Europe continue to recognize just one system: the Carboniferous. Because this book covers world geology, we will follow the European practice, but it will occasionally be helpful to reiterate that the Lower and Upper Carboniferous systems are equivalent to the Mississippian and Pennsylvanian.

Roderick Murchison, who established the Silurian System and coestablished the Devonian, recognized and named the Permian System in 1841. Permian rocks were later identified in Britain and in other regions, but it was in Russia that Murchison founded the system. He named it after Perm, a town on the western flank of the Ural Mountains where an expedition had taken him in 1840.

LIFE

Marine life of the late Paleozoic interval did not differ markedly from that of Late Devonian time except for the absence of several groups of marine organisms that died out in the Late Devonian mass extinction. As the following pages will reveal, the changes that took place on land were far more profound. Numerous insects of remarkably modern appearance evolved during this time, and many new kinds of spore-bearing trees colonized broad swamps, where they formed coal deposits. These plants later dwindled and were replaced by seed-bearing trees, which eventually dominated the land. In parallel fashion, amphibians—which, like spore-bearing plants, were tied to water for reproduction—initially dominated terrestrial habitats but were subsequently replaced by more fully terrestrial reptile groups. By the end of Permian time, the latter displayed a variety of adaptations for feeding and locomotion, many of which resembled those of mammals.

Marine life

Some groups of marine life never recovered from the mass extinction of Late Devonian time. Tabulates and stroma-

toporoids, for example, never again played a major ecologic role. The ammonoids, on the other hand, rediversified quickly and once again assumed an important ecologic position; indeed, ammonoid fossils are widely employed to date late Paleozoic rocks (Figure 14-1). Also persisting from Devonian time as mobile predators were diverse groups of sharks and ray-finned bony fishes. Gone shortly after the start of Carboniferous time, however, were the armored placoderms that had ruled Devonian seas. The absence of armored placoderms and of similar fishes after earliest Carboniferous time reflected a general trend: Although the late Paleozoic is not known for vast changes in the composition of marine life, heavily armored taxa of nektonic (swimming) animals tended to give way to more mobile forms. Apparently, as the Paleozoic Era progressed, the ability to swim rapidly became a near necessity, probably because of the increasingly effective predators that inhabited the seas during this interval. After Devonian time, armored fishes never again dominated marine habitats, and heavy-shelled nautiloids also declined in number. In contrast to these heavy, awkward forms, the sharks, ray-finned fishes, and ammonoids that thrived in late Paleozoic time were mobile swimmers.

We know little about the groups of algae that floated in late Paleozoic seas alongside fishes and ammonoids. Phytoplankton are not well represented in the late Paleozoic fossil record, although some groups must have prospered without leaving recognizable fossils. The record of the readily preserved acritarchs shows that this group persisted but never reexpanded after the great extinction that led to its decline near the end of the Devonian Period.

Following the decline of the tabulate-strome community (page 381), organic reefs remained poorly developed throughout late Paleozoic time because of the scarcity of effective frame-building organisms. Corals of the type that build modern reefs did not evolve until the Triassic Period, and in the interim only low banks and relatively small mounds were formed. Among the groups of animals and plants responsible for these structures were brachiopods, bryozoans, and calcareous algae, all of which will be discussed below in a review of late Paleozoic sea-floor life. Certain groups of animals also contributed vast amounts of skeletal debris to the formation of bedded limestones; these included bryozoans, crinoids, and foraminifers, which will also be described below.

Brachiopods rebounded from the Late Devonian mass extinction to reassume a prominent ecologic role. A group of spiny brachiopods known as **productids** enjoyed particular success. Some immobile productids employed their

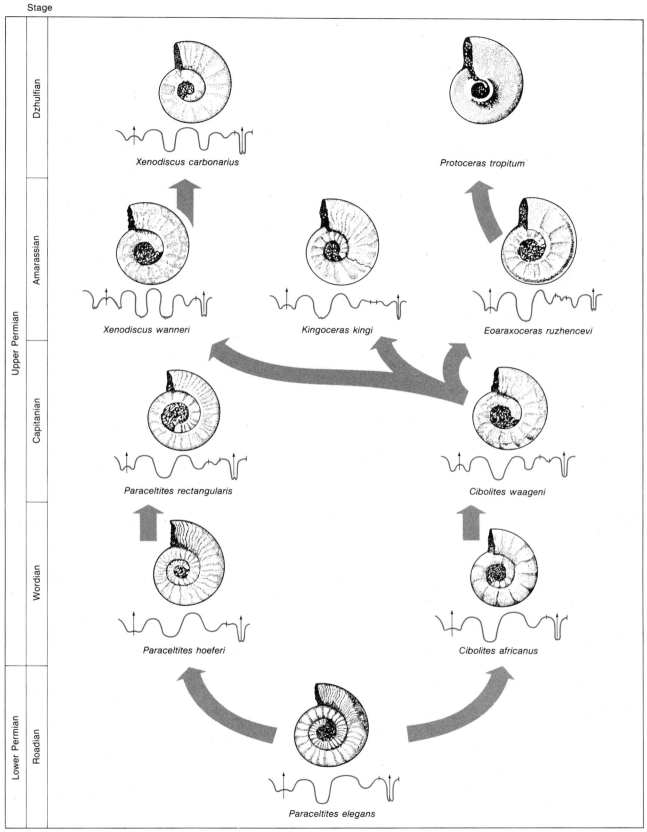

FIGURE 14-1 Species of the ammonoid family Xenodiscidae, arranged according to their evolutionary relationships. A few species of this type were virtually the only ammonoids that survived into the Triassic Period. (*After C. Spinosa et al., Jour. Paleont. 49:239–283, 1975.*)

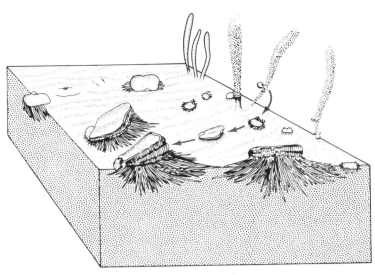

FIGURE 14-2   Modes of life of late Paleozoic spiny brachiopods of the productid group. *A.* Reconstruction showing
changes in the life habits of a mud-dwelling species during its
lifetime (*arrows*). The juvenile brachiopods appear to have been
attached to stalks of algae by curved spines. Then the algae
died, and the small brachiopods came to rest on fine-grained
sediment. As the brachiopods grew, their long spines served as
"snowshoes," preventing the animals from sinking into the sediment. Thus, the brachiopods could pump water in and out between the two halves of their shells in order to obtain food and
oxygen without the danger of being clogged by mud. *B.* A group
of Permian brachiopods of the genus *Prorichthofenia.* The lower
halves of the shells of these coral-like animals were cone-
shaped rather than cup-shaped, and throughout their lives their
spines were attached to hard objects—in this case, the shells of
neighboring brachiopods. The upper halves of the shells were
flattened lids. (*A. After R. E. Grant, Jour. Paleont.
40:1063–1069, 1966. B. The Smithsonian Institution.*)

spines to anchor or support themselves in sediment, and
a group of Permian productids developed cone-shaped
shells that were attached by spines to the frameworks of
solid reefs (Figure 14-2). Like the brachiopods, burrowing
and surface-dwelling bivalves continued to thrive in late
Paleozoic time, as did gastropods.

FIGURE 14-3   Reconstruction of an Early Carboniferous (Mississippian) meadow of crinoids (sea lilies) on the sea floor of
the central United States. Crinoids were animals that employed
their arms for suspension feeding. (*Field Museum of Natural
History.*)

FIGURE 14-4   Early Carboniferous limestone composed largely of skeletal debris from crinoids. The particles that are shaped and stacked like poker chips are segments of crinoid stalks, most of which were flexible. The largest of the stalks shown here are about as thick as a pencil. (*Institute of Geological Sciences, British Crown copyright.*)

Crinoids, suspension feeders that were attached to the sea floor, expanded to their highest diversity early in the Carboniferous Period, forming meadows in many areas of the sea floor (Figure 14-3). During this time, these organisms contributed vast quantities of carbonate debris to the rock record (Figure 14-4), leading to widespread limestone deposition during the Early Carboniferous (Mississippian) Period. Lower Carboniferous limestones of northern France look much like those of Indiana and are quarried in a similar manner (Figure 14-5). Roughly contemporaneous limestones are also conspicuous in the Rocky Mountains (Figure 13-44) and in the Grand Canyon (Figure AII-17). Other animal groups also contributed to the formation of Carboniferous limestones. Prominent among these were lacy bryozoans (Figure 14-6) and the fusulinid foraminifers (Figure 14-7), whose late Paleozoic adaptive radiation made them the primary constituents of some limestones (Figure 14-8).

Lacy bryozoans were sheetlike, colonial suspension feeders that stood above the sea floor. These organisms not only contributed skeletal debris to limestones but also trapped sediment to form reeflike structures. During Early Carboniferous time, lacy bryozoans grew locally in great profusion, trapping fine-grained sediment and growing upward through it to form mounds that usually stood well below sea level in quiet water (Figure 14-9). Around the flanks of the mounds, crinoids often grew in large meadows.

The **fusulinids**, a group of large foraminifers that lived on shallow sea floors, were represented by only a few genera in Early Carboniferous time but underwent an enormous adaptive radiation during the Late Carboniferous and Permian. Some 5000 species have been found in Permian rocks alone. Although they were single-celled, amoebalike

FIGURE 14-5   Quarrying of Lower Carboniferous limestone in the Empire State Quarry near Bloomington, Indiana. (*From D. V. Ager, The Nature of the Stratigraphical Record. John Wiley, New York, 1973.*)

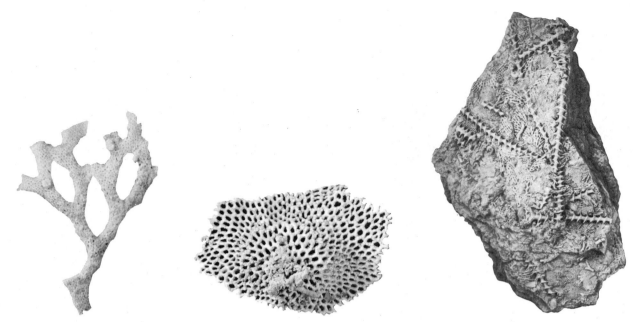

FIGURE 14-6   Lacy, fanlike bryozoans of late Paleozoic age. The specimen on the right shows fossils of the genus *Archimedes* in which the fan-shaped colony was wound into a spi-ral. The central axis of this spiral is usually best preserved. (*Courtesy of R. S. Boardman and the Smithsonian Institution.*)

creatures with shells, fusulinids included species that exceeded 10 centimeters (~4 inches) in length. Owing to their abundance and rapid evolution, fusulinids represent important guide fossils for Upper Carboniferous and Permian strata.

During Late Carboniferous time, certain types of calcareous algae shaped somewhat like large cornflakes assumed a dominant role in trapping fine-grained sediment to form carbonate mounds on the sea floor (Figure 14-10). During the Permian Period, other kinds of algae joined sponges and lacy bryozoans to form reeflike banks. The most spectacular of these banks was a structure in west Texas that will be described later in this chapter. The large size of this bank was unusual for late Paleozoic time, when, as noted above, there were few effective frame builders.

Plant life on land

Plants gave the Carboniferous Period its name, and in no other geologic interval are plant fossils more conspicuous; low-grade coal (Appendix I) from this period typically contains recognizable stems and leaves. Coal deposits developed chiefly in lowland swamps, where fallen tree trunks accumulated in large numbers. Because it takes several

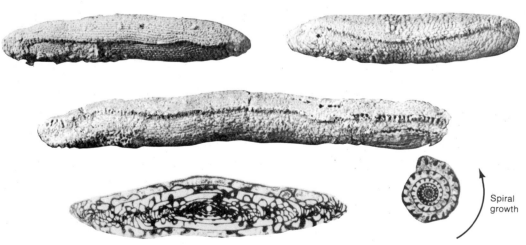

Spiral growth

FIGURE 14-7   Fusulinid foraminifers. These unusually large, single-celled creatures secreted skeletons that were commonly shaped like grains of wheat, which many resembled in size. The lower pictures are of a longitudinal section (*left*) and a cross section (*right*), showing the spiral mode of growth and the com-plex internal structure. Fusulinids are highly useful for dating late Paleozoic rocks. (*External photographs courtesy of the Smithsonian Institution; cross-section photographs courtesy of R. C. Douglass, U.S. Geological Survey.*)

A

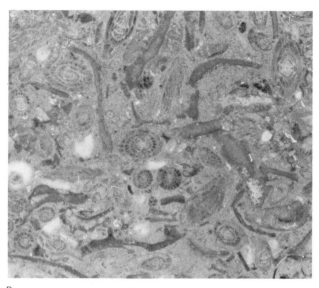

B

FIGURE 14-8 Fusulinids forming the bulk of a specimen of limestone from the Upper Carboniferous of Texas. The photographs show a weathered surface (A), and a polished surface (B) of the same specimen. Compare this illustration to Figure 14-7. (*Courtesy of H. B. Whittington.*)

FIGURE 14-9 A Lower Carboniferous carbonate mound in New Mexico. The mound, which is about 100 meters thick, grew on the sea floor, where lacy, fanlike bryozoans trapped carbonate mud. Bedded, coarse-grained limestones on the flanks of the mound were produced by meadows of crinoids. (*Courtesy of L. Pray.*)

FIGURE 14-10   Upper Carboniferous reeflike mound (*right*) in the Sacramento Mountains of New Mexico. The mound was formed by flakelike calcareous algae, which are shown in cross section in the polished section on the left. These belong to the group known as red algae. The bar on the polished section is 2 centimeters (~0.8 inch) long. The mound is about 40 meters (~130 feet) in diameter. (*Courtesy of S. J. Mazzulo.*)

cubic meters of wood to make one cubic meter of coal, it is evident that the vast coal beds of Late Carboniferous age represent an enormous biomass of original plant material.

Early Carboniferous floras, which formed little coal, resembled floras of Late Devonian time. They included a wide variety of plants, some of which were early representatives of the groups that became conspicuous in Late Carboniferous coal swamps. It appears that considerable evolutionary "experimentation" took place early in the Carboniferous Period, and what emerged as the highly successful late Paleozoic flora of the coal swamps and adjacent habitats consisted of a small number of genera, each represented by large numbers of species. The most important coal-swamp genera were *Lepidodendron* and *Sigillaria,* two types of lycopod trees that contributed many of the logs that were buried and compressed to form coal (Figure 14-11). As we have seen, the lycopod group had been present during Early and Middle Devonian time, but only as small plants (Figure 13-20). Like smaller lycopods, *Lepidodendron* and *Sigillaria* were spore plants that were confined to swampy areas. *Lepidodendron* was the more successful genus; some of its species grew more than 30 meters (~100 feet) tall and measured 1 meter (~3 feet) across at the base.

At the feet of the treelike plants of the Upper Carboniferous Period was an undergrowth that consisted primarily of a wide variety of ferns and fernlike plants. Although some of these were spore plants like the modern fern illustrated in Box 13-1, many others were so-called seed ferns, which, as their name suggests, reproduced by means of seeds (Figure 14-12). Because seed ferns are difficult to distinguish from spore-bearing ferns on the basis of their foliage alone, they were not recognized as a separate group until 1904. Many seed ferns were small, bushy plants, but others were large and treelike.

Glossopteris, the famous plant so abundant in Gondwanaland, was a seed fern. Although it had tonguelike leaves, familiar to most geologists (Figure 7-4), the entire plant had a treelike appearance, with its leaves arranged in clusters (Figure 14-13).

Not all Late Carboniferous vegetation occupied coal swamps. *Calamites,* for example—another prominent Upper Carboniferous genus of trees—seems to have favored higher ground. Because it was also a spore plant, however, we can assume that its habitat was at least seasonally moist. *Calamites* belonged to the **sphenopsids,** a group of plants characterized by branches that radiate from discrete nodes along the vertical stem and by horizontal underground stems that bear roots (Figures 14-14 and 14-15).

A second important group of Late Carboniferous plants that occupied high ground was the **cordaites,** a group of tall trees that often reached 30 meters (~100 feet) in height

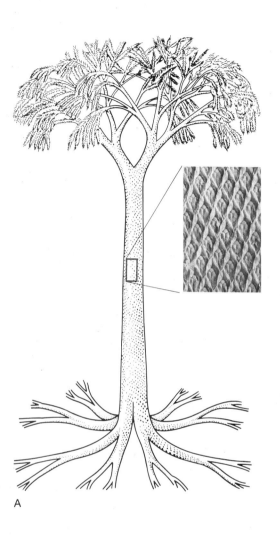

A

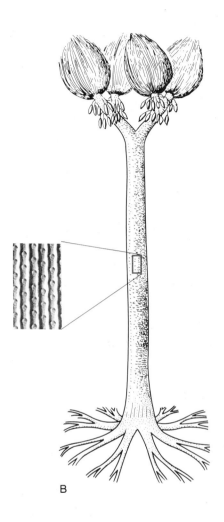

B

C

FIGURE 14-11   The dominant trees of Carboniferous coal swamps, belonging to the lycopod group. The genus *Lepidodendron* (*A*) is characterized by a trunk with a spiral pattern of leaf scars (positions where leaves were formerly attached). Note the similar spiral arrangement of branches in the early vascular plant *Baragwanathia* (Figure 13-18); this small, simple plant may have been an ancestor of the treelike lycopods. *Sigillaria* (*B*) is characterized by vertical columns of leaf scars. The roots of lycopod trees are often preserved as fossils (*C*). (*A and B. Photographs of leaf scars courtesy of the Field Museum of Natural History. C. Courtesy of the Institute of Geological Sciences, British Crown copyright.*)

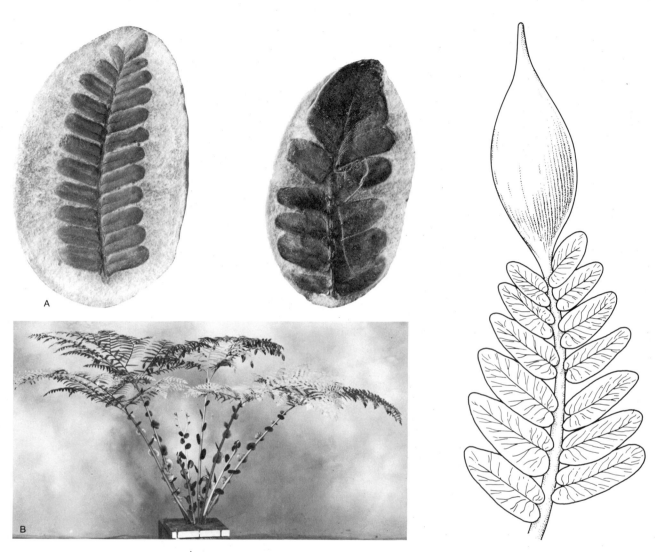

FIGURE 14-12   Late Carboniferous (Pennsylvanian) seed ferns of the genus *Neuropteris*. *A*. Fossil leaves from the Mazon Creek Formation of Illinois and a drawing showing the seed pod at the end of a branch. *B*. Reconstruction of a bushlike plant representing this genus. (*Field Museum of Natural History*.)

(Figure 14-16). As seed plants, cordaites were liberated from moist habitats and probably formed large woodlands that resembled modern pine forests. In fact, cordaites belonged to the group known as **gymnosperms** ("naked-seed plants"), which include the living **conifers,** or cone-bearing plants (pines, spruces, redwoods, and their relatives). The seeds of these plants are lodged in exposed positions on cones or on other reproductive organs and thus differ from the covered seeds of flowering plants, a group that did not emerge until the Mesozoic Era.

The floras that flourished in Late Carboniferous time continued to dominate into the Early Permian Period but subsequently declined. By the end of Permian time, for example, few lycopods or sphenopsids the size of trees remained on earth, while the cordaites, which were all tree-like plants, disappeared altogether. It is interesting to note that nearly all of the lycopods and sphenopsids that survived the Paleozoic Era were small, inconspicuous creeping forms, some of which persist today as "living fossils" (Figures 14-15 and 13-20). During the Permian Period, gymnosperms, including conifers, took over terrestrial environments. Figure 14-17 shows how the foliage of one of these conifers, *Walchia,* resembled the needled branches of certain living conifers. In *Walchia,* as in other conifers, seeds were borne nakedly on cones. Gymnosperm floras, having expanded in Late Permian time, prevailed through-

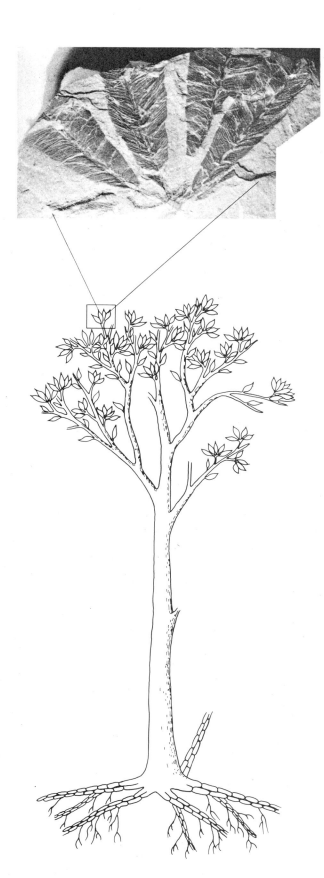

FIGURE 14-13 [Left]   The famous Gondwanaland seed fern *Glossopteris*. The name means "tongue leaf," and the tongue-shaped leaves, which are sometimes found preserved in the clusters in which they grew, were positioned at the top of a large trunk. This is one of many treelike genera of seed ferns. (*After D. D. Pant and R. S. Singh, Palaeontographica 147[B]:42–73, 1974.*)

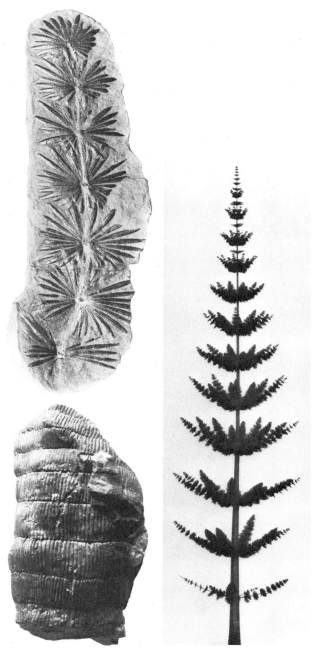

FIGURE 14-14   The Late Carboniferous sphenopsid plant *Calamites*. The branches (*fossil, top left*) were positioned on the segmented trunk (*fossil, bottom left*) at discrete intervals, as shown in the reconstruction on the right. (*Field Museum of Natural History.*)

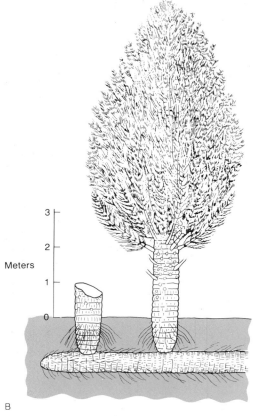

A

B

FIGURE 14-15   Comparison of a treelike Upper Carboniferous species of *Calamites* (*B*) with small living horsetails (*A*) that belong to the same group of plants (the sphenopsids). The 9-meter (~30-foot) height of the *Calamites* plant contrasts dramatically with the small size of the modern horsetails, which are growing in flower pots. (*A. Courtesy of L. Arnold. B. F. and E. M. Gifford, Comparative Morphology of Vascular Plants. W. H. Freeman and Company, New York, 1974.*)

out Triassic, Jurassic, and Early Cretaceous time (Figure 14-18) and thus are often thought of as representing Mesozoic vegetation. Mesozoic vegetation, however, had a head start on other life of the new era. We will learn more about this kind of vegetation in the next chapter, which describes the first portion of the Mesozoic Era—a time in which not only gymnosperms but also dinosaurs came to dominate the land.

Freshwater and terrestrial animals

In late Paleozoic freshwater habitats, aquatic ray-finned fishes continued to diversify and were joined by freshwater sharks that have no close modern relatives. For the first time, mollusks also became conspicuous in freshwater environments; shells of many species of clams are found in freshwater and brackish sediments associated with coal deposits.

On land, a group of invertebrate animals, the insects, assumed a very important ecologic role—one that they have never relinquished. The oldest known insects are of Early Devonian age, but these were wingless forms. Although no Lower Carboniferous insect fossils are known, insects had evolved wings by Late Carboniferous time; in fact, a wide variety of Upper Carboniferous insects have been identified, primarily by means of their preserved wings (Figure 14-19). The earliest flying insects differed from most modern species in that they could not fold their wings back over their bodies; the only two living orders of insects that lack this ability are the dragonflies and the mayflies, both of which are represented in Upper Carboniferous deposits along with several other orders. From the Carboniferous of France have come giant dragonflies—animals with wingspans of nearly half a meter (~18 inches)—and this has given rise to the false impression that Carboniferous landscapes were populated with many kinds of huge insects. In fact, only one giant Carboniferous species is known. The rest were of normal size by modern standards.

The fact that advanced insects with foldable wings are found in younger Upper Carboniferous deposits indicates that this group underwent an extensive radiation before the beginning of the Permian Period. Like many modern insect

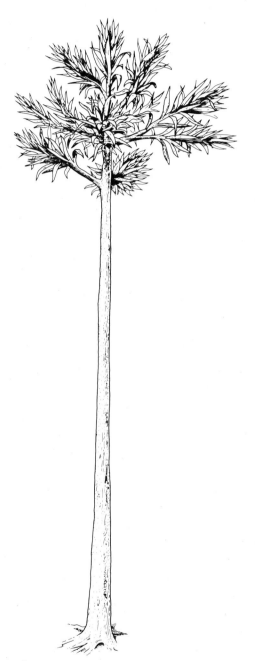

FIGURE 14-16   Reconstruction of a tall cordaite tree of Late
Carboniferous time. Cordaites were seed plants that formed
large forests on dry ground.

FIGURE 14-17   The early conifer *Walchia* of Late Carboniferous age (*top right*) compared to a living species of conifer that
is related to the redwoods (*bottom right*). Like living conifers,
Walchia had needled branches and reproduced by means of
cones. (*From E. B. Blazey, Palaeontographica 146[B]:1–20,
1974, courtesy of J. E. Canright.*)

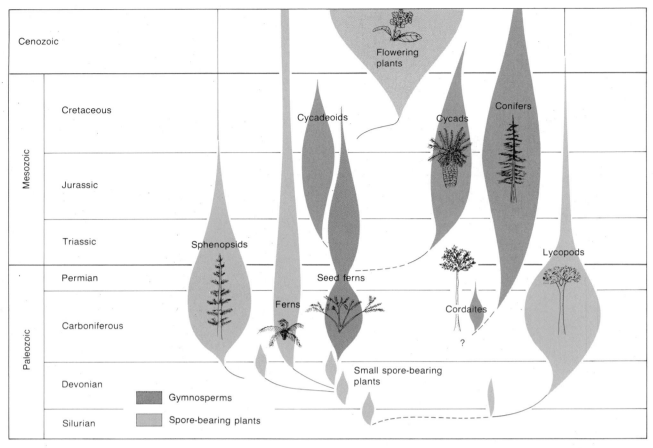

FIGURE 14-18   History of major groups of swamp- and land-dwelling plants. Spore-bearing plants dominated Silurian, Devonian, and Carboniferous floras. Seed ferns are the oldest known seed plants. Gymnosperms (naked seed plants) dominated Mesozoic floras, but the early conifers, which belonged to this group, diversified greatly during the Permian Period, while spore-bearing groups declined. (*Modified from A. H. Knoll and G. W. Rothwell, Paleobiology 7:7–35, 1981.*)

species, some of these Carboniferous forms had eggs that hatched into caterpillar-like larvae, while others possessed specialized egg-laying organs or mouth parts that were adapted to sucking juices from plants—and the legs of still other species were highly modified for grasping prey, leaping, or running. Indeed, many of these insects appear to have been as highly adapted for particular modes of life as are insects of the modern world.

In discussing the vertebrate animals of late Paleozoic time, it is sometimes difficult to distinguish between aquatic and terrestrial life, since many four-legged animals lived along the shores of lakes, rivers, shallow seas, or swamps, dividing their time between the land and the water. Figure 14-20 summarizes the general history of terrestrial and semiaquatic vertebrate animals of late Paleozoic time and also the history of plants, insects, and marine life.

FIGURE 14-19   Wings of a Late Carboniferous insect from the Mazon Creek Formation of Illinois. The color pattern is beautifully preserved in this remarkable specimen. (*F. M. Carpenter, Proc. N. Amer. Paleont. Conv. [I]:1236–1251, 1969.*)

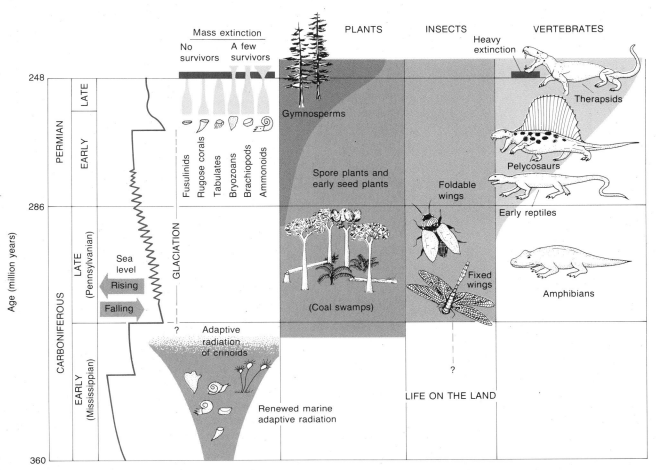

FIGURE 14-20   Major events of late Paleozoic time. Gondwanaland glaciation began during the Carboniferous Period and continued into the Permian Period. Sea level declined from a high level in Early Carboniferous time, dropping sharply three times—first at the end of the Early Carboniferous Epoch. Sea level also rose and fell on a smaller scale during repeated expansions and contractions of glaciers, which locked up water on the land. In the oceans, late Paleozoic time began with continued recovery from the Late Devonian mass extinction and ended with mass extinction in Late Permian time. The Late Permian extinction began before the end of this epoch and eliminated some large groups of animals while causing others to dwindle. During the Permian Period, climates became increasingly arid, and gymnosperms replaced spore-bearing plants as the dominant plants on the land. Late in Carboniferous time, insects with the ability to fold their wings evolved for the first time, and during the Permian Period, reptiles of increasingly high levels of adaptation replaced amphibians as the dominant large land animals. The most advanced mammal-like reptiles, the therapsids, underwent rapid adaptive radiation during Late Permian time and then suffered heavy losses at the close of this epoch.

In Early Carboniferous time, the only vertebrates populating the landscapes were amphibians, many of which retained aquatic or semiaquatic habits throughout their lives. Carboniferous amphibians, however, did not closely resemble their modern relatives. The frogs, toads, and salamanders that comprise most living species of Amphibia are small, inconspicuous creatures; they seem to be the only kind of animals belonging to this class that can thrive in the modern world, in the face of competition and predation by advanced mammals and reptiles. Carboniferous and Early Permian amphibians, in contrast, had the world largely to themselves and thus displayed a much broader spectrum of shapes, sizes, and modes of life; some superficially resembled alligators (Figure 14-21), others were small and snakelike, and a few were lumbering plant eaters. Some Carboniferous amphibians measured 6 meters (~20 feet) from the ends of their snouts to the tips of their tails. The species that were fully terrestrial as adults were covered by protective scales.

The oldest known **reptiles** are found in deposits near

FIGURE 14-21   *Eryops,* a large amphibian from the Lower Permian Series of Texas. This animal, which reached a length of about 2 meters (~6 feet), was probably semiaquatic, occupying the margins of rivers and lakes and preying on fishes. (*Field Museum of Natural History.*)

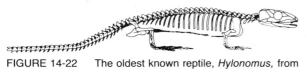

FIGURE 14-22   The oldest known reptile, *Hylonomus,* from the lowest Upper Carboniferous of Nova Scotia. The remains of this animal have been collected from sediments that filled rotted tree stumps. This animal was about 30 centimeters (~1 foot) in length. (*After R. L. Carroll, Jour. Linnean Soc. 45:61–83, 1964.*)

the base of the Upper Carboniferous (Pennsylvanian) System. Specifically, these small animals occur in sedimentary fillings of rotted tree trunks in the Joggins Formation of Nova Scotia (Figure 14-22). Most of the skeletal differences between the earliest reptiles and their amphibian ancestors were minor, relating to such features as the roof of the mouth, the back of the skull, the inner ear, and the vertebrae.

Reptiles also differ from amphibians in their mode of reproduction. The key feature in the origin of the reptiles was the **amniote egg,** which is also employed by modern reptiles and birds. In this egg, the embryo is provided with a nutritious yolk and two sacs. One of these sacs, the amnion, contains the embryo, while the other collects waste products. A durable outer shell protects the developing embryo. The significance of the amniote egg lies in the fact that it allowed vertebrates for the first time to live and reproduce away from bodies of water. The oldest structure suspected of being a reptile egg is of Early Permian age (Figure 14-23), but it is generally assumed that the amniote egg originated in Carboniferous time, when reptiles evolved.

Because the amniote egg was in essence a self-contained pond, it eliminated the need for early life in water and thus enabled reptiles to more fully exploit the land. There is an interesting parallel here with the evolution of the seed in plants. As we have seen, spore plants, like amphibians, require environmental moisture during part of their life cycle. The origin of the more advanced groups— the seed plants and reptiles—represented a transition to a fully terrestrial existence.

Later reptiles developed yet another feature of great importance: an advanced jaw structure that could apply heavy pressure upon closing and could slice food by means of bladelike teeth. Carboniferous amphibians and early reptiles could close their jaws quickly with a snap, but they could apply little pressure. Moreover, they had pointed teeth that could puncture prey in order to kill it but that could not slice or tear food apart; therefore, these animals were forced to swallow their meals whole.

Despite the origin of reptiles in Late Carboniferous time, amphibians continued to prosper into Early Permian time. During the Permian Period, however, reptiles diversified and apparently began to replace amphibians in various ecologic roles, probably because the reptiles had more advanced jaws and teeth as well as greater running speed and agility. Permian rocks of Texas have yielded large faunas of amphibians and reptiles that reveal this pattern. By Early Permian time, the **pelycosaurs—finback reptiles** and their relatives—had become the top carnivores of widespread ecosystems (Figure 14-24). Their strat-

FIGURE 14-23   A specimen alleged to be the world's oldest known fossil egg, from the Lower Permian Series of Texas. It is not known exactly what kind of reptile might have laid this egg, and some experts question its authenticity. Amniote eggs are thought to have appeared in Early Carboniferous time, when reptiles first evolved. (*Courtesy of A. S. Romer.*)

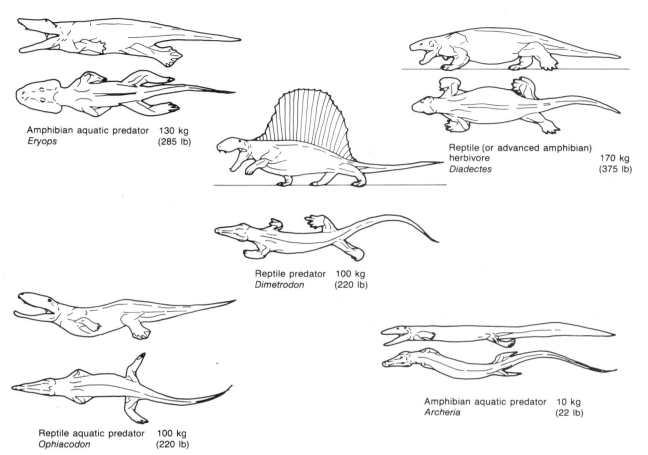

Amphibian aquatic predator   130 kg
Eryops   (285 lb)

Reptile (or advanced amphibian)
herbivore   170 kg
Diadectes   (375 lb)

Reptile predator   100 kg
Dimetrodon   (220 lb)

Reptile aquatic predator   100 kg
Ophiacodon   (220 lb)

Amphibian aquatic predator   10 kg
Archeria   (22 lb)

FIGURE 14-24   The fauna of reptiles and amphibians that oc-cupied parts of Texas during Early Permian time. The large rep-tilian predator *Dimetrodon* (see also Figure 14-25) and the large herbivore *Diadectes* probably roamed the margins of bodies of freshwater inhabited by the more fully aquatic forms. Weights of the animals are indicated in kilograms and in pounds. (*After R. T. Bakker, Amer. Assoc. for the Advancement of Sci. Se-lected Symp. 28:351–463, 1980.*)

FIGURE 14-25   Skeleton of *Dimetrodon,* a fin-backed mam-mal-like reptile of the pelycosaur group from the Lower Permian Series of Texas. The fin, which was supported by long vertebral spines, served an uncertain function. Some workers believe that skin stretched between the spines was used to catch the sun's rays, allowing the animal to raise its temperature to a level above that of its surroundings. From snout to tail, this predator exceeded 2 meters (~6 feet) in length. (*Field Museum of Natu-ral History.*)

FIGURE 14-26   Skeleton of *Diadectes,* a large herbivorous reptile or advanced amphibian from the Lower Permian Series of Texas. This animal was approximately the same length as *Dimetrodon* (Figure 14-25), exceeding 2 meters (~6 feet), but was considerably bulkier. (*Field Museum of Natural History.*)

igraphic occurrence suggests that many lived in swamps and that some may have been semiaquatic. *Dimetrodon* (Figure 14-25) was one such carnivore. It was about the size of a jaguar and had sharp, serrated teeth. *Diadectes,* a large herbivore (Figure 14-26), was probably among its

FIGURE 14-27   Early Permian scene beside a body of water. *Araeoscelis* is climbing a tree of the genus *Cordaites*. The vine *Dimetrodon* (see Figure 14-25) threatens *Eryops* (see Figure is *Gigantopteris* and the small plants are *Lobatannularia*. 14-21). Early insects are in the foreground, and the small lizard (*Drawing by Gregory S. Paul.*)

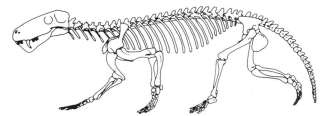

FIGURE 14-28   Skeleton of the therapsid *Lycaenops,* a Late Permian therapsid that was particularly mammal-like. This predator, with its highly differentiated teeth, was about 1.5 meters (~5 feet) long in the posture shown. (*After E. H. Colbert, Amer. Mus. Nat. Hist. Bull. 89:353–404, 1948.*)

prey. (*Diadectes* was either an advanced amphibian or a primitive reptile; its skeleton exhibits features of both classes.) While even the Permian carnivorous amphibians, such as the alligator-like *Eryops,* were forced to swallow small prey whole, *Dimetrodon* could tear large animals to pieces (Figure 14-27).

Dimetrodon belonged to a group known as the **mammal-like reptiles**. The skull structure of these animals in some ways resembled that of mammals, which evolved from them. In mid-Permian time, there evolved one particular group of mammal-like reptiles that was especially similar to mammals. These were the **therapsids** (Figure 14-28), whose legs were positioned more vertically beneath the body than were the sprawling legs of primitive reptiles or even pelycosaurs. In addition, the jaws of therapsids were complex and powerful, and the teeth of many species were differentiated, somewhat like those of a dog, into frontal incisors for nipping, large lateral fangs for puncturing and tearing, and molars for shearing and chopping food.

Many experts believe that the therapsids were **endothermic,** or warm-blooded—which means that by virtue of a high metabolic rate they maintained their body temperatures at relatively constant levels that usually exceeded the temperature of their surroundings. Hair similar to that of modern mammals may have insulated therapsids' bodies (Figure 14-29). Even if they were endothermic, however, therapsids may not have kept their body temperature at levels as constant as those of mammals. In any case, the upright postures and complex chewing apparatuses of advanced Permian therapsids show that these active animals approached the mammalian level of evolution in anatomy and behavior.

The endothermic condition is especially significant in that it allows animals to maintain a sustained level of activity—to hunt prey or to flee from predators with considerable endurance. **Ectothermic** (or cold-blooded) reptiles, in contrast, must rest periodically in order to soak up heat from their environments. Endothermic metabolism, along with advanced jaws, teeth, and limbs, may account not only for the success of the therapsids during Permian time but also for the decline of the pelycosaurs, which were probably ectothermic. In fact, while pelycosaurs declined to extinction during Late Permian time, therapsids underwent a spectacular adaptive radiation. More than 20 families of these advanced animals seem to have evolved in just 5 or 10 million years, and they were the dominant groups of large animals in Late Permian terrestrial habitats (Figure 14-30). Therapsids seem to have represented an entirely new kind of animal—one so advanced that it was able to diversify very quickly. While it is tempting to assume that the therapsids caused a worldwide decline of the pelycosaurs, it is only in the Karroo beds of South Africa that the therapsid record is sufficiently complete for a large enough geologic interval to suggest such a pattern. The sizable Permian therapsid fauna of the eastern Soviet Union, however, includes families that have not been found in South Africa, suggesting that a great variety of therapsids probably populated the Late Permian world. Unfortunately, elsewhere in the world the Permian fossil record of therapsids is rather poor.

The great extinction

The Paleozoic Era ended with what may have been the greatest mass extinction in all of earth history. In the marine realm, the fusulinids, which had been highly successful in mid-Permian seas, disappeared completely, as did the rugose corals and many kinds of brachiopods, bryozoans, and stalked echinoderms. The bivalves and gastropods suffered moderate losses, and the trilobites and tabulates—two groups that were already on the decline—disappeared altogether. The ammonoids nearly died out as well; only a handful of their species seem to have survived into the Triassic Period. It has been estimated that at the end of Permian time the number of marine invertebrate species in the world declined by 70 to 90 percent.

The exact nature of the biotic transition from the Permian to the Triassic is difficult to assess for marine ecosystems because transitional strata with abundant fossils are lacking in most regions. One of the apparent exceptions to this rule is found in the Salt Range of Pakistan, where a fossiliferous rock sequence appears to straddle the

FIGURE 14-29   Late Permian scene in the South African part of Gondwanaland. Therapsids are shown along an ice-covered stream in a snowy environment, where they may have been able to live by virtue of being endothermic. In this reconstruction, they are shown to have hair, which is associated with endothermy in mammals. The largest animal is *Jonkeria*; in the background is *Dicynodon*; animals at lower right are *Trochosaurus*; and the very small form is *Blattoidealestes*. (*Drawing by Gregory S. Paul.*)

boundary (Figure 14-31). Findings at this site have given rise to a debate about the precise location of the boundary between the Permian and Triassic systems—for in the Salt Range as well as in New Zealand, brachiopods that have traditionally been considered to be of latest Permian age occur along with ammonoids that have been thought to be of earliest Triassic age. In fact, the placement of this boundary is arbitrary; the most important goal for stratigraphers is to locate whatever boundary they finally choose as consistently as possible wherever transitional strata are found.

On land, most families of therapsid mammals also died out near the end of the Permian Period. In the section that follows, we will examine the possible causes of the terminal Permian mass extinction.

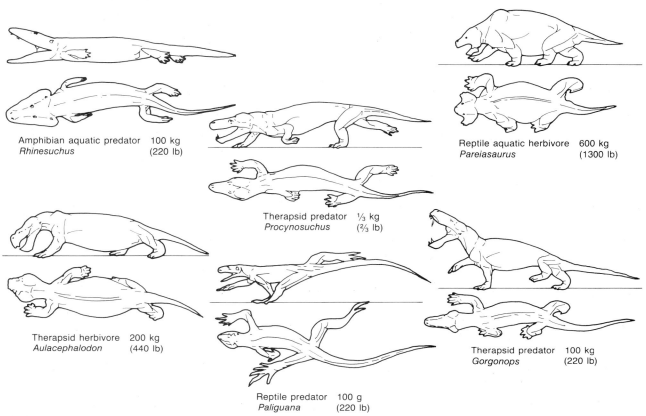

Amphibian aquatic predator   100 kg
Rhinesuchus   (220 lb)

Reptile aquatic herbivore   600 kg
Pareiasaurus   (1300 lb)

Therapsid predator   ⅓ kg
Procynosuchus   (⅔ lb)

Therapsid herbivore   200 kg
Aulacephalodon   (440 lb)

Therapsid predator   100 kg
Gorgonops   (220 lb)

Reptile predator   100 g
Paliguana   (220 lb)

FIGURE 14-30   An assemblage of vertebrates of Late Permian time, when therapsids dominated terrestrial environments after having undergone a spectacular adaptive radiation.

Weights of the animals are indicated in kilograms and in pounds. (*After R. T. Bakker, Amer. Assoc. for the Advancement of Sci. Selected Symp. 28:351–463, 1980.*)

FIGURE 14-31   The contact between the Permian and Triassic systems in the Salt Range of Pakistan. The hammerhead lies at the approximate position of the boundary. (*B. Kummel and C. Teichert, Univ. Kansas Dept. Geol. Spec. Publ. 4:1–110, 1970.*)

PALEOGEOGRAPHY

During late Paleozoic time, the major continents moved closer and closer together until, by early Mesozoic time, they seem to have fused together as the supercontinent Pangaea. Even early in the Carboniferous Period, however, the continents were rather tightly clustered (Figure 14-32*A*). As the period progressed, the sector of Gondwanaland that lay over the South Pole became covered by a large continental glacier that persisted into the Permian Period. Meanwhile, hot conditions prevailed in equatorial regions. Thus, not only was the late Paleozoic a time of vast biotic change, it was a time of changing and extreme climatic conditions. Coal deposits formed most extensively during Late Carboniferous time, accumulating both at low latitudes (for example, in areas now forming central North America and western Europe) and at high latitudes near

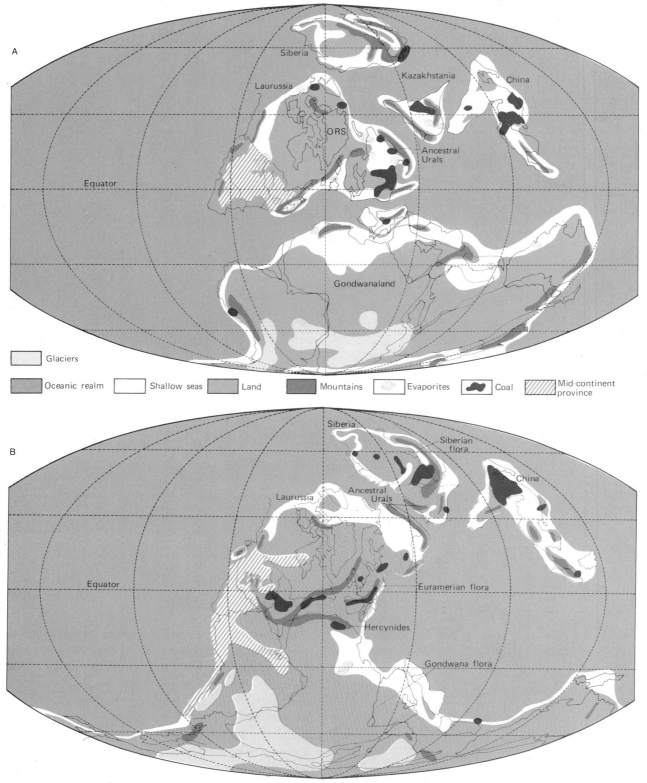

FIGURE 14-32 World geography in Carboniferous time.
A. Map representing Early Carboniferous (Mississippian) time, when the major continents were rather tightly clustered on one side of the earth. The Old Red Sandstone continent (ORS) persisted from Devonian time north of the Tethys Seaway. At low latitudes, coal deposits were formed along the eastern portions of continents, and limestones and evaporites accumulated in many areas, especially in the Midcontinent Province to the west of the Old Red Sandstone continent. Enormous glaciers spread over Gondwanaland near the South Pole. *B*. Map representing

Late Carboniferous time, when the Tethys Seaway was closed in association with Hercynian mountain building. Coal deposits formed over a larger total area at this time than at any other period in earth history; some formed quite far north. In Gondwanaland, continental glaciers spread to remarkably low latitudes and were separated from tropical coal swamps (formed by the Euramerian flora) by steep temperature gradients. The Gondwana and Siberian floras flourished under cooler conditions. (*Modified from R. K. Bambach et al., American Scientist 68:26–38, 1980.*)

the southern ice sheets. We will now consider these events in greater detail by examining precisely what happened in Early Carboniferous, Late Carboniferous, and Permian time.

The Early Carboniferous Period: Widespread limestone deposition

The manner in which the late Paleozoic began gave little hint of what was to come. In many areas, earliest Carboniferous rocks and fossils resemble those of the latest Devonian Period. Soon, however, warm-water limestones became unusually widespread, largely as a result of the great diversification of crinoids.

Limestones accumulated over large areas in Gondwanaland early in Carboniferous time, but even before the beginning of the Late Carboniferous Epoch, it is suspected

FIGURE 14-34   Escarpment of Lower Carboniferous limestones near Llangollen, northern Wales. (*Institute of Geological Sciences, British Crown copyright.*)

that polar areas of the great southern continent were blanketed by the first ice sheets of a major interval of glaciation. Nearer the equator, warm moist conditions prevailed in some continental areas. Coal-swamp floras, for example, which first became established early in Carboniferous time, flourished along the northeastern margin of the Old Red Sandstone continent, which survived from Devonian time (Figure 14-32*A*). Trade winds must have continued to bring this continent moisture from the oceans to the northeast, but this left the western part of the continent in the rain shadow of the Caledonides and the Appalachian Mountains. Here, across what is now central and western North America, evaporites and limestones accumulated in broad, shallow seas (Figure 14-33). Along the southern margin of the Old Red Sandstone continent, Scotland continued to receive nonmarine deposits, but most of England formed a continental shelf where limestones accumulated in shallow water (Figure 14-34). Beyond the edge of the shelf, near what is now the southern coast of England, black shales and other deep-water deposits were laid down.

The later Carboniferous Period: Continental collision and temperature contrasts

In mid-Carboniferous time, the northward movement of Gondwanaland caused that continent to collide with the Old Red Sandstone continent. As we have already learned (page 231), the mountains thus formed in southern Europe are known collectively as the **Hercynides,** and the orogeny as a whole is known as the Hercynian (or Variscan). Hercynian mountains also formed in northwestern Africa, where they became known as the Mauritanides. In North America,

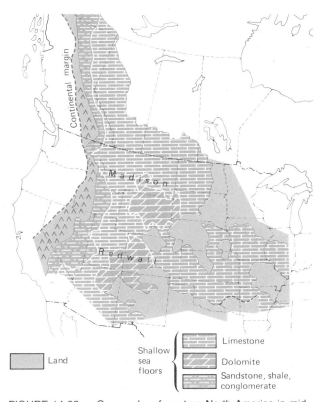

Land

Shallow sea floors

Limestone

Dolomite

Sandstone, shale, conglomerate

FIGURE 14-33   Geography of western North America in mid-Early Carboniferous (mid-Mississippian) time. Among the limestone units deposited at this time were the Redwall in the Grand Canyon region and the Madison farther north. The Antler orogeny had left the continental margin elevated as a ridge of land from which sand and mud was shed into the shallow seas to the east. (*Modified from J. G. Johnson, Amer. Jour. Sci. 274:465–470, 1974.*)

the Hercynian orogeny, known here as the Alleghenian, in effect, continued where the Caledonian orogeny had left off, extending the Appalachian mountain chain southwestward and forming the adjacent Ouachita Belt in Oklahoma and Texas.

Early in Carboniferous time, fusulinid foraminifers and certain other groups that inhabited shallow marine shelves had formed an enormous province that included all areas of the world but one. This exceptional area, which is known as the Midcontinent Province, now forms the central and western United States (Figure 14-32*A*). In Early Carboniferous time, this province was separated from other tropical shelves by the Appalachian Mountains to the east. By Late Carboniferous time, however, North America was welded to Gondwanaland, and the Midcontinent fauna spread into what is now South America—the part of Gondwanaland that had become attached to North America during the Hercynian orogeny.

On the land, latitudinal temperature gradients steepened during Late Carboniferous time—that is to say, there were extreme differences in temperature between the equator and the poles. Continental glaciers pushed northward to within nearly 30° of the ancient equator, a latitude where subtropical conditions have prevailed during most of Phanerozoic time. It seems amazing that tropical coal swamps flourished in North America and western Europe not much farther north than the northernmost Carboniferous glaciers (Figure 14-32*B*).

Recall that coal deposits formed in frigid Gondwanaland as well. Nonetheless, the *Glossopteris* flora that produced the coal deposits in Gondwanaland differed substantially from the so-called **Euramerian** flora of the equatorial region, which was named for Europe and North America. *Lepidodendron* and *Sigillaria*, the dominant Euramerian elements, were represented in Gondwanaland, but many of the plants of Gondwanaland (Figure 7-4) are unknown from northern continents. The *Glossopteris* flora was adapted to the cool climates of the glacial regime in the south. Siberia, which lay near the earth's other pole, also had a distinctive flora adapted to cold conditions.

There is compelling evidence that the fossil floras of Gondwanaland and Siberia grew under cold conditions. Many cold climates are strongly seasonal, and seasonal growth of wood produces distinctive rings in the cross sections of tree trunks. The Upper Carboniferous floras of Gondwanaland in the south and of Siberia in the north are known for their distinctive tree rings (Figure 14-35). In contrast, the Euramerian fossil trees that grew near the Car-

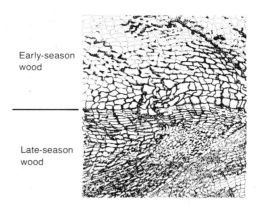

FIGURE 14-35 Section through fossil wood of Late Paleozoic or Triassic age from Antarctica, showing the details of the boundary between two growth rings. Each space in the woody tissue was occupied by a single cell. When the late-season wood of the earlier (lower) layer formed, the cells were small, and the tree grew slowly. During winter, growth was interrupted, but spring then stimulated the growth of early-season wood, resulting in large cells and rapid growth. The number of growth rings indicates the age of a tree, but rings are not well developed in tropical climates. In late Paleozoic time, high-latitude climates were strongly seasonal, producing growth rings in Siberia near the North Pole as well as in regions like Antarctica near the South Pole. (*After a photograph by J. M. Schopf.*)

boniferous equator were of the tropical type: They lacked seasonal rings.

The Permian Period: Climatic complexity

As a result of complex topographic conditions and steep climatic gradients, the floras that characterized Permian time were more provincial than those of any other Phanerozoic period—with the possible exception of the most recent interval, when continents have been widely dispersed and the waxing and waning of continental glaciers has produced much geographic differentiation.

Contributing to the climatic contrasts of the late Paleozoic globe were several mountain chains—including those of the Hercynian, which formed during the suturing of Gondwanaland to the Old Red Sandstone continent. Furthermore, until late in Permian time polar regions remained quite cold and equatorial regions quite hot. As a result of these contrasts, the Late Permian floras of low latitudes remained distinct from the *Glossopteris* flora to the south and also from the flora of Siberia, a continent that was now attached to Europe but remained near the North Pole (Figure 14-36). Southeast Asia was still a separate continent,

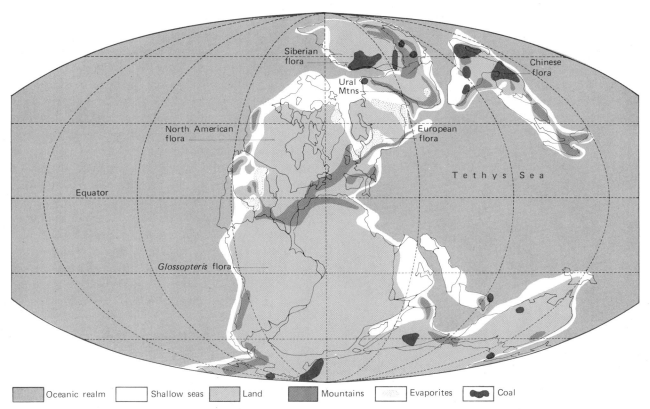

FIGURE 14-36   World geography in Late Permian time, when the ocean separating Europe from Asia was closing along the Ural Mountains to form the supercontinent of Pangaea. The Permian landmasses had a complex geography. Many mountain ranges stood high above lowlands. Five distinctive floras are labeled on the map. Floras produced coal only at high latitudes, while equatorial regions were hot and dry. (*Modified from R. K. Bambach et al., American Scientist 68:26–38, 1980.*)

and its flora had by now become unique. The Euramerian flora had also now broken down into more local floras separated by barriers. The Permian floras east and west of the Appalachians, for example, differed from one another to a significant extent.

The Permian Period was also an interval when climates underwent profound changes within particular regions. Early in the period, for example, glaciation continued in Gondwanaland, and, as indicated by dropstones (stones dropped from melting ice) in marine deposits of northern Eurasia, icebergs were present near the North Pole. As the Permian Period progressed, however, climates at high latitudes warmed up. Shallow-water marine faunas in Australia included increasing numbers of genera derived from equatorial regions, indicating that climates became quite warm after the end of the glacial interval. Then, near the end of Permian time, a new flora spread throughout Gondwanaland. This *Dicroidium* flora, named for a genus of gym-

nosperms with forked, leaf-bearing branches, replaced the *Glossopteris* flora, which had been adapted to cooler conditions. *Dicroidium* first appeared in tropical areas and seems to have migrated into the southern regions of Gondwanaland as climates warmed.

At high latitudes in the Northern Hemisphere, similar changes occurred, but at different times and in different places. Between late Early Permian time and the end of the Permian Period, gymnosperm seed plants replaced many of the spore-bearing plants that had flourished during Late Carboniferous time. At high latitudes in Europe and Asia, this change was apparently a response to climatic warming, as was the floral change in Gondwanaland.

In North America and western Europe, which lay near the ancient equator (Figure 14-36), a different kind of climatic change led to the decline of spore-bearing floras. This was a drying of climates, which killed off coal swamp floras and other spore-bearing plants. As Figure 14-36

FIGURE 14-37   Cross-bedded dune deposits of the Permian System. *A.* The Coconino Sandstone, which crops out in the vicinity of the Grand Canyon in Arizona. *B.* The Dawlish Sandstone in South Devon, England. The latter and associated deposits are sometimes referred to as New Red Sandstone. (*A. U.S. Geological Survey. B. Institute of Geological Sciences, British Crown copyright.*)

illustrates, these new arid conditions resulted in the deposition of great thicknesses of evaporites in the southwestern United States and in northern Europe. There is, in fact, a greater concentration of salt deposits in the Permian than in any other geologic system. Nonmarine sediments and faunas also offer clear evidence of a drying of climates. Throughout the southwestern United States and in parts of Europe, coal deposits decline in importance upward through stratigraphic sections from the Carboniferous into the Permian, and dune deposits, indicative of arid climates, appear in the record locally (Figure 14-37). In Texas, which is the site of a long and rich fossil record of Permian vertebrates, a number of types of aquatic and semiaquatic vertebrates disappear in successively younger Permian strata. Many kinds of plants also vanish; Late Permian floras are dominated by conifers such as *Walchia* (Figure 14-17), which seem to have thrived under dry conditions, as many conifers do today.

The terminal Permian extinction

Because the Late Permian was a time of widespread regression of the seas, many geologists have suggested that the crowding of marine life into narrow continental shelves was the major cause of the terminal Permian mass extinction. One problem with this idea is that there is no evidence of sea level lowering at the very end of Permian time, when many of the extinctions occurred. Another problem was mentioned in Chapter 12: Today a large percentage of shallow-water marine invertebrates occupy narrow sea floors fringing cone-shaped volcanic islands that rise up from the deep sea floor, and obviously require little space for survival. Furthermore, the area of these seas grows when sea level falls, rather than shrinking. In addition, the ammonoids were severely affected by the crisis, and many of them lived in the water column itself. Another twist to the problem is that therapsids also suffered mass extinction on the land; nearly 20 of their families disappeared at or very near the end of Permian time, having evolved just a few million years earlier.

Another hypothesis that has been proposed to explain the mass extinction in the seas is that the salinity of the oceans throughout the world dropped to a level that proved lethal for many animals. According to this idea, salts were removed by the widespread accumulation of marginal marine evaporites (Figure 14-36). The difficulty with such a notion is that not enough salts could have been removed by this means to greatly lower the salinity of the entire ocean.

Because Permian climatic changes caused widespread floral transformations on the land, it is tempting to invoke climatic changes to explain the mass extinction. Furthermore, it appears that many forms of marine life died out at high latitudes in response to a warming of polar seas well before the end of Permian time, while at the same time groups of marine animals that had previously lived at low latitudes moved poleward. As we have seen, the spread of gymnosperm floras over the land during Late Permian time followed a complex pattern that developed in response to climatic changes. The mass extinction in the ocean was similarly complex and may also have resulted from climatic change, although direct evidence of such a connection is lacking. Groups like the fusulinids, rugose corals, bryozoans, brachiopods, and ammonoids have impoverished fossil records even at the base of the final stage of the Permian System. Within that stage, and especially near its top, they decline further; indeed, the fusulinids as well as the rugose and tabulate corals disappear from the rocks altogether. Certain groups, including the fusulinids, died out at high latitudes long before they disappeared near the equator. Conodonts, however, appear to have survived the Permian crisis intact (Figure 12-6). In latest Permian time, the same conodont species lived under cool conditions in Greenland and under warm conditions in the Mediterranean. These species, which were obviously adapted to a wide range of temperature conditions, survived into the Mesozoic Era.

No matter which factor or group of factors caused the Late Permian extinction, the faunas that remained in earliest Triassic time both on land and in the ocean were of very low diversity, and a large percentage of their taxa were distributed over large areas of the globe. Either these taxa were tolerant of an unusually broad range of temperature conditions or latitudinal temperature gradients were so gentle that species of normal temperature tolerance were able to spread far and wide.

REGIONAL EXAMPLES

We will begin our discussion of specific regions of the late Paleozoic world by examining the development of two river deltas just to the west of the Appalachian Mountains, one of which is the forerunner of a major river system of the modern world. Next we will discuss the economically important coal deposits in North America and western Europe; and then we will review tectonic events in the American southwest, the history of an enormous reef complex

in west Texas (the site of large petroleum reservoirs), and Permian orogenic activity along the Pacific margin of North America. Finally, we will examine Late Permian evaporite deposition in western Europe, some of which occurred near the newly forming Ural Mountains.

West of the Appalachians: Black shales, limestones, and deltas

At the end of Devonian time, the Chattanooga Shale, Cleveland Shale, and similar black shales extended over an unusually broad expanse of shallow seas on the North American part of the Old Red Sandstone continent (Figure 13-40). These deposits continued to accumulate in the earliest part of Carboniferous time. Soon, however, the epicontinental seas shallowed, and the carbonate deposition that typified Early Carboniferous (Mississippian) time came to prevail in the western half of what is now North America (Figure 14-33). In the east, with the seas partially withdrawn, an enormous structure known as the Red Bedford delta built southward across central Ohio, carrying with it sediment that was eroding rapidly from the Appalachians of what is now eastern Canada. Other smaller deltas grew to the east of the Red Bedford, at the mouths of rivers that drained the Appalachians (Figure 14-38). In areal extent, the Red Bedford delta roughly equaled the deltas of the modern Nile or the Mississippi. The Red Bedford grew to an unusually great length, however, probably because it lay to the east of an elongate uplift known as the Cincinnati Arch. Protected from storm waves by this arch, the Red Bedford was more successful than most deltas in battling destructive marine forces (Figures 4-2 and 4-6).

Dozens of wells that have penetrated the Red Bedford delta have permitted geologists to map many of its channels. The meandering pattern of these channels is striking (Figure 14-38). The delta grew into such shallow seas that it formed almost entirely of delta-plain beds; there was almost no sloping delta front of the sort that produces the delta-front beds in deltas like the Mississippi (Figure 4-2). Many other ancient deltas that grew into shallow epicontinental seas probably consisted largely of delta-front beds as well. We lack good examples of such deltas today simply because epicontinental seas are rare in the modern world. (The Yellow Sea of China is perhaps the best example.)

Yet another delta formed to the west of the Cincinnati Arch. During Early Carboniferous (Mississippian) time, this so-called Michigan River delta failed to grow very far southward (Figure 14-38). Later, however, it built a bird's-foot delta, like that of the modern Mississippi, along what is now

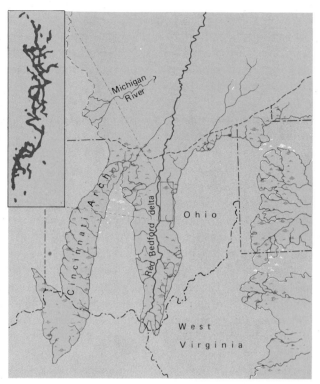

FIGURE 14-38   Paleogeographic map showing the Red Bedford delta of Early Carboniferous age. The delta was formed by sediments eroded from highlands to the northeast. It grew far southward in the protection of the Cincinnati Arch, a gentle uplift centered in western Ohio. The enlargement on the left shows the distribution of channel sandstones in one part of the delta. (*Inset from J. F. Pepper et al., U.S. Geol. Surv. Prof. Paper No. 259, 1954.*)

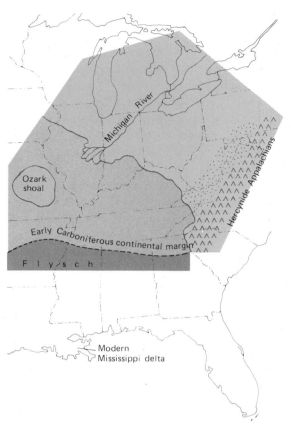

FIGURE 14-39   Paleogeographic map showing the Michigan River delta in Early Carboniferous time. This map represents a later interval than that depicted in the previous figure; during the earlier interval, the shoreline lay farther north, and the Michigan River had not yet built a large delta. During the interval shown here, not only the Cincinnati Arch but all of the state of Ohio stood above sea level. Near the end of the Carboniferous Period, the seas made their final withdrawal from the area to the west of the Appalachians. Subsequently, the North American craton accreted southward, and the area drained by the ancient Michigan River became part of the drainage area of the Mississippi River, whose modern delta is shown in this figure. Sediments carried southward by these river systems form much of the land that has been added between the ancient continental margin shown here and the modern shoreline.

the border of Indiana and Illinois (Figure 14-39). This delta continued to grow well into Late Carboniferous (Pennsylvanian) time. The alluvial and deltaic sandstones of the ancient Michigan River frequently occur as channel deposits (Figure 14-40). Where buried, these sandstones have yielded large quantities of petroleum. The Michigan delta, like the Red Bedford, represents molasse deposits formed of sediments shed from the youthful Appalachians to the northeast.

What is particularly interesting about the ancient Michigan River is not only that its delta resembles that of the modern Mississippi but also that the Michigan River was, in fact, the forerunner of the Mississippi. From the time when the Michigan River came into being some 250 million years ago right up to the present, there has been a river system running southward down the middle of North America. Although this ancient river has shifted in position through time along with the area it has drained, there would be

some justification for calling the ancient river not the Michigan, but the Mississippi!

Coal deposits in North America and Europe

In Late Carboniferous (Pennsylvanian) time, while the Michigan and other rivers flowing from the Appalachians continued to form molasse deposits in eastern North America, coal swamps spread over the floodplains of these rivers and over the margins of epicontinental seas. In North Amer-

FIGURE 14-40   A sand-filled channel (*lower right*) in the Up-
per Carboniferous Tradewater Formation in Muhlenberg County,
Kentucky. This channel was formed as part of the Michigan
River system. (*From J. A. Simon, courtesy of P. E. Potter.*)

ica, these swamps extended far to the west of the mountains over much of the nearly flat midcontinent. Some coal
basins that are now separate were probably connected at
their time of formation. The Michigan Basin, however, formed
in isolation, and the basins in New England and eastern
Canada might have done so as well (Figure 14-41).

We now recognize that most of the coal beds of western
Europe developed as molasse deposits equivalent to those
of North America, but at the other end of the Hercynian
mountain chain (Figure 14-32*B*). In Europe, the Hercynian
orogeny began near the beginning of Late Carboniferous
time. Remnants of the Hercynian mountains there now

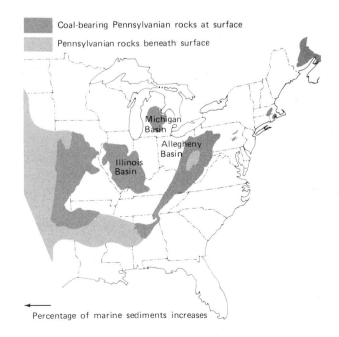

FIGURE 14-41   Distribution of coal-bearing cyclothems of
Pennsylvanian (Upper Carboniferous) age in eastern North
America. Before experiencing erosion, the cyclothems were
even more extensive in the areas east and south of the Illinois
Basin. The main area of coal formation extended from the margin of the early Appalachian Mountains on the east to an area
of fully marine deposition to the west.

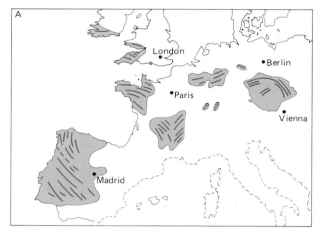

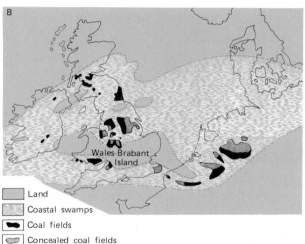

FIGURE 14-42   Late Carboniferous features of Europe. *A*. Geologic structures present today. The colored blocks are massifs, or remnants of Hercynian fold belts, with trends of folding indicated by dark lines. *B*. Paleogeography of northern Europe, showing the locations of major coal fields. Hercynian folding in southern Britain produced the Wales-Brabant Island.

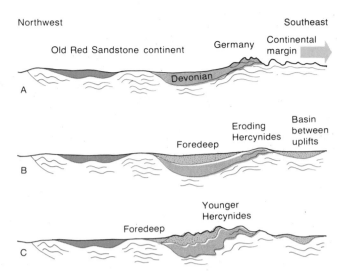

FIGURE 14-43   Stages in the development of coal-bearing molasse deposits in front of the Hercynian mountain chains in Germany. *A*. Folding began at the start of Late Carboniferous time. *B*. During the middle portion of Late Carboniferous time, basins formed between uplifts, and a foredeep developed in front of the Hercynian folding; coal-bearing molasse sequences (*colored areas*) accumulated in both the basins and the foredeep. *C*. Deformation shifted progressively northward, folding older coal deposits and causing the foredeep to advance toward the northeast.

stand as elevated massifs surrounded by flat-lying Mesozoic and Cenozoic sediments (Figure 14-42). The most extensive coal beds formed as part of the molasse deposition in the foredeep that lay to the north of the early folding, but other coal deposits formed within basins between Hercynian uplifts (Figure 14-43). Repeated deformation in Late Carboniferous time incorporated many earlier coal-bearing molasse deposits in folds and uplifts. We have seen that orogeny often shifts in the direction of a foredeep as time passes (Figure 8-3), and the Hercynian orogeny was no exception. In Britain, the main deformation occurred near the end of Carboniferous time, and it affected previously deposited coal beds.

One of the most striking aspects of Late Carboniferous coals is the cyclical nature of their stratigraphic occurrence. Cycles that included coal were of two types: those formed by meandering rivers and those formed by changes in sea level that apparently resulted from the expansion and contraction of continental glaciers.

In cycles produced by meandering rivers, coal beds represent overbank deposits. In other words, a single coal bed was produced by a forest growing in a swamp that was separated from a river channel by a natural levee. Cyclical coal-bearing deposits of this kind are found in Pennsylvania, where they formed along large rivers that drained the youthful Appalachians (Figure 3-24).

Farther west, in the nearly flat midcontinent region of the United States, Late Carboniferous (Pennsylvanian) coal swamps were associated with shallow seas. Here and in similar settings on other continents, coal beds are thin but widespread, occurring within cycles that include marine deposits. Dozens of similar cycles are commonly found superimposed on one another. Such cycles in coal beds are known as **cyclothems** in North America and as **coal measures** in Britain (see Box 14-1). The fact that so many marine and nonmarine habitats are often represented in just a few vertical meters of stratigraphic section indicates

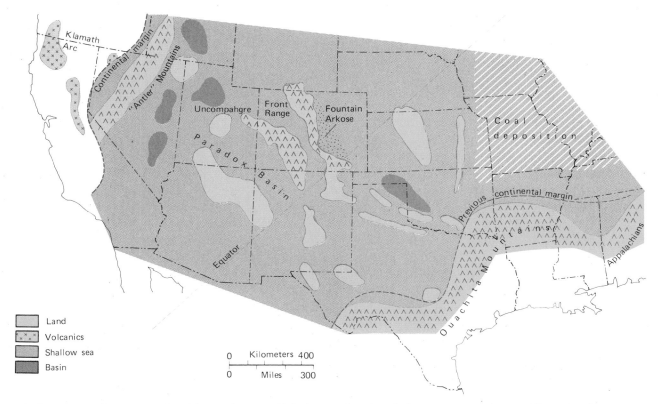

Land

Volcanics

Shallow sea

Basin

0 Kilometers 400

0 Miles 300

FIGURE 14-44   Paleogeography of the southwestern United States during Late Carboniferous (Pennsylvanian) time. Partway through the Late Carboniferous interval, Hercynide deformation formed the early Ouachita Mountains at the southern end of the Appalachians. Most of this deformation occurred offshore, uplifting oceanic sediments and welding them onto the previous continental margin. Coal-bearing cyclothems formed in marginal marine environments in the midcontinent region. To the west, shallow seas covered most of the craton, but many uplifts and basins developed from Texas to eastern Nevada, apparently in association with the Ouachita orogeny. The highest uplifts, the Front Range and the Uncompahgre, are together known as the Ancestral Rocky Mountains. Farther west, mountains produced by the earlier Antler orogeny still bordered the continent, and the Klamath Arc now lay along a subduction zone offshore in what is now California and Oregon.

that the depositional gradient was very gentle. It would appear that only a slight vertical movement of the sea or of the earth's crust accounted for substantial advance or retreat of the water with related shifting of environments.

The origin of the Carboniferous cyclothems has long been debated. It is doubtful that tectonic movements of the crust caused these frequent transgressions and regressions, since such crustal movements should be highly localized, while many cyclothems are widespread (some, for example, can be traced from Pennsylvania to Kansas). Furthermore, there is no apparent reason why the Carboniferous Period alone should have been characterized by such tectonic movements. The most likely explanation for the rapid transgressions and regressions is that the world's oceans rose and fell as the Gondwanaland glaciers repeatedly melted and re-formed. Why, then, do we not find similar cycles representing the Pleistocene interval of the past 1.8 million years, when continental glaciers have expanded and retreated many times? The explanation would seem to be that in recent times—even during interglacial periods of high sea level such as the one in which we live— the continents have remained relatively emergent. The seas are rising and falling over steeply sloping continental margins; they are not invading and receding from vast, almost flat interior lowlands as they did when they formed the cyclothems of Kansas, Illinois, and neighboring regions.

Sedimentary records show that glaciation ceased altogether in South America long before glaciers disappeared from Australia. This observation suggests that glaciers did not wax and wane simultaneously in all parts of Gondwanaland. Thus, the rise or fall of the world's oceans at any time must have depended on the averaging of glacial events within the entire glaciated area. The glacial hypothesis for the control of cyclical deposition gains support from the fact that the interval during which this kind of deposition took place coincided with the mid-Carboniferous to mid-Permian interval of glaciation in Gondwanaland.

Earth movements in the southwestern United States

The late Paleozoic was also a time of mountain building along a zone that extended from Utah across Oklahoma and Texas to Mississippi (Figure 14-44). Here the Ouachita

Box 14-1

Pennsylvanian

cyclothems in

central North America

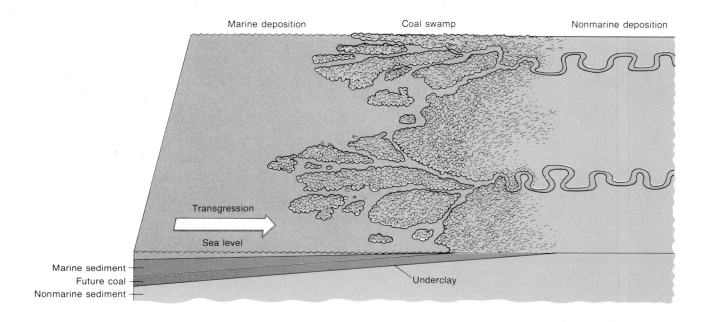

Marine deposition   Coal swamp   Nonmarine deposition

Transgression

Sea level

Marine sediment
Future coal
Nonmarine sediment

Underclay

The drawing above depicts the origin of a single cyclothem; the diagram on page 429 represents two idealized cyclothems. The coal bed in the cyclothem is underlain by a clayey deposit known as the underclay, which is often riddled with root casts (tubes formed by roots but later filled by sediment). Below the underclay are nonmarine deposits, many of which represent meandering-river deposits that cut into sediments of the preceding cycle. Above the coal deposits are shallow marine sediments. Thus, the coal swamps seem to have occupied lowland areas neighboring

the sea. They were fed by the rivers whose deposits lie beneath them. It is possible that the entire swamp was, in effect, a broad river into which inland streams emptied—one that flowed so slowly that its movement could not have been observed with the naked eye. This is what the Everglades swamp of Florida is today. The Everglades "river" also flows over a very flat region—the southern part of the Florida peninsula, which is being partially drowned by the rising sea. The water of the Everglades remains fresh except near the edge of the sea.

Thus, the early part of the typical cycle passes upward from meandering-river deposits through the coals to shallow marine beds and then to the limestones typical of an open marine environment. A reversal—a shallowing of the seas—is then recorded by the presence in the next beds of nearshore faunas, including algae and brackish-water invertebrates. Finally, the cycle comes to an end with river-channel deposits that truncate its upper beds.

Two additional points should be noted. First, the zigzag pattern of transgression and regression shown

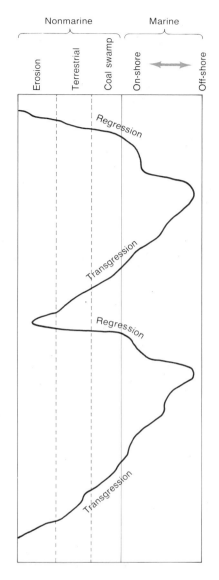

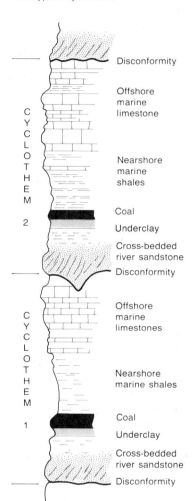

Two typical cyclothems

Nonmarine — Marine

Erosion | Terrestrial | Coal swamp | On-shore ←→ Off-shore

Disconformity

Offshore marine limestone

Nearshore marine shales

Coal

Underclay

Cross-bedded river sandstone

Disconformity

CYCLOTHEM 2

Regression

Transgression

Offshore marine limestones

Nearshore marine shales

Coal

Underclay

Cross-bedded river sandstone

Disconformity

CYCLOTHEM 1

Regression

Transgression

in the accompanying diagram of a stratigraphic section can seldom be fully traced out in the rock record; the record is simply too sparse, and the individual beds are too widespread, for outcrops to display a complete bed from one end to the other. Second, the cycles shown here are ideal ones. Some transgressions were terminated by sudden lowering of sea level (regression) before progressing very far, and some regressions were similarly interrupted; for this reason, many cycles did not proceed through every stage described here ☐

Ss.

Coal

Underclay

Incomplete cyclothem in Illinois. Erosion during regression produced a disconformity on top of the coal bed. Thus, marine rocks are missing, and nonmar-ine sandstone (Ss.) lies above the coal, separated from it by the disconformity. (*Illinois Geological Survey.*)

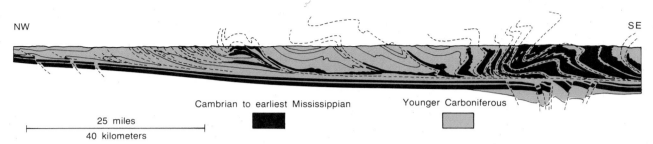

FIGURE 14-45   Cross section through the Ouachita Mountains as they exist today in the southeastern corner of Oklahoma. Here large volumes of deep-water sediment have been thrust northwestward. Note the general similarity between the style of deformation here and in the Appalachian fold-and-thrust belt (Figure 8-19). (*After J. Wickham et al., Geology 4:173–180, 1976.*)

Mountains formed as a westward continuation of the Appalachians. Today, traces of the two mountain chains meet at right angles beneath flat-lying younger deposits, and although the zone of contact is not well understood, the exposed segment of the Ouachitas is a fold-and-thrust belt resembling the Appalachian Valley-and-Ridge Province (Figure 14-45). One difference is that the folded rocks of the Ouachitas, which range in age from Ordovician to Middle Pennsylvanian, consist of deep-water black shale and flysch deposits that have been thrust northward against shelf-edge carbonates of similar age. In other words, deformation took place offshore from the continental margin (Figure 14-44), and after it began, the rate of deposition in the adjacent basin increased. In fact, the deformed region seems to have behaved as an unusually deep foredeep in which enormous volumes of deep-water Carboniferous deposits continued to accumulate on top of already-deformed older deposits. These younger, thicker deposits were, in turn, folded and thrust northward. Most of this deformation was completed prior to the start of the Permian Period.

The Ouachita deformation was part of the general Hercynian sequence of events that united Gondwanaland with the Old Red Sandstone continent (Figure 14-36). Although plate-tectonic events in the vicinity of the Ouachita system were complex and remain poorly understood, it is known that there were several microplates in this region that were similar to those involved in the origin of the Alps. Some of the microplates south of the Ouachita system eventually became parts of Central America (Figure 14-46).

The craton to the north and west of the Ouachita deformation also underwent important tectonic movement. It is not clear just how these cratonic movements were related to the Ouachita orogeny, but they were largely vertical; enormous areas in what is now the southwestern United States became transformed into a series of uplifts and basins (Figure 14-44). Many of these structural features are bounded by high-angle faults. The basins accumulated Late Carboniferous (Pennsylvanian) and, in some cases, Permian deposits. Clastic debris shed from the uplifts was deposited rapidly in nearby basins as coarse arkose. (Re-

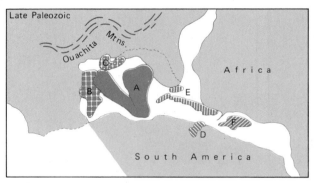

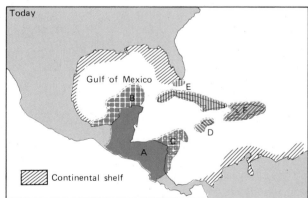

FIGURE 14-46   Approximate positions of microplates south and east of the Ouachita fold-and-thrust belt late in Paleozoic time (*above*) and today (*below*). Since late Paleozoic time, some microplates have shifted southward, leaving the Gulf of Mexico in their place. (*Modified from A. G. Smith and J. C. Briden, Mesozoic and Cenozoic Paleocontinental Maps, Cambridge University Press, Cambridge, 1977.*)

FIGURE 14-47 Black Canyon of the Gunnison River, Colorado, cut into crystalline Precambrian rocks that formed part of the Uncompahgre uplift. This uplift was leveled by erosion during the first half of the Mesozoic Period, and Upper Jurassic and Cretaceous rocks, seen here in the distance, were deposited on top of the erosion surface. (*W. R. Hanson, U.S. Geological Survey.*)

call that arkose is a sedimentary rock that consists largely of feldspar, a mineral that weathers to clay if it is not buried rapidly; see Appendix I.)

Two of the uplifts, the Front Range and Uncompahgre uplifts, are commonly referred to as the **Ancestral Rocky Mountains.** The Ancestral Rockies developed during Late Carboniferous time; they were elevated and then subdued by erosion in an area where portions of the Rocky Mountains stand today. It is estimated that the Uncompahgre uplift rose to an elevation of between 1.5 and 3.0 kilometers (~1 or 2 miles) above the surrounding seas, which flooded much of western North America. This is comparable to the elevation of the modern Rockies above the Great Plains to the east. The Front Range uplift is named for the Front Range of the modern Rockies, which now extends slightly farther east than the late Paleozoic uplift. Growth of the Ancestral Rockies elevated Precambrian basement rocks, which were subsequently leveled by erosion during Permian time. The Precambrian roots of the Ancestral Rockies can still be observed where more recent secondary uplift has caused rivers to cut deep gorges (Figure 14-47).

At places in the basin lying between the Uncompahgre and Front Range uplifts, arkosic sands and conglomerates accumulated to thicknesses exceeding 3 kilometers (~2 miles). The Fountain Arkose, which formed along the eastern flank of the Front Range uplift, was later upturned along the front of the modern Rockies when they were uplifted. Here, through differentiated erosion, the Fountain stands out in central Colorado as a series of spectacular ridges (Figure 14-48). The fact that the orogeny that formed the

FIGURE 14-48 Northward view along the Front Range of the Rocky Mountains near Morrison, Colorado. The core of the Rockies lies to the left. Tilted upward along its margin are the so-called Flatirons. These are formed of Fountain Arkose, which consists of sediment shed from the Ancestral Rocky Mountains, which lay slightly to the west in Late Carboniferous time. The Fountain Arkose was later tilted upward when the modern Rockies formed. (*T. S. Lovering, U.S. Geological Survey.*)

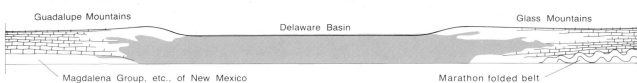

Guadalupe Mountains   Delaware Basin   Glass Mountains

Magdalena Group, etc., of New Mexico   Marathon folded belt

FIGURE 14-49   Aerial view of the Guadalupe Mountains from above the Delaware Basin. The Capitan reef limestone of the Guadalupe Mountains rims the basin. The cross section below, from northwest to southeast, shows the configuration of the basin. (*Courtesy of R. Muldrow III, Kargl Aerial Surveys, Inc.*)

modern Rockies deformed the Fountain Arkose tells us that this orogeny extended slightly to the east of the old Front Range uplift into part of the adjacent basin in which the Fountain was originally deposited.

The Ancestral Rockies lay close to the Late Carboniferous equator, where easterly equatorial winds must have prevailed. It is therefore understandable that the ancient mountains seem to have produced a rain shadow to their west. Here, in the Paradox Basin (Figure 14-44), great thicknesses of evaporites—primarily halite (Figure AI-1)—accumulated. Along the margin of the basin, especially in the southwest, carbonates were deposited in shallow water,

and flakelike algae here and there formed moundlike reefs in waters that were probably, at least some of the time, hypersaline (Figure 14-10).

The Permian System of west Texas

The Delaware Basin of Texas and New Mexico is one of the most famous geologic structures in the world—both because of its economic importance and because it offers spectacular geologic scenery. Although it has not been occupied by the sea for more than 200 million years, it

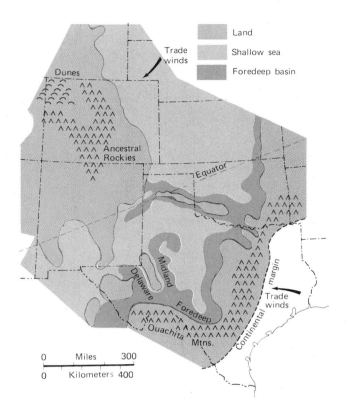

FIGURE 14-50   Paleogeography of Texas and neighboring regions in earliest Permian time. Shallow seas now flooded only parts of western North America, including this region between the Ouachita Mountains and the Ancestral Rockies. Mild deformation of the craton northwest of the Ouachitas continued from Late Carboniferous time, and the Delaware and Midland basins formed in western Texas as part of the Ouachita foredeep. (*Modified from J. M. Hills, Amer. Assoc. Petrol. Geol. Bull. 56:2303–2322, 1972.*)

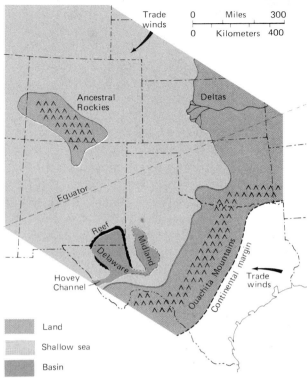

FIGURE 14-51   Paleogeography of Texas and neighboring regions when reefs encircled the Delaware Basin in Late Permian time. During this interval, a narrow passageway (Hovey Channel) connected the Delaware Basin with the open ocean to the west, but the Midland Basin was eventually filled with sediment.

remains a topographic basin. A person can stand in its center today and view ancient carbonates that formed as banks or reefs around the margin during the Permian Period (Figure 14-49). Earlier, during the latter part of Late Carboniferous (Pennsylvanian) time, the shallow seas that had shifted back and forth over the coal basins of the central United States withdrew westward, never to return. In earliest Permian time, they remained only in Texas and in neighboring areas, where they were connected to the seas that still flooded the western margin of North America (Figure 14-50). The Ancestral Rockies still stood high, as a mountainous island, and the young Ouachita Mountains bordered the southern margin of North America as what must have been a rugged and imposing mountain range. To the northwest of this range, detrital material came to rest in a marginal foredeep.

During Early and Middle Paleozoic time, marine de-

posits accumulated in the area that now forms west Texas, which was a broad, shallow basin on the continental shelf. During the uplift of the Ouachita range and other Carboniferous uplands, a small fault block rose up within the west Texas basin, dividing it into the Delaware Basin and the Midland Basin (Figure 14-50). Both of these basins subsequently received large thicknesses of sediment that have yielded enormous quantities of petroleum.

While reefs grew upward around the Delaware Basin, the Midland Basin to the east became filled with sediment (Figure 14-51). Along with surrounding areas, it was then flooded by shallow seas. Although by this time they had been lowered by erosion, the Ancestral Rockies still formed a large island to the northwest. The Delaware Basin lay very close to the Permian equator, and the Ouachita chain must have left the basin in the rain shadow of equatorial winds blowing from the east. The shallow seas that

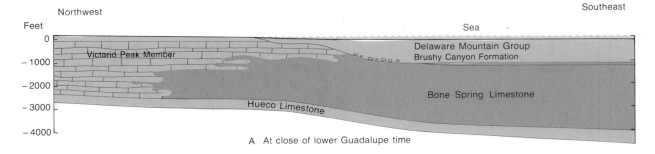

Northwest
Southeast

Feet
Sea

0
Delaware Mountain Group
Brushy Canyon Formation

−1000
Victorio Peak Member

−2000
Bone Spring Limestone

−3000
Hueco Limestone

−4000

A At close of lower Guadalupe time

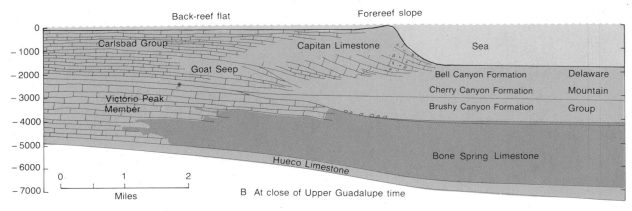

Back-reef flat
Forereef slope

0

−1000
Carlsbad Group
Capitan Limestone
Sea

−2000
Goat Seep
Bell Canyon Formation
Delaware

−3000
Cherry Canyon Formation
Mountain

Victorio Peak
Member
Brushy Canyon Formation
Group

−4000

−5000
Bone Spring Limestone

−6000
Hueco Limestone

0   1   2

−7000
Miles
B At close of Upper Guadalupe time

FIGURE 14-52   Profiles of the reef now forming the Guadalupe Mountains early in Guadalupian time (A) and later in Guadalupian time (B), when the reef had grown rapidly upward while the basin deepened. (The Guadalupian Stage was the next-to-last stage of the Permian Period.) Note that the reef advanced toward the basin center while growing upward. (*After N. D. Newell et al., The Permian Reef Complex of the Guadalupe Mountains Region, Texas and New Mexico, W. H. Freeman and Company, New York, 1953.*)

surrounded the basin were the sites of carbonate and evaporite deposition in what was obviously an arid climate.

As time passed and sea level rose, the reef grew upward more rapidly than the Delaware Basin filled in, and eventually the reef stood high above a basin that was some 600 meters (~1000 feet) deep (Figure 14-52). Although the waters have long since withdrawn, this is the configuration that remains today (Figure 14-49). Early in its history, when the Delaware Basin was relatively shallow, its floor was inhabited by snails, deposit-feeding bivalves, sponges, and brachiopods. Later, when the basin had deepened, these animals decreased in number; the only abundant marine fossil remains in the younger basin deposits are conodonts (Figure 12-6), radiolarians (Figure

FIGURE 14-53   Fossil sponges from the Permian System of Texas. These were among the forms that contributed to the reefs that encircled the Delaware Basin. One of them is attached to a mullusk shell (*lower right*). Like many other Permian fossils of this region, these sponges are extremely well preserved because their skeletons were replaced by durable silica soon after they died. (*Smithsonian Institution.*)

AI-18), and ammonoids (Figure 14-1), all of which lived high in the water column and sank to the bottom after death. Also present are plant spores that were blown into the basin from the land. We can conclude that when the Capitan Limestone formed (Figure 14-52), the floor of the deep basin below was poorly oxygenated; oxygen used up in the decay of organic matter was not replenished, and few bottom-dwelling animals could survive.

The reeflike structure that rims the Delaware Basin was built during the Guadalupian Age, the second-to-last age of the Upper Permian. The reef rock has been extensively altered from limestone to dolomite (Appendix I), and, in the process, much of its original fabric has been destroyed. Nonetheless, it is still possible to see that parts of the structure were formed primarily by sponges (Figure 14-53), algae, and lacy bryozoans (Figure 14-6). The crest of the reef was covered by shallow water that also bathed an extensive back-reef flat (Figure 14-52). Rubble from the reef periodically tumbled down the forereef slope into the basin. The ancient talus slope is present today in bedding that dips at angles as high as 40°. In the rubble of the forereef slope are many beautifully preserved fossils whose originally calcareous hard parts have been replaced by durable silica; among these are shells of fusulinid foraminifers that lived in shallow habitats but periodically washed down to lodge in the slope rubble. Some were swept into the basin by turbidity flows that left conspicuous graded beds in the rocks of the basin; these beds constitute the Delaware Mountain Group. Most sediments of this unit are dark sands and silts that periodically washed into the basin, apparently during low stands of sea level, when the reef surface was exposed to erosion.

When the older bedding surfaces of the carbonates that ring the Delaware Basin are traced laterally, a different configuration becomes apparent. These older bedding surfaces show that the early reefs, known as the Goat Seep Formation, stood in much lower relief above the basin (Figure 14-52). From its earliest days until late in the Permian, the Delaware Basin was connected with the open sea to the southwest through what is called Hovey Channel. Early in the evolution of the basin, when the reefs were low, the connection and the resulting pattern of water circulation permitted oxygen-rich waters to reach the basin floor so that animals could live there (Figure 14-54). Later, the basin deepened, but Hovey Channel remained shallow, a conformation that caused the bottom waters of the basin to become stagnant and poor in oxygen, excluding almost all forms of life.

Eventually, near the end of Permian time, the Delaware

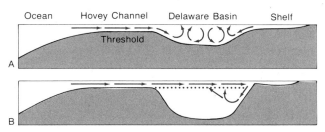

FIGURE 14-54 Patterns of water circulation in the Delaware Basin. Early in its history, the basin was shallow enough that well-oxygenated surface water reached its floor. Later, when the basin had deepened, good circulation was restricted to the upper waters, and the bottom waters thus became stagnant. (*After N. D. Newell et al., The Permian Reef Complex of the Guadalupe Mountains Region, Texas and New Mexico, W. H. Freeman and Company, New York, 1953.*)

Basin filled with evaporites. As we have seen, the climate of Texas and neighboring areas became arid toward the end of the Permian, and this may have increased the rate of evaporation of waters in the Delaware Basin. Hovey Channel may also have become so constricted that the rate at which water evaporated from it occasionally exceeded the rate at which new water was supplied. In any case, the basin ultimately filled with evaporites; distinct layers in some of these evaporite deposits extend over many hundreds of square kilometers (Figure 5-7). It is assumed that this layering reflects seasonal changes similar to those responsible for glacial varves.

The Delaware Basin evaporites remained in place for a long period of geologic time, protecting the magnificent geologic record along the walls and floor of the basin. Freshwater later dissolved the evaporites in many areas, exposing the ancient reef-encircled structure (Figure 14-49).

The western margin of North America

What was happening to the west of the Delaware Basin and the Ancestral Rocky Mountains? During late Paleozoic time, the western margin of the North American craton passed through what is now northwestern Nevada (Figure 14-44). Hundreds of kilometers from the margin of the craton, the Klamath Arc was again active in the area that is now California and slightly to the north. Here coarse clastic deposits were shed from volcanic highlands into the surrounding seas.

There is no evidence of volcanism and tectonic activity farther east along the margin of the craton in the region of Nevada, where the Antler orogeny had taken place during

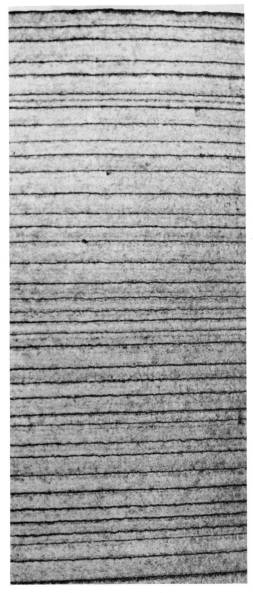

FIGURE 14-55   Uniform laminations in a specimen of the Linien-Anhydrite, an evaporite deposit of Permian age in northern Germany. Individual laminations of this deposit, which range from about 1 to 3 millimeters in thickness, formed over large areas of the Zechstein Basin and can now be traced over substantial distances. (*From G. Richter-Bernburg, courtesy of P. E. Potter.*)

latest Devonian and Early Carboniferous (Mississippian) time (Figure 13-45). Coarse clastic debris spread eastward from the Antler uplift into nearby areas of the craton that were flooded by shallow seas throughout almost all of late Paleozoic time. The deep-sea region lying between the shelf edge and the Klamath Arc received turbidites, black shales, cherts, and pillow lavas.

Latest Permian and Early Triassic times were marked by an orogenic episode in Nevada that was remarkably similar to the Antler orogeny. In this second, Sonoma orogeny, as in the Antler, deep-water deposits of the type just described were thrust upward over the continental margin (Figure 14-56). The Sonoma orogeny was of great significance in that it entailed the complete closure of the basin between the Klamath Arc and the North American continent. Some of the deep-sea deposits of the basin were thrust onto the continent, and others were welded onto the continental margin along with the volcanic terrane of the arc. The result was a considerable westward growth of the North American crust.

While the Sonoma orogeny was getting under way in latest Permian time, the seas shallowed in what is now western North America, as they did in most parts of the world. The combination of this regression and the increasingly arid climate led to the widespread deposition of evaporites and dune deposits.

Continental movement and evaporites in northern Europe

During late Paleozoic time, two northern continental margins converged to form the supercontinent Pangaea, with the resulting suture demarcated by the Ural Mountains (Figures 14-32 and 14-36). Along the western flank of the Urals that bordered what is now Europe, a foredeep subsided to accommodate Lower Permian shallow-water sediments totaling some 5 kilometers (~3 miles) in thickness (Figure 14-57). During Late Permian time, circulation between this foredeep and the open ocean became restricted. The partially landlocked waters then became hypersaline, their faunas were impoverished, and eventually evaporites were deposited. The great thickness of post-Carboniferous nonmarine deposits preserved in the nearby Russian province of Perm gave the Permian System its name.

During Late Permian time, the eastern European foredeep was invaded again, briefly and from the north, by what is known as the Zechstein Sea (Figure 14-58). This sea is most famous for the vast evaporite deposits that it

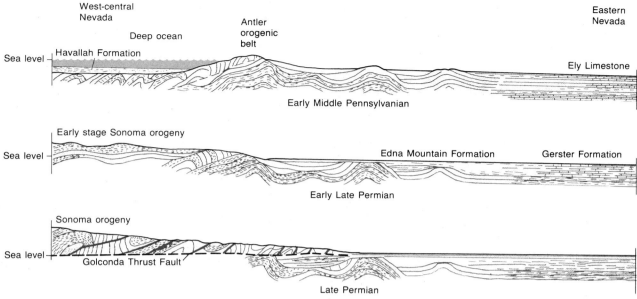

FIGURE 14-56   Sequence of events during the late Paleozoic Sonoma orogeny in Nevada. During Pennsylvanian time, the Havallah Formation formed in deep water beyond the marginal highlands that had been produced by the Antler orogeny. Dur-ing Permian time, the highlands were lowered by erosion, and the Havallah deposits were folded and thrust inland. (*After R. J. Roberts, U.S. Geol. Surv. Prof. Paper No. 459-A, 1964.*)

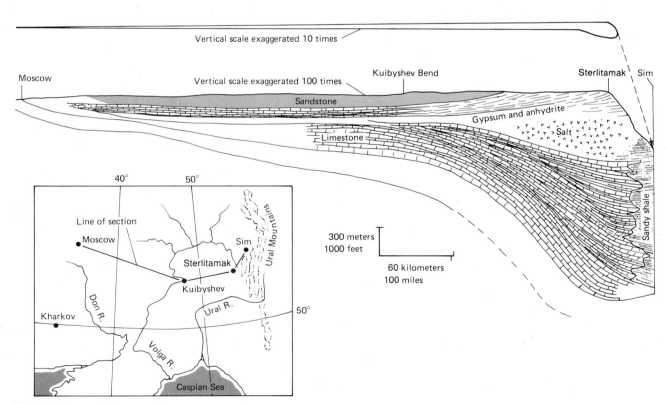

FIGURE 14-57   Diagrammatic cross section showing the Permian deposits between Moscow and the Ural Mountains. These deposits formed in the foredeep west of the Urals. This includes the region (Perm) that gave the Permian Period its name. Note how limestones gave way to evaporites (salt, gypsum, and an-hydrite) near the end of Permian time, reflecting a reduced connection between waters of the foredeep and the ocean to the north. (For a paleogeographic overview of this region, see Figure 14-36.) (*After B. Kummel, History of Earth, W. H. Freeman and Company, New York, 1970.*)

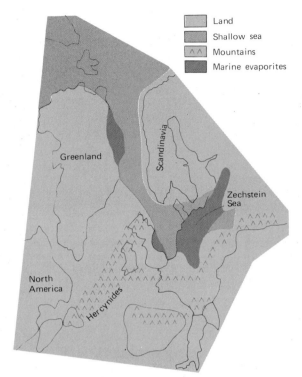

FIGURE 14-58   The Zechstein Sea, which invaded northern Europe during Late Permian time. Four cycles of evaporite deposition are apparent in northern Germany, reflecting four separate episodes in which the seas expanded and then became restricted and hypersaline. The restrictions produced economically important salt deposits.

left behind in northern Germany. The German Zechstein succession begins with a conglomerate that marks the initial transgression of the sea from the northeast at the beginning of Late Permian time. Above this layer, the Zechstein deposits comprise four cycles that reflect expansions and constrictions of the sea. Within each cycle, normal marine deposits underlie evaporites. We can infer that the normal marine deposits represent times of effective communication with the open ocean, while the evaporites that follow represent times of restricted communication. During each cycle, halite was deposited over large areas of the Zechstein basin, and, as the sea became further constricted, commercially valuable potassium salts accumulated on top of the halite in the central region. In northwestern Germany, the Zechstein evaporites reach a thickness of nearly 1 kilometer ($\sim\frac{1}{2}$ mile).

The Zechstein sea flooded not only northern Germany but also the North Sea area and the region now forming east Greenland. The layering of the Zechstein evaporites resembles the layering of the contemporaneous evaporites of the Delaware Basin of west Texas (compare Figures 5-7 and 14-55). Both reflect the warm, dry climate that characterized equatorial regions at the close of the Paleozoic Era.

CHAPTER SUMMARY

1.   Marine life of late Paleozoic time in many ways resembled life of the middle Paleozoic, but the tabulate-strome reef community was gone. In addition, four groups that expanded enormously contributed large volumes of skeletal debris to limestones: first, crinoids and lacy bryozoans and, later, flakelike algae and fusulinid foraminifers.

2.   In Carboniferous time, the coal-swamp floras, which were dominated by trees of the genera *Lepidodendron* and *Sigillaria*, played a major ecologic role, as did seed ferns and, on drier land, sphenopsids and cordaites. During the Permian Period, however, climates in the northern hemisphere became warmer and drier, and these plant groups gave way to conifers and other gymnosperms.

3.   Carboniferous coal swamps produced coal beds that commonly occur within depositional cycles, some of which represent meandering rivers and others alternating transgressions and regressions of shallow seas.

4.   From mid-Carboniferous to mid-Permian time, continental glaciers blanketed the south polar region of Gondwanaland. Throughout late Paleozoic time, large areas of this great southern continent were populated by the cold-adapted *Glossopteris* flora.

5.   Insects originated during the Carboniferous Period and underwent a great adaptive radiation. During the Permian Period, amphibians were displaced from terrestrial habitats by early mammal-like reptiles, including finbacks, but these soon gave way to more advanced mammal-like reptiles, the therapsids.

6.   In mid-Carboniferous time, Gondwanaland be-

came sutured to the Old Red Sandstone continent, forming the Hercynide mountain chains.

7.   In the United States west of the Appalachians, the Early Carboniferous (Mississippian) Period was a time of widespread black-shale deposition, followed by extensive limestone accumulation. The Hercynian orogenic episode affected the central and southern Appalachians and also created the Ouachita Mountains in Oklahoma and Texas. Uplifts and basins formed north and west of the Ouachitas. The Delaware Basin in western Texas became encircled by a reef complex in the Permian System, and, near the end of the Permian, this basin was filled by evaporite deposition. Beginning in the latest Permian time, the Sonoma orogeny resulted in continental accretion along the western margin of North America.

8.   In Europe, widespread limestone deposition in Early Carboniferous time was followed in Late Carboniferous time by Hercynian mountain building and by coal-swamp deposition between Hercynian uplifts. In Late Permian time, evaporite deposition prevailed in the foredeep of the newly forming Ural Mountains as well as in embayments in Germany and neighboring areas.

9.   The latter part of the Permian Period was a time of hot, dry conditions and widespread evaporite deposition in equatorial regions. The Permian Period—and thus the Paleozoic Era as well—ended with an enormous extinction that appears to have eliminated most species of marine invertebrates as well as many species of terrestrial vertebrates.

EXERCISES

1.   What became of the Old Red Sandstone continent during late Paleozoic time?

2.   If you could examine Late Carboniferous (Pennsylvanian) coal-bearing cycles in the field, how would you determine whether the coal deposits formed along a river or a shallow sea? (Hint: Refer to Figure 3-24 and Box 14-1.)

3.   In eastern North America, mountain building progressed from New England to Texas during Paleozoic time. How does this pattern relate to continental movements? (Review the relevant parts of Chapters 7, 12, and 13 as well as the relevant part of this chapter.)

4.   In what way may therapsids have been superior to the amphibians and reptiles that preceded them?

5.   Why were evaporite deposits more widespread in Europe and North America than in most parts of Gondwanaland during Late Permian time?

ADDITIONAL READING

Andrews, H. N., *Studies in Paleobotany,* John Wiley & Sons, Inc., New York, 1961, chaps. 4, 5, 8–12, 16.

Bakker, R. T., "Dinosaur Renaissance," *Scientific American,* April 1975.

Cloud, P., (ed.), *Adventures in Earth History,* W. H. Freeman and Company, New York, 1970, chaps. 49–51.

Colbert, E. H., *Evolution of the Vertebrates,* John Wiley & Sons, Inc., New York, 1980, chaps. 6–10.

PART SIX

The Mesozoic Era

When the Mesozoic Era, or Age of Dinosaurs, began, all of the major continents of the world were joined in the supercontinent Pangaea. As the era progressed, Pangaea separated into many fragments, and tectonic activity created major mountain chains in many parts of the world, including the western margin of the Americas. Marine and terrestrial biotas were impoverished at the start of the era, following the terminal Paleozoic mass extinction. In the oceans, the biotic recovery featured the ascendancy of mollusks, swimming reptiles, and new kinds of fishes. Dinosaurs soon came to dominate the land, while mammals, which also evolved early in the era, remained small and unobtrusive. Huge flying reptiles and primitive birds appeared, and near the end of the era flowering plants replaced conifers and their relatives as the dominant forms of terrestrial vegetation. Mass extinctions punctuated the Mesozoic history of life; the crisis that marked the end of the era resulted in the disappearance of the dinosaurs.

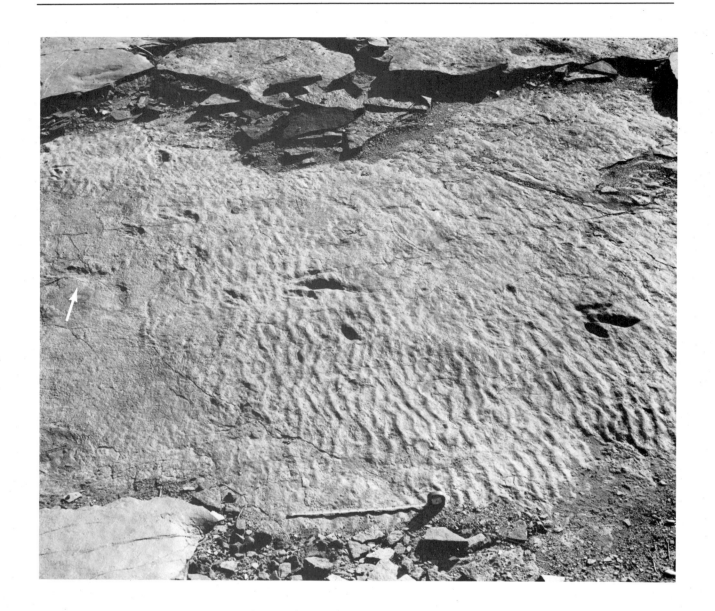

CHAPTER 15
The Early Mesozoic Era

The Mesozoic Era, or the "interval of middle life," began with the Triassic Period. The Triassic and the subsequent Jurassic Period together constitute slightly more than half of the era, and the rocks representing these periods are especially well exposed and well studied in Europe.

Near the transition from the Paleozoic Era to the Mesozoic Era, the great supercontinent Pangaea took its final form, encompassing virtually all the major segments of the earth's continental crust. Pangaea was so large that much of its terrane lay far from any ocean and as a result became arid. During Jurassic time, however, sea level rose, and marine waters spread rapidly over the land, leaving a more extensive record of shallow marine deposition than that of the Triassic System. Then, later in early Mesozoic time, Pangaea began to fragment, and before the end of the Jurassic Period Gondwanaland was once again separate from the northern landmasses.

Life of early Mesozoic time differed substantially from that of the Paleozoic Era. For many groups of animals, recovery from the Late Permian biotic crisis was sluggish, but by the end of the Triassic Period mollusks had reexpanded to become more diverse than they had been during the Paleozoic Era—and their success has continued to the present time. The marine ecosystem was also transformed during Triassic and Jurassic time by the addition of both modern reef-building corals and large reptiles, which joined fishes as swimming predators. On the land, gymnosperm floras that had conquered the land during the Permian Period continued to dominate landscapes, and flying reptiles and birds appeared as well. The most dramatic event in the terrestrial ecosystem, however, was the emergence and diversification of the dinosaurs. Mammals also evolved during this time, but they remained small and relatively inconspicuous throughout the Mesozoic Era, which is informally known as the Age of Dinosaurs.

The Triassic System is bounded by the terminal Permian extinction below and by another lesser extinction above. It was the unique fauna of this system that led Friedrich August von Alberti to distinguish the Triassic in 1934. Von Alberti originally named the system the Trias for its natural division in Germany into three distinctive stratigraphic units. We will examine these later in this chapter.

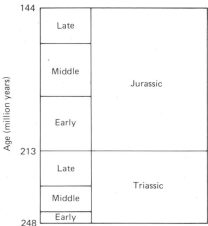

A new kind of animal, the dinosaur, emerged during the Mesozoic Era. The tracks visible on this bedding surface at Rocky Hill, Connecticut, were left by a dinosaur that stepped from firm mud at the left, where imprints of raindrops are visible, into a shallow body of water, where the animal sank deeply into soft mud that had a ripple-marked surface. (*Courtesy of J. H. Ostrom.*)

The Jurassic System also originated with a more abbreviated name—Jura, a label that was borrowed from a portion of the Alps in which the system is especially well exposed (Figure 8-9). The Jurassic was not formally established by a published proposal, however; instead, it gradually came to be accepted as a valid system during the first half of the nineteenth century, when its many distinctive marine fossils were widely investigated.

LIFE IN THE OCEANS: A NEW BIOTA

By the end of the great extinction that brought the Paleozoic Era to a close, several previously diverse groups of marine life had vanished, and others had become rare. Gone were fusulinid foraminifers, lacy bryozoans, rugose corals, and trilobites. Most common in Lower Triassic rocks are mollusks. The ammonoids made a dramatic recovery after almost total annihilation in the terminal Paleozoic extinction; although only two ammonoid genera are thought to have survived the Permian crisis, Lower Triassic rocks have yielded more than 100 genera of ammonoids. The adaptive radiation that produced these genera seems to have issued from the single genus *Ophiceras*, which appears to have been a descendant of *Xenodiscus* (see Figure 14-1). Other groups of marine life were slower to recover, but by Late Triassic time the seas were once again richly populated.

Sea-floor life

Bivalve and gastropod mollusks were less severely affected by the Permian extinction than were many other groups. In fact, bivalves, like ammonoids, are frequently found in Lower Triassic rocks, although their diversity is somewhat limited. Both bivalves and gastropods expanded in number and in variety to become among the most important groups of early Mesozoic marine animals. As in the Paleozoic Era, some of the bivalves burrowed in the sea floor, while others rested on the sediment surface (Figure 15-1).

In addition to ammonoids and bivalves, only brachiopods make a modest showing in Lower Triassic rocks; all other marine invertebrates are rare. The brachiopods also diversified during the Triassic and Jurassic periods, but they subsequently declined, becoming less abundant and diverse in late Mesozoic seas and, ultimately, very sparsely represented in the modern world. As will be described in the next chapter, their decline can probably be attributed to the appearance of new kinds of predators in late Mesozoic time.

Sea urchins, which had existed in limited variety during the Paleozoic Era, diversified greatly during the first half of the Mesozoic Era (Figure 15-2). Some of the new forms that emerged at this time were surface dwellers, like most of the Paleozoic sea urchins, but others lived within the

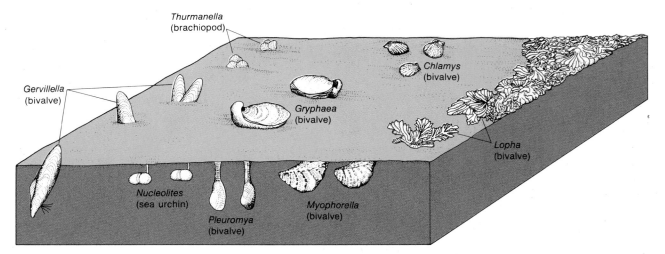

FIGURE 15-1   Life of a Late Jurassic sea floor. As in Paleozoic time, many animals lay exposed on the sea floor, but a number of these were of new types, such as the irregularly shaped oyster *Lopha*, which cemented itself to other shells, and the coiled oyster *Gryphaea*. Other new animals, including sea urchins such as *Nucleolites*, lived within the sediment. (*Modified from F. T. Fürsich, Palaeontology 20:337–385, 1977.*)

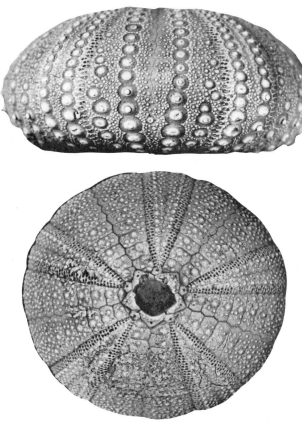

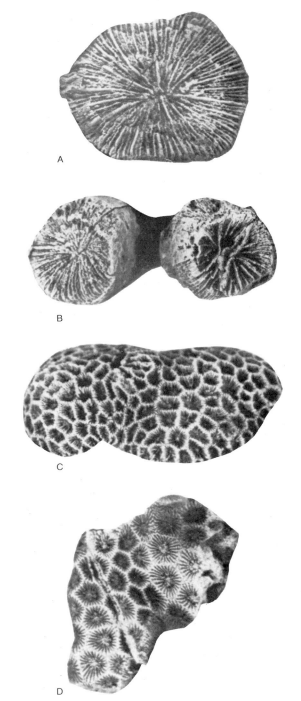

FIGURE 15-2   Side and bottom views of the Jurassic sea urchin *Leioechinus*. Like other nearly circular sea urchins, this animal lived on the surface of the sediment. In life, movable spines were attached to the numerous knobs by ball-and-socket joints. These spines served for defense and for locomotion. (*Courtesy of P. M. Kier, Smithsonian Institution.*)

sediment as actively burrowing deposit feeders (Figure 15-1).

Early in Mesozoic time, the place of the extinct Paleozoic corals was taken by the group that is still successful today—the **hexacorals**. This tentacle-feeding group of corals includes solitary species that resemble the solitary ruguse corals of the Paleozoic Era (Figure 12-16) as well as colonial reef builders, living examples of which are described in Box 2-1. Early reeflike structures of Middle Triassic age were low mounds that stood no more than 3 meters (~10 feet) above the sea floor. Most of these mounds were built by only a few kinds of organisms (Figures 15-3 and 15-4), but by latest Triassic time reefs were larger, some having been constructed by more than 20 different species.

Some of the early coral mounds grew in relatively deep waters, which suggests that, unlike the corals that form large tropical reefs today, the earliest hexacorals did not

FIGURE 15-3   Early hexacorals from the Triassic System of Idaho. Tentacle-bearing polyps rested in the cuplike depressions in the skeleton (Box 2-1). Specimens *C* and *D* were colonial, housing many interconnected polyps. As the name *hexacoral* suggests, the partitions within each cup number six or some multiple of six. *A* and *B*. *Montlivaltia*. *C*. *Elysastraea*. *D*. *Actinastrea*. (*G. D. Stanley, Univ. Kansas Paleont. Contrib. No. 65, 1979.*)

A

FIGURE 15-4   Structure of an early coral reef of Triassic age in the Pilot Mountains of Nevada. In these reefs, platy corals grew on a foundation of sponges; later these mounds were colonized by branching corals. Ultimately, a bed of oysters grew over the corals. Scales in the photographs are 15 centimeters (~6 inches) long. This reef was not well developed by modern standards, and it may have grown in deep water without the benefit of symbiotic algae. (*Modified from G. D. Stanley, Univ. Kansas Paleont. Contrib. No. 65, 1979.*)

B   C

live in association with symbiotic algae. Perhaps it was not until latest Triassic or early Jurassic time, when hexacorals began to form large reefs, that this important symbiotic relationship was established. Living with corals in Jurassic reefs were the earliest coralline algae of modern types. As described on page 91, these plants encrust the surface of reefs, thus protecting them from destructive forces.

Because of the success of bivalve and gastropod mollusks, sea urchins, and reef-building hexacorals, sea-floor life of the Late Jurassic world looked much more like that of today than had sea-floor life of the Paleozoic Era. Still missing were many kinds of modern arthropods, but the group that includes crabs and lobsters got off to a modest evolutionary start during the Jurassic Period (Figure 15-5).

Pelagic life

Although many kinds of planktonic organisms in Triassic and Jurassic seas left no fossil record, a few types are well represented in early Mesozoic rocks. Acritarchs left a meager record, which indicates that they had not yet recovered from the great extinction that had taken place near the end of the Devonian Period. In contrast, the **dinoflagellates**— which had existed even in the Paleozoic Era—underwent extensive diversification during mid-Jurassic time (Figure 15-6) and remain an important group of producers in mod-

FIGURE 15-5   An early lobsterlike animal belonging to the Jurassic genus *Cycleryon*. Note how weak the claws of this animal were in comparison to those of a modern lobster or crab. (*Field Museum of Natural History.*)

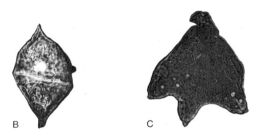

A

B   C

D

FIGURE 15-6   Early Mesozoic dinoflagellates, a group that remains very important in the modern world. *A. Dapcodinium* (Late Triassic; height 32 microns). *B. Rhaetogonyaulax* (Late Triassic; height 96 microns). *C. Nannoceratopsis* (Early Jurassic; height 77 microns). *D. Valvaeodinium* (Early Jurassic; diameter 39 microns). (*Courtesy of W. Wille.*)

ern seas (Figure 2-31). The **coccolithophores**, another important group of living algae, made their first appearance in earliest Jurassic time (Figure 15-7). Today, these floating algae are concentrated in tropical seas (Figures 2-31 and 2-35), and their armor plates (**coccoliths**) rain down on the deep sea to become important constituents of deep-sea sediments.

Higher in the food chain, the ammonoids and the belemnoids played major roles as swimming predators. The ammonoids' evolutionary recovery following the Permian crisis led to great success throughout the Mesozoic Era. Individual ammonoid species, however, survived for relatively brief intervals, often a million years or less—and this has made them extremely useful index fossils for Mesozoic rocks (Figure 15-8). The belemnoids, which were squidlike relatives of the ammonoids, also pursued prey by jet propulsion (Figure 15-9). They evolved in late Paleozoic time but remained inconspicuous until the Mesozoic Era, at which

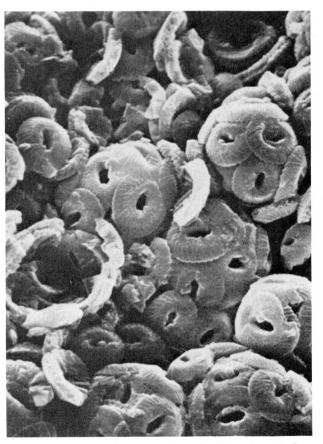

FIGURE 15-7   Coccolithophores from the Upper Jurassic Series of England. Many of the calcareous discs shown here are united just as they were when they formed armored shields around spherical algae cells. (*Courtesy of D. Noël.*)

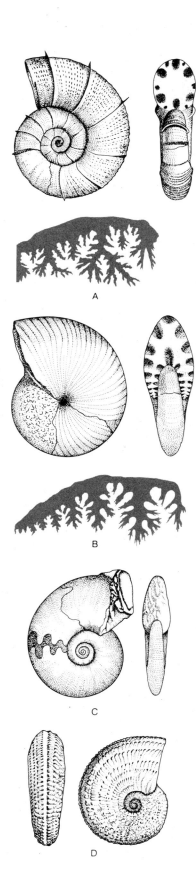

A

B

C

D

FIGURE 15-8   Representative Triassic and Jurassic ammonoids. *A* and *B. Lytoceras* and *Phylloceras* are Jurassic forms. The suture pattern for each is shown below the shell. (The suture is the juncture between the convoluted internal partitions, or septa, and the coiled outer shell.) *C.* The earliest Triassic genus *Ophiceras. D.* The Late Triassic genus *Protradryceras.*

5 centimeters

2 inches

FIGURE 15-9   Belemnoids existed in the Triassic Period, but they populated Jurassic seas in especially large numbers. Belemnoids were squidlike cephalopod mollusks that were related to ammonoids but lacked external shells. Like ammonoids, they were predators that swam by jet propulsion. The most commonly preserved part of a belemnoid is the cigar-shaped counterweight shown above. This heavy structure acted to offset the buoyant effect of gas within the shell, thereby maintaining balance. (*Field Museum of Natural History.*)

time many types evolved. Conodonts, the toothlike structures that are now known to have belonged to eel-like animals (Figure 12-6), have also proved useful in the correlation of Triassic rocks, but by Jurassic time conodont-bearing animals no longer existed.

Paleozoic ray-finned bony fishes gave rise to forms that were successful in early Mesozoic time but were still more primitive than most of their modern-day descendants. The scales that covered the bodies of these fishes, for example, were diamond-shaped structures that overlapped slightly or not at all (Figure 15-10)—contrasting sharply with the circular, strongly overlapping scales of nearly all living bony fishes. Presumably the primitive, diamond-shaped scales were less protective than the modern kind. Other features that distinguished early Mesozoic bony fishes from their modern counterparts were skeletons that consisted partly of cartilage rather than entirely of bone; relatively simple, primitive jaws; and tails that were highly asymmetrical, like those of all Paleozoic bony fishes (Figure 15-10). Some early Mesozoic bony fishes had teeth shaped like rounded pegs, which served to crush durable items of food—probably small shellfish. Bony fishes underwent many changes

during the Mesozoic Era, and by the end of the era, few species with the aforementioned primitive traits remained. One especially useful feature that developed during this time was the swim bladder, a sac of gas that allows advanced fishes to regulate their buoyancy. The swim bladder evolved from the lung, which was present in Paleozoic fishes and was ultimately passed on to land-living vertebrates (Figure 13-26).

Sharks were also well represented in Early Mesozoic seas. One particularly prominent group, the **hybodonts** (Figure 15-11), had already diversified by the end of Paleozoic time. Their teeth, like those of the bony fishes shown in Figure 15-10, were adapted for crushing shellfish. The living Port Jackson shark of Australia, a descendant of the hybodonts, has similar teeth and feeds on mollusks. While

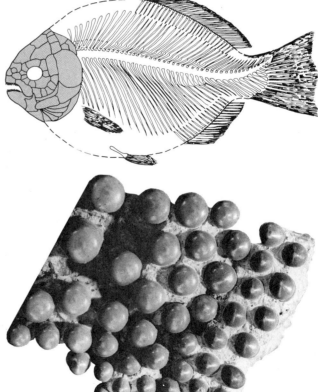

FIGURE 15-10 Two fishes of Jurassic age (*Dapedius* above and *Lepidotus* below). Unlike modern fishes, these early forms had scales that barely overlapped one another. They also had asymmetrical skeletal supports within their tails; note the upturned vertebral column in the upper right-hand drawing. These

particular genera possessed knobby teeth for crushing shellfish, as illustrated by the roof of a *Lepidotus* fish's mouth in the lower right-hand photograph. (*Museum Hauff, Holzmaden, West Germany.*)

FIGURE 15-11   A hybodont shark (*Hybodus*) from the Lower Jurassic Series of Germany. This animal, which was about 2.2 meters (~7 feet) in length, resembled the bony fishes of Figure 15-10 in that it possessed teeth adapted for crushing shellfish. (*Museum für Geologie und Paläontologie, Tübingen, West Germany.*)

the hybodonts declined in diversity late in Mesozoic time, more modern sharks appeared. Mackerel sharks, for example, evolved during the Jurassic Period, as did the family that includes the modern tiger shark.

Many of the reptiles that emerged in early Mesozoic seas were creatures that resembled the popular conception of sea monsters. Among these were the **placodonts**, which, like many early Mesozoic fishes, were blunt-toothed shell crushers (Figure 15-12). The broad, armored bodies of placodonts gave these animals the appearance of enormous turtles. Cousins of the placodonts were the **nothosaurs** (Figure 15-13), which have been found in Early Triassic deposits and seem to have been the first reptiles to invade

the marine realm. Nothosaurs had paddlelike limbs resembling those of modern seals, and it seems likely that, like modern seals, they were not fully marine but instead lived along the seashore, periodically plunging into the water to feed on fishes. Although the placodonts and nothosaurs did not survive the Triassic Period, a group of more fully aquatic reptiles evolved from the nothosaurs in mid-Trias-

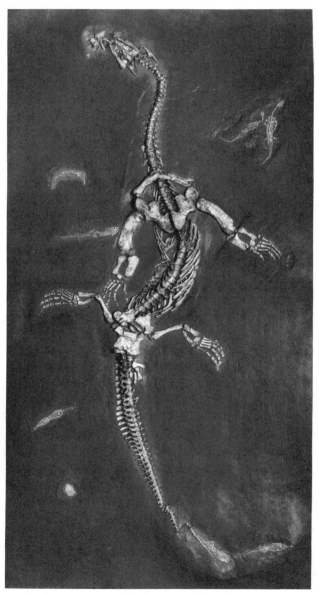

FIGURE 15-13   The nothosaur *Ceresiosaurus,* which was about 2.2 meters (~7 feet) long, preserved with small nothosaurs of a different family in the Middle Triassic Muschelkalk of Germany. Like modern seals, these animals probably fished along the shore. (*Museum für Geologie und Paläontologie, Tübingen, West Germany.*)

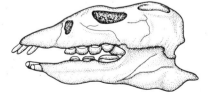

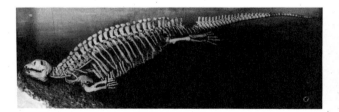

FIGURE 15-12   A placodont reptile (*Placodus*) from the Middle Triassic Muschelkalk of Germany. The skeleton has been reconstructed in a feeding posture. As shown in the drawing of the skull, this reptile, like the fishes of the preceding two figures, possessed teeth for crushing mollusks. (*Museum für Geologie und Paläontologie, Tübingen, West Germany.*)

FIGURE 15-14   Late Jurassic (Oxfordian) plesiosaurs from England mounted in swimming position. Note the paddlelike limbs. These two animals illustrate the two body types of plesiosaurs. *Cryptoclidus*, above, has a long neck and a short head, whereas *Peloneustes*, below, has a short neck and a long head. *Cryptoclidus* is about 3 meters (~10 feet) long. (*Museum für Geology und Paläontologie, Tübingen, West Germany.*)

sic time, and these reptiles, known as **plesiosaurs**, played an important ecologic role for the remainder of the Mesozoic Era. Plesiosaurs apparently fed on fishes and, in Cretaceous time, attained the proportions of modern predatory whales, reaching some 12 meters (~40 feet) in length. The limbs of plesiosaurs were winglike paddles that propelled these animals through the water in much the same way that birds fly through the air (Figure 15-14).

By far the most fishlike reptiles of Mesozoic seas were the **ichthyosaurs**, or "fish-lizards," many of whose species must have been top predators in marine food webs. Superficially, ichthyosaurs bear a closer resemblance to modern dolphins, which are marine mammals, than to fishes; outlines of skin preserved in black shales under low-oxygen conditions show the dolphinlike profiles of some ichthyosaurs (Figure 15-15). The ichthyosaurs, however, had upright tail fins rather than the horizontal pair of rear flukes that propel dolphins through the water. In addition, the extension of the backbone into the ichthyosaur tail bent downward, contrasting with the upward curve that characterized early Mesozoic bony fishes (Figure 15-10). Large eyes supplemented other adaptations of ichthyosaurs for rapid swimming in the pursuit of prey. Ichthyosaurs were fully marine and thus could not easily lay eggs; instead, they bore live young. In fact, skeletons of ichthyosaur embryos have been found within the skeletons of adult females.

Surprising as it may seem, the last important group of early Mesozoic marine reptiles to evolve were the early **crocodiles**, which, as we will see, were related to the dinosaurs. Although crocodiles evolved in Triassic time as terrestrial animals, some were adapted to the marine environment by earliest Jurassic time (Figure 15-16). In fact, certain crocodiles became formidable ocean-going predators whose tail fins were well adapted for rapid swimming.

A

B

FIGURE 15-15   The ichthyosaur *Stenopterygius*. A. The animal preserved with the outline of its skin in the Lower Jurassic Posidonia Shale, Holzmaden, West Germany. Note how the vertebral column bends downward to support the lower portion of the tail fin. B. Reconstruction of the animal in pursuit of belemnoids, two of which are emitting black ink as they escape by jet propulsion. (*A. Field Museum of Natural History. B. Drawing by Gregory S. Paul.*)

TERRESTRIAL LIFE

The presence of dinosaurs during the Mesozoic Era gave the biotas of large continents an entirely new character, but Mesozoic land plants were also distinctive. Because these plants were positioned at the bases of the food webs to which dinosaurs belonged, we will review them first.

Land plants: The Mesozoic gymnosperm flora

Unlike terrestrial animals, land plants do not appear to have undergone a dramatic mass extinction at the close of the Paleozoic Era. As we learned in the preceding chapter, the decline of the late Paleozoic floras began long before the end of the Permian Period. In effect, the transition from the late Paleozoic kind of flora to the Mesozoic kind of flora began before the start of the Mesozoic Era.

Among the groups that decreased in diversity long before the end of Permian time were the lycopod trees, which formed coal swamps, and the sphenopsid and cordaite trees, which inhabited higher ground. Persisting into the Mesozoic Era in greater numbers were ferns and seed ferns. The seed ferns, however, were reduced in abundance and apparently failed to survive into Jurassic time.

FIGURE 15-16   A marine crocodile, *Steneosaurus*, in the Lower Jurassic Posidonia Shale of Holzmaden, West Germany. The animal was more than 2.5 meters (~8 feet) in length. Adjacent to the skeleton in the left part of the diagram is a cluster of stones that were located in the gut of the animal, where they served to grind up food. (*Museum für Geologie und Paläontologie, Tübingen, West Germany.*)

FIGURE 15-17   A fossil fern from the Upper Triassic Chinle Formation of New Mexico (×0.7). (*U.S. Geological Survey.*)

FIGURE 15-18   The trunk of a fossil cycad, *Cycadeoidea*, from the Potomac Formation of Maryland (about one-third of natural size). The numerous scars mark the positions of branches. Figure 2-20 shows a living cycad tree. (*G. R. Wieland, American Fossil Cycads, Carnegie Institution, Washington, D.C., 1906.*)

For ordinary ferns, the story was quite different. As we all know, ferns are well represented in the modern world, but they are nowhere near as prevalent today as they were in Triassic time. Ferns, in fact, dominate Triassic fossil floras (Figure 15-17).

Although some of the trees that stood above Triassic ferns were sphenopsids and lycopods, most belonged to three groups of gymnosperms that had already become established during the Permian Period. The most diverse of these three groups were the **cycads**, followed by the conifers, which we have already discussed (page 406), and the **ginkgos**. All three of these groups survive to this day, but the cycads are rare, and there is only one living species of ginkgo remaining on earth.

The trees that belonged to these three dominant groups are united as gymnosperms because they were all characterized by exposed seeds. The seeds of pines and other conifers, for example, rest on the projecting scales of their cones. There is a reason for this configuration. Whereas flowering plants, which did not evolve until Cretaceous time, can attract insect and bird pollinators, most gymnosperms rely primarily on wind to carry their pollen from tree to tree. With the possible exception of the pine family, all of the modern conifer families were present in early Mesozoic time. Cycads, as noted, are less familiar plants. The few modern species of cycads are tropical trees that superfi-

cially resemble palms (Figure 2-20). Cycad trunks are well known as early Mesozoic fossils (Figure 15-18). The single living species of ginkgo looks more like a hardwood tree (that is, like an oak or a maple) than a conifer, and, like hardwoods, it sheds its leaves seasonally. This surviving species of ginkgo is a true living fossil whose record extends back some 60 million years to the Paleocene Epoch (Figure 15-19), early in the Cenozoic Era.

FIGURE 15-19   Leaves of the living ginkgo, *Ginkgo biloba*, compared to a similar leaf of early Cenozoic age. (*Photograph of fossil courtesy of the Smithsonian Institution.*)

FIGURE 15-20   Reconstruction of a Mesozoic landscape. Cycads and ferns appear in the foreground and conifers along the horizon. (*Drawing by Z. Burian under the supervision of Professor J. Augusta.*)

The cycads, conifers, and ginkgos formed the forests of the Jurassic Period, but the cycads dominated to such an extent that the Jurassic interval might well be called the Age of Cycads. Ferns of the Jurassic Period were less conspicuous as undergrowth than those of Triassic time. Both Triassic and Jurassic landscapes, however, would have looked more familiar to us than Paleozoic landscapes, largely because of the presence of conifers that closely resembled modern evergreens (Figure 15-20). Even so, the absence of flowering plants such as grasses and hardwood trees would have made Mesozoic floras appear archaic to a modern observer.

Terrestrial animals:
The Age of Dinosaurs begins

During Late Permian and Early Triassic time, much of the supercontinent of Pangaea stood above sea level. Under such circumstances, one would think that terrestrial deposits spanning the boundary and recording the history of land animals from Late Permian time into the Triassic would be abundant and easy to find. Unfortunately, this is not the case. Continuous fossil records have been found in only two regions: in the Karroo Basin of South Africa and in the Soviet Union near the Ural Mountains. Significantly, the fossil records of these two regions tell the same story. Just below the Permian-Triassic boundary in both regions, most of the dozens of genera of Late Permian mammal-like reptiles disappeared suddenly from the fossil record, marking a major mass extinction. What remained at the start of Triassic time were a few predatory genera and the large herbivore *Lystrosaurus,* which is famous for its fossil occurrence on many of the now widely dispersed fragments of Gondwanaland (Figure 7-19).

Although the mammal-like reptiles rediversified during the Triassic Period to play an important ecologic role once more, they barely survived into the Jurassic Period. None-

FIGURE 15-21   Mid-Triassic animals of the genus *Lagosu-chus* intimidating an early mammal of smaller body size. *Lago-suchus,* which was about 30 centimeters (~1 foot) tall, was either a primitive dinosaur or a thecodont that closely resembled the earliest dinosaurs. Thecodonts were the ancestors of dino-saurs. (*Drawing by Gregory S. Paul.*)

theless, they left an important legacy in the form of the true mammals, which evolved from them near the end of the Triassic Period. **Mammals**, which are endotherms with hair and which suckle their young, are the dominant large an-imals of modern terrestrial habitats, but they remained small and peripheral throughout the Mesozoic Era. Apparently no species grew larger than a house cat. Their problem seems to have been that the **dinosaurs** also evolved dur-ing the Late Triassic interval and quickly rose to dominance. (Mammals will be discussed at length in the following chap-ter. The fossil record for mammals is better in the Creta-ceous System than in the Triassic or Jurassic.)

Dinosaurs were also small at first (Figure 15-21), but their advantage over primitive mammals may have been that they were extremely agile. They inherited their loco-motory capacity from their ancestors, the **thecodonts**, which evolved during the Triassic Period. Some thecodonts were adapted for speedy two-legged running in the fashion of ostriches and other flightless birds, but all thecodonts probably spent much time standing or walking on all fours. The upper portion of the legs of many thecodonts stood beneath their bodies rather than sprawling slightly out to the side as they did in mammal-like reptiles. This feature, which facilitated running, was passed on to the dinosaurs and seems to have been the key to the dinosaurs' success.

The first dinosaurs resembled bipedal thecodonts (that is, thecodonts that traveled on their hind legs), but the dinosaurs had skulls of a different form as well as teeth that were more highly developed. Dinosaurs (formally known as the **Dinosauria**) did not become gigantic before the end of the Triassic, but during Triassic time they did reach lengths of more than 6 meters (~20 feet). Figure 15-22 shows a moderately large species along with a thecodont that stood on all fours and a primitive, long-legged croc-odile. The crocodiles, like the dinosaurs, evolved from the-codonts in Late Triassic time, and all three groups are united formally in a single taxonomic category. Thus, the living crocodiles are the dinosaurs' closest living relatives.

FIGURE 15-22 Scene representing Late Triassic life of Argentina. In the foreground, the thecodont *Riojasuchus* (*left*) contests the long-legged crocodile *Pseudohesperosuchus* for a meal of a small reptile. In the background, the prosauropod dinosaur *Massospondylus* feeds on *Dicroidium*, a land plant that characterized Gondwanaland floras of Late Triassic age. (*Drawing by Gregory S. Paul.*)

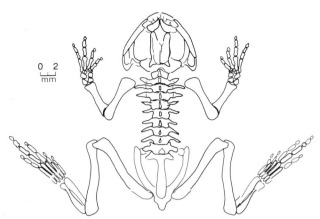

FIGURE 15-23   Drawing of a partly restored skeleton of a frog belonging to the genus *Vieraella* from the Lower Jurassic Series of Argentina. This is one of the oldest true frogs known from the fossil record. (*After R. Estes and O. A. Reig, in J. L. Vial [ed.], Evolutionary Biology of the Anurans, Columbia University Press, New York, 1973.*)

Two other important groups, both of which are familiar in the modern world, also appear to have become established in Triassic time. One was the **frogs**, which were and remain amphibians of small body size. The oldest known fossil displaying the form of a modern frog is of earliest Jurassic age (Figure 15-23), but froglike skeletons have also been found in Triassic rocks. The other modern group was the **turtles**, although the earliest turtles lacked the ability to pull their heads and tails fully into their protective shells.

By Late Triassic time, mammal-like reptiles had declined in diversity. Those that remained lived alongside increasing numbers of dinosaurs, the still-diverse thecodonts, and other, mainly smaller amphibians and reptiles. A few kinds of large amphibians persisted as well. The stage was thus set for the ascendancy of the dinosaurs to a dominant position for the remainder of the Mesozoic Era.

The rise of the dinosaurs   Unfortunately, the known fossil record of Early Jurassic time is too poor to permit us to piece together the details of the dinosaur's great rise to dominance. There are, nonetheless, fossil remains of huge dinosaurs even in Lower Jurassic rocks, indicating that the dinosaurs evolved rapidly; the oldest dinosaur giants are found in Australia.

Dinosaurs fall into two groups, which are characterized by different pelvic structures (Figure 1-7). The "bird-hipped" (**ornithischian**) dinosaurs were all herbivores, whereas the "lizard-hipped" (**saurischian**) group included both herbi-

vores and carnivores. In both groups, there were species that traveled on two legs and others that moved about on all fours. The largest of all the dinosaurs were the **sauropods**, lizard-hipped herbivores that moved about on all fours. Some interesting aspects of dinosaur biology are presented in Box 15-1.

By Late Jurassic time both bird-hipped and lizard-hipped dinosaurs were quite diverse. The most spectacular Jurassic assemblage of fossil dinosaurs in the world is found in the Upper Jurassic Morrison Formation, which extends over a large area of the western United States, from Montana to New Mexico. At Como Bluff, Wyoming, where the Morrison has been extensively quarried for fossils, dinosaur bones are so common that a local sheep herder constructed a cabin of them because they were the most readily available building materials! Later in this chapter, we will examine the nature of the Morrison deposits in greater detail. The dinosaurs of the Morrison Formation, which include more than a dozen genera, are representative of the kinds of dinosaurs that lived throughout the world during Late Jurassic time. Several of the common Morrison species are shown in Figures 15-24 through 15-26.

FIGURE 15-24   Herbivorous dinosaurs from the Morrison Formation of Utah. A. *Diplodocus*, an animal that stretched to a length of more than 26 meters (~87 feet) from nose to tail. B. Skeleton of *Camarasaurus*, mounted in the position in which it was found preserved in the rock. (*Smithsonian Institution.*)

Box 15-1

Who were the dinosaurs?

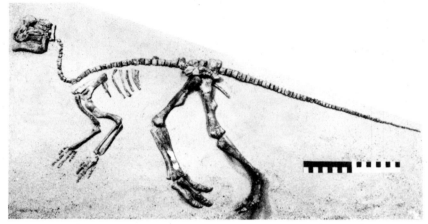

Cross section of the skull of a duck-billed dinosaur (*left*), showing chambers. Juvenile dinosaur (*right*) found in a nest in Cretaceous sediments of Montana; scale 20 centimeters (~8 inches) long. (*Drawing after J. A. Hopson, Palebiology 1:21−43, 1975. Photograph from J. R. Horner and R. Makela, Nature 282:296−298.*)

As researchers bring increasing knowledge and experience to bear on their examination of dinosaur remains, many earlier notions of how these creatures looked and behaved are being debunked. It is now evident, for example, that not all dinosaurs were of massive proportions; many, in fact, were less than 1 meter (~3 feet) long. Moreover, while dinosaurs have often been portrayed as hulking, sluggish creatures lacking in both speed and agility, it is now known that many were as mobile as ostriches, which are famous for their great speed. The orientations of dinosaur limbs in their sockets show that dinosaur legs were positioned almost vertically beneath the body, and fossil trackways of dinosaurs such as that shown on page 442, which was found in the Upper Triassic Newark Group of Connecticut, confirm that this posture was typical. Note

that the left and right tracks here are nearly in line, signifying that both feet were positioned almost beneath the center of the body. Most fossil trackways also reveal long strides for the size of the individual tracks, suggesting that dinosaurs tended to move rapidly.

Perhaps even more surprising, however, is evidence suggesting that dinosaurs were social animals. The Late Cretaceous duck-billed dinosaurs, for example, had tall, crested skulls that may have been resonating chambers that functioned like those of a trumpet. Whether dinosaurs communicated by sound or not, some of their well-preserved track-

ways show that large groups of animals sometimes traveled together in herds. Moreover, a nest of baby dinosaurs has been found in the Upper Cretaceous of Montana. It has long been known that dinosaurs laid eggs. Some of these have remained in circles, just as they were carefully laid; often several circles of eggs were buried on top of one another. The nest of baby dinosaurs—a cluster of juvenile skeletons with broken eggshells in a depression on top of a 3-meter-wide (10-foot-wide) hill—shows that the eggs were not abandoned. Like living birds and crocodiles, dinosaurs cared for their young.

Two questions have often been

Eggs laid by a Cretaceous dinosaur in Mongolia. (*American Museum of Natural History, 1979.*)

posed with regard to the largest dinosaurs (Figure 15-24). The first has to do with nutrition: How did the relatively small jaws of these enormous animals chew up enough food to live on? The answer is that these giant herbivores used their mouths and jaws only for gathering and swallowing plant food. In the animals' intestinal tracts were "gizzard stones," like those of birds but much larger, which helped the animals grind up coarse material after it had been swallowed.

Another question is whether the largest dinosaurs could have supported their own weight on land. In the past, it was widely believed that dinosaurs lived a semiaquatic life, like modern hippopotamuses, ordinarily depending on water to buoy up their enormous bodies. It now appears that this conclusion was erroneous. The giant dinosaurs did not have the broad feet that would have been necessary for support on the muddy floors of lakes and rivers; moreover, the animals are as commonly preserved in nonaquatic environments as are smaller dinosaurs. Apparently, the largest dinosaurs walked on all fours on dry land.

The greatest controversy about dinosaurs, however, has related to the issue of metabolism. Dinosaurs have traditionally been classified as reptiles, and it has thus been assumed that they were ectothermic (or cold-blooded). It has recently been argued, however, that dinosaurs were actually endothermic, or warm-blooded. The newer idea stems in part from evidence that dinosaurs were very active animals and in part from additional evidence indicating that they were more successful during the Mesozoic Era than were warm-blooded early mammals or mammal-like reptiles, which had had a long evolutionary head start in the race for dominance on the land. The argument here is that dinosaurs could not have competed with mammals if they had not been capable of sustaining a high rate of activity for long intervals of time while hunting or fleeing from predators. Ectothermic animals, including modern reptiles, have little endurance (page 415). Additional evidence derives from the fact that in communities of dinosaurs, predators usually made up less than 10 percent of the volume of living species, as is the case for living and fossil mammal communities. In communities of ectothermic animals, predators commonly represent 40 percent or so of the volume of living tissue. Having low metabolisms, they need little food, and many can sustain themselves on small populations of prey animals. The low percentage of predators in many dinosaur communities argues for a closer resemblance to mammal communities than to reptile communities. The microscopic structure of bone has also entered the debate. Endothermic animals differ from ectothermic animals in bone structure, and it has been contended that the dinosaurs' bone structure was of the endothermic type.

Whatever the details of dinosaurs' metabolism may have been, we at least recognize that even large dinosaurs were active, highly adapted animals rather than simple, lumbering beasts. For all we know, many dinosaurs might have fared well in the modern world had they been given the chance. The sudden extinction of the dinosaurs at the end of the Cretaceous Period appears to have been more of an unfortunate accident than an indication of biological inferiority.☐

FIGURE 15-25   Dinosaurs from the Morrison Formation of Utah. *A. Ceratosaurus*, a carnivorous form with large teeth; the horn on the animal's nose may have been used in combat with other animals of the same species. *B. Stegosaurus*, an armored herbivore with spikes on its tail. (*Smithsonian Institution.*)

Late in the Triassic Period, vertebrate animals invaded the air for the first time as the **pterosaurs** came into being. These reptiles had long wings and tails (Figure 15-27) and were probably capable of powered flight, but the flapping of their wings did not propel them very efficiently. Like birds, however, pterosaurs had many hollow bones that served to facilitate flight. It is not known what pterosaurs did when

they were not in the air. Their small hind legs may have been incapable of supporting them, and thus it has been suggested that they might have hung upside-down while resting, in the manner of modern bats.

Birds appeared near the end of the Jurassic Period. The first clue to their time of origin was a feather discovered in 1861 in the fine-grained Solnhöfen Limestone of Ger-

FIGURE 15-26   Reconstruction of the dinosaur fauna of the Morrison Formation, featuring three of the animals whose skeletons are shown in the previous two figures. These are *Camarasaurus*, walking toward the right in the background; *Diplodocus*, drinking water; and *Stegosaurus* on the right. Reclining on the left is *Allosaurus*, a large carnivorous dinosaur, about 6 to 7 meters (~20 to 25 feet) long. In the foreground are dragonflies, pterosaurs, turtles, and a crocodile. The trees are conifers. (*Drawing by Gregory S. Paul.*)

A

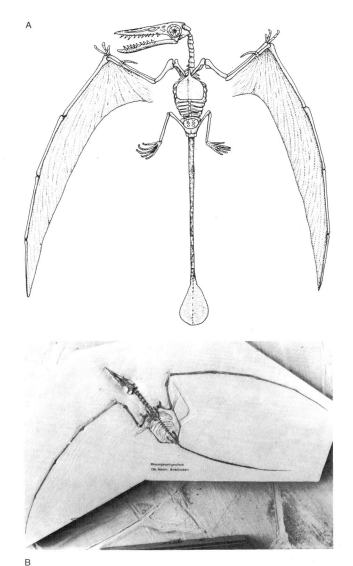

B

FIGURE 15-27   The pterosaur, or "flying lizard," *Rhamphor-hynchus*. *A.* Skeleton of the animal preserved intact in the Solnhöfen Limestone of Germany; it measures about 50 centimeters (~18 inches) long. Many of the bones are hollow like those of a bird. *B.* Reconstruction of the animal, showing the skin that stretched along the margins of the forelimbs to form wings. The fourth fingers were enormously elongated to support the wings. (*A. Museum für Geologie und Paläontologie, Tübingen, West Germany. B. After S. W. Willistan, The Osteology of the Reptiles, Harvard University Press, Cambridge, Massachusetts, 1925.*)

many, followed a few months later by the discovery of an entire skeleton of the species to which the feather belonged (Figure 15-28). This feathered animal was given the name *Archaeopteryx*, which means "ancient wing." *Archaeopteryx* had a skeleton so much like that of a dinosaur that

it would be regarded as such were it not for its feathery plumage. This represents a classic discovery of a missing link—in this case, the link between birds and their flightless ancestors. The teeth, large tail, and clawed forelimbs of *Archaeopteryx*, which are absent from advanced birds, reflect its dinosaur ancestry. *Archaeopteryx* lacks a breast bone, and is thus assumed to have possessed weak flying muscles. It was probably a clumsy flier by the standards of modern birds. Unfortunately, the hollow bones of birds are fragile, and no bird bones but those of *Archaeopteryx* have been found within the Jurassic System.

PALEOGEOGRAPHY

At about the time the Mesozoic Era began, all of the major landmasses of the world became united as the supercontinent Pangaea (Figure 15-29). Near the end of Triassic time Pangaea began to break apart, but continental movement is so slow that even by the end of Jurassic time, the newly forming continental fragments were hardly separated. Thus, throughout the interval of time covered in this chapter, the earth's continental crust was concentrated on one side of the globe. Figure 15-30 illustrates that sea level, which changed little during the Permian-Triassic transition, rose slightly during Early Triassic time. As in Late Permian time, however, the bulk of the continental crust during the Triassic Period stood above sea level, forming one vast continent. At the start of Triassic time, the Tethys, sometimes called the Tethyan Seaway, was an embayment of the deep sea projecting into the portion of equatorial Gondwanaland that today constitutes the Mediterranean. Later in Triassic and Jurassic time, rifting extended the Tethys between Eurasia and Africa and all the way westward between North and South America in the Pacific.

Pangaea during the Triassic Period

Although the dominant land plants of the Triassic Period differed from those of the Permian, the distributional pattern of floras on Pangaea remained much the same; a Gondwana flora existed in the south, and a Siberian flora existed in the north (Figure 15-29). The Euramerian flora grew under warmer, drier conditions at low latitudes; in fact, unusually extensive deposition of evaporites attests to the presence of arid climates far from the equator. This condition may have resulted in part from the sheer size of Pangaea, which was such that many regions of the supercontinent lay far from the low-standing oceans.

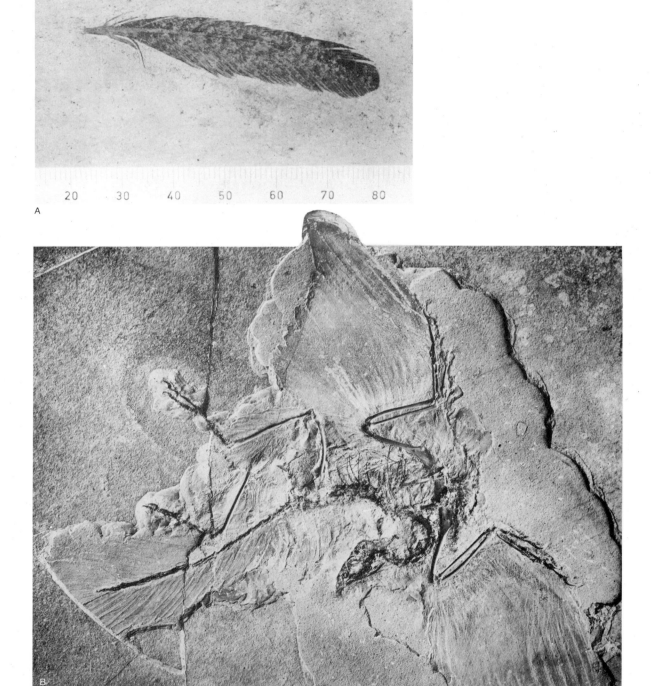

FIGURE 15-28 Fossil remains of *Archaeopteryx lithographica*, the oldest known bird, from the Upper Jurassic Solnhöfen Limestone of Germany. The existence of a bird during Solnhöfen deposition was first indicated by the discovery of a feather (*A*). The asymmetry of the feather suggests that it aided in flight; flightless living birds have feathers that are symmetrical about the central shaft. The bird itself was soon found, and impressions of long feathers are clearly visible around it in the fine-grained limestone (*B*). Despite the fact that it possessed feathers, *Archaeopteryx* had a skeleton and teeth similar to those of dinosaurs. (*A. Courtesy of J. H. Ostrom. B. Smithsonian Institution.*)

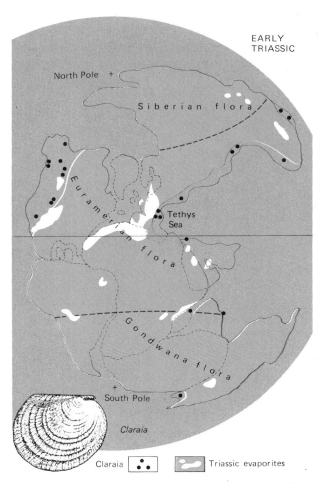

FIGURE 15-29 World geography of Early Triassic time. The Euramerian flora occupied a broad, warm belt across the middle of Pangaea, and the Siberian and Gondwana floras occupied regions to the north and south. The bivalve mollusk genus *Claraia* (*lower left*) was broadly distributed along both the eastern and western borders of Pangaea. The widespread evaporites illustrated here represent the entire Triassic Period, which was relatively warm and dry even at high latitudes. (*Modified from G. E. Drewry et al., Jour. Geol. 82:531–553, 1974.*)

The survival of only a small percentage of Permian species into the Triassic Period resulted in some striking biogeographic distributions. In the oceans, the scalloplike bivalve genus *Claraia* ranged over an enormous area (Figure 15-29) and is assumed to have occupied the sea floor even in deep water. On land, the large herbivorous therapsid *Lystrosaurus* ranged over large areas of the globe; *Lystrosaurus* has been found in the fossil records of several continents that represent fragments of Gondwanaland (Figure 7-19). It appears that at the start of the Triassic Period, few large therapsids preyed on *Lystrosaurus*, and as a

consequence, its populations were vast. As described above, the therapsids underwent renewed radiation during the Triassic Period, and partway through the period thecodonts and dinosaurs began their expansion. Even when these vertebrate groups radiated to high diversities, most of their species were also wide-ranging. Many families of Triassic terrestrial vertebrates are found as fossils on several modern continents. In fact, it has been remarked that the Triassic is the only period whose fossil land vertebrates clearly indicate that all the continents were connected.

The breakup of Pangaea

The most spectacular geographic development of the Mesozoic Era was the fragmentation of Pangaea, an event that began in the Tethyan region. As the Triassic Period progressed, the Tethyan Seaway spread farther and farther inland, and eventually the craton began to rift apart. The Tethys subsequently became a deep, narrow arm of the ocean separating what is now southern Europe from Africa. Then, during the Jurassic Period, this rifting propagated westward, ultimately separating North and South America. South America and Africa, however, did not separate to form the South Atlantic until the Cretaceous Period; in fact, all of the Gondwanaland continents remained attached to one another until Cretaceous time. North America began to break away from Africa in mid-Jurassic time. Interestingly, this rifting generally followed the old Hercynian suture. Rifting occurred as some of the arms of a series of triple junctions joined, tearing Pangaea in two (Figure 15-31).

The rifting that formed the Atlantic had another important consequence. When continental fragmentation begins in an arid region near the ocean, evaporite deposits often form in that region (page 198). Thus, as rifting began in Pangaea, extension produced normal faults between Africa and the northern continents (Figure 15-31); zones bounded by such faults sank, and water from the Tethys to the east periodically spilled into the trough and evaporated. Evaporites that were precipitated in this trough are now located on opposite sides of the Atlantic, both in Morocco and offshore from Nova Scotia and Newfoundland.

During Middle and Late Jurassic time, one arm of rifting passed westward between North and South America, giving rise to the modern Gulf of Mexico. The early intermittent influxes of sea water into this rift, apparently from the Pacific Ocean, caused great thicknesses of evaporites to accumulate. Today these evaporites, which are known as the Louann Salt, lie beneath the Gulf of Mexico and in the

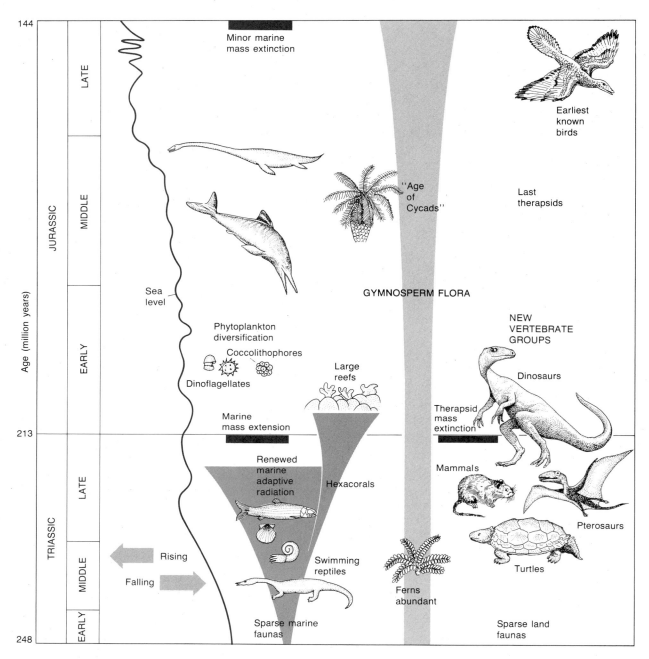

FIGURE 15-30   Major events of early Mesozoic time. Sea level underwent a general rise during this interval. Ferns flourished during Triassic time, but gymnosperms continued to dominate terrestrial environments. After the Late Permian crisis, marine and terrestrial faunas were impoverished, but adaptive radiations soon replenished animal life. Mass extinctions weaker than the terminal Permian crisis struck at the end of the Triassic and Jurassic periods. Several new groups of vertebrate animals, including dinosaurs, occupied Late Triassic terrestrial environments, and birds were present before the end of Jurassic time. In the oceans, adaptive radiation gave mollusks a major ecologic role. Hexacorals evolved in mid-Triassic time and by Early Jurassic time had produced reefs resembling those of the modern world. Large predatory reptiles evolved in the Triassic Period and flourished in the Jurassic Period as well.

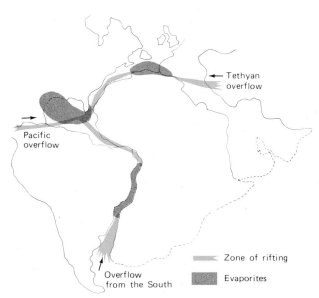

FIGURE 15-31   Early Mesozoic evaporites that accumulated during the early stages of rifting that formed the Atlantic Ocean as oceans overflowed intermittently into newly forming fault basins. (*After K. Burke, Geology 3:613–616, 1975.*)

subsurface of Texas. Because its density is low, the Louann Salt has in some places pushed up through younger sediments to form salt domes (Figure 15-32), many of which are overlain by valuable reservoirs of petroleum. The rifting that forms the South Atlantic did not begin until Early Cretaceous time, when the record shows that salt deposits formed after sea water spilled inland from the south (Figure 15-31).

The Jurassic world

Although sea level underwent only minor changes during Late Triassic and Early Jurassic time, it subsequently rose, with minor oscillations, until Late Jurassic time. Then, very late in the Jurassic Period, it underwent more rapid oscillations but remained at a high level, causing epicontinental seas to flood large areas of North America and Europe throughout Late Jurassic time (Figure 15-33). Long before the advent of plate-tectonic theory, it was recognized that there were two biogeographic provinces of marine life in Europe during the Jurassic Period: the southern province,

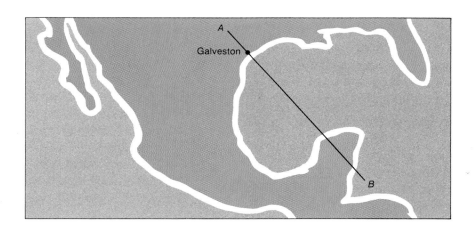

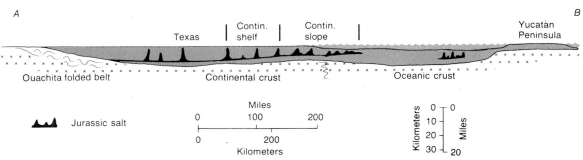

FIGURE 15-32   Position of Jurassic salt within the sediments beneath the Gulf of Mexico. The salt deposits, which rise into domes in some places because of their low density, accumu-

lated when the Gulf of Mexico began to form by continental rifting. (*Modified from O. Wilhelm and M. Dwing, Geol. Soc. Amer. Mem. 83, 1972.*)

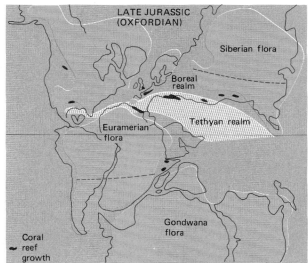

FIGURE 15-33   Late Jurassic (Oxfordian) world geography. The three floras persisted from Triassic time. The Tethyan marine realm, which was characterized by tropical life including reef corals, extended all the way from the eastern Pacific across the newly forming Mediterranean and the newly forming Gulf of Mexico to the western Pacific. The cooler Boreal realm lay to the north.

which was centered in the Tethys and was designated the Tethyan realm, and the northern province, which was named the Boreal realm. Figure 15-33 shows the positions of these provinces when sediments of the Oxfordian Stage were laid down at the time of maximum worldwide transgression for the Jurassic Period.

Coral reefs were largely restricted to the Tethyan realm, as were limestones and certain groups of mollusks. These observations suggest that the Tethyan was largely a tropical province. The transition from the Tethyan to the Boreal province resembled that which we see today from the tropical conditions of southern Florida, with its carbonates and coral reefs, to the subtropical conditions of northern Florida, where siliciclastic sediments prevail and reefs are lacking.

There is no doubt that temperature gradients from equator to pole were gentle throughout the Jurassic Period. Plants that appear to have required warmth (the Euramerian flora of Figure 15-33) occupied a broad belt extending to about 60° north latitude. Even the Gondwana flora to the south and the Siberian flora to the north included groups of ferns whose modern relatives cannot tolerate frost. The high-latitude floras do not seem to have been tropical, however; they contained few cycads, and modern cycads are restricted to warm regions (Figure 2-20). Thus, it appears that the large landmass that began to fragment during the Jurassic Period was bathed in tropical climates well

to the north and south of the equator, even though it experienced somewhat cooler temperatures toward the poles.

Modest mass extinctions

Both the Triassic and Jurassic periods ended with mass extinctions, but neither of these extinctions approached that of the Late Permian event in severity. Nearly all major taxa of marine animals, for example, survived the early Mesozoic crises. The conodonts (Figure 12-6) and the placodont reptiles (Figure 15-12), however, disappeared in the terminal Triassic event. On land, the therapsids barely survived this crisis and then disappeared during the Jurassic Period. Bivalves and ammonoids also suffered major losses at the end of Triassic time, but they recovered during the Jurassic Period.

Although the terminal Jurassic extinction severely affected the ammonoids, these organisms recovered rapidly, just as they had following the Triassic extinction. However, the Jurassic event did eliminate most of the ichthyosaurs (Figure 15-15) and plesiosaurs (Figure 15-14) in the ocean, and on land the most noteworthy disappearances were those of the stegosaurian dinosaurs (Figure 15-25B) and the larger sauropods (Figures 15-24A and 15-26). None of the herbivorous dinosaurs of the Cretaceous Period was as large as the largest Jurassic sauropods.

Unfortunately, the possible causes of these extinctions have not been extensively investigated. It is interesting to note, however, that coral reefs disappeared from the Tethyan region of Europe during the terminal Triassic mass extinction, suggesting that the Tethyan waters may have cooled. In addition, cooling in Asia as the Jurassic Period came to a close is suggested by a decline in abundance of fossil pollen of conifer trees adapted to warm conditions.

REGIONAL EXAMPLES

In viewing regional events of early Mesozoic age, we will first focus on depositional basins in eastern North America—structures that illustrate events associated with the initial opening of the Atlantic Ocean. We will then examine the origin of the tectonic episodes that ultimately produced the mountains now standing in the American West. Finally, we will review the history of Europe—first for Triassic time, when nonmarine deposition was widespread, and then for Jurassic time, when marine deposition expanded and when Africa and Europe split apart to form a deep-water connection between the Tethys and the young Atlantic Ocean.

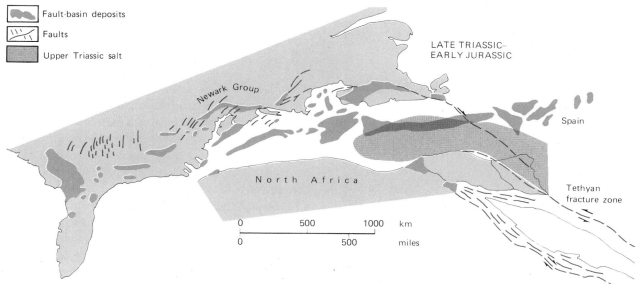

LATE TRIASSIC–
EARLY JURASSIC

Newark Group

North Africa

Spain

Tethyan
fracture zone

| 0 | 500 | 1000 | km |
| 0 | 500 | | miles |

FIGURE 15-34   Geologic features in eastern North America and nearby regions during Late Triassic and Early Jurassic time. In eastern North America, block faulting produced elongate depositional basins, most of which paralleled the enormous rift that eventually formed the Atlantic Ocean. Salt deposits ac-cumulated from the sporadic westward spilling of sea water from the Tethyan Seaway, where the Mediterranean was form-ing as a result of movement of Africa relative to Europe. (*After W. Manspeizer et al., Geol. Soc. Amer. Bull. 89:901–920, 1978.*)

Atlantic fault basins

During Early and Middle Triassic time, erosion subdued the Appalachian Mountains, which were centrally located in Pangaea, but in Late Triassic time, long, narrow depositional basins bounded by faults developed on the gentle Appalachian terrane (Figure 15-34). These basins formed when Pangaea was splintered by normal faults on either side of the great rift that began to divide the continent, forming the Atlantic Ocean. One of the largest of these basins extended from New York City to northern Virginia and received sediments known as the Newark Group. Here, during a Late Triassic and Early Jurassic interval of subsidence, the nonmarine sediments of the Newark Group accumulated to a thickness of nearly 6 kilometers (~4 miles). Early Mesozoic basins resembling those of eastern North America are also found in Africa and South America, but these contain thick evaporite deposits. It was in the early phase of rifting that water from the Tethys periodically spilled into these southern basins to form vast salt deposits (Figure 15-31).

The best locations for investigating the basin sediments are in the eastern United States. One particularly well studied basin passed through present-day Connecticut and Massachusetts, bounded on the east by a large normal fault along which the basin subsided continually while sediments accumulated from an eastern source area (Figure 15-35A). Several types of depositional environments occupied this basin. Coarse conglomerates wedging out to the west represent alluvial fans that spread from the eastern fault margin. Many sand-sized sediments of the basin

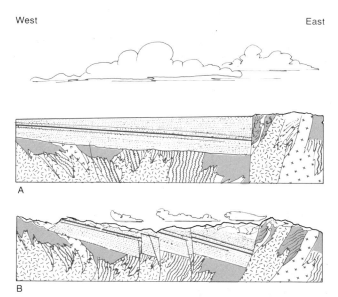

West   East

A

B

FIGURE 15-35   Diagrammatic cross sections of the large early Mesozoic fault basin that passes through central Connecticut, where the Newark Group was deposited (Figure 15-34). A. The basin late in its depositional history, when great thicknesses of sediment had accumulated. As the basin subsided, lavas welled up periodically, forming dikes and sills, and gravels from the uplands to the east spread into the basin as alluvial fans. B. The eventual destruction of the basin by extensive faulting.

are stream deposits. The fact that most of these deposits are composed of red arkose suggests that deposition in this area was rapid in that there was little time for feldspars to disintegrate to clay. In the basin center, well-laminated

mudstones floored the lakes (Figure 15-36). Cycles now visible in the sediments reflect expansion and contraction of the lakes, which for the most part must have been quite shallow. During some dry intervals, evaporite minerals were precipitated from the shrinking waters, but abundant fossil fish remains indicate that at other times the waters were hospitable to life.

Although dinosaur tracks are common in rocks representing lake margins (page 442), conditions in the basins seldom favored the preservation of dinosaur skeletons. Some of the ancient soils in the basin that passes through Connecticut and Massachusetts contain caliche nodules, indicating that the climate here was warm and seasonally arid. Apparently bones decayed rapidly under these conditions, so that relatively few were preserved as fossils.

Periodically, basic lavas welled up through faults, forming dikes and widespread sills within the basin. One of the largest of these sills forms the Palisades along the Hudson River near New York City (Figure 15-37). At least some of the North American basins continued to subside until Early

FIGURE 15-36   Lake sediments of East Berlin, Connecticut, found in the fault basin represented in the previous figure. This is a typical lake deposit, consisting primarily of well-laminated mudstone. (*J. F. Hubert et al., Amer. Jour. Sci. 276:1183–1207, 1976.*)

FIGURE 15-37   The famous Palisades along the western shore of the Hudson River west of New York City. These imposing structures are formed by vertical joints (see Appendix II) in a widespread igneous sill. The sill represents lava that flowed horizontally after rising from the mantle during Early Mesozoic block faulting. (*UPI/Bettmann Archive.*)

Jurassic time, when deposition ended with a final episode of faulting. After this time, the basins apparently moved so far westward along with the North American plate that they were no longer affected by the mid-Atlantic rifting. The fact that some of the basins are located several hundred kilometers from the present margin of North America (Figure 15-33) indicates how imprecise the fracturing of a large continent can be; in most such instances, many small breaks and ruptures occur rather than a clean separation from sea to sea.

Western North America

Throughout the Triassic Period, much of the American West was the site of nonmarine deposition. Shallow seas expanded and contracted along the margin of the craton but for the most part remained west of Colorado. During Middle and Late Triassic time, western North America was especially free of marine influence.

As in the Permian Period, the climate here remained largely arid. At times, however, there was sufficient moisture to permit the growth of large trees belonging to the Euramerian flora. The series of river and lake sediments in Utah and Arizona that are collectively known as the Chinle Formation, for example, erodes spectacularly in some places to reveal the well-known Petrified Forest of Arizona (Figure 15-38). In southwest Utah, the Chinle is overlain by the Wingate Sandstone, a desert dune deposit (Figure 15-39). This, in turn, is followed by a river deposit called the Kayenta Formation, on top of which rests the Navajo Sandstone. The Navajo, also a desert dune deposit, ranges upward in the stratigraphic sequence from approximately the position of the Triassic-Jurassic boundary. The Navajo is famous for its large-scale cross-bedding in the neighborhood of Zion National Park (Figure 15-40).

During Middle and Late Jurassic times, as sea level rose globally (Figure 15-30), waters from the Pacific Ocean spread farther inland in a series of four transgressions, each more extensive than the last. The first such transgression went no farther than British Columbia and northern Montana, but the last, which is known as the Sundance Sea, spread eastward to the Dakotas and southward almost to the Mexican border (Figure 15-41). Eventually, as mountain building progressed along the Pacific Coast in Late Jurassic time, the Sundance Sea retreated. We will now examine this tectonic episode and analyze how it caused marine sedimentation in the Sundance Sea to give way to nonmarine deposition.

FIGURE 15-38   Silicified logs that have weathered out of the Triassic Chinle Formation in the Petrified Forest of Arizona. (*National Park Service.*)

After the end of the Sonoma orogeny early in the Triassic Period, there was a brief interlude of tectonic quiescence along the west coast of North America. Then, in mid-Triassic time, the continental margin once again came to rest against a subduction zone, thus marking the beginning of an orogenic episode that extended from Alaska all the way to Chile. Mountain building along the Pacific coast of North America during the Mesozoic Era resembled the growth of the Andes to the south, which has continued to the present day (page 217).

Subduction of the oceanic plate beneath the margin of North America thickened the continental crust by leading to the accumulation of intrusive and extrusive igneous rocks. The oldest intrusives of the Sierra Nevada Mountains were emplaced during Jurassic time (Figure 15-42), although larger volumes were added later in the Mesozoic Era.

The Mesozoic history of the Pacific coast of North America is highly complex (Figure 15-43). At times, more than one subduction zone lay offshore, and exotic slivers of crust were added to the continental margin. Near the end of the Jurassic Period, the continent accreted westward when the Franciscan sequence of deep-water sediments and volcanics was forced against the craton along a subduction zone after having been metamorphosed at high pressures and at low temperatures. The Franciscan sediments

FIGURE 15-39   Nonmarine Triassic and Jurassic sediments exposed along the Virgin River in Zion National Park, Utah. The Triassic-Jurassic boundary probably lies near the base of the Navajo Sandstone. Figure 15-38 provides a more detailed view of the Chinle Formation, and Figure 15-40, of the Navajo Sandstone. (*National Park Service.*)

FIGURE 15-40   Enormous cross-beds within the Navajo Sandstone in Zion National Park, Utah. These Jurassic sediments almost certainly represent dune deposits. (*National Park Service.*)

include graywackes and mudstones together with smaller amounts of chert and limestone. As shown in Figure 7-51, these sediments became a mélange (page 200) that was piled up against the continent, with a segment of ocean floor (the Great Valley Ophiolite) squeezed in between. This Late Jurassic event approximately coincided with eastward folding and thrusting from the Sierra Nevada uplift. These tectonic events of Jurassic age are collectively known as the **Nevadan orogeny**, although related orogenic activity that is not generally assigned this name continued well into the Cretaceous Period.

The eastward thrusting and folding of Late Jurassic time greatly altered patterns of deposition as far east as Colorado and Wyoming. The Sundance Sea spread over a large area of the western United States, representing the most extensive marine incursion since late Paleozoic time (Figure 15-41). In latest Jurassic time, however, the folding and thrust faulting that extended over Nevada, Utah, and Idaho produced a large mountain chain. The elevation of the land and the shedding of clastics eastward from the mountains drove back the waters of the Sundance Sea, leaving only a small inland sea to the north (Figure 15-44).

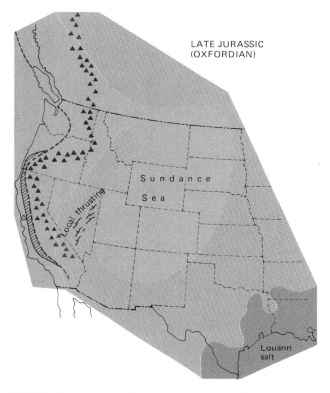

FIGURE 15-41   Geologic features of western North America during Late Jurassic (Oxfordian) time, when the Sundance Sea flooded a large interior region from southern Canada to northern Arizona and New Mexico. In mid-Triassic time, the western margin of the continent had ridden up against a subduction zone. As a result, during the Jurassic Period, a belt of igneous activity extended for hundreds of kilometers parallel to the Pacific Coast. At this time, thrust faulting was largely limited to the state of Nevada.

What remained in Colorado, Wyoming, and adjacent regions was a nonmarine foredeep in which molasse deposits accumulated. Apparently, on the gentle profile of the foredeep even the lowest depositional environments were above sea level, since there was no initial deposition of marine flysch. The molasse of the foredeep was deposited in rivers, lakes, and swamps, creating the famous Morrison Formation, which has yielded the world's most spectacular dinosaur faunas (Figure 15-45; see also Figures 15-24 and 15-25). The dinosaur skeletons in this formation are usually disarticulated, but often as many as 50 or 60 individuals are found together in one small area, indicating that these fossils may have accumulated during floods.

The Morrison Formation consists of sandstones and multicolored mudstones deposited over an area of about 1 million square kilometers. Caliche soil deposits indicate that the climate was seasonally dry during at least part of the Morrison depositional interval, while the scarcity of crocodiles, turtles, and fishes suggests that many of the lakes may have been saline at this time. The dinosaurs are found in deposits representing all of the Morrison environments—rivers, lakes, and swamps. This broad environmental distribution suggests that none of the species—not even the huge sauropods (Figure 15-24A)—was adapted specifically for a life of wading in large bodies of water. The Morrison Formation spans the last 10 million years or so of the Jurassic Period and is overlain by the nonmarine Cloverly Formation of Early Cretaceous age, which contains a completely different fauna of dinosaurs, apparently because of major extinctions at the end of the Jurassic Period.

Europe: The Triassic Period

In Europe, as in North America, the Triassic interval was a time in which nonmarine deposition prevailed. It was also a period in which the pattern of marine sedimentation underwent a major change. Early in the Triassic Period, as in Permian time, seas from the north periodically spread into northern Europe. Soon, however, the northern seas withdrew, and marine waters instead encroached on Europe from the south as a result of the major plate movements that were gradually carrying Africa away from Europe. As a consequence of this separation, the Tethys Seaway spread westward across what is now the Mediterranean region. As the sea expanded here, it was rising throughout the world (Figure 15-30), and thus it flooded southern Europe (Figure 15-46). To view these events in a broader geographic perspective, keep in mind that the Atlantic Ocean had not yet opened; Europe, Greenland, and North America remained connected (Figure 15-33). Given this perspective, we will view the effects of these events on deposition in northern and central Germany, Great Britain, and the Tethyan region of Europe.

Northern and central Germany   Although Europe as a whole was not extensively flooded during the Triassic Period, the seas did encroach on northern Germany and adjoining areas, and the shift of influence from the northern seas to the Tethys Seaway can be seen there. Even in this broad depositional basin, however, most intervals of Triassic time were characterized either by nonmarine deposition or by the absence of deposition. The Triassic deposits of Europe indicate that hot, dry conditions persisted

FIGURE 15-42   The Sierra Nevada Mountains at Yosemite National Park, California. Many of the granitic rocks that formed the Sierra Nevada were emplaced during the Jurassic Period. (*U.S. Geological Survey.*)

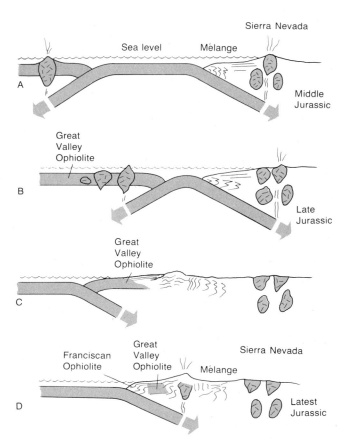

here from Permian times, just as they did in western North America.

The sediments of the Germanic basin of north-central Europe reveal three divisions: a marine sequence, the Muschelkalk (or "mussel limestone"), sandwiched between two predominantly nonmarine sequences, the Bunter deposits below it, and the Keuper deposits above. This is the threefold ("tri-") division that gave the Triassic Period its name.

In northern Germany, the lower part of the Bunter sequence is a continuation of the pattern of deposition of the Permian Zechstein sequence (Figure 14-58). During Early Triassic time, under hot, dry conditions, evaporites accumulated here from marine waters that encroached from the

FIGURE 15-43   Diagram of the inferred history of the Sierra Nevada during the Jurassic Period. During Middle and early Late Jurassic time (*A* and *B*), there seem to have been two opposing subduction zones. Later (*C*), the offshore subduction zone apparently ceased to be active, and its igneous products and neighboring sea floor were attached to the continent as the Great Valley Ophiolite. Finally, near the end of the Jurassic Period (*D*), the Franciscan Ophiolite, a mélange of deep-water sediments that formed along the subduction trench, was crumpled against the Great Valley sequence. (*After R. A. Schweichert and D. S. Cowan, Geol. Soc. Amer. Bull. 86:1329–1326, 1975.*)

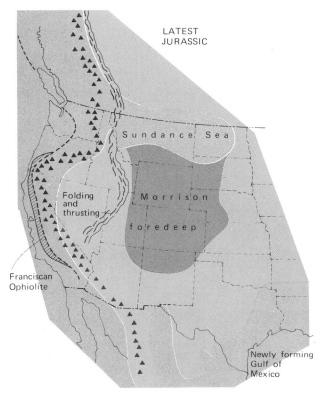

FIGURE 15-44   Geologic features of western North America during latest Jurassic time. A fold-and-thrust belt now extended for hundreds of kilometers roughly parallel to the coastline but far inland. Tectonic activity had driven the Sundance Sea from the western interior, leaving a foredeep where the nonmarine Morrison Formation accumulated.

north. Elsewhere in central Europe, nonmarine siliciclastics, including channel-filling conglomerates, are the dominant Bunter sediments (Figure 15-47).

Near the end of Early Triassic time, the pattern of marine encroachment shifted. A marine incursion now spread northward from the expanding Tethyan Seaway to form the

FIGURE 15-45   The Morrison Formation, which is well known for its large fauna of fossil dinosaurs. *Above.* Excavation of dinosaur bones from the Morrison. *Below.* A vista of Morrison outcrops near Dinosaur National Monument, Utah. (*U.S. Geological Survey; National Park Service.*)

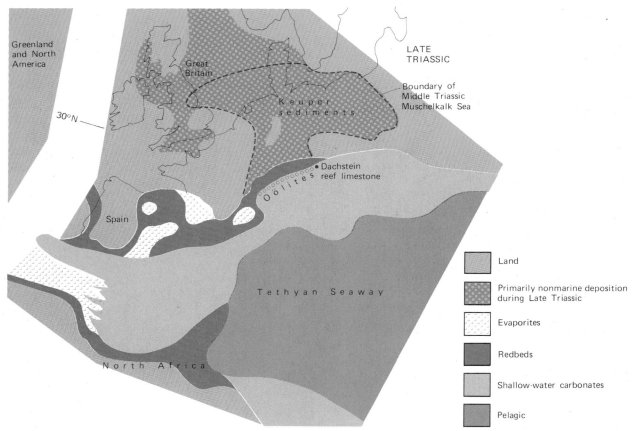

FIGURE 15-46   Depositional patterns of Europe and the Mediterranean area late in the Triassic Period. As rifting began, the Tethyan Seaway spread westward across Pangaea between Spain and North Africa. Shallow-water limestone deposition bordered the central deep-sea portion of the seaway, and in many areas, red beds and evaporites were deposited along the land. The nonmarine Keuper sediment accumulated over a larger region than had the marine Muschelkalk of Middle Triassic age.

shallow Muschelkalk Sea (Figure 15-46). Even though this sea spread from the south, it occupied the same broad basinal area that had been invaded by the Permian Zechstein Sea during its maximum expansion from the north (Figure 14-58).

The Muschelkalk deposits include limestones, evaporites, and marginal sandstones. The marine fauna consists of only a few species, but these were present in great numbers—a pattern suggesting that the shallow Muschelkalk Sea was characterized by unusual salinities (page 55). These waters may well have been brackish at times, but the occasional occurrence of evaporites suggests that at other times they were hypersaline. The remains of placodont reptiles and fish-eating nothosaurs are well preserved in the Muschelkalk (Figures 15-12 and 15-13), indicating that these predators ranged near the shore. At one time, a forest of crinoids populated much of the Muschelk-

FIGURE 15-47   Channel-filling conglomerates of the Bunter (Lower Triassic) of Germany. These deposits were laid down by rivers at a time when nonmarine deposition prevailed in Germany. (E. Backhaus, Guidebook, VIII Internat. Sedim. Congress, pp. 105–124, 1971.)

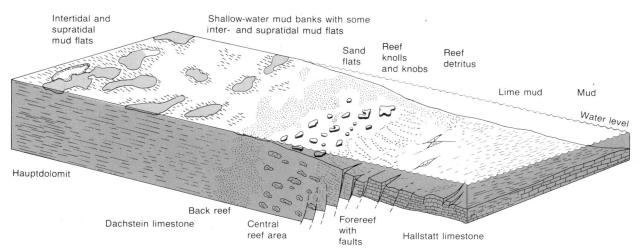

FIGURE 15-48   Reconstruction of reefs that were formed by a community of corals, algae, sponges, and bryozoans in southern Germany late in the Triassic Period. The location of the Dachstein Limestone, in which these reefs are found, is shown in Figure 15-46. The cyclical back-reef deposits are illustrated in Figure 15-49. The block faulting in the forereef was associated with the movement of Africa relative to Eurasia. Pelagic sediments accumulated in fairly deep water in front of the reef, and, as shown in Figure 15-50, some of these settled into fissures produced by the faulting. (*After H. Zankl, Guidebook, VIII Internat. Sedim. Congress, pp. 147–185, 1971.*)

alk sea floor. The transition from the late Middle Triassic Muschelkalk to the Late Triassic Keuper reflects a reduction of marine influence in the central European basin. Marine sediments and fossils form part of the Keuper, but they are often associated with sediments that yield land plants together with bones of terrestrial vertebrates.

Great Britain   Throughout the Triassic Period, Great Britain remained largely inland from marine incursions of Tethyan waters. In Late Permian time and early in the Triassic Period, a nonmarine sedimentary unit known as the New Red Sandstone accumulated across a large portion of the British Isles. (The adjective *new* was added to this term in the nineteenth century to distinguish the red beds of this unit from "Old Red" deposits of middle Paleozoic age; see page 385.) The New Red Sandstone includes spectacular dune deposits that indicate the presence of a warm, arid climate (Figures 3-14 and 14-37). Triassic deposits in England also include evaporites, but there are river deposits as well; thus, it appears that climatic conditions alternated between moist and dry. The only Triassic marine deposition in Britain appears to have resulted from a single, localized transgression of the Middle Triassic Muschelkalk Sea—an event that left telltale marine microfossils in central England (Figure 15-46).

The Tethys   To the south, as we have seen, the Tethyan Seaway began to spread westward as Africa split away from Europe. The records of this progression are extensive. Along the northern margin of the Tethys, the development of a carbonate platform resembling the modern Bahama Banks (Figure 4-22) but much longer, followed an Early Triassic interval of clastic deposition. By Middle Triassic time, there were many organic reefs, which were built both by the newly evolved hexacorals and by organisms such as algae, sponges, and bryozoans. These reefs are now magnificently exposed in the Alps. Figure 15-48 illustrates Late Triassic reef growth in southern Germany. The Dachstein reef limestone depicted here reaches thicknesses of 1.2 kilometers (~¾ mile). The individual reefs within the Dachstein, however, were not large; they were patch reefs separated by broad stretches of bedded sediment. In Late Triassic time, the newly evolved hexacorals had not yet diversified to a significant extent (Figure 15-3) and had not developed the ability to form elongate barrier reefs like those of the modern world (page 93). Behind the belt of reef growth in southern Germany were subtidal, intertidal, and supratidal environments represented by cyclic deposits that reflect changes in the relative positions of land and sea (Figure 15-49).

Fine-grained limestones accumulated in front of these reefs. Among the fossils in these rocks are those of pelagic organisms, including ammonoids, conodont animals, and single-celled calcareous algae. Block faulting disturbed the forereef area (Figure 15-48), and ammonoids settled along with fine-grained pelagic sediment in some of the large

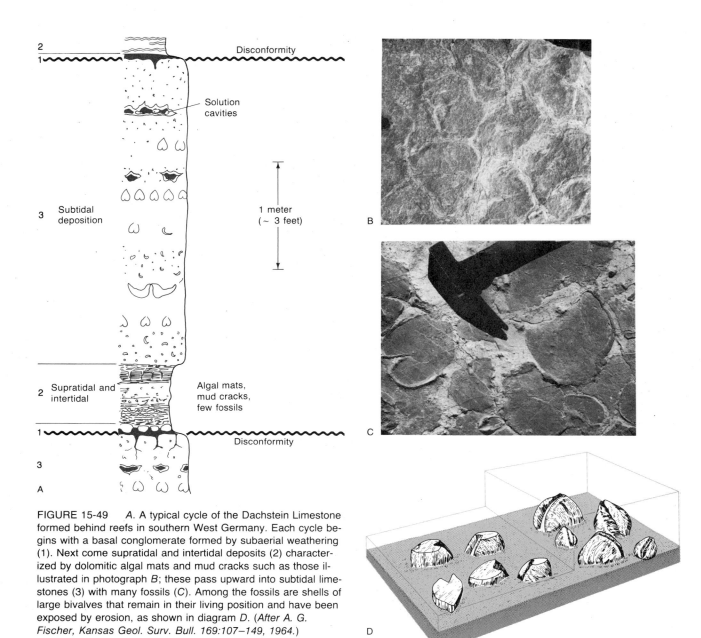

FIGURE 15-49   A. A typical cycle of the Dachstein Limestone formed behind reefs in southern West Germany. Each cycle begins with a basal conglomerate formed by subaerial weathering (1). Next come supratidal and intertidal deposits (2) characterized by dolomitic algal mats and mud cracks such as those illustrated in photograph B; these pass upward into subtidal limestones (3) with many fossils (C). Among the fossils are shells of large bivalves that remain in their living position and have been exposed by erosion, as shown in diagram D. (After A. G. Fischer, Kansas Geol. Surv. Bull. 169:107–149, 1964.)

fissures that formed (Figure 15-50). This faulting, which marked the early stages of Tethyan rifting, became more extensive during Jurassic time as Africa separated entirely from Eurasia. Volcanic activity was associated with the faulting, as is often the case during continental rifting, and it, too, continued into the Jurassic Period.

Europe: The Jurassic Period

It is appropriate to begin a discussion of Jurassic events in Europe with a continuation of the Triassic story for the Tethyan region, because events initiated here in Triassic time persisted into the Jurassic Period and had major effects on the European continent.

The Tethys   Carbonate deposition continued in shallow water around the margins of the Tethys, and in most areas there was no break in deposition during the Triassic-Jurassic transition. Deposits similar to those of the modern Bahamas characterized many of the shallow-water platforms of southern Europe. Many of these deposits, like the

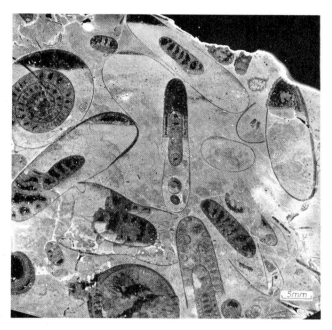

FIGURE 15-50   Fine-grained limestone filling a fissure in the forereef facies of the Dachstein Limestone, southern West Germany. The ammonite shells preserved in the limestone fell into the fissure where the lime mud accumulated. (*H. Zankl, Guidebook, VIII Internat. Sedim. Congress, pp. 147–185, 1971.*)

Triassic reefs, are now impressively exposed at high altitudes, having been elevated tectonically as part of the Alps. Among the limestones are oolites. Figure 15-46 illustrates a spectacular band of oolitic limestones that stretches 150 kilometers (~90 miles). It represents a chain of oolitic shoals that formed parallel to the latest Triassic and earliest Jurassic shoreline.

At the same time, however, block faulting persisted and even increased, reaching its greatest intensity in the latter part of Early Jurassic time. The tectonic pattern of block faulting was complicated by the fact that Africa not only was separating from Eurasia but was also moving eastward with respect to the northern landmass. Thus, some crustal blocks were elevated by compression before dropping downward. Nonetheless, by Early Jurassic time, many areas in which shallow-water carbonates had been accumulating were submerged to sufficient depths to receive pelagic sediments. Carbonate banks all around the margins of the Tethys were now foundering (Figure 15-51A). The region of accumulation of the Dachstein reef limestone (Figure 15-46) subsided unusually early, and superimposed on it are pelagic carbonates that represent the beginning of Early Jurassic time (Figure 15-51B). Associated with some

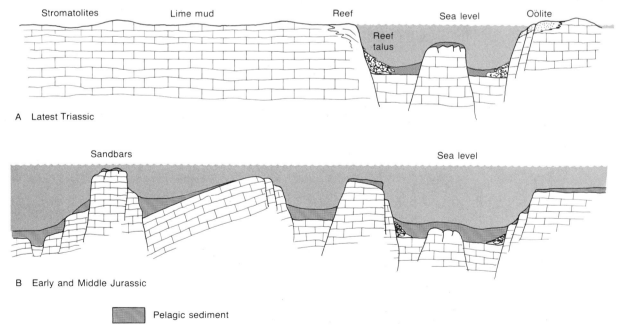

A   Latest Triassic

B   Early and Middle Jurassic

▨ Pelagic sediment

FIGURE 15-51   Diagrammatic illustration of the beginning of block faulting and deep-sea sedimentation in the Tethyan region. *A.* At first, reefs and oolite shoals formed on some of the shallow block margins, and coarse talus was shed into deep water along some fault scarps, grading into pelagic carbonates. *B.* By Jurassic time, most of the fault blocks had foundered to depths so great that pelagic sedimentation prevailed. This sequence of events represented the first step in the origin of the Mediterranean Sea. (*After D. Bernoulli and H. C. Jenkyns, Soc. Econ. Paleont. and Mineral. Spec. Publ. No. 19:129–160, 1974.*)

pelagic carbonates of the Mediterranean region are **radiolarites**—rocks that consist of alternating layers of chert and red shale (Figure 15-52). The chert is formed of densely packed, partly recrystallized skeletons of radiolarians, from which the rocks' name is derived. There is some disagreement as to the depths at which the pelagic limestones and radiolarites formed, but certainly these sediments accumulated on sea floors that were more than a few meters deep, and it is possible that the sea floors were hundreds of meters below sea level.

Central and northern Europe   Unlike the areas bordering the Tethys, central and northern Europe experienced major changes in sedimentary patterns during the Triassic-Jurassic transition. Most important was that the prevalence of nonmarine deposition ended with a transgression that flooded much of western Europe before the end of Early

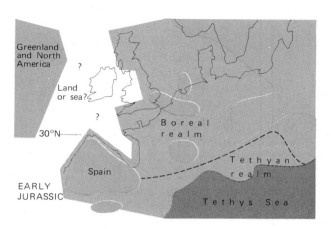

FIGURE 15-53   Paleogeography of Europe during late Early Jurassic (Pliensbachian) time, showing the effect of the transgression that brought marine conditions to a large area. By this time, the Boreal realm was differentiated from the tropical Tethyan realm.

FIGURE 15-52   An Upper Jurassic radiolarite in northern Italy. The light-colored beds are chert formed of radiolarians, and the darker ones are red shale. These beds formed on the Tethyan sea floor in the rift zone between Eurasia and Africa. (*Courtesy of R. L. Folk and E. F. McBride.*)

Jurassic time (Figure 15-53). The marine faunas that occupied Europe as a result of this transgression were separated into the Boreal and Tethyan biogeographic realms, which persisted, with some migration, throughout the remainder of the Jurassic interval (Figure 15-33). At all times, there remained some highlands, most of which were massifs that had been produced during the Hercynian orogeny (Figure 14-42).

In central Europe, the Jurassic System has three informal divisions, which correspond in an approximate way to the Lower, Middle, and Upper Jurassic. These divisions, from bottom to top, are often referred to as the black, brown, and white Jurassic.

The black Jurassic, or Liassic, consists of black and blue-gray rocks. The most famous of these belong to the Posidonia Shale of West Germany, notable for its spectacularly well preserved vertebrate fossils (Figures 15-10, 15-11, 15-15, and 15-16). These vertebrate fossils underwent little decay and escaped dismemberment by scavengers, suggesting that the sea floor was starved of oxygen at the time they were preserved. Such conditions prevailed at the same time in Yorkshire, England, where organic-rich black muds accumulated as the Jet Rock.

The brown Jurassic, or Dogger, corresponds to the Middle Jurassic System in central Europe and consists of iron-rich sandstones and limestones. In England, as in central Europe, a prevalence of shaly deposits in the Lower Jurassic System gives way to a greater variety of sediment types in the Middle Jurassic. Limestone deposition pre-

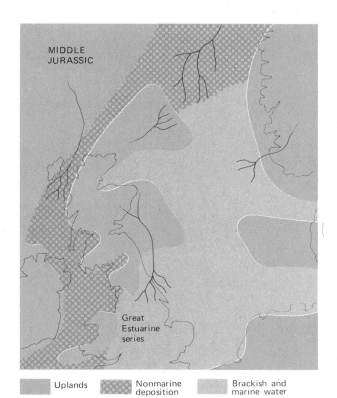

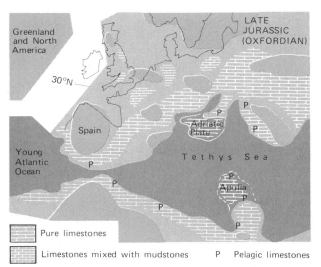

FIGURE 15-55   Late Jurassic (Oxfordian) paleogeography of Europe and the Mediterranean region. Limestones were deposited over broad areas during this interval of maximum transgression of Jurassic seas. Fine-grained pelagic limestones accumulated in areas of deep water adjacent to the Tethys. (*Modified from A. Hallam, Palaeontology 12:1–18, 1969.*)

Uplands Nonmarine deposition Brackish and marine water

FIGURE 15-54   Middle Jurassic paleogeography of Great Britain and the North Sea. Rivers drained uplands in the north, depositing sediment on deltas and in the shallow brackish and marine waters. Thus, the siliciclastic Great Estuarine series accumulated over a large area of central Britain. (*After P. Kent, Jour. Geol. Soc. London. 131:435–468, 1975.*)

FIGURE 15-56   Cliffs of Late Jurassic Portland Limestone capped by the younger Purbeck Beds on the Isle of Portland along the southern coast of England. Faunas of the Portland comprise Tethyan species that appear to have lived under tropical conditions. (*Institute of Geological Sciences.*)

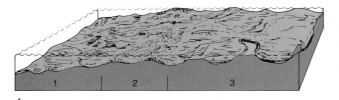

A

B

FIGURE 15-57   Unusual preservation in the fine-grained Solnhöfen Limestone of southern West Germany. *A*. A reconstruction of the sea floor on which the Solnhöfen formed: (1) living
sponge reefs well below the sea surface; (2) shallower living
reefs; (3) uplifted dead sponge reefs in the lagoon behind the
living reefs, with the Solnhöfen sediments accumulating in the
basins between dead reefs. *B*. Bedding plane in the Solnhöfen
showing how an ammonite settled in an upright position, leaving
a mark on the sediment, and then keeled over on its side. *C*.
Final tracks of a horseshoe crab, *Mesolimulus* (*arrow*), and
scratch marks made by the dead or dying animal as it slid over
the sediment. (*A. K. W. Barthel, News Jahrb. Geol. Paläont.
135:1–18, 1970. C. H. Leich, Aufschluss 1:5–7, 1965.*)

vailed in the south of England near the warm Tethys, producing the deposits known as the Inferior and Great Oolite
Series. To the north, great thicknesses of deltaic and
lagoonal siliciclastic deposits accumulated as the Great
Estuarine Series. These predominantly brackish-water
shoreline deposits formed along the southeastern border
of a vast upland area that now occupies eastern Greenland
and North America (Figure 15-54).

The white Upper Jurassic, or Malm, formed in central
Europe at a time when limestone deposition was taking
place sporadically as far north as northern England (Figure
15-55). It prevailed in southern England, where reef corals
and other tropical organisms lived just as they did in southern Eurasia along the margin of the Tethys (Figure 15-56).
In southern Germany, the Solnhöfen Limestone accumulated and preserved beautiful fossils, including the famous
bird *Archaeopteryx*, some rare flying reptiles that must have
fallen into the water when they were dying, and the remains
of creatures that appear to be ancestors of modern crabs
(Figures 15-5, 15-27, and 15-28).

The Solnhöfen is an extremely fine grained limestone
that is partially composed of coccoliths. The story of its
accumulation begins with the formation of sponge reefs.
During Late Jurassic time, sponge reefs that had grown in

moderately deep waters in Bavaria and in neighboring areas
were subjected to tectonic tilting. The northernmost of these
reefs began to die as they were elevated to positions above
sea level or to shallow depths where they could not flourish.
Coral reefs then colonized a band of dead sponge reefs in
shallow water, leaving a quiet lagoon in the nearshore area
behind them—and it was in this lagoon that the Solnhöfen
deposits accumulated (Figure 15-57). At one time, it was
widely believed that the Solnhöfen had formed intertidally
and that the animals fossilized in it had died after being
stranded as the sea retreated. Evidence against this lies
in the absence of typical intertidal structures such as tidal
channels, mud cracks, and bird's-eye holes (Figure 4-29).
Moreover, the Solnhöfen fossils represent species that are
more likely to have lived in a subtidal environment. The
ammonoid shown in Figure 15-57, for example, obviously
settled from standing water before falling sideways to its
final resting place; it was not washed up onto a tidal flat.
It appears that the bottom waters of the Solnhöfen lagoon
were stagnant and anoxic, at least in some places and at
some times. The only trails left by living benthic animals
belong to the horseshoe crab *Mesolimulus,* but these markings are not characteristic of the trails left by healthy crabs.
The animal shown in Figure 15-57 died in its tracks.

CHAPTER SUMMARY

1.   Triassic and Jurassic seas lacked important Paleozoic groups such as fusulinid foraminifers, rugose corals, and trilobites. Important groups of marine life that were present during this time included bivalve gastropods, ammonoid mollusks, brachiopods, sea urchins, hexacorals, bony fishes, sharks, and swimming reptiles.

2.   Gymnosperms were the dominant group of plants in Triassic and Jurassic landscapes.

3.   Although mammals originated in Triassic time, dinosaurs were much more successful than mammals during the Mesozoic Era. Flying reptiles evolved in the Triassic Period, and birds emerged in Jurassic time.

4.   Mass extinctions that were less severe than the terminal Permian event marked the ends of the Triassic and Jurassic periods.

5.   Very early in the Triassic Period, nearly all of the earth's continental crust was consolidated in the supercontinent Pangaea, and even at the end of Jurassic time, all of the continents remained close together. Evaporites mark zones where Pangaea began to rift apart early in the Mesozoic Era.

6.   Fault-block basins that formed in eastern North America received thick deposits of sediment during the rifting episode that eventually formed the Atlantic Ocean between this continent and Africa.

7.   The Sundance Sea invaded western North America during the Jurassic Period but was expelled by uplifting and sediment influx associated with the Nevadan orogeny. Dinosaur fossils are abundantly preserved in the molasse deposits that were then deposited in the vicinity of Utah.

8.   The Triassic and Jurassic systems are well developed in Germany, where both were named, as well as in nearby areas. Along the Tethys Sea, which flanked southern Europe, hexacorals formed reefs in both Triassic and Jurassic time. Simultaneously, faulting occurred as Africa was rifted away from Eurasia, leaving a deep-sea basin in between.

EXERCISES

1.   How did the marine life of Late Jurassic time differ from that of the earliest Triassic?

2.   How did reefs formed by hexacorals during the Jurassic Period differ from those formed during Triassic time?

3.   What observations suggest that dinosaurs had relatively advanced behavior of the sort that might explain their evolutionary success?

4.   Which two deposits in Germany have yielded many of the exceptionally well preserved vertebrate fossils shown in this chapter? Why?

5.   In what areas is there evidence that large oceans started to form in Triassic and Jurassic time? What is that evidence?

6.   What plate-tectonic change might explain the eastward migration of the zone of igneous activity in western North America during the Jurassic Period?

ADDITIONAL READING

Bakker, R. T., "Dinosaur Renaissance," *Scientific American,* April 1975.

Hallam, A., *Jurassic Environments,* Cambridge University Press, New York, 1975.

Kurtén, B., *The Age of Dinosaurs,* McGraw-Hill Book Company, New York, 1968.

CHAPTER 16
The Cretaceous World

Throughout most of Cretaceous time, the seas stood much higher than they do today. Widespread shallow marine deposits on continental surfaces, together with nonmarine and deep-sea sediments, reveal that the Cretaceous Period was in many ways an interval of transition. Some Cretaceous sediments are lithified, like nearly all those of older systems; many others, however, consist of soft muds and sands, like most deposits of the younger Cenozoic Era. Fossil biotas of the Cretaceous Period also display a mixture of archaic and modern features. They include members of important extinct taxa, such as dinosaurs and ammonoids (groups that failed to survive the Cretaceous Period), as well as important modern taxa, such as flowering plants and the subclass of fishes that is the most diverse in the world today. It was during the Cretaceous Period that continents moved toward their modern configuration. At the start of the period, the continents were tightly clustered, and Gondwanaland was prominent in the south—but by the end of Cretaceous time the Atlantic Ocean had widened, and Gondwanaland had fragmented into most of its daughter continents.

The Cretaceous System was first described formally in 1822 by the Frenchman d'Omalius D'Halloy. For many years before that, however, Cretaceous rocks had been recognized as constituting a stratigraphic interval distinct from the Jurassic rocks below and the sediments above, now labeled Cenozoic. The name *Cretaceous* derives from *creta*, the Latin word for chalk, which is a soft, fine-grained kind of limestone that accumulated over broad areas of the Late Cretaceous sea floor. This remarkably white rock is composed predominantly of the armorlike plates of planktonic coccolithophores, which flourished during Late Cretaceous time to a greater degree than ever before or since (Figure 2-31).

LIFE

Life of the Cretaceous Period, both in the seas and on the land, was a curious mixture of modern and archaic forms. In the marine realm, for example, strikingly modern types of bivalve and gastropod mollusks populated Late Cretaceous seas along with enormous coiled oysters and other now-extinct sedentary bivalves. Diverse fishes of the

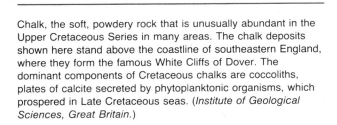

Chalk, the soft, powdery rock that is unusually abundant in the Upper Cretaceous Series in many areas. The chalk deposits shown here stand above the coastline of southeastern England, where they form the famous White Cliffs of Dover. The dominant components of Cretaceous chalks are coccoliths, plates of calcite secreted by phytoplanktonic organisms, which prospered in Late Cretaceous seas. (*Institute of Geological Sciences, Great Britain.*)

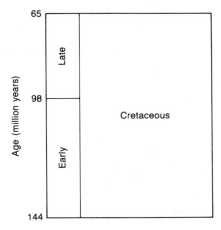

modern kind occupied the same waters as a variety of ammonoids, belemnoids, and reptilian sea monsters—none of which have any close living relatives. On the land, floras changed from the Mesozoic type, which were dominated by gymnosperms, to the modern type, in which flowering plants predominate. Many groups of vertebrate animals that are still extant also evolved at this time: snakes and modern types of turtles, lizards, crocodiles, and salamanders. Dinosaurs, however, continued to rule the terrestrial ecosystem. Of all the modern groups of terrestrial vertebrates present during Cretaceous time, only the crocodiles approached the dinosaurs in bodily proportions. Mammals, in particular, remained very small by modern standards.

Pelagic life in the oceans

As a result of the appearance of new groups of single-celled organisms, marine plankton had acquired a thor-oughly modern character by the end of Cretaceous time. The primary change among the phytoplankton was the evolutionary expansion of the diatoms, solitary cells that lived within siliceous structures resembling pillboxes (Figure 16-1). Diatoms may have existed during the Jurassic Period, but they did not radiate extensively until mid-Cretaceous time. Together with dinoflagellates (Figure 15-6) and, in warm seas, coccolithophores (Figure 15-7), diatoms must have accounted for most of the photosynthesis that occurred in the pelagic realm of Cretaceous seas (page 53). Recall that today diatoms are the dominant contributors to the siliceous oozes of the deep sea (page 109), and their accumulation in deep-sea sediment was well under way before the end of the Cretaceous Period.

Higher in the pelagic food web, the modern planktonic foraminifers diversified greatly for the first time. This group, known as the **globigerinaceans**, has a meager fossil record in Jurassic rocks; not until the upper part of the Lower

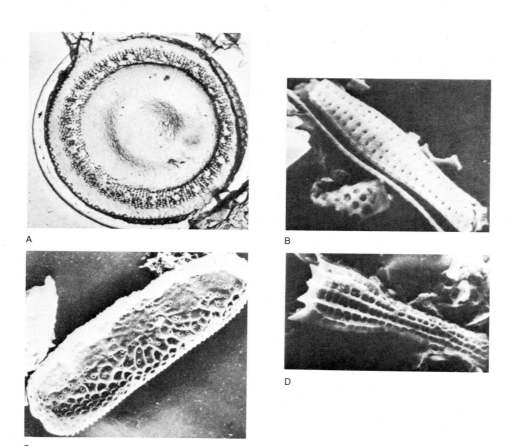

A

B

C

D

FIGURE 16-1   Cretaceous diatom tests (skeletons) from South Pacific deep-sea cores. These skeletal structures are composed of hydrated silica like that which forms the semiprecious stone opal. They belonged to single-celled members of the plankton and fell to the sea floor when their inhabitants died. A. *Pseudopodosira* (×3200). B. *Pterotheca* (×2080). C. *Xanthiopyxis* (×3300). D. *Sceptroneis* (×2990). (*Courtesy of M. Hajoś.*)

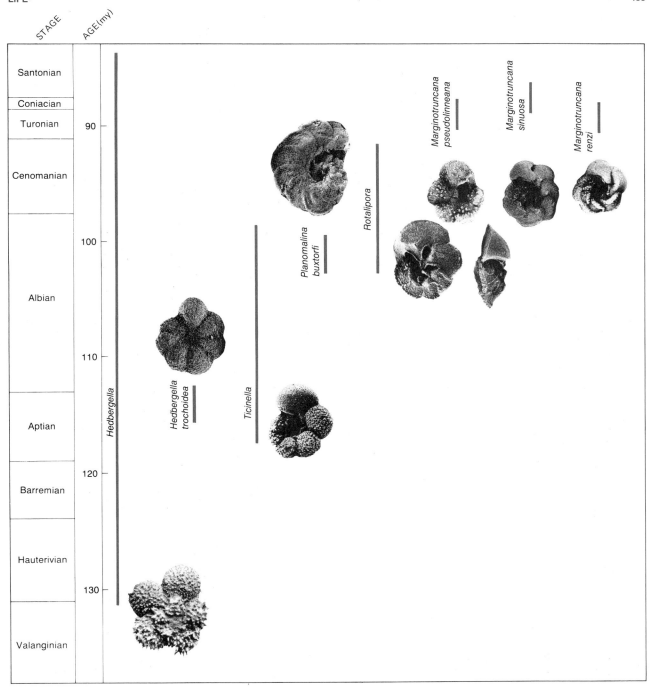

FIGURE 16-2   Early planktonic foraminifers (Globigerinacea). Many individual species lived for relatively short geologic intervals and are useful index fossils in offshore deposits (Creta-ceous stages are shown at the left). These species average about $\frac{1}{2}$ millimeter ($\frac{1}{50}$ inch) in diameter. (*After A. Boersma.*)

Cretaceous System is it well enough represented to be of great value in biostratigraphy (Figure 16-2).

Late Cretaceous adaptive radiations of two of the single-celled planktonic groups altered depositional patterns in the pelagic realm: Since mid-Cretaceous time, both the globigerinid foraminifers and the coccolithophores have contributed vast quantities of calcareous sediment in oceanic areas, whereas before about 100 million years ago, little or no calcareous ooze was present on the deep-sea floor.

During Late Cretaceous time, the coccolithophores blossomed in warm seas to such an extent that cocco-liths—the small plates that armored their cells—accumu-

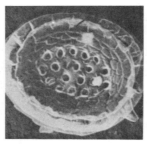

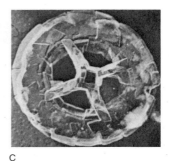

A   B   C

FIGURE 16-3   A piece of Cretaceous chalk (*A*) and coccoliths like those of which it is composed (*B* and *C*). The coccoliths represent the genera *Cribrosphaerella* (*B*, ×9750) and *Deflan-* *drins* (*C*, ×7000). (*A. Institute of Geological Sciences, Great Britain. B and C. Courtesy of M. Black.*)

lated in huge volumes as fine-grained limestone commonly known as chalk (Figure 16-3). Cretaceous chalk is, in fact, widely used for writing on blackboards. The most famous chalk deposits in the world crop out along the southeastern coast of England, where they are known formally as the Chalk (page 482). Similar but less massive chalks formed in shallow seas in Kansas and nearby regions and along the Gulf Coast of the United States. Why coccolithophores were so productive in the Cretaceous Period is uncertain, but never before or since have they yielded chalk deposits of such great extent or thickness.

Still higher in the pelagic food web of Late Cretaceous time, the ammonoids and belemnoids persisted as major swimming carnivores (Figure 16-4). The ammonoids serve as valuable index fossils for the Cretaceous System, just as they do for the Triassic and Jurassic. Among the Cretaceous ammonoids were many species with straight, cone-shaped shells and others with strangely coiled shells (Figure 16-5).

New on the scene in Cretaceous time were the **teleost fishes**, a subclass that is today the dominant group of marine and freshwater fishes. Teleosts are characterized by such features as symmetrical tails, round scales, specialized fins, and short jaws that are often adapted to take particular kinds of food. By Late Cretaceous time there were already a wide variety of teleosts, including the biggest species known from the fossil record (Figure 16-6). This group also included close relatives of the modern sunfish, carp, and eel as well as members of the families that include salmon, pompanos, and the vicious South American piranha. Similarly, Cretaceous sharks resembled present-day forms; in fact, all of the living families of sharks had evolved by the end of the Cretaceous Period.

Most of the top carnivores of Cretaceous pelagic habitats, however, were not at all modern. Whereas whales of

FIGURE 16-4   Reconstruction of the Late Cretaceous shallow sea represented by fossil faunas at Coon Creek, Tennessee. A large coiled ammonoid rests on the sea floor, surrounded by numerous swimming and feeding members of the straight-shelled ammonoid genus *Baculites* (see also Figure 16-5). (*Field Museum of Natural History.*)

one kind or another have occupied the "top carnivore" adaptive zone during most of the Cenozoic Era, reptiles were the largest marine carnivores until the end of Cretaceous time. Ichthyosaurs and marine crocodiles were rare by this time, but plesiosaurs still thrived, some exceeding 10 meters (~35 feet) in length. Plesiosaurs are depicted in Figure 16-7 along with other members of the Late Cretaceous pelagic community of the western interior of the United States. Huge marine lizards known as

FIGURE 16-5   Cretaceous ammonoids. *A. Neogastroplites,* a form with heavy nodes on its shell, in the approximate life position (rear and side views, ×2). *B.* Similar views of *Tragodesmoceras* (×1). *C. Nostoceras,* a strangely coiled form (×1).

D. Baculites, a straight-shelled form, shown in the reconstruction depicted in Figure 16-4. The specimen on the left is an internal filling that shows the complex folded septa, or partitions, within the shell (×1). (*U.S. Geological Survey.*)

FIGURE 16-6   *Xiphactinus,* a Cretaceous fish that, at about 5 meters (~16 feet) in length, is the largest known teleost. A careful look will reveal that the animal shown here died with a good-sized fish in its belly. A reconstruction of *Xiphactinus* in life can be found in Figure 16-7. (*Smithsonian Institution.*)

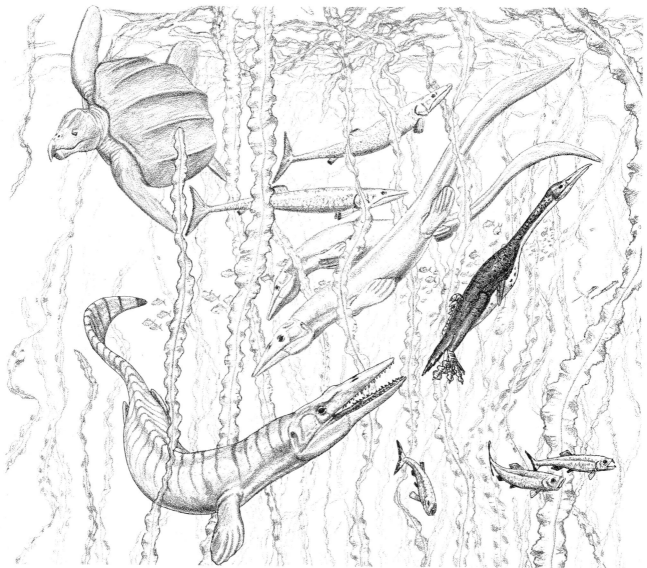

FIGURE 16-7   A reconstruction of marine life preserved in the Upper Cretaceous Pierre Shale of the western interior of the United States. The animals are shown swimming in a bed of kelp, which are algae of large proportions. The giant turtle on the left is *Archelon*, which reached a length of almost 4 meters (~12 feet). The striped animal in the lower left is the mosasaur *Clidastes,* and beyond it is a pair of mosasaurs of the genus *Platecarpus. Clidastes* is in pursuit of the diving bird *Hesperor-*

mosasaurs were probably the most formidable marauders of Cretaceous seas; some grew to be longer than 15 meters (~45 to 50 feet). Although there is direct evidence that mosasaurs attacked ammonoids (Figure 16-8), the reptiles' pointed teeth were not well adapted for shell crushing, and ammonoids probably did not form a major part of their diet.

    Other, less important animals in the marine ecosystem included the flightless diving bird *Hesperornis* and the marine turtles that evolved during the Cretaceous Period (Fig-

ure 16-7). The large feet and small wings of *Hesperornis* (Figure 16-9) were adapted for swimming, and the sharp, backward-directed teeth were adapted for catching slippery fishes. The swimming turtle *Archelon* (Figure 16-10) grew to a length of nearly 4 meters (~12 feet).

Sea-floor life

On the sea floor, life began to take on a modern appear-

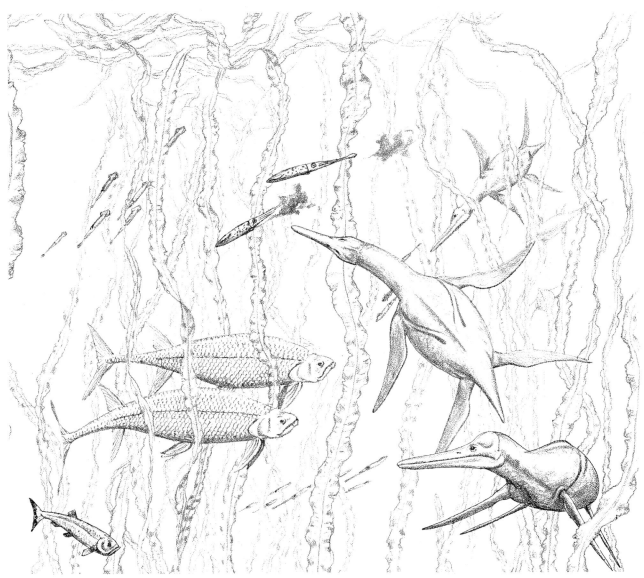

nis. On the right are three plesiosaurs of the genus *Trinacromerum.* Above them and to the left are belemnoids. The teleost fishes in this drawing are *Cimolichthyes* (pikelike pair near the turtle), *Enchodus* (small fishes in the lower center), and *Xiphactinus* (pair of large fishes on the right). (*Drawings by Gregory S. Paul.*)

ance during the Cretaceous Period. One noteworthy feature was the decline of the brachiopods, which had suffered greatly in the mass extinction at the end of the Paleozoic Era but had experienced a moderate expansion again early in Mesozoic time. Certain other groups of sea-floor life that had been successful in the Jurassic Period continued to hold their own during Cretaceous time. Among these were the sea urchins and the hexacorals, which diversified but underwent no startling adaptive changes. Still other major groups retained many of their Jurassic families and genera but also produced new representatives that have survived to the present. Some of these are:

Foraminifers (Figure 16-11). A large percentage of the families of bottom-dwelling foraminifers in existence today appeared during the Cretaceous Period, giving this group a modern aspect.

Bryozoans (Figure 16-12). The most common modern bryozoans are the **cheilostomes**, which typically encrust

FIGURE 16-8   A Late Cretaceous ammonoid that was bitten by a mosasaur. This fossil, which was collected in South Dakota, displays 16 separate bites. (*University of Michigan.*)

marine surfaces, including the hulls of boats, in the form of low-growing mats. Cheilostomes originated in Jurassic time but did not enjoy success until the Late Cretaceous, when they expanded to include more than 100 genera.

Burrowing bivalve mollusks (Figure 16-13). Early Cretaceous burrowing bivalves resemble those of the Jurassic, but by the end of the period new genera had appeared as well, including some that pumped water in and out of their shells through fleshy siphons.

Gastropod mollusks, or snails (Figure 16-14). During the Cretaceous Period, the aptly named **neogastropoda**, or "new snails," produced many modern families and genera. Unlike most earlier snails, these animals are generally carnivorous, feeding on such prey as worms, bivalves, and other snails. Some live in the sediment, while others live on the sediment surface. Many modern seashells popular with collectors belong to neogastropod species.

Crabs. A more or less modern type of crab had evolved during the Jurassic Period, but other, even more modern groups appeared during the Cretaceous Period.

Sea grass. **Sea grass** is not true grass, which evolved during the Cenozoic Era, but is instead a grasslike plant that carpets the sea floor and originated during the Cretaceous Period (Figure 16-15). Today, many kinds of ma-

FIGURE 16-9   The Late Cretaceous diving bird *Hesperornis*. The wings of this flightless animal had been almost entirely lost, but large, paddlelike feet were present, as were teeth for catching fish. *Hesperornis* was about 1.2 meters (~4 feet) long. (*Smithsonian Institution.*)

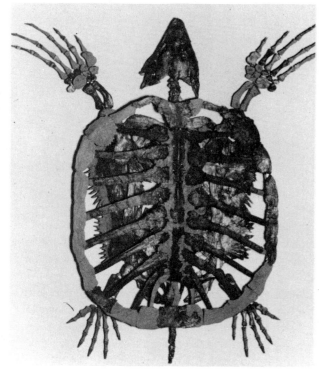

FIGURE 16-10   The Late Cretaceous marine turtle *Archelon*, which reached a length of almost 4 meters (~12 feet). (*Smithsonian Institution.*)

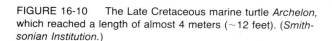

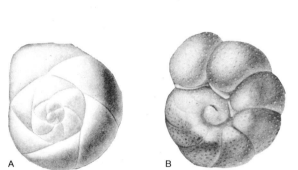
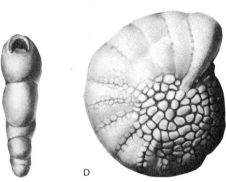
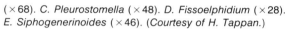

A   B   C   D   E

FIGURE 16-11   New genera of bottom-dwelling foraminifers of Cretaceous age. *A. Alabamina* ($\times$140). *B. Anomalinoides* ($\times$68). *C. Pleurostomella* ($\times$48). *D. Fissoelphidium* ($\times$28). *E. Siphogenerinoides* ($\times$46). (*Courtesy of H. Tappan.*)

rine life live only on or in beds of sea grass. Presumably, this was true in Cretaceous time as well.

Some uniquely Mesozoic groups of considerable importance also colonized Cretaceous sea floors. Among bivalve mollusks living on the surface of the substratum, for example, coiled oysters and other groups that had existed during Jurassic time evolved species of enormous size (Figure 16-16). Of these, the **rudists** were of special significance because they lived like corals, forming large tropical reefs. Of the bivalve groups represented in Figure 16-16, only the small scallops are alive today; the rudists and other large surface-living groups did not survive the Cretaceous Period.

Before their demise, the rudists apparently flourished at the expense of reef-building corals. Shallow-water reefs built in Early Cretaceous time, like those built during the Jurassic Period and in the modern world, were formed primarily by hexacorals and coralline algae. In mid-Cretaceous time, however, rudist bivalves assumed the dominant role in tropical reef growth. The rudists were mollusks with a cone-shaped lower shell and a lidlike upper shell (Figure 16-16C). These curious animals attached

A   B

FIGURE 16-12   Two genera of Cretaceous cheilostome bryozoans: (*A*) *Rhiniopora* ($\times$25) and (*B*) *Onychocella* ($\times$15). Both grew over the substratum as crusts. A colony forming a crust consisted of many interconnected individuals, each of which emerged from a small opening to feed. (*Courtesy of E. Voigt.*)

FIGURE 16-13   Burrowing bivalves of Cretaceous age. Like similar living animals, they obtained food and oxygen from water currents. *A. Scabrotrigonia* (×0.7) lacked siphons for channeling their water currents. *B* and *C. Panope* (×0.7) and *Aphrodina* (×1) possessed fleshy, tubelike siphons; the deflection of the line of muscle attachment within the shell allowed the siphons to be withdrawn. The fossils are all shown in living orientations in the line drawing (*D*). (*Smithsonian Institution.*)

FIGURE 16-14 Cretaceous snails of the carnivorous neogastropod group. The opening of the neogastropod shell is extended to house a long siphon that senses the direction of prey. *A. Morea* (×2). *B. Odontobasis* (×3). *C. Buccinopsis* (×1.5). *D. Ornopsis* (×2). *E. Sargana* (×3). (*U.S. Geological Survey.*)

themselves to hard objects (often other rudists) and grew upward, some reaching heights of more than 1 meter. Nearly all shallow-water reefs of Late Cretaceous age are dominated by rudists (Figure 16-17), who seem to have defeated the corals temporarily in competition for space. Most likely the rudists, like reef-building corals, grew rapidly by feeding on symbiotic algae that lived and multiplied in their tissues (Box 2-1). Only after the end of the Cretaceous Period, when rudists, like dinosaurs, became extinct, did corals and coralline algae prevail on reefs once more.

The rise of modern marine predators

Many of the general changes that occurred in bottom-dwelling marine life during Jurassic and Cretaceous time seem to have been related to the great expansion of modern types of marine predators. Among the new predators were the advanced teleost fishes (Figure 16-6), modern kinds of crabs, and carnivorous gastropods (Figure 16-14). Many

FIGURE 16-15   Living and fossil sea grass. *A*. Sea grass growing in the Mediterranean Sea off the coast of Spain. Currents have scoured out a channel along the margin of the grass bed. *B*. Fossil imprint of a blade of Cretaceous sea grass (×20). This imprint formed on the underside of a colony of bryozoans that encrusted the sea grass. (*A. Courtesy of E. Shinn. B. Courtesy of E. Voigt.*)

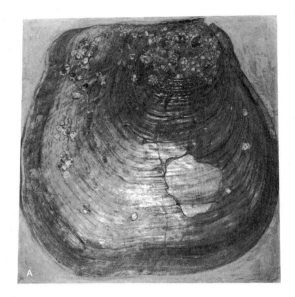

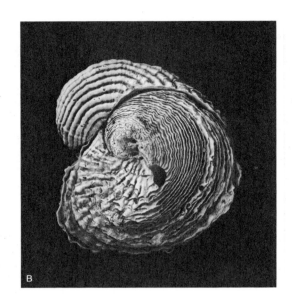

FIGURE 16-16   Cretaceous bivalve mollusks that lived on the surface of sediment. *A. Inoceramus,* a giant animal that was nearly 1 meter (~3 feet) long; some members of this species measuring more than half again this size are found in Kansas. *B. Exogyra,* a coiled oyster (×0.5). *C. Sphaerulites,* a rudist that stood upright on the elongate lower valve; the upper valve was a cup-shaped lid (×0.6). *D. Chlamys,* a scallop (×2). (*A. Sternberg Memorial Museum, Fort Hays State University. B. and D. Smithsonian Institution. C. Photograph by the author.*)

of these new predators were efficient at penetrating shells: fishes by biting, crabs by crushing or peeling with their claws, and some of the gastropods by drilling holes.

The contrast between Paleozoic and Mesozoic predation on the sea floor is exemplified by the absence during Paleozoic time of large arthropods with crushing claws and by the virtual absence in fossilized Paleozoic brachiopods and bivalve shells of holes drilled by predators. The appearance of new groups of shell-crushing reptiles, sharks, and bony fishes in Triassic and Jurassic seas must have represented an important step toward the situation existing today; most of their members, however, were sizable animals that could only have seized large and conspicuous prey. Many of the younger, smaller Mesozoic predator groups—including the teleost fishes, the crabs, and the snails—were better able to attack shelled animals that were small or partially hidden in the substratum.

The decline of the brachiopods and the stalked crinoids, both of which were moderately well represented in early Mesozoic seas, can probably be attributed to the diversification of modern predators. The few species of stalked crinoids that survive today live in deep water; in shallow waters, predation by fishes is probably too severe to permit their existence. Similarly, more species of brachiopods today live in temperate seas than in tropical seas, where predation by crabs, fishes, and snails is severe. By the end of the Mesozoic Era, relatively few immobile species of animals lived on the surface of the sea floor in the mode typical of many groups of Paleozoic brachiopods (Figure 12-16). It would appear that the ability to swim or to burrow actively was the best defense against predation, except for species which had defensive spines or unusually heavy protective shells.

Flowering plants conquer the land

The greatest change in terrestrial ecosystems during the Cretaceous Period was the ascendancy of the **flowering plants (angiosperms)**, although gymnosperm floras resembling those of Triassic and Jurassic age continued to dominate the land well into Cretaceous time. The most conspicuous change during Early Cretaceous time was in the types of gymnosperms that dominated: Conifers became the most numerous species, and the Age of Cycads came to a close. This development, too, was short-lived. About 100 million years ago, midway through the Cretaceous Period, the first angiosperms made their appearance on earth, and during Late Cretaceous time, they came to surpass the conifers in diversity. Today, there are some 200,000 species of flowering plants, including many types of grasses, weeds, wildflowers, and hardwood trees. In contrast, there are only about 550 modern conifer species, although some, including pines, firs, and spruces, are conspicuous in the modern landscape. The success of the flowering plants is one of the most fascinating chapters in the history of life. As we will see, this story includes several important episodes, some of which unfolded during the Cretaceous Period and others which took place during Cenozoic time.

The term *flowering plant* can be misleading, since not all so-called flowering plants have showy flowers; all of them do, however, possess the kinds of reproductive structures that are found within showy flowers. The key repro-

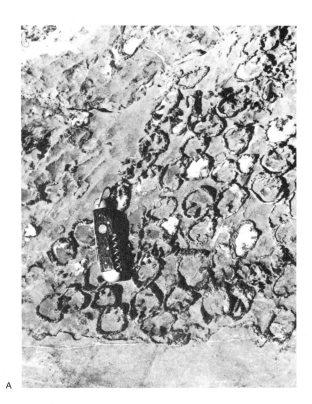

A

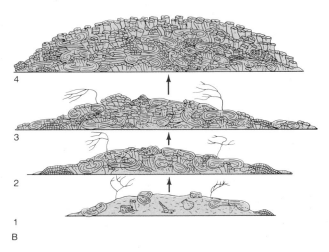

FIGURE 16-17   Rudist reefs. *A.* Erosion surface exposing rudists of the genus *Hippurites* that formed a reef in the Upper Cretaceous Cuantla Formation near San Lucas, Mexico. *B.* Stages in the development of a Cretaceous rudist reef of the Caribbean region. Some of the first rudists to colonize the sea floor were unusual forms that looked like buffalo horns stretched out on the substratum; later, upright cone-shaped rudists grew crowded together, dominating the reef surface. (*A. U.S. Geological Survey. B. After E. G. Kauffman and N. F. Sohl, Verhandl. Naturf. Ges. Basel 84:339–467, 1974.*)

FIGURE 16-18   Fossils of two Cretaceous members of the flowering plant group that are very similar to species of the modern world. The leaf represents *Platanus,* the genus of modern sycamores. The fruit is a fig that may belong to the modern genus *Ficus.* (*Smithsonian Institution.*)

ductive feature that distinguishes angiosperms (flowering plants) from gymnosperms (naked-seed plants) is the enclosure of the seed within a hollow ovary (Box 13-2). This ovary is actually a modified leaf of the spore-bearing plant.

In the early part of this century, many Upper Cretaceous fossil leaves were mistakenly assigned to genera of modern flowering plants. We now know that the resemblance between the leaves of nearly all Cretaceous genera and those of modern taxa is only superficial; all but a few Cretaceous genera belong to extinct groups. Similarly, many living genera and families do not extend as far back in geologic time as was once believed. Of some 500 living families of flowering plants, only about 50 are now believed to be represented by Upper Cretaceous fossils. Some of these surviving Cretaceous groups are quite familiar (Figure 16-18); examples include the sycamore trees (*Platanus*), the hollies, the palm trees, the oak family, the walnut family, and the family that today includes birches and alders. To a modern observer lacking a detailed knowledge of botany, forests of latest Cretaceous time would look relatively familiar. The open, unforested areas, however, would look quite strange; missing altogether would be grasses—the kind of vegetation that now characterizes meadows, prairies, and savannahs.

An early phase of the adaptive radiation of flowering plants is documented by fossils of the Atlantic Coastal Plain in Maryland. Here, within a sedimentary interval representing only about 10 million years of mid-Cretaceous time, both fossil leaves and fossil pollen increase in variety and in complexity of form (Figure 16-19). The early leaves have simple, smooth outlines, and their supporting veins branch in irregular patterns. Later leaves include varieties with many marginal lobes and veins that follow more regular geometric patterns. Probably the more regular patterns gave the leaves greater strength to withstand tearing. The oldest pollen is also of simple form, with only one germination furrow (the structure through which an egg is fertilized). Later forms of pollen show more variation and typically have three germination furrows, making successful germination more likely because of the requirement that a germination furrow be in contact with the female portion of the flower.

Before the adaptive radiation depicted in Figure 16-19 was understood, the impressive variety of fossil angiosperms in Upper Cretaceous sediments was a great puzzle to paleontologists. To Charles Darwin, what appeared to be geologically instantaneous diversification was "an abominable mystery." Assuming that the flowering plants must have gone through a long period of evolutionary di-

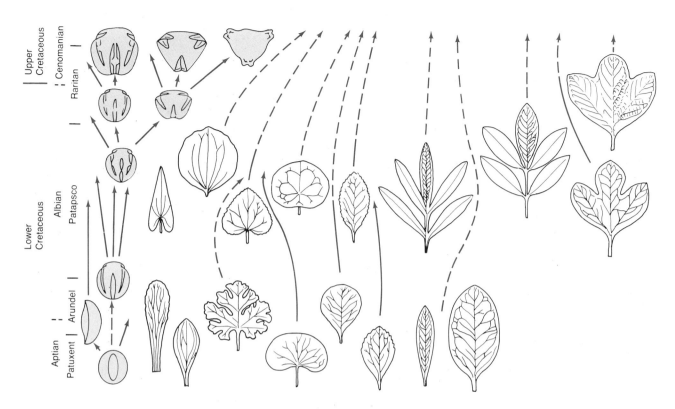

FIGURE 16-19   The pattern of initial adaptive radiation of flowering plants, as illustrated by fossil leaves and pollen in formations (Patuxent through Raritan) of the Cretaceous Potomac Group of Maryland. Both pollen (*left*) and leaves exhibit an in-crease in complexity and variety of form through time. (*After J. A. Doyle and L. J. Hickey, in C. B. Beck [ed.], Origin and Early Evolution of the Angiosperms, Columbia University Press, New York, 1976.*)

versification, specialists searched the early Mesozoic record unsuccessfully for primitive flowering plants. The sequence of leaves and pollen shown in Figure 16-19 indicates that flowering plants did not undergo this initial diversification until the Cretaceous Period and that this process was indeed rapid, but not as sudden as was once believed. By the end of the Cretaceous Period, angiosperms had overtaken gymnosperms as the dominant group of higher plants.

The reasons why angiosperms diversified during the Late Cretaceous while gymnosperms declined are quite evident. One great advantage of the flowering plants is their provision of nutritional food for their seeds. By a process known as double fertilization, one fertilization event produces a seed within the ovary and a second fertilization event, also within the ovary, produces a supply of stored food for the seed: the nutritional part of a kernel of corn or a grain of wheat is an example. The rapid manufacture of this food supply allows for the quick release of a well-fortified seed. Because gymnosperms lack this double-fertilization mechanism, it takes much longer for the parent plants to supply their seeds with enough food to enable the progeny to survive on their own. As a result, most gymnosperms have reproductive cycles of 18 months or more. In contrast, thousands of species of flowering plants grow from a seed and then release seeds of their own in just a few weeks. *Herbaceous* is a term used to describe plants whose stems and foliage grow up from the ground during a single growing season, produce seeds, and then wither to the ground again. Many flowering plants have this property. Most angiosperm plants that are not herbaceous are woody shrubs and trees, but even some species of these are capable of occupying bare ground quickly. Rapid colonization has been one of the secrets of the ecologic success of the flowering plants.

A second reproductive mechanism of flowering plants that has contributed enormously to their success is the flowers' attraction of insects. The insects benefit from the nutritious nectar that the flowers provide, and the flowers benefit because the insects unknowingly carry pollen from one flower to another, fertilizing the plants on which they feed. This attraction is often specialized: A particular kind

FIGURE 16-20   The Late Cretaceous duck-billed dinosaur *Edmontosaurus.* Duck-billed dinosaurs of many species lived together in the American West (*Smithsonian Institution.*)

FIGURE 16-21   The Late Cretaceous horned dinosaur *Triceratops.* This rhinolike genus appears to have been the last of the dinosaurs to disappear from the earth, having survived to the very end of the Cretaceous Period. (*Smithsonian Institution.*)

of insect feeds upon a particular kind of plant, providing a unique mechanism for speciation. If a flower of a new shape, color, or scent develops within a small population of plants, the flower may attract a different kind of insect than the one that visited its ancestors. The plants with the new kind of flower will thus be reproductively isolated from their ancestral species; in other words, the new forms will become a new species. In general, new kinds of insects create opportunities for the development of new species of plants (with new kinds of flowers); and, similarly, new kinds of plants create feeding opportunities for new species of insects. This reciprocity has apparently accelerated rates of speciation in both flowering plants and insects. High rates of speciation have permitted the frequent development of new adaptations. Thus, the symbiotic (mutually beneficial) relationship between flowering plants and insects has played a major role in the great success of both groups since mid-Cretaceous times.

Large dinosaurs and small mammals

Owing to a patchy fossil record, Early Cretaceous vertebrate faunas are not well known, but the available fossils suggest the continued dominance of the dinosaurs. The lowest Cretaceous faunas from the Wealden beds of southern England and Germany include few sauropods (Figure 15-24*A*), but various new kinds of dinosaurs are represented; these appear to be precursors of the Late Cretaceous dinosaurs that are well known from fossil-bearing deposits of Wyoming, Montana, Alberta, and Asia.

In the American West, dinosaurs formed a community that has been compared to the modern mammal faunas of the African plains. Instead of antelopes, zebras, and wildebeests, there were many species of duck-billed dinosaurs (Figure 16-20). These fast-running herbivores probably traveled in herds and may have trumpeted signals to one another by passing air through complex chambers in their skulls (see Box 15-1). In place of rhinoceroses, there were horned dinosaurs with beaks and teeth for cutting harsh vegetation (Figure 16-21). Sharing the Late Cretaceous plains with these herbivores were fearsome predators, including the largest carnivorous land animals of all time, *Albertosaurus* (Figure 16-22) and *Tyrannosaurus.* Here, too, were terrestrial crocodiles (Figure 16-23) that grew to the remarkable length of 15 meters (~45 to 50 feet). Huge predatory dinosaurs of all types were typically represented by far fewer species and individuals than the herbivorous dinosaurs; presumably, these predators played an ecologic role resembling that of modern African lions and hyenas, which are also much less numerous than their hoofed prey.

A general evolutionary trend toward large body size is evident in the Upper Cretaceous fossil record of the American West: Not only among the carnivorous group but also among the duckbills and horned dinosaurs, the species of largest body size were some of the last to evolve. Along with these giants, however, smaller species remained, so

FIGURE 16-22 *Albertosaurus,* a carnivorous dinosaur that lived during Late Cretaceous time. This animal, which stood about 4.2 meters (~14 feet) tall, was typical of large carnivorous dinosaurs in that its small forelimbs apparently were not used for grasping prey; instead, the enormous jaws must have served this function. (*Field Museum of Natural History.*)

casses, to which their size would have been appropriately scaled.

Living a less conspicuous existence on the ground were several groups of vertebrate animals with a future: amphibians (frogs and salamanders), reptiles (snakes, lizards, and turtles), and mammals. Frogs and salamanders, you will recall, had been present since the Jurassic Period, and by the end of the Cretaceous some of their modern families had evolved. Lizards and turtles were present even earlier in the Mesozoic Era, and by the time the Cretaceous Period drew to a close, many of their modern families had come into being as well.

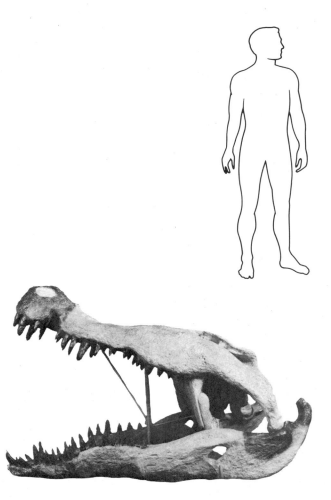

FIGURE 16-23 Head of a huge terrestrial crocodile, *Phobosuchus,* which probably fed on Late Cretaceous dinosaurs of small and medium size. The length of the head equaled the height of a large man, as indicated by the outline, and the entire crocodile measured about 15 meters (~45 to 50 feet). (*British Museum of Natural History.*)

that both the number and the variety of species were highly diverse toward the end of the period.

The skies above the plains where the dinosaurs roamed were populated—perhaps sparsely—by the two groups of flying vertebrates that had evolved earlier in the Mesozoic: the birds and the flying reptiles (Figure 16-24). Most of the Cretaceous birds were probably wading birds and shorebirds that lived like modern herons and cranes; there were no songbirds of the kind that surround us today. The flying reptiles were among the most spectacular of all Cretaceous animals. While they may have relied heavily on passive transport, soaring on the wind, it appears that many if not all species at times flapped their wings in flight. The largest known species, represented by fossils from the uppermost Cretaceous of Texas (Figure 16-25), is estimated to have had a wingspan of 15.5 meters (~50 feet). Members of this species, like modern vultures, may have soared through the sky in search of carrion in the form of dinosaur car-

FIGURE 16-24   Reconstruction of a Late Cretaceous fauna of Alberta. On the left is the armored herbivorous dinosaur *Edmontonia*. The duck-billed herbivores to its right belong to the genus *Corythosaurus*. The ferocious carnivore to the right of center is *Tyrannosaurus*; it confronts horned dinosaurs of the genus *Chasmosaurus*. Passing overhead in the foreground are

Snakes are a younger group. These limbless reptiles did not originate until the Cretaceous Period, and their great evolutionary expansion did not come about until the Cenozoic Era; all Cretaceous snakes were of the primitive group that includes the present-day constrictors and pythons.

Another Cretaceous vertebrate group with an especially bright future were the mammals, which are contrasted with reptiles in Box 16-1. Mammals of the Cretaceous Period, like those of Triassic and Jurassic time, were of small body size, but during this period a wider variety of species evolved. Several primitive mammal groups died out, and the two large modern groups came into being. The first comprises the marsupials, or pouch-bearing forms, which today are especially diverse in Australia; the second consists of the placentals, a group that includes most living species of mammals (Box 16-1). We will see much more of these two

pterosaurs (flying reptiles) of the genus *Quetzalcoatlus,* and flying in the distance are water birds. The latter have feathered wings, in contrast with the naked wings of the pterosaurs. (*Drawing by Gregory S. Paul.*)

groups of mammals in the following chapters on the Cenozoic Era, when they expanded rapidly both in body size and in variety of adaptation.

PALEOGEOGRAPHY

Because it has undergone less metamorphism and erosion than older geologic systems, the Cretaceous System is represented on modern continents by an extensive record of shallow marine and nonmarine sediments and fossils. In addition, Cretaceous sediments and fossils are widespread in the deep sea, in contrast to the sparse deep-sea records for the Triassic and Jurassic periods; this difference reflects the fact that movements of plates across the earth's surface are rapid enough that a large percentage of deep-sea sediments older than the Cretaceous Period have by now been swallowed up along subduction zones. The rel-

Box 16-1

The early mammals

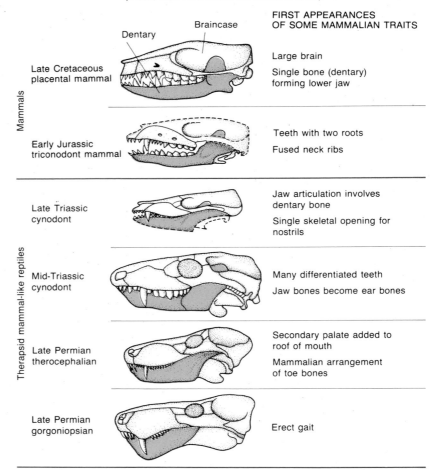

STAGES IN THE EVOLUTION OF MAMMALS
(skulls, after R.E. Sloan, not drawn to scale)

FIRST APPEARANCES
OF SOME MAMMALIAN TRAITS

Mammals

Late Cretaceous placental mammal — Dentary, Braincase
- Large brain
- Single bone (dentary) forming lower jaw

Early Jurassic triconodont mammal
- Teeth with two roots
- Fused neck ribs

Therapsid mammal-like reptiles

Late Triassic cynodont
- Jaw articulation involves dentary bone
- Single skeletal opening for nostrils

Mid-Triassic cynodont
- Many differentiated teeth
- Jaw bones become ear bones

Late Permian therocephalian
- Secondary palate added to roof of mouth
- Mammalian arrangement of toe bones

Late Permian gorgoniopsian
- Erect gait

The oldest known mammals are of Late Triassic age, but until the beginning of the Cenozoic Era some 150 million years later, species of mammals remained small. Few Mesozoic mammals were larger than a house cat, and most were no larger than a rat. Mammals differ from reptiles in a number of ways. For one thing, the brain case of mammals is larger for the size of the animal. For another, the lower jaw of mammals is formed of only one bone, but it bears cheek teeth of greater complexity than those of reptiles. Other physiological and behavioral features that separate the two groups are listed in the accompanying table.

It is debatable whether dinosaurs were reptiles (Box 15-1). Scientists who argue that dinosaurs maintained nearly constant body temperature conclude that they had evolved beyond the reptilian condition. Mammal-like reptiles may also have been endothermic (page 415), in which case they also may deserve taxonomic status separate from the true reptiles.

The fossil record, though incomplete, illustrates many stages of evolution in the transition from mammal-like reptiles to modern mammals. Three of the most conspicuous and important changes marking this transition were: (1) a general enlargement of the braincase and brain, (2) reduction of the number of bones forming the lower jaw, and (3) modification of two posterior jaw bones to become two of the three small bones of the middle ear that form part of an advanced hearing mechanism. This origin of the advanced ear represents one of the most remarkable examples of the evolutionary alteration of a biological feature to perform an entirely new function. It was in many ways comparable to the transformation of one of the gill arches of primitive fishes to form the vertebrate jaw (Figure 13-12).

Numerous groups of small mammals that are now extinct existed in Triassic and Jurassic time. One of these groups, the multituberculates, was well represented in the Cretaceous Period. The name *multituberculate* refers to the many tubercles

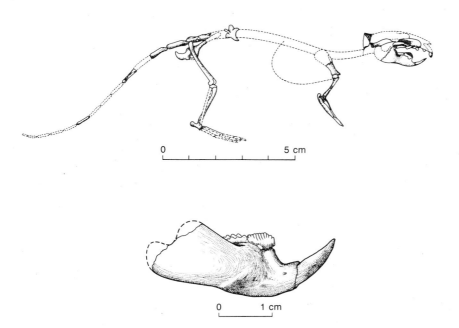

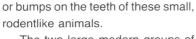

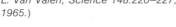

The Cretaceous multituberculate mammal *Stygimys,* with an enlargement of its lower jaw. (*After R. E. Sloan and L. Van Valen, Science 148:220–227, 1965.*)

North American opossum.

or bumps on the teeth of these small, rodentlike animals.

The two large modern groups of mammals also became established during the Cretaceous Period. One of these, the placentals, includes most living mammal species. The other, the marsupials, includes the native species of Australia, among which are the kangaroo, wombat, and koala, as well as several species in the Americas, including the opossum. Whereas an embryonic placental is nurtured in the uterus by means of a placenta attached to the mother, a marsupial is born at an early stage of development and transferred to the protection of a pouch, where the mother's teats are located. Like other mammals of the period, Cretaceous placentals and marsupials were of small body size. Some of the marsupials belonged to the opossum family, which can thus be considered a "living fossil" group.

Although primitive mammals were present before the end of the Triassic Period, it was not until Late Cretaceous time that placental mammals evolved. □

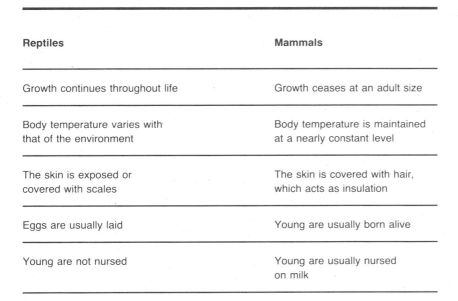

Reptiles	Mammals
Growth continues throughout life	Growth ceases at an adult size
Body temperature varies with that of the environment	Body temperature is maintained at a nearly constant level
The skin is exposed or covered with scales	The skin is covered with hair, which acts as insulation
Eggs are usually laid	Young are usually born alive
Young are not nursed	Young are usually nursed on milk

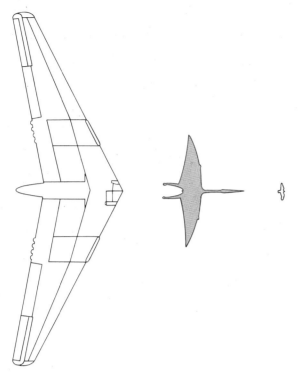

FIGURE 16-25   *Quetzalcoatlus,* a Texas-sized pterosaur from the Latest Cretaceous System of the Big Bend area of Texas (*center*), compared with a Northrop YB-49 "flying wing" and a modern condor. The Texas pterosaur had a wingspan of about 15.5 meters (~50 feet). It is shown in flight in Figure 16-24. (*After D. A. Lawson, Science 187:947–948, 1975.*)

ative abundance of Cretaceous sediments in the ocean basins and on the continents assists us in interpreting paleogeographic patterns and global events of the period. Additional information is drawn from the Upper Cretaceous fossil record of flowering plants. As was described in Chapter 2, these organisms are particularly sensitive to climatic conditions.

Sea level, climates, and ocean circulation

In the course of the Cretaceous Period, there was a global rise of sea level, with only minor interruptions (Figure 16-26). As a result, throughout most of Late Cretaceous time, sea level stood perhaps as high as at any other time in the Phanerozoic history of the earth, and continents were extensively blanketed with marine deposits.

During the Cretaceous Period, temperatures changed in different ways in different places, but both oxygen isotopes and fossil plant occurrences suggest that climates grew generally warmer during the first part of the period; at the end of Early Cretaceous time, about 100 million years ago, the average temperature on earth reached a level perhaps higher than it has ever been since. Temperatures then generally declined throughout Late Cretaceous time.

The final stage of the Cretaceous Period—the Maastrichtian—represents the coolest Cretaceous interval. There may also have been a marked cooling trend during the Maastrichtian itself, as evidenced by a decline in the percentage of flowering plants whose leaves had entire (smooth) margins (page 45). As shown in Figure 16-27, the Maastrichtian decline in this percentage from about 65 to 30 percent for floras near the border between Wyoming and Montana represents a drop in mean annual temperatures from about 20 to about 10°C (68 to 50°F)!

During the middle part of the Cretaceous Period, there were intervals when black muds covered large areas of shallow sea floor (Figure 16-28). There is an apparent connection between global temperature, water circulation, and the presence or absence of black muds in the sedimentary record. These muds are known to form where bottom waters are depleted of oxygen: It appears that extensive black muds of this kind accumulated in shallow seas when unusually poor circulation within ocean basins led to the stagnation of much of the water column. As shown in Figure 16-29, these waters may at times have spilled over from oceanic areas into shallow seas, leading to the epicontinental deposition of black muds. At other times in the earth's history, including the present, cold waters in polar regions have sunk to the deep sea and spread along the sea floor toward the equator, carrying with them oxygen from the atmosphere (page 47). The light color of the sediments representing these intervals in cores taken from the deep-sea floor indicates the presence of oxygen during deposition. The depositional records of these intervals show frequent interruptions, because the flow of the bottom water toward the equator scoured the sediment from the sea floor. Extensive black mud deposition occurred when polar regions were too warm for oxygen-rich surface waters to descend and spread toward the equator. Thus, the widespread accumulation of black muds provides still more evidence that the middle portion of the Cretaceous Period was a particularly warm interval; not even the waters of the deep sea were cold.

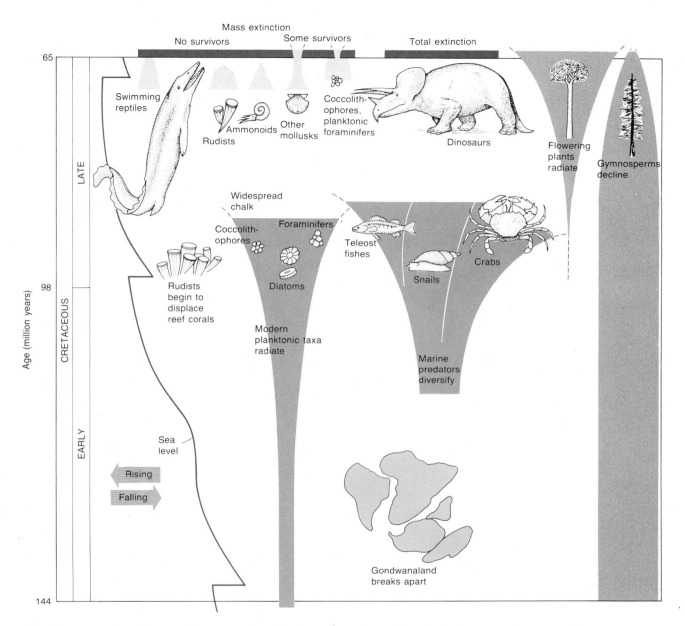

FIGURE 16-26   Major events of Cretaceous time. Gondwana-land fragmented during this interval, forming the South Atlantic Ocean. On the land, while dinosaurs continued to reign in the animal world, flowering plants (angiosperms) expanded at the expense of gymnosperms. Sea level rose throughout most of the period, and when it stood high in Late Cretaceous time, coccoliths rained down on the sea floor, producing widespread chalk deposits. The Late Cretaceous adaptive radiation of coc-colithophores yielded the chalk-producing species, and two other modern planktonic groups, diatoms and foraminifers, di-versified at the same time. Teleost fishes originated in mid-Cretaceous time and radiated along with two carnivorous groups that had originated earlier in the Mesozoic Era: the crabs and the predatory snails. In mid-Cretaceous time, rudist bivalves became the dominant builders of organic reefs, but they died out at the close of the Cretaceous Period along with numerous other marine groups, including the swimming reptiles and the dinosaurs.

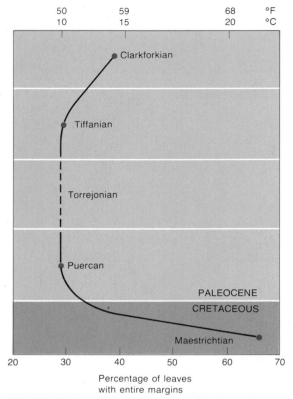

Estimated mean annual temperature

FIGURE 16-27   Reduction of temperature in northwestern Wyoming during the transition from the Cretaceous to the Paleocene. Temperature is estimated from the percentage of fossil leaves having entire (not jagged) margins (see Figure 2-21). Even floras of the Puercan Stage (first stage of the Paleocene in this region) show that temperatures had declined markedly. (*After L. J. Hickey, Univ. Michigan Papers in Paleont. 24:33–49, 1980.*)

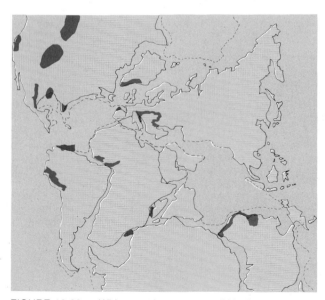

FIGURE 16-28   Widespread occurrence of black shales and muds of Aptian and Albian age on modern continents and continental shelves. (*After A. G. Fischer and M. A. Arthur, Soc. Econ. Paleont. and Mineral. Spec. Publ. No. 25:19–50, 1977.*)

New continents and oceans

Although Pangaea had begun to break apart early in the Mesozoic Era, the smaller continents that had formed from the supercontinent remained tightly clustered at the beginning of the Cretaceous Period. The continued fragmentation of Pangaea and the dispersion of its daughter continents were among the most important developments in global geography during Cretaceous time. Especially dramatic was the breakup of Gondwanaland, two stages of which are shown in Figure 16-30 and Figure 16-31. At the start of the Cretaceous Epoch, Gondwanaland, still intact, was barely attached to the northern continent. By the end of the period, however, South America, Africa, and peninsular India had all become discrete entities; of the present-day continents that represent fragments of Gondwanaland, only Antarctica and Australia remained attached to one another.

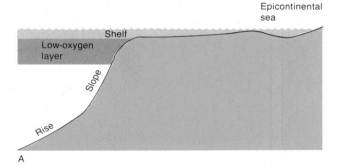

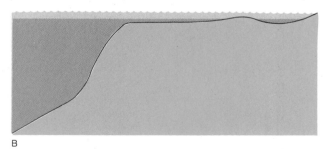

FIGURE 16-29   Expansions of the low-oxygen layer in the ocean at a time when the deep sea is warm. At times like the present, when cold water at the poles is sinking to the deep sea, it supplies deep waters with oxygen, so that the low-oxygen layer is relatively thin (*A*). When polar waters are warmer and are thus of low density, they do not sink; hence, they fail to supply oxygen to the deep sea. Under these conditions, the warm waters below the depth of wave activity are relatively stagnant and the low-oxygen layer thickens, extending even into some epicontinental seas (*B*). (*Modified from A. G. Fischer and M. A. Arthur, Soc. Econ. Paleont. Mineral. Spec. Publ. No. 25:19–50, 1977.*)

The fragmentation and separation of continents during Cretaceous time caused new oceans to form and narrow oceans to grow wider. Most notable were the Early Cretaceous openings of the South Atlantic, the Gulf of Mexico, and the Caribbean Sea. As we have seen, evaporites had formed during the Jurassic Period when marine waters spilled into the rifts that later widened to form the Gulf of Mexico and South Atlantic (Figure 15-31). Early in the Cretaceous Period, these basins were connected with the rest of the world's oceans, but the connections were narrow, and evaporites accumulated along the basin margins in the restricted bodies of water that resulted (Figure 16-30).

In fact, evaporites were deposited over a large portion of the Early Cretaceous globe (Figure 16-30), a condition that reflected the warm, equable climates of the interval. These climates also resulted in the growth of coral reefs as far as 30° from the equator. Further evidence of warm

temperatures at high latitudes is the presence of fossil leaves of breadfruit trees in Cretaceous deposits of Greenland (Figure 16-32); similarly, warm-adapted plant species are also found in the fossil record of northern Alaska. Latitudinal temperature gradients were so gentle that the trade-wind belts probably extended farther north and south of the equator than they do today, resulting in the widespread accumulation of evaporites as their dry air swept over the land (page 39).

A dominant feature of the Cretaceous world was the great Tethys Seaway, where the trade winds drove surface waters westward without obstruction by large landmasses. Animals largely confined to the fully tropical Tethyan region included reef-forming corals and rudists, certain types of bottom-dwelling foraminifers, and several groups of ammonoids. As in Jurassic time, the Tethys was an essentially tropical belt where carbonate deposition prevailed. During

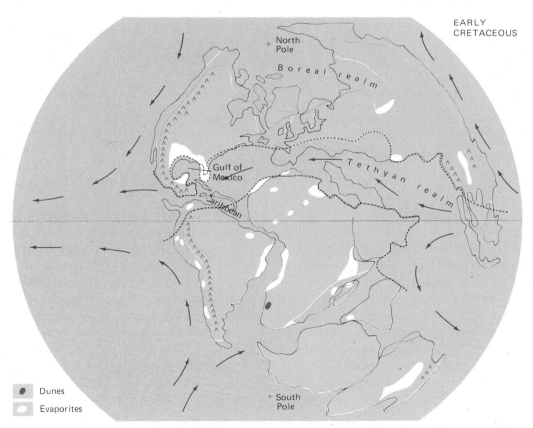

FIGURE 16-30   Global geography during Early Cretaceous time. Gondwanaland was beginning to break apart, but North America, Greenland, and Eurasia remained connected to one another. Climates were warm, and evaporites accumulated even far to the north and south of the tropical Tethyan realm.

(*Partly after A. G. Smith and J. C. Briden, Mesozoic and Cenozoic Paleocontinental Maps, Cambridge University Press, Cambridge, England, 1977, and G. E. Drewry et al., Jour. Geol. 82:531–553, 1974.*)

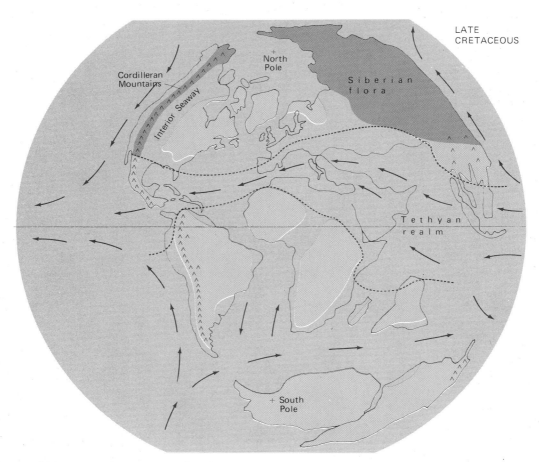

FIGURE 16-31   Global geography during Late Cretaceous time. The Siberian flora occupied not only Siberia but also the northern part of the narrow island formed by the Cordilleran mountain chain. Sea level stood higher in relation to most land areas than it had during Early Cretaceous time. The tropical Tethyan realm remained. (*Partly after A. G. Smith and J. C. Briden, Mesozoic and Cenozoic Paleocontinental Maps, Cambridge University Press, Cambridge, England, 1977.*)

Late Jurassic time, however, only shallow Tethyan seas connected Caribbean waters with those of the Pacific (Figure 15-33); during the Cretaceous Period, the separation of North and South America provided a deep oceanic passage for these waters (Figures 16-30 and 16-31). The great strength of the westward-flowing Tethyan current through this passage and beyond is demonstrated by the fact that Cretaceous faunas resembling those of the Tethyan Seaway have been found on submerged seamounts as far as 1750 kilometers (~1000 miles) west of Hawaii.

The equatorial position of Africa deflected the Tethyan current northward, producing a steeper temperature gradient toward the cool North Pole than that found in the Southern Hemisphere. As it had in the Jurassic Period, the largely nontropical Boreal realm lay to the north of the tropical Tethys. Throughout the Cretaceous Period, the North Atlantic Ocean probably remained too small for the Coriolis force to produce a fully circular clockwise gyre like that of the modern Atlantic (Figure 2-22).

Early in Cretaceous time, the Arctic Ocean remained largely isolated from the Atlantic and supported a distinct marine fauna. Later in the period, this isolation ended when an important episode of continental rifting split the huge landmass of the Northern Hemisphere into the modern continents of North America, Greenland, and Eurasia (Figure 16-33).

It was not only rifting that connected the Atlantic and Arctic oceans during Late Cretaceous time. The progressive elevation of sea level that characterized most of the Late Cretaceous Period formed a narrow seaway that spread from the Gulf of Mexico all the way across Alaska to the Arctic Ocean (Figure 16-31). Later in the chapter, we will examine patterns of deposition within this great body of water, which is known as the Cretaceous Interior Seaway.

FIGURE 16-32   A fossil breadfruit leaf (*A*) from the Cretaceous System of Greenland, a continent that was much warmer during Cretaceous time than it is today, and a modern breadfruit leaf (*B*) in a tropical garden in the Caribbean. (*A. The Swedish Natural History Museum. B. Courtesy of R. A. Howard.*)

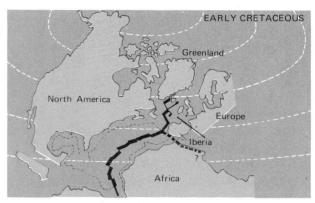

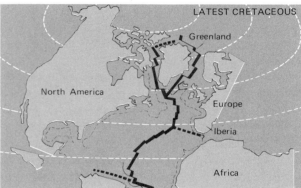

FIGURE 16-33   The opening of the North Atlantic during the Cretaceous Period. As North America and Africa moved apart, the Mid-Atlantic Rift was propagated northward along two branches, leaving Greenland as a separate landmass between North America and Eurasia. (*After J. G. Sclater et al., Jour. Geol. 85:509–552, 1977.*)

At present, we will simply note that the seaway divided North America into two land areas. The western segment was the long, narrow Cordilleran mountain belt, where tectonic activity continued from Jurassic time. This narrow highland supported a flora similar to that of Siberia and distinct from that of eastern North America.

Cretaceous mass extinctions

The most famous mass extinction in all of earth history occurred at the end of the Cretaceous Period. Although it was less devastating in the marine realm than the Late Permian crisis, the terminal Cretaceous event eliminated the dinosaurs, and because of this it has attracted a great deal of interest. Earlier in Late Cretaceous time, about 90 million years ago, another, less severe mass extinction in the marine realm had eliminated many species and genera but relatively few families. Because the terminal Cretaceous mass extinction was more severe and has been much more intensively studied, it will be the focus of the following discussion.

The cause of the mass extinction that ended the Age of Dinosaurs remains a mystery. Nonetheless, because the event was relatively recent, geologists have been able to accumulate more relevant evidence than is available for the evaluation of earlier mass extinctions, and it is possible that we will soon at least partly understand its cause or causes. Over the years, many explanatory hypotheses have been proposed, some ingenious and others outlandish, some testable and others purely speculative. Among the many suggested causes have been cosmic radiation from a supernova, a sudden lowering of sea level, collapse of the marine planktonic ecosystems resulting from an absence of nutrient supply, a burst of volcanic activity that temporarily screened the earth from the sun, a sudden change in the earth's magnetic field, and a spillover of cold, fresh water into the Atlantic Ocean from the previously isolated Arctic basin.

If we are to evaluate such hypotheses, we must first review the kinds of organisms that were affected.

The victims Whereas the dinosaurs disappeared entirely at the end of the Mesozoic Era, the much smaller mammals suffered only moderate losses. In the sea, the ammonoids and large marine reptiles (plesiosaurs and mosasaurs) died out, and other groups also declined sharply: In the water column, few species of coccolithophores, planktonic foraminifers, radiolarians, or belemnoids survived into the Cenozoic, and on the sea floor the reef-building rudists disappeared along with many smaller taxa of bivalves, snails, corals, bryozoans, and large foraminifers.

Temporal patterns In studying the Late Cretaceous mass extinction, scientists have asked the question, Did the lethal agent or agents of the Cretaceous extinction strike suddenly and only once, or were their effects spread over a substantial interval of time? As it turns out, the pattern differs for different groups of animals and plants.

Of the many taxa that disappeared, some, like the inoceramid bivalves (Figure 16-16A) and the ammonoids, declined in diversity over most of the Maastrichtian Age and disappeared before its end. This final age of the Cretaceous Period lasted perhaps 8 million years.

Interestingly, the dinosaurs also appear to have died out gradually, with different species affected at different times. By the end of Maastrichtian time, the dinosaur faunas of western North America were not at all like the diverse faunas described on page 498: The duck-billed forms had disappeared, and the most conspicuous survivors were the

FIGURE 16-34 Tilted sequence of deep-sea sediments at Zumaya, Spain, spanning the Mesozoic-Cenozoic boundary. Plankton preserved within the sequence experienced a severe extinction at the boundary during an interval of approximately 10,000 years. (*Courtesy of S. F. Percival.*)

FIGURE 16-35   A coccolith of the species *Braarudisphaera
bigelowi* (×3135). This living species has survived for more
than 150 million years. It is remarkably hardy, and after the
terminal Cretaceous extinction, it spread throughout the
oceans in great abundance, apparently because in the barren
waters there were few animals to eat it and few species of
phytoplankton to compete with it. (*From B. U. Haq, in B. U.
Haq and A. Boersma [eds.], Introduction to Micropaleontology,
Elsevier, New York, 1978.*)

horned dinosaurs (Figure 16-21). For a brief time, until they
too died out, these huge herbivores had the world largely
to themselves in a situation similar to that of the mammal-
like reptile *Lystrosaurus* after the latest Permian mass extinction, when this large herbivore was the dominant vertebrate genus in Gondwanaland (Figure 7-19).

Other forms of Late Cretaceous life, especially groups
of marine plankton, died out more suddenly. Although there
are extremely few marine stratigraphic sections spanning
the Mesozoic-Cenozoic boundary, the few uplifted sections
that do represent deep-sea deposits of the time offer excellent fossil records of pelagic and deep-sea life. One of
these sections is found at Gubbio, Italy; another is located
at Zumaya on the northern coast of Spain (Figure 16-34).
In each of these areas, changes in the tropical plankton
are revealed in a narrow stratigraphic interval representing
perhaps only 10,000 years at the very end of the Maastrichtian. First, planktonic foraminifers began to die out.
Then, unusual coccolithophores appeared in the open ocean,
among them the remarkable species *Braarudosphaera bigelowi* (Figure 16-35). This is the oldest living species of
coccolithophore, with a fossil record extending back to the
Upper Jurassic. It lives mainly in brackish bays today, but
at times in its history it has spread widely over the oceans
in great abundance; in fact, it seems to flourish when other
types of plankton are suffering extinction. Slightly above
the stratigraphic level where it first appears in latest Maas-

trichtian sediments of Italy and Spain, this species and a
small number of others dominate the fossil phytoplankton.

There are other major groups that experienced relatively sudden extinction during Maastrichtian time but well
before the end of the interval. Perhaps the most important
early victims were the various members of the rudist reef
community. Fossil faunas in the Caribbean region and in
Spain show that nearly all of the rudists and other tropical
marine animals associated with them died out during a very
brief interval at the end of the middle segment of the Maastrichtian, perhaps 2 million years before the decline of the
plankton. Late Maastrichtian reef faunas are known, but
these typically consist of only a few species.

In summary, some of the groups that died out during
the Maastrichtian underwent a gradual decline to extinction, others disappeared suddenly at the end of the epoch,
and still others disappeared suddenly but before the end
of the Maastrichtian.

Regression of the seas   Until very recently, it was widely
believed that sea level fell markedly at the end of the Cretaceous Period and that this contributed to the mass extinction. North American geologists in particular held this
belief based on evidence that the seas shrank back from
the western interior of North America as an orogeny began
in the Rocky Mountain region at the end of the Cretaceous
Period. This was not a time of global regression, however;
in fact, a major lowering of the seas took place in mid-
Maastrichtian time a few million years *before* the end of
the Cretaceous Period, and the seas then rose again during
the transition to the Cenozoic Era. Thus, regression of the
seas can be related to few, if any, of the terminal Cretaceous extinctions.

Global refrigeration   As we have seen, the Maastrichtian
was the coolest interval of the Cretaceous Period, and land
floras reveal that a pronounced cooling took place during
the Maastrichtian itself (Figure 16-27). There is also indirect
evidence of a cool interval in the deep sea near the close
of Cretaceous time. Deep-sea deposits straddling the
boundary tend to be greenish-gray or reddish rather than
black, suggesting that cool polar waters were sinking and
carrying oxygen to the deep sea, just as they do today.

Is it possible that a final pulse of cooling, or a series of
pulses, was simply too much for many groups of animals
and plants to withstand? If so, we might expect to see
latitudinal patterns of extinction—and, indeed, such patterns do exist. In the seas, Tethyan marine life, which was
tropical in character, was decimated; faunas of cooler re-

gions, on the other hand, experienced only modest losses. The rudist reef community, which was Tethyan in distribution, disappeared altogether, just as the tabulate-strome reef community had vanished in the Late Devonian mass extinction. Other victims that lived on Tethyan sea floors included certain tropical groups of snails, nonrudist bivalves, and large foraminifers. The major groups of calcareous plankton that were nearly eliminated altogether—the coccolithophores and the planktonic foraminifers—were also centered in the tropics.

In contrast, northern marine faunas suffered only modest losses. One indication of this is the historical debate over whether the Danian Stage, named for Denmark, where the stage was recognized, should be included in the Cretaceous System or in the initial (Paleocene) series of the Cenozoic Erathem: The debate arose because the Danian includes many fossil species closely related to unquestioned Cretaceous forms, together with others that are related to unquestioned Paleocene forms. The Danian is now assigned to the Paleocene Epoch, primarily because fossils of Danian age in the Tethyan region to the south are very different from older fossils of Cretaceous age; here, mass extinction is evident at the base of the Danian. Similarly, the Cannonball Sea, which occupied the cool northern region of North Dakota in Early and Middle Paleocene time, harbored many species that survived from the Cretaceous Period. In fact, roughly half of the Cannonball species are holdovers from the Cretaceous, even though a gap of 2 or 3 million years separates it from the youngest Cretaceous faunas of the region.

The fact that the mass extinction struck marine faunas hardest in the tropics suggests that temperature change may have been the primary lethal agent. We must wonder whether particular faunas suffered extinction because they were unable to migrate to warmer regions. The loss of Tethyan marine species, including those of oceanic islands of the Pacific, may have resulted from the disappearance of tropical seas. Under conditions of cooling, faunas of nontropical climates could have migrated toward the equator along with the climatic zones to which they were adapted. On the other hand, tropical faunas would have had nowhere to go. The history of marine snails suggests that this situation may indeed have developed. Tethyan snail faunas disappeared during the crisis; afterward, in early Cenozoic time, groups of snails that had occupied the Boreal realm during the Cretaceous Period ranged from high latitudes to as far south as northern Africa, which had previously been Tethyan (tropical).

The idea that a lowering of temperature was a lethal agent is further supported by the observation that global temperatures declined progressively, and perhaps fluctuated, during the Maastrichtian. Unlike hypotheses that suggest one sudden event as the cause of extinction, this argument accounts for a complex temporal pattern of extinction by theorizing that temperatures declined beyond the tolerances of different taxa at different times as additional agents of cooling came into play.

The temperature hypothesis also offers a convenient explanation for the fact that extinction occurred both on land and in the sea. However, it should be noted that we have no direct evidence that the temperature levels in late Maastrichtian time were lethal. Furthermore, we do not know what factor or factors might have caused widespread refrigeration.

The arrival of an asteroid or comet   It has recently been noted that in many regions there is a thin layer of sediment right at the Mesozoic-Cenozoic boundary which contains an unusually high concentration of iridium, an element that is extremely rare on earth (Figure 16-36). This so-called iridium anomaly probably reflects one of three conditions or events: (1) low rates of sediment deposition during the critical interval, causing the element, which is constantly reaching the earth in extraterrestrial dust, to become concentrated; (2) the impact on earth of a group of meteorites or a large foreign body, such as an asteroid or comet, that contained a high concentration of iridium; or (3) the eruption of a huge volcano on earth that emitted an unusually large quantity of iridium.

Two facts seem to argue against the idea that low rates of sediment accumulation concentrated the iridium. One is the occurrence of the "iridium anomaly" in both deep-sea and nonmarine deposits, and we would not necessarily expect rates of deposition to slow down for both simultaneously. The second has to do with the level of iridium contained in normal cosmic dust that settles on the earth. It has been estimated that the amount of iridium in the anomaly represents the amount that would normally settle to the earth in about 10 million years, which is much longer than the time that the iridium-rich layer could represent in most areas. Intensive volcanism on earth might have yielded a heavy fallout of iridium from the atmosphere during a much shorter interval, but there is independent evidence of a meteorite impact: within the iridium-rich layer in Montana are fragments of quartz riddled with microscopic fractures of a type caused by severe impacts like those of large meteorites.

The notion that the earth might have sustained a dev-

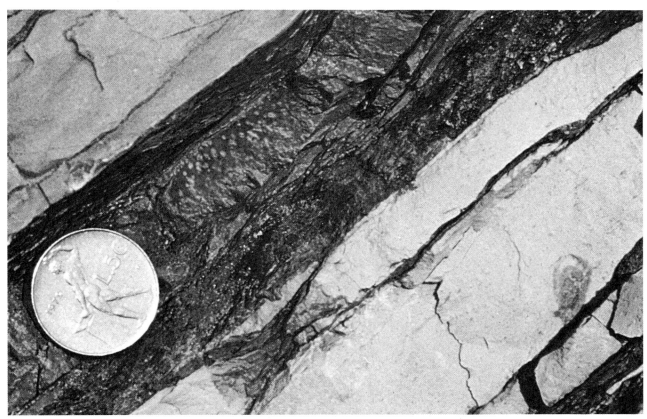

FIGURE 16-36   The band of clay at the Cretaceous-Cenozoic boundary near Gubbio, Italy. The concentration of the rare element iridium is about 30 times higher in this clay than in sediments above and below. A rich assemblage of fossil coccoliths is present below the clay, but there is only a sparse assemblage above. (*Courtesy of D. A. Russell.*)

astating meteorite impact at the end of the Cretaceous Period has generated great interest in the scientific community. The dust cloud associated with the impact of a large extraterrestrial body might for a brief time, geologically speaking—a few years or perhaps a few thousand years—have screened out a large fraction of the sunlight that normally reaches the earth, killing off certain plants and perhaps cooling the planet as well. Still unexplained, however, would be the decline or disappearance before the very end of the Cretaceous of groups like rudists, inoceramids, ammonoids, and dinosaurs. It may be that a meteorite impact was the last in a series of events that led to excessive extinction near the end of Cretaceous time.

REGIONAL EXAMPLES

The great worldwide elevation of the seas that began near the end of Early Cretaceous time produced much of the Cretaceous record exposed on modern continents. This record tells many of the regional stories that follow. We will first examine the continuation of mountain building in western North America, a process that produced an enormous foredeep that became flooded by the seaway that extended from the Gulf Coast to the Arctic. We will see that the Gulf Coast itself was fringed by rudist reefs and that a rudist-rimmed carbonate bank also stretched along a large segment of the adjacent Atlantic coast until midway through the Cretaceous Period, when it gave way to the deposition of mud and sand that continues today. On the other side of the Atlantic, we will observe how siliciclastic deposition early in the Cretaceous Period was followed by the widespread accumulation of chalk in Europe. Finally, we will review the development of a vast interior sea in Australia.

Cordilleran mountain building continues

During Cretaceous time, an important change took place in the pattern of igneous activity in western North America. Subduction of the Franciscan complex along the western

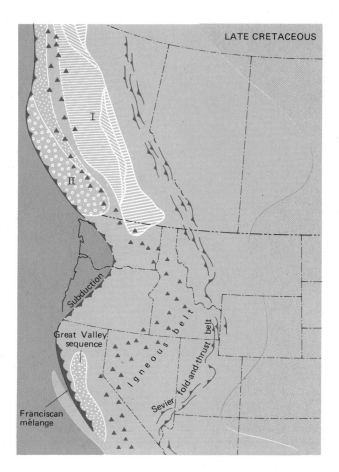

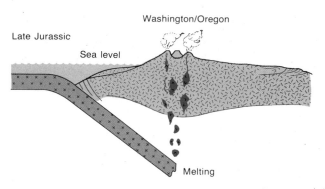

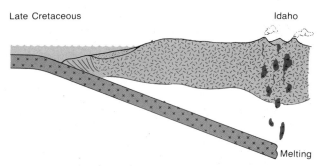

FIGURE 16-38   Diagram illustrating a likely explanation for the eastward migration of igneous activity in the Cordillera during Cretaceous time. As suggested here, the subducted plate began to pass downward at a reduced angle, so that it reached the depth of melting only after passing far to the east.

FIGURE 16-37   Late Cretaceous geologic features of western North America. Subduction produced the Franciscan mélange in California. North of California, igneous activity resulting from subduction was located far to the east of the continental margin; this activity, together with the folding and thrusting to the east, represented the latter part of the Sevier orogeny. In Canada, the margin of the continent consisted of two blocks of exotic terrane (I and II) that had been sutured to North America earlier in the Mesozoic Era; each of these blocks consisted of two or more slivers of crust that were welded together to form the block before it was attached to North America.

margin of the continent continued, as did the associated igneous activity. However, by Late Cretaceous time, although volcanic and plutonic activity persisted in the Sierra Nevada region, the northern igneous activity had come to be concentrated to the east, in Nevada and Idaho (Figure 16-37). This pattern contrasted with that of the Late Jurassic Epoch, when igneous activity in the north had been centered near the coast, in northern California and Oregon (Figure 15-41). The likely explanation for the eastward mi-

gration of igneous activity in the northern United States is that the angle of subduction there had changed. In mid-Cretaceous time, the subducted plate in this region probably began to pass downward beneath the continent at a low angle; the subducted crust therefore failed to sink deep enough to melt until it had extended far inland (Figure 16-38). The fold-and-thrust belt in front of the mountainous igneous regions also shifted inland in the northern United States. By Late Cretaceous time, folding and thrusting extended eastward as far as the Idaho-Wyoming border (Figure 16-37).

A major episode of igneous activity and eastward folding and thrusting coincided approximately with the Cretaceous Period; although this episode was not entirely separate from earlier and later tectonic activity, it has become known as the **Sevier orogeny**. East of the Sevier orogenic belt lay a vast foredeep; during almost all of Late Cretaceous time, this foredeep was occupied by a narrow seaway stretching from the Gulf of Mexico to the Arctic Ocean. The history of this seaway will be discussed in the section that follows.

The orogenic belt that occupied western North America during the latter half of Cretaceous time was unusually broad, apparently because of low-angle subduction—but in its development of a foredeep and certain other features, the orogenic belt was otherwise typical. Like the modern Andes (Figure 8-16), for example, it was symmetrical: The Franciscan deformation at the continental margin (Figure 16-37) was mirrored on a larger scale by the Sevier folding and faulting east of the belt of igneous activity.

The Mesozoic history of western Canada is far more complicated. For one thing, the crust that now forms most of British Columbia is made up of exotic terranes that actually became attached to North America during the Mesozoic Era. Subduction welded one large block of foreign terrane to the continental margin during the Jurassic Period and another during the Cretaceous Period (Figure 16-37). The evidence for these events comes from three sources: (1) paleomagnetic measurements show that the two blocks (I and II in Figure 16-37), were previously located far from North America; (2) paleontological evidence shows that block I and block II and adjacent North America belonged to different biogeographic provinces in Permian and early Mesozoic times; and (3) the three terranes are structurally incompatible. Furthermore, the same kinds of evidence reveal that blocks I and II are themselves composite terranes; before becoming attached to North America, each existed as a microcontinent that was formed by the coalescing of several smaller slivers of crust. Thus, the westward growth of Canada after the Paleozoic Era was achieved less by gradual accretion than by the stepwise suturing of elongate microcontinents to the continental margin.

The Gulf Coast and North American interior seaway

During the latter part of Early Cretaceous time, a spectacular array of marine environments developed in the Cordilleran foredeep. This foredeep extended from the Gulf Coast all the way to the Arctic Ocean, which was probably much warmer than today, judging by nearby Alaskan floras. We will now trace the events that led to the formation of this Cretaceous Interior Seaway.

Shortly before the end of the Early Cretaceous, during Albian time, Arctic waters spread southward, flooding a large area of western North America with the Mowry Sea (Figure 16-39). This body was named for the Mowry Formation that accumulated within it, and that consists mostly

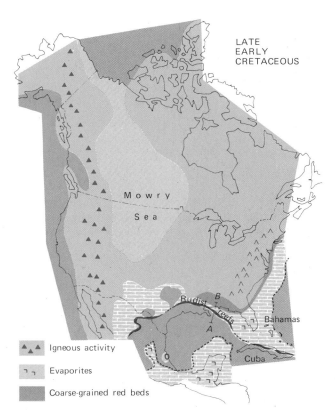

FIGURE 16-39   Geography of North America later in Early Cretaceous time. The Mowry Sea, which was the site of black mud deposition, spread southward from the Arctic Ocean. A carbonate bank bordered by rudist reefs encircled the Gulf of Mexico, and carbonate deposition extended far to the north along the East Coast. A cross section along the line A–B is shown in Figure 16-40.

of oil shale containing large numbers of fish bones and scales. The Mowry Sea formed as a part of the great mid-Cretaceous marine transgression that resulted in the deposition of black shales on many continents (Figure 16-28).

To the south, the Gulf of Mexico was originally part of the tropical Tethyan realm (Figure 16-31). Reefs flourished around its margin, especially in Albian time, when the Gulf was almost entirely encircled by rudist-dominated barrier reefs (Figure 16-39). Behind the reefs, lime muds accumulated in protected lagoons. A broad carbonate bank encompassed the areas now occupied by peninsular Florida and the Bahamas and extended south to Cuba. Farther south, another broad carbonate bank extended northward from the Yucatán region, leaving only a narrow channel for communication between the Gulf and the Atlantic Ocean.

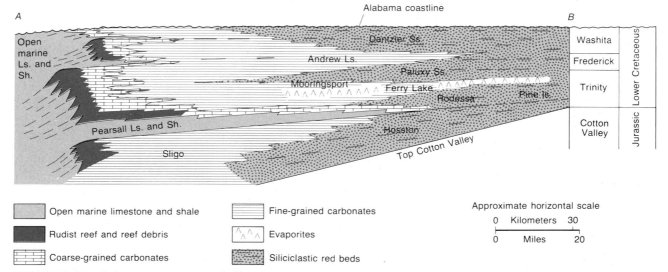

FIGURE 16-40 Schematic section through Alabama along the line *A–B* shown in Figure 16-39. Here, the Lower Cretaceous rock units range from coarse-grained red beds along the coastline to lagoonal carbonates to bank-margin rudist reef deposits to deeper water limestones and shales. The coarse-grained sediments along the coastline were shed from the southern Appalachians. (*Courtesy of R. C. Vernon, Shell Oil Company.*)

At various sites behind the rudist reefs, evaporites formed in lagoons and tidal flats. Fossil cycad stumps in marginal marine sediments of central Texas attest to the presence of tropical conditions some 30° north of the equator. The only interruption in the belt of carbonate deposition was located in the area to the east of the present Mississippi delta. Here, carbonates gave way, landward, to coarse siliciclastic sediment shed from the Appalachian highlands (Figure 16-40).

The Mowry Sea made brief and intermittent contact with the Gulf of Mexico before the end of Early Cretaceous time, but an enduring connection was established at the start of the Late Cretaceous. The result of this contact was an enormous inland sea, the Cretaceous Interior Seaway, that occupied the foredeep in front of the Sevier orogenic belt. Until just before the end of the Cretaceous Period, this seaway extended from the Gulf of Mexico to the Arctic Ocean (Figure 16-41). Most of the sediments deposited here were shed from the Cordilleran Mountains that formed to the west. The history of the seaway is especially well understood because of excellent stratigraphic correlations based on abundant fossil ammonoids; in addition, ashfalls from volcanic eruptions to the west provided numerous marker horizons that can often be dated radiometrically (Figure 16-42).

Barrier islands bounded much of the seaway, with lagoons standing behind them. The lagoons were bordered by broad swamps; on the western margin of the seaway, these gave way to alluvial plains, which were succeeded near the mountains by alluvial fans. As we saw in Chapter 4, the record of the barrier island–lagoon complexes is

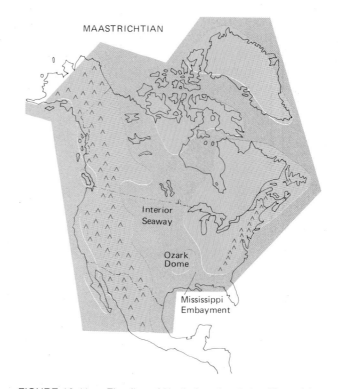

FIGURE 16-41 Flooding of North America during Maastrichtian time, just 2 or 3 million years before the great regression at the end of the Cretaceous Period. Prior to the regression, the Cretaceous Interior Seaway stretched from the Gulf of Mexico to the Arctic Ocean. The Mississippi Embayment occupied the area now occupied by the southern segment of the Mississippi River; at times this embayment was connected to the seaway by a passage along the east side of the Ozark Dome, a highland centered in Arkansas. (*After G. D. Williams and C. R. Stelck, Geol. Soc. Canada Spec. Paper No. 13, 1975.*)

FIGURE 16-42   The Upper Cretaceous x-bentonite (*arrow*), a thin bed of clay (the mineral bentonite) formed by the alteration of a layer of ash that settled after a volcanic eruption. Shown here in Russell County, Kansas, the x-bentonite is present across a broad area of the western interior of the United States and serves as a key bed for correlation. (*Courtesy of D. B. Macurda.*)

remarkably well displayed in rocks of Colorado and nearby states (Figure 4-10).

The western shoreline of the seaway shifted back and forth, primarily in response to the rate of sediment supply. At all times, conglomeratic sediments were shed eastward from the neighboring mountains as clastic wedges, but at times of particularly active thrusting or uplift, these wedges prograded especially far to the east (Figure 16-43). Because of the great weight of sediments on the western side of the seaway, subsidence was more rapid there than farther east. Along the western margin, in nonmarine environments, Late Cretaceous dinosaurs left a rich fossil record.

The Upper Cretaceous strata of the interior seaway represent two large depositional cycles, one of which is illustrated in Figure 16-44. Each cycle consists of an interval of transgression followed by an interval of regression. In addition to changing rates of sediment supply, global changes in sea level and the changing rate of subsidence of the seaway floor must have influenced these patterns of transgression and regression. At times of low supply and maximum lateral expansion of the seaway, chalks were laid down in the center. The most famous of these is the Niobrara Chalk, which occupies the middle of the upper

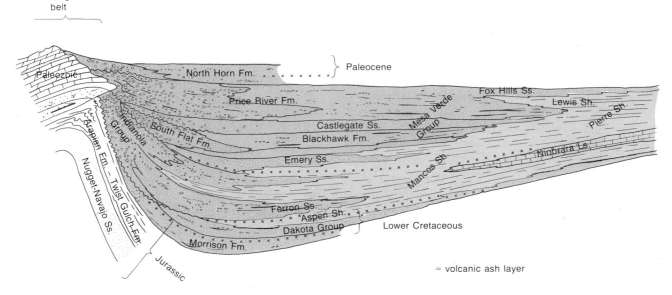

FIGURE 16-43   Diagrammatic cross section of Upper Cretaceous sediments in central Utah. These sediments were deposited in the foredeep east of the Sevier orogenic belt. Clastic wedges in the west pass eastward into finer-grained marine sediments. (*After R. L. Armstrong, Geol. Soc. Amer. Bull. 79:429–458, 1968.*)

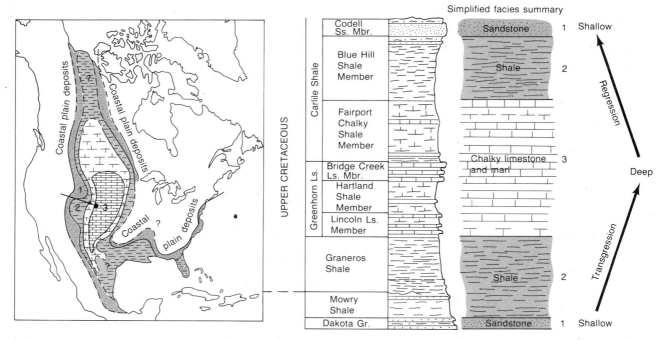

FIGURE 16-44   The early Late Cretaceous "Greenhorn" depositional cycle of the North American interior seaway. The three facies in the simplified facies summary are shown in map view on the left. The cross section represents the vicinity of eastern Colorado (*arrow on map*), where the cycle developed by an oscillation of the shoreline (transgression and regression). During the transgression, the area of chalky limestone and marl (limey clay) deposition in the center of the basin (facies 3) expanded; first facies 2 and then facies 3 spread into eastern Colorado. During regression, the area of deposit of facies 3 contracted, and facies 2 and then facies 1 shifted into eastern Colorado. (*Modified from E. G. Kauffman, The Mountain Geologist 6:227–245, 1969.*)

transgressive-regressive cycle. The Niobrara has yielded beautifully preserved fossil vertebrates (Figures 16-6, 16-7, 16-9, and 16-10).

To the south, Late Cretaceous seas spread from the Gulf of Mexico into the Mississippi Embayment, which was partially separated from the interior seaway by the Ozark Dome (Figure 16-41). Near the end of the period, siliciclastic sediments were shed into the eastern and western extremities of the Mississippi Embayment—into the region of Georgia from the ancestral eastern Appalachians and into the region of Tennessee from the ancestral Mississippi River. Between these two areas of siliciclastic deposition, large volumes of chalk accumulated in what is now central Alabama (Figure 16-45).

Just before the end of the Cretaceous Period, the seas retreated southward from the interior seaway and the Mississippi Embayment, and a new pulse of mountain building began. This orogenic event, which is known as the **Laramide orogeny**, continued well into the Cenozoic Era and will be discussed in the next chapter. Except for a brief and less extensive incursion just after the beginning of the Ce-

nozoic Era, the seas have never returned to the western interior of North America.

The east coast: Development of the modern continental shelf

In eastern North America, seismic studies reveal a great thickness of sediments beneath the continental shelf (Figure 16-46). These sediments consist of deposits laid down during the early Mesozoic episodes of rifting that formed the modern Atlantic Ocean. At the base are fault basin deposits like those of the Newark Series that are exposed on the continent to the west (page 467). Next come large thicknesses of Jurassic carbonates that accumulated in the narrow, young Atlantic Ocean. Above the Jurassic carbonates are more carbonates from the Early Cretaceous interval, when, under much warmer climatic conditions than exist today, reef-rimmed carbonate banks bordered the ocean from Florida to New Jersey (Figure 16-39).

Before the end of Early Cretaceous time, reef growth

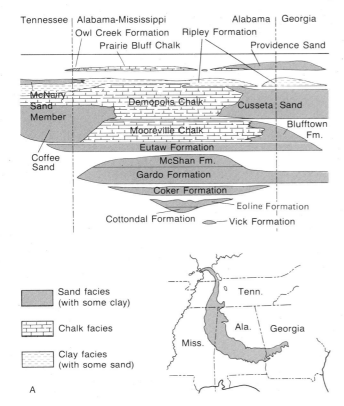

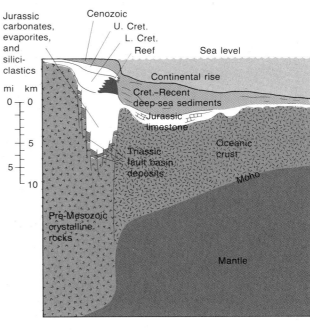

FIGURE 16-46 Cross section of the continental shelf and deep sea off New Jersey. Here the opening of the Atlantic is recorded by early Mesozoic sediments deposited within fault-bounded basins. During Early Cretaceous time, a reef-rimmed carbonate bank extended this far north under the influence of Tethyan ocean currents (see Figure 16-39). During Late Cretaceous time, carbonate deposition gave way to siliciclastic deposition, which has predominated to the present day. (*After R. E. Sheridan et al., The Geology of Continental Margins, Springer-Verlag, New York, 1974.*)

FIGURE 16-45 Upper Cretaceous deposits of the Mississippi Embayment. *A.* Siliciclastic deposition prevailed at first, but as the embayment deepened, siliciclastics became limited to marginal areas and chalk accumulated in the center of the embayment. *B.* An outcrop of chalk along the Tombigbee River in Alabama. (*A. After W. H. Monroe, Amer. Assoc. Petrol. Geol. Bull. 31:1817–1824, 1947. B. U.S. Geological Survey.*)

gave way to deposition of predominantly siliciclastic sediments. This change marked the beginning of the growth of the large clastic wedge that forms the modern continental shelf. The clastic wedge consists largely of sands and muds from the Appalachian region laid down in nonmarine and shallow marine settings. Off the coast of New Jersey (Figure 16-46), the total thickness of Cretaceous and Cenozoic sediments is approximately 3 kilometers (~2 miles). Only to the south, in southern Florida, did carbonate deposition persist to the present. Here, about 3 kilometers of sediments, consisting mainly of carbonates, accumulated during the Cretaceous Period, and another 2 kilometers or so were added during the Cenozoic Era (Figure 4-22).

The clastic wedge apparently began to develop because of a renewed lifting of the Appalachian Mountain belt to the west, after it had been largely leveled during the

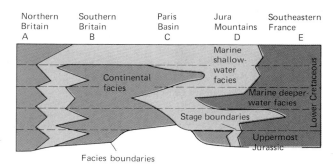

FIGURE 16-47   Pattern of deposition in Britain and France during Early Cretaceous time. Boreal seas flooded eastern Britain and southern Scandinavia, and the warm waters of the Tethys bathed southern Europe. The nonmarine and marginally marine Wealden beds (C) accumulated in southern England and in northern France. (*Upper diagram modified from T. R. Owen, The Geological Evolution of the British Isles, Pergamon Press, Oxford 1976.*)

Jurassic. Cretaceous sediments of the wedge lie mostly beneath the ocean or beneath younger deposits on the land; some, however, are exposed in two belts of the Coastal Plain Province, one stretching from New Jersey to Maryland and Virginia and the other positioned in the Carolinas. Like the younger Coastal Plain sediments, these belts remain largely unconsolidated.

The Chalk Seas of Europe

To many geologists, the term *chalk* refers specifically to the soft, fine-grained limestones of western Europe, although, as we have seen, rocks of a chalky nature are found elsewhere, including the Cretaceous System of North America. The Cretaceous chalks of Europe, which are spectacularly displayed as the White Cliffs of Dover and other coastline cliffs of southeastern England (page 482), accumulated throughout almost all of Late Cretaceous time, when high sea levels flooded much of western Europe.

Early in the Cretaceous Period, European geography was quite different from what it is today. A largely nonmarine basin extended from southern England into northern France and Germany (Figure 16-47). Here, sediments accumulated in lakes, rivers, and deltas to form the dinosaur-bearing Wealden beds. The landmass on which they formed separated the Boreal seas to the north from the Tethys to the south.

Late in Early Cretaceous time, a transgression inundated not only the Wealden Basin, but much of Britain and western Europe as well. In the first stage of this transgression, siliciclastic sediments were deposited in shallow water, forming units like the Folkstone beds (Figure 16-48) and the Greensands of England. Later, when shallow seas spread inland from the Atlantic as far as Poland, siliciclastics were restricted to shallow, marginal environments, and chalk accumulated over a large region in epicontinental seas that were remarkably deep—commonly 200 to 300 meters (~600 to 1000 feet).

Except in the newly forming Alps to the south (Figure 8-11), tectonic activity was largely absent from western Europe during Late Cretaceous time, when the chalk accumulated. Here and there, massifs stood relatively high as islands or, during transgressions, as shallow sea floors (Figure 16-49). Chalk accumulated almost continuously in the basins that surrounded the massifs. The lower portions of the chalk often contain clay, but the remainder is relatively pure calcium carbonate, consisting of minute skeletal debris that is about 75 percent planktonic in origin. In fact, the rock remains soft because coccoliths are the dominant components; these are secreted as the mineral calcite, which is less easily dissolved and reprecipitated as cement than aragonite, which is the form of calcium carbonate secreted by many other organisms.

Although the Chalk Seas were unusually deep for epicontinental bodies of water, the general presence of oxidizing conditions along the sea floor is demonstrated by

FIGURE 16-48 Lower Cretaceous cross-bedded sandstones (the Folkstone beds) of southeastern England. These were the first sediments to be laid down during the transgression that flooded the Wealden Basin. (*Institute of Geological Sciences, Great Britain.*)

the widespread occurrence of a fauna of bottom-dwelling species including bryozoans, ostracods, foraminifers, brachiopods, bivalves, echinoids, and soft-bodied burrowers.

Knowing the typical thickness of the chalk and the total time elapsed during its deposition, we can estimate that it accumulated at a rate perhaps as great as 15 centimeters (~6 inches) per 1000 years. The coccoliths of the Chalk were produced for the most part by species not found in sediments of nearby oceanic areas, suggesting that the biotic environment of the Chalk Seas was unusual, or at least partly isolated by ocean currents. The productivity of the coccolithophores here and in the interior seaway of North America was apparently greater than that of coccolithophores in any region of the oceans today.

Seas flood Australia

Almost as spectacular as the mid-Cretaceous spread of seas over North America and Europe was the simultaneous inundation of much of the Australian continent. The primary site of the marine invasion of Australia was the Great Artesian Basin. During most of Jurassic time, this lowland area in the eastern half of the continent was already receiving nonmarine sediments, including plant remains that later formed beds of coal (Figure 16-50). Seas began to flood the basin early in the Cretaceous Period and spread to

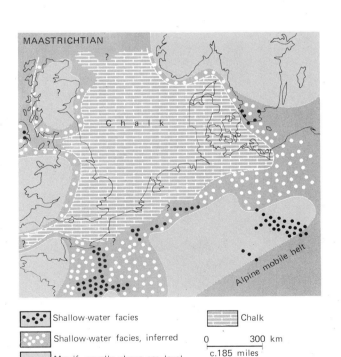

FIGURE 16-49 Paleogeography of northwestern Europe during Maastrichtian time. Chalk deposition was centered in the North Sea basin. Several stable blocks (massifs) formed islands around which marginal facies developed; these facies consisted mainly of coarse limestones but included siliciclastics as well. (*Modified from E. Hakansson et al., Spec. Publ. Internat. Assoc. Sedimentol. 1:211–233, 1979.*)

LATE JURASSIC

Great Artesian Basin

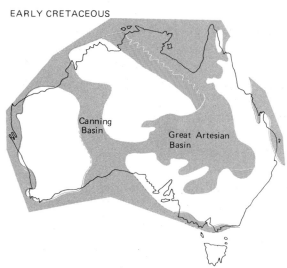

EARLY CRETACEOUS

Canning Basin

Great Artesian Basin

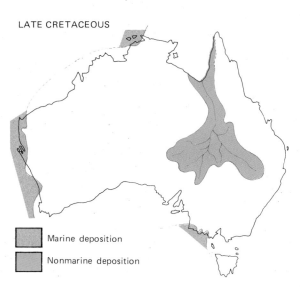

LATE CRETACEOUS

Marine deposition

Nonmarine deposition

their maximum extent in Aptian and Albian times, during the worldwide elevation of sea level (page 504). Flooding of the Great Artesian Basin and of the Canning Basin to the west was largely responsible for the Cretaceous rocks that appear today in surface or subsurface areas over about one-third of the Australian continent. Late Cretaceous time was marked by regression and a resulting transition from marine to nonmarine deposition in the Great Artesian Basin (Figure 16-50). Just as in North America, the seas never again spread widely over Australia. Cenozoic marine sediments fringe a few segments of the continent, but nowhere are such sediments found far inland.

CHAPTER SUMMARY

1.   With the evolutionary expansion of the dinoflagellates, diatoms, and coccolithophores during the Cretaceous Period, the phytoplankton assumed a modern character. Similarly, the diversification of the planktonic foraminifers contributed to the modernization of the zooplankton.

2.   Coccoliths raining down on the sea floor produced thick deposits of chalk in Western Europe and elsewhere.

3.   On the sea floor, predators such as crabs, teleost fishes, and carnivorous snails diversified markedly during Cretaceous time, thereby altering the nature of the marine ecosystem.

4.   In mid-Cretaceous time, rudist bivalves temporarily displaced corals as the primary builders of organic reefs.

5.   On the land, angiosperms, or flowering plants, displaced gymnosperms as the most diverse group of plants.

FIGURE 16-50   Flooding of Australia during the worldwide elevation of sea level of Aptian and Albian (late Early Cretaceous) time. The largest area to be flooded was the great Artesian Basin, which had already become an eastward-draining non-marine depositional basin in Late Jurassic time. During Late Cretaceous time, the Australian continent stood almost entirely above sea level, and the great Artesian Basin was again the site of nonmarine deposition, though it now drained to the north. (*After D. A. Brown et al., The Geological Evolution of Australia and New Zealand, Pergamon Press, Oxford, 1968.*)

6. Gondwanaland broke apart during the Cretaceous Period, forming the South Atlantic and other oceans.

7. Sea level rose throughout most of Cretaceous time, flooding much of Europe and spreading over western North America from the Gulf of Mexico to the Arctic Ocean.

8. In western North America, orogenic activity shifted eastward from its Jurassic position.

9. World climates cooled in Late Cretaceous time, but before this happened they were so warm that rudist reefs flourished in southern Europe and extended from the Gulf of Mexico northward along the Atlantic coast to New Jersey.

10. At the end of the Cretaceous Period, mass extinction eliminated the ammonoids, rudists, marine reptiles, and dinosaurs and devastated many other groups of organisms.

EXERCISES

1. What were the most important groups of swimming predators in Cretaceous seas?

2. What kinds of gymnosperms and flowering plants of Cretaceous age survive to the present?

3. What evidence do land plants offer about the nature of Late Cretaceous climates?

4. What conditions may account for the formation of widespread black shales on continental surfaces at certain times during the Cretaceous Period?

5. What modern continents that were once part of Gondwanaland remained attached to each other at the end of the Cretaceous Period?

6. Why did thick siliciclastic deposits accumulate in the western interior of North America during Cretaceous time?

ADDITIONAL READING

Bakker, R. T., "Dinosaur Renaissance," *Scientific American,* April 1975.

Kurtén, B., *The Age of Dinosaurs,* McGraw-Hill Book Company, New York, 1968.

Kurtén, B., "Continental Drift and Evolution," *Scientific American,* March 1969.

Russell, D. A., "The Mass Extinctions of the Late Mesozoic," *Scientific American,* January 1982.

PART SEVEN

The Cenozoic Era

The Cenozoic Era earned its nickname, The Age of Mammals, because of the rapid diversification of these animals after the dinosaurs had disappeared. Modern snakes, frogs, and songbirds made their appearance during Cenozoic time, as did grasses and weedy plants, while marine life underwent only modest changes during the era—a notable exception being the origin and diversification of whales. Youthful mountains such as the Alps, Himalayas, and Rockies that stand tall today had Cenozoic origins.

Mid-Cenozoic time was characterized by climatic cooling in many areas, refrigeration of the deep sea, and a mild mass extinction in the oceans. Climates and deep sea waters have since remained cooler than they were early in the era, and about 3 million years ago, continental glaciers formed in the Northern Hemisphere, initiating a glacial age that has continued to the present. This glacial age has witnessed the appearance of the human species and the evolution of its various modern cultures.

CHAPTER 17
The Paleogene World

The close of the Cretaceous Period marked a major transition in earth history. Never again did coccolithophores abound to the degree that they formed massive chalk deposits; scarcely any belemnoids survived, and ammonoids, rudists, and marine reptiles were gone from the seas. What remained were marine taxa that persist as familiar inhabitants of modern oceans, including bottom-dwelling mollusks, teleost fishes, and other groups that will be described in this chapter. On the land, the flowering plants of the Paleogene resembled those of latest Cretaceous time in many ways, but animal life differed dramatically from that of the previous period. Taking the place of the dinosaurs were the mammals, which were universally small and inconspicuous at the start of the Paleogene interval but in many ways resembled modern mammals by the period's end.

The most profound geographic change during Paleogene time was a refrigeration of the earth's polar regions, which resulted in a chilling of the deep sea and, ultimately, in continental glaciation, a phenomenon that will be discussed in the next chapter. Of great importance in North America were Paleogene mountain-building events in the west that foreshadowed subsequent Neogene uplifts of ranges like the Sierra Nevada and Rocky Mountains. In southern Europe, the Paleogene elevation of the Alps was of comparable significance.

The sediments that record these and other Cenozoic events are for the most part unconsolidated, or soft, although most Neogene carbonates and some siliciclastics are lithified. Cenozoic deposits are widespread in Europe, and the distinction between their fossil biotas and those of Mesozoic deposits was recognized early in the history of geologic investigation. Subdivision of the Cenozoic Era is less clear-cut; today, many geologists divide the era into two periods: the **Paleogene Period**, which includes the **Paleocene**, **Eocene**, and **Oligocene** epochs, and the **Neogene Period**, which includes the **Miocene, Pliocene, Pleistocene**, and **Recent** (or **Holocene**) epochs. This Paleogene-Neogene classification has become increasingly popular during the past two decades and has thus been

Scotts Bluff in western Nebraska. This landmark of the high plains is the erosional remnant of a thin blanket of sediments that were shed eastward during Oligocene time from an area of mountain building to the west. (*Photograph by W. S. Keller, National Park Service, U.S. Department of the Interior.*)

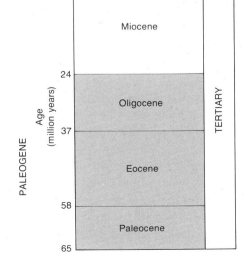

FIGURE 17-1   Teleost fishes from the Eocene System of
Monte Bolca, Italy, illustrating the great variety of body forms
present by this time. (*Field Museum of Natural History.*)

adopted in this book. Traditionally, however, the Cenozoic Era has been divided into two periods of quite different lengths: the **Tertiary**, which encompasses the interval from the Paleocene through the Pliocene, and the **Quaternary**, which includes only the Pleistocene and Recent epochs (an interval of less than 2 million years). The advantage of the Paleogene-Neogene division is that it yields two periods of more similar duration.

Fortunately, the question of how to divide the Cenozoic into periods is not of great consequence, because geologists tend to focus less on the periods of this era than on its epochs. The reason is that the Cenozoic epochs, though shorter than Paleozoic or Mesozoic periods, are so recent that their sedimentary records are well displayed and can be analyzed in great detail—especially now that the history of deep-sea sediments and their fossil biotas have come to light through the use of coring techniques.

The first formally recognized Paleogene epoch was the Eocene, which was established in 1833 by Charles Lyell on the basis of deposits found in the Paris and London basins. (These deposits will be discussed later in the chapter.) Lyell named and described the Eocene Series in his great book *Principles of Geology,* a work that also served to popularize the uniformitarian view of geology (page 3). As it was initially designated, the Eocene Series constituted a thick stratigraphic interval that corresponded to what is now recognized as the entire Paleogene System. In other words, the Eocene Series originally included the sediments now assigned to the Paleocene and Oligocene series. It was not until 1854 that Heinrich Ernst von Beyrich distinguished the Oligocene Series from the Eocene in Germany and Belgium on the basis of fossils, and then, in 1874, W. P. Schimper established the Paleocene Series on the basis of distinctive fossil assemblages of terrestrial plants in the Paris Basin.

WORLDWIDE EVENTS

In this chapter, we will depart from the format of the preceding five chapters by discussing life of the Paleogene Period under the heading of "Worldwide Events." The logic underlying this approach is that Paleogene life is so familiar to us that it requires no special introduction; its most interesting features are the major expansions and contractions of certain taxonomic groups on a worldwide scale.

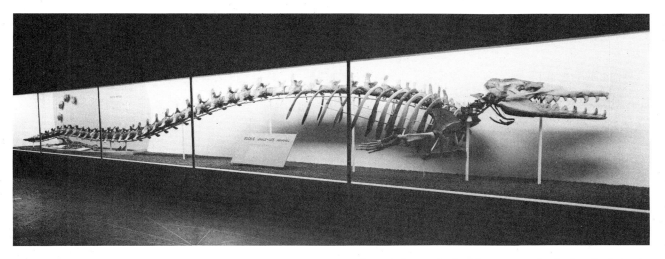

FIGURE 17-2   The Eocene whale **Basilosaurus**, a carnivore endowed with large teeth. Although it was not enormous by modern whale standards, this genus attained a length of about 14 meters (~55 feet). (*Smithsonian Institution.*)

Nor will we devote a special section to Paleogene paleogeography, since the Paleogene lasted no more than 41 million years—a span in which paleogeographic changes occurred on such a small scale that they are best considered under the heading "Regional Events."

Evolution of marine life

The present marine ecosystem is for the most part populated by groups of animals, plants, and single-celled organisms that survived the extinction at the end of the Mesozoic Era to expand during the Cenozoic. Many benthic foraminifers, sea urchins, cheilostome bryozoans, crabs, snails, bivalves, and teleost fishes survived in sufficiently large numbers to assume prominent ecologic positions in Paleogene seas (Figure 17-1). Perhaps the biggest beneficiaries of the terminal Cretaceous extinction, however, were the reef-building corals, which had relinquished their dominant reef-building role to the rudists in mid-Cretaceous time but reclaimed it following the rudists' extinction. Unfortunately, few Paleocene coral reefs have been found, partially because of the relative rarity of exposed Paleocene strata representing tropical conditions. The rarity of Paleocene coral reefs also seems to indicate that corals did not recover quickly from the terminal Cretaceous mass extinction. During the warm Eocene Epoch that followed the Paleocene, however, corals were again widespread.

The calcareous phytoplankton, including the coccolithophores, suffered severe losses at the end of the Cretaceous Period but rediversified somewhat during the Paleogene. These forms as well as the diatoms and dinoflagellates, which were not as adversely affected, have accounted for most of the ocean's productivity throughout the Cenozoic Era, just as they did in Cretaceous time.

Although many elements of Paleogene marine life closely resembled those of Late Cretaceous age, some forms were dramatically new. Perhaps the most distinctive marine organisms of this period were the **whales**, which during the Eocene Epoch evolved from carnivorous land mammals and quickly achieved success as large marine predators (Figure 17-2). Joining the whales as replacements for the reptilian "sea monsters" that were the top carnivores of the Mesozoic Era were enormous sharks (Figure 17-3). Unlike the whales, however, sharks descended from similar creatures that lived during Cretaceous time.

It would appear that the marine ecosystem expanded during Paleogene time to include new niches along the fringes of the oceans. **Sand dollars**, for example, which are the only sea urchins able to live along sandy beaches, evolved at this time from biscuit-shaped ancestors (Figure 17-4), and new kinds of bivalve mollusks also invaded exposed sandy coasts. Both of these new groups have successfully inhabited shifting sands by virtue of their ability to reburrow quickly when washed from the sand by waves. Other newcomers to the ocean margins were the penguins, a group of swimming birds of Eocene origin, and possibly the **pinnipeds**, a group that includes walruses, seals, and sea lions. It is widely believed that the pinnipeds evolved

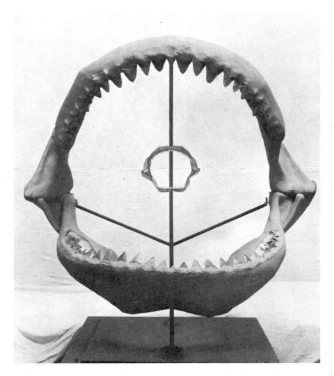

FIGURE 17-3   Jaws of the enormous fossil shark **Carcharodon** from the Eocene System of South Carolina—compared here with the jaws of a modern shark. The jaws of the fossil shark were more than 2 meters (>6.5 feet) across. (*Field Museum of Natural History.*)

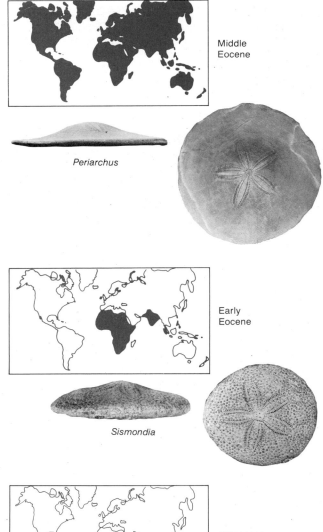

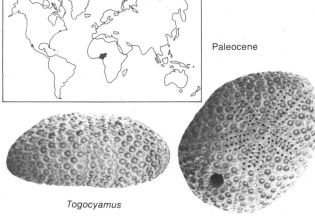

before the beginning of the Neogene Period, although this group left no known Paleogene fossil record.

Evolution of terrestrial plants

For land plants, the story is quite different. The transition to the Paleogene was apparently not marked by any drastic change in the character of terrestrial floras; instead, the great radiation of flowering plants simply continued. In the process, modern families of flowering plants evolved. By the beginning of Oligocene time, some 37 million years ago, about half of all genera of flowering plants were ones that are alive today, and although many modern plant genera had not yet evolved, forests had taken on a distinctly modern appearance. The oldest known rose species, for example, is of Late Eocene age (Figure 17-5).

One major evolutionary event that did take place during the Paleogene interval was the origin of the **grasses**. Al-

FIGURE 17-4   Specimens illustrating the evolution of the sand dollars from biscuit-shaped Paleocene ancestors (*Togocyamus*), which are known only from Africa. True sand dollars (*Periarchus*) were present throughout the world by Middle Eocene time. *Sismondia* is an intermediate genus that has been found in Africa and India. (*Courtesy of P. M. Kier, Smithsonian Institution.*)

Another interesting evolutionary point to consider is that, with the enormous numbers of individual grass plants populating marshes and grasslands, effective reproduction would be difficult if these plants required pollination by insects. Thus, it comes as no surprise that grasses are wind-pollinated plants. It should be noted, however, that grasses can also carpet large areas by means of budding, a nonsexual reproductive process in which individual plants put out underground runners that sprout new plants, thus rapidly and densely colonizing the soil.

Early Paleogene terrestrial and freshwater animals

In Chapter 6, we saw that the mammals, having inherited the world from the dinosaurs, underwent a remarkably rapid adaptive radiation during the early part of the Cenozoic Era. It was probably through both competition and predation that the dinosaurs prevented the mammals from undergoing any great evolutionary expansion during Mesozoic time.

At the start of the Paleocene Epoch, when they first had the world to themselves, most mammals were small creatures that resembled modern rodents; at this time, it would appear that no mammal was substantially larger than a good-sized dog. Furthermore, most mammal species tended to remain generalized in both feeding and locomotory adaptations; those that dwelt on the ground generally retained a primitive limb structure in which the heels of the hind feet and the "palms" of the front feet touched the ground during movement. Perhaps 12 million years later, however, by the end of Early Eocene time, mammals had diversified to the point that most of their modern orders were in existence (Figure 17-6). Bats already fluttered through the night air (Figure 17-7), for example, and, as we have seen, large whales swam the oceans.

Also included among the Paleocene mammals were groups that survived from Cretaceous time (Box 16-1), such as marsupials, multituberculates, and placental mammals called insectivores. Other Paleocene mammals have been assigned by some experts to the **Primates**, the order to which modern humans belong. Although these small animals were quite different from monkeys, apes, or humans in many respects, by Early Eocene time they did climb with grasping hindlimbs and forelimbs that foreshadowed our own hands and feet (Figure 17-8). By mid-Paleocene time, true mammalian carnivores—members of the living order

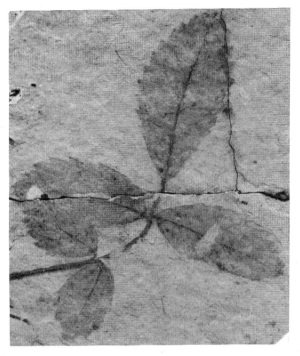

FIGURE 17-5　One of the oldest known rose plants. This specimen was preserved in fine-grained sediments of the Florissant Formation of Colorado. Fossils of the same species, *Rosa hilliae*, have also been found in the Upper Eocene Green River Formation of Utah. It probably produced insect-attracting flowers, although these would not have been as beautiful, by our standards, as those developed in modern times by artificial breeding. (*Smithsonian Institution.*)

though these usually low-growing flowering plants were present before the end of the Paleocene, they did not reach their full ecologic potential until Late Oligocene and Miocene times. Early grasses were apparently confined to wooded or swampy areas. Like the modern sedges that form marshlands along continental coastlines, the mode of growth of early grasses did not allow their leaves to grow continuously and thus to recover from heavy grazing by animals of the sort that inhabit open country in large numbers. It was only by virtue of an adaptive breakthrough— the origin of continuous growth, the process that forces us to cut our lawns every week or two—that grasses were ultimately able to invade open country with great success. Once they were able to survive the effects of heavy grazing by animals, grasses quickly spread over vast expanses of the earth to form grasslands.[1]

---

[1] This idea has been proposed informally, but not yet published, by Leo J. Hickey of Yale University.

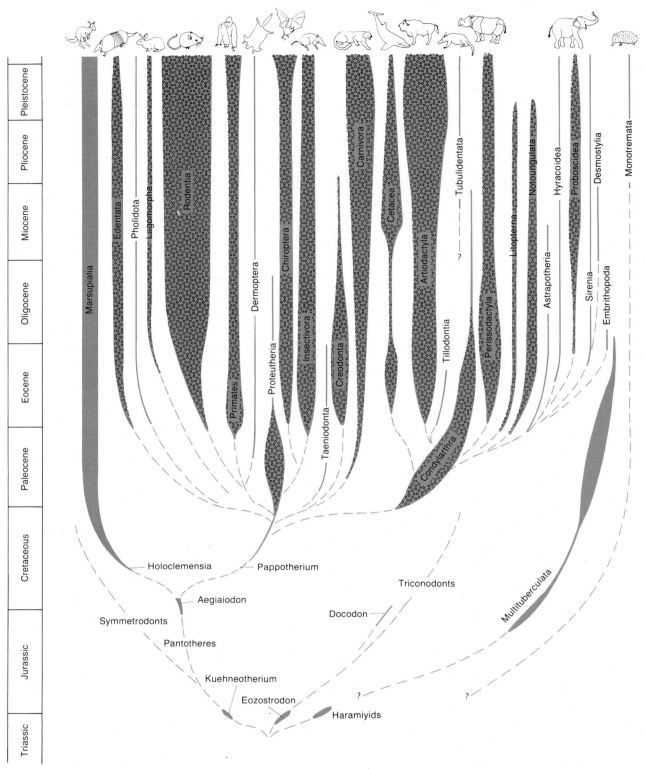

FIGURE 17-6 The pattern of adaptive radiation of mammals that gave the Cenozoic Era its informal name, the Age of Mammals. It is thought that the two large groups of modern mammals, the placentals (*solid color*) and the marsupials, shared a common ancestry in the Cretaceous Period. The multituberculates, which failed to survive into Neogene time, apparently evolved separately—as did the Monotremata, primitive survivors of which include the Australian platypus, which lays eggs instead of giving birth to live young. For the placental mammals, each vertical bar represents an order of mammals. Note that most of the new Cenozoic orders had already evolved by the beginning of the Eocene Epoch, about 10 million years after the disappearance of the dinosaurs. (*After P. D. Gingerich, in A. Hallam [ed.], Patterns of Evolution, Elsevier Publishing Company, Amsterdam, 1977.*)

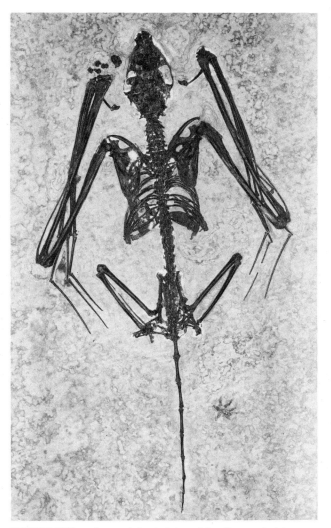

FIGURE 17-7   A complete skeleton of the Eocene bat *Icaronycteris*, preserved in the Green River Formation of Wyoming. This bat had tiny, sharp claws that have no counterparts in living bats. (*Courtesy of W. Starks, Princeton Museum of Natural History.*)

FIGURE 17-8   Reconstruction of the early primate *Cantius*, a small genus of Early Eocene time. This arboreal animal had large toes on its hind feet and nails much like our own, and it apparently jumped from limb to limb. (*Courtesy of R. T. Bakker.*)

Carnivora—had emerged (Figure 17-9). This is the order to which nearly all living carnivorous placental mammals belong. By the end of Paleocene time, the earliest members of the horse family had evolved as well; these animals were no larger than small dogs (Figure 17-10), but by the end of the epoch larger herbivorous mammals, some the size of cows, had also appeared.

The Eocene Epoch was marked by a continued increase in the variety of mammals. The number of mammalian families doubled to nearly 100, approximating that of the world today. In addition, more modern varieties of hoofed herbivores began to appear. Most animals of this

kind are known as **ungulates** and these are divided into **odd-toed ungulates** (living horses, tapirs, and rhinos) and **even-toed ungulates**, or cloven-hoofed animals (cattle, antelopes, sheep, goats, pigs, bisons, camels, and their relatives). The odd-toed ungulates expanded before the even-toed group did, but primitive even-toed ungulates were also present early in the Eocene (Figure 17-11); of the modern even-toed types, camels and relatives of the present-day chevrotain (the Oriental "mouse deer") evolved before the end of the epoch. The earliest members of the **elephant** order also appeared during Eocene time; The oldest known genus, *Moeritherium*, was a bulky animal

FIGURE 17-9 Reconstruction of a mid-Paleocene biota of New Mexico. The large trees represent the sycamore genus *Platanus*, which has survived to the present time. The early arboreal carnivore (*Criacus*) climbing the tree on the left watches a small insectivore (*Deltotherium*) on a small branch in the center of the drawing. On the ground, mesonychid carnivores of the genus *Ancalagon* feed on the small crocodile *Allognathosuchus*. Ferns and sable palms (or fan palms) constitute the undergrowth in the background. (*Drawing by Gregory S. Paul.*)

about 3 meters (~10 feet) long, that possessed rudimentary tusks and a short snout rather than a fully developed trunk (Figure 17-12). The **rodents**, which had originated in the Paleocene, continued to diversify as well, but their success may have been attained at the expense of the archaic multituberculates, which were also specialized for gnawing seeds and nuts. As the rodents expanded during

FIGURE 17-10 *Hyracotherium* ("eohippus"), the earliest genus of the horse family. This animal, which was present in Late Paleocene and Eocene time, was no larger than a small dog. It had four toes on each front foot and three on each hind foot. (*Field Museum of Natural History.*)

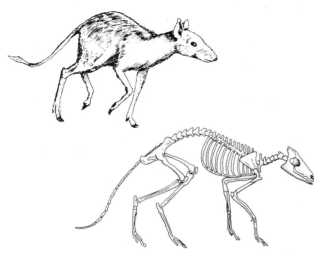

FIGURE 17-11   *Diacodexis*, an early even-toed ungulate, or cloven-hoofed herbivore. *Hyracotherium* (Figure 17-10), in contrast, was an early odd-toed ungulate. The limb structure of *Diacodexis* shows that it was an unusually adept runner and leaper for Early Eocene time. (*After K. D. Rose, Science 216:621–623, 1984.*)

FIGURE 17-12   *Moeritherium*, an early member of the elephant group. This elongate animal stretched to a length of about 3 meters (~10 feet). During the Eocene Epoch, it probably wallowed in shallow waters and grubbed for roots or other low-growing vegetation. (*After B. Kurtén, The Age of Mammals, Columbia University Press, New York, 1971.*)

the Eocene, the multituberculates declined, finally becoming extinct early in the Oligocene Epoch.

Among the animals that preyed on the herbivorous mammals described above were groups that had their evolutionary origin in the Paleocene Epoch. These include the superficially doglike **mesonychids** and the **diatrymas**, which were huge flightless birds with powerful clawed feet and enormous slicing beaks (Figure 17-13). The diatrymas disappeared toward the end of the Eocene Epoch. At that time the mesonychids were joined by primitive members of three familiar modern carnivore groups: the **dog**, **cat**, and **weasel** families.

FIGURE 17-13   Large terrestrial predators that had evolved by Early Eocene time. The animals that superficially resemble dogs are giant mesonychids of the genus *Pachyhyaena*, which were the size of small bears. The flightless birds guarding their chicks are members of the genus *Diatryma*, which stood about 2.4 meters (~8 feet) tall. (*Drawing by Gregory S. Paul.*)

FIGURE 17-14   The long-legged Eocene duck *Presbyornis*. Fossils of this wading bird are found in large numbers in Green River sediments of Wyoming and Utah, which suggests that it lived in enormous colonies. *Presbyornis* left tracks revealing that webbed feet at the end of its long legs supported it when it walked in mud. Along with the tracks are lines of probing marks made by the beak while searching for food. The legs of *Presbyornis* were too long to allow the bird to swim, but modern ducks have inherited its webbed feet and employ them as paddles in swimming. (*After drawing by J. P. O'Neill. Photograph by the author.*)

The monstrous diatrymas were not the only birds of the Eocene, but flying birds were much less diverse then than they are today. Most species were shore birds that waded in shallow water when they were not in flight (Figure 17-14). Not yet present were many other modern kinds of birds, including the song birds, which are especially numerous today.

As for other forms of vertebrate life, reptiles and amphibians were relatively inconspicuous during Paleogene time. The first record of the Ranidae, the largest family of living frogs, is in the Eocene Series (Figure 17-15), but the fossil record of this group of fragile animals is not good, and thus we do not know precisely when the Ranidae originated or attained high diversity.

There is no question that the insects took on a modern appearance with the origin of several modern families in Paleogene time (Figure 17-16). Several Oligocene forms,

FIGURE 17-15   Complete skeleton of an Eocene frog that belongs to a still-living family. This specimen comes from the Green River Formation of Wyoming. (*Courtesy of L. Grande.*)

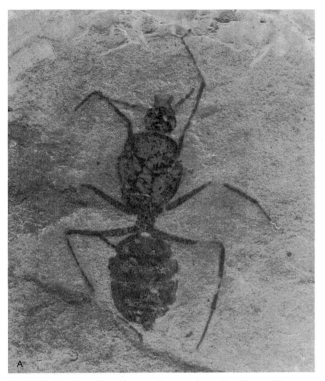

FIGURE 17-16   Two Eocene insects from the Green River Formation, both of which belong to families that have undergone large adaptive radiations since Eocene time. *A.* An ant, *Liometopum,* of the family Formicidae, which was about 1.5 centimeters—or more than $\frac{1}{2}$ inch—in length. *B.* A crane fly of the family Tipulidae, whose length was about 2 centimeters, or $\frac{4}{5}$ inch. (*A. Courtesy of L. Grande. B. Field Museum of Natural History specimen PE 39254.*)

which have been preserved with remarkable precision in amber, closely resemble living species.

Mammals of the Oligocene Epoch

During the Oligocene Epoch, the modernization of mammals continued. Many Eocene families died out, and there were important expansions of many living (Recent) groups ranging from the tiny rodents to the massive elephants. The horse family had disappeared from Eurasia during the Eocene Epoch but survived in North America by way of a small number of species, including members of the three-toed genus *Mesohippus* (Figure 17-17). Other odd-toed ungulates that enjoyed greater success during the Oligocene Epoch than in previous intervals were the **rhinos,**

FIGURE 17-17   Reconstruction of an Oligocene horse belonging to the genus *Mesohippus.* This three-toed animal stood about $\frac{1}{2}$ meter (~20 inches) tall at the shoulder. (*Field Museum of Natural History.*)

FIGURE 17-18   *Indrichotherium* (formerly *Baluchitherium*), which, as far as we now know, was the largest mammal ever to walk the earth. This Oligocene giant, which belonged to the rhi- noceros family, stood about 5.5 meters (~18 feet) at the shoul- der. This is the height of the top of the head of a good-sized modern giraffe. (*Drawing by Gregory S. Paul.*)

FIGURE 17-19   The titanothere *Brontops*, which belonged to a family of odd-toed ungulates that were related to the rhinos but had blunt horns rather than sharp ones. Titanotheres were abundant in the American West during the Oligocene Epoch. (*Field Museum of Natural History.*)

which included the largest land mammal of all time (Figure 17-18), and the rhinolike **titanotheres** (Figure 17-19). As these animals illustrate, there were many more large mammals during the Oligocene Epoch than during the Eocene.

As the Oligocene Epoch progressed, odd-toed ungulates were outnumbered for the first time by even-toed ungulates, including deerlike animals (Figure 17-20) and pigs, which became especially diverse.

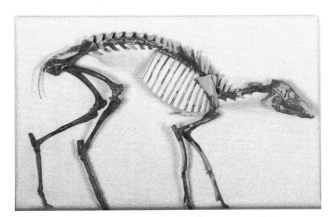

FIGURE 17-20   *Leptomeryx*, a deerlike, even-toed (cloven-hoofed) ungulate of Oligocene age. This animal was only about 0.6 meters (~2 feet) in length. (*Smithsonian Institution.*)

Among the carnivores, the dog, cat, and weasel families, which had their origins in Eocene time, radiated during the Oligocene Epoch and produced more advanced forms, including large saber-toothed cats (Figure 17-21), bearlike dogs, and animals that resembled modern wolves.

An especially important aspect of the modernization of mammals during the Oligocene Epoch was the appearance of **monkeys** and apelike primates. The genus *Aegyptopithecus,* an arboreal animal the size of a cat, had teeth resembling those of an ape but possessed a head and a tail resembling those of a monkey (Figure 17-22). In Neogene time, before the arrival of humans, apes attained considerable importance in the terrestrial ecosystem.

Climatic change and mass extinction

Flowering plants (angiosperms) are commonly viewed as the thermometers of the past 100 million years. As we have seen, their value derives in part from the strong correlation between mean annual temperature and percentage of species within a flora that have leaves with entire margins (Figure 2-21). Recall that evaluation of this relationship indicates that there was a 10°C (18°F) temperature drop between Maastrichtian and Paleocene time in Wyoming and Montana (page 504). Younger fossil floras indicate that relatively cool temperatures prevailed here until Late

FIGURE 17-21   The Oligocene cat *Dinictis*, which was approximately the size of a modern lynx. This animal possessed advanced adaptations for running and springing upon prey. Its canine teeth were elongated for stabbing, but *Dinictis* was not a member of the true saber-toothed cat group. (*Drawing by Gregory S. Paul; photograph from the Field Museum of Natural History.*)

Paleocene time, when a warming trend began that persisted into the Eocene. As shown in Figure 17-23, data for floras of the Mississippi Embayment show a similar pattern.

Leaf-margin data for several North American localities also reveal that two cool episodes took place before the end of the Eocene Epoch (Figure 17-23). During Middle and Late Eocene time, however, there were warm intervals as well. This was a time when sea level stood relatively high (Figure 17-24), and so there were many warm, shallow seas. The composition of the Middle Eocene flora of southern Alaska suggests a mean annual temperature of about 22°C (~72°F), which is comparable to that of lowland areas in southern Mexico today.

Near the end of the Eocene Epoch, a dramatic episode of refrigeration occurred on a global scale. Changes in the compositions of floras indicate that along the west coast

of North America, the mean annual temperature at this time declined by about 12°C (~22°F). Here there was a large extinction of marine mollusks. Some taxonomic groups, including the planktonic foraminifers, suffered heavy extinction throughout the world. There is evidence, too, that a pulse of extinction struck land mammals, although the details of this episode have yet to be worked out.

The events that took place in the deep sea provide an explanation for the episode of cooling and extinction that occurred in Late Eocene time. Study of deep-sea cores reveals not only a major extinction of abyssal benthic foraminifers and ostracods but also an increase of the ratio of oxygen 18 to oxygen 16 in the skeletons of the foraminifers in equatorial regions and near Antarctica. This increase apparently reflected the first growth of glacial ice on and adjacent to Antarctica, with the lighter oxygen iso-

FIGURE 17-22   The Oligocene primate *Aegyptopithecus*, whose name reflects its discovery in Egypt. The skull (*right*) resembles that of a monkey, but the teeth are apelike. The brain of *Aegyptopithecus* was unusually large for the size of the animal, perhaps reflecting a high level of intelligence for the Oligocene world. (*Drawing by S. F. Kimbrough. Photograph courtesy of R. F. Kay.*)

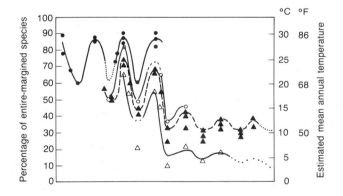

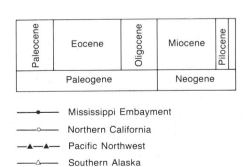

FIGURE 17-23   Curves showing estimates of Cenozoic temperature changes in four areas of North America. The curves are based on percentages of entire-margined leaves in fossil floras. The fact that the curves follow parallel paths where they overlap in time suggests that the trends hold for large areas of the earth's surface. (*After J. A. Wolfe, American Scientist 66:694–703, 1978.*)

tope (^{16}O) accumulating preferentially in this ice (page 55). The growth of glacial ice on Antarctica was a prelude to the greater expansion of high-latitude ice sheets that took place near the end of the Cenozoic Era during what is informally known as the recent Ice Age.

It was during the Late Eocene cooling episode that the psychrosphere—the cool bottom layer of the ocean (page 55)—came into being as cold, dense polar water began to sink to the deep sea. The short interval of isotopic change that characterizes deep-sea cores suggests that the psychrosphere formed in less than 100,000 years; the temperatures of bottom waters over large areas of the deep sea dropped by perhaps 4 to 5°C (~7 to 8°F).

It is assumed that the upwelling of psychrospheric waters affected climates in many parts of the world. Nonetheless, it is not known to what extent this widespread cooling was an effect of the Antarctic glacial buildup and to what extent it was a cause.

A modest global lowering of the sea occurred near the Eocene-Oligocene boundary, probably in response to the growth of Antarctic glaciers (Figure 17-25). In contrast, the oceans of Early Oligocene time stood high in relation to the land—probably as a result of partial glacial melting. Floral evidence in many areas, however, indicates that climates remained cooler during this period than during the early part of the Eocene Epoch, when glaciers had not yet formed. In fact, global temperatures may never again have risen to the levels that they attained in Eocene time.

LATE EOCENE

FIGURE 17-24   World geography during Late Eocene time, an interval in which substantial areas of continental crust were inundated. Relatively warm climates prevailed even as far north as Ellesmere Island, where a diverse fauna of terrestrial plants and vertebrate animals flourished in Middle Eocene time. The Bering land bridge, a neck of continental crust, stood above sea level throughout the Eocene interval, permitting faunal interchange between North America and Eurasia.

A sudden drop in sea level occurred during the latter part of the Oligocene Epoch. This event, whose origin is not known, has been shown by seismic stratigraphy to have produced one of the lowest ocean levels in all of Phanerozoic time. This appears to have been the only time during the entire Paleogene Period when sea level on a global scale was lower than it is today. No heavy marine extinctions occurred at this time, but several groups of land mammals, including the titanotheres (Figure 17-19), died out, probably in response to changes in climate and vegetation (Figure 17-23).

The Eocene Epoch was a time when moist tropical and subtropical forests cloaked much of North America and Eurasia, but the cooling events of late Eocene and Oligocene time altered this condition forever. Paleobotanical studies show that during the Oligocene Epoch, in response to cooler and drier conditions, savannahs—grassy plains with scattered trees and shrubs (page 41)—spread across large areas of major continents. At the same time, moist subtropical and tropical forests came to be confined largely to low latitudes, where they remain to this day in the form of jungles and rain forests. Because many of the evolutionary changes associated with this climatic transition took place in Miocene time, we will pick up this story in the next chapter.

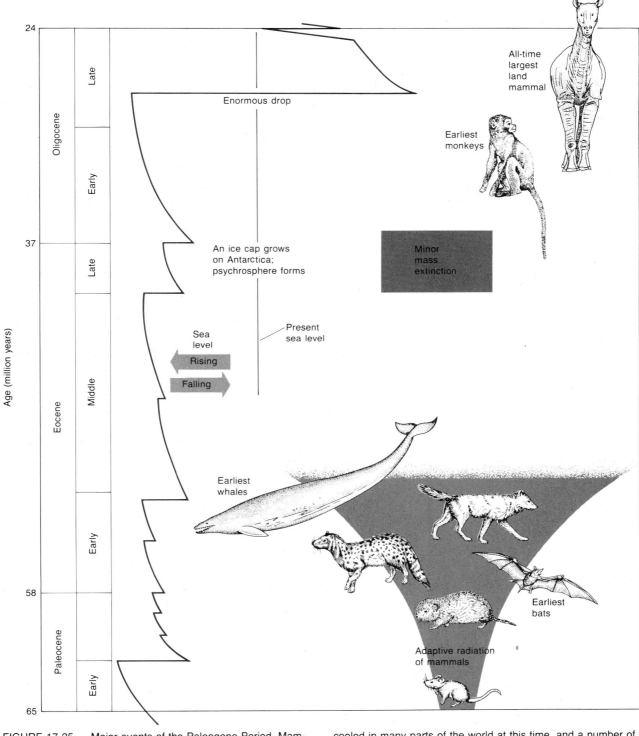

FIGURE 17-25   Major events of the Paleogene Period. Mammals, which were small and of limited diversity early in the Paleocene Epoch, radiated rapidly. Near the end of the Eocene Epoch, an ice cap formed on Antarctica, and cold water began to settle to the deep sea, forming the psychrosphere. Climates cooled in many parts of the world at this time, and a number of taxa became extinct. Sea level stood considerably higher than it stands today until Late Oligocene time, when it dropped to a very low level.

REGIONAL EVENTS

In examining important regional events of the Paleogene Period, we will travel to the ends of the earth—first to the South Pole, where Antarctica became separated from Australia and developed its icy cover, and then toward the North Pole, where land areas of North America and Eurasia were more closely connected than they are today. Next we will look at the history of the Cordilleran region of North America, where the Laramide orogeny yielded many structures that remain conspicuous in the Rocky Mountains today. Before leaving North America, we will examine the nature of deposition along the Gulf Coast, where large volumes of petroleum are trapped in Paleogene sediments. Our final stop will be Europe, where many fossiliferous deposits accumulated and the modern structure of the Alps began to form.

Antarctica and the origin of the psychrosphere

The only major event of continental rifting that took place exclusively during the Cenozoic Era was the separation of Australia from Antarctica, which occurred during the Eocene Epoch. With the inception of this rifting early in Eocene time, Australia began to move northward, and the resulting alteration of wind and water circulation near the South Pole had consequences that extended over much of the planet.

Even before its separation from Australia, Antarctica had been centered over the South Pole, but it remained warm because its shores were bathed in relatively warm waters from lower latitudes (Figure 17-26A). Oxygen isotope analyses of foraminifers from deep-sea cores suggest, however, that sea temperatures adjacent to Antarctica dropped during Eocene time as Australia drifted northward. A major cause of this cooling was the formation of the circum-Antarctic current that, up to the present time, has flowed clockwise around Antarctica partly because of the prevailing winds that result from the earth's rotation in the opposite direction (Figure 2-23). During Oligocene time, despite the fact that narrow continental connections remained with Australia and South America, the circum-Antarctic current was established near the sea surface (Figure 17-26B), and the apparent consequence was that warm waters from low latitudes failed to reach Antarctica. As a

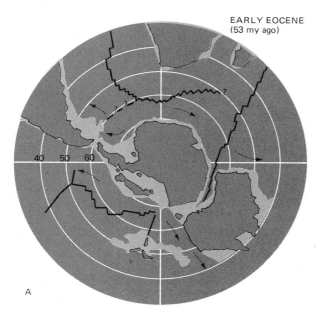

EARLY EOCENE
(53 my ago)

A

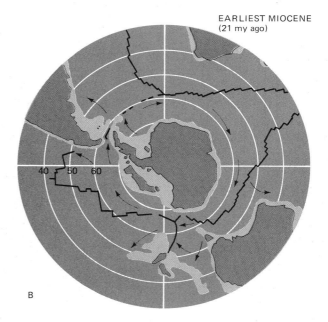

EARLIEST MIOCENE
(21 my ago)

B

FIGURE 17-26   Separation of Antarctica and Australia, remnants of Gondwanaland near the South Pole. Before the two continents separated (A), Antarctica was warmed by currents that reached it from lower latitudes. After separation, however, the earth's rotation set up a clockwise gyre of water around Antarctica, thereby isolating the newly formed continent from warm currents and allowing it to cool in its polar position (B). As a

result of this cooling, sea ice began to form around Antarctica at the very end of the Eocene interval; cold water plunged downward and spread throughout the deep sea, contributing to the origin of the psychrosphere, which still exists today. (*After J. Kennett, Marine Geology, Prentice-Hall, Inc., Englewood Cliffs, New Jersey, 1982.*)

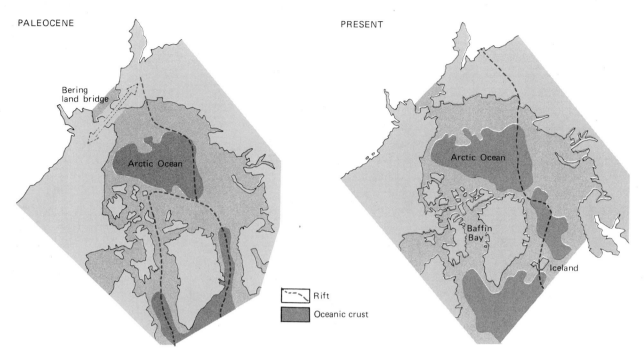

FIGURE 17-27   Changes in the pattern of rifting in the North Atlantic during the Cenozoic Era. In mid-Paleogene time, rifting ceased between Greenland and North America but continued between Greenland and Eurasia, where it persists today. (*After E. M. Herron et al., Geology 2:227–280, 1974.*)

result, the continent and the waters adjacent to it, being largely isolated in a polar position, cooled down.

In response to this cooling—and perhaps also in response to worldwide cooling resulting from some other, unknown factor—the first sea ice began to form around Antarctica near the end of the Eocene, and waters of near-freezing temperature sank to the deep sea and spread northward, forming the psychrosphere. The direct result was an extinction of bottom-dwelling foraminifers and other larger inhabitants of the deep-sea floor. At the same time—and perhaps as a result of the Arctic cooling—the earth experienced more widespread climatic deterioration. As we have seen, floras reveal that temperatures plummeted at this time even in the terrestrial habitats of the Northern Hemisphere (Figure 17-23).

Changes in oxygen isotopes within sea water, as recorded in fossil foraminifers, reveal that a slow buildup of ice occurred on Antarctica during Oligocene time. Once the Antarctic "cooling system" was in place, the mean annual temperatures at high latitudes never again reached their Eocene levels; in fact, the climatic deterioration at the end of the Eocene fostered further deterioration later in Cenozoic time. The recent Ice Age of the Northern Hemi-

sphere, which will be described in Chapter 18, would never have occurred if global temperatures had not already been lowered.

Australia, on its journey northward during the Paleogene Period, experienced only minor marine encroachment along its margins—except during Late Eocene time, when global transgression reached its Paleogene peak. In the next chapter, we will review the history of Australia during Neogene time, when the migration of that continent caused it to experience major climatic modifications.

The top of the world: Changing positions of land and sea

Major geologic and geographic changes also took place during the Paleogene Period within 30° latitude of the North Pole (Figure 17-27). Many of these changes can be read from the ages of various segments of the deep-sea floor. The Arctic deep-sea basin was in existence during the Cretaceous Period, but until nearly the end of Cretaceous time, it remained separated from the Atlantic, except by way of shallow seas, because North America, Greenland, and Eurasia were still united as a single landmass. Then, as we saw in the previous chapter, the Mid-Atlantic Rift proceeded northward along two forks, splitting Greenland from North America on the west and Eurasia on the east. Finally, in mid-Paleogene time, the pattern of plate movement simplified as the western arm of rifting ceased to be active, ending the relative movement of Greenland away from North America. Since that time, the rifting of the northern Atlantic has been limited to the area between Greenland and Scandinavia. It has been suggested that plate movements in this area, like those in the south polar region, contributed to the origin of the psychrosphere. The suggestion is that early in Paleogene time the Arctic Ocean was isolated from larger oceans to the south and thus retained its frigid waters. Then, late in Eocene time, rifting between Greenland and Scandinavia proceeded far enough to allow a channel to open up between the Arctic basin and the Atlantic, causing dense, frigid Arctic waters to spill into the larger ocean as part of the psychrosphere.

Meanwhile, throughout the Cenozoic Era, continental crust has continued to separate the Arctic and Pacific basins; Alaska and Siberia are connected by a stretch of continental crust despite the fact that this connecting segment now lies submerged beneath the Bering Sea. During much of the Cenozoic Era, this segment, which is known as the **Bering land bridge**, stood above sea level, serving as a land corridor between North America and Eurasia. Throughout Paleogene time, this corridor remained open, allowing mammals and land plants to migrate between Asia and North America. As we will see, however, during the Neogene Period, more frequent inundations of the sea prevented the exchange of species between the old and new worlds.

Tectonics of western North America

Mountain-building activity in the Cordilleran region of North America continued into the Paleogene Period, but with a number of changes. Figure 17-28, which summarizes the orogenic history of the eastern Cordilleran region, shows that the Sevier episode occupied almost the entire Cretaceous Period. In latest Cretaceous time, however, a new style of tectonic activity was initiated, and it persisted through the Paleocene and well into Eocene time. The episode characterized by this new style is known as the Laramide orogeny.

The Laramide orogeny was not unusual in the north or in the south. In the north, extending from the United States into Canada, there remained an active belt of igneous activity and, inland from it, an active fold-and-thrust belt (Fig-

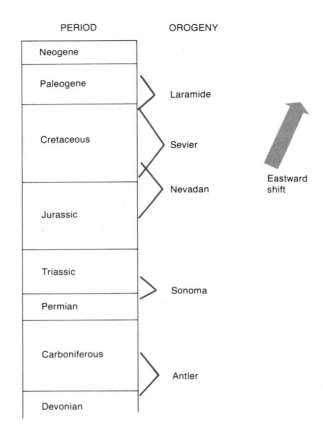

FIGURE 17-28   Summary of major orogenic events in the eastern Cordilleran region. Between Jurassic and Paleogene time, orogenic activity migrated eastward (see Figure 15-41).

ure 17-29). Thrust sheets of enormous proportions are spectacularly exposed in the Canadian Rockies (Figure 17-30). A similar pattern of tectonism persisted both in the southern United States and in Mexico.

The unusual features of the Laramide orogeny were in the central part of the western United States, where a broad area of tectonic quiescence extended from central Utah through Nevada to California. East of this inactive region there was a strange pattern of tectonism in which large blocks of crystalline basement rock were uplifted in a belt extending from Montana to Mexico. The largest of these blocks were centered in Colorado, where the Ancestral Rocky Mountains had formed as basement uplifts more than 200 million years earlier, late in the Paleozoic Era (page 431).

What created the unusual tectonic pattern that characterized the central part of the Cordilleran region? Note that the Paleogene basement uplifts were for the most part positioned well to the east of Sevier orogenic activity (Fig-

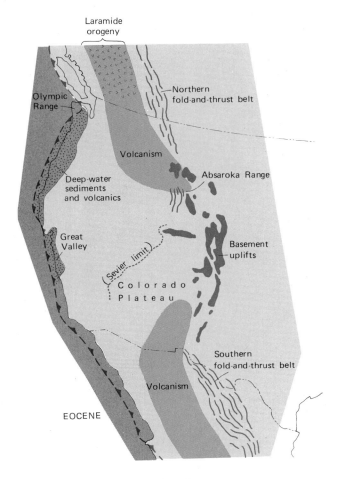

EOCENE

ure 17-29), and recall, in addition, that an eastward migration of orogenic activity culminating in the Sevier orogeny took place during the Mesozoic Era. The widely favored explanation for the earlier eastward shift applies to the Laramide shift as well: A central segment of the subducted plate that passed beneath North America began to penetrate the mantle at a still lower angle, extending a great distance eastward before becoming deep enough to melt and to create igneous activity within the overlying crust (page 514).

Farther west, along the coast, the Great Valley of California continued to receive marine sediment, while north-

FIGURE 17-29   Geologic features of western North America during Eocene time. Subduction continued along the west coast. Marine sediments were deposited in the Great Valley region of California, and deep-water sediments and volcanics accumulated in a forearc basin to the north in Washington and Oregon. Farther inland in the north and south, the Laramide orogeny produced a band of volcanism and, inland from this, a belt of folding and thrusting. In Colorado and adjacent regions, however, the orogeny was expressed as a series of uplifts of crystalline basement that extended far to the east of the Cretaceous Sevier orogenic belt. These may have formed by a slight clockwise rotation of the Colorado Plateau relative to the continental interior.

FIGURE 17-30   The Lewis Thrust Fault in the Laramide fold-and-thrust belt of the northern Rockies. In this view, the fault is exposed in the side of Summit Mountain, Glacier National Park, Montana. The upper half of the mountain is formed of Precambrian rocks that were thrust over the lighter colored rocks below, which are of Cretaceous age. (U.S. Geological Survey.)

ern California and the Sierra Nevada region remained as highlands. A separate basin in Washington and Oregon received deep-water sediments and layers of pillow lava. Here the Olympic Range began to form along a sharp inward bend of the subduction zone (Figures 17-29 and 17-31).

Having reviewed the general pattern of Paleogene orogenic activity in the Cordilleran region, let us now focus on the eastern belt of uplifts, which stretches from Montana to New Mexico. Because this is the region in which the central and southern Rocky Mountains developed during Neogene time, Paleogene events that preceded the uplift of this segment warrant special attention. Deformation here began in latest Cretaceous time with the origin of north-south trending ranges and basins (Figure 17-32); in Utah and Wyoming, these structures lay along the eastern margin of the northern fold-and-thrust belt. The basement uplift farthest to the east formed the Black Hills of South Dakota (page 300). In Colorado, many ranges were formed by the elevation of large basement uplifts along thrust faults. It has been suggested that these uplifts were produced by a slight clockwise rotation of the Colorado Plateau, which behaved as a rigid crustal block, absorbing some of the convergence along the subduction zone to the west (Figure 17-29). It is not known why there was little volcanism in this central region.

Today, of course, many peaks and ridges of the central and southern Rocky Mountain uplifts stand at very high elevations; the Front Range uplift, for example, rises far above the high plains of eastern Colorado (Figure 14-48). It is inappropriate, however, to evaluate the effects of the Laramide orogeny simply by viewing the elevations of the Rocky Mountains today, since, as we will see, the high elevations of the modern Rockies reflect post-Laramide uplift. During the Laramide orogeny, erosion nearly kept pace with uplift in most areas of the United States, resulting in a regional topography that was less rugged than that

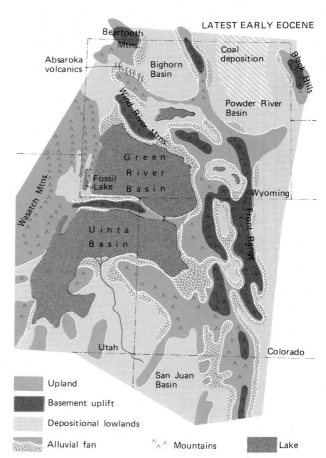

FIGURE 17-32   Geologic features associated with the Laramide basement uplifts of Colorado and adjacent regions at the end of Early Eocene time. The cores of major uplifts consist of Precambrian crystalline rock. The Green River Formation, which is well known for its oil shales and splendid fossils, was accumulating in the Green River and Uinta basins. To the north, volcanism formed the Absaroka Mountains, where Yellowstone Park is now located. The Black Hills of South Dakota represent the easternmost uplift.

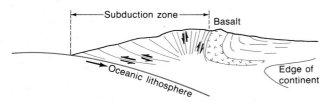

FIGURE 17-31   Formation of the Olympic Range in an embayment of the Pacific border of the state of Washington. Subduction piled sediments against basaltic rocks. (*After D. E. Kari and G. F. Sharman, Geol. Soc. Amer. Bull. 86:377–389, 1975.*)

which characterizes the area today. Basins in front of elevated areas were receiving large volumes of rapidly eroding material.

By the end of Early Eocene time, the regional north-south pattern had weakened, and individual basins were experiencing independent histories. Most of these basins received alluvial and swamp deposits with abundant fossil mammal remains, and some were at times occupied by lakes. Near the beginning of Eocene time, sediments of the Wasatch Formation were laid down in and around rivers and swamps within the Bighorn, Green River, and Uinta basins. Then, later in Early Eocene time, lakes came to occupy most of the areas within the basins, and these lakes

FIGURE 17-33   The transition from the Wasatch Formation to the Green River Formation (*arrow* A) near the border between Wyoming and Utah. The Wasatch is famous for its fauna of terrestrial mammals, while the Green River is noted for fossils of many kinds. Arrow B points to the base of the Wilkins Peak Member of the Green River, depicted in Figure 3-13. (*U.S. Geological Survey.*)

survived, sometimes in restricted areal extent, throughout much of the Eocene Epoch (Figure 17-32). The famous Green River deposits accumulated in and around the margins of these lakes. The transition from Wasatch sediments to those of the Green River is well displayed today in many poorly vegetated areas of Wyoming and Utah (Figure 17-33). Plant remains in these sediments reveal that Eocene climates in this region were quite different from those of today, having been warm (perhaps subtropical) at times.

In Chapter 3 (page 71) we saw that the Wilkins Peak Member of the Green River Formation includes alluvial-fan, braided-stream, salt-flat, and lake deposits. These lake deposits, which are extremely well laminated, have commonly been termed "oil shales" because algal material within them has broken down to yield vast quantities of petroleum. Unfortunately, this petroleum is disseminated throughout the rock and has thus proved difficult to extract. Nonetheless, the Green River deposits form the largest body of ancient lake sediments known, and they may eventually serve as a valuable source of fuel. The fine undisturbed lamination of the Green River lake deposits accounts for the remarkable preservation of a host of animal and plant fossils, including delicate creatures such as frogs (Figure 17-15) and insects (Figure 17-16).

Another interesting group of rocks found in the region of Paleogene basins and uplifts are the volcanics that form the Absaroka Range in western Wyoming and Montana (Figure 17-34). A large portion of Yellowstone National Park lies within the Absarokas; here, the still-active geysers and hot springs serve as evidence that igneous activity has not ceased completely. Recall that Yellowstone seems to represent a hot spot in which igneous activity is localized (page 191). During Eocene time, this area stood at the eastern margin of the volcanic belt of the Pacific northwest (Figure 17-29). Volcanism at this time was episodic, and each episode yielded a variety of rocks, including volcanic

FIGURE 17-34   Table Mountain in the Absaroka Range of northwestern Wyoming. The Absarokas, which include most of Yellowstone National Park, consist of Eocene volcanic rocks of various kinds. The episodic nature of volcanic accumulation is evidenced by the layering visible on the side of Table Mountain. (*U.S. Geological Survey.*)

igneous units as well as sedimentary units composed of volcanic debris. All of these volcanic episodes, however, were catastrophic, destroying entire forests and, we must assume, the animal life within them. Fossilized leaves, needles, cones, and seeds reveal the presence of lowlands with subtropical vegetation, like that simultaneously occupying the Green River Basin, and of slightly cooler uplands. Today, the remnants of the Eocene forests can be seen at high elevations in Yellowstone National Park, where trees are preserved upright as stumps buried by lavas, mud flows, and flood-deposited volcanic debris (Figure 17-35). More than 20 successive forests, all of which were killed in this way, can be identified along the Lamar Valley in Wyoming (Figure 17-36).

As the Eocene Epoch drew to a close, the level of volcanism declined sharply in every part of the northwestern United States except Oregon and Washington. In a few areas to the east, including the Absarokas, volcanic activity persisted but was weak and sporadic. In addition, most of the depositional basins that lay between Montana and New Mexico were filled with sediment by the end of Eocene time. The Laramide orogeny was completed, and the up-

lands that it had produced were largely leveled; thus, as the Oligocene Epoch dawned, a monotonous erosion surface stretched across a broad expanse of western and central North America, interrupted by only a few isolated hills.

Where, then, did the modern Rocky Mountains come from? As we will see in the following chapter, the Rockies are the product of a renewed uplift that took place during Neogene time. In many areas in the Rockies today, a person can look into the distance and view a flat surface formed by the tops of mountains (Figure 17-37). This is what remains of the broad erosional surface that existed at the end of the Eocene Epoch, but the surface now stands high above the Great Plains as a result of Neogene uplift.

During the Oligocene Epoch, after most of the Laramide uplifts had been leveled, sediment moved eastward at a reduced rate, but a thin veneer of deposits spread as far east as South Dakota. The Badlands of South Dakota consist of rugged terrane carved from Oligocene deposits (Figure 17-38). These deposits, which were laid down largely in rivers and lakes, have yielded rich faunas of fossil mammals.

FIGURE 17-35   Three prominent petrified stumps standing upright in the Absaroka volcanics at Specimen Ridge, Yellowstone National Park. Some logs here have diameters of about 1 meter (~3 feet). (*National Park Service, U.S. Department of the Interior.*)

The Gulf Coast

Unlike the Cordilleran region, the Gulf Coast of North America has remained an area of tectonic quiescence throughout the Cenozoic Era. The Gulf Coast did, however, experience a number of changes just before the close of the Mesozoic Eon as a result of the worldwide regression of the seas. As the Paleocene Epoch progressed, marine waters spread far inland for the last time during the Phanerozoic Eon. In fact, North Atlantic waters spread all the way to the Dakotas in the form of the Cannonball Sea, a body of water named for a stratigraphic unit known as the Cannonball Formation. Subsequently, the seas retreated to the Mississippi Embayment, where a thick sequence of Eocene marine sediments accumulated (Figures 17-39 and 17-40); and finally, during the Oligocene Epoch, the seas withdrew to the approximate position of the present shoreline. During Oligocene time, the waters rose again, but they

FIGURE 17-36   Succession of 27 petrified forests now exposed along the Lamar Valley in Yellowstone National Park. Each of these Eocene forests was destroyed by a volcanic eruption. (*After E. Dorf, Scientific American, April 1964.*)

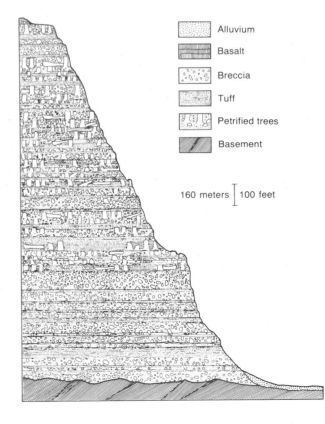

Alluvium

Basalt

Breccia

Tuff

Petrified trees

Basement

160 meters ⏐ 100 feet

FIGURE 17-37   The so-called subsummit surface formed by the flat-topped Rocky Mountains as it appears northwest of Colorado Springs. Pikes Peak is a rare peak that rises high above this surface. The surface developed near the end of the Eocene Epoch, when nearly all of the highlands produced by the Laramide orogeny had been leveled. The subsummit surface has since been dissected by erosion following secondary uplift during Neogene time. (*U.S. Geological Survey*.)

never spread as far inland as they had been during most of Eocene time. The total thickness of Paleogene sediments near the present coastline of the Gulf of Mexico exceeds 5 kilometers (~3 miles), largely because of the enormous quantities of sediment carried to the region by the Mississippi River system. The Cenozoic clastic wedge in this region, though largely buried, has been studied in great detail during the highly successful search for petroleum there.

Europe

The separation of northern Europe from Greenland during the Paleogene Period (Figure 17-27) had consequences

FIGURE 17-38   Oligocene deposits forming the Badlands of South Dakota. These units, which were deposited in and around rivers, are much thinner than those that formed far to the west during Eocene time in subsiding basins near the mountains. The Oligocene beds nonetheless yield rich faunas of fossil mammals.(*National Park Service, U.S. Department of the Interior*.)

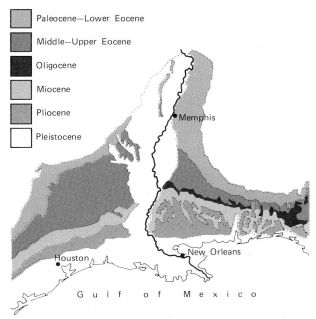

FIGURE 17-39   Map of the segment of the Gulf coastal plain known as the Mississippi Embayment. Here marine sediments of Paleogene age were deposited farther inland than those of Neogene age.

that extended far to the east. Northern Ireland and western Scotland experienced intensive igneous activity, which appears to have been related in some way to the presence of the new rift that separated Europe and Greenland. Lava flows more than 1.5 kilometers thick (~1 mile thick) accumulated in these parts of the British Isles. The most scenic of these flows form Giant's Causeway in northern Ireland (Figure 17-41).

Another phenomenon that was probably related to rifting was a sagging of the North Sea basin, in the center of which thousands of meters of sediment accumulated during Paleogene time. The North Sea also lapped upon the neighboring continents, contributing extensive marine deposits to southeastern Britain and to the European mainland between northern France and Denmark (Figure 17-42).

Figure 17-43 illustrates how Paleogene transgressions and regressions produced an alternation of marine and nonmarine units in the London and Paris basins. Although it is a marginal marine unit, the London Clay of late Early

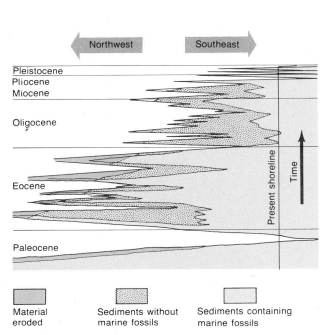

FIGURE 17-40   Transgressions and regressions recorded in the Cenozoic deposits of the Mississippi Embayment by the shifting position of the shoreline (that is, the boundary between marine and nonmarine sediments).

FIGURE 17-41   Giant's Causeway in northern Ireland. The columnar structure resulted from jointing that developed when lava cooled. (Northern Ireland Tourist Bureau.)

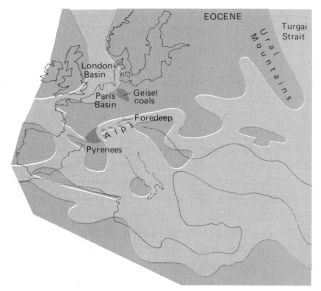

FIGURE 17-42   Europe and the Mediterranean region during Eocene time. Africa had by this time moved close to Spain, leaving little connection between what remained of the Tethyan Seaway (the newly formed Mediterranean Sea) and the Atlantic Ocean. The seas stood relatively high, flooding southeastern England, the Davis Basin, and the margins of the Mediterranean. The Turgai Strait separated the land area of Europe from that of Asia. Large-scale folding began in the Alps, and flysch was deposited in the Alpine foredeep. The Adriatic plate, the microplate that later formed peninsular Italy (*dashed outline*), was moving northward toward its present position (see also Figure 17-44).

faunas of the European Eocene—and perhaps of the entire Phanerozoic record—are found in the lignites, or brown coals, of the Geisel Valley in Germany, which lies close to the North Sea (Figure 17-42). These units are the product of decaying vegetation in Middle Eocene swamps, where even fragile flowers are preserved, fossil leaves often retain their green chlorophyll, beetles display their original iridescent colors, and frogs are so well mummified that their skin is preserved with the original cells intact and still bearing nuclei! Here the dense organic layers of lignite have excluded oxygen and decay-causing oxygen-dependent bacteria, offering us a unique glimpse of biological detail for a biota more than 40 million years old.

Farther east, during much of Eocene and Oligocene time, a great seaway, the **Turgai Strait**, extended along the eastern margin of the Ural Mountains from the Tethys to the Arctic Sea (Figures 17-24, 17-42, and 17-44). The presence or absence of land connections across the Turgai Strait determined whether mammals could migrate between Asia and Europe during the Paleogene Period, just as the presence or absence of the Bering land bridge governed interchange between North America and Asia and the rifting of Greenland governed migration between Europe and land areas to the west. During Late Paleocene and Early Eocene time, mammal faunas of Europe were remarkably similar to those of North America. Many genera, including the earliest horse, *Hyracotherium* (Figure 17-10),

Eocene age yields a rich terrestrial flora that contains about 350 species, many of which are related to species now found in the jungles of the Malay Peninsula. Also present are the fossil remains of marine crocodiles. Unlikely as it may seem considering its frigid nature today, the North Sea warmed up in Eocene time to become more or less tropical. Southeastern England appears to have had an average annual temperature of about 25°C (~78°F) compared to the modern figure of 10°C (~50°F). There were two causes: global warming during the Paleocene-Eocene transition (Figure 17-23) and the establishment of a connection, in the area of the present English Channel, with warm waters to the south (Figure 17-42).

It was largely through his studies of the Paleogene mammal faunas in nonmarine beds of the Paris Basin that Baron Georges von Cuvier established the science of vertebrate paleontology at the beginning of the nineteenth century. Nonetheless, the most remarkable terrestrial fossil

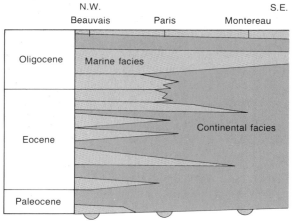

FIGURE 17-43   Diagrammatic cross section showing the intertonguing of marine and nonmarine Paleogene facies in the Paris Basin. The Late Oligocene migration of marine fossils eastward represents the major transgression that connected the North Sea with the Turgai Strait (Figure 17-44). The major worldwide regression that followed is represented by the return of nonmarine deposition to the entire Paris Basin.

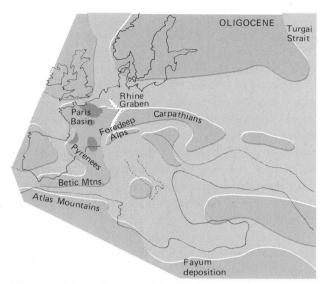

FIGURE 17-44   Europe and the Mediterranean region in the latter part of Oligocene time, when a worldwide high stand of sea level caused a seaway to spread from the North Sea to the Turgai Strait. Mountains were forming in a broad belt from the Gilbraltar region across southern Europe to the Himalayas. The Alps were now a well-developed mountain range, and in the Alpine foredeep, flysch deposition was giving way to molasse deposition. The Adriatic plate, the microplate that later formed peninsular Italy (*dashed outline*), approached its present position. In northern Africa, the waters of the newly formed Mediterranean were pushed back by the uplift of the Atlas Mountains in the west and by the progradation of Fayum deposits near the position of the modern Nile delta.

inhabited both continents. Apparently, at this time the rifting depicted in Figure 17-27 had not yet fully separated Europe, Greenland, and North America, and passage was still possible. Europe and Asia, on the other hand, were parts of a single craton but nonetheless had distinctive mammalian faunas because of the presence of the Turgai Strait. During the Oligocene Epoch, the similarities between the faunas of Asia, Europe, and North America fluctuated with the coming and going of corridors for migration.

A major transgression that occurred in Late Eocene and Oligocene time ushered in an invasion of North Sea waters that extended across all of Europe to the Turgai Strait (Figure 17-42). This invasion, which flooded northern France more extensively than had earlier Paleogene marine incursions, was responsible for an especially thick accumulation of marine deposits in the **Rhine Graben** (Figure 17-44), a fault-block basin that actually represents one arm of a triple junction of rifts (page 191). The history of the Rhine Graben began in late Mesozoic time, when it appears that a mantle plume (a hot spot) became stationed beneath

the crust, forming a dome. The typical three-pronged rift developed early in the Paleogene Epoch, and the Rhine Graben, which was the longest of these prongs, subsided during the Cenozoic Era to receive thousands of kilometers of alluvial-fan and river deposits as well as marine sediments that resulted from the great Oligocene transgression.

The Oligocene transgression in Europe was ended by a regression that permanently drained the Turgai Strait before the end of Oligocene time. This regression, which left the European continent standing above sea level, was the global event that caused the world's oceans to drop to one of the lowest levels of the entire Phanerozoic (Figure 17-25).

The Mediterranean

In the Mediterranean region, many events of Paleogene time were associated with the westward movement of Africa in relation to Eurasia. Early in Paleogene time, the Adriatic plate began to approach the European continent (Figure 8-11). As we saw in Chapter 8, the subduction beneath the Adriatic plate produced extensive igneous activity and folding-and-thrusting, and the Alps began to assume their present form (Figure 8-10). A foredeep developed north of the Alpine suture; here, flysch accumulated until mid-Oligocene time, when molasse deposition began. The Alpine orogenic events were only one part of widespread mountain building that extended beyond the Caucasus to the Himalayas (Page 209).

Meanwhile, along the southwestern border of the Mediterranean, the Atlas Mountains began to form as a result of the movement of the African plate in relation to Eurasia. Early in the Paleogene Period, marine deposition had prevailed in this part of North Africa (Figure 17-42), but it ceased during the latter part of the Eocene Epoch, when folding and thrusting formed the structures known as the High Atlas. In contrast, Libya and Egypt, which lay to the east, experienced an interval of tectonic quiescence at this time. Here, sedimentary units accumulated throughout Paleogene time. During the Eocene Epoch, thick bodies of limestone were deposited. Many of these consist primarily of the remains of large, disc-shaped foraminifers called nummulitids (Figure 17-45). The Sphinx and the great pyramids of Egypt were made of blocks of nummulitic limestone, which is relatively resistant to chemical weathering in the arid climate of the Sahara (Figure 17-46). This marine deposition was short-lived, however. As the Eocene progressed, the pattern of sedimentation here changed, owing

FIGURE 17-45   Giant nummulitid foraminifers of the genus *Nummulites*. These Eocene specimens are from Pyramids Plateau, Giza, Egypt. The scale is in centimeters; the largest specimen in the photograph has a diameter of about 2.5 centimeters (~1 inch). (*Courtesy of T. Aigner.*)

to the northward transport of vast quantities of siliciclastic sediment from the interior of Africa. This pattern of sedimentation continues to the present day through the growth of the Nile delta. The thick Fayum deposits of Egypt record progradation of siliciclastics around the mouth of the ancestral Nile. The older Fayum sediments, which are of Late Eocene age, yield fossil marine faunas that include sharks and early whales. By latest Eocene time, however, progradation had progressed to the point at which marine faunas had given way to nonmarine biotas. These comprised huge tree trunks and a rich variety of mammals that constitute one of the world's greatest mammalian faunas of Oligocene age.

As brief as it was, the marine deposition that prevailed in the Mediterranean region during much of Eocene time was uncharacteristic of Africa. This continent, which now stands unusually high above sea level (page 312), also experienced only very minor marginal flooding by marine waters throughout the Paleogene Period.

FIGURE 17-46   The Sphinx and the pyramids of Egypt, which were built with blocks of Eocene limestone that consisted largely of nummulitid foraminifers. Herodotus, the Greek historian of the fifth century B.C., interpreted the disc-shaped nummulitids as lentils that had been fed to the slaves who built the pyramids and that had accidentally been spilled and had turned to stone. (*Courtesy of B. Kummel.*)

CHAPTER SUMMARY

1. Marine life of the Paleogene Period resembled that of the modern world.

2. Plants on land also resembled those of the present time, except that grasses were rare until late in Paleogene time.

3. Terrestrial vertebrate animals were more primitive than those of the present. Although most of the mammalian orders alive today existed in Paleogene time, many Paleogene families and genera are now extinct. Most species of birds were wading animals that resembled modern storks or herons.

4. Late in Eocene time, polar regions cooled, while cold, dense waters began to descend to the deep sea to form the psychrosphere—an ocean layer that has persisted to the present time.

5. Climates cooled in at least some regions during Late Eocene time, and this cooling and drying persisted into Oligocene time. Partly as a result of these trends, grasslands replaced forests in many parts of the world.

6. Both north and south polar regions underwent plate-tectonic changes in mid-Paleogene time. In the south, Australia broke away from Antarctica, which was left isolated over the South Pole. In the north, Europe and North America moved further away from Greenland.

7. In western North America, the Laramide orogeny produced northern and southern fold-and-thrust belts that were separated by a curious zone of uplifts in Colorado, Wyoming, and neighboring regions. By the end of the Eocene Epoch, the Laramide orogeny had ended, and the mountains that it had produced had been subdued by erosion.

8. Paleogene fossils in western Europe record the presence of warm climates during most of Eocene time. Marine sediments and faunas reveal that seas flooded large areas of Europe until they receded during a global lowering of sea level late in Oligocene time.

9. North of the Mediterranean, the Alps and other mountain chains formed as the African plate and nearby microplates shifted with respect to Eurasia.

EXERCISES

1. Which group changed more from the beginning of the Paleogene Epoch to the end—marine or land animals? Explain your answer.

2. In Paleogene oceans, which animals took the place of marine Mesozoic reptiles?

3. How does the fossil record of flowering plants reveal that climates changed during Paleogene time?

4. What may have caused the psychrosphere to form?

5. How did the location of the Laramide orogeny differ from that of the Sevier orogeny of the Cretaceous Period? What might explain this change?

ADDITIONAL READING

Kennett, J., *Marine Geology,* Prentice-Hall, Inc., Englewood Cliffs, New Jersey, 1982.

Kurtén, B., *The Age of Mammals,* Columbia University Press, 1971.

Pomerol, C., *The Cenozoic Era,* John Wiley & Sons, Inc., New York, 1982.

Wolfe, J., "A Paleobotanical Interpretation of Tertiary Climates in the Northern Hemisphere," *American Scientist* 66:694–903, 1978.

CHAPTER 18
THE NEOGENE WORLD

Because it includes the present, or Recent Epoch, the Neogene Period holds special interest for us. In examining the Neogene, we will learn how the modern world took shape—that is, how the ecosystem acquired its present configuration and how important topographic features assumed the forms we are familiar with today.

The boundary between the Paleogene and Neogene systems has no great historical significance inasmuch as no mass extinction marks the close of one and the start of the other. During the Neogene itself, however, major changes in life and in the physical features of the earth have occurred even though the period has spanned only about 24 million years. The most far-reaching biotic changes were the spread of grasses and weedy plants and the modernization of vertebrate life. In addition, snakes, songbirds, frogs, rats, and mice expanded dramatically, and humans evolved from apes. In the physical world, the Rocky Mountains and the less rugged Appalachians took shape during Neogene time, as did the imposing Himalayas. The Mediterranean Sea also dried up and rapidly formed again during the Neogene. The most widespread physical change on earth, however, was climatic. Continental glaciation began in the latter part of Neogene time and had especially profound effects in the Northern Hemisphere. Although this interval is commonly thought of as corresponding to the Pleistocene Epoch, it actually began late in the Pliocene Epoch and presumably will continue long into the future. The spreading of glaciers from polar regions is episodic,

and what is formally referred to as the Recent Epoch is almost certainly only the latest of numerous intervals between pulses of severe glaciation. Today, ice caps in the far north remain poised to spread southward, as they have done repeatedly during the past 2 million years or so.

All four Neogene epochs—the Miocene, Pliocene, Pleistocene, and Recent—were named by Charles Lyell, who introduced them in the 1833 volume of his *Principles of Geology*—although the Pleistocene was known as the "Newer Pliocene" until 1837. Lyell distinguished the epochs of the Neogene Period on the basis of his observations of marine strata and fossils in France and Italy, noting that Pleistocene strata were characterized by molluscan faunas in which about 90 percent of the species are still alive in modern oceans. Lyell found that Pliocene strata contained fewer surviving species and that Miocene strata contained fewer still. It was not until later in the nineteenth century,

A perched boulder that was transported to its present position in Yorkshire, England, by a Pleistocene glacier. (*British Institute of Geological Sciences.*)

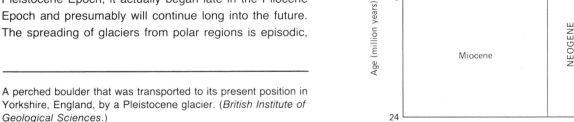

however, that glacial deposits were recognized on land and found to correlate with the marine Pleistocene record.

WORLDWIDE EVENTS

In this chapter, as in the preceding one, we will review major biological events in the context of global environmental changes. The reason once again is that the animals and plants that currently inhabit the earth are, in general, representative of more ancient Cenozoic life, which thus requires no special introduction. Instead, we will examine Neogene forms of life in the context of the world ecosystem and will learn how this ecosystem changed as a result of the climatic fluctuations that were so profound during the Neogene Period. Also, as in the previous chapter, we will not devote a special section to paleogeography, which has undergone little change during the brief Neogene interval.

FIGURE 18-1   Reconstruction of the marine fauna represented by fossils of the Middle Miocene Calvert Formation of Maryland. Representing the whale family were early baleen whales (*Pelocetus*), which strained minute zooplankton from the sea (*upper left*); long-snouted dolphins (*Eurhinodelphis—lower left*); short-snouted dolphins (*Kentriodon—upper center*); and carnivorous sperm whales (*Orycterocetus—right*). Sharks include the hammerhead *Sphyrna* (*far right*), shown here under

Life in aquatic environments

Charles Darwin observed long ago that invertebrate life tends to evolve less rapidly than vertebrate life; thus, the short Neogene Period has produced only modest evolutionary change in the marine invertebrate fauna, which as a consequence will receive little attention in this chapter. Less is known about freshwater Neogene life, which has consisted largely of soft-bodied species not readily pre-

served in the fossil record. We will, however, note the expansion of one major group of freshwater phytoplankton that has produced preservable hard parts: the diatoms.

Adaptive radiations in the sea   Not surprisingly, the most dramatic evolutionary development in Neogene oceans has been the expansion of a group of vertebrate animals, the whales. During the Miocene Epoch, a large number of whale species came into existence (Figure 18-1); among them

attack by a sperm whale, sand sharks of the genus *Odontaspis* (*bottom center*), the six-gilled shark *Hexanchus* (*left of center*), and the giant white shark *Carcharodon*, one of which is killing a whale. Among the inhabitants of the sea floor were large bi-

valve mollusks of the scallop family (*lower right*). (*Drawing by Gregory S. Paul.*)

were the earliest representatives of the group that includes modern sperm whales, which are carnivores with large teeth, and also the group that includes modern baleen whales, which feed by straining zooplankton from sea water. Dolphins, which are specialized whales, also made their first appearance early in Miocene time.

At the other end of the size spectrum for pelagic life, the globigerinacean foraminifers, which barely survived the terminal Eocene mass extinction, expanded again early in the Miocene Epoch. Remarkably, the new array of genera resembled the globigerinacean genera that had perished in the Eocene mass extinction (Figure 6-17). Neogene globigerinaceans serve as valuable index fossils for oceanic sediments (Figure 18-2).

On the sea floor, evolutionary changes from Paleozoic time were relatively minor. One important development of Miocene time, however, was the appearance of the first **algal ridges** on coral reefs (page 93). For millions of years, coralline algae had contributed to reef growth but lacked the capacity to form a substantial ridge facing powerful waves. Protected by algal ridges, coral reefs have since Miocene time thrived along coasts pounded by heavy surf.

Expansion of freshwater diatoms   Unlike the marine diatoms, which began their evolutionary expansion in the Mesozoic Era, the dominant group of freshwater diatoms, the **Pennales**, did not evolve until early in Cenozoic time. By Miocene time, the Pennales comprised about 2000 known species (Figure 18-3) and had already assumed the role that they play today as primary freshwater producers both in the planktonic realm and on lake and river bottoms.

Life on the land

Climatic changes exerted a profound influence over Neogene terrestrial biotas, and the geographic and evolutionary modifications of biotas represented in the fossil record help us reconstruct these changes. As in Late Cretaceous and Paleogene times, flowering plant fossils represent our best gauge of climatic shifts.

Flowering plants: Climatic deterioration and an explosion of herbs   In the world of plants, the Neogene Period might be described as the Age of Herbs. **Herbs**, or herbaceous plants, as we have seen, are small, nonwoody plants that die back to the ground after releasing their seeds.

(Defined in this way, herbs include many more plants than the few that we use to season our food.) The recent success of herbs is primarily the result of the worldwide climatic deterioration that took place during Oligocene and Miocene times when cooler, drier conditions caused forests to shrink and opened up new environments to plants such as herbs and grasses, which prefer open habitats and can withstand low rainfall (Figure 18-4). Today, there are some 10,000 species of grasses alone.

The **Compositae**, an important family of herbs that includes such seemingly diverse members as daisies, asters, sunflowers, and lettuces, appeared near the beginning of the Neogene Period only 20 or 25 million years ago, and yet today this family contains some 13,000 species, including the plants that ecologists refer to as "weeds" (page 34). As any gardener knows, weeds are exceptionally good

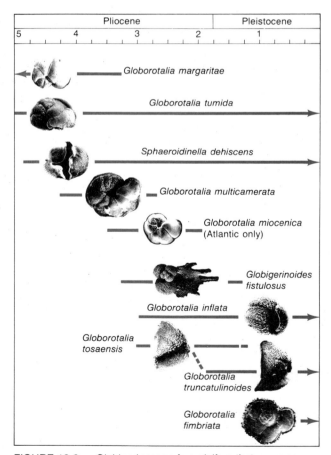

FIGURE 18-2   Globigerinacean foraminifers that serve as valuable index fossils in Cenozoic sediments. (*Courtesy of A. Boersma*).

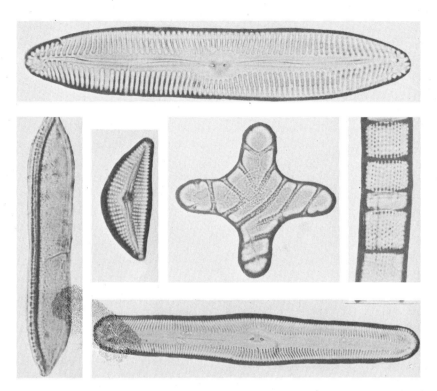

FIGURE 18-3   Miocene species of freshwater diatoms. This important group of single-celled algae underwent a major adaptive radiation during the Cenozoic Era, having evolved from marine diatoms. (*G. W. Andrews, U.S. Geol. Surv. Prof. Paper No. 683A, 1970.*)

invaders of bare ground. They may not compete successfully against other plants to retain the space that they invade, but they soon disperse their seeds to other bare areas created by such destructive agents as fires, floods, or droughts, and spring up anew.

Neogene climatic changes were in part the result of local tectonic events that caused areas in eastern Africa, western North America, and South America to receive less rainfall than they had during the Paleocene Epoch. The regional changes were also partly due to global climatic trends. The presence in the Southeast Pacific of ice-rafted, coarse sediments of Early Miocene age shows that by this

FIGURE 18-4   A remnant of true prairie in Pottowatomie County, Kansas. This is the kind of grassland that greeted early settlers of the American West. (*National Park Service, U.S. Department of the Interior.*)

time Antarctic glaciers had begun to flow to the sea. Cores of deep-sea sediment also reveal that the belt of siliceous diatomaceous ooze that encircled Antarctica (Figure 4-38) simultaneously expanded northward at the expense of carbonate ooze, which tends to accumulate where climates are warmer. Because of this polar cooling, the Early Miocene Age marked the beginning of strong latitudinal gradients in the distribution of oceanic plankton. Since that time, assemblages of species at high latitudes have differed greatly from those at low latitudes.

Analyses of leaf margins from terrestrial floras of North America reveal only modest fluctuations during Neogene time, a trend that contrasts sharply with the dramatic drop in temperatures that marked the end of the Eocene Epoch (Figure 17-23). There was, however, a slight cooling trend from early in the Oligocene Epoch until late in the Pliocene Epoch, when the Ice Age began. At this time, climates not only became slightly cooler but also grew drier and more seasonal. As a result, dry grasslands expanded into areas that had once been open woodlands and dense forests.

Modernization of terrestrial vertebrates   Because we are naturally interested in the origins of large mammals, we often ignore the great success of smaller creatures—or we assume that because these creatures are small, they are also primitive and have changed relatively little in the more recent geologic periods. In fact, the Neogene Period might well be labeled the Age of Frogs, the Age of Rats and Mice, the Age of Snakes, or the Age of Songbirds, since all four of these groups have undergone tremendous adaptive radiations over the past few million years.

Many species of rats and mice dig burrows in dry terrane and eat the seeds of grasses and herbs. To some extent, the success of these small rodents during Neogene time may be related to that of both the grasses and the Compositae and also, more fundamentally, to the drying and cooling of climates that favored these plants. Perhaps we can attribute the success of modern frogs and toads, whose species number about 2000, to their remarkable ability to catch insects through the quick protrusion of their long tongues. In any case, the snakes have obviously flourished largely because of the proliferation of frogs and rodents, since few other predators can pursue small rodents down their burrows without digging. Before the start of the Neogene Period, there were few snakes except members of the primitive boa constrictor group. Today, however, the more advanced snakes of the family Colubridae include about 1400 species, many of which are poisonous.

Also poorly represented before Neogene time were the **passerine** birds, or songbirds and their relatives, which are highly conspicuous today. Presumably, these birds have also benefited from the diversification of seed-bearing species of herbs, but like frogs, they owe much of their success to the fact that they are well equipped to capture flying insects. It is probable that many types of flying insects were not heavily preyed upon until groups of passerine birds took to the air.

Of course, groups of large animals also developed their modern characteristics during of the Neogene Period. Among the herbivores, for example, the horse and rhinoceros families dwindled after mid-Miocene time in a continuation of the general decline of the odd-toed ungulates. Meanwhile, the even-toed, or cloven-hoofed, ungulates expanded, especially through the adaptive radiation of both the **deer** family and the family called the **Bovidae**, which includes cattle, antelopes, sheep, and goats. The **giraffe** family and the **pig** family also radiated during the Miocene Epoch, but the number of species in these families has since declined. Similarly, many types of elephants, including those characterized by long trunks, experienced great success during the Miocene and Pliocene intervals but later declined. Today, there are only two elephant species: the large-eared African elephant and the smaller, more docile Indian elephant that is commonly trained to perform in circuses.

The carnivorous mammals also assumed their modern character in the course of the Neogene Period; this group included the dog and cat families, both of which had appeared during Paleogene time. The **bear** and **hyena** families were also important Miocene additions to the carnivore group.

Many of the Neogene mammal groups expanded successfully because of the spread of open country in the form of savannahs and open woodlands (Figure 18-5). Several radiating herbivore groups, such as the antelopes and cattle, included many species that were well adapted for long-distance running over open terrane and that grazed on harsh grasses with the aid of high-crowned, continuously growing teeth. Many grasses contain tiny fragments of silica that offer resistance to grazing, and only continuously growing teeth can tolerate the resulting wear. Also on the increase were the groups of rodents that are adapted for burrowing in prairies and the elephants, which require open country simply to move around. As we might expect, the diversification of herbivores in savannahs and woodlands in turn fostered the success of groups of carnivores that were well adapted for attacking herbivores in open country—groups such as hyenas, lions, cheetahs, and long-legged dogs.

Ultimately, however, the greatest change in the terrestrial ecosystem was wrought by primates, simply because the ecologically disruptive humans belong to this group. As a group, primates tend to favor forests over savannahs; in fact, most live in trees. As we have seen, monkeys were present by Oligocene time; the oldest group includes the so-called **Old World Monkeys**, which now live in Africa and Eurasia. Before the end of the Oligocene interval, however, a distinctive group of monkeys reached South America. These **New World monkeys**, which differed from their Old World counterparts in that most possessed prehensile (or grasping) tails, probably had a separate evolutionary origin. In any event, monkeys on both sides of the Atlantic underwent adaptive radiations during Neogene time.

Apes, which evolved in the Old World, flourished for a time but have since declined in number of species. We will discuss apes and apelike animals when we examine the origins of humans, which belong to the same superfamily, the Hominoidea (Figure 1-6). The most recent phases of human evolution have taken place within the climatic context of the recent Ice Age (Pleistocene Epoch); therefore, before we discuss the Hominoidea, it is appropriate that we examine the major global events of this fascinating interval of geologic time and of the Pliocene Epoch that preceded it.

Late Neogene climatic change

The cooling and drying of climates that characterized Oligocene and Miocene time caused grasses and weedy plants to expand their coverage of the land, and Early Pliocene time saw modest climatic changes as well. In contrast, Late Pliocene and Pleistocene time were marked by strong, rapid climatic fluctuations in the Northern Hemisphere—changes that characterized the modern Ice Age.

Pliocene equability Stratigraphic unconformities in various parts of the world indicate that at the very end of the Miocene Epoch, between 6 and 5 million years ago, global sea level fell by perhaps as much as 50 meters (Figure 18-6). This so-called **Messinian Event**, as we will see, isolated the Mediterranean Sea from other oceans, causing the Mediterranean to dry up temporarily. It is generally believed that the Messinian Event can be attributed to re-

moval of water from the ocean by the sudden expansion of glaciers in Antarctica. There is, in fact, evidence that climates became cooler in the Southern Hemisphere at this time. Siliceous, rather than calcareous, sediments were suddenly deposited over larger areas of southern oceans than before. The success of diatoms, which produced these sediments, indicates that upwelling had increased, apparently because of steepening temperature gradients. This cooling was not worldwide, however; terrestrial floras and marine microfossils near the Pacific coast of North America show a slight warming trend across the Miocene-Pliocene boundary.

As the Pliocene Epoch got underway, about 5 million years ago, sea level rose again (Figure 18-6), and the seas continued to stand well above their present level between about 4.5 and 3.5 million years ago, leaving marine deposits inland of coastlines in areas like California, eastern North America, and countries bordering the North Sea and the Mediterranean. Fossil faunas and floras also reveal that global climates during this time were more equable than they are today; pollen analyses, for example, indicate that southeastern England was subtropical, or nearly so, and that northern Iceland enjoyed a temperate climate. Especially in the Northern Hemisphere, however, this warm interval came to a sudden close with the start of the modern Ice Age slightly more than 3 million years ago. Almost certainly, the glacial interval that ensued has not yet ended.

Pleistocene continental glaciation The evidence supporting the existence of a recent ice age is now so conclusive that it is difficult for us to understand why so many competent naturalists during the last century were skeptical that continental glaciation had indeed occurred. Early in the century, naturalists generally invoked the rushing waters of the biblical flood to explain the appearance throughout much of Europe of large boulders far from their sources (see page 558). This idea gave way to a second hypothesis: that the boulders had been rafted by icebergs floating on floodwaters. Not until the 1830s, in large part because of the efforts of the great geologist Louis Agassiz, was it widely recognized that the so-called **erratic boulders** and other coarse sedimentary debris had been transported great distances and deposited by glaciers of continental proportions. We now know that the recent Ice Age has consisted of many intervals of glacial expansion that have been separated by warmer interglacial intervals, including the one in which we almost certainly live today. There is no reason to believe that the Ice Age has ended.

FIGURE 18-5   A reconstruction of the so-called *Hipparion* fauna of Asia. This diverse fauna occupied open country in Asia about 10 million years ago, in Late Miocene time. *Hipparion* is the galloping horse (*center*). The elephant on the left, with downward-directed tusks, is *Dinotherium*, and the straight-tusked elephants on the right are members of the genus

(*Trilophodon*). A saber-toothed cat trots in front of *Trilophodon*, and, in the foreground, short-legged hyenas of the genus *Percrocuta* look on from their den. (*Drawing by Gregory S. Paul*.)

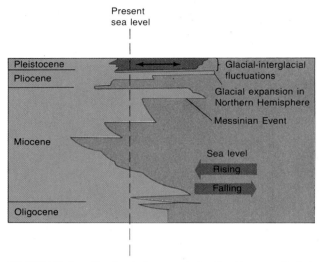

FIGURE 18-6   Sea-level changes during the Neogene Period, a time when sea level has been controlled primarily by the volume of glacial ice. These changes have been determined primarily by the application of seismic stratigraphic techniques to sediments underlying continental shelves (page 126). (*After P. R. Vail et al., Amer. Assoc. Petrol. Geol. Mem. 26:83–97, 1977.*)

FIGURE 18-7   Reconstruction of glaciers as they existed during a typical glacial interval of the Pleistocene Epoch. Large continental glaciers were centered in North America, Greenland, and Scandinavia. (*After a drawing by A. Sotiropoulos in J. Imbrie and K. P. Imbrie, Ice Ages, Enslow Publishers, Short Hills, New Jersey, 1979.*)

The marks of recent glaciation tell of ice sheet development primarily in the Northern Hemisphere (Figure 18-7). They take many forms. Moraines (page 66) and other glacial deposits are scattered over large areas of Eurasia and North America. Some, like the moraines that form much of Cape Cod, Massachusetts, extend into the marine realm. Terminal moraines often flank depressions in the earth that were left after glaciers retreated, and many of these depressions became the beds of modern lakes, including the Great Lakes of North America. Hudson Bay is not bounded by a terminal moraine, but it is an arm of the ocean that spread into a region where the earth's crust was depressed by the thickest continental glaciers in North America and has not yet rebounded fully. Figure 18-8 shows the amount of crustal uplift that has taken place in Scandinavia, which was similarly burdened with a large ice cap until the crust began to rebound about 10,000 years ago.

Some elevated regions, such as Mount Monadnock, New Hampshire, provide clues to the maximum thickness of the Pleistocene glaciers that flowed past them. Like some mountains in Antarctica today (Figure 18-9), Mount Monadnock once stood partially above surrounding ice sheets. The maximum elevation reached by the neighboring glaciers can be determined from the position of the line that separates the lower part of the mountain, which was smoothed by glacial flows, from the still-rugged upper part, which the glacier didn't reach.

Even in areas far removed from continental glaciation, smaller glaciers—which are known as Alpine, or mountain, glaciers—modified the landscape; these glaciers flowed from high elevations, where climates are cold, through valleys between mountains to lower elevations. Their most spectacular Pleistocene productions are the U-shaped valleys sculpted by glaciers flowing through valleys that were originally more nearly V-shaped in cross section (Figure 11-5). Even in the tropical Hawaiian Islands, small moraines attest to the growth and movement of glaciers at high elevations during glacial intervals.

The alternations of Pleistocene glacial and interglacial intervals have caused climatic belts and the floras and faunas that occupy them to shift over distances measured in hundreds of kilometers. Thus, fossils of mammals such as the muskrat, which today does not range south of Georgia, reveal that climates in Florida were cool when glaciers pushed southward into the northern United States. Other fossil occurrences, such as those of hippopotamuses in Britain, show that during at least some interglacial intervals, climates were warmer than they are today.

One of the most useful fossil indicators of Pleistocene

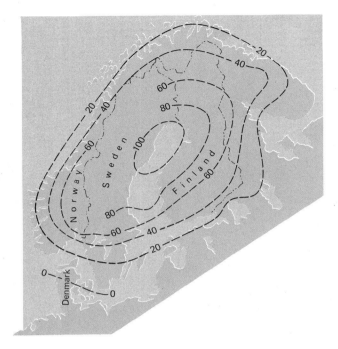

climates is the pollen of terrestrial plants. Pollen assemblages reveal climatic change by indicating the shifting of floras to the north or south. Figure 18-10 shows the southward movement of floras in Europe by about 20° latitude during the most recent glacial interval there. Beetles are also extremely valuable indicators of Pleistocene climates, since some are adapted to cool climates and others to warm climates. Beetle species can be readily identified from their genitalia, which survive well in fine-grained sediments, and virtually all beetle species known from the Pleistocene are alive today, so their ecologic requirements are well understood.

During glacial intervals, some mountaintops that stood above ice sheets served as refuges for plant and animal

FIGURE 18-8   The amount of uplift (in meters) in Scandinavia since the disappearance of the Pleistocene continental glacier that was centered here. The depressed crust is still rebounding some 10,000 years after the glacier disappeared.

FIGURE 18-9   Dark rock of the Shackleton Mountain Range projecting above the surface of the modern Antarctic ice cap.

Many North American mountains were partly buried in ice during the Pleistocene Epoch. (*U.S. Navy.*)

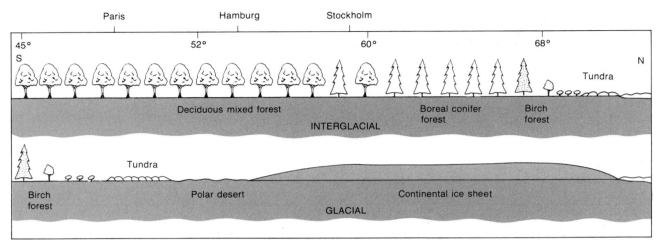

FIGURE 18-10   North-south migration of vegetation in Europe during the Pleistocene Epoch. During glacial intervals, continental glaciers spread southward to the vicinity of Hamburg, and tundra shifted to the latitude of Paris. (*After T. Van Der Hammen, in K. K. Turekian [ed.], The Late Cenozoic Glacial Ages, Yale University Press, New Haven, 1971.*)

species that could not live on the ice. Some species that were stranded during the most recent glacial advance could not migrate across warmer lowlands when the glaciers melted back and thus today remain isolated from other members of their species, which generally live farther north. In his book *On the Origin of Species*, Charles Darwin discussed the isolation, during the Ice Age, of these still-marooned populations, which thus served as early evidence of continental glaciation.

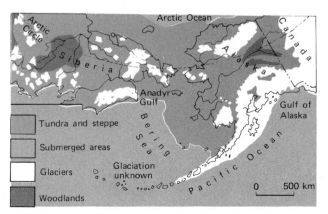

FIGURE 18-11   The geography of Beringia during the most recent glacial interval (the Wisconsin, or Riss-Würm, Stage). Though positioned at a high latitude, Beringia was dry and free of large glaciers, and animals migrated across this land bridge between Eurasia and North America. (*After D. M. Hopkins, in D. M. Hopkins [ed.], The Bering Land Bridge, Stanford University Press, Stanford, California, 1967.*)

Larger refuges between major centers of ice accumulation also existed during glacial intervals. The most famous of these refuges was an area known as **Beringia**, which was so designated because it included the region of the Bering Strait. During glacial episodes, regression of the seas turned the Bering Strait into a land corridor between Asia and North America by means of which not only lower mammals but also the first humans entered the New World. Ironically, Beringia, which was hospitable to terrestrial mammals during the height of glaciation (Figures 18-11 and 18-12), included portions of Siberia and Alaska—areas that we now view as inhospitable to most species but that happened to remain unglaciated when other parts of the Northern Hemisphere were covered with ice.

Today, there remain only two ice caps of the sort that many times during the Pleistocene Epoch expanded to cover broad areas. One of these modern ice caps covers much of Greenland (Figure 2-17), while the other covers nearly all of Antarctica (Figure 18-9). Today, about three-quarters of the world's freshwater is locked up in glacial ice, and most of this belongs to the Antarctic ice cap. It may seem impressive that glaciers now contain about 25 million cubic kilometers of ice, but it has been estimated that during glacial advances of the Pleistocene Epoch, the volume of ice was nearly three times as great, with larger ice sheets measuring an average of about 2 kilometers (~1.2 miles) in thickness. Shelves of ice projected into the sea, and these shelves, together with the icebergs and pack ice that broke loose from them, spread over half the world's oceans.

FIGURE 18-12   Reconstruction of the mammalian fauna that occupied the steppe in the Alaskan portion of Beringia about 12,000 years ago during the most recent glacial interval. Of 61 species depicted here, 11 are extinct; among these are the woolly mammoth, the American mastodon, the long-horned bison, a lionlike cat, and a saber-toothed cat. (*Mural by J. H. Matternes in the Smithsonian Institution.*)

One important effect of each major expansion of ice sheets during the Pleistocene Epoch was a partial draining of the oceans, which took place as great quantities of water were locked up on the land. During major glacial expansions, most of the surfaces that now form continental shelves stood above sea level. Rivers cut rapidly downward through the soft sediments of many continental shelves to form valleys that exist today as submarine canyons, following further excavations by submarine turbidity currents (Figure 4-35). During some glacial episodes, sea level dropped slightly more than 100 meters (~330 feet) below its present position.

The chronology of glaciation   The microplankton fossil record of the North Atlantic suggests that rapid cooling occurred slightly more than 3 million years ago, apparently in response to the initial formation of continental ice sheets. The beginning of continental glaciation at this time is confirmed by the first appearance of coarse, ice-rafted sedimentary debris in the North Atlantic and also of glacial deposits in Iceland above well-dated lava flows. It was somewhat later that the first glaciers spread southward into the central United States, and the beginning of the Pleistocene Epoch is now placed at 1.8 million years before the present. The sequence of glacial tills in the United States and Europe suggests that there were four major glacial intervals during the Pleistocene Epoch; these intervals are formally recognized as glacial ages, and the intervening intervals are known as interglacial ages. In accordance with normal stratigraphic practice, each age corresponds to a stage in the geologic column (Table 18-1).

In recent years, it has become evident not only that the major glacial ages were characterized by smaller pulses of glaciation but also that minor glacial advances occurred during interglacial ages as well. These smaller fluctuations are reflected in sediments obtained from deep-sea cores. The crucial data are the oxygen-isotope ratios of foraminifer skeletons preserved in deep-sea sediments. Oxygen, as we have seen, is present in sea water as the stable isotopes ^{16}O and ^{18}O, and these isotopes are incorporated into the growing skeletons of foraminifers in the same proportions in which they occur in the surrounding sea water.

TABLE 18-1   Glacial and interglacial stages of the Pleistocene Epoch

Those recognized in North America are thought to correspond to those recognized in Europe as shown here.

North America	Europe
Wisconsin (Sangamon interglacial)	Riss-Würm (interglacial)
Illinoisan (Yarmouth interglacial)	Mindel (interglacial)
Kansan (Aftonian interglacial)	Günz (interglacial)
Nebraskan	Danube

As it turns out, the relative proportions of these two isotopes in fossil foraminifer skeletons fluctuate greatly when traced downward through cores of deep-sea sediments (Figure 18-13). When these fluctuations were first discovered, some scientists erroneously assumed that they reflected variations in water temperature during the Pleistocene Epoch. Soon, however, it was found that similar fluctuations were exhibited not only by planktonic foraminifers but also by foraminifers that lived on the sea floor, an area that has remained at near-freezing temperatures since the origin of the psychrosphere. As we have previously noted (page 540), we now know that temperature has only a minor effect on the oxygen-isotope ratio within a foraminifer skeleton and that this ratio tends to conform more closely to the ratio in the sea water where the foraminifer lived. For the ocean as a whole, the ratio varies greatly with the size of continental ice sheets.

Once it was recognized that oxygen-isotope ratios have varied with the volume of glacial ice, scientists faced the task of establishing a time scale for fluctuation. The scale in Figure 18-13 is based on correlation of isotopic fluctuations for a small number of cores with reversals in the earth's magnetic polarity as recorded in the same cores. The Pleistocene Epoch is often arbitrarily defined as beginning with the Olduvai Event, a brief reversal of the earth's magnetic polarity that occurred about 1.8 million years ago. In fact, there have been about 18 minor glacial expansions during the Pleistocene Epoch—one every 100,000 years

or so—with the intensity of the glaciations increasing toward the latter part of the epoch. Some major glaciations have been compound events represented by two or more peaks on the curve, and there have been many minor advances as well. The difference between "major" and "minor" advances, however, is largely a matter of individual judgment.

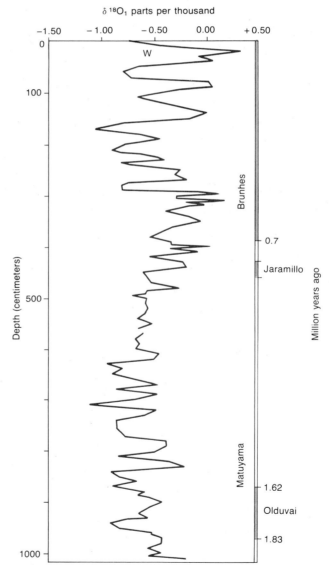

FIGURE 18-13   Oxygen-isotope fluctuations in planktonic foraminifers within a deep-sea core. The index of oxygen-isotope composition plotted here records decreases in temperature toward the right and increases toward the left. On the right is a paleomagnetic time scale for the core. W = the Wisconsin, or Riss-Würm, Stage (the most recent glacial interval). (*After J. van Donk, Geol. Soc. Amer. Mem. 145:147–163, 1976.*)

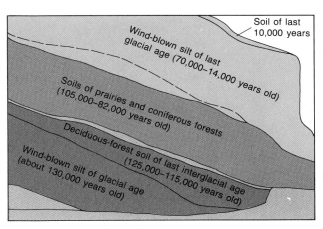

FIGURE 18-14   Sediments in a Czechoslovakian brickyard representing the two latest glacial intervals and the intervening interglacial period. The interglacial deposits record an interval of cooling, with forests of conifer trees replacing forests of decidu-ous trees. The glacial intervals, in contrast, appear to have ended abruptly; the stratum directly above each one was deposited in a markedly warmer environment. (*Courtesy of G. J. Kukla.*)

The high probability that the glaciers will eventually flow southward again has engendered curiosity about how suddenly glacial episodes begin and end. Unfortunately, the curves displayed in Figure 18-13 are inexact: errors have been introduced by the activities of burrowing animals that disturb the deep-sea sediments and mix up fossil foraminifers of different sedimentary levels. Although the curves seem to reveal rates of transition from glacial to interglacial intervals and vice versa, the evidence is inconclusive.

Studies of Pleistocene deposits on land, however, indicate that glacial intervals have in general, begun more slowly than they have ended. Many years ago, counts of glacial varves suggested that the continental ice sheets of the last glacial interval melted back between about 15,000 and 8000 years ago. More recently, carbon 14 dating has confirmed this timing and has further revealed that the maximum glacial advance took place only 18,000 years ago, at the time the southernmost moraines formed. Thus, the last major glacial interval, which began more than 100,000 years ago, ended not only recently but also suddenly, on a geologic scale of time.

Cyclical Pleistocene deposits in Czechoslovakia and Austria suggest that this pattern is typical. Each cycle includes deposits of one major interglacial and glacial interval (Figure 18-14). The lowest interglacial unit in each of these cycles is a forest soil that contains pollen from a humid interglacial climate slightly warmer than the present one. Above this are soils representing progressively cooler conditions that supported prairies and conifer forests. Sometimes soils representing even cooler but nonglacial Arctic climates follow. The entire interglacial sequence represents a long, gradual episode of cooling. Above each interglacial sequence are loess deposits formed during a glacial interval. **Loess** is windblown silt that accumulated in front of the ice sheets, in the frigid Pleistocene deserts that developed there. Most of this silt was derived from glacial meltwaters.

The Pleistocene deposits of Czechoslovakia and Austria were not themselves glaciated, which is fortunate in that glaciation would have damaged their excellent stratigraphic record. These areas were, however, close enough to the periodically advancing glaciers to record their presence through the accumulation of loess. In contrast to the depositional evidence for a normally slow onset of glacial conditions, an abrupt transition from loess to forest soil typically indicates the sudden ending of a glacial interval (Figure 18-14).

The nature of glacial and interglacial intervals   Because of their relative recency, the last glacial interval (the Wisconsin or Riss-Würm interval) and the interglacial interval that preceded it have left the best record of Pleistocene climatic fluctuations. A major international project called CLIMAP was recently formed to reconstruct the oceanographic conditions and climates that characterized this glacial interval. Many of CLIMAP's conclusions have been based on studies of the geographic distributions of fossil coccoliths and planktonic foraminifers that have been obtained from deep-sea cores but represent species that remain alive today. Knowledge of the environmental re-

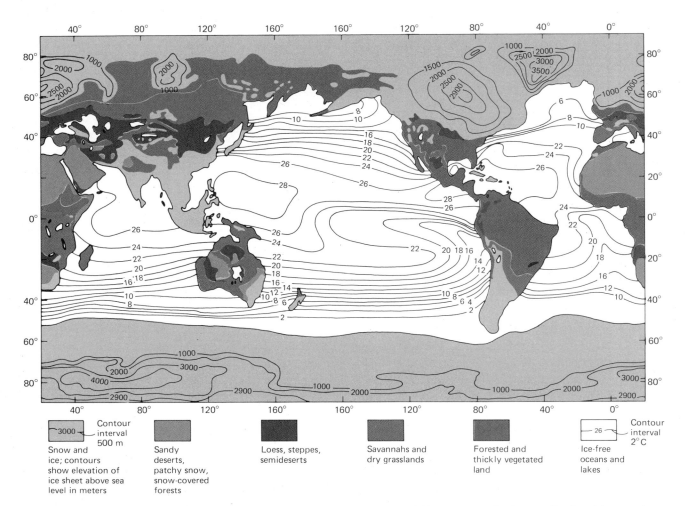

FIGURE 18-15   Geographic features reconstructed from the height of the most recent glacial interval, about 18,000 years ago, when sea level was about 85 meters (~280 feet) below its present level. Temperatures (in degrees Centigrade) are estimates for the month of August. (*Modified from CLIMAP, Science 191:1131–1144, 1976.*)

quirements of these planktonic species has permitted paleontologists to reconstruct the distributions of water masses associated with particular temperatures, and these temperatures, combined with information on the distribution of terrestrial floras and faunas, have made it possible to map the distribution of Pleistocene vegetation and environments on land (Figure 18-15).

During the Wisconsin glacial interval, north-south temperature gradients steepened in the Northern Hemisphere both in shallow seas and on land. Winter temperatures fell only slightly in most tropical areas but plummeted at latitudes north of 30° in the Northern Hemisphere. Other patterns of climatic change, however, were more complex. Some areas that lay only a few degrees south of continental glaciers, for example, became much wetter than they are

today. Among these was the Great Basin in the American West, which drew rainfall from glaciated terranes to the north and accumulated numerous lakes in areas that are now arid (Figure 18-16). Subtropical deserts, such as portions of the Sahara near the Mediterranean, also became wetter in their northern regions. In contrast, environments of low rainfall—steppes, semideserts, savannahs, and dry grasslands— stretched in a broad belt across Eurasia south of glaciers. At the same time, the tropical rain forests of South America and Africa, which lay relatively far from the lowered oceans, shrank back in areas where drying conditions came to prevail, eventually giving way to vegetation that required less moisture.

In Europe, the Alps and the Pyrenees developed glaciers of their own under the cold climatic conditions, and

Mountain glaciers

Principal lakes of Wisconsinan Age

—··— Southern limit of Wisconsinan continental ice

FIGURE 18-16   Locations of glaciers and lakes in the United States during the most recent glacial interval. Lake Agassiz and the ancestral Great Lakes formed to the south of the continental glacier as it retreated northward near the end of the glacial interval. (*C. B. Hunt, Natural Regions of the United States and Canada, W. H. Freeman and Company, New York, 1974.*)

FIGURE 18-17   Modern ice floes in the sea adjacent to northern Canada. Similar floes occupied large areas of the Atlantic Ocean during Pleistocene glacial intervals (see Figure 18-18). (*National Air Photo, Library of Canada.*)

the north-south trend of these cold barriers blocked the southward migration of many species that were moving ahead of continental glaciers advancing from the north. Many forms of life, such as magnolia trees, were caught in this glacial "trap" and disappeared from Europe but still managed to survive in North America, where southward migration to Florida and Central America remained unimpeded.

Antarctica, which was positioned over the South Pole, accumulated a continental glacier long before the beginning of the Pleistocene Epoch, but the isolation of this island continent following the breakup of Gondwanaland prevented its glaciers from spreading over a vast continental area in the manner of the late Paleozoic glaciers of Gondwanaland. Nonetheless, snow and ice accumulated over the southern tip of Africa, an eastern segment of Australia, and a large part of southern South America (Figure 18-15). A large lake occupied central Australia, and much of this continent was wetter and more heavily vegetated than today.

Three large glacial centers developed in the Northern Hemisphere—one in North America, one in Greenland, and one in Scandinavia (Figure 18-15). The northern Atlantic Ocean, being adjacent to these three glacial centers, was more profoundly affected by the Pleistocene glaciers than any of the world's other major oceans except the Arctic, but land areas adjacent to the Atlantic also suffered

marked climatic change. During glacial intervals, pack ice similar to that which now occupies bays adjacent to northern Canada (Figure 18-17) must have choked large areas of the North Atlantic (Figure 18-18). Farther south, along the east coast of North America, glaciers flowed southward to New Jersey, and tundra occupied what is now Washington, D.C. Today, the Gulf Stream swings far northward, maintaining a relatively warm climate in the British Isles and northwestern continental Europe at about 55° north latitude (Figure 18-18), but planktonic foraminifers in deep-sea cores have revealed that the Gulf Stream has periodically shifted from a northern interglacial position, like that of the present, to a glacial position that carried its warm waters more directly eastward, toward Spain (Figure 18-18). As a consequence of this pattern, northern Europe remained frigid during glacial intervals.

Another geographic change compounded this cooling trend in the northern Atlantic. Because major wind systems are driven by convection produced by temperature contrasts, the steepening of north-south temperature gradients during glacial intervals strengthened the trade winds of the Northern Hemisphere. As Figure 18-18 illustrates, this caused the winds to blow more strongly along their diagonal paths toward the equator—as a result of which the west-

PRESENT   GLACIAL INTERVAL

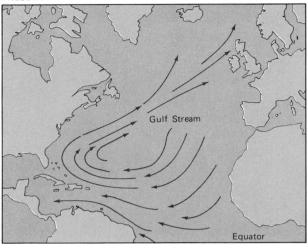

FIGURE 18-18   The Atlantic Ocean today and during the most recent Pleistocene glacial interval. During the glacial interval, ice floes occupied much of the North Atlantic, and the Gulf Stream was pushed almost directly eastward toward Spain.

Strong latitudinal temperature gradients strengthened the trade winds of the Northern Hemisphere, pushing the warm equatorial currents into the Southern Hemisphere.

ward-flowing equatorial currents must have been pushed southward. Today, these currents are deflected northward by the large hump of South America at Brazil, feeding warm water to the Northern and Southern Hemispheres, with the portion of the Atlantic north of the equator capturing more than its share of the warm equatorial currents. It is assumed that when strong northeasterly trade winds pushed the equatorial currents southward during recent glacial intervals, less of the warm water of these currents was captured by the hump of South America, causing the northern Atlantic gyre to remain cooler than it would have been otherwise.

The causes of glaciation   What factors have brought about Pleistocene glaciation? This is a double question. On the one hand, we must know why the Ice Age began in the first place. On the other, we must determine the reason, once the Ice Age had begun, that the glaciers waxed and waned at frequent intervals. In considering these questions, it is important to understand that both the expansion and the contraction of glaciers are unstable processes; once either of these processes has begun, it usually accelerates automatically. This happens because ice reflects a much higher percentage of warming sunlight than does land or water: its albedo is higher. Thus, if a small amount of climatic cooling generates glacial ice, this ice reflects a higher percentage of sunlight than was reflected before the glacier developed, and this leads to further cooling and

glacial expansion, which leads to even less absorption of sunlight, and so on. In the same manner, climatic warming melts ice, resulting in exposure of light-absorbing land, which produces even more warming. These conditions tell us that the question of why glaciers expand and contract is to a large extent a question of why a new trend (expansion or contraction of ice) gets started. A small change in climatic conditions can ultimately have an enormous effect.

The first question—What might have initiated the Ice Age?—is one that geologists have not been able to resolve with any degree of certainty. The formation of ice earlier in geologic history can be partly attributed to the movement of continents across poles; whereas polar seas are often warmed by the exchange of water with warmer regions, a large landmass centered over a pole tends to become quite cold. Apparently it is no coincidence that the major ice caps of both the Ordovician Period (Figure 12-37) and late Paleozoic time (Figure 14-32) developed at times when large continents were passing over the South Pole. The Pleistocene Ice Age, in contrast, was focused in the Northern Hemisphere, with the North Pole itself centered in the Arctic Ocean. It is true that the North Pole was partly isolated from warmer oceans by neighboring landmasses, but this condition existed for millions of years before ice sheets formed and does not in itself provide the answer.

It has been suggested that the sudden elevation of mountain ranges might have triggered the Ice Age by cre-

FIGURE 18-19   Reef-built terraces fringing the Huon Peninsula of New Guinea. These high terraces represent interglacial epochs when sea level stood relatively high throughout the world. (*Courtesy of A. Bloom.*)

ating high-altitude mountain glaciers which expanded by reflective cooling and eventually lowered global temperatures. It has also been hypothesized that a reduction in the sun's energy output might have led to the Ice Age. These theories are neither supported nor refuted by existing evidence. Perhaps the most reasonable idea yet put forward to explain the onset of glaciation is that when the Isthmus of Panama formed about 3.5 million years ago, the Gulf Stream was strengthened by the northward deflection of the equatorial current. Moisture that the warm Gulf Stream waters thus supplied to northern regions led to an increase in snowfall here and to a buildup of ice caps. We will return to this idea when we review Neogene events in the Atlantic Ocean.

More readily evaluated is the notion that changes in the relationship of the earth to the sun may cause glacial fluctuations. Supporters of this theory point to the observation that the earth's orbit periodically changes shape as a result of the movements of other planets, which exert a gravitational pull on the earth. The most pronounced orbital changes of the earth follow a 92,500-year cycle, which approximates the periodicity of the oxygen-isotope cycle recorded in deep-sea cores.

Astronomical arguments have been carried even further in attempts to explain smaller-scale climatic cycles, such as changes in sea level. Thorium dating of coral reefs that now stand about 6 meters (~20 feet) above sea level has revealed that about 125,000 years ago the world's oceans stood much higher than they do today. Sea level subsequently dropped but rose quickly about 105,000 years ago, and it then dropped and rose once more about 82,000 years ago. The second and third high stands were lower than the first, but it is not known precisely how far sea level fell in between the high stands. Terraces formed by reef growth are found at many localities, including New Guinea (Figure 18-19), Barbados (an island in the West Indies), and the Florida Keys. They appear to record minor oscillations that occurred during a single larger oscillation, which is the last major interglacial interval, known as the Sangamon or Eem (Table 18-1).

It has been suggested that the minor oscillations recorded in the fossil-reef terraces represent an astronomical cycle known as the **precession of the equinoxes**. Precession is a slight wobble in the axis of the earth's rotation caused by the gravitational pull exerted by the sun and the moon on this planet. The wobble causes each hemisphere of the earth to experience its longest and shortest days at a different position in its elliptical orbit each year. Winter in the Northern Hemisphere is likely to be coldest when, on the longest day of the year (June 21), the earth happens to be positioned at the point in its elliptical orbit that lies farthest from the sun. The earth is in this position once every 22,000 years, the time required for completion of one precession cycle. Those who favor the theory of astronomical control of glacial cyclicity point to the fact that the earth's 125,000-, 105,000- and 82,000-year-old high stands of sea level were separated by intervals of approximately 22,000 years. They argue that precession of the equinoxes imposes small-scale cycles on larger (92,500-year) cycles that reflect changes in the shape of the earth's orbit. Skeptics question the certainty of this temporal correspondence maintaining that the effects of the astronomical cycles have not been proven strong enough to influence continental glaciation. Instead, some scientists favor the notion that fluctuations in the intensity of solar radiation have governed Pleistocene climatic fluctuations.

Climatic cycles, which are reflected in glacial movements, have also occurred on smaller scales. Between about 1500 and 1850 A.D., for example, the world experienced a minor episode of refrigeration known as the **Little Ice Age**, which reached its peak about 1700. During this period, areas like Scandinavia and New England suffered bitter winters and short summers that caused numerous major crop failures. George Washington and his troops at Valley Forge were actually fortunate, in 1777 and 1778, to have experienced a winter that, for the times, was relatively mild. Studies of glacial moraines show that over the last 10,000 years (since the last major glacial interval), the world has experienced three other cold intervals comparable to the Little Ice Age. The warm interval preceding the first of these peaked about 7000 years ago, at which time global temperatures were warmer than they have been at any time since.

On a still smaller scale, numerous cycles are evident in climatic records of the last hundred years or so (Figure 18-20). Beginning around the turn of the century, the average annual temperature at the earth's surface showed a rising trend, reaching a maximum rate of increase in 1938. Since that time, the rate of change has followed a zigzag course downward, and the average annual temperature has dropped many times. Does this indicate that we are

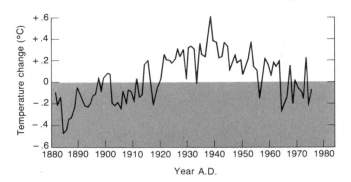

FIGURE 18-20   Climatic changes over the past 100 years. The graph represents changes in average annual temperature from year to year. We cannot be certain whether the cooling trend of the past few years will soon end or whether it will continue and produce another Little Ice Age. (*After J. M. Mitchell, in Energy and Climate, National Academy of Sciences, Washington, D.C.*)

entering another Little Ice Age? Or will the trend reverse itself so that we will climb further out of the cold period that reached a climax about 1700? Unfortunately, attempts to answer these questions represent little more than conjecture.

One factor that concerns many scientists is the possibility that the burning of wood and fossil fuels for energy, a process that adds carbon dioxide to the atmosphere, may eventually cause the earth's climate to warm up considerably as a result of the so-called greenhouse effect (page 36). If the temperature of the earth were to rise by only a few degrees, the partial melting of glaciers would elevate the oceans to such an extent that the major port cities of the world would be flooded. Unfortunately, we cannot yet assess the risk that we are incurring by burning carbon compounds to obtain energy.

We have been discussing how humans may be influencing the climate without first having discussed where our species came from to begin with. Having reviewed global events of Neogene time, we can now step back and review what is known about human origins.

Human evolution

In addition to the single species that now constitutes the human family, the superfamily Hominoidea currently consists of just four species of the ape family—the common chimpanzee, the pigmy chimpanzee, the gorilla, and the orangutan—together with six species of the gibbon family (Figure 1-6). The human family, Hominidae, did not evolve from the modern ape family, Pongidae; instead, the two

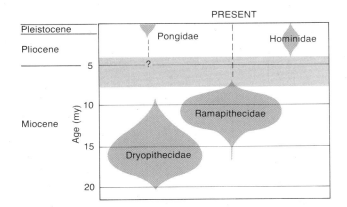

FIGURE 18-21   Families of advanced primates. Modern apes (Pongidae) have a very poor fossil record, and the fossil record of humans (Hominidae) extends back only about 4 million years to the early part of the Pliocene Epoch. The relationship of modern apes and humans to the apelike Dryopithecidae and Ramapithecidae remains uncertain.

families have followed independent lines of evolution. Although it is possible that Hominidae and Pongidae evolved independently from a single family of primitive apes, the case is not clear. Thus, we will first review the evolution of the Pongidae and then outline that of the Hominidae.

Early apes in Africa and Asia   Although an extensive Plio-Pleistocene fossil record has been uncovered for the Hominidae, very few known fossil remains of any age represent the Pongidae. Furthermore, the fossil record of the superfamily Hominoidea in latest Miocene and earliest Pliocene time (8 to 4 million years ago) is almost entirely barren. Sediments representing this interval in Africa, where apes and humans may have evolved, are rare and poorly studied.

Farther back in the Miocene Series, numerous fossils represent two families of apelike species that have both been considered potential ancestors of modern apes and humans. These early families are the **Dryopithecidae** and the **Ramapithecidae** (Figure 18-21). The record of the older group, the Dryopithecidae, begins in Africa at about the base of the Miocene Series, extending upward to sediments that are of mid-Miocene age (about 14 million years old). In Eurasia, dryopithecids first appear in sediments about 14 million years old, and their last occurrence is in sediments 8 or 9 million years old. The fact that dryopithecids did not spread from Africa to Eurasia until about 14 million years ago is not surprising in that the African plate did not collide with Asia until about 20 million years

ago. As a result of this collision, not only hominoids but also many other previously isolated groups of African mammals, including elephants and giraffes, spread into Eurasia. Until recently, dryopithecids were thought to have resembled and given rise to modern apes, but closer examination has revealed that the dryopithecids were actually not so similar to apes; their tooth proportions resemble those of Old World (African and Eurasian) monkeys, and their limb structure suggests a life of four-legged motion in the trees, perhaps with occasional forays onto the ground.

In Asia, ramapithecid fossils have not been found in sediments older than about 14 million years, but a jaw fragment from Africa extends their range back to about 17 million years. The ramapithecids were generally assumed to have evolved from the dryopithecids, from which they differ in their reduced canine teeth (or eyeteeth) and in their large, relatively flat cheek teeth, which in advanced forms had thick enamel (Figure 18-22). Unfortunately, few ramapithecid limb or pelvic bones are known, so we have little information about modes of locomotion within the group. Fragmentary limb bones suggest that these early apes ranged in weight from 20 kilograms (~44 pounds) to 70 kilograms (~154 pounds). *Gigantopithecus*, which is probably the largest of the ramapithecids, moved about on the ground like a gorilla, but the smaller forms may have been arboreal.

The ramapithecids became so diverse that by Late Miocene time the Old World was much more heavily populated by apes than Africa is today. These animals virtually disappeared before the end of Miocene time, however.

Species of Australopithecus: *Ape people of Africa*   The fossil record of the Hominidae is so poor for the latter part of the Miocene Epoch that little is known of the history of the family during this interval. By comparison, the Pliocene fossil record is a treasure trove, yielding numerous remains of the genus *Australopithecus,* the oldest known genus of the human family, which has been found in African deposits ranging in age from about 1.3 to 4.0 million years.

Most experts currently recognize the existence of three species of *Australopithecus* (Figure 18-23). The oldest of these species is *Australopithecus afarensis*, which includes the famous skeleton known as "Lucy." East African fossils reveal that this species averaged about 1.2 meters (~4 feet) in height, but its average brain capacity was larger than that of a chimpanzee (ranging from 380 to 450 centimeters, compared to 300 to 400 for a chimp). By comparison, the modern human brain averages about 1330 cubic centimeters. Our larger body requires a somewhat

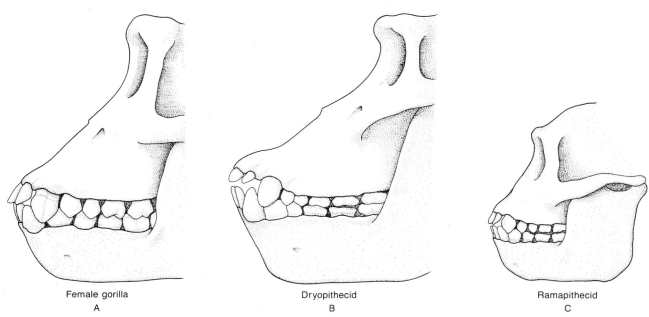

Female gorilla
A

Dryopithecid
B

Ramapithecid
C

FIGURE 18-22   Skulls of apelike animals. The skull of the dryopithecid (*Sivapithecus*) in the center in many ways resembles that of a modern ape (*left*). The ramapithecid skull, represented by *Ramapithecus* on the right, has reduced canine eyeteeth and broad molars, and the muzzle is less elongate than in the other forms. (*After E. L. Simons, Scientific American, May 1977.*)

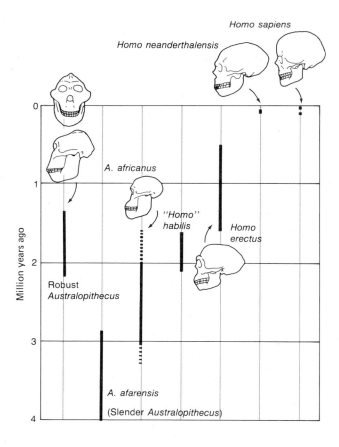

larger brain simply for control of bodily functions. Even in relation to body size, however, the human brain is much larger than that of *Australopithecus afarensis*, which for simplicity we will call "Afarensis."

On the other hand, the pelvis of Afarensis is remarkably like that of humans (Figure 18-24). Unlike the elongate pelvis of apes, the pelvis of Afarensis was adapted to support an upright body. This upright posture is also evidenced by a spectacular set of tracks left by either Afarensis or a close relative in soft volcanic ash in Tanzania slightly more than 3 million years ago (Figure 18-25).

Although its pelvis was similar to that of a human, Afarensis had a skull that retained many apelike features, including massive brow ridges, a low forehead, and a projecting, muzzlelike mouth. A reconstruction of the ani-

FIGURE 18-23   Stratigraphic ranges of species of the Hominidae (human family) as currently recognized from fossil data. Two slender species and one robust species of *Australopithecus* are represented on the left. Species of the genus *Homo* are represented on the right. Some workers place the species labeled "*Homo*" *habilis* in the genus *Australopithecus* rather than in the genus *Homo*. Some workers classify Neanderthal as a subspecies of our own species, *Homo sapiens*.

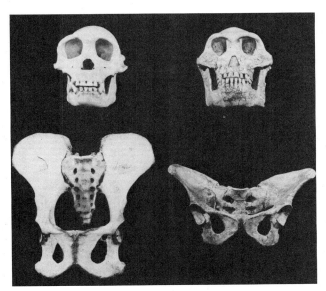

FIGURE 18-24   Skeletal features of *Australopithecus afarensis* (*right*) compared to those of a modern chimpanzee (*left*). The skulls of the two animals are more similar than the pelvises. The wide pelvis of *Australopithecus* resembles that of a modern human, where it serves as an adaptation for upright posture. (*Courtesy of D. C. Johanson.*)

FIGURE 18-25   Tracks made in volcanic ash at Laetolil, Tanzania, more than 3 million years ago. These tracks, which represent some species of *Australopithecus*, record upright walking behavior much like that of modern humans. (*Photograph by J. Reader courtesy of the National Geographic Society.*)

mal in side view illustrates these features especially well (Figure 18-26).

The teeth of Afarensis are in many respects intermediate between those of apes and humans. Compared to humans, apes have large incisors (or front teeth) in relation to the size of their molars (Figure 18-27). Apes also have large projecting canines (eyeteeth) and a space between their incisors and canines in the upper jaw to accommodate the upward-projecting canine of the lower jaw. Humans do not have this space; Afarensis does, but it is reduced in size, in keeping with the size of the canine teeth of Afarensis, which are larger than a human's but smaller than an ape's. The incisors of Afarensis are smaller for the size of the jaw than those of an ape, but the molars are larger, and so the relative proportions of the teeth more closely resemble those of modern humans.

Fossils clearly assignable to *Australopithecus afarensis* range in age from about 4 to about 2.8 million years, thus spanning more than 1 million years. About 3 million years

FIGURE 18-26   Reconstruction of the oldest known species of the human family, *Australopithecus afarensis*. The low forehead, heavy brow, and projecting lower face of this species are all evident here. (*After a reconstruction by J. H. Matternes.*)

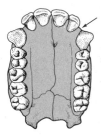
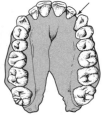
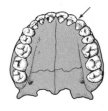

Chimpanzee | Australopithecus afarensis | Modern human

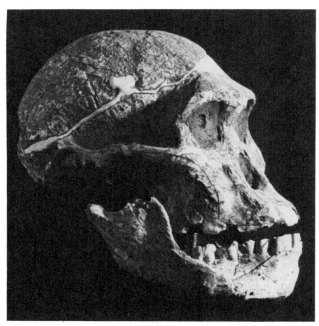

FIGURE 18-27   Comparison of the upper teeth of *Australopithecus afarensis* with those of a chimpanzee and a modern human. The chimpanzee has larger canine (eyeteeth) than humans, with a gap between each canine and the adjacent incisor (*arrow*). The incisors of the ape are also large in relation to the molars. In these important features, *Australopithecus afarensis* is intermediate between apes and humans. (*After D. Johanson and M. Edey, Lucy: The Beginnings of Humankind, Simon and Schuster, New York, 1981.*)

FIGURE 18-28   The skull of *Australopithecus africanus* in side view, revealing a much lower forehead and a more projecting lower face than those found in modern humans. (*Courtesy of J. T. Robinson.*)

ago, Afarensis was joined by *Australopithecus africanus*, and the two species appear to have shared the African continent for a time, with Africanus surviving until about 1.6 million years ago in southern and eastern Africa. *Australopithecus africanus* was slightly taller than Afarensis, averaging perhaps 1.4 meters ($\sim$4½ feet). Well-preserved Africanus skulls exhibit the heavy brow and the long, sloping facial region below the eyes that characterize all member species of the genus *Australopithecus* (Figure 18-28).

The third species of *Australopithecus* was *Australopithecus robustus*, whose geologic "lifetime" overlapped with that of Africanus, spanning an interval from about 2.3 to 1.3 million years ago. Robustus lived up to its name in many ways—its head and teeth were heavily constructed (Figure 18-29); the crown of its skull was elevated into a bony crest similar to that of a gorilla, thereby providing for the attachment of strong jaw muscles; and its wide, flange-like cheekbones and massive lower jaws provide further evidence of powerful chewing muscles. Despite its presumably formidable appearance, the robust *Australopithecus* was not necessarily a fierce creature. Nor was it large in its overall proportions; it is thought to have averaged not much more than 1.5 meters ($\sim$5 feet) in height. Set in its huge jaws were exceptionally broad, flat cheek teeth. The power of the Robustus jaw was apparently directed toward grinding up coarse plant food.

It is widely argued that the specialization of *Australopithecus robustus* for chewing coarse food makes it an unlikely ancestor of modern humans. It is almost universally

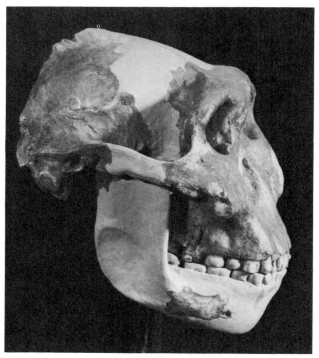

FIGURE 18-29   *Australopithecus robustus*, a heavy-browed species with massive jaws and teeth that served for grinding up vegetable matter. (*Smithsonian Institution.*)

FIGURE 18-30   Reconstruction of *Homo erectus*, a species that existed from about 1.6 million years ago to perhaps 400,000 years ago and employed a distinctive stone culture. (*Painting by Z. Burian under the supervision of Professor J. Augusta, Professor J. Filipa, and Dr. J. Moleho.*)

Homo erectus resembled modern humans and differed from *Australopithecus* in several ways, but the recent discovery in Africa of the 1.6 million year old skeleton of a twelve-year-old boy indicates that *Homo erectus* rivaled us in size. The cranial capacity of *Homo erectus* skulls that are currently available for study ranges from about 800 to 1300 cubic centimeters. The maximum size approximates the average for modern humans (about 1300 cubic centimeters), but the minimum size is much lower. There is some evidence, however, that the brain size of *Homo erectus* increased somewhat over the course of the species' existence. The brain case of *Homo erectus* was also relatively longer and lower than ours; furthermore, its forehead was low and interrupted by brow ridges that were even bigger than those of the two slender species of *Australopithecus* (Figure 18-31). These features reflect derivation from *Australopithecus*, as do the projecting mouth and heavy lower jaw—but the mouth and jaw as well as the cheek bones were less highly developed than those of *Australopithecus*. The teeth of *Homo erectus* were smaller than those of *Australopithecus*, and though larger than the teeth of modern humans, nevertheless bear a close re-

agreed, however, that one or both of the more slender australopithecine species have a place in our evolutionary history.

The human genus makes its appearance   We do know that our own genus, *Homo*, was alive in Africa about 1.6 million years ago in the form of *Homo erectus* (Figure 18-30). In contrast to *Australopithecus*, which is unknown from localities outside Africa, *Homo erectus* was a widely traveled species. Known as *Pithecanthropus* before its similarities to modern humans were fully acknowledged, *Homo erectus* lived not only in Africa and Europe but also in China, where it has been referred to as "Peking man," and in Java, where it has been called "Java man." Fossil skulls of *Homo erectus* represent a long interval of time, extending from about 1.6 million years ago to perhaps 400,000 years ago.

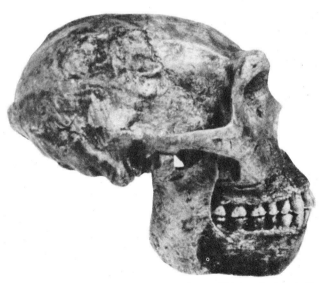

FIGURE 18-31   A skull of *Homo erectus*. This particular skull has been informally labeled "Java man." In comparison to the skulls of modern humans, the brain case is long and low, and the brow, jaw, and teeth are robust. Except in size, the teeth resemble those of modern humans. (*W. W. Howells, Scientific American, November 1966. Photograph by C. H. R. von Koenigswald.*)

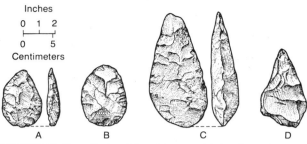

Inches
0 1 2

0   5
Centimeters

A   B   C   D

FIGURE 18-32   Tools of the widespread Acheulian culture attributed to *Homo erectus*. *A*. Twisted oval tool from Saint-Acheul near Amiens, France. *B*. Oval hand ax from south of Wadi Sidr, Israel. *C*. Large hand ax from Orgesailie, Kenya. *D*. Hand ax from Hoxne, Suffolk, England. (*After K. P. Oakley, Man the Toolmaker, British Museum of Natural History, London, 1950.*)

Inches
0   1   2

0   5
Centimeters

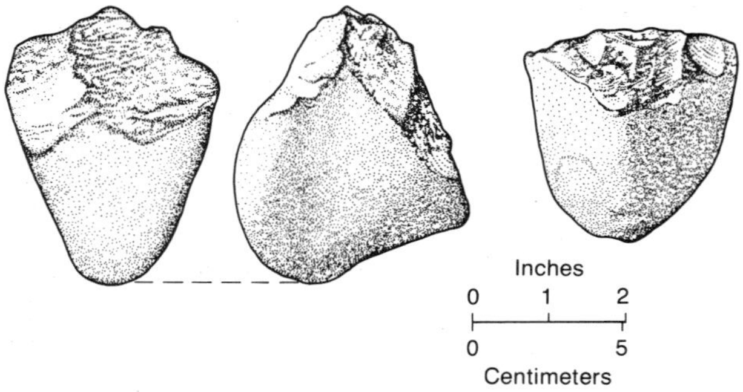

FIGURE 18-33   Simple stone implements from Olduvai Gorge, Tanzania, representing the Oldowan culture that existed about 2.5 million years ago. These are "core" implements from which flakes were struck. (*After K. P. Oakley, Man the Toolmaker, British Museum of Natural History, London, 1950.*)

semblance to them. *Homo erectus* was also a toolmaker. Its stone culture, known as **Acheulian**, included the production of magnificent hand axes (Figure 18-32).

One problem in tracing the evolution of our genus concerns the classification of a species that made its earliest known appearance below *Homo erectus*, in strata about 2 million years old. This problematic species has traditionally been designated "*Homo*" *habilis*. (The generic name is in quotation marks because some workers consider the skull of this species to be of the *Australopithecus* type.) The brain capacity of Habilis (650 to 800 cubic centimeters) was larger than that of other *Australopithecus* species but smaller than that of most specimens of *Homo erectus*. Part of the reason that Habilis has been placed in the genus *Homo* is that stone tools of the so-called Oldowan type (Figure 18-33) have been found in associated sediments of Olduvai Gorge, the locality that gave the stone culture its name—and it was once believed that only true humans, or members of the genus *Homo*, could fabricate tools. This belief has no real foundation and is damaged by recent discoveries of Oldowan tools in sediments about 2.5 million years old—about half a million years older than the oldest known Habilis skull. The Oldowan tools comprise sharp flakes of stone and many-sided "core" implements that represent pieces of stone left after the removal of the flakes. The flakes may have served for slicing and the core implements for pounding and chopping. Members of the genus *Australopithecus* remain strong suspects in our search for the producers of these implements, which are the oldest of all known tools.

Was Neanderthal one of us?   Just as there are gaps in our knowledge of the evolutionary pathway that led to *Homo*

erectus, so are there gaps in the evolutionary pathway connecting this species to our own, *Homo sapiens*. Unfortunately, sediments in Africa representing the interval above the youngest specimens of *Homo erectus* have yielded a poor suite of fossil remains. As a result, we are not certain exactly what evolutionary pathway led from *Homo erectus* to our species, *Homo sapiens*. The interval of poor fossil remains extends from about 400,000 to 100,000 years ago, and most of the partial skulls found within this interval have robust brow ridges and low foreheads reminiscent of *Homo erectus*.

In sediments about 100,000 years old, the hominid fossil record improves with the appearance of well-preserved fossils of the creature known as Neanderthal (Figure 18-34). Some scientists regard Neanderthal as a variety or subspecies of our species (*Homo sapiens neanderthalensis*), while others regard it as a separate species (*Homo neanderthalensis*). The record of this humanoid creature extends from Spain to central Asia, ranging in time up to about 35,000 years ago. Enough of its bones and artifacts have been found in caves to suggest that Neanderthal was frequently a cave dweller.

Neanderthal is so designated because its bones were first found in the Neander Valley of Germany in 1856, three years before Charles Darwin wrote *On the Origin of Species*. This creature differed from modern humans in a number of skeletal features. Neanderthal resembled *Homo erectus* in that its skull was long and low with prominent brow ridges, a projecting mouth, and a receding chin (Figure 18-34). On the other hand, its brain was quite large— slightly larger, on average, than that of modern humans. The large brain is by no means indicative of superior intellect, however; Neanderthal's body was somewhat more

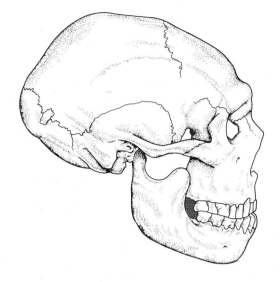

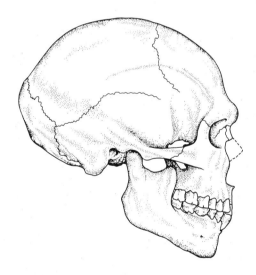

FIGURE 18-34   A Neanderthal skull (*above*) and a skull of the modern human type (*below*). The Neanderthal brain case is larger and lower in the forehead region. The brow is also heavier, as is the lower jaw. A characteristic of Neanderthals is the gap (colored here) between the rear teeth and the jawbone. (*After E. Trinkaus and W. W. Howells, Scientific American, December 1979.*)

massive than ours, though slightly shorter, and may have required some extra brain cells simply for the purpose of motor control.

The reputation of Neanderthal has been damaged by generalizations that have been unfairly gleaned from the remains of a single individual found at Lachapelle-aux-Saints, France, which attracted much attention early in the twentieth century. The stooped posture of this individual was taken as evidence that Neanderthals as a group were primitive creatures, lacking fully erect posture. The fact is that the French specimen was poorly reconstructed. Neanderthal normally stood as erect as we do, although his bones were more dense, and he probably moved more slowly.

Neanderthal also differed from modern humans in the presence of a gap between its cheek teeth and the back of its jaw (Figure 18-34) and in certain distinctive features of the shoulder blade, pelvis, and hand. The finger bones of Neanderthal have stronger attachment areas for tendons than those found in modern humans, suggesting that this creature possessed a grip stronger than our own. Those who favor the idea that Neanderthal was a species separate from *Homo sapiens* cite these features along with the *Homo erectus*–like shape of its skull as evidence. Those who prefer to include Neanderthal in our species emphasize similarities in teeth and brain size.

The distribution of the two species in time and space suggests that *Neanderthal* and *Homo sapiens* may, in fact, have been reproductively isolated. The interval during which Neanderthals lived represented the most recent glacial stage, the Riss-Würm Stage, but Neanderthals disappeared before the glaciers melted back. Radiocarbon dating tells us that Neanderthals vanished from eastern Europe about 40,000 years ago and from western Europe perhaps 5000 years later. At this time, the Cro-Magnon people, who were undisputed members of the species *Homo sapiens* and were anatomically almost identical to modern humans, spread throughout Europe (Figure 18-35). The fact that Cro-Magnon suddenly replaced Neanderthal in Europe suggests that the two did not interbreed successfully and thus may well have been distinct species.

Where did the biologically modern populations of *Homo sapiens* known as Cro-Magnon originate? One possibility is that they evolved in Europe from a population of Neanderthals. Another is that they migrated north from Africa about 40,000 years ago. There are, in fact, incomplete African skulls dated at approximately 100,000 years of age that seem to have a modern appearance. Perhaps descendants of *Homo erectus* in Africa were the predecessors of our species, while Neanderthal originated from *Homo erectus* descendants in Europe.

In any event, Neanderthals had developed a distinctive stone culture known as **Mousterian** in which flakes of stone were fashioned into knives, scrapers, and projectile heads—

FIGURE 18-35 A reconstruction of a Cro-Magnon camp in Europe. Cro-Magnon people were anatomically almost identical to modern Europeans. (*Painting by Z. Burian under the super-* *vision of Professor J. Augusta, Professor J. Filipa, and Dr. J. Moleho.*)

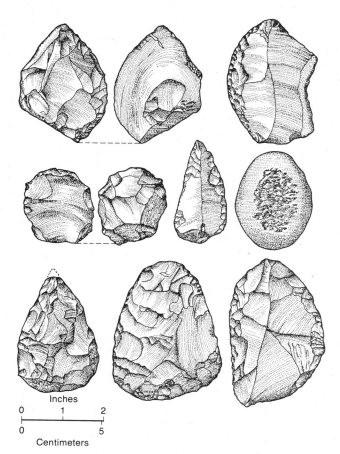

Inches

0   1   2

0   5

Centimeters

far more sophisticated tools than the Acheulian implements of *Homo erectus*. The Cro-Magnon culture that followed at first incorporated Mousterian elements but then emerged as a more sophisticated culture known as **Late Neolithic** in which a variety of specialized tools were invented (Figure 18-36).

Even more innovative was the artwork of the Cro-Magnon people, which seems to reflect a new use of the powers of imagination. Their magnificent cave paintings, primarily of animals, can still be admired in France and Spain (Figure 18-37), while other artifacts show that their artistic efforts included modeling in clay, carving friezes, decorating bones, and fabricating jewelry from teeth and shells.

Thus, we are left with several intriguing questions. Did modern humans, in the form of Cro-Magnon, evolve from a population of Neanderthals? There is no question that the two groups occupied Europe simultaneously for a brief time between about 40,000 and 35,000 years ago. Did early Cro-Magnon people retain Neanderthal culture briefly because they evolved from Neanderthals, or did they as-

FIGURE 18-36 Tools of the Cro-Magnon people. Among these are scrapers (*top row*) and an anvil or hammerstone (*middle row, at the right*). (*After K. P. Oakley, Man the Toolmaker, British Museum of Natural History, London, 1950.*)

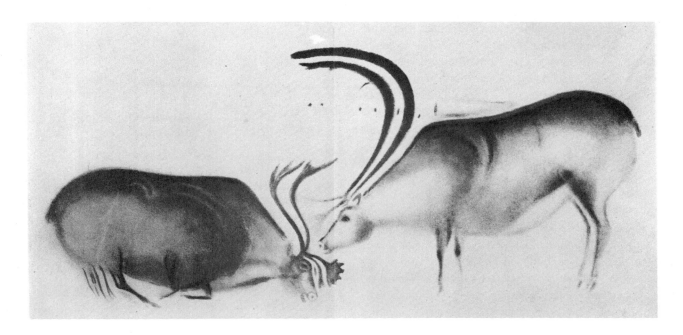

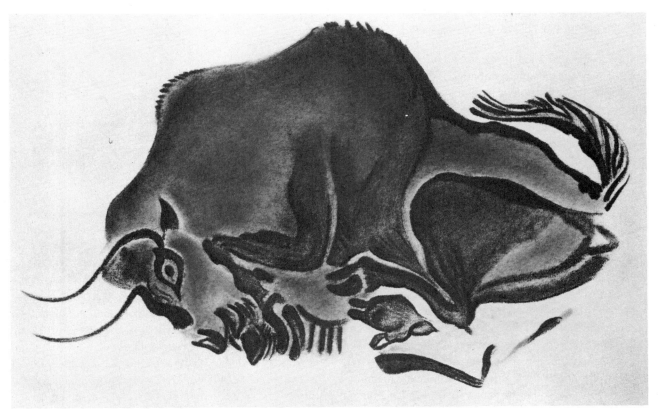

FIGURE 18-37   Cave paintings of the Cro-Magnon people. *Below*: Wisent (European bison), Altamira Cave, Spain. (*American Museum of Natural History*.)
Above: Male and female reindeer, Fonte de Gaume, France.

similate the culture when they came into contact with Neanderthals following a separate origin? Were the two groups members of the same species, or did they represent different species? Did *Homo sapiens* exterminate Neanderthal through warfare in Europe between 40,000 and 35,000 years ago? And finally, does the advanced art and technology of Cro-Magnon, together with the phenomenal cultural expansion that followed Cro-Magnon's ascendancy, reflect the fact that modern humans have a substantially more powerful intellect than that of Neanderthals? Although we know that our brain case differs in shape from the long, low structure of the Neanderthal, we do not know whether our brains actually function in a different manner.

One intriguing fact is that, whatever his mental powers may have been, Neanderthal did have religion. Burial sites reveal that Neanderthals sometimes prepared their dead for a future life by interring them with flint tools and cooked meat (Figure 18-38). In the Zagros Mountains of Iraq, a Neanderthal man who died following a skull injury was buried in a bed of boughs and flowers that can be identified from the pollen they left beneath the skeleton!

Human expansion and the extinction of large mammals
Anthropologists continue to debate not only how the various races of *Homo sapiens* spread throughout the world, but also whether all of these races evolved from Cro-Magnon or simply share with Cro-Magnon features characterizing all older populations of *Homo sapiens*. One particularly noteworthy question regarding the geographic spread of *Homo sapiens* is, When did our species first reach the Americas? This issue is intertwined with another: the cause of an extinction of many large terrestrial mammals at the end of the last glacial advance about 10,000 years ago. As a result of this extinction event, the world today is impoverished with regard to large mammals; only the faunas of the African savannahs and open woodlands give us a hint of what the rich Pliocene terrestrial ecosystems must have been like, not only in Africa but also in Eurasia and North America. In fact, the modern African fauna, which is rich by today's standards, is impoverished in comparison with the diverse faunas that characterized the Pliocene and Pleistocene epochs. Climatic deterioration led to a reduction of faunal diversity in Eurasia during the Pleistocene Epoch, but even the fossil record of that epoch has yielded a fauna of more than 50 genera of Bovidae (cattle, antelope, sheep, goats, and related hoofed animals).

Within the rich Plio-Pleistocene faunas were many species of immense proportions. Because increased body size

FIGURE 18-38   Reconstruction of the burial of a young Neanderthal whose skeleton shows that the head was cradled on one arm. Also present are the charred bones of animals that were apparently left as food for the dead individual in an afterlife. (*Painting by Z. Burian under the supervision of Professor J. Augusta, Professor J. Filipa, and Dr. J. Moleho.*)

reduces heat loss (big animals have smaller surface areas for their body volume than do small animals), the large size of some Pleistocene species has been considered to have been an adaptation to cold climates. Among the huge Pleistocene species were an elephant-sized bison of the American plains whose horns spread more than 2 meters (~7 feet), or three times the span of the living species; a North American beaver (*Castoroides*) that was nearly as large as a black bear; horses of the modern genus *Equus*, which evolved to the size of our artificially bred Clydesdale draft horses; and mammoths (members of the elephant genus *Mammuthus*) that stood about 4.5 meters (~15 feet) at the shoulder, about 30 percent taller than an average African elephant!

Near the end of the Pleistocene Epoch, about 11,000 years ago, at the time the glaciers were receding from Eurasia and North America, numerous species of large

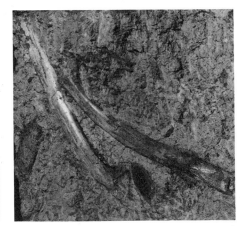

FIGURE 18-39 *A*. Reconstruction of "Folsom man" hunting giant bison. *B*. The original Folsom point, found between the fossil ribs of a giant bison. (*Denver Museum of Natural History.*)

mammals disappeared throughout the world. Some have argued that these extinctions occurred at the hands of humans armed with newly developed weapons for which large, conspicuous mammals were easy marks. Because they were represented by few individuals to begin with, such species were eventually hunted out of existence. This theory suggests that a wave of humans swarmed into North America about 11,000 years ago, by way of the Bering land bridge, and were such successful big-game hunters that their populations expanded and spread rapidly—perhaps by about 16 kilometers (~10 miles) per year—in their efforts to find new game. Radiocarbon analyses of wood associated with stone weapons reveals that humans at this time employed sophisticated projectile systems in the New World; they were known to have launched flint tips of many types at the ends of lances and may also have propelled darts with throwing sticks. The famous Clovis and Folsom projectile points represent this stage of weaponry development (Figure 18-39).

REGIONAL EVENTS

Because many shifts in climate and in biogeographic distribution have been reconstructed in detail for the Neogene Period, geologists have been able to relate many regional geologic events to these phenomena. This relationship represents a recurring theme in the review of Neogene regional events that follows.

We will begin our regional tour of the Neogene Period by reviewing the history of the western United States, which is highlighted not only by the elevation of imposing mountains that form part of our scenery today—the Cascade Range, the Sierra Nevada, and the Rocky Mountains—but also by climatic changes that resulted from the uplifting of these mountains. Next, we will examine events in and around the Atlantic Ocean; among our topics here will be the origin of the modern mountainous topography of the Appalachians, the uplift of the Isthmus of Panama, the origin of the Caribbean Sea, and the cooling of the Atlantic with the onset of the Ice Age. We will then review the Neogene history of the African continent and its famous rift valleys, which have contributed valuable fossil remains of human ancestors. Following this, we will learn how the northward movement of Africa against Eurasia closed the venerable Tethys Sea and, in the process, formed the Mediterranean. At one point, this remnant of the Tethys suddenly dried up, but it was soon refilled with sea water. Finally, we will observe how the northward movement of the Australian plate has brought biotas of the Southern Hemisphere into contact with animals and plants of Eurasian ancestry.

Development of the American West

The pre-Neogene history of mountain building in the Cordilleran region was described in earlier chapters and is summarized in Figure 17-28. By late Paleozoic time, uplifts resulting from the final mountain-building episode of the western interior, the Laramide orogeny, had been largely subdued by erosion, thereby setting the stage for the Neogene events that produced the Rocky Mountains. In the broad region west of the Rockies, the Neogene Period was a time of widespread tectonic and igneous activity, which built most of the mountains standing there today.

Provinces of the American West   Lying between the great plains of the United States and the Pacific Ocean are several distinctive physiographic provinces that have taken shape largely in Neogene time, primarily as a result of uplift and igneous activity. Let us briefly review the present characteristics of these provinces before considering how they have come into being (Figure 18-40).

The lofty, rugged peaks of the **Rocky Mountains**, some of which stand more than 4.5 kilometers (~14,000 feet) above sea level, could only be of geologically recent origin. We have seen that the widespread subsummit surface of the Rockies was all that remained of the Laramide uplifts by the end of the Eocene Epoch about 40 million years

ago (Figure 17-37). One question we must answer, then, is how the Rocky Mountain region became mountainous once again.

Centered adjacent to the Rockies in the "four corners" area where Colorado, Utah, New Mexico, and Arizona meet is the oval-shaped **Colorado Plateau**, much of which stands about 1.5 kilometers (~ 1 mile) above sea level. The Phanerozoic sedimentary units here are not intensively deformed. Some, however, are gently folded in a steplike pattern, and others, especially to the west, are offset by block faults (Figure 18-41). Cutting through the plateau is the spectacular Grand Canyon of the Colorado River (see Figure AII-17), but about 10 million years ago there was neither a Colorado Plateau nor a Grand Canyon. The origin of these features forms another part of our story.

West of the Rockies and the Colorado Plateau, within the belt of Mesozoic orogeny, lies the **Basin and Range Province** (Figure 18-42). This is an area of north-south trending block-fault valleys and intervening ridges (Figure 18-43)—features of Neogene origin. A large area of this

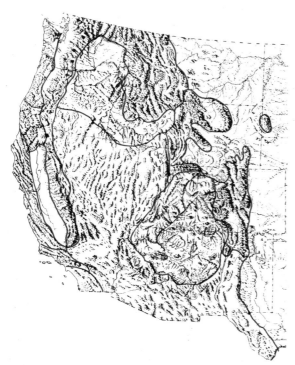

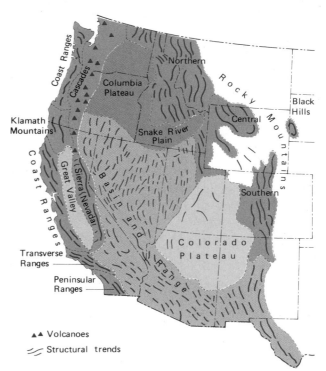

FIGURE 18-40   Major geologic provinces of western North America. The map on the left shows the relationships of the provinces to topographic features. (*Topographic map based on* *the United States Geological Survey, National Atlas of the United States of America.*)

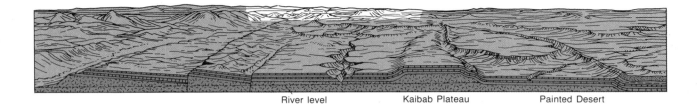

River level   Kaibab Plateau   Painted Desert

Kilometers

0   80

FIGURE 18-41   Block diagram of the western part of the Colorado Plateau north of the Grand Canyon. This high-standing region is characterized by block faulting (*left*) and by gentle, steplike folds (*right*). (*After P. B. King, The Evolution of North America, Princeton University Press, Princeton, New Jersey, 1977.*)

province forms the **Great Basin**, an arid region of interior drainage (page 69). Volcanism has been associated with some faulting episodes, and sediment eroded from the ranges blankets the valleys to depths ranging from a few hundred meters to about 3 kilometers (~2 miles). The thickness of the earth's crust in the Basin and Range Province ranges from about 20 to 30 kilometers compared to thicknesses of 35 to 50 kilometers in the Colorado Plateau. The thinning and block faulting in the Basin and Range Province point to extension of the crust by at least 65 percent and perhaps by as much as 100 percent.

Farther north, centered in Oregon, is a broad area covered by volcanic rocks of the **Columbia River and Snake River plateaus** (see Figure AI-8). Today the climate here

FIGURE 18-42   Fault-block mountains and basins of the Basin and Range Province. The view is from the Panamint Mountains. The eastern fault scarp of the Sierra Nevada can be seen in the distance. (*Courtesy of W. B. Hamilton, U.S. Geological Survey.*)

FIGURE 18-43   Diagrammatic cross section showing the possible pattern of the block faulting in the Basin and Range Province that might have been responsible for lateral extension of the crust.

is cool and semiarid; only about one-quarter of the plateau area is cloaked in forest and woodland, while sagebrush and drier conditions characterize about half of the terrane. In Oligocene time, however, lavas had not yet blanketed the region, and, as revealed by fossil plant remains, a large forest of redwood trees grew there.

Along the western margin of the Columbia Plateau stand the lofty peaks of the **Cascade Range** (Figure 18-44). These are cone-shaped volcanoes that represent the volcanic arc associated with subduction of the Pacific plate along the western margin of the continent. Volcanism began here in Oligocene time and continues to the present, as manifested by the recent eruptions of Mount St. Helens (Figure 1-1).

The Cascade volcanic belt passes southward into the **Sierra Nevada Range**, a mountain-sized fault-block of granitic rocks. The plutons forming the Sierra Nevada were emplaced in east-central California during Mesozoic time, before igneous activity at this latitude shifted inland. As we will see, however, the present topography of the Sierra Nevada is of Neogene origin. This mountain range is unusual in that throughout its length of some 600 kilometers (~350 miles), it is not breached by a single river. This is why it represented such a formidable obstacle to early pioneers attempting to reach the Pacific.

The Sierra Nevada Range stands between the Basin

FIGURE 18-44   Mount Hood, one of the high peaks of the Cascade Range in Oregon. (*Oregon State Highway Commission*.)

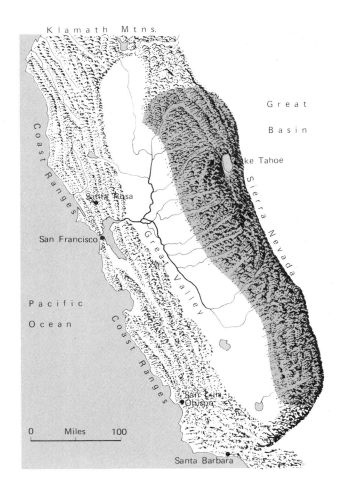

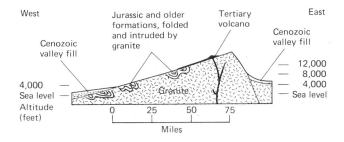

FIGURE 18-45　Map and diagrammatic cross section of the Sierra Nevada fault block of California and Nevada. Sediments of the Great Valley lap onto the gentle western slope of the Sierra Nevada. (*After C. B. Hunt, Natural Regions of the United States and Canada, W. H. Freeman and Company, New York, 1974.*)

and Range Province to the east and the **Great Valley** of California to the west (Figure 18-45). The Great Valley is an elongate basin containing large volumes of Mesozoic sediment (the Great Valley Sequence) eroded from the plutons of the Sierra Nevada region long before the modern Sierra Nevada formed by block faulting (Figure 15-43). Above these are Cenozoic deposits, some of which accumulated during marine invasions of the Great Valley and others during times of nonmarine sedimentation.

West of the Great Valley are the California **Coast Ranges**, which consist of slices of crust that include crystalline rocks representing Mesozoic orogenic activity, Franciscan rocks of deep-water origin (Figure 18-46), and Tertiary rocks. To the south, the **Transverse** and **Peninsular ranges** are formed of similarly faulted and deformed rocks, but these ranges lie inland of the main belt of Franciscan rocks in the region of intensive Mesozoic igneous activity. Striking features of all of these mountainous terranes are the great faults—including the Garlock Fault (Figures 18-47 and 18-48)—that divide the crust into sliver-shaped blocks. The longest and most famous of these faults is the San Andreas (facing page 1), which extends for about 1600 kilometers (1000 miles). Until the great San Francisco earthquake of 1906, it was not widely recognized that movement along the San Andreas persisted. The earthquake of 1906 was produced by a sudden movement of up to 5 meters (~16 feet) along the fault. Study of geologic features cut by the San Andreas Fault shows that its total movement during the past 15 million years has amounted to about 315 kilometers (190 miles). Continued movement at this rate for the next 30 million years or so would bring Los Angeles northward to the latitude of San Francisco—through which the fault passes. As we will see, the faulting and uplifting of the Coastal Ranges of California are probably causally related not only to the Neogene uplift of the Sierra Nevada but also to the origins of the Basin and Range topography to the east.

The **Olympic Mountains** of Washington have quite a different history. These relatively low mountains, which lie to the west of the Cascade Volcanics, consist of oceanic sediments and volcanics that were deformed primarily during Eocene time in association with subduction along the continental margin (Figure 17-31).

Development of the American West: The Miocene Epoch　We will begin our story of the development of the modern Cordilleran provinces described above with a summary of major events of the Miocene Epoch. Then we will

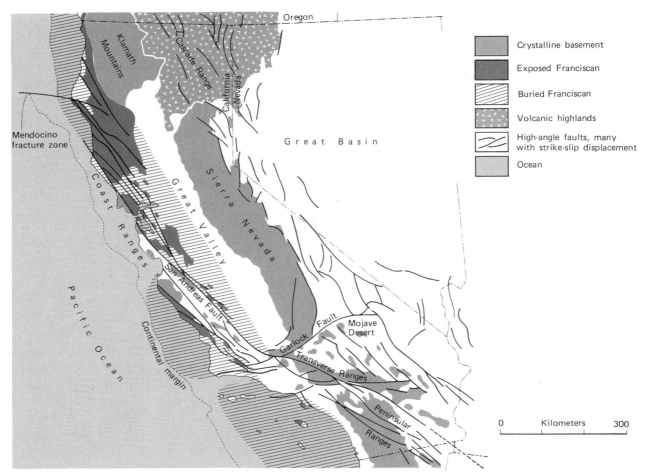

FIGURE 18-46   Major geologic features of California. The
Franciscan terrane was attached to the coast late in the Jurassic Period (Figures 15-43 and 15-44). The many faults depicted

here, including the famous San Andreas, are of Neogene Age.
(*Modified from P. B. King, The Evolution of North America,
Princeton University Press, Princeton, New Jersey, 1977.*)

examine the Pliocene events that followed, and we will conclude by analyzing the tectonic mechanisms that may account for events of both epochs.

Geologic features of the far west in Miocene time are
shown in Figure 18-49. Subduction continued beneath the
continental margin in the northwestern United States, and
the resulting volcanic arc produced peaks in the Cascade
Range, where volcanism continues today. To the south, in
California, the mid-Miocene interval was a time of faulting
and mountain building; elements of the modern Coast
Ranges and other nearby mountains were raised, and the
seas were driven westward. Meanwhile, as in Paleozoic
time, the Great Valley remained a large embayment, and
during Miocene time it received great thicknesses of siliciclastic sediments, most of which were shed from the region of the modern Sierra Nevada. Although the block

faulting that eventually produced the Sierra Nevada did
not begin until Pliocene time, volcanoes of the southern
end of the Cascade Island Arc were shedding sediments
westward, from the position where the Sierra Nevada later
stood.

Before the Sierra Nevada rose appreciably, the Basin
and Range Province began forming to the east. During
Paleogene time, light-colored rhyolitic (felsic) ashfall deposits were occasionally spread across the province, but
near the beginning of the Miocene Epoch basaltic volcanism predominated—and it was at this time or slightly
earlier that the Basin and Range topography began to form.
Since the beginning of block faulting, most lavas of the
region have welled up along faults rather than reaching the
surface through cylindrical vents. To the north of the Basin
and Range Province, basalt also spread from fissures in

FIGURE 18-47   Aerial view of the Garlock Fault in the Mojave Desert of California, looking westward. The trace of this strike-slip fault is clearly visible. (*J. S. Shelton, Geology Illustrated, W. H. Freeman and Company, New York, 1966.*)

much greater volume. Most of the great Columbia Plateau formed in this way between about 16 and 13 million years ago; here, individual basalt flows range in thickness from 30 to 150 meters (~ 100 to 500 feet), and in places, the total accumulation reaches about 5 kilometers (~ 3 miles).

One of the most important Miocene events in the Cordilleran region was a broad regional uplift that affected the Basin and Range Province, the Colorado Plateau, and the Rocky Mountains. Today, even the basins of the Basin and Range Province nearly all stand at least 1.3 kilometers

FIGURE 18-48   Movements of crustal blocks along faults in southern California. The San Andreas Fault bends at the southern end of the Great Valley and is met by the Garlock Fault. (*From D. L. Anderson, "The San Andreas Fault." Copyright © 1971 by Scientific American, Inc. All rights reserved.*)

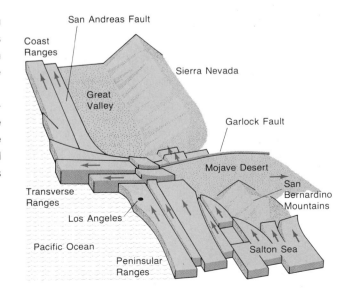

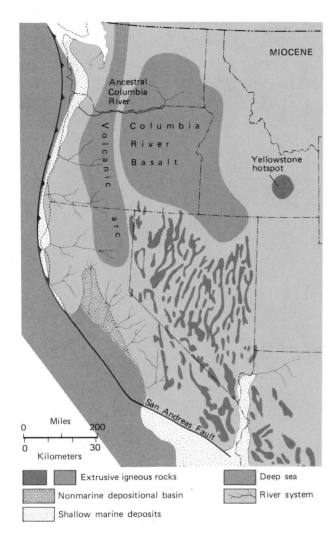

FIGURE 18-49   Geologic features of western North America in Miocene time. West of the San Andreas Fault, coastal southern California lay farther south than it does today. The area currently occupied by the Great Valley of California was for the most part a deep-water basin from which a nonmarine depositional basin extended to the north. Volcanoes of the early Cascade Range formed along a volcanic arc inland from the subduction zone along the continental margin. Igneous rocks were extruded along north-south trending faults in the Great Basin, and farther north the Columbia River Basalts spread over a large area. (*Modified from J. M. Armentrout and M. R. Cole, Soc. Econ. Paleont. and Mineral. Pacific Coast Paleogeog. Symp. 3:291–323, 1979.*)

($\sim\frac{3}{4}$ mile) above sea level. Thus, it is remarkable that mid-Cenozoic fossil floras of this region are characterized by species that could only have lived at low altitudes. Geologists have reconstructed even more precise histories of uplift for the Colorado Plateau and Rocky Mountains by studying the time at which rivers have cut through well-dated volcanic rocks. Many of the rivers of these regions existed before uplift began in the Miocene Epoch, and these cut rapidly downward as the land rose, producing deep gorges. A large part of the Grand Canyon, for example, was incised during the rapid elevation of the Colorado Plateau between about 10 and 8 million years ago.

Uplift in the Rockies began slightly earlier, in Early Miocene time, and terrane that now forms the Southern Rockies has since risen between 1.5 and 3.0 kilometers ($\sim$1 to 2 miles).

Development of the American West: Plio-Pleistocene time   In both the Colorado Plateau and the Rockies, the Miocene pulse of uplifting was followed by a lull and then by an episode of renewed elevation; in fact, large-scale uplifting was the dominant process in the Cordilleran region during Pliocene and Pleistocene time. We will discuss some of the details of this uplifting after noting several other key Plio-Pleistocene events.

During the Pliocene and Pleistocene epochs, igneous activity continued in the volcanic provinces of Oregon, Washington, and Idaho (Figure 18-50). Many of the scenic volcanic peaks of the Cascades, including Mount St. Helens (Figure 1-1), have formed within the past 2 million years or so. Beginning in Late Miocene time and continuing sporadically to the present, the flowing of basalt from fissures has produced the Snake River Plain, which amounts to an eastward extension of the Columbia Plateau (Figure 18-50).

In California, faulting and deformation continued during the Pliocene and Pleistocene epochs. Since the beginning of the Pliocene Epoch about 5 million years ago, the sliver of coastal California that includes Los Angeles has moved northward on the order of 100 kilometers ($\sim$60 miles). The Great Valley has, of course, remained a lowland to the present day, but during Pliocene and Pleistocene time it became transformed from a marine basin into a terrestrial one (Figure 18-51). Early in the Pliocene Epoch, seas flooded the basin from both the north and the south. The sedimentary sequence of the basin reveals that several transgressions and regressions occurred during Pliocene time, but as the epoch progressed, uplift associated with

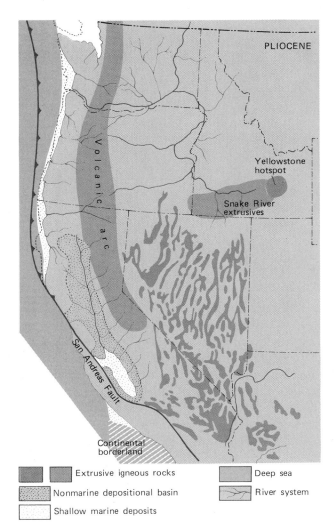

FIGURE 18-50   Geologic features of western North America in Pliocene time. The Great Valley of California was a shallow basin that received nonmarine sediments except where a shallow sea flooded its southern portion. The volcanic arc continued to form volcanoes of the Cascade Range, and igneous rocks continued to be extruded along faults in the Great Basin. The Snake River extrusives spread over a large area of southern Idaho west of Yellowstone. (*Modified from J. M. Armentrout and M. R. Cole, Soc. Econ. Paleont. and Mineral. Pacific Coast Paleogeog. Symp. 3:297–323, 1979.*)

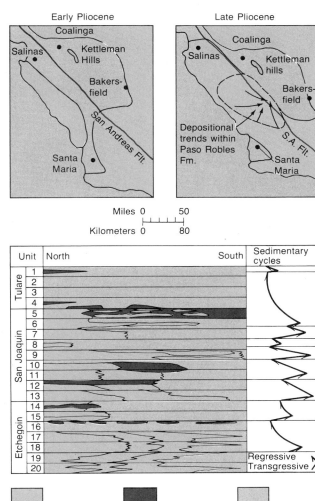

FIGURE 18-51   Decreasing marine influence in the Great Basin of California during the Pliocene Epoch as the sliver of crust west of the San Andreas Fault moved northwestward. The marine channel passing through Santa Maria was closed during the Pliocene Epoch; the Paso Robles Formation is of nonmarine origin. The lower diagram displays the history of deposition in the area now forming the Kettleman Hills; marine deposition dominated early in the Pliocene Epoch but gave way to freshwater deposition. (*After R. J. Stanton and J. R. Dodd, Jour. Paleont. 44:1092–1121, 1970.*)

movement along the San Andreas Fault eliminated the southern connection. Eventually, nonmarine deposition prevailed throughout the Great Valley, which is now one of the world's richest agricultural areas as well as the site of large reservoirs of petroleum.

Meanwhile, areas to the east of the Great Valley underwent major uplift. The Sierra Nevada had experienced con-

siderable tilting during Miocene time, but fossil plants of Miocene age preserved on the crest of the Sierra Nevada were still types that could not have lived as much as a kilometer above sea level. Thus, it was not until Pliocene time that the Sierra Nevada (Figure 18-52) became elevated to its present height, which exceeds 4.3 kilometers

FIGURE 18-52   The eastern face of the Sierra Nevada. This is a fault scarp that is partly dissected by youthful valleys. The view is from the Owens Valley, Inyo County, California. (*Courtesy of W. C. Mendenhall, U.S. Geological Survey.*)

(~14,000 feet). The consequences of this uplift were enormous for the Basin and Range Province to the east, where uplift continued from Miocene time on a smaller scale. Sitting in the rain shadow of the Sierra Nevada (Figure 2-16), this area—which in Miocene time had been covered by evergreen forests—came to be carpeted by savannah vegetation. The trend toward increasing aridity, which was compounded by the global trend toward drier climates, continued into very late Neogene time, when the Great Basin became a desert.

The Colorado Plateau and Rocky Mountains, where uplift had slowed in Late Miocene time, experienced rapid elevation once again. Streams that had been established millions of years earlier cut rapidly downward as the uplift proceeded. The Rockies attained most of their modern elevation during Pliocene time, and the result was the origin of deep canyons that now carry streams through tall mountain ranges (Figure 18-53). One of the most scenic of these canyons is Royal Gorge near Canon City, Colorado (Figure 18-54). The Colorado Plateau also rose again during Pliocene and Pleistocene time, and the Colorado River responded by cutting swiftly downward. In fact, it appears that much of the Grand Canyon of the Colorado (see Figure AII-17) formed during just the past 2 or 3 million years!

The renewed elevation of the Rocky Mountains left the Great Plains to the east in a partial rain shadow. Sediments derived from the rejuvenated Rockies spread eastward, forming the Pliocene Ogallala Formation. Caliche nodules are abundant in many parts of the Ogallala, indicating the presence of seasonally arid climates (page 63). The Ogallala is a thin, largely sandy unit that lies buried under the Great Plains from Wyoming to Texas, and serves as a major source of ground water. Unfortunately, this is ancient water that is not being renewed as rapidly as it is drawn from the earth. As a result, severe water shortages may one day strike many areas of the central United States.

It is important to recognize that the rain-shadow effects of both the Sierra Nevada and the Rockies were superimposed on the larger global trend toward drier, cooler climates that has characterized the post-Eocene world. Then, with the onset of the Pleistocene Epoch, frigid conditions brought glaciation to mountainous regions of the western United States just as they foster glaciation in Alaskan mountains today (Figure 2-18). The Sierra Nevada, for example, was heavily glaciated, as were portions of the Rocky Mountains (Figure 18-16). Today, broad U-shaped valleys in both mountain systems testify to the scouring activity of Pleistocene glaciers (Figure 11-5).

Possible mechanisms of uplift and igneous activity in the American West   What has led to the many tectonic and igneous events of Neogene time in the American West? It seems likely that the secondary uplift of the Colorado Plateau and the Rocky Mountains, which took place long after the Laramide orogeny, may to a large extent represent simple isostatic adjustment (Figure 1-19). When uplifts in these areas were largely leveled during Eocene time, they left behind felsic roots of low density. With the weight of the mountains removed, these roots were apparently out of isostatic equilibrium, and hence they began to rise up toward positions of equilibrium during the Neogene Period.

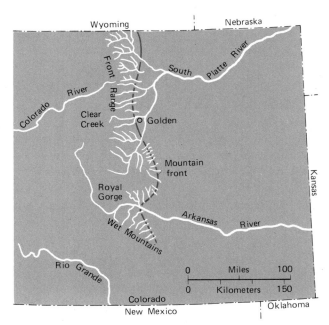

FIGURE 18-53   Large canyons cut during Pliocene time along the eastern flank of the Front Range and Wet Mountains of Colorado when these mountains underwent considerable elevation. (*After G. R. Scott, Geol. Soc. Amer. Mem. 144:227–248, 1975.*)

FIGURE 18-54   Royal Gorge, spanned by a bridge in the foreground. This is one of the deep Pliocene canyons of Colorado depicted in Figure 18-53. (*W. G. Pierce, U.S. Geological Survey.*)

The elevation of the Basin and Range, with its block faulting and relatively thin crust, requires a different explanation. Basin and Range events, the spreading of the Columbia River Basalt, and the extensive faulting and folding along the California coast all began in Miocene time and seem to be related in some way to plate-tectonic movements, including those that simultaneously occurred along the Pacific Coast. One idea is that the **East Pacific Rise**, a large oceanic rift that passes into the Gulf of California, breaks up as it passes inland through thick continental crust (Figure 18-55) and branches out to become the many normal faults of the Basin and Range Province that have caused thinning and elevation of the crust. The high heat flow that is associated with such a spreading zone is also alleged to have caused the Neogene elevation of the Basin and Range Province.

Objections have been raised against the rifting hypothesis for Basin and Range development based on the belief that spreading should have ceased when the East Pacific Rise came into contact with the subduction zone at the western boundary of the North American plate; instead, it is argued, movement must have been propagated along one or more transform faults such as the San Andreas,

passing along the continental margin. This argument leads to an alternative hypothesis of Great Basin development (Figure 18-55B). The crux of this hypothesis is that crustal shearing adjacent to a strike-slip fault like the San Andreas will automatically cause extensional faulting similar to that of the Great Basin. This second hypothesis is deficient in one regard: It fails to account for the broad elevation of the Basin and Range Province during Neogene time.

Still another hypothesis is that heat from the subducted Pacific plate elevates the crust in the Basin and Range regions, placing it under tension that results in block faulting and in thinning of the crust. At present, we have no actual evidence that this mechanism has operated.

We are more certain, however, about the general pattern of tectonism along the Pacific coast. As Figure 18-55B illustrates, North America became attached to the Pacific plate near the beginning of the Miocene Epoch. Movements along the San Andreas and other faults that have formed

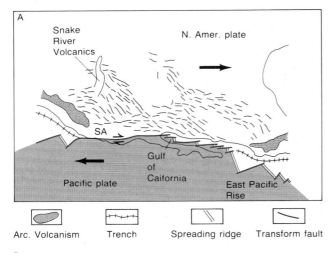

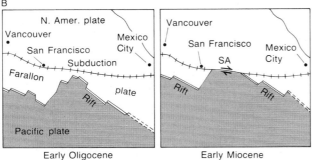

Early Oligocene   Early Miocene

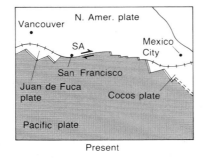

Present

FIGURE 18-55   Plate-tectonic features that may account for the Basin and Range structure in western North America. *A.* The present situation, with the spreading ridge known as the East Pacific Rise passing into the Gulf of California. One idea is simply that the spreading ridge passes beneath western North America, thinning and widening the crust by extension along normal faults. An alternative idea is that the East Pacific Rise abuts against the North American continent and is offset westward along the San Andreas Fault (SA); according to this idea, shearing forces resulting from relative movement of terrane on either side of the San Andreas Fault (*heavy arrows*) pull the crust apart, producing the north-south trending faults of the Basin and Range Province. *B.* Illustration of the second alternative. The rift zone between the Pacific and Farallon plates encountered the thick crust of North America along the subduction zone that bordered the continent. Unable to pass inland, the rift was divided along a strike-slip fault (the San Andreas). (*Based on J. H. Stewart, Geol. Soc. Amer. Mem. 152:1–31, 1975. After Atwater.*)

since that time account for the complex slivering and deformation in the Coast Ranges and neighboring areas (Figure 18-46).

The Atlantic Ocean and its environs

Although the margins of the Atlantic Ocean were relatively quiescent during the Neogene Period, they did experience mild vertical tectonic movements—and these movements, together with more profound changes in sea level, had major effects on water depths and shoreline positions. We will review events in these regions first for Miocene and then for Plio-Pleistocene time.

Miocene events   Global sea level has never stood as high during the Neogene Period as it did during much of Cretaceous or Paleogene time. For this reason, along the Atlantic Ocean, Neogene marine sediments stand above sea level in only a few low-lying areas. Among the most impressive of the Miocene deposits found here are those of the Chesapeake Group, which form cliffs along the Chesapeake Bay in Maryland (Figure 18-56). The Chesapeake Group accumulated during a worldwide high stand of sea level between about 16 and 14 million years ago, and they were deposited in the Salisbury Embayment, one of several downwarps of the American continental margin (Figure 18-57). Inhabiting the waters of the Salisbury Embayment was a rich fauna that included many large vertebrates, especially whales, dolphins, and sharks (Figure 18-1). Most of the fossils of baleen whales represent juvenile animals, which suggests that the embayment may have been a calving ground. Perhaps sharks were numerous because the young whales were especially vulnerable prey. Land-mammal bones are also found here and there in the Chesapeake Group, indicating that the waters of the embayment were shallow. Pollen from nearby land plants settled in the Salisbury Embayment, leaving a fossil record that shows a warm, temperate flora near the base of the Chesapeake Group slowly giving way upward in the sedimentary sequence to a slightly cooler but still temperate flora.

The Chesapeake Group and other buried deposits of earlier age to the south consist primarily of siliciclastic sediments shed from the Appalachians to the west. Erosional features associated with the Appalachians reveal that these ancient mountains have a complex history. Like the modern Rockies, the existing topographic mountains that we call the Appalachians are the product of secondary uplift. The Appalachian orogenic belt was largely leveled by erosion

FIGURE 18-56   Cliffs formed by the Middle Miocene Choptank Formation along the western shore of the Chesapeake Bay
in Maryland. A thick shell bed containing a rich fauna of mollusks is visible in the cliff. The vertebrate fauna of the Calvert
Formation, which underlies the Choptank, is depicted in Figure
18-1. (*Courtesy of R. E. Gernant.*)

by the end of the Mesozoic Era, as a result of which the
Schooley Peneplain, or "almost plain," formed the surface of the land on top of the deformed Appalachian rocks.
As shown in Figure 18-58, three intervals of uplift and erosion followed. None of the intervals of erosion leveled the
region, but each left erosional surfaces that can be identified in the present complex topography (Figure 18-58E).
The pulses of secondary uplift in the Appalachians presumably reflect isostatic response to the persistence
of low-density roots of the original mountain system.
Unfortunately, geologists have found it impossible to correlate well-dated deposits of the Atlantic Coastal Plain of
North America with pulses of uplift and erosion in the Appalachians. These pulses remain poorly dated.

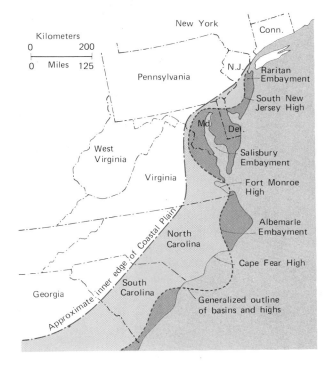

FIGURE 18-57   Elevated regions and depositional embayments that existed along the mid-Atlantic coast during
the Miocene Epoch. (*After J. P. Menard et al., Geol. Soc.
Amer. Northeast-Southeast Sections, Field Trip Guidebook 7a,
1976.*)

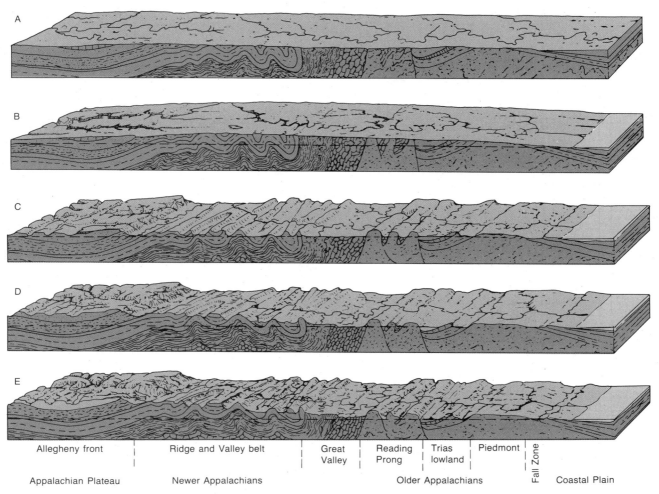

| Allegheny front | | Ridge and Valley belt | | Great Valley | Reading Prong | Trias lowland | Piedmont | Fall Zone | Coastal Plain |

Appalachian Plateau   Newer Appalachians   Older Appalachians

FIGURE 18-58   Erosional history of the Appalachian Mountains. *A.* Schooley erosion surface truncating older, folded rocks of the Appalachians and Cretaceous formations of the coastal plain. *B.* Arching of the Schooley erosion surface. *C.* Dissection of the Schooley erosion surface by acceleration of erosion and development of the Harrisburg erosion surface. *D.* Uplift and dissection on the belts of weakest rocks. *E.* Uplift and dissection of the Somerville erosion surface, producing the present condition. (*After D. W. Johnson, Stream Sculpture on the Atlantic Slope, Columbia University Press, New York, 1931.*)

On the other side of the Atlantic, Miocene seas also encroached only modestly beyond the present shorelines. The North Sea did extend inland beyond the present coastline, and the Rhine River mouth lay in the area now occupied by the city of Cologne, where a large swamp formed in the area where the Rhine met the sea (Figure 18-59). During Early and Middle Miocene time, logs of swamp-dwelling trees accumulated to produce lignite, an immature brown coal that is now of considerable economic value.

In Late Miocene time sea level declined to some extent, and fewer marine deposits formed on either side of the Atlantic inland of the present coastlines.

Plio-Pleistocene events   After the Messinian Event at the end of the Miocene Epoch, the Antarctic ice cap melted back, and sea level rose throughout the world (Figure 18-6). This trend, which reached a peak about 4 million years ago, resulted in marine deposition in areas that are now exposed above sea level—on both sides of the North Sea, for example, as well as on the other side of the Atlantic from Virginia to Florida and the Caribbean. The Caribbean region deserves special mention here because it underwent major changes late in the Neogene Period. Although the Caribbean Sea is now an embayment of the Atlantic Ocean, it was once connected to the Pacific (Figure

FIGURE 18-59   Coastal swamps of Middle Miocene age in the lower part of the Rhine Bay along the North Sea, where large deposits of brown coal accumulated. (*R. Teichmuller, Fortschr. Geol. Rheinl. u. Westf. 2:721, 1958.*)

18-60). During the Cretaceous Period, the floor of the Caribbean, which consists of oceanic rocks, was a small segment of the Pacific plate that was pushing toward the Atlantic, but during the Cenozoic Era the Caribbean sea floor has lain along the north coast of South America while the Atlantic plate has been subducted beneath it. The Caribbean plate became a discrete entity late in Cenozoic time, when a new subduction zone came to connect the subduction zone bordering North America with the one bordering South America. The Greater Antilles—Cuba, Puerto Rico, Jamaica, and Hispaniola—represent an ancient mountain belt that is actually the southern end of the North American Cordillera (Figure 18-61). The Lesser Antilles represent an island arc west of the subduction zone, together with islands formed by deformation associated with subduction. The Yucatán Peninsula is a broad carbonate platform that lies to the west of the Caribbean, and, as we have seen, the Bahamas are an ancient carbonate platform lying to the north (Figure 4-22).

During the Pliocene Epoch the waters of the Caribbean became separated from those of the Pacific when tectonic activity formed the narrow **Isthmus of Panama**. Deep-sea cores reveal that different planktonic foraminifer species lived on opposite sides of the isthmus about 3.5 million years ago, suggesting that the isthmus had formed by this time.

The most profound effect of the newly formed isthmus lay in its role as a land bridge that allowed mammals to

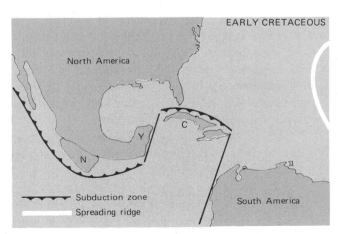

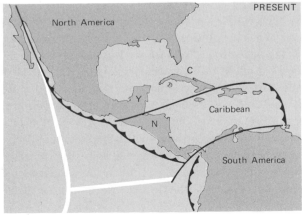

FIGURE 18-60   Tectonic development of the Caribbean Sea as a segment of Pacific oceanic crust that has overridden Atlantic crust along a subduction zone that now extends northward from South America east of the Lesser Antilles. N = Nicaragua; Y = Yucatán; C = Cuba. (*After G. W. Moore and L. Del Castillo, Geol. Soc. Amer. Bull, 85:607–618, 1974.*)

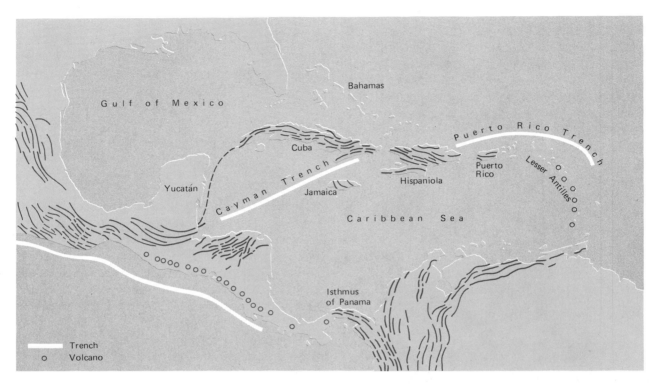

FIGURE 18-61   Major geologic features of the Caribbean region. Structural features of the Greater Antilles, from Cuba to Puerto Rico, represent a continuation of the Cordilleran mountain belt of North America. Subduction along the eastern margin of the Caribbean Sea has created the volcanic islands that form the Lesser Antilles.

migrate between North and South America. Previously, in Neogene time, the terrestrial faunas of North and South America had remained largely separate from each other—although a few species had passed from one continent to the other early in the Neogene Period, perhaps by swimming or floating on logs. South America had been a great island continent and, like Australia, was populated by many marsupial mammals. The marsupials of Australia and South America share common ancestors that populated these continents as well as Antarctica when all three were part of a single landmass late in the Mesozoic Era; in fact, Eocene mammal faunas resembling those of South America occur in Antarctica. By Pliocene time, however, when the isthmus developed, South American marsupials differed greatly from Australian marsupials, and the South American fauna included several groups of placental mammals whose ancestors had reached the continent from the north at the beginning of Cenozoic time or even earlier. Among the South American marsupials present when the

land bridge formed were members of the opossum family, and among the placentals were sloths and armadillos that dwarf their relatives in the modern world (page 137).

More North American species invaded South America than vice versa (Figure 18-62). Among those that reached South America were members of the camel, pig, deer, horse, elephant, tapir, rhino, rat, skunk, squirrel, rabbit, bear, dog, raccoon, and cat families. Migrating in the opposite direction were monkeys, anteaters, armadillos, porcupines, opossums, and other, less familiar animals.

The formation of the Isthmus of Panama was associated with a Plio-Pleistocene pulse of orogeny that elevated the Andean mountain ranges to the south by 2 to 4 kilometers (~1.5 to 2.5 miles). Large areas of South America thus came to lie in the rain shadow of the Andes. Dry conditions intensified during Pleistocene glacial episodes, at which time parts of the great Amazonian rain forest were fragmented into small pockets separated by grassland barriers or by poorly wooded areas. Even today, certain groups of

FIGURE 18-62   Animals that took part in the great faunal interchange between North and South America when the Isthmus of Panama was elevated, connecting the continents. The animals shown in North and Central America are immigrants from the south; these include armadillos, sloths, porcupines, and opossums. The animals shown in South America are immigrants from the north; among these are rabbits, elephants, deer, camels, and members of the bear, dog, and cat families. More animals migrated southward than northward. (*Drawing by M. Hill Werner, courtesy of L. G. Marshall.*)

rain-forest–dwelling birds and lizards remain clustered in the areas where they were concentrated during the Pleistocene Epoch.

The formation of the Isthmus of Panama also had profound oceanographic effects. Recall the suggestion that it may even indirectly have triggered the Ice Age in the Northern Hemisphere by deflecting warm waters into the Gulf Stream, which then increased humidity and snowfall at high latitudes (page 577). The Caribbean Sea was also affected. Today the Caribbean is isolated from other tropical areas and consequently harbors a unique fauna of reef-building corals. Early in the Cretaceous Period, however, the coral fauna here closely resembled that of the Tethyan Seaway in what is now the Mediterranean region, since the spreading of the Atlantic had not yet separated the two regions by a vast expanse of ocean. Then, in Late Cretaceous time, the width of the Atlantic became so great that very few coral larvae were able to cross it, and the fauna of the Caribbean began to diverge from that of the Tethys Sea. Finally, the uplift of the Isthmus of Panama separated the Caribbean coral fauna from that of the Pacific. Since Miocene time, both the restricted area of the isolated Caribbean and a progressive climatic deterioration have caused the coral fauna here to dwindle. Especially during glacial episodes of the Pleistocene Epoch, the Caribbean has been near the lower temperature limit for coral-reef growth. What remains today is a small coral fauna consisting of about 60 species, whereas nearly 10 times as many species inhabit the vast biogeographic province formed by the western Pacific and Indian oceans.

In fact, marine life in and around the Atlantic Ocean suffered greatly during the expansion of glaciers and nearby land areas slightly more than 3 million years ago. During the Pliocene high stand of sea level just before the glacial interval began, enormous molluscan faunas occupied the borders of the Atlantic under relatively warm conditions. Seas that invaded Virginia supported a subtropical marine fauna, and even northern Iceland was bordered by a temperate fauna. In mid-Pliocene time, the global high stand of sea level, together with temperate climates, permitted a wave of Pacific mollusks to spread over the polar region into the Atlantic (Figure 18-63). One of the immigrants was *Mya arenaria*, the "steamer clam" that is widely consumed in eastern North America.

Pliocene strata in Iceland have been dated by the application of radiometric and paleomagnetic methods to interbedded lavas associated with the Mid-Atlantic Rift, which passes through Iceland (Figure 7-30). Consequently, geologists can estimate with reasonable accuracy when the

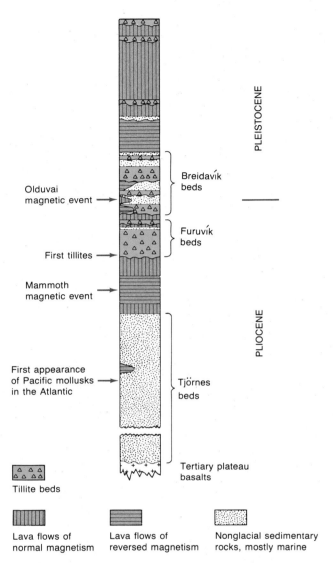

FIGURE 18-63   Stratigraphic section of the Tjörnes Peninsula of northern Iceland. This sequence has been dated by paleomagnetic correlation. Two magnetic reversal events are labeled; the Olduvai Event marks the beginning of the Pleistocene Epoch. A high stand of sea level in mid-Pliocene time flooded the north polar region, allowing Pacific species of mollusks to cross into the Atlantic. Temperate climates then gave way to colder conditions, and about 3 million years ago glacial tills were spread over the region. (*After T. Einarsson et al., in D. M. Hopkins [ed.], The Bering Land Bridge, Stanford University Press, Stanford, California, 1967.*)

richly fossiliferous Pliocene beds in Iceland suddenly gave way to tillites—a transition marking the expansion of ice sheets across the island.

The refrigeration that accompanied continental glaciation reached far southward. We have seen, for example, that during glacial episodes of the Pleistocene Epoch, the lowering of sea level and the accumulation of pack ice reduced the size of the clockwise gyre of ocean currents north of the equator and, at the same time, pushed the Gulf Stream diagonally across the Atlantic toward Spain (Figure 18-18). Recall, too, that warm equatorial waters, which the hump of South America had deflected into the Caribbean, now passed southward under the influence of stronger trade winds (page 576). Studies of planktonic foraminifers in deep-sea cores suggest that February sea-surface temperatures in the central Caribbean during the most recent glacial advance (the Wisconsin or Riss-Würm Stage) were about 4°C (9°F) cooler than they are today. Long before this time, during the early phases of glacial cooling, thousands of species of marine mollusks disappeared from the Atlantic Ocean and neighboring seas in the Northern Hemisphere. Along the Atlantic coast of North America, for example, about 70 percent of these species disappeared, and the Caribbean fauna was similarly decimated. Faunas of the North Sea and the Mediterranean experienced less severe but still noteworthy pulses of extinction. Tropical species suffered especially severe losses.

Africa, southern Europe, Asia, and Australia

In Chapter 11, we noted that the landmass of Africa has remained unusually stable for hundreds of millions of years, in part because most of its area lay well within Gondwanaland for a long interval of time. Currently, however, the strength of the African craton is being tested by continental rifting, and it is not certain whether this great continent will survive or be torn asunder. We will now examine this situation and also the geographic consequences of the movements of two lithospheric plates: the northward movement of the African plate, which is associated with the uplift of the Alps and other mountain chains, and the northward movement of the Australian plate, which produced the Himalayas.

Africa: Rifting and climatic change   Near the beginning of Miocene time, two large domes, the Ethiopian Dome and the Kenyan Dome, began to rise up in eastern Africa—a process that was followed by volcanism and crustal rifting (Figure 18-64). The Ethiopian Dome in the north was the site of the large three-pronged rift that split the Arabian Peninsula from the rest of Africa, allowing the Red Sea and the Gulf of Aden to open in Early Miocene time (page 168). The southern prong was crossed by the western rift, which eventually propagated all the way to the south-

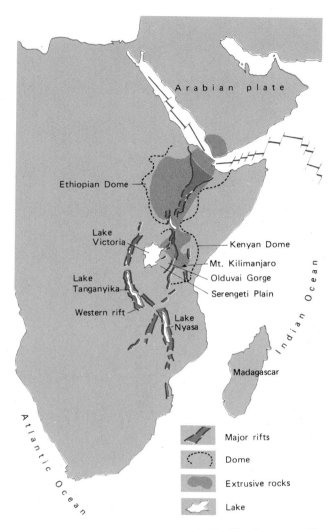

FIGURE 18-64   Features associated with rifting in eastern Africa. The Arabian plate is splitting off from Africa along two prongs of a three-pronged rift centered in an area of crustal doming, part of which forms the Ethiopian Dome, where extrusive rocks are widespread. The third prong extends into continental Africa, forming the rift valleys, some of which harbor large lakes. (*After E. R. Oxburgh, in W. W. Bishop [ed.], Geological Background to Fossil Man, Scottish Academic Press, Edinburgh, 1970.*)

western margin of Africa. Since early in the Miocene Epoch, the ever-deepening valleys of this rift system have received large quantities of sediment and have also harbored some large lakes.

Neogene tectonism in Africa has had important biotic consequences that were enhanced by trends toward more arid climatic conditions—as were similar changes in western North America and in South America. With respect to

FIGURE 18-65   Olduvai Gorge, a locality in East Africa famous for fossil remains and artifacts of members of the human
family. (For its location, see Figure 18-64.) (*Courtesy of R. L. Hay.*)

the moist winds from the South Atlantic, much of eastern
Africa now lies in the rain shadow of highlands of the Kenyan Dome east of the western rift. East of the eastern rift,
tall volcanic mountains like scenic Mount Kilimanjaro trap
moisture from trade winds moving inland from the Indian
Ocean.

The emergence of wind barriers, together with the global
trend toward cooler, more arid conditions, caused grassy
terrane to replace most forests in eastern Africa—a change
analogous to that in western North America. Fossil mammal faunas reveal that early in the Miocene Epoch, eastern
Africa was cloaked in equatorial rain forest. By mid-Miocene time, however, there was some open terrane in this
area, as evidenced by the presence of ostriches and of
herbivorous mammals that possessed continuously growing molar teeth adapted to grinding up hard grasses. By
Late Miocene time, the savannah fauna resembled that of
the modern Serengeti Plains in that there was a prepon-

derance of large ground-dwelling mammals, including many
species of gazelles and antelopes.

Another geologic development of great biotic consequence was the collision of the Arabian section of Africa
with Eurasia, between 20 and 18 million years ago. This
contact allowed for biotic interchange between Africa and
Eurasia; for example, elephants for the first time escaped
from their African birthplace and, in the course of the Miocene Epoch, spread throughout Eurasia and across the
Bering land bridge to the Americas. As we have seen, the
attachment of Africa to Eurasia also afforded apes their
first opportunity to emigrate northward from Africa.

Rifting and volcanism have continued to the present
day in Africa without fully separating eastern Africa from
the remainder of the continent. Like rift valleys of other
regions, those of Africa have subsided rapidly and have
received large volumes of sediment interlayered with volcanic rocks. It is because of the almost continuous sedi-

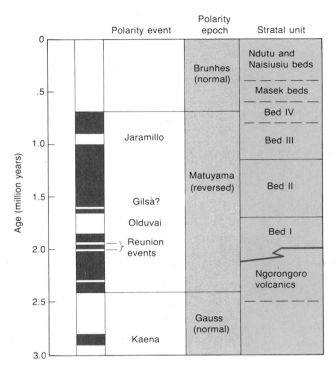

FIGURE 18-66 Magnetic stratigraphy of Olduvai Gorge, where an unusually complete sequence of Pleistocene deposits is exposed. (*R. L. Hay, Geology of Olduvai Gorge, University of California Press, Berkeley, 1976.*)

mentation here that the stream-cut gorges subsequently cut down through the layers of deposits in African rift basins have proven to be fruitful collection sites for hominid remains. Olduvai Gorge, the most famous of these sites, is incised into the Serengeti Plain at the margin of the eastern rift (Figure 18-65). From Olduvai have come not only fossils representing the human family but also their very early stone implements (Figure 18-33). The relatively complete depositional sequence at Olduvai for the past 2 million years has made the stratigraphic section here a standard against which other African sequences are often compared. The presence of volcanic rocks that can be dated by radiometric and fission-track techniques (page 121) and that can be correlated paleomagnetically further enhances the value of the Olduvai section as a standard for worldwide comparison (Figure 18-66).

Closure of the Tethyan Seaway   The collision of the African plate and southern India with Eurasia during the Cenozoic Era destroyed what remained of the Tethyan Seaway. Today, only vestiges of the seaway remain in the

form of the isolated Mediterranean, Black, Caspian, and Aral seas.

We have seen that between 18 and 14 million years ago, Africa and Eurasia collided, creating a land corridor across which terrestrial animals began to migrate. This event marked the end of the eastward connection between the western Tethys and the waters of the Indo-Pacific region (Figure 18-67). By Early Miocene time, the eastern Tethys was divided into a northern and a southern arm; thus, during the collision in which the Dinaride and Hellenide ranges of Yugoslavia and Greece as well as the Taurus Mountains of southern Turkey rose up as barriers, the two arms became separate seas. Consequently, the Mediterranean Sea was born and also, to the north of it, an isolated body of water known as the **Paratethys**. The Paratethys received freshwater from rivers in Eurasia and survived as a brackish inland sea from the middle to the end of Miocene time.

At the end of the Miocene Epoch, both the Mediterranean and the Paratethys underwent spectacular changes. The first strong hint that geologists had of these changes was the discovery in 1961 of pillar-shaped structures in seismic profiles of the Mediterranean sea floor (Figure 18-68). These structures looked very much like the salt domes of Jurassic age in the Gulf of Mexico (Figure 15-32), but if the strange features were indeed salt domes, the salt could only have formed by evaporation of the Mediterranean. In 1970, the presence of evaporites here was confirmed by drilling that brought up anhydrite in cores representing the latest Miocene Epoch. Also present here was gravel that represented shallow marine or nonmarine conditions. The idea that the Mediterranean had somehow turned into a shallow evaporite basin was confirmed by the discovery of halite (rock salt) near the center of the eastern Mediterranean. Because it is highly soluble and does not precipitate until after the other salts in a solution have precipitated, halite is usually found in the center of an evaporite basin.

Further evidence that the Mediterranean dried up at the end of Miocene time was the discovery of deep valleys filled with Pliocene sediments lying beneath the present beds of rivers like the Rhone in France, the Po in Italy, and the Nile in Egypt. Rivers like the Rhone and the Nile were already flowing into the Mediterranean earlier in the Miocene Epoch, and when the waters of the sea fell, the rivers cut deep canyons to the basin floor. In attempting to find solid footing for the Aswan Dam, Soviet geologists discovered the canyon buried beneath the present Nile delta and

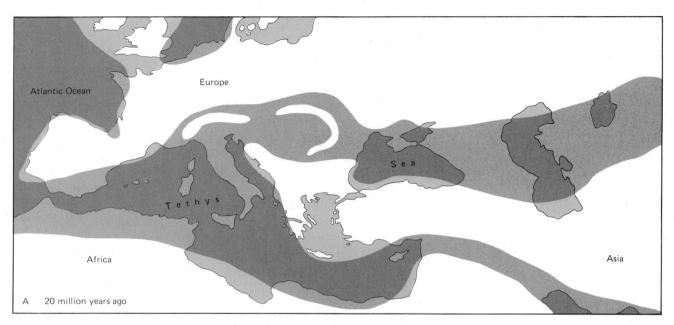

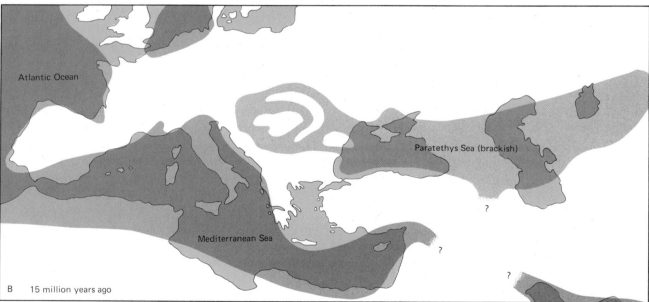

FIGURE 18-67   Closure of the Tethys Sea. Early in Miocene time, the Tethys was open, and a long northern arm of the sea spread eastward into Asia (A). In mid-Miocene time, this arm, the Paratethys, was isolated by mountain building in Asia Minor as Africa collided with Eurasia (B). Near the end of the Miocene Epoch, the Atlantic connection near the remaining southern arm of the Tethys (the Mediterranean Sea) became severed, and the Mediterranean dried up (C). Streams at the eastern end of

judged it to rival the modern Grand Canyon of Arizona in size!

Clearly, what happened at the end of the Miocene Epoch was that the single narrow connection between the Mediterranean Sea and the Atlantic Ocean closed—probably as a result of a lowering of sea level in the Atlantic (page 565). Rates of evaporation similar to those of the Mediterranean region today would dry up an isolated sea as deep as the Mediterranean in a mere thousand years. Not only evaporites but also fossils of freshwater (or brack-

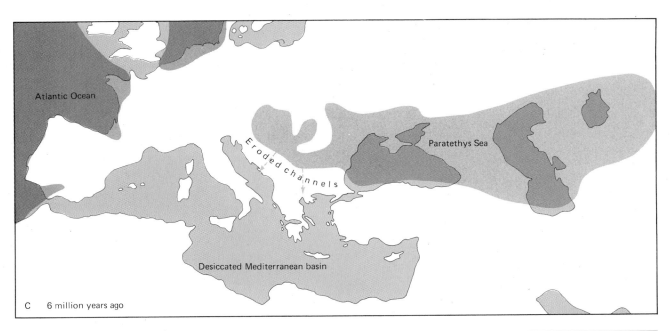

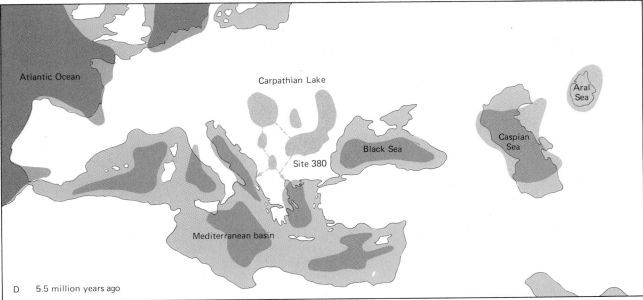

the Paratethys then cut downward, draining much of the Para- tethys into the desiccated Mediterranean basin and separating the remaining Paratethyan waters into a series of lakes, one of which became the Black Sea, another the Caspian Sea, and a

third the Aral Sea (*D*). (*From K. J. Hsu, "When the Black Sea Was Drained." Copyright © 1978 by Scientific American, Inc. All rights reserved.*)

ish-water) ostracods occur in the latest Miocene record of the Mediterranean sea floor, suggesting that some of the rivers that flowed into the evacuated Mediterranean basin not only cut rapidly downward to reach the basin floor but also cut northward, capturing freshwater from the Para-

tethys and carrying it to the Mediterranean basin to form scattered lakes. Soon the Paratethys disintegrated into several smaller bodies of water, including the ancestral Black, Caspian, and Aral seas (Figure 18-67).

All of this happened between about 6 million years ago,

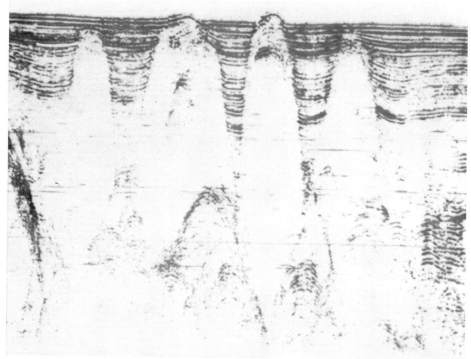

FIGURE 18-68   Pillarlike salt domes in the floor of the western Mediterranean as recorded by seismic profiling. These structures offered the first hint that salt lay beneath the present Mediterranean sea floor. Since it is less dense than ordinary sediment, salt tends to rise and form pillar structures. (*K. J. Hsu, Scientific American, December 1972.*)

when the eastern passage to the Atlantic closed, and 5 million years ago, when the Mediterranean basin refilled with deep water. Five-million-year-old deep-water microfossils in sediments on top of evaporites attest to the refilling. The brief nonmarine interlude corresponded to most of the Messinian Stage (the final stage of the Miocene Series in Italy, where the disappearance of the Mediterranean obviously had special significance), and it is for this reason that the temporary demise of the sea is often referred to as the Messinian Event.

Apparently the Mediterranean-Atlantic connection was reestablished when the natural dam at Gibraltar was suddenly breached. It has therefore been suggested that the first Atlantic waters must have been carried into the deep basin by a waterfall that would have dwarfed Niagara Falls.

As for the remnants of the Paratethys, many simply dried up. Fossils show that the Black Sea, which survived, remained isolated from marine waters as a freshwater lake until very recently, when erosion finally connected it with the Mediterranean by way of the narrow Bosporus. Thus, the Black Sea became a brackish basin once again.

Farther east, the Tethyan Seaway was interrupted during Miocene time by the attachment of the Indian Peninsula to the Eurasian plate (page 216), where molasse that shed southward from the newly forming Himalayas produced a lengthy and relatively complete fossil record for mammals. The famous Siwalik beds of Pakistan and India, for example, provide a nearly continuous record for the interval from 11 to 1 million years ago, documenting the composition of the rich faunas that occupied the spreading savannahs of Late Miocene and Pliocene age (Figure 18-5). The great rivers of eastern Asia, which flow from the Himalayas to the sea, also formed in Miocene time during the uplift of the Himalayan region. Because of the high relief and abundant rainfall of the region, the Indus and Ganges of India and the several large rivers of Indochina contribute huge volumes of sediment to the ocean each year (Figure 8-12).

Australia's northward journey   After Australia broke away from Antarctica, Antarctica remained positioned at the South

Pole, but Australia moved northward along with New Zealand and New Guinea, both of which are part of the same lithospheric plate—the plate that moved against Asia to form the Himalayas (page 216). Paleomagnetic data enable us to track the progress of the Australian plate on its journey (Figure 18-69).

One of the consequences of Australia's move to lower latitudes is that its climate became warmer and drier. Today, much of Australia lies in the latitudinal belt extending from 20 to 30° south latitude. Deserts tend to form within this belt because it is where equatorial air descends after rising, cooling, and losing much of its moisture (page 41). Therefore, deserts now occupy a large area in the central and western parts of Australia, while easterly trade winds bring rain to the east coast of the continent (Figure 2-13). Pollen analyses show that quite different climatic conditions prevailed during the Miocene Epoch, when Australia lay largely within the belt of westerly winds that were laden with moisture. Rain forests cloaked the eastern and southeastern parts of the continent, and in central Australia bordered rivers and streams, although some open grassland was also present.

The migration of Australia toward the equator has also

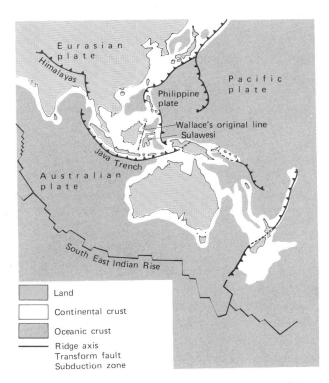

FIGURE 18-70 Wallace's line, a biogeographic division that was recognized in the nineteenth century by Alfred Russel Wallace. Late in his life, Wallace relocated the line southeast of Sulawesi, and this revision is accepted by most modern experts. We now know that the faunas north and south of the line differ because they evolved on two different plates (the Eurasian and Australian plates) that collided only about 15 million years ago.

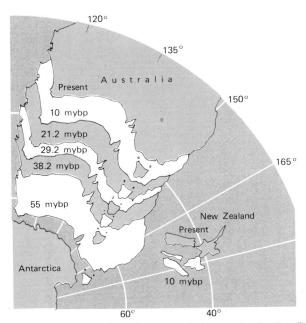

FIGURE 18-69 Northward movement of Australia after it split off from Antarctica early in the Cenozoic Era. (*After J. P. Kennett, Marine Geology, Prentice-Hall, Inc., Englewood Cliffs, New Jersey, 1982.*)

permitted the Great Barrier Reef to flourish (Box 2-1). It was only late in Oligocene time that the northeastern shore of Queensland, which is now bordered by the reef, came to lie fully within the latitudinal zone that accommodates luxuriant reef growth.

One of the most interesting effects of the northward drift of Australia was that it brought into contact two entirely distinct terrestrial biotas: the biota of Australia and that of Asia. The zone of contact between the two biotas, located in the Malay Archipelago, was first recognized by Alfred Russel Wallace, an English naturalist who, working independently from Charles Darwin, also conceived of the idea of evolution by natural selection. In fact, it was in 1859—the very year that Darwin published *On the Origin of Species*—that Wallace, who was traveling in the Malay Archipelago, published a paper on the biogeographic line of demarcation, which then became known as **Wallace's line**.

Wallace originally placed his line north of the island of Sulawesi (Figure 18-70), but later in life he decided that the Sulawesi fauna was dominated by Asian elements, and, accordingly, he shifted the line to a position south of the island. This decision is accepted by most modern biogeographers, since only one marsupial mammal from the Australian region has reached Sulawesi. Still, of the diverse Asian fauna of mammals to the northwest, only a few species of shrews, two monkeys, a deer, a pig, and a porcupine have reached Sulawesi. Migration in both directions has been difficult because of the presence of water barriers. Plants migrate more easily than mammals, however, and Australian and Asian floras intermingle over a broader zone.

Until the advent of plate-tectonic theory, there was no explanation for the narrow zone of overlap between Asian and Australian biotas in the vicinity of Sulawesi. If landmasses were immobile, why had there not been greater intermingling in the long course of Cenozoic time? Geologic evidence now shows that the Australian plate, which was bounded on the north by New Guinea, collided with the Asian plate to which Sulawesi belongs in Middle Miocene time, only about 15 million years ago. Furthermore, it was only in Late Miocene and Pliocene time that the islands between New Guinea and Sulawesi emerged to form stepping stones for the limited interchange of species.

CHAPTER SUMMARY

1.   Invertebrate life in the oceans underwent only minor changes during Neogene time, but whales radiated rapidly.

2.   On land, grasses and herbs diversified and occupied more territory during Miocene time, benefiting from a cooling of climates.

3.   Terrestrial mammals assumed their modern character during Neogene time, and frogs, rats and mice, snakes, and songbirds diversified markedly.

4.   Apes radiated during the Miocene Epoch but have since dwindled in number of species.

5.   *Australopithecus*, the oldest known genus of the human family, evolved at least 4 million years ago. It possessed a small body, a small brain relative to its body size, and apelike teeth. *Homo*, the modern human genus, evolved at least 1.6 million years ago or earlier, and at least 40,000 years ago humans of the modern type were present.

6.   The modern Ice Age began about 3 million years ago with the origin of continental glaciers in the Northern Hemisphere.

7.   During the past 1.8 million years, continental glaciers have expanded and contracted about 18 times, with expansions lowering sea level by as much as 100 meters. Glaciers are now in a contracted state, but they will probably expand again.

8.   During Neogene time, the American West has been affected by extensive volcanic activity, block faulting, and major uplifts that have produced the Sierra Nevada, the Rocky Mountains, and the Colorado Plateau.

9.   In the western Atlantic region, the Caribbean Sea, which is bounded on the east by an island arc, developed its modern configuration during Neogene time, and the Isthmus of Panama was uplifted, permitting extensive biotic interchange between North and South America.

10.   East of the Atlantic, the Mediterranean Sea became sealed off from this larger ocean and briefly dried up about 6 million years ago before renewed connection allowed it to fill again.

11.   During Neogene time, Africa has been fragmenting along spreading zones known as rift valleys, and in Early Miocene time it moved northward into contact with Eurasia, allowing biotic interchange between the two continents. The Australian plate also moved northward, forcing the Indian Peninsula against Asia to form the Himalayas and bringing terrestrial biotas of Australia into contact with those of Asia.

EXERCISES

1.   In what ways did mammals become modernized during the Neogene Period?

2.   How do modern humans differ from members of the genus *Australopithecus*?

3.   What factors influence the isotopic composition of oxygen in skeletons of marine organisms? Which has been the dominant factor during the Pleistocene Epoch?

4.   Describe three changes in the geographic distribution of animals produced by tectonic events during Neogene time.

5.   How did the Rocky Mountains develop their present configuration in the course of the Neogene Period?

6.   How did the Appalachian Mountains develop their present configuration in the course of Neogene time?

ADDITIONAL READING

Imbrie, J., and K. P. Imbrie, *Ice Ages*, Enslow Publishers, Short Hills, New Jersey, 1979.

Kennett, J., *Marine Geology*, Prentice-Hall, Inc., Englewood Cliffs, New Jersey, 1982.

Kurtén, B., *The Age of Mammals*, Columbia University Press, New York, 1971.

Pilbeam, D., *The Ascent of Man: An Introduction to Human Evolution*, The Macmillan Company, New York, 1972.

Pomeral, C., *The Cenozoic Era*, John Wiley & Sons, Inc., New York, 1982.

Epilogue

On our voyage through geologic time, we have seen continents grow as they have crept over the asthenosphere, and we have seen them break apart along great rifts where new oceans have formed. We have also seen how mountain systems have risen up and worn down, how seas have flooded continents and receded, and how from humble Archean beginnings life expanded to form a complex marine ecosystem and then to conquer the land. Through the operation of these and other processes, the earth has taken its present shape and its inhabitants have assumed their modern character.

Turning our gaze far into the future, we can predict major geologic events from what has gone before. About 30 million years from now, the movement of the narrow sliver of crust that joins the rest of North America along the San Andreas Fault will have brought Los Angeles to a position alongside San Francisco. In 10 million years, barring renewed uplifts, the rugged Rocky Mountains and Alps will have been subdued by erosion; in contrast, the Himalayas will probably still stand tall, owing to continued wedging of the Indian Peninsula beneath them. As we have seen, Australia is moving northward into the dry zone of the trade winds; during the next 10 million years or so, southern Australia will become arid, and much of its terrestrial biota will suffer extinction. Africa, though characterized by unusually thick crust, may fragment along its present rift valleys in another several million years; alternatively, convective cells in the mantle may shift, leaving the continent intact. Whether our species will observe these events is uncertain. It is sobering to contemplate that few species of our class (the Mammalia) have survived for more than 3 or 4 million years. While it would be nice to believe that our intelligence and cultural flexibility will earn us an exceptionally long stay on earth, the enormous destructive potential of our nuclear weaponry must dampen our optimism.

In the shorter term, the implications of the Ice Age pique our curiosity. The Northern Hemisphere emerged from the last Pleistocene glacial episode about 10,000 years ago. We do not know exactly how long major interglacial intervals of the Pleistocene have lasted, but the average duration has been many times 10,000 years. This means that our species, which endured the most recent advance of continental glaciers, need not fear another major glacial advance for many millennia. Nonetheless, climatic changes on a smaller scale could quickly disrupt our civilization. During the present interglacial interval, humans have built cities far to the north, which are vulnerable to even a minor glacial advance. Moreover, a cool interval comparable to the Little Ice Age of the past few centuries would impair agriculture in large areas of the Soviet Union, Scandinavia, and Canada. On the other hand, as we have already noted, a warming climatic trend could melt enough polar ice to flood coastal cities; such warming could result from natural causes or from our civilization's fuel-burning practices that add carbon dioxide to the atmosphere, accentuating the greenhouse effect.

A question that warrants less concern but nonetheless attracts much interest is what the evolutionary future of our species might be, however long we may ward off extinction and whatever climatic changes we may face. Our species has experienced very little anatomical change, at least in Europe, during the past 40,000 years. Will this stability continue? In Chapter 6 we noted that some paleontologists believe that most species, in fact, change relatively little once they have originated by rapid evolution from populations of preexisting species; other paleontologists believe that most species evolve at substantial rates throughout their existence. This controversy, however, has little bearing on the future of biological evolution of the human species. Our species, unlike all others, evolves culturally as well as biologically, and our cultural evolution has interfered with our biological evolution, making our biological evolution different from that of all other species. For example, many physically impaired individuals who would not survive in nature are sustained by society—especially by modern medicine—and given the opportunity to produce children. On the other hand, our society discourages reproduction of certain individuals who carry genetic features that impair their health and might impair the health of their offspring. Selection pressures in modern societies not only differ from those of other species, but also from those that characterized more primitive human societies long ago. Physical strength and speed are not now at a premium to the extent that they were when members of our species, using only simple weapons, frequently battled wild animals and other humans. *Homo sapiens* actually has the potential to direct its own evolution as it sees fit by practicing artificial breeding of the sort that has produced such divergent breeds of domestic dogs as the Saint Bernard, Chihuahua, and Greyhound. And if the prospect of selective breeding in our species were not disturbing enough in itself, we must also contemplate the ethical questions raised by modern advances in genetic engineering. By these techniques, scientists can introduce entirely new genetic features to the human species.

We can speculate about yet another evolutionary possibility for *Homo sapiens*. This is the fanciful prospect that someday a human population isolated in outer space might evolve into a new species. Today, early in the Space Age, we have no way of assessing the likelihood that such an event will occur. We can, however, conclude that there is little chance that a similar speciation event will take place here on earth, as long as our populations are as widespread and mobile as they are today. Under these circumstances, it is highly improbable that any population will remain isolated long enough to diverge into a new species without interference from other populations of the human species.

Most of the lessons of this book illuminate less perplexing issues. We have traced the geologic histories of many regions and have reviewed the major developments in the history of life. In the perspective thus gained, all the features of our planet's landscape, from small-scale local outcrops to jagged mountain chains and sandy coastal plains, take on new meaning. Fossils, which are abundant in nearly all regions of the world except crystalline Precambrian shields, also come to life. You can supplement your new understanding of these phenomena by making use of the many regional libraries and state and federal agencies from whom information about local rocks and fossils is available. Museums and parklands also offer special glimpses of geology and paleontology. In summary, the world around you will be a richer store of information now that you comprehend in broad outline the evolution of the earth and life through time.

Minerals and Rocks

The rocks and soil on which we stand are aggregates of mineral grains. As defined in Chapter 1, a mineral is a naturally occurring solid element or compound whose atoms are organized in a particular configuration. (A compound is a substance that consists of chemically combined elements.) This appendix introduces some characteristics of important minerals and rocks, beginning with a review of the fundamental properties of minerals.

MINERAL PROPERTIES

A **chemical element** is a fundamental substance made up of a particular kind of atom that, in its most stable state, has equal numbers of electrons and protons. The electrons have a negative charge, while the protons have a positive charge. Electrons surround the atom's nucleus, where protons reside along with neutrally charged neutrons. Some kinds of atoms tend to lose one or more of their outermost electrons to other kinds of atoms. Atoms that lose their electrons are said to become positively charged **ions**, whereas those that gain electrons are called negatively charged ions. Positively and negatively charged ions are attracted to each other in the way that hair is sometimes attracted to a synthetic fabric that is drawn over it. The attachment that results, which is called an **ionic bond**, is essential to the formation of mineral compounds. Uncharged atoms can also combine in other ways; for instance, they can share electrons in a process called **covalent bonding**. When different kinds of atoms bond together, they form compounds whose characteristics sometimes differ dramatically from those of their constituent elements in pure form.

Stable chemical configurations such as those that characterize most minerals cannot exist unless two conditions have been met: first, positively and negatively charged ions must be combined in the proper proportions so that there is no charge imbalance; and second, these positive and negative ions must be of relative sizes that enable them to fit snugly together to form a solid structure. The geo-

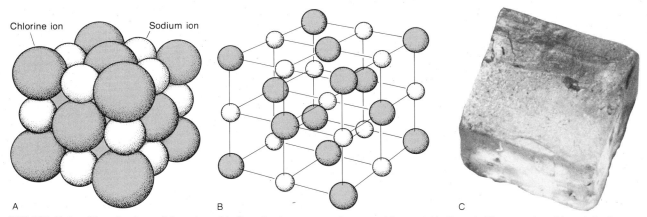

FIGURE AI-1   The structure of the mineral halite, also known as rock salt. Halite is composed of equal numbers of positive sodium ions and negative chloride ions, which fit together to form a cubic crystal lattice. A. The actual position of the ions. B. The cubic pattern is evident in the expanded lattice. C. Photograph of a cubic crystal. (C. Photograph by the author.)

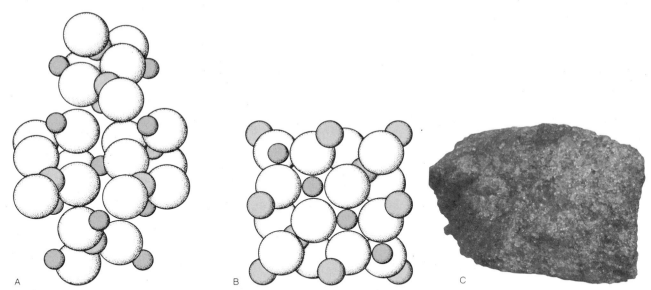

FIGURE AI-2   The structure of olivine (a mixture of Mg$_2$, SiO$_4$, and Fe$_2$SiO$_4$), which is an important mineral of the earth's mantle. The less dense crystal form of the mineral (A) exists under lower pressures (or at shallower depths) than the more dense crystal form (B). The large ions are oxygen, the small ions are silicon, and the medium-sized ions are magnesium (or iron). C. A rock consisting of many olivine crystals. (C. Photograph by the author.)

metrically consistent internal structure of a mineral is known as a **crystal lattice** (Figure AI-1).

The physical properties of minerals, including their hardness, density, external form, and pattern of breakage, are determined by their chemical composition and crystal lattice structure. The strength of a mineral's chemical bonds, for example, is the primary determinant of that mineral's hardness. Thus, diamond, which is composed solely of the element carbon, is the hardest mineral on earth because all of its carbon atoms are strongly bonded to one another. The mineral graphite also consists of pure carbon, but because its carbon atoms are arranged in layers that are weakly attached to one another, it is very soft—soft enough to be used as the "lead" in pencils.

The density of a mineral—or its mass per unit volume—is partially determined by the types of atoms of which it is constituted. An iron atom, for example, is heavier than an aluminum atom because it contains more protons and neutrons, and compounds of iron are therefore more dense than compounds of aluminum. The density of a mineral is also determined to some extent by the degree to which its atoms are packed together. Figure AI-2, for example, shows two crystal structures of the mineral **olivine**, which is thought to be the primary component of the earth's mantle. At a depth of about 400 kilometers (~250 miles) beneath the earth's surface, seismic waves abruptly accelerate, indi-

cating that the mantle abruptly increases in density at this point. Laboratory experiments further reveal that as temperature and pressure are increased to the values that are thought to exist at this depth, olivine alters from its less dense form to its more dense form as a result of changes in the mineral's crystal lattice. This phenomenon probably explains the seismic acceleration that takes place within the earth's mantle at a depth of 400 kilometers.

Minerals may vary slightly in their chemical composition, but only within specified limits. Such variation usually results from the substitution, within a mineral's crystal lattice, of one element for another whose atoms are roughly equivalent in size. Iron, for example, frequently substitutes for magnesium in some crystal lattices, and strontium often substitutes for calcium in others.

A crystal's external form and the manner in which it breaks also reflect its internal structure, as illustrated by the mineral **halite** (NaCl), which forms from the evaporation of natural waters (Figure AI-1). In halite, which is informally known as rock salt, sodium and chlorine ions form natural cubic units within the crystal, maintaining the cubic shape as the crystal grows. Halite also breaks along planes of weakness known as **cleavage planes**, which lie at right angles to one another following zones of weak attachment between planes of atoms. In other minerals, characteristic external forms and breakage patterns also reflect internal

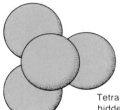

Tetrahedron of four oxygens with a silicon hidden in the center

Amphibole (single chain)

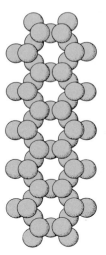

Pyroxene (double chain)

FIGURE AI-3   Rock-forming silicate minerals. Diagrams show the arrangements of tetrahedra in five important rock-forming mineral groups. (Other ions are omitted for simplicity.)

In amphiboles and pyroxenes, the silicate tetrahedra are assembled into long chains that are bonded together by ions of iron, calcium, or magnesium positioned between them. The iron and magnesium make these minerals relatively dense and also make many of them dark in color.

In mica and clay minerals, the silicate tetrahedra are more fully connected to form two-dimensional sheets that are bonded together by sheets of aluminum, iron, magnesium, or potassium. Because the bonds between these sheet silicates are weak, micas and clays cleave into thin flakes. Clay minerals are especially weak and almost always occur naturally as small flakes.

The feldspar minerals are the most common group of minerals in the earth's crust, and quartz is second in abundance. Quartz is the simplest of all silicate minerals in chemical composition, consisting of nothing but interlocking silicate tetrahedra. In this structure, each oxygen is shared by two adjacent tetrahedra. This means that, for the mineral as a whole, there are only two times rather than four times as many oxygens as silicons (the ratio in a single tetrahedron). Hence, the chemical formula of quartz is SiO_2. Quartz is very hard because its silicons and oxygens are bonded tightly together.

Feldspars differ from quartz in that their structure includes both silicate tetrahedra and tetrahedra in which aluminum takes the place of silicon. Ions of one or more additional types (potassium, sodium, and calcium) also fit into the framework in differing proportions. Unlike quartz, feldspars display good cleavage. Feldspars are also slightly softer than quartz, and they decay much more readily in nature. In fact, chemical weathering quickly alters feldspars into clay in the presence of water at the earth's surface. (*Photographs by the author, except clay [kaolite], courtesy of W. D. Keller.*)

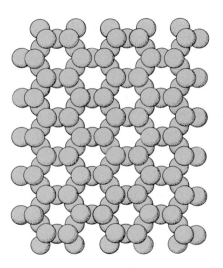

Mica, *left*, and clay, *right* (sheet)

Quartz

Finely crystalline quartz

Feldspar

crystal structure. Unlike halite, however, many minerals do not ordinarily break along cleavage planes; moreover, cleavage planes in other minerals do not parallel crystal surfaces.

A few groups of minerals are especially abundant on earth because they are important rock-forming minerals. We will now review several of these groups, whose properties and roles in the formation of rock are summarized in Table AI-I.

FAMILIES OF ROCK-FORMING MINERALS

The mineral groups known as **silicates** are the most abundant minerals in the earth's crust and mantle. In silicates, four negatively charged oxygen atoms form a tetrahedral structure around a smaller, positively charged silicon atom. Figure AI-3 illustrates how silicate tetrahedra unite in var-

ious ways, usually with other atoms, to form the most important silicate minerals of the earth's crust. Not only are silicates the primary constituents of the igneous rocks of the crust, but they also represent the most important minerals of crustal sedimentary and metamorphic rocks. Silicates tend to form at the high temperatures that exist deep within the earth's crust and mantle and that also characterize molten rock that reaches the surface from great depths.

Carbonate and **sulfate** minerals are also important rock formers—but unlike silicates, most of these minerals form at low temperatures near the earth's surface. Carbonates are constructed of one or more positive ions, such as calcium, magnesium, and iron, bonded to the negative carbonate ion. The carbonate ion is a composite ion formed of one carbon linked to three oxygens (CO_3). **Calcite** ($CaCO_3$) is the most abundant carbonate mineral (Figure AI-4). **Aragonite**, another important carbonate mineral, has the same chemical composition as calcite but differs in its crystal structure; thus, aragonite's physical and chemical properties also differ from those of calcite. The shells of corals, mollusks, and many other marine organisms consist of calcite or aragonite. **Dolomite** resembles calcite, but half of the calcium ions are replaced by magnesium. Sulfates are formed of positive ions (such as calcium, iron, or strontium) that are attached to negative sulfate ions, each of which consists of sulfur bonded by four oxygens (SO_4). Many sulfates also form at low temperatures near the earth's surface.

Although **oxides** make up only a small percentage of the large bodies of rock on earth, these minerals form many important ore deposits. Rocks whose primary components are the oxides magnetite (Fe_3O_4) and hematite (Fe_2O_3), for example, yield most of the iron that is put to human use. Many other ore minerals are **sulfides**, which are compounds of metals such as copper or iron in combination with sulfur.

TYPES OF ROCKS

Rocks are classified on the basis of their size, their composition, and the arrangement of their constituent grains. We will now examine the main types of rocks that belong to each of the fundamental groups of rocks—igneous, sedimentary, and metamorphic.

Igneous rocks

When igneous rocks are classified on the basis of their chemical composition, they fall into two major groups—

Table AI-1   **Major mineral groups**

	Chemical properties	Physical properties	Rock-forming contribution	Comments
Silicates	SiO_4 tetrahedra are the basic units	Mostly hard, except for mica and clay minerals; most have a glassy or pearly luster	Dominant mineral group in igneous, sedimentary, and metamorphic rocks	Most crystallize at high temperatures and occur in sediments only as detritus
Carbonates	Positive ions attached to CO_3	Soft, light-colored	Mostly sedimentary, but also form marble, a metamorphic rock	Include calcite, aragonite, and dolomite
Sulfates	Positive ions attached to SO_4	Soft, light-colored, water-soluble	Most rock-forming varieties are sedimentary	Form large sedimentary evaporite deposits, including gypsum and anhydrite
Halides	Positive ions attached to negative ions of elements such as chlorine (Cl) or bromine (Br)	Soft, light-colored, water-soluble	Most rock-forming varieties are sedimentary	Form large sedimentary deposits, including halite (rock salt)
Oxides	Metallic ions combined with oxygen	Soft to hard	Mostly sedimentary, but many varieties are present in igneous and metamorphic rocks	Some, including the iron minerals magnetite and hematite, are major ore minerals
Sulfides	Metallic ions combined with sulfur	Soft to medium hard, often with a metallic luster	Have a minor role in rock forming	Many are important ore minerals that form at high temperatures

felsic and **mafic**. The term *felsic,* which is derived from the first three letters of *feldspar,* the silicon-rich mineral introduced in Figure AI-3, is used as a general designation for all silicon-rich igneous rocks. Such rocks are generally light-colored and of low density. Granite is the most abundant felsic igneous rock; two kinds of feldspar constitute about 60 percent of its volume (Figure AI-5). Not surprisingly, granite and other felsic rocks also contain high percentages of quartz. Continental crust (Figure 1-18) is predominantly felsic, containing large volumes of granite-like rocks. Mafic rocks, in contrast, are relatively low in silicon and contain no quartz. Such rocks are, however,

FIGURE AI-4   The mineral calcite (CaCO₃). Above are crystals, and below is a block of the mineral produced by fracturing a crystal along cleavage planes. (*Photographs by the author.*)

rich in magnesium and iron; hence their name, which derives from the *ma* in *magnesium* and the *f* in *ferrous*. This abundance of magnesium and iron makes mafic rocks, like **gabbro** (Figure AI-5), darker and heavier than felsic rocks. Mafic rocks form most of the oceanic crust, while **ultramafic** rocks, which are even lower in silicon, form the mantle below the crust. Olivine (Figure AI-2) is one of the primary constituents of ultramafic rocks.

Igneous rocks can also be classified according to grain size. This classification system is especially useful in that grain size reflects the rate at which igneous rocks cooled from a molten state. If a rock cools slowly, its crystals can grow large, thus producing a coarse-grained rock, whereas rapid cooling "freezes" molten rock into small crystals that yield a fine-grained rock. The fine-grained equivalent of granite is called **rhyolite**, and the fine-grained equivalent of gabbro is called **basalt**. The dense rock that forms the oceanic crust consists primarily of basalt.

Most molten rock that cools within the crust or at the earth's surface comes from the mantle; as this molten rock rises, in the form of a blob or plume, it melts the crustal rock with which it comes into contact. When molten rock is found within the earth, it is known as **magma**, but when it appears at the earth's surface through an opening called a **vent**, it is called **lava**. Some magma cools within the earth and thus never reaches the surface. Because cooling here usually takes place slowly, the result is nearly always an igneous rock of coarse grain size, such as granite or gabbro. Lava, in contrast, usually cools rapidly at the earth's surface, and the result is usually a fine-grained igneous rock such as rhyolite or basalt.

Magma that cools within the earth forms bodies of rock that are sometimes referred to as **intrusions**—so called because they often displace or melt their way into preexisting rocks—or **plutons**. **Sills**, on the other hand, are sheetlike or tabular plutons that have been injected between sedimentary layers, and **dikes** are similarly shaped plutons that cut upward through sedimentary layers or crystalline rocks (Figures AI-6 and AI-7).

In contrast to intrusive, or plutonic, rocks, which form from magma at depths within the earth, extrusive, or volcanic, rocks form when lava cools at the earth's surface. Some volcanic rocks form simply by flowing out of cracks called fissures, from which they spread over large areas. The volcanic rocks that spread widely from fissures are almost always mafic, since felsic lavas are more viscous than mafic lavas. Mafic extrusive rocks that have flowed widely are often referred to as **flood basalts** (Figure AI-8).

Other lavas erupt from tubes or fissures to build the cone-shaped structures that we call **volcanoes**. Volcanoes that form oceanic crust seldom produce tall, narrow cones, but instead form broad cones called **shield volcanoes**. This shape results from the mafic composition of oceanic lavas, which gives the lavas a low viscosity that allows them to spread rapidly in all directions (Figure AI-9). Lava that emerges from the crust beneath the sea cools rapidly in a way that gives its surface a hummocky configuration, creating rock known as **pillow basalt** (Figure AI-10). On continents, hot magma that rises toward the earth's surface

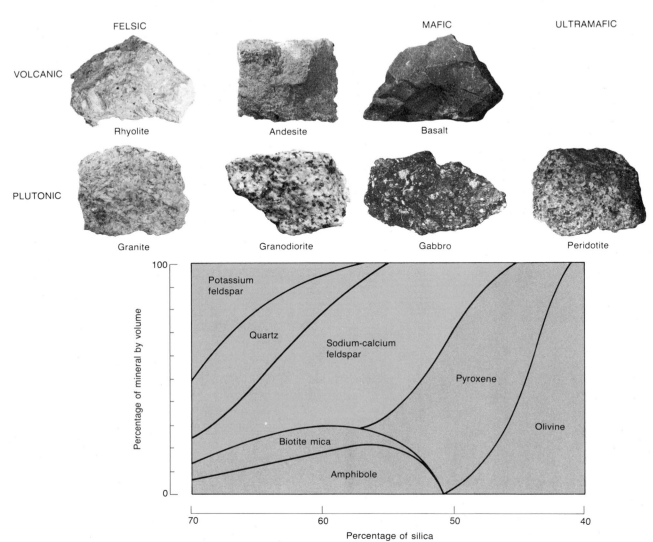

FELSIC   MAFIC   ULTRAMAFIC

VOLCANIC

Rhyolite   Andesite   Basalt

PLUTONIC

Granite   Granodiorite   Gabbro   Peridotite

FIGURE AI-5 The composition of igneous rocks. In contrast to felsic rocks, mafic rocks contain no quartz and not much more than 50 percent silica. Note that plutonic rocks, which crystallize slowly within the earth, are more coarse-grained than volcanic rocks, which crystallize more rapidly at the earth's surface. All of the specimens shown here would fit in the palm of your hand. (*Photographs by the author.*)

often melts surrounding felsic rock, thereby mixing with the felsic magma. Because felsic lava is more viscous than mafic lava, volcanoes on the surfaces of continents tend to take the form of tall cones. An example is Mount St. Helens, which erupted in the state of Washington in 1980 (Figure 1-1). At the summit of most volcanoes is a hollow

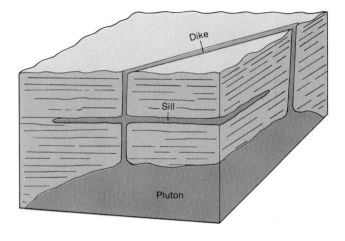

FIGURE AI-6 A block diagram showing the configurations of intrusive bodies of igneous rock. A pluton is a large body of irregular shape; a dike is a body of tabular shape that cuts across layered rocks; and a sill is a tabular body that has been intruded between layers.

crater that forms after an eruption, when the unerupted lava sinks back down into the vent and hardens.

Volcanic rocks can also form in ways that do not involve the cooling of flowing lava. Some volcanic eruptions, for example—including that of Mount St. Helens in 1980—are explosive, hurling solid fragments of previously formed volcanic rock great distances. The finest material produced by these eruptions is termed **volcanic dust**; slightly coarser material is called **volcanic ash**. Chunks larger than about 6 centimeters ($\sim2\frac{1}{2}$ inches) are called **bombs**; bombs weighing many tons are sometimes thrown several kilometers from a vent. Loose debris of various sizes settles to form rock known as **tuff**. Although tuff is deposited in the same manner as sedimentary rock, it is usually classified as volcanic simply because it consists of volcanic particles. In fact, some tuffs form from hot grains that melt together as they settle, but others harden following the precipitation of cement from water percolating through them.

FIGURE AI-7   A dike of basalt cutting through the Old Red Sandstone Formation south of Wemyss Point, Scotland. (*British Institute of Geological Sciences.*)

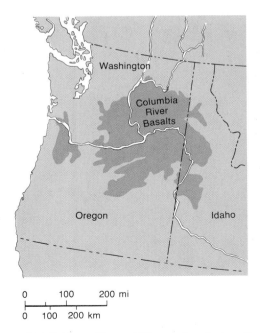

FIGURE AI-8   Flows of Miocene basalt along the Columbia River near Vantage, Washington. The cliff is about 330 meters ($\sim$1100 feet) high. The map shows the widespread distribution of these so-called Columbia River Basalts. (*Washington Department of Conservation and Development.*)

Rocky material, however, is not all that issues from volcanic vents. Gases of many kinds are also emitted from volcanoes along with rock fragments, and in volcanic areas such as Yellowstone Park, steamy geysers shoot up from the earth at sites where underground water is heated against magma and vaporized.

Sediments and sedimentary rocks

Most sediments that form rocks belong to one of three categories—siliciclastic, chemical, or biogenic.

To understand what **siliciclastic sediments** are, one must first understand what a clast is. A **clast** is a particle of rock that has been transported; **clastic rocks** are aggregates of clasts. Clasts are the solid products of erosion

FIGURE AI-9   Basalt formed by cooling of lava on Fernandina, one of the Galápagos Islands. The lava flowed from the low-shield volcano that can be seen in the background. The ropy configuration in the foreground developed when the surface of the flowing lava wrinkled like dough; the large cracks in the center of the picture formed as the lava cooled. (*Photograph by the author.*)

FIGURE AI-10   Pillow basalts that formed at Española, one of the Galápagos Islands, more than 3 million years ago, when lava cooled beneath the sea. Most of the "pillows" are less than 1 meter (~3 feet) across. (*Photograph by the author.*)

(see below) and are also referred to as **detritus**, and the adjective **detrital** is often applied to clastic rocks that have been transported and deposited by wind, water, or ice. Siliciclastic sedimentary rocks, then, are detrital rocks that are composed of silicate clasts. Clasts composed of quartz and clay are particularly common at the earth's surface because they are supplied in vast quantities by the breakdown of preexisting rocks under conditions in which many other solid materials are destroyed. Often, however, grains that are mineral aggregates and grains consisting of feldspars and other individual minerals also survive this breakdown process to become components of siliciclastic rocks.

Chemical sediments are created by precipitation from natural waters; most accumulate near the site where they originally formed. Many chemical sediments form by the evaporation of bodies of water.

Biogenic sediments consist of mineral grains that were once parts of organisms. Some of these grains are pieces of organic skeletons, such as snail shells or "heads" of coral, and others are the tiny, complete skeletons of single-celled creatures. Most biogenic sediments, however, consist of the skeletal remains of a variety of organisms rather than just one or two.

We will soon examine these three types of sedimentary rocks in greater detail, beginning with the siliciclastic group. First, however, it is necessary to consider where the clasts of siliciclastic rocks come from.

Erosion and sediment production   As we have seen, rocks are not indestructible; those that are located at or near the earth's surface can be transformed by chemical reactions or broken by external forces. The products of this decay ultimately move away from the site of origin under the influence of gravity, wind, water, or ice. As noted in Chapter 1, the term *erosion* is used to describe all of the processes that cause rock to loosen and move downhill or downwind. Erosion refers both to the chemical processes that loosen particles of rock and to the physical processes that break rock apart and transport the resulting particles.

The term **weathering** is applied to those aspects of erosion that take place before transport. There are physical as well as chemical weathering processes, and the two kinds of processes interact constantly. Fragmentation of rock, for example, accelerates chemical weathering by increasing the surface area exposed to destructive chemical processes. Similarly, chemical weathering weakens rock, making it more vulnerable to physical breakage. Thus, when one kind of process begins to weather a rock, the other usually starts as well.

Ice, snow, water, and earth movements are the primary agents of physical weathering. Water expands when it freezes, and when it does so in cracks and crevices within rocks, it exerts tremendous pressure that can often split rocks apart. Movements of the earth, which will be discussed in Appendix II, can also fracture rock.

Destructive chemical processes constitute the most pervasive kind of weathering, and water and watery solutions act as the primary agents of this weathering. Water at the earth's surface, for example, readily converts feldspar to clay, carrying away some ions in the process. Because feldspars are the most abundant minerals in the igneous rocks of continents (Figure AI-3), clay is also very abundant at the earth's surface. Like micas, clay minerals are sheet silicates (Figure AI-3). Clay differs from mica, however, in that the molecular structure of its sheets is weak. Thus, clay minerals do not form large flakes such as those that typify mica, but instead form minute flakes, which give clay sediments their fine-grained texture.

Quartz, in contrast to feldspar, is quite resistant to weathering, a characteristic that accounts for the abundance of sand on the earth's surface. Most sand consists of quartz grains that are similar in size to, or slightly smaller than, the quartz grains of granites and similar crystalline rocks.

When the feldspar grains of a crystalline rock weather to clay, the rock crumbles, releasing both the flakes of clay and the grains of quartz as sedimentary particles (Figure AI-11). Rainfall or the meltwaters of snow or ice wash many of these particles into streams, from which they are carried to larger bodies of water. Eventually, many of the particles settle from the waters of rivers, lakes, or oceans as sediment that, in time, may become hard sedimentary rock.

Both water and oxygen take part in the weathering of mafic rocks, converting these iron-rich minerals to iron oxide minerals that resemble rust. In the process, silica is carried away in solution. Because mafic rocks form at higher temperatures than felsic rocks, mafic minerals at the earth's surface are generally less stable than felsic minerals and therefore undergo more rapid chemical weathering. Mafic crystals, in other words, grow from magma or lava before it cools to a sufficient degree to allow quartz, mica, and other crystals of felsic rock to grow as well. Thus, we can think of mafic minerals as "high-temperature" minerals that tend to disintegrate at the earth's surface because they are so far from their temperatures of formation. Mafic minerals are rarely seen in the beach sands that fringe oceans, since most of these minerals weather to crumbly iron oxide either within the rock where they formed or close to the area in

which that rock was exposed to air and water. In the same manner, feldspars are rarely found on sandy beaches, since most turn to clay either within or close to their parent rocks.

Siliciclastic sedimentary rocks   Siliciclastic sedimentary particles and rocks are classified according to grain size, as Figure AI-12 illustrates. In this definitional system, the term **clay** denotes particles that are smaller than $\frac{1}{256}$ millimeter. Although this quantitative definition may seem redundant and contradictory, because clays also constitute

FIGURE AI-11   Weathering of a large boulder of gabbro in Mesa Grande, San Diego County, California. As minerals in the gabbro undergo chemical weathering, layers of rock "spall off" and crumble to clay and other minerals. (*Courtesy of W. T. Schaller, U.S. Geological Survey.*)

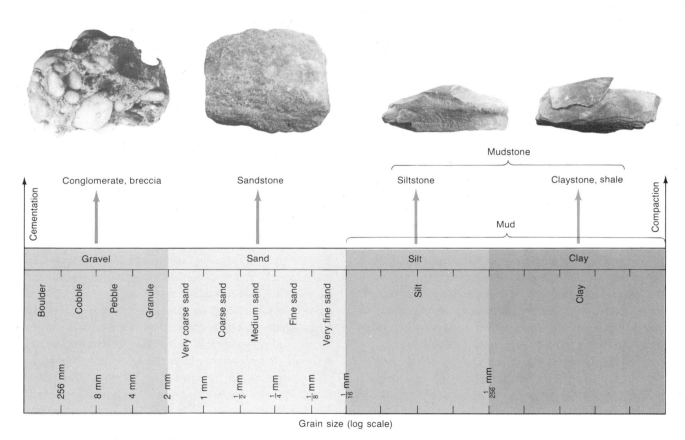

FIGURE AI-12   Classification of sedimentary rocks according to grain size. Sediments range in size from clay to silt, sand, and gravel; gravel is divided into granules, pebbles, cobbles, and boulders. Gravelly rocks are called conglomerates when their pebbles and cobbles are rounded and breccias when they are angular; these rocks normally contain sand as well. Rocks in which sand dominates are called sandstones. Rocks in which silt dominates are called siltstones. Rocks formed of clay are called claystone if they are massive and shale if they are fissile (or platy). Siltstones, claystones, and shales are all varieties of mudstone. All of the specimens shown here would fit in the palm of your hand. (*Photographs by the author.*)

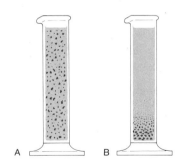

FIGURE AI-13   The settling pattern of sediment after it is suspended in water (A). The coarsest sediment settles most quickly and therefore ends up at the bottom of the deposit; the finest sediment settles last (B).

a family of sheet silicate minerals, the use of the term *clay* in reference to both mineralogy and particle size seldom creates contradictions, since nearly all clay mineral particles are of clay size and nearly all particles of clay size are clay minerals. **Mud** is a term that embraces aggregates of clay, silt, or some combination of clay and silt. Material of **sand** size ranges from $\frac{1}{16}$ millimeter to 2 millimeters, and **gravel** includes all particles larger than sand, including **granules**, **pebbles**, **cobbles**, and **boulders**.

When sediment settles from water, coarse-grained particles settle faster than fine-grained particles, as can be seen when sediments of various grain sizes are mixed with water in a tall glass container and allowed to settle (Figure AI-13). Clay settles so slowly in water that in nature very little of it falls from rapidly moving water such as that of a river channel or a wave-ridden sandy shore of the ocean. Instead, most of it is deposited in calm waters such as those of lakes, quiet lagoons, and the deep sea.

Not only do coarse sediments settle more quickly from water than do fine sediments, they are also less easily picked up or rolled along surfaces by moving water. **Gravel**, for example, tends to be deposited near its source area (the area in which it was originally produced by erosion); thus, gravel usually accumulates along the flanks of the mountains from which it is eroded, seldom reaching the center of large oceans.

Siliciclastic rocks are classified according to composition as well as grain size. Rocks formed largely of clay-sized particles, for example, consist primarily of flakes of clay minerals, since few other minerals are abundant in nature at such a small size. When rocks such as these exhibit a tendency to break along bedding surfaces, they are said to be **fissile** and are called **shale**. Not all rocks composed primarily of clay are fissile, however, since fis-

sility results from the horizontal alignment of flakes of clay during deposition. Thus, **claystone**, a rock that is identical in composition to shale, exhibits little or no fissility because of the irregular orientation of its clay particles.

Most sedimentary grains of silt and sand size are composed of quartz. Even though quartz is very hard, quartz grains suffer abrasion as they bounce and slide downstream along the floors of rivers. In the process, they tend to become smaller and more rounded. Because so many sand-sized grains are quartz, the word *sandstone* is sometimes automatically interpreted to mean **quartz sandstone**. There are, however, several other siliciclastic rocks that consist largely of sand-sized grains. One of these is **arkose**, whose primary constituents, grains of feldspar, often give the rock a pinkish color. Arkose usually accumulates only in proximity to its parent rock, soon after the feldspar grains are released by partial weathering and erosion.

Another important rock in which sand-sized particles normally predominate is **graywacke**, which is so designated because it is usually dark gray. Graywacke consists of a variety of sedimentary particles, including sand-sized and silt-sized grains of feldspar and dark rock fragments and also substantial amounts of clay. It is now believed that most of the clay in graywacke was not carried to the environment of deposition in its present state, but was formed by the disintegration of larger grains within the rock. Chemically unstable grains of feldspar and other minerals were initially present in graywacke because most graywackes were deposited by rapidly moving water currents that carried particles from the source area to the place of deposition, leaving little time for disintegration before burial. Conversion of particles to clay took place after deposition.

A rock that contains large amounts of gravel is termed a **conglomerate** if the gravel is rounded and a **breccia** if it is angular. In both cases, however, sand or granules nearly always fill the spaces between the pieces of gravel. The phrase **poor sorting** refers to this mixing of grain sizes, and its presence implies that moving water did not separate the grains according to size before deposition. Sand along a beach, in contrast, has usually been washed and transported by water currents and waves, and thus it tends to be **well sorted**—that is to say, it tends to consist of particles whose size range is very narrow. Most of the particles in a handful of beach sand are likely to be either medium- or fine-grained.

Siliciclastic grains vary not only in their size and chemical composition but also in the manner in which they are arranged within rocks. When the cross section of a sedi-

FIGURE AI-14   Sedimentary structures. *A.* A block of sand-stone showing laminations. *B.* A graded bed, with gravel at the base (5-centimeter [~1-inch] scale in *B* also applies to *A*). *C.* Cross-stratification in a sandstone. *D.* Ripples on a bed of sand-stone. (*Photographs by the author.*)

mentary rock is examined, distinct divisions are revealed. These divisions, which are called **beds** if they are thicker than 1 centimeter and **laminations** if they are thinner (Figure AI-14*A*), usually represent discrete depositional events. Often the grains at the base of one bed are either coarser or darker and heavier than those of the bed below—or else these grains differ conspicuously in some other way that reflects a later and slightly different origin. As described in Chapter 3, bedding patterns, which are termed **sedimentary structures**, reflect modes of deposition and provide useful tools for interpreting the environments of deposition of ancient sedimentary rocks.

A **graded bed** is one in which grain size decreases from the bottom to the top (Figure AI-14*B*). This pattern usually results from the normal settling process that characterizes sediments of mixed grain size, with the coarser sediment settling more rapidly than finer sediment (Figure AI-13).

When beds or laminations were deposited at an angle from the horizontal, the pattern that they form is referred to as **cross-stratification** (Figure AI-14*C*). Cross-stratification forms by deposition of sediment along the slope of a dune or sand bar (Figure 1-20). In many cases, one set of parallel cross-strata in a rock cuts across another, reflecting a shift in the position of the depositional slope accompanied by the erosion of previous deposits. Because cross-stratification reflects the general direction of sediment transport, measurement of the orientation of cross-stratification in ancient rocks often provides a strong indication of the orientation of an ancient river or shoreline.

Ripples, which are small elongate dunes that form at right angles to the direction of wind or water movement, can also reveal the direction of water movement in ancient environments. Ripples often produce cross-stratification, as well (Figure AI-14*D*).

A variety of processes transform soft siliciclastic

sediments into hard rock. These processes, which affect sediments after deposition but before metamorphism, are collectively referred to as diagenesis. Both physical and chemical diagenetic processes harden, or lithify, sediment. The primary physical process of lithification is **compaction**, a process in which grains of sediment are squeezed together beneath the weight of overlying sediment. The extent of compaction varies with grain size. Sand grains, for example, are deposited in a closely packed state; thus, sand-sized sediments experience only a small amount of compaction. When clay particles are deposited, on the other hand, they are often separated by films of water that represent as much as 60 percent of the total sediment volume. This is why muddy sediments usually experience a great deal of compaction after burial, as water is squeezed from them. In the process, the sediments become hardened.

Chemical diagenesis takes many forms, one of the most important of which is the lithifying process of **cementation**. In this process, minerals crystallize from watery solutions that percolate through the pores between grains of sediment. The cement that is thus produced may or may not have the same chemical composition as the sediment. Sandstone, for example, is often cemented by quartz but more commonly by calcite. Cement is most easily studied by examining thin sections, which are slices of rock ground so thin that they transmit light and can thus be examined by microscope (Figure AI-15). Cementation is less extensive in clayey sediments than in clean sands because after they undergo compaction, clays are relatively impermeable to mineral-bearing solutions. Clean sands are initially much more permeable than clays, but their porosity is reduced—and sometimes virtually eliminated—by pore-filling cement.

Cement sometimes gives sedimentary rocks their color. This is often the case for red siliciclastic sedimentary rocks called **red beds**. Some red beds are mudstones, some are sandstones, and some are conglomerates, but almost all derive their color from iron oxide, which acts as a cement.

Chemical and biogenic sedimentary rocks　The grains that form chemical and biogenic sedimentary rocks were not produced by erosion but were instead precipitated within the bodies of water in which they accumulated. Some

FIGURE AI-15　Cement bordering a quartz grain in sandstone as seen microscopically in a thin section. The border of the original grain is outlined by dark foreign materials called inclusions, beyond which cement has grown outward, locking the grain to neighboring grains. (*Courtesy of I. E. Odom.*)

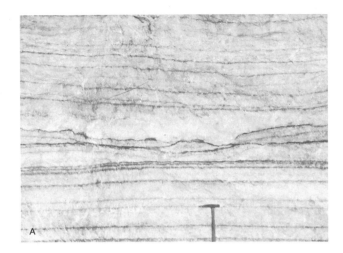

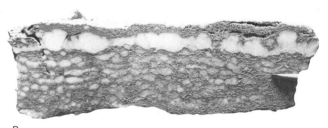

FIGURE AI-16   Evaporite deposits. *A.* White bands of halite alternate with thin, dark bands of clayey anhydrite in New York State; this unit, which is of Silurian age, is mined for rock salt. *B.* Nodular gypsum rock of Miocene age from Sicily. (*A. Courtesy of L. F. Dellwig. B. Courtesy of L. A. Hardie.*)

chemical and biogenic sediments are difficult to distinguish from one another, so we will consider both types in this section.

The most common chemical sedimentary rocks are **evaporites**, which form from the evaporation of sea water or other natural water. Many evaporites are massive deposits that consist of vast numbers of crystals (Figure AI-16). Among the most important evaporites are **anhydrite** (calcium sulfate, $CaSO_4$) and **gypsum** (calcium sulfate with water molecules attached, $CaSO_4 \cdot H_2O$). It is important to note, however, that the terms *anhydrite* and *gypsum* refer both to the minerals with these names and to rocks that are composed largely of these minerals. *Halite* is another term that refers both to a mineral and to an evaporite rock. It is the presence of sodium chloride in large amounts that makes sea water salty, so it is no surprise that halite (Figure AI-1) should accumulate in large quantities when sea water evaporates. Halite deposits have great economic value, providing us both with table salt and with the rock salt that is used to melt ice on highways.

Evaporites are readily precipitated from water but are also readily dissolved; hence, they do not survive for long periods of time at the surface of the earth except in arid climates. When evaporites are buried far beneath younger deposits, however, they are protected from potentially destructive ground water and can thus survive for long geologic intervals.

FIGURE AI-17   Chert. *A.* A chert nodule. *B.* Bedded chert of the Paleozoic Caballos Formation of Texas. (*A. Photograph by the author. B. Courtesy of E. F. McBride.*)

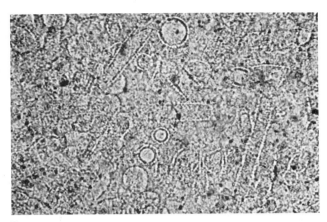

A

FIGURE AI-18   Siliceous oozes. *A*. Siliceous skeletons, mostly of radiolarians, contributing to siliceous ooze dredged from the northern equatorial Pacific Ocean. *B*. A photomicrograph of liquefied ooze forming the Caballos Formation of mid-

B

Paleozoic age. "Ghosts" of sponge needles and radiolarians are still visible. (*A. Courtesy of J. R. Hein. B. Courtesy of E. F. McBride.*)

Other types of chemical sediments are less abundant than evaporites. Among the most important of these are chert, phosphate rocks, and iron formations. **Chert**, which is also called **flint**, is composed of extremely small quartz crystals that have been precipitated from watery solutions. Some cherts occur as bedded rocks, while others appear as irregular, rounded masses called **nodules** (Figure AI-17). Typically, impurities give chert a gray, brown, or black color. Chert also breaks along curved, shell-like surfaces; American Indians took advantage of this feature when they fashioned chert into arrowheads. Some cherts form within bedded rock such as limestone when nodules or layers grow from silica-rich solutions that have moved through the rock. On the other hand, some bedded cherts are thought to form by direct precipitation of silica (SiO_2) from sea water, and others are biogenic deposits that result from the accumulation on the sea floor of the microscopic skeletons of single-celled organisms. These skeletons consist of silica of a type that differs from quartz in that it is amorphous (or noncrystalline). During diagenesis, water percolates through deposits of these skeletons, converting them to very hard chert that consists of minute interlocking quartz crystals (Figures AI-17 through AI-19). Cherts older than 100 million years or so have suffered extensive chemical diagenesis; thus, many are difficult to identify as biogenic or chemical.

Phosphate rocks consist primarily of calcium phosphate. Although they are of marine origin, their precise mode of formation is not known. It would appear that these rocks form in areas where marine life is abundant, since this is thought to be the source of their phosphate.

Iron formations are complex rocks that usually consist of oxides, sulfides, or carbonates of iron, often in association with chert. Iron formations are widespread only in very old Precambrian rocks, and many form important iron ore deposits. They are examined in greater detail in Chapter 11.

Limestones include both chemical and biogenic bodies of rock. Because they are not as soluble in water as evap-

FIGURE AI-19   The Monterey Formation near Santa Barbara, California. This unit is composed largely of siliceous diatoms and of chert derived from these diatoms. The light-colored intervals consist of soft diatomaceous sediment. The dark-colored intervals consist of dense chert in which diatoms are difficult to discern. (*M. N. Bramlette, U.S. Geological Survey.*)

orites, limestones are much more common at the earth's surface, where they are quarried extensively for the production of building stone, gravel, and concrete. Although ancient limestones consist primarily of the mineral calcite (Figure AI-4), many of their grains were initially composed of the mineral aragonite. (You will recall that aragonite has the same chemical composition as calcite but differs in its crystal structure.) Aragonite can form at the temperatures

and pressures that exist at the earth's surface, but it is relatively unstable under these conditions and in time becomes transformed into calcite. Little or no aragonite remains in most limestones older than a few million years.

Dolomite is a carbonate mineral that is relatively uncommon in modern marine environments but common in many ancient rocks. As mentioned earlier, it differs from calcite in that half of the calcium ions of the crystal lattice

0   0.4 mm

FIGURE AI-20   Limestones. *A.* A weakly lithified limestone that consists of fragments of organic skeletons. *B.* A block from a bed of well-lithified Paleozoic limestone formed largely of shells and shell fragments, some of which can be distinguished on the surface of the bed (5-centimeter [1-inch] scale applies to both *A* and *B*). *C.* Aragonite needles from a modern tropical sea floor ($\times 65$); such needles are the primary components of most carbonate muds. *D.* Section of an ancient fine-grained limestone that was carbonate mud prior to lithification. (*A and B. Photographs by the author. C. Courtesy of P. E. Cloud, U.S. Geological Survey. D. Courtesy of P. A. Scholle.*)

are replaced by magnesium ions. In fact, much dolomite has formed by the chemical alteration of calcite. When dolomite is the dominant mineral of an ancient rock, the rock is also called dolomite. Because limestones and dolomites are often intimately associated and frequently intergrade with one another in mineral composition, they are sometimes referred to collectively as **carbonate rocks**. Similarly, unconsolidated sediments consisting of aragonite, calcite, or both minerals are often called **carbonate sediments**.

Carbonate sediments form in two ways: by the direct precipitation of aragonite needles from sea water or through the activities of organisms. Many types of marine life, including corals and most mollusks, grow shells or other kinds of skeletons that consist of aragonite or calcite. At death or even earlier, these organisms contribute skeletal material to the sea floor as sedimentary particles. Some of these particles retain their original sizes, while others diminish in size through breakage or wear. The product of this biological contribution is an array of carbonate particles that are similar to siliciclastic grains and, like siliciclastic grains, can be classified according to size. Thus, we speak of **carbonate sands** and **carbonate muds**, and we find that carbonate sands display many of the sedimentary structures that we also see in siliciclastic sands, including cross-stratification. Most carbonate particles that are sand-sized or larger can be seen to be skeletal particles (Figure AI-20), but it is often difficult to determine the origin of mud-sized material. Even **aragonite needles**, which are the primary components of carbonate muds, are produced both by direct precipitation and by the collapse of carbonate skeletons, especially of algae. In ancient fine-grained limestones, aragonite needles have been transformed to tiny calcite grains. The resulting granular texture reveals little about the configuration or mode of origin of the original carbonate particles.

Calcium carbonate is precipitated only from sea water that contains relatively little carbon dioxide, and carbon dioxide is less soluble in warm water than cold water. Thus, carbonate sediments accumulate primarily in tropical seas, where winter water temperatures seldom drop below 18°C (~64°F). This condition not only explains the direct precipitation of carbonate minerals in tropical or near-tropical seas but also accounts for the fact that few organisms living in cold water secrete massive skeletons of calcium carbonate. Carbonate sediments can also form in freshwater habitats, usually as a result of the carbonate-secreting activities of certain algae.

Unlike siliciclastic clays, carbonate rocks do not compact greatly after burial. Instead, they harden primarily by cementation, in the same manner as siliciclastic sands. Carbonate sediments are nearly always cemented by carbonate minerals simply because rich sources of such cements are close at hand. Where carbonate sediments accumulate, cement is sometimes immediately available from the surrounding sea water, and some carbonate sediments are partially or completely lithified right on the sea floor. When carbonate sediments are elevated above sea level and exposed to rain water, they become cemented especially quickly, since carbon dioxide dissolved in rain water forms weak carbonic acid that dissolves carbonate grains (especially aragonite). When a solution formed in this way evaporates, calcium carbonate is precipitated between grains as cement.

Metamorphic rocks

Metamorphic rocks form by the alteration of other rocks at temperatures and pressures that exceed those normally found at the surface of the earth. Metamorphism alters both the composition and the texture of all kinds of other rocks—igneous rocks, sedimentary rocks, and rocks that are already metamorphic.

There are three fundamental types of metamorphism: dynamic, regional, and contact. **Dynamic metamorphism**, which results from earth movements, causes rock to fracture and even pulverize along zones of movement and then welds the resulting fragments together in new textures. In this type of metamorphism, pressure plays a greater role than temperature. **Regional metamorphism** transforms deeply buried rocks at high temperatures and pressures; as its name implies, this process operates over areas whose dimensions are measured in hundreds of kilometers. **Contact metamorphism** is caused by igneous intrusion, which "bakes" surrounding rock. This is usually a local phenomenon that may occur deep within the earth or near the surface. High temperature usually plays a larger role in contact metamorphism than high pressure.

We will take a closer look at the three classes of metamorphism and the rocks that they produce.

Dynamic metamorphism When movements of the solid earth or of magma cause rocks to bend intensively or to break apart, the rocks may be shattered or their grains squeezed into new shapes and orientations. The resulting rocks are called **cataclastics**. Some cataclastic rocks are coarse-grained, resembling sedimentary breccias (Figure AI-21), while others are fine-grained rocks that consist of

FIGURE AI-21   A cataclastic breccia formed by the fracturing of rock during an igneous intrusion. The large fragments consist of the metamorphic rock hornfels, which was produced by local heating of rocks by magma. Later, pressure from the igneous intrusion fragmented the hornfels to form the breccia. (*Courtesy of H. C. Granger, U.S. Geological Survey.*)

A

B

C

FIGURE AI-22   Foliated metamorphic rocks. Slate (*A*) is a low-grade metamorphic rock, schist (*B*) is a medium-grade rock, and gneiss (*C*) is a high-grade rock. (*Photographs by the author.*)

pulverized material the size of sedimentary silt or very fine sand. Most cataclastic rocks become lithified at the time of deformation, when heat and especially pressure are so intense that they cause the particles of these rocks to grow together and interpenetrate. Nonetheless, dynamic metamorphism is primarily a physical rather than a chemical process.

Regional metamorphism   For reasons that are discussed in Chapter 8, igneous activity usually extends along the length of an actively forming mountain chain. Along each side of such an igneous belt is a zone of regional metamorphism produced by high temperatures and pressures extending outward from the igneous belt. Most rocks in zones of regional metamorphism display a texture known as **foliation**, which is an alignment of platy minerals caused by the pressures applied during metamorphism (Figure AI-22). **Gneiss** is a coarse-grained metamorphic rock whose intergrown crystals resemble those of igneous rock but

FIGURE AI-23   Metamorphic rocks that lack foliation because they consist largely of single minerals that do not form platy or flaky grains. *A.* Marble, a rock that usually consists of calcium and magnesium carbonate. *B.* Quartzite, a rock consisting almost exclusively of the mineral quartz. (*Photographs by the author.*)

whose minerals tend to be segregated into wavy layers. **Schist** consists largely of platy minerals such as micas, which lie roughly parallel to one another in such a way that the rock tends to break along parallel surfaces. **Slate** is a finer-grained rock in which aligned platy minerals produce fissility much like that of shale. The mineral alignment of slate, however, results from deformational pressures rather than from depositional orientation of the sort that produces the bedding of shale.

Metamorphic rocks of a particular texture can form over a wide range of temperature and pressure conditions. Mineral assemblages of metamorphic rocks, however, serve as critical "thermometers" and "barometers," varying with the temperature and pressure of metamorphism. White micas, for example, occur only in low- and medium-grade metamorphic rocks. (**Grade** is the word used to indicate level of temperature and pressure.) The green micalike mineral **chlorite** is more restricted in its occurrence, since it is exclusively a low-grade mineral. In contrast, **pyroxenes** (Figure AI-3) are restricted to rocks of intermediate and high grade. Typically, grade of metamorphism in a regional metamorphic zone declines away from the neighboring belt of igneous activity that supplied the heat for metamorphism.

Not all rocks in regional metamorphic zones are foliated. Some have homogeneous, granular textures, which indicates not only that their interlocking mosaics of crystals lack preferred orientations but also that certain minerals are not segregated into bands. **Marble** and **quartzite**, for example, are usually homogeneous, granular metamorphic rocks (Figure AI-23). Marble consists of calcite, dolomite, or a mixture of the two, and it forms from the metamorphism of sedimentary carbonates. Quartzite consists of nearly pure quartz, and it forms from the metamorphism of quartz sandstone. The simple mineralogical composition of marble and quartzite, together with their lack of platy minerals, prevents both of these rocks from exhibiting foliation even when they form under great pressure.

Contact metamorphism   Contact metamorphism is a more localized phenomenon than regional metamorphism, but like the latter, it displays a gradient in which the grade of metamorphism declines away from the heat source. **Hornfels** usually forms in areas adjacent to local igneous intrusions. This is a fine-grained granular metamorphic rock of varying composition that is formed at high temperatures (Figure AI-21). Farther away are lower-grade metamorphic rocks that may be either granular or foliated (Figure AI-24).

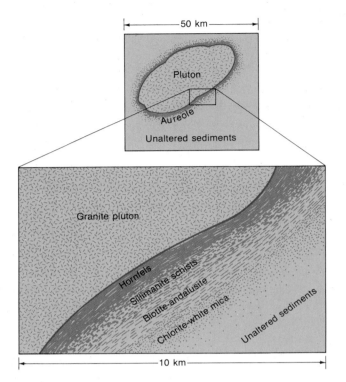

The upper limit of metamorphism   When conditions become so hot that a rock melts, metamorphism ceases. Rocks that form when the resulting molten material cools are then classified as igneous. Very high grade metamorphic rocks that nearly pass through a molten state before cooling resemble coarse-grained igneous rocks in that they are granular rocks that display little or no foliation. Those that consist largely of quartz and feldspar are known as **granulites**. High-grade metamorphic rocks are difficult to distinguish from igneous rocks, and the origins of many high-temperature rocks are the subject of debate.

FIGURE AI-24   Contact metamorphism adjacent to a granite pluton. The heat from the magmatic intrusion that formed the pluton metamorphosed the sediments that were intruded. The sediments closest to the pluton were metamorphosed to hornfels, a fine-grained metamorphic rock that forms at high temperatures. A series of metamorphic zones with other mineral-assemblages extends away from the pluton, with assemblages formed at progressively lower temperatures away from the pluton. (*After F. Press and R. Siever, Earth, W. H. Freeman and Company, New York, 1982.*)

Appendix II

Deformation Structures in Rocks

It has long been known that the rocks of the earth's crust and the fossils within them can be fractured and deformed on a very large scale. Especially where mountains have been uplifted, pieces of crust many square kilometers in areal extent have been transported great distances, and large bodies of rock have been contorted in complicated ways. In Chapter 1 we noted that the study of these movements is called tectonics. Many of the forces that have caused such movements result from the motions of large plates of the lithosphere—motions that are considered in Chapters 7 and 8 under the label of plate tectonics. This appendix summarizes the kinds of geologic structures that result from the application of large forces to rock. It also illustrates how these visible structures allow us to decipher earth movements of the past.

BENDING AND FLOWING OF ROCKS

We can see from geologic outcrops that rocks have been warped, twisted, and folded, and that they have even flowed. Appendix I makes the statement that most rocks consist of discrete mineral grains. When forces are applied to a rock, the component grains may be affected in various ways; they may slide past one another, change shape, or break along parallel planes. (In the last case, a grain deforms in the way that a deck of cards on a table changes shape when we push on one end of the deck near the top.) Internal deformation of a large body of rock by any of these mechanisms takes place very slowly, but when many of the grains are affected, the entire body of rock can, in the course of millions of years, undergo radical changes in shape. Such changes usually take place at great depth within the earth's crust.

One common type of large-scale rock deformation is referred to as **folding**. Compressive forces can shorten the earth's crust, creating folds (Figure 1-14A) that come in many sizes. When folded sedimentary rocks are viewed with their oldest beds at the bottom and their youngest beds on top, the folds that are concave in an upward direction, with their vertexes at the top, are termed **synclines**, while those that are concave in a downward direction, with their vertexes at the bottom, are termed **anticlines** (Figure AII-1). Many rocks do not simply bend when folded; instead, material is displaced from one part of a bed toward another. When a large, complex body of rock is subjected to an external force, certain weak beds are often more intensely folded than other, more durable beds. Shales, for example, tend to be weak and to deform quite severely in comparison to massive sandstones and limestones that are subjected to the same forces. Igneous and metamorphic rocks are relatively durable, but they, too, can be folded.

To describe shapes of folds in greater detail, geologists have developed special terminology. A tilted bed, for example, is said to have a **dip**—a term that describes the angle that the bed forms with the horizontal plane. Put another way, the dip is the direction that water would run down the surface of the bed. The **strike** of a bed, in contrast, is the compass direction that lies at right angles to the dip (Figure AII-2); strikes are always horizontal. It is sometimes stated that the **regional strike** for a given area is in a particular geographic orientation—for example, north-south. This does not mean, however, that every strike in this area has the same orientation. What it means for a folded terrane is that most of the fold axes trend north-south so that most beds have strikes with the same general orientation.

A fold is said to have an **axial plane**, which is an imaginary plane that cuts through the fold and divides it as symmetrically as possible. In actuality, many folds are asymmetrical, with one limb (or flank) dipping more steeply than the other. If either limb is rotated more than 90° from

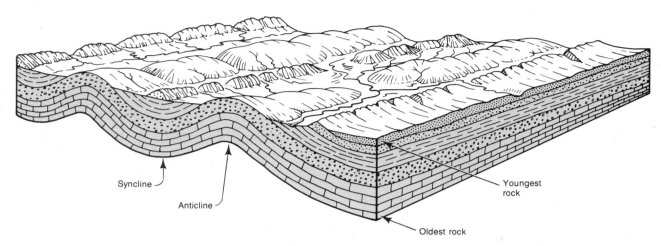

FIGURE AII-1   Block diagram showing humplike folds, or anticlines, and troughlike folds, or synclines. (*After F. Press and*

*R. Siever, Earth, W. H. Freeman and Company, New York,
1982.*)

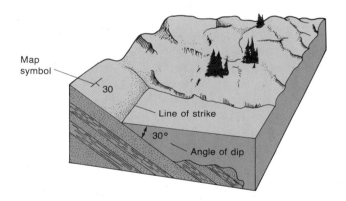

FIGURE AII-2   A geologist measuring the dip of inclined
beds. The dip is the angle between the edge of the instrument
and the horizontal plane. The horizontal plane is found by rotating a mechanism within the instrument until a bubble floating in
liquid is centered. (*U.S. Geological Survey.*)

its original position, the fold is said to be **overturned** (Figure AII-3), and a fold lying on one limb, with the axial plane
nearly horizontal, is said to be **recumbent**.

The **axis** of a fold is the line of intersection between
the axial plane and the beds of folded rock. Often the axis
plunges, which means that it lies at an angle to the horizontal (Figure AII-4A). When a **plunging fold** is truncated
by erosion, its beds form a curved outcrop pattern (Figures
AII-4B and AII-5). A series of plunging folds then produces
a scalloped surface pattern (Figure AII-6). Because most
folds plunge (we could not expect many to be perfectly
horizontal), this scalloped pattern is characteristic of regions where sedimentary rocks have been extensively folded.

Stratified rocks do not always bend into folds such as
those just described. Local uplift of a sedimentary sequence can form a **dome**, and local depression can form
a **basin**. Both of these structures, when eroded, yield concentric, circular bands of outcrop for stratified rocks. There
is a simple way to distinguish the outcrop pattern of a dome
from that of a basin. In a dome, the oldest beds lie in the
center, whereas in a basin, it is the youngest beds that are
centrally positioned (Figure AII-7). Some basins are enormous. The entire state of Michigan, for example, forms the
central part of a structural basin (Figure AII-8). It is important to understand, however, that some structural domes

Overturned folds

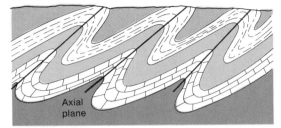

Recumbent folds

FIGURE AII-3   Cross-sectional views of overturned and recumbent folds. Both limbs of an overturned fold dip in the same
direction. One limb of a recumbent fold is upside-down. (*After
F. Press and R. Siever, Earth, W. H. Freeman and Company,
New York, 1982.*)

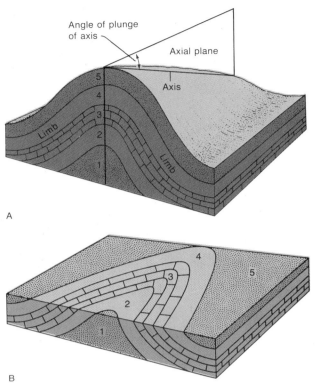

FIGURE AII-4   A plunging fold before (*A*) and after (*B*) truncation resulting from erosion. Note how erosion produces a
curved outcrop pattern.

FIGURE AII-5   Aerial view of a plunging fold. The view is
along the axis of the Virgin Anticline of southwestern Utah.

Compare to Figure AII-4. (*J. S. Shelton, Geology Illustrated,
W. H. Freeman and Company, New York, 1966.*)

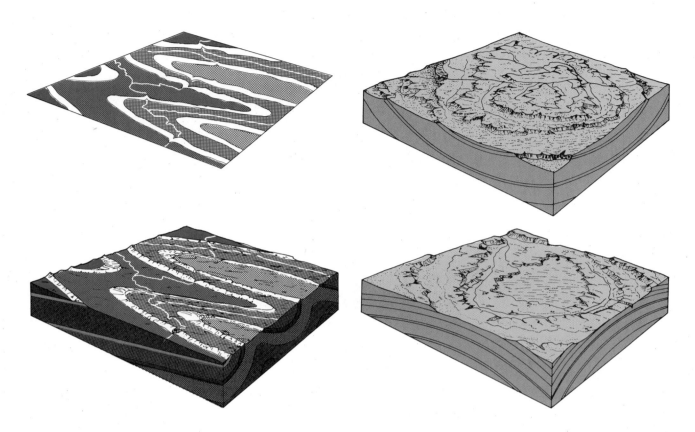

FIGURE AII-6   Block diagram (*below*) and map (*above*) showing the scalloped outcrop pattern of a series of plunging folds. Note that the resistant rock units form ridges on the eroded surface. (*After W. K. Hamblin and J. D. Howard, Exercises in Physical Geology, Burgess Publishing Company, Minneapolis, Minnesota, 1975.*)

FIGURE AII-7   Block diagram (*below*) showing the concentric outcrop pattern of a structural basin; the rocks in the center are relatively durable and have thus remained at a high elevation. Block diagram (*above*) showing the concentric outcrop pattern of a structural dome; the rocks in the center are relatively weak and have been deeply eroded to form a topographic basin. (*After W. K. Hamblin and J. D. Howard, Exercises in Physical Geology, Burgess Publishing Company, Minneapolis, Minnesota, 1975.*)

form topographic basins. This happens when the oldest (central) beds happen to be weak so that they erode easily. This is true of the basin shown in Figure AII-9, in which many of the beds forming the flanks stand well above the heavily eroded "core." Similarly, the centers of structural basins sometimes stand at high elevations.

Figure AII-9 also illustrates that a dome can be oblong rather than nearly circular. An extremely elongate dome amounts to an anticline that plunges in two directions. Similarly, a very long basin amounts to a syncline that has upturned ends.

BREAKING OF ROCKS

Rocks do not always bend or flow when stressed heavily. Sometimes they behave in a more brittle fashion and simply break. **Joints** are fractures in rock that can result from various kinds of stress. In some bodies of rock, joints are oriented randomly, but a particular source of stress often forms a set of nearly parallel joints. Two or more intersecting sets give many outcrops a blocky or columnar appearance (Figure AII-10). Joints are simple fractures along which there is no appreciable movement of the opposing rock surfaces. **Faults**, on the other hand, are fractures along which measurable movement has occurred. A fault, like a bed of rock, has an orientation that can be described by a strike and a dip. Various kinds of faults are illustrated in Figure AII-11. A **normal fault** has a steep dip—one that is closer to vertical than horizontal—and the rocks above the fault plane move downward in relation to those below. This type of fault is frequently formed when extensional stresses spread rocks apart; the rocks break and then the upper block slides downward under the influence of gravity. A **reverse fault**, in contrast, usually results from compression, which causes the rock to fracture and the block above the fracture to slide up over the lower block. **Strike-slip faults** result from horizontal shearing forces—forces that break the rock at a high angle and move the blocks on each side of the break horizontally past each other. Rocks may move only a few centimeters along a strike-slip fault, or they may move hundreds of kilometers. An **oblique-slip fault** is intermediate in character between a strike-slip fault and a normal or reverse fault.

Low-angle reverse faults, termed **thrust faults**, can account for enormous earth movements when mountains form. Large, tabular segments of crust known as **thrust sheets** may be transported for tens or even hundreds of kilometers along fault planes that lie just a few degrees from horizontal. A cross section through a much smaller and higher-angle thrust fault is shown in Figure AII-12.

Thrust faults are causally related to overturned folds. A force that compresses bedded rocks to form an overturned fold may eventually break the bedded rocks and move the upper limb of the fold along a thrust fault (Figure AII-13). Thus, it is not surprising that belts of deformation within many mountain chains have been formed by a combination of folding and thrusting. Elongate zones of such structures are known as **fold-and-thrust belts** (Figure AII-14).

Movement along most faults is sporadic. Between

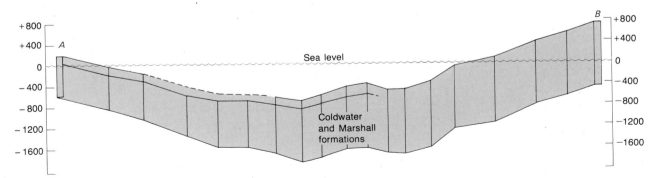

FIGURE AII-8   A simplified geologic map of the state of Michigan. The fact that the youngest rock units are in the center indicates that this is a structural basin. The cross section from *A* to *B* shown below illustrates the configuration of the Coldwater and Marshall Formations, which occur throughout most of the state but are buried beneath younger deposits in the center. This cross section has been constructed from information obtained by drilling. (*After V. Brown Monnett, Amer. Assoc. Petrol. Geol. Bull. 32:629–688, 1948.*)

FIGURE AII-9   Structural dome near Rawlings, Wyoming. Here weak rocks in the center have been eroded deeply to form a topographic basin. (*J. S. Shelton, Geology Illustrated, W. H. Freeman and Company, New York, 1966.*)

movements, stress within the earth builds up against friction along the fault plane. Finally, the stress overcomes the friction, and a sudden movement takes place. After stress builds up again, the process is repeated, and movement thus occurs in small steps. Sudden movement along a fault causes the earth to shudder, and such movements create most earthquakes. It is difficult to know just when stress will overcome friction along a fault, and this is why we cannot accurately predict the occurrence of earthquakes.

New faults have originated during recorded history (Figure 1-14*B*), and measurable movement has also taken place recently along old faults. Active faults often produce visible linear features at the earth's surface (facing page 1). The reason for this is that rocks are often fractured along fault planes, and fracturing tends to invite erosion where faults intersect the earth's surface. Thus, faults may

FIGURE AII-10   Vertical joints in the Portage Formation, Cayuga Lake, New York. Two sets of joints are present here. The fact that they are oriented nearly parallel to each other gives the outcrop a blocky appearance. Joints are simply fractures in the rock, not planes along which rocks move. (*U.S. Geological Survey.*)

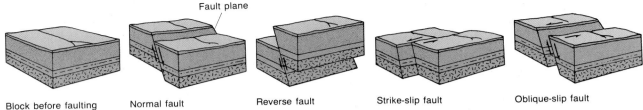

Block before faulting   **Normal fault**   **Reverse fault**   **Strike-slip fault**   **Oblique-slip fault**

FIGURE AII-11   Block diagrams illustrating several types of faults. (*After F. Press and R. Siever, Earth, W. H. Freeman and Company, New York, 1982.*)

FIGURE AII-12   Thrust fault in thin-bedded limestone. As indicated by the arrows, the body of rock above the fault (hanging wall) has moved toward the right with respect to the body of rock below the fault (foot wall). (*Photograph by K. Segerstrom, U.S. Geological Survey.*)

FIGURE AII-13   Diagram (*right*) showing how an overturned fold can give rise to a thrust fault as force continues to be applied. (*After F. Press and R. Siever, Earth, W. H. Freeman and Company, New York, 1982.*)

be expressed topographically as valleys. The Great Glen Fault, for example, runs the full width of Scotland and is generally marked by narrow valleys, one of which cradles Loch Ness, the lake that is famous for its apparently mythical monster (Figure AII-15).

STRUCTURAL CROSS SECTIONS

Sequences of folding and faulting leave records that can be deciphered to reveal earth movements of the past. An example is presented in Figure AII-16, which illustrates an idealized cross section of rocks within the Basin and Range

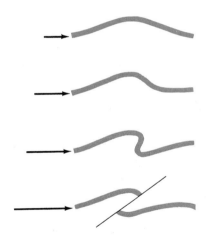

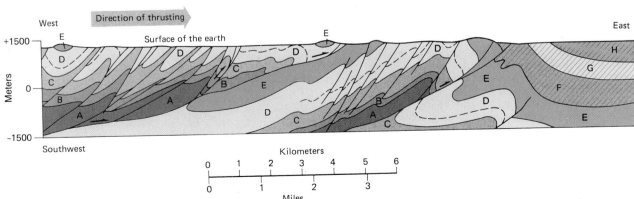

FIGURE AII-14   A fold-and-thrust belt of the Rocky Mountains southwest of Calgary, Alberta, Canada. Thrust faults, shown as dark lines, are intimately associated with overturned folds. Thrusting has been toward the east, and folds are overturned in the same direction. The oldest beds (*A*) are Lower Carboniferous, and the youngest (*H*) are Paleogene. (*After P. B. King, The Evolution of North America, Princeton University Press, Princeton, New Jersey, 1977.*)

Province of western North America. Here the sequence of events began with deposition of horizontal sediments on top of crystalline basement (A). This was followed by folding and thrusting from the west (B). Next, an erosion surface developed, truncating the folds (C). Then a sheet of lava was spread over the erosion surface (D). Finally, normal faults broke the terrane into a series of blocks (E). We can reconstruct the sequence of these events by working backwards. The normal faulting had to be the last event because the normal faults cut through all other features—through the sheet of lava, the folds, and the thrust faults. The next-to-last event must have been the extrusion of the lava

FIGURE AII-15   The Great Glen of Scotland, a large valley formed by erosion along the Great Glen Fault. Loch Lochy occupies the valley in the foreground, and Loch Ness is visible in the distance. The Great Glen Fault passes across the full width of Scotland. (*Institute of Geological Sciences, Great Britain.*)

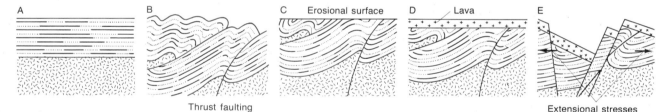

FIGURE AII-16   Stages in the development of the Basin and Range Province of western North America, according to W. M. Davis. A. Sedimentary units are deposited. B. The sedimentary deposits are folded and thrust faulted by compressive forces. C. An erosion surface develops. D. Sheets of lava are spread

over the erosion surface. E. Block faults break up the terrane. The final configuration (E) is all that geologists see, and from this they reconstruct the earlier events. (*P. B. King, The Evolution of North America, Princeton University Press, Princeton, New Jersey, 1977.*)

FIGURE AII-17   The history of the Grand Canyon region. Here the sequence of Paleozoic and Triassic rocks rests right-side-up on Precambrian rocks and is separated from these rocks by an angular nonconformity. In some places the lowest Cambrian unit, the Tapeats Sandstone, overlies the Precambrian sedimentary rocks, and in others it overlies a metamorphic unit, the Vishnu Schist. The Vishnu itself is occasionally intruded by granite rocks. The intimate relationship between the intrusions and the metamorphic structures of the Vishnu Schist indicates that intrusion and metamorphism took place at the same time. A summary of the events recorded in the Grand Canyon region is as follows:

STEP 1.   The earliest event that we can recognize is the deposition of the Vishnu sediments, although we cannot reconstruct the properties of these sediments in detail.

STEP 2.   Next came the metamorphism of the Vishnu sediments to schist and the accompanying intrusion of the sediments by granitic rocks.

STEP 3.   The Precambrian sedimentary unit numbered 1 lies unconformably above the Vishnu. Step 3, then, is the development of an erosion surface on the Vishnu.

STEP 4.   The deposition of units 1 through 4.

STEP 5.   In places, faults have elevated units 1 through 4 so that they are in contact with the older Vishnu Schist. The normal, or block, faulting that had this effect must therefore have followed the deposition of units 1 through 4.

STEP 6.   After the block faulting, erosion planed off many irregularities in the topography.

STEP 7.   On the erosional surface thus formed, the Paleozoic and Triassic sediments were laid down, with a major unconformity separating Cambrian and Devonian deposition (the Muav and Temple Butte limestones) and with minor unconformities separating some younger units.

LATER STEPS.   The section of rocks exposed in the Grand Canyon does not in itself reveal the history of the region after deposition of the Triassic sediments that cap the sedimentary sequence. Regional studies, however, show that Cenozoic uplift of the plateau over the past 2 million years or so led to the formation of the present topography.

sheet, since it was laid down on an erosion surface that truncated all other features except the normal faults. This implies that the folding and thrusting was the first mode of deformation. By using this kind of logic, we can unravel local geologic history.

Paleozoic and Triassic sedimentary rocks that form the upper walls of the mile-deep Grand Canyon of Arizona provide us with a more complex exercise in the analysis of regional geologic history. Rocks of every Paleozoic period but the Ordovician and Silurian are present along this great valley (an unconformity separates Devonian rocks from Cambrian rocks). Figure AII-17 illustrates how the regional history can be read from the Grand Canyon walls.

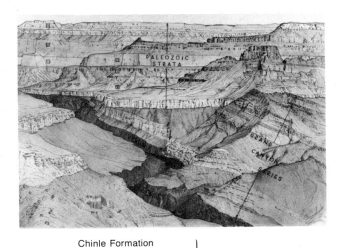

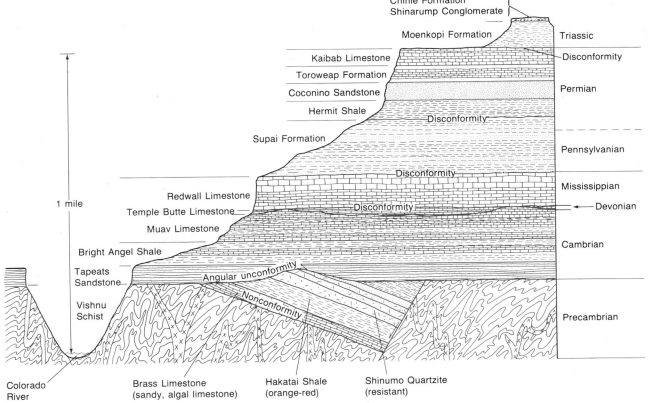

Chinle Formation
Shinarump Conglomerate
Moenkopi Formation — Triassic
Kaibab Limestone — Disconformity
Toroweap Formation
Coconino Sandstone — Permian
Hermit Shale
Disconformity
Supai Formation — Pennsylvanian
Disconformity — Mississippian
Redwall Limestone
Temple Butte Limestone — Disconformity — Devonian
Muav Limestone
Bright Angel Shale — Cambrian
Tapeats Sandstone
Angular unconformity
Vishnu Schist
Nonconformity
1 mile
Precambrian
Colorado River
Brass Limestone (sandy, algal limestone)
Hakatai Shale (orange-red)
Shinumo Quartzite (resistant)

Classification of Major Fossil Groups

This is a compilation of important fossil taxa. Listed with most groups are figure numbers directing the reader's attention to illustrations of one or more fossil or living representatives. An examination of these figures and their captions in the context of this appendix will provide a comprehensive overview of fossil life. The reader should bear in mind that because the fossil record is incomplete, many taxa undoubtedly lived longer than the time ranges indicated for them here. Also, classifications are subjective, and some workers employ different classifications than the one adopted in this text.

Kingdom MONERA

Includes all **prokaryotes**, which are single-celled organisms whose DNA is not organized into chromosomes or housed in a nucleus. Mitochondria and chloroplasts are lacking. Some experts divide the Monera into two kingdoms, one primitive and the other more advanced. Figures 9-35 and 10-15

Phylum "**Fermenting bacteria**" Obtain energy by breaking down organic compounds. Cannot tolerate oxygen. (*Archean–Recent*) Figure 9-36

Phylum **Thiopneutes (sulfate-reducing bacteria)** Obtain energy by converting sulfate to hydrogen sulfide (H_2S), which smells like rotten eggs. Live in mud. Cannot tolerate oxygen. (*Archean–Recent*) Figure 9-36

Phylum "**Anaerobic photosynthetic bacteria**" The most primitive photosynthetic organisms. Some types tolerate very high or low temperatures, but few tolerate oxygen. (*Archean–Recent*) Figure 9-36

Phylum **Cyanobacteria (blue-green algae**, also called blue-green bacteria or cyanophytes) Photosynthetic cells that are threadlike or nearly spherical. Form algal mats and stromatolites. (*Archean–Recent*) Figures 4-28, 9-34, 9-36, 10-5, 10-13, 10-14, and 12-28; also pp. 272 and 324

Kingdom PROTOCTISTA

Single-celled and simple multicellular eukaryotes, including algae.

Phylum **Dinoflagellata** Floating algae, mostly single-celled. Characterized by two whiplike flagella. Important phytoplankton in lakes and in the ocean. Some species live as symbiotic individuals in the tissues of corals or other animals. (*Triassic–Recent*) Figures 2-31 and 15-6

Informal Group **Acritarcha** An artificial group of hollow, organic-walled fossils that apparently represent algae. (*Proterozoic–Recent*) Figures 10-16 and 12-9

Phylum **Haptophyta (coccolithophores)** Photosynthetic, single-celled floating algae whose nearly spherical cells are armored by shieldlike plates of calcium carbonate. Important phytoplankton in the tropics, where their plates accumulate on quiet sea floors. (*Paleozoic? Triassic–Recent*) Figures 2-31, 15-7, and 16-3

Phylum **Bacillariophyta (diatoms)** Photosynthetic, single-celled algae, protected by pillboxlike skeletons composed of silica. Some species are planktonic and some are benthic; some are marine and others are lake dwellers. Important phytoplankton in lakes and cool seas. (*Cretaceous–Recent*) Figures 2-31, 16-1, and 18-3

Phylum **Rhodophyta (red algae)** Photosynthetic multicellular algae in which the chlorophyll is contained in reddish intracellular structures. Predominantly marine. Coralline red algae produces the massive algal ridge of modern coral reefs. (*Proterozoic? Paleozoic–Recent*) Figures 4-13 and 14-10

Phylum **Chlorophyta (green algae)**   Photosynthetic single-celled and multicellular algae whose chlorophyll gives them a green color. Some are calcareous and contribute vast quantities of calcium carbonate to the sedimentary record. (*Proterozoic? Paleozoic–Recent*)   Figure 10-23(?)

Phylum **Foraminifera (foraminifers)**   Single-celled animal-like marine forms with skeletons (tests) through which projections of protoplasm extend. Some (superfamily **Globigerinacea**) are planktonic but most are benthic. Some experts unite them with the radiolarians and with *Amoeba* and its relatives in the phylum **Sarcodina.** (*Cambrian–Recent*)   Figures 2-30, 4-40, 6-17, 14-7, 14-8, 16-2, 16-11, 17-45, and 18-2; also p. 114

Phylum **Actinopoda (radiolarians** and their relatives)   Single-celled animal-like marine forms with radially symmetrical skeletons (tests) through which projections of protoplasm extend. The skeleton consists of silica or strontium sulfate. (*Cambrian–Recent*)

Kingdom PLANTAE

Plants. Multicellular, sexually reproducing eukaryotes with chloroplasts. More complex than algae. All plants except some mosses and liverworts have vascular tissue for fluid conduction. (Phyla marked with asterisks are **gymnosperms,** or vascular plants with naked seeds.)

Phylum **Bryophyta (mosses** and **liverworts)**   Small plants that lack well-developed fluid-conducting tissues and live in moist environments. (*Late Paleozoic–Recent*)

Phylum **Psilophyta**   Small vascular plants with simple stems. Mostly extinct. (*Middle Paleozoic*)   Figures 12-29(?), 13-17, 13-19, and 13-20

Phylum **Lycopodophyta (lycopods—club mosses** and their relatives)   Spore-bearing plants in which the leaves are arranged in a helical pattern around the stem. All living species are small and inconspicuous, but many extinct species, including the dominant plants of Carboniferous coal swamps, were large trees. (*Silurian–Recent*)   Figures 13-18, 13-21, and 14-11; also p. 396

Phylum **Sphenophyta (sphenopsids—horsetails** and their relatives)   Spore-bearing plants in which the stem is divided into nodes that bear whorls of branches. Living members are small plants but some Carboniferous species were trees. (*Devonian–Recent*)   Figures 14-14, 14-15, and 14-18

FIGURE AIII-1   Ferns (phylum Filicinophyta). Unlike many ferns, these particular species grow upright as small trees. (*A. S. Foster and E. M. Gifford, Comparative Morphology of Vascular Plants. W. H. Freeman and Company, New York, 1974.*)

Phylum **Filicinophyta (ferns)**   Spore-bearing plants with branching leaves that have spore-bearing organs on their lower surface. (*Devonian–Recent*)   Box 13-1; Figures 14-18, 15-17, and AIII-1

*Phylum **Cycadophyta**   Primitive seed plants, most of which have the form of shrubs or trees.

   Class **Pteridospermales (seed ferns)**   Seed-bearing plants with fernlike foliage. Some were the size of bushes or trees. (*Middle Paleozoic–Jurassic*) Figures 14-12, 14-13, and 14-18

   Class **Cycadeoidales** (extinct **cycads)**   Cycad plants of a primitive type. (*Triassic–Cretaceous*)   Figures 14-18 and 15-18

   Class **Cycadales** (modern **cycads)**   Plants with columnlike trunks and featherlike leaves. (*Triassic–Recent*)   Figures 2-20, 14-18, and 15-20

   Class **Cordaitophyta (cordaites)**   Trees with tall trunks and elongate leaves. (*Late Paleozoic*)

*Phylum **Ginkgophyta (ginkgos)**   Deciduous trees with small, fan-shaped leaves that have regularly branching veins. Only a single species survives today. (*Early Mesozoic–Recent*)   Figures 14-18 and 15-19

*Phylum **Coniferophyta (conifers—cone-bearing gym-nosperms,** including **pines, spruces,** and **firs)**   Shrubs and trees with cones and needle-shaped leaves. (*Late Paleozoic–Recent*)   Figures 6-28, 14-17, 14-18, and 15-20

Phylum **Angiospermophyta (flowering plants,** including

grasses and **hardwood trees**)   Seed plants in which the gamete-bearing generation is hidden within the flower. (*Cretaceous–Recent*)   Box 13-2; Figures 14-18, 16-15, 16-18, 16-19, 17-5, 17-9, and 18-4

Kingdom ANIMALIA

Animals. Multicellular organisms that obtain their nutrition by consuming other organisms.

Phylum **Porifera (sponges)**   Very simple sedentary aquatic animals that strain food from water that they pump through pores in their walls. Supporting the tissues of most are spinelike spicules of calcite or silica that can be preserved as fossils. Include the **Stromatoporoidea**, which were important reef builders during the Paleozoic. (*Cambrian–Recent*)   Figures 2-32, 12-1, 12-20, 14-53, and AIII-2

Phylum **Archaeocyatha**   Vase-shaped, double-walled organisms that attached to hard substrata to form reef-like structures. They were probably suspension feeders that sieved food from water that passed through holes in their skeletons. (*Cambrian*)   Figures 12-13 and 12-14

Phylum **Coelenterata** or **Cnidaria**   Saclike animals having inner and outer body layers with jellylike material in between. Most are carnivores that capture food with stinging cells.

Class **Scyphozoa (jellyfishes)**   Floating marine forms with limited swimming capacity. Most capture food with dangling tentacles. (*Precambrian–Recent*) Figure 10-27

Class **Anthozoa** (**corals**, **sea anemones**, and **sea whips** and **fans**)   Benthic forms whose tentacles extend upward.

Order **Rugosa** (**rugose corals**, or **tetracorals**) Corals with fourfold symmetry. Most species were solitary "cup corals." (*Cambrian–Recent*) Figures 12-16, 12-18, and 13-3

Order **Tabulata** (**tabulates**)   Colonial, often reef-building corals with horizontal platforms (tabulae) in their skeletons. Some forms assigned to this group may actually have been sponges. (*Ordovician–Recent*)   Figures 12-16, 12-20, and 13-3 through 13-5

Order **Scleractinia** (**hexacorals**)   Solitary or colonial corals with septa (vertical partitions) of the skeleton present in multiples of six. Most colonial species have symbiotic dinoflagellates in their tissues that play a role in the building of reefs. (*Ordovician–Recent*)   Box 2-1; Figures 4-17, 6-13, 15-3, and 15-4

Phylum **Bryozoa** or **Ectoprocta** (**moss animals**)   Colonial aquatic animals in which the connected individuals are tiny structures with tentacles used for suspension feeding. Many Paleozoic types grew upright as fanlike or branching stony colonies. Most post-Paleozoic types have less heavily calcified skeletons and encrust hard substrata. (*Ordovician–Recent*)   Figures 12-16, 12-19, 12-25, 14-6, 16-12, and AIII-3

Phylum **Brachiopoda** (**lamp shells**)   Double-valved marine animals that suspension feed by means of a loop-shaped structure bearing tentacles.

Class **Inarticulata**   Brachiopods that lack teeth. Since the Cambrian they have included fewer species than the Articulata. (*Cambrian–Recent*)   Figure 12-4

Class **Articulata**   Brachiopods with hinge teeth, that is, teeth that lock their valves together. Most attach to the substratum by means of a fleshy stalk called a pedicle. (*Cambrian–Recent*)   Figures 1-9, 6-19, 12-4, 12-16, 12-17, 12-19, 13-1, and 14-2

FIGURE AIII-2   A sponge (phylum Porifera). This is a marine species of simple form. Note the vaselike shape. Water passes through pores in the wall of the sponge and exits through the central opening at the top. (*Courtesy of W. Sacco.*)

FIGURE AIII-3   A single bryozoan individual (phylum Bryozoa or Ectoprocta). The feeding tentacles are clearly visible. Individuals like this bud from others and remain attached to form colonies. (*Courtesy of T. S. Wood.*)

Phylum **Mollusca**   A diverse group of invertebrates, most of which have a foot and a flaplike mantle that secretes a shell.

Class **Monoplacophora**   Primitive crawling forms with cap-shaped shells. (*Cambrian–Recent*)   Figure 12-1*B*

Class **Gastropoda (snails)**   Mostly crawling forms with coiled shells and bodies that are twisted into a U shape. Some are marine, some freshwater, and some terrestrial. (*Cambrian–Recent*)   Figures 2-32, 12-16, 12-21, 13-2, and 16-14

Class **Bivalvia (bivalves—clams, mussels, oysters, scallops,** and their relatives)   Mollusks with two shell halves (valves). Some are infaunal and others epifaunal. Some are suspension feeders and others deposit feeders. (*Cambrian–Recent*)   Figures 2-32, 6-20, 6-25, 12-16, 12-23, 13-2, 16-13, 16-16, 16-17, and 18-1

Class **Cephalopoda (octopuses, squids, nautiloids,** and **ammonoids)**   Carnivores that swim by jet propulsion. They capture prey with tentacles and tear it apart with a beak.   Figures 12-11 and 12-16

Subclass **Nautiloidea**   Possess a chambered shell in which the partitions (septa) are attached to the wall along slightly curved lines. *Nautilus* is the only living genus. (*Cambrian–Recent*)   Figure 13-5

Subclass **Ammonoidea**   Descendants of nautiloids in which the septa are attached to the shell wall along wavy lines. (*Devonian–Cretaceous*) Figures 1-8, 6-15, 13-5, 13-6, 14-1, 15-8, 16-4, and 16-5

Subclass **Coleoidea (squids, octopuses,** and their relatives, including the extinct **belemnoids)** Groups in which the shell is reduced or lacking. (*Late Paleozoic–Recent*)   Figures 1-13, 15-9, and 16-4

Phylum **Annelida**   Segmented worms, some of which are marine, others freshwater, and still others (including earthworms) nonmarine. They are responsible for many of the burrows visible in marine sediments younger than about 700 million years. (*Late Precambrian–Recent*) Figures 2-32 and 10-29

Phylum **Echinodermata**   Marine animals having fivefold radial symmetry (most starfishes have five arms); internal skeletons that consist of plates of calcite; and small, elongate suction cups (tube feet) for feeding or locomotion.   Figure 12-5

Class **Echinoidea (sea urchins)**   Globe-shaped and disc-shaped species with moveable spines. Most are grazers or deposit feeders. (*Ordovician–Recent*) Figures 1-8, 2-32, 12-22, 15-1, 15-2, and 17-4

Class **Stelleroidea (starfishes)**   Mostly carnivorous forms that can pry open bivalves with their tube feet. (*Ordovician–Recent*)   Figures 2-32, 12-16, and 12-24

Class **Crinoidea (sea lilies)**   Animals that use branching arms for suspension feeding. Some attach to the sea floor by a stalk, but others are free-living and can swim awkwardly. (*Cambrian–Recent*) Figures 1-13, 12-16, 12-18, 14-3, and 14-4

Phylum **Onychophora**   Creatures that resemble annelids in having segmented, flexible bodies and arthropods in having jaws derived from legs. They have unjointed legs and are probably evolutionary intermediates between annelids and arthropods. Living species are terrestrial, but Cambrian fossils occur in marine deposits. (*Cambrian–Recent*)   Figures 12-8 and AIII-4

Phylum **Arthropoda**   Forms with an external skeleton and jointed appendages.

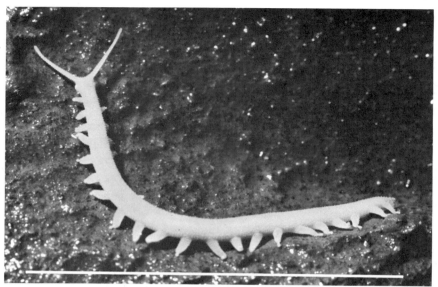

FIGURE AIII-4   An onychophoran (Onychophora). These liv-ing-fossil animals, which require moist conditions, are intermedi-ate in form, and apparently in evolutionary position, between annelid worms and arthropods. Compare this living form to the Cambrian onychophoran shown in Figure 12-8E. The latter is more than 500 million years old. (*Courtesy of R. Norton.*)

Subphylum **Trilobitomorpha (trilobites)**   Primitive marine arthropods with a three-lobed body (a central lobe and two lateral lobes) and hard skeletons. (*Cam-brian–Permian*)   Figures 8-35, 12-2, 12-3, 12-15, 12-16, and 13-5; also p. 324

Subphylum **Chelicerata (spiders, scorpions, horse-shoe crabs**, and their relatives)   Arthropods whose six front segments are united into a head and which possess a pair of jointed pincers. (*Cambrian–Recent*) Figure 13-7; also p. 322

Subphylum **Crustacea (crabs, lobsters, ostracods**, and numerous other groups, including important mem-bers of the zooplankton)   A varied group of arthro-pods, mostly aquatic. (*Cambrian–Recent*)   Figures 12-7 and 15-5

Subphylum **Insecta (insects)**   Includes more living species than all other animal taxa combined. (*Devo-nian–Recent*)   Figures 1-10, 14-19, and 17-16

Phylum **Conodonta (conodonts)**   These animals, long known only from their teeth, are now known from rare fossils of their bodies. They were apparently eel-shaped swimmers. (*Cambrian–Recent*)   Figure 12-6

Phylum **Hemichordata**   An inconspicuous group in the modern world, but it includes the Paleozoic graptolites (class **Graptolithina**). These were colonial animals, some of which floated and some of which were attached to the sea floor. (*Cambrian–Recent*)   Box 5-1

Phylum **Chordata**   Animals with a dorsal nerve chord and embryonic features that include gill slits in the throat region and a rod of cartilage (notochord) that in some groups becomes the vertebral column.

Class **Agnatha (jawless fishes)**   Include the living lamprey and extinct Paleozoic **ostracoderms**. (*Cambrian–Recent*) Figures 12-12, 13-8, and 13-10

Class **Acanthodii**   Primitive jawed fishes of small size with numerous spiny fins. (*Middle and late Paleozoic*)   Figure 13-8

Class **Placodermi**   Early jawed fishes. Most were heavily armored and some were very large. (*Middle Paleozoic*)   Figures 13-8, 13-13, and 13-14

Class **Chondrichthyes (sharks** and their relatives) Fishes with skeletons composed of cartilage. (*Middle Paleozoic–Recent*)   Figures 13-8, 13-14, 15-11, 17-3, and 18-1

Class **Osteichthyes (bony fishes)**

Subclass **Actinopterygii (ray-finned fishes)** Fishes with fins supported by numerous slender, raylike bones.

Infraclass **Chondrostei**   Primitive ray-finned fishes with lungs, a primitive jaw, diamond-shaped scales, an asymmetrical tail, and a partly cartilaginous skeleton. Only a few species, including sturgeons, survive today. (*Devonian–Recent*)   Figures 13-8, 13-11, and 13-15

Infraclass **Holostei**   Ray-finned fishes resembling the Chondrostei, but with the lungs transformed into a swim bladder and a slightly more advanced jaw. Only a few species, including the garpike and bowfins, survive today. (*Triassic–Recent*)   Figures 6-26 and 15-10

Infraclass **Teleostei**   The advanced group that includes most living species of ray-finned fishes. The jaws are highly developed, the scales are rounded and overlapping, the tail is symmetrical, and the skeleton is entirely bony. (*Cretaceous–Recent*)   Figures 6-14, 16-6, 16-7, and 17-1

Subclass **Sarcopterygii**   Characterized by fleshy fins and the ability to breathe air.

Order **Crossopterygii (lobe-finned fishes)**   Fishes with teeth and fin bones resembling those of primitive amphibians, to which they were ancestral. There is only one living genus. (*Devonian–Recent*)   Figures 13-8, 13-16, 13-25, 13-26, and AIII-5

FIGURE AIII-5   A coelacanth fish (order Crossopterygii). This primitive type of lobe-finned fish was thought extinct until discovered living deep in the Indian Ocean southwest of Africa in 1938. The coelacanth is a living fossil, resembling middle Paleozoic lobe-finned fishes. (*Courtesy of P. H. Dudley.*)

Order **Dipnoi (lungfishes)**   Freshwater fishes with cylindrical bodies. Some can survive buried in mud for months. (*Devonian–Recent*)   Figure 13-8

Class **Amphibia**   The oldest terrestrial vertebrates, although they require water for reproduction.

Subclass **Labyrinthodontia**   Early amphibians with solid skulls and complex internal tooth structure. Some were quite large. (*Devonian–Triassic*)   Figures 13-25, 13-26, 14-21, 14-26, and 14-27

Subclass **Salientia (frogs, toads,** and their relatives)   Swimming and hopping groups. (*Early Mesozoic–Recent*)   Figures 15-23 and 17-15

Subclass **Caudata (salamanders** and their relatives)   Small amphibians that possess a tail. Some remain in aquatic habitats as adults. (*Early Mesozoic–Recent*)   Figure 6-21

Class **Reptilia (reptiles)**   Vertebrate animals that have scales or armor and reproduce by amniote eggs.

Subclass **Anapsida**   Reptiles whose skull has a solid roof.

Order **Cotylosauria** (stem reptiles)   The earliest reptiles. (*Late Paleozoic–Triassic*)   Figure 14-22

Order **Chelonia (turtles)**   Reptiles with few or no teeth and a protective shell. (*Early Mesozoic–Recent*)   Figures 6-6 and 16-7

Subclass **Euryapsida**   Reptiles with a large opening in the skull behind the eye. Mostly marine.

Order **Protorosauria**   Ancestral euryapsids that lived on land. (*Permian–Mesozoic*)

Order **Ichthyosauria (ichthyosaurs)**   Swimming reptiles, shaped like dolphins, that bore live young. (*Triassic–Cretaceous*)   Figure 15-15

Order **Placodontia (placodonts)**   Large, turtle-shaped animals that crushed mollusks with rounded teeth. (*Triassic*)   Figure 15-12

Order **Sauropterygia**   Swimming reptiles with paddlelike limbs and small heads, often on long necks. The primitive **nothosaurs** probably lived along the shore like seals, but the advanced **plesiosaurs** were fully aquatic.   Figures 15-13, 15-14, and 16-7

FIGURE AIII-6   The tuatara (infraclass Lepidosauria). This is a living-fossil reptile, representing a primitive group of lizards that flourished during the Triassic Period. (*Courtesy of St. Louis Zoological Park.*)

Subclass **Diapsida**   Reptiles with two openings in the skull behind the eye.

Infraclass **Lepidosauria**   Primitive diapsids and their descendants, including **lizards**, **snakes**, and the huge Cretaceous marine lizards called **mosasaurs.**   Figures 16-7, 16-8, and AIII-6

Infraclass **Archosauria**   Advanced diapsids.

Order **Crocodilia** (**crocodiles** and **alligators**). Aquatic and terrestrial carnivores with flattened skulls. (*Triassic–Recent*)   Figures 15-16, 15-22, and 16-23

Order **Pterosauria** (**flying reptiles**)   Winged reptiles with long skulls and short bodies. (*Mesozoic*)   Figures 15-27, 16-24, and 16-25

Order **Thecodontia**   A varied group that was ancestral to the dinosaurs. Some types were able to run on two legs. Others superficially resembled crocodiles. (*Triassic*)   Figure 15-21

(Group uncertain) **Dinosauria**   Although traditionally classified as reptiles, dinosaurs may have been distinct enough to form a separate class, especially if they were endothermic.

Order **Saurischia** (**lizard-hipped dinosaurs**)   This group included carnivores that traveled on two legs and often attained giant proportions, and even bigger herbivorous sauropods that walked on all fours. (*Triassic–Cretaceous*) Figures 1-7, 1-11, 15-22, 15-24, 15-25*A*, 15-26, 16-22, and 16-24; also p. 440

Order **Ornithischia** (**bird-hipped dinosaurs**) Herbivorous forms, some of which traveled on two legs (for example, duck-billed dinosaurs) and others that traveled on all fours. (*Triassic–Cretaceous*)   Figures 1-7, 15-25*B*, 15-26, 16-20, 16-21, and 16-24; Box 15-1

Subclass **Synapsida** (**mammal-like reptiles**) Terrestrial animals with an opening in the skull behind the eye and with highly differentiated teeth. Like the dinosaurs, they may deserve a taxonomic position apart from the Reptilia, with which they have traditionally been classified.

Order **Pelycosauria**   Early mammal-like reptiles, including forms in which the vertebral spines were elongated to support a huge fin or sail of uncertain function. (*Late Paleozoic*)   Figures 14-25 and 14-27

Order **Therapsida**   Advanced mammal-like reptiles, with legs positioned more fully beneath the body than the legs of pelycosaurs. The lower jaw was also formed largely of a single bone. May have been endothermic or partly so. (*Permian–early Mesozoic*)   Figures 7-19, 14-28, 14-29, and 15-22; Box 16-1

FIGURE AIII-7   The duckbill platypus (subclass Prototheria). This unusual Australian animal resembles other mammals in that it suckles its young, but the young hatch from eggs, like young reptiles. (*Australian Information Service.*)

Class **Aves (birds)**   Endothermic flying vertebrates with feathers. (*Jurassic–Recent*)

Subclass **Archaeornithes**   Primitive toothed birds, of which only *Archaeopteryx* is know. Closely resembled dinosaurs in skeletal form. (*Jurassic*) Figure 15-28

Subclass **Neornithes**   Modern birds. (*Late Mesozoic–Recent*)   Figures 6-4, 6-7, 16-7, 16-9, 16-24, 17-13, and 17-14

Class **Mammalia**   Vertebrates that have hair and suckle their young.

Subclass **Eotheria**   Small, primitive mammals that evolved from mammal-like reptiles. (*Early Mesozoic*)   Figure 15-21

Subclass **Prototheria**   Egg-laying mammals, of which only the living echidna and platypus are known. (*Recent*)   Figure AIII-7

Subclass **Allotheria (multituberculates)**   Small mammals with complex teeth for grinding food. (*Jurassic–Paleogene*)   Box 16-1

Subclass **Theria**   The group that includes nearly all modern mammals.

Infraclass **Pantotheria**   Forms whose dentition suggests that they were ancestral to marsupial and placental mammals. (*Mesozoic*)

Infraclass **Metatheria (marsupials)**   Mammals that rear their offspring in a pouch. (*Cretaceous–Recent*)   Figure 6-16; Box 16-1

Infraclass **Eutheria (placentals)**   The dominant group of modern mammals except in Australia, where marsupials prevail. (*Cretaceous–Recent*) See Figure 17-6 for a summary of the stratigraphic distribution of placental mammal orders; other placentals are represented in Figures 2-6, 3-3, 6-1, 6-5, 6-23, 6-24, 17-2, 17-7 through 17-13, 17-17 through 17-22, 18-1, 18-5, 18-12, 18-22 through 18-39, 18-62, and AIII-8; Box 16-1

FIGURE AIII-8   Tarsiers (class Mammalia). These primitive members of the human order Primates display two key adaptations of this order for life in trees, forward-directed eyes (for stereoscopic vision) and grasping toes. (*Zoological Society of London.*)

In many parts of the world, the geologic record has been divided into stages. As discussed in Chapter 5, stages are time-stratigraphic units. For the most part, the stages recognized in Europe have become the standard stages with which stages that have been defined elsewhere are correlated. Correlations remain imperfect, however, as do estimates of the absolute ages of stage boundaries. This appendix is meant to serve as a reference for students who encounter unfamiliar stage names in their studies. Figure AIV-1 lists important Paleozoic and Mesozoic stages that were first defined in Europe and shows how a number of

North American stages are currently believed to correlate with them. Figure AIV-2 presents the same kind of information for Cenozoic stages, showing how European stages are thought to relate to American stages that are based on biostratigraphic zones for fossil land mammals.

FIGURE AIV-1   Major Paleozoic and Mesozoic stages of Europe and North America. Through correlation and absolute dating, efforts are being made to extend the European stages to all parts of the world.

	AGE (million years)	SYSTEM		SERIES	STAGE (European)	STAGE (North American)
CENOZOIC			QUATERNARY	PLEISTOCENE		
		NEOGENE		PLIOCENE		
				MIOCENE	*(See Figure AIV-2)*	
	24			OLIGOCENE		
		PALEO-GENE	TERTIARY	EOCENE		
				PALEOCENE		
	65					
MESOZOIC				UPPER	Maastrichtian	
					Campanian	
					Santonian	
					Coniacian	
					Turonian	
					Cenomanian	
		CRETACEOUS			Albian	
					Aptian	
				LOWER	Barremian	
					Hauterivian	
					Valanginian	
					Berriasian	

AGE (million years)	SYSTEM	SERIES	STAGE (European)	STAGE (North American)	(SERIES)
144					
	JURASSIC	UPPER	Tithonian		
			Kimmeridgian		
			Oxfordian		
		MIDDLE	Callovian		
			Bathonian		
			Bajocian		
			Aalenian		
		LOWER (LIAS)	Toarcian		
			Pliensbachian		
			Sinemurian		
			Hettangian		
213					
	TRIASSIC	UPPER	Rhaetian		
			Norian		
			Carnian		
		MIDDLE	Ladinian		
		LOWER	Anisian		
			Scythian		
248					
	PERMIAN	UPPER	Tatarian	Ochoan	
			Ufimian/Kazanian	Guadalupian	
		LOWER	Kungurian	Leonardian	
			Artinskian		
			Sakmarian	Wolfcampian	
			Asselian		
286					
	CARBON-IFEROUS	PENNSYLVANIAN UPPER	Stephanian	Virgilian	
				Missourian	
			Westphalian	Desmoinesian	
				Atokan	
				Morrowan	
320					
		MISSISSIPPIAN LOWER	Namurian	Springerian	
				Chesterian	
			Visean	Meramecian	
				Osagean	
			Tournaisian	Kinderhookian	
360					
	DEVONIAN	UPPER	Famennian	Chautauquan	
			Frasnian	Senecan	
		MIDDLE	Givetian	Erian	
			Eifelian		
		LOWER	Emsian	Ulsterian	
			Siegenian		
			Gedinnian		
408					
	SILURIAN	UPPER	Ludlovian	Cayugan	
		LOWER	Wenlockian	Niagaran	
			Llandoverian	Medinan	
438					
	ORDOVICIAN	UPPER	Ashgillian	Richmondian	Cincinnatian
				Maysville	
			Caradocian	Edonian	
		LOWER	Llandeilian	Chazyan	Champlanian
			Llanvirnian	Whiterockian	
			Arenigian		Canadian
			Tremadocian		
505				Trempealeauan	Croixan
	CAMBRIAN	UPPER	Dolgellian	Franconian	
			Maentwrogian	Dresbachian	
		MIDDLE	Menevian		Albertan
			Solvan		
		LOWER	Lenian		
			Atdabanian		Waucoban
			Tommotian		
590					

MESOZOIC — PALEOZOIC

AGE (million years)	EPOCH		STAGE (European)	STAGE (North American land mammal)
1.8	PLEISTOCENE		(See Table 18-1)	RANCHOLABREAN / IRVINGTONIAN
	PLIOCENE	UPPER	PIACENZIAN	BLANCAN
5		LOWER	ZANCLEAN	
	MIOCENE	UPPER	MESSINIAN	HEMPHILLIAN
			TORTONIAN	CLARENDONIAN
		MIDDLE	SERRAVALLIAN	BARSTOVIAN
			LANGHIAN	
		LOWER	BURDIGALIAN	HEMINGFORDIAN
24			AQUITANIAN	
	OLIGOCENE	UPPER	CHATTIAN	ARIKAREEAN
		LOWER	RUPELIAN	WHITNEYAN
37				CHADRONIAN
	EOCENE	UPPER	PRIABONIAN	
		MIDDLE	BARTONIAN	DUCHESNEAN
			LUTETIAN	UINTAN
				BRIDGERIAN
		LOWER	YPRESIAN	WASATCHIAN
58	PALEOCENE	UPPER	THANETIAN	CLARKFORKIAN
			(UNNAMED)	TIFFANIAN
				TORREJONIAN ?
		LOWER	DANIAN	?
65				PUERCAN

FIGURE AIV-2   Major Cenozoic stages of Europe and North America. The North American stages, which are currently only crudely correlated with the European stages, are based largely on fossil occurrences of land mammals in the Midwest and West. (Figure 5-18 illustrates the Cenozoic pattern of reversals of the earth's magnetic polarity—a pattern that is also widely employed for global correlation.) The most recent epoch, not listed in this diagram, is known as the Recent or Holocene. This brief epoch began at the end of the Pleistocene, roughly 10,000 years ago.

Glossary

Terms that are used in these definitions and that are also defined in this glossary are in many instances *italicized* for the reader's convenience. For descriptions of important minerals and more extensive discussions of rock types, the reader is referred to Appendix I. Similarly, a more detailed review of rock deformation and its terminology can be found in Appendix II. Taxonomic names of organisms are not included in this glossary; Appendix III outlines the major biological taxa, with references to illustrations in the text.

A

A-horizon　(See *Horizon, Soil*)

Abyssal plain　The broad expanse of sea floor lying between about 3 and 6 kilometers ($\sim$ 2 to 4 miles) below sea level.

Acheulian culture　The tool culture of *Homo erectus*, a species of the human family that lived during the Pleistocene Epoch.

Active lobe, of a delta　The site on a *delta* where functioning distributary channels cause the delta to grow seaward.

Actualism　The interpretation of ancient rocks by applying the results of analyses of modern-day geologic processes in accordance with the principle of *uniformitarianism.*

Adaptation　A feature of an organism that serves one or more functions useful to the organism.

Adaptive breakthrough　An evolutionary innovation that affords a group of organisms a special ecologic opportunity and often leads to the *adaptive radiation* of that group.

Adaptive radiation　The rapid origins of many new species or higher *taxa* from a single ancestral group.

Age, geologic　The division of geologic time smaller than an *epoch.*

Albedo　The percentage of solar radiation reflected from the earth's surface. This percentage is higher for ice than for land or water, and usually higher for land than for water.

Algal ridge　The durable structure formed by coralline algae that buttresses the front of a modern coral reef.

Allele　One of two or more types of *genes* that may occur at a given position on a strand of *DNA.*

Alluvial fan　A low, cone-shaped structure that forms where an abrupt reduction in slope—for example, the transition from a highland area to a broad valley—causes a stream to slow down.

Amino acid　One of the chemical building blocks of a protein. There are 20 amino acids, each a unique combination of carbon, hydrogen, oxygen, and nitrogen.

Amniote egg　The type of egg laid by reptiles and birds, having a nutritious yolk and a hard outer shell to protect the embryo from the dry environment. The amniote egg is named for the amnion, a sac that contains the embryo.

Andesite　A find-grained extrusive igneous rock, intermediate in composition between *rhyolite*, which is *felsic*, and *basalt,* which is *mafic.*

Angular unconformity　An *unconformity* separating horizontal strata, above, from older strata that had been tilted and eroded.

Anhydrite　The mineral that consists of calcium sulfate ($CaSO_4$), or the rock composed of this mineral.

Anomaly, magnetic　A local increase or decrease in the strength of the earth's magnetic field caused by the magnetism of nearby sediments or rocks.

Anticline　A fold that is concave in a downward direction—that is, the vertex is the lowest point.

Apparent polar wander　A hypothetical migration of the earth's magnetic pole that would account for the changing orientation with age of paleomagnetism in rocks at a particular fixed location. Most geologists believe that it is the continents that have moved, and not the magnetic pole—that polar wander is indeed only apparent, not real.

Aragonite needles　Slender crystals of the mineral aragonite that constitute most *carbonate* muds in the modern ocean. Some of the needles form by direct precipitation from sea water and some by the collapse of the skeletons of organisms.

Arkose　A rock consisting primarily of sand-sized particles of feldspar. Most arkose accumulates close to the source area of the feldspar, because feldspar weathers quickly to clay and seldom travels far.

Asthenosphere　The *ultramafic* layer of the earth lying below the *lithosphere.* The asthenosphere is marked by low seismic velocities, suggesting that it is partly molten.

Atmosphere The envelope of gases that surrounds the earth.

Atoll A circular or horseshoe-shaped organic *reef* growing on a submerged volcano.

Axial plane An imaginary plane that cuts through a fold, dividing it as symmetrically as possible.

B

B-horizon (See *Horizon, Soil*)

Backswamp A broad vegetated area that lies adjacent to a *meandering river* and becomes covered with water when the river overflows its banks during floods.

Banded iron formation An *iron formation* that consists of alternating iron-rich and iron-poor layers. Most rocks of this type are older than about 2 billion years.

Barrier island An elongate island composed of sand heaped up by waves that lies approximately parallel to the shoreline of an ocean.

Barrier island–lagoon complex The set of marginal marine environments that consists of a barrier island, the *lagoon* behind it, and (usually) *tidal flats, marshes,* and sandy beaches.

Barrier reef An elongate organic *reef* that parallels a coastline and is large enough to dissipate ocean waves, leaving a quiet-water *lagoon* on its landward side.

Basalt A fine-grained, *extrusive, mafic* igneous rock. The dominant rock of oceanic *crust.*

Basement rocks Rocks beneath a large geologic feature (such as the fold-and-thrust belt of a mountain system) which are genetically unrelated to the overlying feature.

Basin, structural A roughly circular depression of stratified rocks.

Bed A distinct sedimentary layer (stratum) thicker than 1 centimeter.

Bedding The arrangement of a sedimentary rock into discrete layers (strata) thicker than 1 centimeter (*beds*).

Bedding surface The surface between two sedimentary *beds.*

Benthic (or **Benthonic**) **life** (See *Benthos*)

Benthos The bottom-dwelling life of an ocean or freshwater environment.

Biogenic sediment Sediment consisting of mineral grains that were once parts of organisms.

Biogeography The study of the distribution of organisms on a geographic scale.

Biostratigraphic unit A body of rock, such as a *zone,* defined on the basis of its fossil content and having approximately time-parallel upper and lower boundaries.

Biota A collective term for all the animals and plants of an *ecosystem.*

Bird's-eye limestone Layered *limestone* that is full of holes, some of which may have been secondarily filled by cementing minerals. The holes were produced by burrowing animals or by gas bubbles. Most of these limestones are of intertidal or supratidal origin.

Borer A bottom-dwelling (*benthic*) animal that excavates living space in a hard substratum such as a rock or a coral skeleton.

Boulder A piece of gravel larger than 256 millimeters (~10 inches) in size.

Bouma sequence The characteristic vertical sequence of sedimentary features in a *turbidite.* From bottom to top, it includes a massive graded bed of coarse sediment, a zone of laminated sand, a zone of rippled sand and silt, a zone of laminated silt, and a zone of mud.

Brackish water Water that is lower in salinity than normal sea water and higher in salinity than freshwater, ranging from 30 to 0.5 parts salt per 1000 parts water.

Braided stream A stream that has many intertwining channels separated by bars of coarse sediment. Braided streams develop where sediment is supplied to the stream system at a very high rate—on an *alluvial fan,* for example, or in front of a melting *glacier.*

Breccia A rock that resembles *con-*

glomerate in consisting of *clasts* of *gravel* surrounded by sand, but in breccia the clasts are angular, whereas in conglomerate they are rounded.

Burrower A bottom-dwelling (*benthic*) animal that digs into or moves through soft sediment.

C

C-horizon (See *Horizon, Soil*)

Calcrete (See *Caliche*)

Caliche Nodular calcium carbonate that accumulates in the *B-horizon* of soils in warm climates that are dry part of the year. Also called calcrete.

Carbonaceous chondrite A stony *meteorite* that contains carbon compounds.

Carbonate mineral A mineral in which the basic building block is a carbon atom linked to three oxygen atoms. Calcite, aragonite, and *dolomite* are the most abundant examples found in sediments and sedimentary rocks.

Carbonate rock A sedimentary rock that consists primarily of carbonate minerals. The dominant mineral is nearly always either calcite, in which case the rock is *limestone,* or dolomite, in which case the rock is *dolomite.*

Carbonate sediment Unconsolidated sediment that consists primarily of carbonate minerals, usually aragonite or calcite.

Carbonization The mode of *fossilization* in which liquids and gases escape, leaving a residue of carbon on the surface of an *impression* of the organism.

Carnivore An animal that feeds on other animals or animal-like organisms.

Cataclastic rock Metamorphic rocks that resemble *breccias* or poorly sorted sandstones and that form by dynamic metamorphism, which breaks and reorients grains.

Catastrophism The doctrine that sudden, violent, and widespread events caused by supernatural forces formed most of the rocks that are visible at the earth's surface.

Cementation The *lithification* of sediment by the precipitation of minerals from watery solutions percolating through the sediment.

Chemical sediment A sediment created by precipitation of one or more minerals from natural waters.

Chemosynthesis The breakdown of simple chemical compounds within a cell for the production of energy. *Sulfate-reducing bacteria* exemplify this process.

Chert An impure rock, often gray in color, that consists primarily of extremely small quartz crystals precipitated from water solutions.

Chloroplast A body within a plant cell or plantlike cell that serves as the site of photosynthesis within the cell. Chloroplasts are apparently evolutionary descendants of blue-green algae that became trapped in other single-celled organisms.

Chromosome One of several elongate bodies in which *DNA* is concentrated within the nucleus of a cell.

Circumpolar current The circular flow of water around Antarctica, resulting from the juncture of the *west-wind drifts* of the Atlantic, Pacific, and Indian Oceans.

Clast A solid product of *erosion.* Clasts are sometimes referred to as *detritus* or detrital material.

Clastic rock A rock that is an aggregate of detrital material, or *clasts.*

Clastic wedge A wedge-shaped body of *molasse.*

Clay A member of the clay mineral family, which includes silicates that resemble micas.

Clay-sized sediment Sediment in which particles are smaller than $\frac{1}{256}$ millimeter. Most sediment of this size belongs to the *clay* mineral family.

Claystone A sedimentary rock that consists primarily of *clay* but that is not *fissile* like *shale.*

Coal A metamorphic rock formed from stratified plant remains. It contains more than 50 percent carbon and burns readily.

Coal measure The British term for a *cyclothem.*

Cobble A piece of *gravel* between 8 and 256 millimeters in size.

Coccolith One of the many armorlike plates that surround single-celled marine algae known as coccolithophores.

Community, ecologic A group of populations belonging to several species and living in the same habitat.

Compaction, of sediment The process in which grains of sediment are squeezed together beneath the weight of overlying sediment.

Competition, ecologic The condition in which two species vie for an environmental resource, such as food or space, that is in limited supply.

Components, principle of The principle stating that a body of rock is younger than any other body of rock from which any of its components are derived.

Concretion A hard, nodular structure formed in sediment or in a sedimentary rock by *diagenesis.*

Conglomerate A rock consisting of rounded *clasts* of *gravel* surrounded by sand.

Consumer, ecologic Animals or animal-like organisms, which feed on other organisms.

Contact metamorphism Local metamorphism caused by igneous intrusion that "bakes" nearby rocks.

Continental accretion The marginal growth of a continent along a *subduction zone* by mountain building or by addition of a *microplate.*

Continental drift The movement of continents with respect to one another over the surface of the earth.

Continental margin The submarine edge of the *continental shelf,* from which the *continental slope* descends. In the modern world this margin generally lies about 200 meters (~600 feet) below sea level.

Continental rise A more gently sloping region along the base of the *conti-nental slope.* The continental rise is a depositional feature, formed of sediment transported down the slope, often by *turbidity currents.*

Continental shelf A continental margin flooded by the sea.

Continental slope The sloping submarine portion of a continent, extending from the *continental margin* to the *continental rise* or the *abyssal plain.*

Convection Rotational flow of a fluid resulting from imbalances in density. This often occurs because the fluid below is heated and becomes less dense than the fluid above or because the fluid above is cooled and becomes more dense than the fluid below.

Convective cell One of a number of rotational units believed to operate within the earth's *mantle* as a result of *convection.*

Convergence, evolutionary The evolution of similar features in two or more different biological groups, or *taxa.*

Cope's rule The tendency for body size to increase during the evolution of a group of animals.

Core of the earth The central part of the earth below a depth of 2900 kilometers. It is thought to be composed largely of iron and to be molten on the outside with a solid central region.

Coring The process of inserting a tube into sediments or rocks and then extracting the tube along with a core, or plug, of material for study.

Coriolis effect The tendency of a current of air or water flowing over the surface of the earth to bend to the right in the Northern Hemisphere and to the left in the Southern Hemisphere.

Correlation The use of fossils to establish that spatially separated stratigraphic sections are the same geologic age.

Craton The portion of a continent that has not experienced *tectonic* deformation since Precambrian or early Paleozoic time.

Cross-bedding *Cross-stratification* in

which the individual strata exceed 1 centimeter in thickness.

Cross-cutting relationships, principle of The principle stating that when one fault is offset by another, the second is younger than the first.

Cross-lamination *Cross-stratification* in which the individual strata are thinner than 1 centimeter.

Cross-stratification A sedimentary structure in which groups of *strata* lie at angles to the horizontal.

Crust The outermost layer of the *lithosphere,* consisting of *felsic* and *mafic* rocks less dense than the rocks of the *mantle* below.

Crystalline rocks *Igneous* or *metamorphic rocks.*

Cycle, sedimentary A composite sedimentary unit which is repeated many times in succession within a given region. The unit includes two or more characteristic beds or groups of beds arranged in a characteristic vertical sequence that often reflects *Walther's law.*

Cyclothem Sedimentary *cycles* that include coal beds. Most cyclothems are of Late Carboniferous (Pennsylvanian) age.

D

Declination, magnetic The angle that a compass needle makes with the line running to the geographic North Pole, reflecting the fact that the magnetic pole (to which the compass needle points) does not coincide with the geographic pole.

Decollement The basal surface along which thrust sheets slide in the *fold-and-thrust* belt of a mountain system.

Deep-focus earthquake An earthquake produced by earth movements along or within a subducted slab of *lithosphere* more than 300 kilometers (~190 miles) below the surface of the earth.

Deep-sea floor The *continental slope* and *abyssal plain.*

Degassing The loss of gases by the earth early in its history, when it became liquefied.

Delta A depositional body of sand, silt, and clay formed when a river discharges into a body of standing water so that its current dissipates and drops its load of sediment. This structure takes its name from the Greek character △, which it resembles in shape.

Delta, river-dominated A *delta* that projects far out into the ocean because its construction from river-borne sediment prevails over the destructive forces of the sea.

Delta, wave-dominated A *delta* situated along a coast where wave activity is so strong that it prevents an *active lobe* of the delta from growing out into the ocean.

Delta front The submarine slope of a *delta* extending downward from the delta plain. The delta front is usually the site of accumulation of silt and clay.

Delta plain The upper surface of a *delta,* characterized by *distributary channels* and their levees and intervening swamps.

Deposit feeder An aquatic animal that feeds by consuming sediment, from which it digests organic matter.

Desert A terrestrial environment that receives less than about 25 centimeters (10 inches) of rain per year and consequently supports only a few kinds of plants.

Detritus, sedimentary Loose material produced by *erosion.*

Diagenesis The set of processes, including solution, that alter sediments at low temperatures after burial.

Dike A sheetlike or tabular body of igneous rock that cuts through sedimentary layers or crystalline rocks.

Dip The angle that a tilted bed or fault forms with the horizontal.

Disconformity An *unconformity* above rocks that underwent erosion before the beds above the unconformity were deposited. The strata above and below a disconformity are horizontal.

Distributary channel Channels on a *delta plain* that radiate out from the mainland, carrying river water to the ocean in several directions.

Diversity, biotic The number of kinds of organisms living in an ecosystem.

DNA Deoxyribonucleic acid, the "double helix" molecule that carries chemically coded genetic information and is passed from generation to generation.

Dollo's law The rule that any substantial evolutionary change is virtually irreversible because genetic changes are not likely to be reversed in an order exactly opposite to the order in which they originally developed.

Dolomite A mineral that consists of calcium magnesium carbonate, with calcium and magnesium present in nearly equal proportions, or the sedimentary rock that consists largely of this mineral.

Dome, structural A blisterlike uplift of stratified rocks.

Doppler effect A shift toward longer wavelengths for waves reaching an observer when the source of the waves is moving away from the observer.

Dropstone A stone dropped to the bottom of a lake or ocean from a melting body of ice afloat on the surface.

Dune, sand A hill of sand piled up by the wind. The sand within a dune is characterized by *trough cross-bedding.*

Dynamic metamorphism Metamorphism that entails the shattering of rocks and the deformation or reorientation of their grains.

E

Ecology The branch of biology concerned with the factors that govern the distribution and abundance of organisms in natural environments.

Ecosystem The organisms of an ecologic *community* together with the physical environment that they occupy.

Ectothermic "Cold-blooded": the physiological condition in which an animal's body temperature is controlled by the external environment.

Embryo sac The female gamete-bearing generation of a seed-bearing plant. This generation lives like a parasite on the spore-bearing generation.

Endothermic "Warm-blooded": the physiological condition in which an animal's body temperature is controlled internally and maintained at a more-or-less constant level.

Energy gradient, reef Dissipation, due to friction, of the energy of motion of water as it passes over a *reef flat*.

Eon The largest formal unit of geologic time. There are three eons: the Archean, Proterozoic, and Phanerozoic.

Epicontinental sea A shallow sea formed when ocean waters flood an area of a continent far from the continental margin.

Epoch, geologic A division of geologic time shorter than a *period*.

Equatorial countercurrent The eastward-flowing global ocean current that carries the water that has been piled up by the *equatorial current*.

Equatorial current The global ocean current pushed westward along the equator by the trade winds.

Era, geologic A division of geologic time shorter than an *eon* but including two or more *periods*.

Erathem A time-stratigraphic unit consisting of all the rocks that represent a geologic *era*.

Erosion The group of processes that loosen rock and move pieces of loosened rock downhill.

Erratic boulder A boulder that a *glacier* has transported from its place of origin to a distant location.

Eukaryote (sometimes spelled **Eucaryote**) An organism that consists of one or more *eukaryotic cells*. All organisms except bacteria and blue-green algae are of this type.

Eukaryotic cell (sometimes spelled **Eucaryotic**) A cell characterized by a nucleus with *chromosomes, mitochondria*, and other complex internal structures. This is the kind of cell that forms higher organisms (all organisms but bacteria and blue-green algae).

Eustatic sea-level change A sea-level change that is worldwide and affects all oceans.

Evaporite A mineral or rock formed by precipitation of crystals from evaporating water.

Evergreen coniferous forest A high-latitude forest, often growing adjacent to *tundra* and always dominated by coniferous trees such as spruce, pine, or fir.

Exposure (See *Outcrop*)

Exterior drainage A drainage pattern in which lakes and rivers carry runoff from a region to beyond the borders of that region.

Extinction The total disappearance of a species or higher taxon.

Extrusive igneous rock An igneous rock that has been erupted onto the surface of the earth.

F

Facies The set of characteristics of a rock that represents a particular local environment.

Fault A surface along which rocks have broken and moved past one another.

Fauna A collective term for all the animals of an *ecosystem*.

Felsic rock A silicon-rich igneous rock that contains only a small percentage of iron and magnesium. Granite is the most abundant example. Felsic rocks dominate the crust of continents.

Fissile Having the property of fissility, or tending to break along *bedding surfaces*. This is a property of some sedimentary rocks, especially *shales*.

Fission-track dating The dating of a rock according to the number of fission tracks produced by the decay of uranium 238. In the process of decaying, uranium 238 atoms eject subatomic particles that leave microscopic tracks in the surrounding rock.

Fissure A crack in a body of rock, often filled with minerals or *intrusive igneous rock*.

Flood basalt *Extrusive rocks* of *mafic* composition that have flowed widely over the earth's surface.

Flora A collective term for all the plants of an *ecosystem*.

Flysch Black shales and *turbidites* that accumulate in deep water within a *foredeep* bordering an active mountain system.

Focus, earthquake The point within the earth at which a rupture occurs, causing an earthquake.

Fold-and-thrust belt The tectonic zone of a mountain chain characterized by folds and *thrust faults* and positioned adjacent to the *metamorphic belt* and farther away than the metamorphic belt from the igneous core of the mountain chain.

Foliation The alignment of platy minerals in metamorphic rocks, caused by the high pressures applied during metamorphism.

Food chain The sequence of nutritional steps in an ecosystem, with *producers* at the bottom and *consumers* at the top.

Food web The nutritional structure of an ecosystem in which more than one species occupies each level. Thus, there are usually several *producer* species and several *consumer* species in a food web.

Foredeep An elongate basin lying in front of an active mountain system and receiving sediment (primarily *flysch* and *molasse*) from the mountains.

Formation The fundamental *rock-stratigraphic unit*. A body of rock characterized by a particular set of *lithologic* features and given a formal name.

Fossil The remains or tangible trace of an ancient organism preserved in sediment or rock.

Fossil fuel Condensed and altered organic matter that can be burned to supply energy for human use. Examples are coal, petroleum, and natural gas.

Fossil succession The vertical ordering of fossil *taxa* in the geologic record, reflecting the operation of evolution and extinction.

Fossilization The formation of a *fossil*.

Fringing reef   An elongate organic *reef* that fringes a coastline and has no lagoon on its landward side.

G

Gabbro   A *mafic* igneous rock; the coarse-grained, *intrusive* equivalent of *basalt*.

Gamete   A sex cell (egg or sperm) which carries half the normal complement of *chromosomes* and combines with another sex cell to produce a new individual possessing the normal complement.

Gene   A unit of inheritance consisting of a segment of *DNA* that performs a particular function. Many genes provide coded information for the synthesis of particular *polypeptides*.

Gene, structural   A *gene* that codes for the formation of a segment of a protein molecule.

Gene pool   The sum total of the genetic components of a population.

Geobiology   (See *Paleobiology*)

Geochemistry   The study of the ways in which chemical reactions produce rocks and alter them.

Geophysics   The study of the physics of the earth, including movements of lithospheric *plates*, motions of rock deep within the earth, and the structure of the earth's *magnetic field*.

Glacier   A large mass of ice that creeps over the earth's surface.

Gneiss   A coarse-grained metamorphic rock resembling an igneous rock but with its component minerals segregated into wavy layers.

Graben   A fault-block basin produced by the depression of a block of crust bounded by *normal faults*.

Grade, metamorphic   A classification system based on the level of temperature and pressure responsible for metamorphism. Metamorphic rocks are divided into high-grade, medium-grade, and low-grade categories.

Granite   A coarse-grained *felsic* igneous rock consisting of feldspar and quartz with minor *mafic* components. The

most common *intrusive* rock of continental crust.

Granule   A piece of *gravel* of small size (between 2 and 4 millimeters).

Granulite   A high-*grade* metamorphic rock that nearly became molten during metamorphism and as a result resembles an igneous rock in being granular and lacking foliation.

Grassland   (See *Savannah*)

Gravel   Sediment larger than sand (larger than 2 millimeters).

Graywacke   A *siliciclastic* rock that is poorly sorted but consists primarily of sand-sized grains. Some of the grains are dark, giving the rock its characteristic dark-gray color.

Grazer   A *benthic herbivore* that feeds on plantlike forms, especially algae that grow on hard surfaces.

Greenstone belts   Podlike bodies of rock characteristic of Archean terranes. They consist of volcanic rocks and associated sediments that have commonly been metamorphosed so that they have a greenish color.

Group   A *rock-stratigraphic unit* of higher rank than a *formation*.

Guide fossil   (See *Index fossil*)

Guyot   A flat-topped volcanic seamount in the deep sea. It appears that guyots form in shallow water when wave action truncates the upper part of a volcano and that they are transported to deeper water by lateral *plate* movement.

Gypsum   A mineral that consists of calcium sulfate with water molecules attached ($CaSO_4 \cdot H_2O$), or the rock that consists primarily of this mineral.

H

Habitat   An environment that supports life.

Half-life   The time required for a particular radioactive *isotope* to decay to half its original amount. This time is consistant for any isotope, regardless of the amount of the isotope present at the outset.

Halite   The mineral that consists of so-

dium chloride (NaCl), popularly known as rock salt, or the rock that consists primarily of this mineral.

Herb   A small, nonwoody terrestrial plant that dies back to the ground before releasing its seeds.

Herbaceous plant   (See *Herb*)

Herbivore   An animal that feeds on plants or plantlike organisms.

Heterogeneous accretion   A possible mode of origin of the earth in which materials would have condensed in a sequence determined by their density, causing concentric layers to form with the most dense materials at the center.

Homogeneous accretion   A possible mode of origin of the earth in which materials of many different densities would have aggregated haphazardly, without becoming layered according to density.

Homology   The presence, in two different animal or plant groups, of organs that have the same ancestral origin but serve different functions.

Horizon, soil   A layer of soil. The A-horizon, or topsoil, consists mainly of sand and clay mixed with *humus*. The B-horizon underlies the A-horizon and contains less humus. The C-horizon underlies the B-horizon and consists of slightly altered and broken bedrock mixed with some sand and clay.

Hornfels   A fine-grained, granular metamorphic rock of varying composition formed by *contact metamorphism* under conditions of high temperature and low pressure.

Hot spot   A small area of heating and igneous activity in the earth's crust where a *thermal plume* rises from the mantle. The Hawaiian Islands represent a hot spot.

Humus   Organic matter in soils, formed largely by the decay of leaves, woody tissues, and other plant materials.

Hydrological cycle, global   The continuous passage of water from the atmosphere to the land and to bodies of water, and back again to the atmosphere.

Hypersaline water   Water that is higher in salinity than normal sea water (contains more than 40 parts salt per 1000 parts water).

Hypsometric curve A graph that displays the proportions of the earth's surface that lie at various altitudes above and various depths below sea level.

I

Igneous rock A rock formed by the cooling of molten material.

Impression The mode of *fossilization* in which the flattened imprint of an organism forms on a *bedding surface.*

Index fossil A species or genus of fossils that provides for especially precise *correlation.* An ideal index fossil is easily distinguished from other taxa, is geographically widespread, is common in many kinds of sedimentary rocks, and is restricted to a narrow stratigraphic interval.

Industrial melanism An increase in the incidence of dark-colored moths in industrial areas, where the trees on which the moths rest became coated with soot. This is a phenomenon that resembles natural selection but results from human alteration of the environment.

Interior drainage A drainage pattern in which lakes and rivers fail to carry runoff from a region to beyond the borders of the drainage area. The rainfall is so light that streams and rivers are temporary, drying up at intervals.

Intertidal zone The belt that is alternately exposed and flooded as the *tide* ebbs and flows along a coast.

Intrusion A body of igneous rock that formed within the earth rather than at the earth's surface.

Intrusive igneous rock A rock formed by the cooling of *magma* within the earth.

Intrusive relationships, principle of The principle stating that an *intrusive igneous rock* is always younger than the rock that it invades.

Iron formation Complex sedimentary or weakly metamorphosed rock that usually consists of oxides, sulfides, or carbonates of iron, often in association with chert.

Iron meteorite A *meteorite* in which iron is the primary component.

Island arc An arcuate chain of islands produced by volcanism at a site where magma rises through the *lithosphere* from a *subducted* plate.

Isostatic adjustment The mechanism whereby areas of the earth's crust rise or subside to keep the *crust* in gravitational equilibrium as it floats on the *mantle.* Thus, a mountain is balanced by a root of crustal material.

Isotope One of two or more varieties of an element that differ in number of neutrons within the atomic nucleus.

Iterative evolution The evolution of the same general type of organism more than once from a single ancestral group.

K

Key bed A sedimentary bed, such as a volcanic ash bed, which is of nearly the same age everywhere and is thus useful for *correlation.*

L

Lagoon A ponded body of water along a marine coastline, usually landward of a barrier island or organic reef.

Lamination A distinct sedimentary layer *(stratum)* thinner than 1 centimeter.

Late Neolithic The tool culture of early modern humans, the Cro-Magnon people of Europe.

Lava Molten rock *(magma)* that has reached the surface of the earth.

Life habit The mode of life of an organism, or the way in which it functions within its niche—how it obtains nutrients or food, reproduces, and stations itself or moves about within its environment.

Limestone A sedimentary rock that consists primarily of calcium carbonate. Limestones may be *biogenic* or chemical in origin.

Limiting factor, ecologic An environmental condition, such as temperature, that restricts the distribution of a species in nature.

Lithification The consolidation of loose *sediment* by compaction, precipitation of mineral cement, or a combination of these processes to form a sedimentary rock.

Lithology The physical and chemical characteristics of a rock.

Lithosphere The outer rigid shell of the earth, situated above the *asthenosphere* and consisting of the *crust* and upper *mantle.* The lithosphere is divided into *plates.*

Loess The wind-blown silt that accumulates in the frigid desert in front of a *glacier.* It is deposited by the *meltwaters* of the glacier.

Longshore current An ocean current that flows along a coast, often sweeping sand in a direction parallel to the coastline.

Longshore drift The zigzag movement of sand grains along a *coast* as a result of *waves* striking the shore diagonally. Particles of sand are carried diagonally shoreward, but gravity causes them to slide back toward the sea, this time in a direction perpendicular to the coast. The next wave moves them diagonally shoreward again, and so on.

M

Mafic rock A dense, dark-colored igneous rock that is relatively poor in silicon and rich in iron and magnesium. Basalt, the characteristic igneous rock of oceanic *crust,* is an example.

Magma Naturally occurring molten rock.

Magnetic field, earth's The field of magnetism that results from motions of the iron-rich outer core of the earth; these motions cause the earth to behave like a giant bar magnet, with a north and south pole.

Magnetic reversal A reversal of the polarity of the earth's magnetic field, as recorded in rocks magnetized by the field.

Mantle The zone of the earth's interior between the *core* and the *crust,* ranging from depths of approximately 40 to 2900 kilometers. It is composed of dense ultramafic silicates and divided into concentric layers.

Marble A homogeneous, granular metamorphic rock that consists of calcite, dolomite, or a mixture of the two and forms by the metamorphism of *limestone* or *dolomite.*

Maria (singular, **Mare**) The large craters on the surface of the moon.

Marker bed   (See *Key bed*)

Marsh, intertidal   Habitats along the seashore that are dominated by low-growing plants and alternately flooded by the tide and exposed to the air. The remains of the plants usually accumulate to form *peat*.

Mass extinction   An episode of large-scale extinction in which large numbers of species disappear in a few million years or less.

Massif   A large, durable remnant of a mountain system or ancient body of *crystalline rocks*, usually standing above the surrounding terrane.

Meandering river   A river that winds back and forth like a ribbon, depositing sediment on the inside of each curve and eroding sediment on the outside.

Mediterranean climate   A climate characterized by dry summers and wet winters, often found along coasts lying about 40° from the equator. Much of California and much of the Mediterranean region of Europe have this kind of climate.

Mélange   A chaotic, deformed mixture of rocks such as often forms where subduction occurs along a deep-sea trench.

Meltwaters   The waters that issue from the front of a melting *glacier*.

Member   A *rock-stratigraphic unit* of lower rank than a *formation*.

Metamorphic belt   The metamorphic zone parallel to the long axis of a mountain chain and near the igneous core of the mountain chain. This is a zone of *regional metamorphism*.

Metamorphic rock   A rock formed by *metamorphism*.

Metamorphism   The alteration of rocks within the earth under conditions of high temperature and pressure.

Meteorite   An extraterrestrial object that crashed into the earth after being captured by the earth's gravitational field.

Microfossils   Fossils so small that they must be studied under a microscope.

Micropaleontology   The study of *microfossils*.

Microplate   A small lithospheric plate, usually of predominantly *felsic* composition.

Midocean ridge   The ridge on the ocean floor where oceanic *crust* forms and from which it moves laterally in each direction.

Mineral   A naturally occurring inorganic solid element or compound with a particular chemical composition or range of compositions and a characteristic internal structure.

Mitochondrion (plural, **Mitochondria**)   A body within a *eukaryotic cell* in which complex compounds are broken down by oxidation to yield energy and, as a by-product, carbon dioxide. The mitochondrion is apparently an evolutionary descendant of a small bacterium that became trapped within a larger one.

Moho   (See *Mohorovičić discontinuity*)

Mohorovičić discontinuity   The boundary between the crust and mantle, marked by a rapid increase in *seismic wave* velocity.

Molasse   Nonmarine and shallow marine sediments—representing such environments as alluvial fans, river systems, and barrier island–lagoon complexes—that accumulate in front of a mountain system after heavy sedimentation from the mountains has driven deep marine waters from the *foredeep* there.

Mold   A *fossil* that consists of a three-dimensional imprint of an organism or part of an organism.

Moraine, glacial   A ridge of till ploughed up in front of a glacier.

Mousterian culture   The tool culture of Neanderthal, a Late Pleistocene human or humanoid creature.

Mud   An aggregate consisting of *silt*- or *clay-sized sediment* or a combination of the two.

Mutation   Chemical changes in genetic features. These changes provide much of the variability on which *natural selection* operates.

N

Natural levee   A gentle ridge bordering a *meandering river* or a *distributary* of a *delta* and composed of sand and silt deposited by the river or distributary when it overflows its banks during floods.

Natural selection   The process recognized by Charles Darwin as the primary mechanism of evolution. The selection process, which operates on heritable variability, results from differences among individuals in longevity and in rate of production of offspring.

Nebula   The dense cloud of matter that remains after a *supernova* explodes.

Niche, ecologic   The ecologic position of a species in its environment, including its requirements for certain kinds of food and physical and chemical conditions and its interactions with other species.

Nodule   An irregular rounded mass of *chert* or some other mineral imbedded in a sedimentary rock. Most nodules formed by diagenetic processes where they are found.

Nonconformity   An *unconformity* separating bedded rocks, above, from *crystalline rocks*, below.

Normal fault   A fault whose dip is steeper than 45° and along which the rocks above have moved downward relative to the rocks below.

O

Oceanic realm   The portion of the ocean that lies above the *deep-sea floor*.

Oil shale   A well-laminated shale in which dark, organic-rich layers alternate with thicker, lighter-colored layers. Petroleum can be removed from such rock, but not at a cost that is at present economically profitable.

Ontogeny   The sequence of development of an individual organism, from its origin by the fertilization of an egg to its death.

Ooid   A nearly spherical sand-sized grain that forms on the sea floor when water movement causes a small particle (or nucleus) to roll around and accumulate needles of aragonite in concentric layers.

Oolite   A clean sand consisting primarily of *ooids*.

Ooze, deep-sea   Fine-grained sediment in the deep sea that consists of carbonate or siliceous skeletons of dead *planktonic* organisms.

Ophiolite   A segment of sea floor that is elevated so as to rest on continental crust. An ophiolite usually includes *turbidites*, black shales, cherts, and *pillow basalts* along with *ultramafic rocks* from the mantle.

Opportunistic species   A species that specializes in the rapid invasion of newly vacated habitats, where there is little competition from other species.

Original horizontality, principle of   The principle enunciated in the seventeenth century which states that all strata are horizontal when they form. (A more accurate statement would be that almost all strata are initially more nearly horizontal than vertical.)

Original lateral continuity, principle of   The principle that states that similar strata found on opposite sides of a valley or some other erosional feature were originally connected.

Orogenesis   The process of mountain building.

Orogenic belt   A belt of mountain building.

Orogeny   An episode of mountain building.

Outcrop   A portion of a body of rock that is visible at the earth's surface. (Some geologists restrict this term to rocks laid bare by natural processes and apply the term exposure to artificially exposed areas of rock.)

Outwash, glacial   Well-stratified glacial sediment deposited by a stream of meltwater issuing from a melting glacier.

Overturned fold   A fold in which at least one limb has been rotated more than 90° from its original position.

Oxide   A mineral in which metallic atoms are bonded to oxygen atoms.

Oxygen cycle, global   The continuous flow of oxygen through the atmosphere. It is supplied to the atmosphere by the breakdown of H_2O in the upper atmosphere and by photosynthesis and is removed from the atmosphere by organic

respiration and by the oxidation of various materials.

Oxygen sink   A reservoir consisting of a chemical element or compound that combines readily with oxygen and thus removes it from the *atmosphere*. During the early part of Precambrian time, sulfur, iron, and other elements and compounds served as important oxygen sinks, preventing oxygen from accumulating in the atmosphere.

P

Paleobiology   The study of biological processes of the past by analysis of the *fossil* record, sedimentary rocks, and biological facts and principles.

Paleogeography   The geography of the *geologic* past.

Paleomagnetism   The magnetism of a rock, developed from the earth's magnetic field when the rock formed.

Paleontology   The study of ancient life, based largely on the nature and occurrence of *fossils*. Three important subdivisions are the study of ancient invertebrate animals, vertebrate animals, and plants, known respectively as invertebrate paleontology, vertebrate paleontology, and paleobotany.

Parasite   An organism that derives its nutrition from other organisms without killing them.

Particulate inheritance   The presence of hereditary factors called *genes* that retain their identity while being passed on from parent to offspring.

Patch reef   A small mound or pinnacle-like reef growing in the lagoon behind a barrier reef.

Peat   Sediment composed primarily of stratified plant debris that has not metamorphosed to form *coal*.

Pebble   A piece of *gravel* between 4 and 8 millimeters in size.

Pellet, fecal   An oval grain formed of fine sediment that has passed through the gut of an animal, usually a deposit feeder.

Peneplain   Literally, an "almost plain"—

a regional erosion surface that is almost flat.

Period, geologic   The most commonly used unit of geologic time, representing a subdivision of an *era*.

Permineralization   The mode of *fossilization* in which porous spaces within part of an organism (such as bony or woody tissue) become filled with mineral material.

Petrology   The study of the compositions and origins of rocks.

Photic zone   The upper layer of the ocean where enough light penetrates the water to permit *photosynthesis*.

Photosynthesis   The process in which plants and single-celled plantlike organisms employ the compound chlorophyll to convert carbon dioxide and water from their environment into energy-rich sugar, which fuels essential chemical reactions.

Photosynthetic bacteria   Bacteria that employ photosynthesis, but with hydrogen sulfide taking the place of water in the production of sugar. Such bacteria are poisoned by oxygen and probably flourished in Archean time when the atmosphere lacked free oxygen.

Phylogeny   A segment of the tree of life that includes two or more evolutionary branches.

Phytoplankton   *Plankton*, or floating aquatic life, that is photosynthetic. Most phytoplankton species are single-celled algae.

Pillow basalt   Basalt with a hummocky surface formed by rapid cooling of *lava* beneath water.

Pinnacle reef   (See *Patch reef*)

Plankton   Life that floats in the ocean or in lake waters.

Plate   A segment of the *lithosphere* that moves independently over the earth's interior.

Plate tectonics   The study of the movements and interactions of lithospheric *plates*.

Playa lake   A temporary lake in a region of interior drainage. When such a lake dries out, *evaporite* deposits form.

Plunging fold A fold whose axis plunges (lies at an angle to the horizontal) so that the beds of the fold have a curved outcrop pattern if truncated by erosion.

Pluton An *intrusion*, or massive body of *intrusive igneous rock.*

Point bar An accretionary body of sand on the inside of a bend of a *meandering river.*

Pollen A tiny grainlike structure that is the male gamete-bearing generation of a seed-bearing plant. Pollen yields *gametes* for fertilization of eggs within the *embryo sac* (the female gamete-bearing generation).

Polypeptide An organic molecule containing a large number of *amino acids.* A *protein* is a large polypeptide.

Population A group of individuals that live in the same area and interbreed.

Precambrian shield The Precambrian portion of a *craton* exposed at the earth's surface.

Precession of the equinoxes A slight wobble in the axis of rotation of the earth caused by the gravitational pull of the sun and moon.

Predation The eating of one animal by another.

Prodelta The gently seaward-sloping area of a *delta front* where *clay* accumulates in deep water.

Producer, ecologic Plant or plantlike organism, which manufactures its own food.

Prograde To grow seaward by the accumulation of sediment or sedimentary rocks. *Deltas* often prograde, as do organic *reefs.* Progradation produces *regression*, or seaward migration, of the shoreline.

Prokaryote (sometimes spelled **Procaryote**) An organism that consists of a *prokaryotic cell:* A member of the kingdom Monera, which includes bacteria and blue-green algae.

Prokaryotic cell (sometimes spelled

Procaryotic) A primitive cell in which there are no *chromosomes* and no nuclei, *mitochondria*, or certain other internal structures characteristic of the cells of higher organisms.

Protein (See *Polypeptide*)

Pseudoextinction The disappearance of a species not by dying out but by evolving to the point that it is recognized as a different species.

Psychrosphere The deepest zone of the modern ocean, characterized by waters that are nearly freezing. Convection causes the cold waters in polar regions to descend to the deep sea and form the psychrosphere.

Q

Quartzite A metamorphic rock formed by the metamorphism of quartz sandstone and consisting of almost pure quartz.

R

Radioactive decay The spontaneous breakdown of certain kinds of atomic nuclei into one or more nuclei of different elements, with a release of energy and subatomic particles.

Radiocarbon dating Radiometric dating using carbon 14, a radioactive isotope with a *half-life* so short that its decay can be used to date materials younger than about 70,000 years.

Radiometric dating The use of naturally occurring radioactive materials to date rocks by measuring the amounts of these materials relative to their daughter products (products of radioactive decay) in rocks.

Rain forest, tropical A jungle that develops in an equatorial region where heavy, regular rainfall results from the cooling of air that has ascended after being warmed and picking up moisture near the earth's surface.

Rain shadow A region located on the downwind side of a mountain and receiving little rain because the winds rise as they pass over the mountain, cooling and dropping most of their moisture before they reach the other side.

Recumbent fold A fold that rests on

one of its limbs, with the axial plane nearly horizontal.

Red beds Sediments of any grain size that are reddish in color, usually because of the presence of iron oxide cement.

Reef, organic A solid but porous limestone structure standing above the surrounding sea floor and constructed by living organisms, some of which contribute skeletal material to the reef framework.

Reef flat The flat upper surface of a *reef*, usually standing close to sea level (often intertidal).

Regional metamorphism The creation of metamorphic rocks over areas whose dimensions are measured in hundreds of kilometers. Usually this happens in association with mountain building.

Regional strike In a deformed terrane, the prevailing orientation of fold axes or of the lines of outcrop of tilted beds.

Regression A seaward migration of a marine shoreline and of nearby environments.

Relict distribution The localized occurrence of a taxonomic group after it has died out throughout most of the geographic area that it previously occupied.

Remobilization Regional metamorphism and deformation that affect a segment of crust previously altered by similar processes.

Reverse fault A fault above which rocks have moved uphill relative to rocks below.

Rhyolite A fine-grained, *felsic* extrusive igneous rock equivalent in mineralogical composition to granite.

Rift A juncture between two plates where *lithosphere* forms and the plates diverge.

Ripples Small dunelike structures formed on the surface of sediment by moving water or wind.

Rock An aggregate of interlocking or attached grains, each of which is typically composed of a single *mineral.*

Rock flour Mud-sized material ground from coarse grains by the movement of a *glacier.*

Rock-stratigraphic unit A body of characteristic lithology that is formally recognized as a *formation. member. group*, or *supergroup*.

Root, mountain The body of crustal rock that extends into the more dense mantle beneath a large mountain and balances the mountain—that is. keeps it in *isostatic* equilibrium.

S

Sabka A flat *supratidal zone* in which the sediments are heavily laden with evaporites. Sabkas form in arid climates.

Salinity The saltiness of natural water. The salinity of normal seawater is 35 parts salt per 1000 parts water.

Sand Sediment ranging in size from $\frac{1}{16}$ to 2 millimeters.

Sandstone A *siliciclastic* sedimentary rock consisting primarily of sand. usually sand that is predominantly quartz.

Savannah A broad grassland. which typically forms where there is enough rainfall to sustain grass but not enough to sustain the trees that form woodlands or forests.

Scavenger An organism that feeds on other organisms after they are dead.

Schist A metamorphic rock that consists mainly of platy minerals such as micas that lie nearly parallel to one another so that the rock tends to break along parallel surfaces.

Section, stratigraphic A local outcrop or series of adjacent outcrops displaying a vertical sequence of layered rocks.

Sediment Material deposited at the earth's surface by water. ice. or air.

Sedimentary rock A rock formed by the consolidation of loose *sediment* or by precipitation from a watery solution.

Sedimentary structure A distinctive bedding pattern in a sedimentary rock.

Sedimentology The study of the compositions and origins of *sedimentary rocks.*

Seed A reproductive structure of a plant—produced by the union of *gametes* and then released from the plant—that has the potential to grow into a new plant. The seed is actually a juvenile stage of the *spore*-bearing generation of the plant.

Seismic wave An oscillatory movement of the earth. caused naturally or by artificial means such as the setting off of an explosion.

Series A time-stratigraphic unit consisting of all the rocks that represent a geologic *epoch.*

Sexual recombination The mixing of *chromosomes* from generation to generation that continually creates new genetic combinations and. hence. new kinds of individuals upon which *natural selection* can operate.

Shale A *fissile* sedimentary rock consisting primarily of clay.

Shield volcano A low cone-shaped volcano of the type that often forms in oceanic areas. where lavas tend to be *mafic* and hence of low viscosity. so that they spread rapidly when they erupt.

Silicates The mineral group that includes the most abundant minerals in the earth's crust and mantle. The basic building block of silicates is a tetrahedral structure consisting of four oxygen atoms surrounding a silicon atom.

Siliciclastic sediment Detrital sediment consisting of *silicate* minerals. This is the most abundant kind of sediment on earth.

Sill A sheetlike or tabular body of intrusive igneous rock injected between sedimentary layers.

Silt Sediment in which articles range in size from $\frac{1}{256}$ to $\frac{1}{16}$ millimeter.

Siltstone A *siliciclastic* sedimentary rock consisting primarily of silt.

Slate A fine-grained metamorphic rock that is *fissile*. like the sedimentary rock *shale*, but whose fissility results from alignment of platy materials by deformational pressures rather than by depositional orientation of particles.

Soil Loose sediment that accumulates in contact with the atmosphere.

Solar nebula theory The theory that the solar system formed from a cloud of cosmic dust.

Speciation The origin of a new species from two or more individuals of a preexisting species.

Species A group of individuals that interbreed or have the potential to interbreed in nature and that are reproductively isolated from other such groups.

Spore A reproductive structure. not produced from *gametes*, that is released from a plant and has the potential to grow into a new plant.

Stabilization, orogenic Consolidation of the sedimentary rocks of an *orogenic belt* along the margin of a continent by *metamorphism*, folding. and *thrust faulting*.

Stage The time-stratigraphic unit ranking below a *series* and consisting of all the rocks that represent a geologic *age.*

Stony meteorite A *meteorite* of rocky composition.

Stony-iron meteorite A *meteorite* that consists of a mixture of metallic material (chiefly iron) and stony material.

Stratification The arrangement of a sedimentary rock into discrete layers (or *strata*).

Stratigraphy The study of the relationships of *strata* in time and space.

Stratum (plural, **Strata**) A distinct sedimentary layer.

Strike The compass direction that lies at right angles to the dip of a tilted bed or fault (that is. the compass direction of a horizontal line lying in the plane of a tilted bed or fault).

Strike-slip fault A high-angle fault along which the rocks on one side move horizontally relative to rocks on the other side. with a shearing motion.

Stromatolite An organically produced sedimentary structure that consists of alternating layers of organic-rich and organic-poor sediment. The organic-rich layers have usually been formed by sticky threadlike algae. which have trapped the sediment of the organic-poor layers.

Structural geology   The branch of geology concerned with the ways in which rocks bend, break, and flow under stress.

Subduction   Descent of a slab of *lithosphere* into the *asthenosphere* along a deep-sea *trench*.

Subduction zone   The region where *subduction* of the lithosphere occurs.

Substratum   The surface—sediment, rock, or another organism—on which or beneath which a *benthic* aquatic organism lives.

Subtidal   Positioned seaward of the *intertidal zone*.

Subzone, biostratigraphic   Subdivision of a *zone*.

Sulfate mineral   A mineral in which the basic building block is a sulfur atom linked to four oxygen atoms. Most sulfate minerals are highly soluble in water and form by the evaporation of natural waters.

Sulfate-reducing bacteria   Fermenting bacteria that obtain energy by converting sulfate compounds into sulfide compounds. These bacteria, which cannot tolerate oxygen, are common in the muds of swamps, ponds, and lagoons. Bacteria of this type seem to have flourished in Archean time, when there was little atmospheric oxygen.

Supergroup   A *rock-stratigraphic unit* of higher rank than a *group*.

Supernova   An exploding star, which casts off matter of low density.

Superposition, principle of   The principle which states that in an undisturbed sequence of strata, the oldest strata lie at the bottom and progressively younger strata are successively higher.

Supratidal zone   The belt along a coast just landward of the *intertidal zone* and flooded only occasionally, during storms or unusually high *tides*.

Surf zone   The zone along the beach where *waves* break.

Suspension feeder   An aquatic animal that feeds by straining *phytoplankton* or plant debris from water.

Suture   The juncture between two continents that have been united along a *subduction zone*.

Suturing   The unification of two continents along a *subduction zone*.

Syncline   A fold that is concave in an upward direction—that is, the vertex is the highest point.

System   A time-stratigraphic unit consisting of all the rocks representing a geologic *period*.

T

Talus, reef   The pile of rubble sloping seaward from the living surface of a *reef*.

Taxon (plural, **Taxa**)   A formally named group of related organisms of any rank, such as phylum, class, family, or species.

Taxonomy   The study of the composition and relationships of *taxa* of organisms.

Tectonic cycle   The characteristic cycle of deposition associated with an episode of mountain building. Shallow-water deposition along a continental shelf gives way to deep-water *flysch* deposition when a *foredeep* forms, and this, in turn, gives way to nonmarine *molasse* deposition when heavy sedimentation drives deep marine waters from the foredeep.

Tectonics   The segment of *structural geology* concerned with large-scale features such as mountains.

Temperate forest   A forest dominated by deciduous trees (trees that lose their leaves in winter). This kind of forest typically grows under slightly warmer climatic conditions than *evergreen coniferous forests*.

Thermal plume   A column of *magma* rising from the *mantle* through the *lithosphere*.

Thrust fault   A low-angle *reverse fault*. During many mountain-building episodes, large slices of rock *(thrust sheets)* move hundreds of kilometers over rigid *basement rocks*.

Thrust sheet   A large, tabular segment of the *crust* that moves along a thrust fault during mountain building.

Tidal delta   The body of sandy sediment that forms in a lagoon where a tidal inlet slows down, dropping some of its sediment.

Tidal flat   A surface where mud or sand accumulates in the *intertidal zone*.

Tide   Major movements of the ocean that result primarily from the gravitational attraction of the moon. Tides ebb and flow in particular regions as the earth rotates beneath a bulge of water created by the pull of the moon.

Till, glacial   Heterogeneous sediment ploughed up and deposited by a *glacier*.

Tillite   Lithified *till*.

Time-stratigraphic unit   A formally named body of rock representing a particular interval of time—an *erathem, system, series,* or *stage*.

Top carnivore   A carnivorous animal positioned at the top of a *food chain* or *food web*.

Topsoil   (See *Horizon, Soil*)

Trace fossil   A track, trail, or burrow left in the geologic record by a moving animal.

Trade winds   Winds that in each hemisphere blow diagonally westward toward the equator within about 30° latitude of the equator. They result from a zone of high pressure forming between about 20 and 30° from the equator, where air that has risen near the equator builds up.

Transform fault   A *strike-slip fault* along which two segments of lithosphere move relative to one another. Many transform faults offset *midocean ridges*.

Transgression   A landward migration of a marine shoreline and of nearby environments.

Transpiration   The emission of water by plants into the atmosphere, primarily through the leaves.

Trench, deep-sea   A deep, elongate trough along which a lithospheric *plate* is *subducted*.

Triple junction   A point where three lithospheric *plates* meet.

Tropical climate   A climate in which the average annual temperature is in the range of 18 to 20°C (64 to 68°F) or higher. Most tropical climates lie within about 30° of the equator.

Trough cross-stratification *Cross-stratification* of sediments in which one set of beds is truncated by erosion in such a way that the next set of beds laid down accumulates on a curved surface.

Tuft A rock deposited as a sediment but consisting of volcanic particles.

Tundra A terrestrial environment where air temperatures rise above freezing during the summer, but where a layer of soil beneath the surface remains frozen. Tundras are characterized by low-growing plants that require little moisture.

Turbidite A graded bed, often with poorly sorted sand at the base and mud at the top, formed when a *turbidity current* slows down.

Turbidity current A flow of dense, sediment-charged water that moves down a slope under the influence of gravity.

Type section A formally designated local stratigraphic section where a *rock-stratigraphic unit* such as a formation was originally defined.

U

Ultramafic rock A very dense rock that is even poorer in silicon and richer in iron and magnesium than a *mafic rock*. Ultramafic rocks characterize the earth's *mantle*.

Unconformity A surface between a group of sedimentary strata and the rocks beneath them, representing an interval of time during which erosion, rather than deposition, occurred.

Uniformitarianism The principle that there are inviolable laws of nature that have not changed in the course of time.

Upwelling Ascent of cold water from the deep sea to the *photic zone*, usually providing nutrients for rich growth of *phytoplankton*. Upwelling is most common where ocean currents drag surface waters away from continental margins.

V

Varves Alternating layers of coarse and fine sediment that accumulated in a lake in front of a *glacier*. A coarse layer forms each summer, when streams of meltwater carry sand and silt to the lake. A fine layer forms each winter, when the surface of the lake is covered by ice, so that only clay and organic matter settle to the bottom.

Vascular plant A plant that has vessels in its stem for transporting water, nutrients, and food.

Vestigial organ An organ that serves no apparent function but was functional in ancestors of the organism that now possesses it.

Volcanic ash Fine-grained material ejected from a volcano. This material settles in the same manner as sedimentary deposit.

Volcanic rock (See *Extrusive igneous rock*)

W

Walther's law The principle that when depositional environments migrate laterally, sediments of one environment come to lie on top of sediments of an adjacent environment.

Wave, ocean surface A wave on the surface of the ocean produced by the circular movement of water particles under the influence of the wind.

Weathering The various aspects of erosion that take place before transport. Some weathering is physical in character and some is chemical.

West-wind drift The eastward-flowing ocean current near the North or South Pole that is created by the major ocean gyres and the reinforcing *westerly* winds.

Westerlies Winds that flow toward the northeast in each hemisphere between about 30 and 60° from the equator. These result from the same high-pressure zone that produces the *trade winds*, which flow in the opposite direction.

Z

Zone, biostratigraphic The body of rocks characterized by the presence of one or more *index* or *guide fossils*. The upper and lower boundaries of a zone are approximately the same ages everywhere.

Zooplankton *Plankton*, or floating aquatic life, that is animal-like in mode of nutrition (feeds on other organisms).

Index

A

A-horizon (soil), 63
aardvark, 160
Absaroka Range, 549, 550
Abitibi Belt, 254
absolute age (rock), 123, 132, 651; radioactivity and, 119–121
abyssal plain, 48, 49, 52, 108
Acadian orogeny, 227, 228, 230, 232–233, 359, 386
acanthodians, 367–368
Acheulian stone culture, 584
acritarchs, 284–286, 288, 330–332, 361, 383, 398, 446
actualism, 2
adaptations, 136
adaptive breakthrough(s), 148, 152, 378–379, 531
adaptive innovations, 374
adaptive radiation(s), 148–151, 152; Cambrian, 346; Cretaceous Period, 486–487, 496–497; Devonian, 367–368, 373; early Mesozoic, 463; first, 326; flowering plants, 496–497, 530–531; insects, 408–410; local, 150; mammals, 531; marine taxa, 334, 335–343, 381, 401–402, 561–562; middle Paleozoic, 361, 362; mollusks, 444; Neogene Period, 561–562, 564, 565; Ordovician, 335–543, 360; reptiles, 415; terrestrial vertebrates, 564; trilobites, 334
Adelaide mobile belt, 316, 351, 356
Adelaide Trough, 316
adenine, 145, 146
Adirondack uplift, 304
Adriatic plate, 213, 214, 231, 555
Aegyptopithecus (genus), 539
Afar Triangle, 191–194, 198
Africa, 34, 212, 259, 311, 369, 555, 556, 589; ape people, 579–583; in breakup of Pangaea, 463; climate, 39, 607–609; collision with Eurasia, 608, 609; in continental drift theory, 169, 607–609; continental rifting, 191–194, 195; cratonization, 260; early apes, 579; fauna, 170, 588;

glacial "pavement," 66; glaciation, 279, 575; greenstone belts, 254; position of, 234, 317, 471, 508; Precambrian record, 312–313; rift valleys, 174, 191, 196–198, 589, 608, 616; rifting, 195, 196–198, 607–609, 616; savannahs, 41–42; separated from Europe, 475–476, 477, 506
African plate, 209, 213, 215; northward movement, 607–609
Agassiz, Louis, 141, 565
Age of Dinosaurs, 19, 443, 454–461, 510; see also Mesozoic Era
Age of Mammals, 19, 216; see also Cenozoic Era
Alaska, 14, 34, 507, 508, 540, 570
albedo, 37, 576
Albertosaurus, 498
Albian Age, 515, 522
Aldan massif, 311, 319
Alexandra reef complex, 390
algae, 284, 330, 332, 370, 395; filamentous, 101; late Paleozoic, 398; multicellular, 289–290; reef-building, 56, 435, 446, 475; single-celled, 50, 52 (see also coccolithophores); symbiotic, 148, 446, 493; see also blue-green algae; calcareous algae; coralline algae; plankton
algal mats, 99–101, 102, 103, 290, 341–342
algal ridges, 562
Alleghenian orogeny (Appalachian orogeny), 227, 228, 230–231, 235, 420
allele, 144
alligators, 160
alluvial fans, 70, 71, 73, 77, 83, 107, 176; Cretaceous Interior Seaway, 516; in moist climates, 75–76; in mountain building, 207
alluvial plains, 198, 516
Alpine glaciers, 568
Alpine orogeny, 212, 214–215
Alps, 219, 220, 231, 317, 430, 444, 475, 477, 520, 527; future of, 616; glaciers, 574–575; history of,

209–215; origin of, 212–215, 544, 555, 607; structure and composition of, 209–212
aluminum oxides, 63
Amargosa failed rift, 309
Amazon Basin, 313
Amazon River, 195, 313
Americas: Homo sapiens in, 588–589
amino acids, 264, 266, 269, 279
ammonia (NH₃), 242, 264, 269
ammonites, 381
ammonoids, 364, 370, 383, 395, 416, 480, 516; Cretaceous Period, 483, 484, 486, 488, 507; Delaware Basin, 435; early Mesozoic, 444, 447, 466; European Mesozoic deposits, 475–476; extinction of, 415, 422, 423, 510, 513, 527; as index fossil, 151; rediversification of, 398
amoebas, 115, 288
Amphibia (amphibians), 155–156, 378, 379, 397, 398, 412; Carboniferous, 411; Cretaceous Period, 499; early Mesozoic, 457; Paleogene Period, 536
Anabar massif, 311, 319
Ancestral Rocky Mountains, 431–431, 433, 546
Andean region, 381
Andes, 206, 208, 219, 313, 314–315, 316, 515, 604; history of, 209, 217–219, 469
andesites, 199
Andros Island, 98, 99–101
angiosperms; see flowering plants (angiosperms)
angular momentum, 242, 246
anhydrite, 70, 103, 385, 609, 632
Animalia (kingdom) (animals), 5, 43, 650–655; adaptive features of, 135, 136; common ancestry of, 141–142; desert, 41; early Mesozoic, 443; early Paleozoic, 325–342; freshwater and terrestrial: late Paleozoic, 408–415; geographic distribution of, 170, 179–180; large, highly specialized, 154–155; migrations across land bridges, 604 (see also biotic interchange); move onto land, 360,

Animalia (*cont.*)
376–380; Neogene Period, 560; new,
397; Ordovician, 341–342;
Paleogene Period, 527, 531–537;
Proterozoic, 290–294; reef-building,
341; similarities among widely
separated, 174, 175; swimming,
364–370, 398; terrestrial: early
Mesozoic, 454–461, 463; Tethyan
region, 507
Animikie sequence, 301
annelids, 294
Antarctica, 311, 356, 564, 568, 604; in
continental drift theory, 171, 175,
177, 178, 180–181; continental
glacier, 66, 540–541, 575; ice cap,
29, 42, 570, 602; and origin of
psychrosphere, 544–545; position of,
312, 313, 506, 612–613;
Precambrian record, 316
antelopes, 42, 533, 564, 588, 608
anticlines, 639, 642
Antilles, 96
Antler orogeny, 393, 435–436
apes, 170, 539, 559, 565, 578–579,
608
Apollo space program, 248
Appalachian orogenic belt, 123–124,
205, 210, 273, 275, 276, 278,
299–300, 304, 305, 311, 344, 347,
351, 356, 381, 383–384, 385, 418,
420, 421–424, 426, 430, 559;
erosion, 467; history, 209, 219–235,
600–601; leveling and renewed
uplifting, 519–520; modern
topography, 589; provinces,
219–220; tectonic cycles of, 219,
220–228
Appalachian plateaus, 219
Appalachian province, 381
apparent polar wander, 181–182
Appalachian system, 205
apparent polar wander, 181–182
Aptian time, 522
aquatic habitats, 30, 31
aquatic life: Neogene Period, 561–562;
see also marine life
Arabian Peninsula, 317, 607
aragonite, 520, 621, 634, 635
aragonite needles, 98, 635
Aral Sea, 609, 611
archaeocyathids, 332–334, 341
Archaeopteryx, 460–461, 480
Archean Eon (Precambrian time), 19,
239–271, 273, 279, 282, 300
Archean rocks, 251, 253–259;
Antarctica, 316; Eurasia, 319–320;
India, 317; South America, 313;
Australia, 317
Archelon, 488
Arctic Ocean, 508, 516, 545–546, 575
argon, 121, 251
arid basins, 69–74
arkose, 308, 430–431, 467, 629
armadillos, 137, 604
Arthropoda (phylum) (arthropods), 294,
327, 329, 332, 376–378, 446, 494
artificial selection, 142
Arunta Belt, 315
asbestos, 3

Ascension Island, 184
Asia, 17, 41, 217, 231, 316, 498, 579
asteroid belt, 248
asteroids, 241, 243, 248, 250, 251; as
agent of mass extinction, 512–513
asthenosphere, 16–17, 189, 190, 195,
205, 206, 616
asthenosphere–lithosphere boundary,
187
astronomical cycles: as cause of
glaciation, 577–578
Aswan Dam, 609
Atlantic Coastal Plain, 219, 496
Atlantic fault basins, 467–469
Atlantic Ocean, 175, 187, 234, 471,
483, 508, 515, 545–546; clays in,
108; and closure of Tethyan Seaway,
610, 611; cooling of, 589; effect of
Pleistocene glaciers on, 575;
formation of, 169, 173, 195, 199, 463,
467, 518; Neogene Period, 589–607;
opening of, 466; seafloor spreading
in, 184
Atlantic plate, 602
Atlantis, 170
Atlas Mountains, 555
atmosphere, 34–39, 268, 279–280;
chemical composition of, 35–36;
formation of, 250–251; as habitat, 30;
oxygen in, 64, 279–283; of planets,
243–244, 262; temperatures and
circulation in, 36–39
atolls, 93–95, 390
atom(s), 618, 619
ATP (adenosine triphosphate), 267
Australia, 171, 174, 259, 311, 343, 351,
369, 421, 604; climate, 44; collision
with Eurasia, 614; continental
accretion, 356; in continental drift
theory, 178, 180; Devonian Great
Barrier Reef, 383, 393–395;
glaciation, 66, 279, 427, 575;
greenstone belts, 254; interior sea,
513, 521–522; mass extinction in,
345; northward migration, 612–614,
616; position of, 506; Precambrian
record, 315–316; separation from
Antarctica, 544, 545, 612–613
Australian plate, 216; northward
movement, 589, 607
Australopithecus, 584; species of,
579–583
Australopithecus afarensis, 579–582
Australopithecus africanus, 582
Australopithecus robustus, 582–583
Austria, 573
Avalon Peninsula (Newfoundland),
232–233
Avalon terranes, 235
axial plane, 639–640
axis (fold), 640
axolotl, 157

B

B-horizon (soils), 63
backswamps, 77–79
bacteria, 263, 266, 270, 284, 327;
anaerobic, 279; in ocean, 54;
fermenters, 267–268; oxygen toxic
to, 279; purple and green, 268;
single-celled, 32; sulfate-reducing,
268
Badlands National Monument, 64, 550
Baffin Bay, 300
Bahama Banks, 96, 222, 475
Bahamas, 98, 99, 102, 259, 349, 476
Baikal (rift system), 217
baleen whales, 562, 600
Balkans, 213, 214
Baltic Sea, 348
Baltic Shield, 259, 305, 320
Baltica, 229, 230, 231, 320, 388; effect
of movement of, on climate,
347–349; positon of, 343, 351;
sutured with Laurentia, 359, 381,
385, 386
banded iron formations, 257, 262,
280–282; Australia, 315; North
American craton, 301
Barbados, 577
Barberton Mountainland greenstone
sequence, 260
barrier-island–lagoon complex, 86–90,
91, 93, 103, 516–517
barrier islands, 48, 86–90; Cretaceous
Interior Seaway, 516–517
barrier reefs, 93, 475
barriers: as limiting factor, 34
basalt(s), 623; Gondwana sequence,
176, 177; North American craton,
303–304; western North America,
594–595, 596, 599
basement: Appalachians, 220
basement rocks, 207
basement uplift: Laramide orogeny,
546–547, 548
bases, 266
Basin and Range Province, 590–591,
592–593, 594, 595–596, 597, 599
basins, 640–642; arid, 69–74; eastern
North America, 466, 467–469;
Europe, 471–476; marine, 259; South
America, 313–314; southwestern
U.S., 430–435, 436; western North
America, 308–310, 548–549, 550
bats, 30, 138–139, 141, 148, 531
bays, 48; salinity, 55, 58
beaches, 85, 125
Beagle (ship), 137–140, 218
Bear Province, 274
bears, 30, 564, 604
Beartooth Mountain, 300
Beaufort Group, 176
beaver (*Castoroides*), 64, 588
Becquerel, Antoine Henri, 119
bed(s), 630, 642
bedding, 4
bedding surfaces, 125–129
beetles, 158; fossil, 554, 569
belemnoids, 447–448, 484, 486, 510,
527
Belt Supergroup (Montana), 290, 291,
309, 310, 311
benthic foraminifers, 158, 529;
extinction of, 540
benthic life, 50, 52
benthic species: graptolites, 117
benthos, 50
Bering land bridge, 546, 554, 589, 608

Bering Sea, 546
Bering Strait, 570
Beringia, 570
Beyrich, Heinrich Ernst von, 528
big bang theory, 242
Bighorn basin, 548–549
Bikini Island, 93
biogenic sedimentary rocks, 631–635
biogenic sediments, 626, 627
biogeographic distribution: shifts in, and
    climatic change, 589, 598, 606, 608
biogeographic provinces: Europe,
    465–466
biogeography, 34, 55
biostratigraphic system, 131–132
biota(s), 31, 42
biotic change: late Paleozoic, 417;
    Neogene, 559, 562
biotic interchange: Africa/Eurasia, 608,
    609; Australia/Asia, 613–614
birds, 30, 443, 449, 460–461, 536;
    flightless, 137, 535; see also
    songbirds
bird's eye limestone, 101, 102
bisons, 42, 533, 588
bivalve mollusks, 10, 157, 327,
    338–339, 361, 362, 389, 408, 510,
    512, 521; burrowing, 490; Cretaceous
    Period, 483, 491; Delaware Basin,
    434; early Mesozoic, 444, 446, 466;
    extinction rate, 151; late Paleozoic,
    401; mass extinctions, 415;
    Paleogene, 529
Black Hills (S. Dak.), 300, 548
black muds/shales, 201; Appalachians,
    222; Cretaceous Period, 504, 515;
    late Paleozoic, 418, 423–424, 436;
    mountain building, 207, 211, 212; Old
    Red Sandstone continent, 389
Black Sea, 609, 611, 612
block faulting, 196, 198; Europe,
    475–476, 477; in origin of Alps,
    212–213
blue-green algae, 71, 99, 101, 270,
    282, 283, 284, 291;
    Archean rocks, 263; photosynthesis,
    268; similarity to chloroplasts, 288–289
Blue Ridge Mountains, 220, 304
Blue Ridge Province (Appalachians),
    219, 220
Blue Ridge uplift, 300
body size: and brain size, 579–580,
    584–585; change in, 153–155; and
    heat loss, 588
bombs, igneous, 625
bony fishes, 494; Cretaceous, 486;
    early Mesozoic, 448–449, 451
Boreal realm, 466, 478, 508
Boreal seas, 520
borers, 52
Bosporus, 612
Botacatu Sandstone, 177
boulders, 629
Bouma sequence, 107–108
Bovidae, 564, 588
bowfin fishes, 160
Braarudosphaera bigelowi, 511,
brachiopods, 155, 327, 334, 337,
    343, 347–48, 362, 383, 389, 395,

416, 494, 495, 521; articulate, 335,
    361; decline of, 489; Delaware Basin,
    434; early Mesozoic, 444; extinction
    of, 415–416, 423; late Paleozoic,
    398–401; in reef-building, 398;
    Silurian, 381
brackish water, 55–58, 86
braided-stream deposits, 70, 73, 76–77,
    376
braided streams, 70, 71, 76
brain: mammals, 503
brain case: human, 583, 588;
    mammals, 502
brain size: human, 157, 579–580, 583,
    . 584–585
Brazil, 175, 177
Brazilian shield, 313, 314
Brazilide orogen, 313
breccias, 629, 635
Bridger Basin (Wyo.), 71, 73
British Columbia, 352–353, 515
British Isles, 305–308, 369, 386,
    387–389, 575; see also England;
    Great Britain; Ireland; Scotland;
    Wales
bromeliad(s), 135
bryozoans, 337, 339–340, 343, 361,
    381, 389, 521; Cretaceous Period,
    489–490; extinction of, 415, 423,
    510; reef-building, 398, 475
budding, 531
Bulawayan Group (Rhodesia), 263
Bunter deposits, 472–473
Burgess Shale, 327–330, 332,
    352–354, 389
burrow structures, 90
burrowers, 52, 86, 90, 101; soft-bodied,
    521
burrows, 11; in ancient-soil
    identification, 64

C

C-horizon (soils), 63, 64
cactuses, 31, 32, 41
Calamites, 404
calcareous algae, 98, 398, 402, 475
calcareous ooze, 485–486
calcareous plankton, 512
calcareous sediment; see carbonate
    sediments
calcite, 520, 621, 631, 634, 635, 637
calcium, 619, 621
calcium carbonate, 56, 62, 71, 90, 95,
    99, 520; sea-floor sediment,
    108–109; in soil formation, 63
calcium phosphate, 633
Caledonian orogeny, 230, 306, 307,
    359, 386, 420
Caledonide mountain belts, 359
Caledonides, 418
caliche, 64, 71
caliche nodules (calcrete), 63; Ogallala
    Formation, 598; Old Red Sandstone
    continent, 387
California, 44, 46, 391, 514, 548
Cambria, 22
Cambrian Period, 19, 22, 222, 316,

325; early, 221–222; marine fauna,
    326, 327–334, 335, 337, 338–339,
    340–341; paleogeography, 343–346
Cambrian rocks, 23, 119
Cambrian System, 115, 131, 291, 325,
    327; formally designated, 22, 359
Canada, 390, 515
Canadian Rockies, 546
Canadian Shield, 239, 253–254, 273,
    276, 298–299, 306, 320; basalt
    deposits, 303; continental accretion,
    302, 304; greenstones, 300; low-lying
    area, 313
Canning Basin (Australia), 393, 394,
    395, 522
Cannonball Formation, 551
Cannonball Sea, 512, 551
canyons: submarine, 106–107, 571
Cape Cod, Mass., 568
Cape Fold Belt, 178
Capitan Limestone, 435
carbon, 251, 262, 264
carbon 12, 121
carbon 14, 120
carbon 14 dating, 84, 121, 573
carbon dioxide (CO$_2$), 36, 56, 244, 250,
    251, 269, 282, 287, 578, 616, 635
carbon monoxide (CO), 250, 264, 282
carbonaceous chondrites, 264
carbonate ooze, 564
carbonate muds, 635
carbonate platform(s), 95–102,
    383–385, 390, 402, 423, 435, 621;
    Appalachian mountains, 229, 305; in
    Archean rocks, 257, 258, 259;
    Caribbean, 603; east coast (U.S.),
    518–519; Gulf Coast (U.S.), 513,
    515, 516; Laurentia, 345, 352–353;
    Tethyan Seaway, 475; Wopmay
    orogen sequence, 275–276
carbonate rocks, 635
carbonate sands, 635
carbonate sediments, 90, 91, 96, 99,
    103, 108–109, 635; in mountain
    building, 211, 226; Tethyan region,
    476–478
Carboniferous cycles, 79
Carboniferous Period, 397–398, 423,
    424, 426–427, 433; Age of
    Amphibians, 379; animal life,
    408–415; continental collision and
    temperature contrasts, 419–420;
    Dwykatillite, 176; late, 230, 234;
    limestone deposition, 418; marine life,
    398–402; mountain building, 227;
    orogenic activity, 436; plant life,
    402–406, 420; plate movements,
    233–235; position of continents, 417;
    tectonic events, 427–432
Carboniferous-Permian boundary, 175
Carboniferous System, 359; formal
    recognition of, 397–398
carbonization, 11
Caribbean plate, 603
Caribbean region, 602–603
Caribbean Sea, 17, 218, 402–403, 507,
    508, 589, 606
Carnivora (order) (carnivores), 31, 34,
    42, 533, 535; bottom-dwelling, 52;

Carnivora (*cont.*)
Cambrian, 327, 332; Cretaceous Period, 486–488; dinosaurs, 457, 498; mammalian, 531–533; marine, 52, 562; Neogene Period, 564–565; Paleogene Period, 539; pelagic, 383; reptiles, 412–415; swimming, 486; in tundras, 43; zooplankton as, 50
Cascade Range, 589, 592, 594, 596
Caspian Sea, 609, 611
cat family, 135, 535, 539, 564, 604
cataclastics, 635–636
catastrophism, 2, 20
Catskill clastic wedge, 230, 389
Catskill delta, 227, 389
cattle, 533, 564, 588
Caucasus, 555
cave paintings, 586
cell(s), 5; early, 266–267, 269, 279
cementation, 4, 631, 635; agents of, 91, 92, 93
Cenozoic Era, 19, 116, 126, 150, 160, 159, 191, 299, 317, 453, 483, 500, 519; adaptive radiations, 151, 152, 531; Age of Mammals, 148; carbonate platforms, 96; climate, 44, 545; continental rifting, 306, 544, 546; deltas, 85; flowering plants, 495; Gulf Coast in, 551; Ice Age, 541; index fossils, 115, 119; magnetic reversals, 130, 186; mammals, 501; marine life, 529; mountain building, 209, 212, 215, 216, 218, 219, 231, 235, 518; Pangaea in, 179–180; plate movement, 191, 195–196; sea-level changes, 126; subdivision of, 527–528
Cenozoic Erathem, 512
Central Province (N. Amer. craton), 299
Cephalopod mollusks, 332, 370
Ceratodus, 369
Chalk (the) (England), 486, 521
chalk deposits, 486, 513, 517–518, 520–521, 527
Chalk Seas, Europe, 520–521
channels, 76, 77, 83, 85; in lagoons, 87; migrating, 79
chaparral vegetation, 44
Chattanooga Shale, 389, 423
cheilostome bryozoans, 489–491, 529
Cheirolepis, 368
chemical element(s), 618
chemical evidence: early life on earth, 264–267
chemical sedimentary rocks, 631–635
chemical sediments, 626, 627
chemosynthesis, 268, 269, 270
chert(s), 10, 109, 210, 211, 633; in Archean rocks, 257; late Paleozoic, 436
Chesapeake Bay, 219, 600
Chesapeake Group, 600
chevrotain, 533
chimpanzee, 578
China, 160, 343; *Homo erectus* ("Peking man"), 582
Chinle Formation, 469
chlorine, 619
chlorite, 253, 637

chlorophyll, 36, 268, 289
chloroplasts, 288–289, 648
chromosomal mutations, 145–147
chromosomes, 143–147; genetic messages in, 164–166
Churchill Province (N. Amer. craton), 302, 304, 311
cichlid group (fishes), 150, 151
ciliates, 288
Cincinnati Arch, 423
circumpolar current, 46–47, 544–545
Cladoselache, 368
Claraia, 463
class(es), 5
clast(s), 4, 626–627
clastic alluvial-fan deposits: Gondwana sequence, 176
clastic rocks, 626–627
clastic wedge: Appalachians, 223–226, 227; Catskill, 230, 389; continental shelf–east coast (U.S.), 519–520; Cretaceous Interior Seaway, 517; Gulf Coast region, 552; North America, 383–384, 386; Precambrian, early Paleozoic, 376; Taconic orogeny, 351
clay(s), 3, 251, 254, 627, 628–629; in deltas, 83, 84, 90; in deep sea, 108, 109; in turbidite deposits, 108
claystone, 629
clean sands, 211, 226, 227, 257, 355–356
cleavage planes, 619–621
Cleveland Shale, 423
CLIMAP (project), 573–74
climate(s), 29, 36, 65, 96; arid, 86, 198; Australia, 613, 616; Cretaceous Period, 504–505, 507, 516; drying of, 378, 379, 421–422; early Mesozoic, 461, 466, 468, 469, 471–472, 475; effect of distribution of organisms on, 39; effect of movement of continents on, 347–351; Eocene Epoch, 549; fossil plants as indicators of, 44; glaciers as indicators of, 65–66; late Paleozoic, 397, 417, 420–422, 435, 438; middle Paleozoic, 381, 386; Miocene Epoch, 608, 613; moist, 75–76, 86; Neogene Period, 559, 560, 562–564, 565–578, 591–592, 598, 613; Paleogene Period, 545; of planets, 243–244; Pleistocene time, 604; Proterozoic Eon, 276, 279; soils as indicators of, 63, 65; and vegetation, 39–44; *see also* cooling
climatic belts: shifting of, 568–569
climatic change, 5, 23–24; Africa, 607–609; and future of human beings, 616; and mass extinctions, 423, 539–543; Neogene Period, 559, 560, 562–564, 598; Pleistocene Epoch, 573–576, 588; and shifts in biogeographic distribution, 589, 598, 606, 608; *see also* cooling
climatic cycles, 577–578
Cloverly Formation, 471
Clovis projectile point, 589
coal deposits, 1, 13, 47, 62, 79, 83, 89, 373, 397, 398, 402–404, 408; brown,

554, 602; Gondwana sequence, 177; late Paleozoic, 417–419, 422, 423, 424–427, 428–429; Old Red Sandstone continent, 386; sedimentary environment and, 61; as system boundary, 23
coal measures, 426–427
coal-swamp floras, 404, 418, 421
coal swamps, 397, 420, 452; late Paleozoic, 424–425, 426, 428
Coalsack Bluff (Antarctica), 180–181
Coast Ranges (Calif.), 593, 594, 600
Coastal Plain (N. Amer.), 219, 305
Coastal Plain Province, 520
cobbles, 309, 629; in Archaen rocks, 254–257
coccolithophores, 50, 108, 447, 483, 510, 511, 512, 521, 527; Cretaceous Period, 484, 485–486; geographic distribution, 55; rediversification, 529
coccoliths, 108, 447, 480, 485–486; in Chalk Seas, 520, 521; fossil, 573–574
coelacanths, 369
Coelenterata (Cnidaria) (phylum), 292, 294
coiled oysters, 157, 158–159, 483, 491
Colorado, 548
Colorado Plateau, 548, 590, 596, 598
Colorado River, 598
Colubridae, 564
Columbia Plateau, 591–592, 594–595, 596
Columbia River basalt, 599
comet(s), 512–513
communities of organisms, 31–34
Como Bluff, Wyoming, 457
compaction, 4, 631, 635
competition, 31, 34, 334, 346, 411, 531; absence of, 378; as limiting factor, 147, 148, 150
components, principle of, 18
Compositae, 562–563, 564
compound(s), 618
concretion(s), 13
conglomerate(s), 629, 631; in Archean rocks, 254, 257, 259; in Witwatersrand sequence, 260
conifers, 406, 422, 466, 495; early Mesozoic, 453, 454
Connecticut rift, 198
Connecticut Valley (U.S.), 64
conodonts, 327, 335, 336, 395, 434; and mass extinctions, 423, 466; Mesozoic, 448, 466, 475
consumers, 31, 287, 291; early forms of cell life as, 266, 267–268, 269; eukaryotic, 291; in freshwater environments, 58; in ocean, 52
contact metamorphism, 635, 637–638
continental accretion, 297; Antarctica, 316; Australia, 315, 316, 356, 393; Eurasia, 320; Gondwanaland, 311; Laurentia, 299, 311; North American craton, 299; Precambrian provinces, 299–300; South America, 314–315
continental collision, 208, 317, 381; Africa/Asia, 579; late Carboniferous, 418–419; and mountain building,

continental collision (*cont.*)
205, 206, 212–217, 230; North
American craton, 305
continental crust, 15, 29, 35, 178, 205,
622; age of, 251; in anatomy of
mountain chain, 206, 207; doubled
from wedging, 205, 208, 214, 216;
"machine" for production of, 251,
254; new, 299; origin of, 251
continental drift, 182, 189, 195, 205,
311; Cretaceous Period, 483; history
of opinion about, 169–182; northern
Europe, 436–438; in plate tectonics,
191
continental glaciation, 66, 126;
Carboniferous Period, 176–178, 420,
426–427; Neogene Period, 527, 559;
Pleistocene, 565–571; Pliocene,
606–607
continental margin(s), 48, 61, 312, 600;
migration of, 126; North America,
297, 299, 300, 304–305, 309–310,
311; western North America, 436,
469–470, 593, 594, 599
continental rise, 48
continental seas, 48
continental shelf, 29–30, 48, 52, 90,
211, 571; deposits rare in Archean
rocks, 257–259; east coast (U.S.),
518–520
continental slope, 48; 104–108
continents, 239, 258–259, 427;
convergence of, 351; formation of
large, 169, 259–262, 616;
fragmentation of, 169, 184–185, 616;
fused into supercontinent, 228–235,
417 (*see also* Pangaea); jigsaw
puzzle fit of, 169, 173; middle
Paleozoic, 381; migration of, 24, 169;
modern, 311; new: Cretaceous
Period, 506–509; origin of, 297–321;
position of, 5, 343, 420; stationary,
195; *see also* rifting; suturing
convection, 37–38, 184, 190–191, 195,
251, 575
convective cells, 184, 185
convergence: of continents, 351; of
plate margins, 199, 200, 201, 202
convergence, evolutionary, 151–152
cooling, 347, 545, 577; as agent of
mass extinction, 346, 350–351, 383,
510, 511–512; Eocene Epoch,
540–541; episodes of, 578; and Late
Devonian mass extinction, 383;
Paleogene Period, 545; Pliocene,
606–607; rate of, 21, 119; *see also*
climate(s)
Cope, Edward Drinker, 153
Cope's rule, 153
copper deposits, 303
coral reefs, 90, 103, 148; Cretaceous
Period, 507; early Mesozoic, 466,
480; Neogene Period, 562;
Paleogene Period, 529; thorium
dating of, 577–578; in tropics,
56–57
coralline algae, 91, 93, 446, 491, 492
corals, 61, 91, 292, 389, 390, 445;
adaptive radiations, 148; Cretaceous
mass extinction, 510, 529; primitive,

337; *see also* reef-building corals
cordaites, 404–406, 452
Cordilleran Mountains, 509, 516, 603
Cordilleran orogen, 278
Cordilleran orogenic belt, 299
Cordilleran provinces, 595–596
Cordilleran region, 544, 546–550;
mountain building, 393, 589–590
Cordilleran system, 205, 209
core (earth), 14, 15, 129, 240; in
concentric layering, 248–251;
mountain chain, 206, 207–208,
209
coring (technique), 61, 528
Coriolis effect, 38–39, 45, 508
correlation, 115–133, 283, 539–540;
651; accuracy of fossils vs.
radioactivity in, 121–123
cosmic radiation, 510, 512
covalent bonding, 618
crabs, 52, 294, 446, 480, 490,
493–494, 495, 529
cratons, 205, 273, 297–298, 343;
appearance of large, 259–262;
Precambrian, 239, 240; Proterozoic,
297–321
Cretaceous Interior Seaway, 508–509,
514, 515–518, 521
Cretaceous mass extinction, 509–513,
529
Cretaceous Period, 150, 463, 470, 527,
545, 606; carbonate platforms, 96;
continental fragmentation, 300, 317;
dinosaurs, 466; early, 408, 465, 471;
fossils, 182; late, 44, 213, reptiles,
451; sea level, 600; tectonic activity,
546
Cretaceous System, 89, 455, 486, 501;
formal description of, 22, 483
Cretaceous World, 483–523
crinoids (sea lilies), 337, 361, 362, 389,
395, 398, 401, 402, 418, 474
crocodiles, 451, 455, 471, 484; marine,
486, 554; terrestrial, 498
Cro-Magnons, 585, 586–588
cross-bedding, 87–89, 99, 107
crosscutting relationships, principle of,
18
cross-stratification, 630
crust, 14, 15–16, 29, 103; in Archean
Eon, 240, 300; concentric layering of,
248–251; continuous alteration of,
241; and great meteorite shower,
253; siliciclastic rocks in, 4; *see also*
continental crust; oceanic crust; plate
tectonics
crust/mantle boundary, 187
crustaceans, 395
crustal blocks: Appalachian system,
231–235
Cryptozoic rocks, 123
crystal lattice(s), 619, 634–635
crystalline rocks, 239
cultural evolution, 616–617
Cuvier, Baron Georges von, 147, 554
cycads, 44, 453, 454, 466, 495
cyclothems, 426–427; Pennsylvanian,
428–429
cytosine, 145, 146
Czechoslovakia, 573

D

Dachstein reef, 475, 477
Dalradian sequence, 308, 344, 351
Daly, Reginald, 180
Dana, James Dwight, 191, 210
Danian Stage, 512
Darwin, Charles, 93, 136–143, 147,
150, 157, 218, 269, 496, 561; *On the
Origin of Species by Natural
Selection*, 137, 143, 210, 570, 584,
613
dating of rock record, 13, 115–133
dawn redwood, 160
Death Valley (Calif.), 69–71, 73, 309,
310, 343
decay systems; *see* radioactive decay
deciduous trees, 43
declination, 181
decollement, 207, 208, 209–212, 219
deep sea: chilling of, 527, 540–541
deep-sea fans, 107
deep sea floor; *see* sea floor
deep-sea trenches, 185, 199, 205;
subduction at, 189–190
deer family, 564, 604, 614
deformation, 13, 14, 21, 199, 200, 297,
639–647; Africa, 312; Alps, 209, 212,
214; Appalachians, 219, 220;
Antarctica, 316; British Isles, 307;
glaciers, 42; mechanism of, 208–209;
mountain building, 206, 207, 209;
North American craton, 299–300,
301, 302, 310, 386; Ouachita
Mountains, 430; and plate tectonic
movement, 169; western North
America, 548, 596, 600; Wopmay
orogen, 274
degassing, 250–251
Delaware Basin, 432–435, 438
Delaware Mountain Group, 435
delta(s), 83–85, 86, 87, 92, 131;
abandoned lobes, 85; active lobes,
84, 85, 89
delta-front deposits, 83–84, 90, 423
delta plain deposits, 83, 423
deltaic cycles, 85
density, 14, 619; in concentric layering
of earth and its fluids, 248–250; of
planets, 243–244
deposit feeders, 52
deposit sequence: coarse-to-fine, 79,
84, 106
deposition, 4, 21, 66, 573; Baltica,
348–349; cycles of, 79, 376; and
dating of rock record, 115; pattern of:
Cambrian, 343, 344–345; *see also*
glacial deposits; marine deposition;
nonmarine deposition; sedimentary
deposition
depositional environments: Atlantic fault
basins, 467–469; east coast
continental shelf (U.S.), 518–519;
Europe, 475–476; lakes, 65; shifting
of boundaries in, 124–125
desert soils, 63
deserts, 32, 41, 61, 574; Australia, 613;
sedimentary deposits in, 69–74
detrital rocks, 627
detritus, 627

"devil's corkscrews," 64
Devonian Great Barrier Reef (Australia), 393–395
Devonian Period, 64, 230, 359; Age of Fishes, 364–366; deltaic cycles, 85; early, 227; land animals, 376–380; land plants, 370–376, 404; marine fauna, 361–370; mass extinctions, 381–383, 398; mountain building, 218, 226–227; paleogeography, 381–395; reef building, 92, 341; sea level, 359; upper, 381
Devonian System (England), 79, 398; founding of, 359
Dharwar Shield, 317
Diadectes, 413–415
diagenesis, 4, 10, 13, 63, 64, 631, 633
diamond, 619
diatomaceous earth, 109–110
diatoms, 50, 158, 484, 529, 561; on deep-sea floor, 109–110; freshwater, 562, 565; geographic distribution of, 55
diatrymas, 535, 536
Dicroidium, 421
dikes, 259, 623; mafic, 196, 198, 259
Dimetrodon, 413, 415
Dinaride range, 609
dinoflagellates, 50, 284–285, 446–447, 484, 529
Dinosauria (dinosaurs), 6–8, 61, 179, 408, 463, 517, 527, 531; bird-hipped, lizard-hipped, 457; body size, 498–499; Cretaceous Period, 483, 484, 498–501; early Mesozoic, 443, 452, 454–461; extinction, 2, 147, 148, 150, 364, 459, 466, 501–511, 513; fossil, 468, 471; mobility, 455, 458; as reptiles, 502; rise of, 457–460; size of, 455, 458, 459
dip(s), 639
disconformity, 22; see also unconformity
distributary channels, 83, 84, 85
divergence: evolutionary, 150, 151; of plate margins, 201–202
diversity, 32–34, 150
DNA (deoxyribonucleic acid), 5, 143–147, 264–266, 287, 648; structure of, 146
dog family, 535, 539, 564, 565, 604
Dogger (brown Jurassic), 478–480
Dollo, Louis, 162
Dollo's law, 162
dolomite, 103, 385, 621, 634–635, 637; in Wopmay orogen sequence, 275–276
dolphins, 451, 562, 600
dome, doming, 194, 195, 640–642
Doppler effect, 242
dragonflies, 408
drainage: exterior, 69, 70, 74–75; interior, 69
drilling, 95, 103–104, 126, 219
dropstones (glacial), 66, 421; Gowganda Formation, 276
Dryopithecidae, 579
duck-billed dinosaurs, 458, 498, 510
Duluth Gabbro, 304
dune deposits: Europe, 475; Gondwana

sequence, 176; late Paleozoic, 422, 436; North America, 469
dunes, 103, 630; submarine, 99
Dunkleosteus, 368, 389
dust cloud, 512, 513
du Toit, Alexander, 169, 174, 175, 177–178
dynamic metamorphism, 635–636

E

ear: mammal, 503
Early Paleozoic world, 76, 325–357
earth, 262, 616; age of, 20, 119, 189, 241–242; change in, 29, 239, 240; compared to other planets, 242, 243–244; concentric layering of, and its fluids, 248–251; history of, 1–25, 245–248, 527; relationship to sun, 577, 578; thermal history of, 250, 251, 259, 276, 279
earth movements, 13–17, 21, 643–644; and fracturing of rock, 627, 635; southwestern U.S., 427–432; see also tectonism
earthquakes, 14–15, 169, 184, 187, 218, 644; deep-focus, 189–190; in Himalayas, 217
earthworms, 174, 294
East Pacific Rise, 191, 599
Ecca Group, 176
Echinodermata (phylum) (Echinoderms), 31, 327, 334
echinoids, 338, 521
ecologic niche, 30
ecological communities, 31–34
ecology: principles of, 30–34
ecosystem(s), 31, 291; aquatic, 364; altered by photosynthesis, 268; marine, 325, 340–341, 443; modern, 559; Neogene Period, 560; terrestrial, 63, 588
ectotherms, 415, 459
Ediacara fauna (Australia), 292–294, 316, 356
Ediacara Sandstone (Australia), 327
eggs: amniote, 412; dinosaur, 458; plant, 372, 374–376
Egypt, 555, 556
electrons, 619
elements: heavy, 246, 250; light, 248; volatile, 246, 251
elephants, 154–155, 170, 533–534, 537, 564, 579, 604, 608
Elsonian orogenic interval, 302, 311
embryo sac, 374, 375
embryos: similarity of, 141
emus, 137
endotherms, 415, 455, 459, 502
energy: in atmosphere, 36; sources of, for single-celled organisms, 267–268, 269
energy gradient(s), 93
England, 18, 418, 480, 565; position of, 229, 230, 347–348, 351; study of natural selection in, 142–143; see also, British Isles; Great Britain
English Channel, 554
environment(s), 13; anaerobic,

353–354; and evolution, 150; and life, 24, 29–59; as limiting factor, 147; Pleistocene, 574; see also depositional environments
enzymes, 146
Eocene Epoch, 71, 527, 528, 529, 533, 544; animal life, 531, 533, 535, 536, 537, 539; buried forests, 64; climate, 540, 541, 542, 545, 549, 564; depositional basins, western North America, 548–549; Gulf Coast, 552; magnetic reversals, 130; marine deposition, 555, 556; mountain building, 209, 214, 216, 590; plate movement since, 191; sedimentation, 555–556; tectonic activity, 546; volcanism; 550; water temperature
Eocene mass extinction, 562
Eocene Series, 528, 536
epicontinental seas, 48–49; inland, 90
Epworth Basin, 273–275
equability, 565
equator, 37, 395; Devonian, 386; Late Carboniferous, 432; temperature at, 39, 41
equatorial countercurrents, 45
equatorial currents, 45, 576
Equus, 588
erathems, 131
erosion, 3, 13, 17, 21, 22, 65, 85, 195, 241, 251, 253, 297, 616; Alps, 209–212; Appalachian Mountains, 226, 227, 235, 467, 600–601; clasts produced by, 626–627; Cordilleran region, 590; Cretaceous System, 501; deserts, 69; effect on continental size, 298; in environmental migration, 89; by glaciers, 66; of mountain chains, 207, 209–212, 219; North and South America, 313; Precambrian rocks, 239; rapid, 376; Rocky Mountains, 431, 433, 548; and sediment production, 258–259, 627–628; western North America, 550
erosional surfaces: and dating of rock record, 115
erratic boulders, 565
Eryops, 415
Ethiopia, 191
Ethiopian Dome, 607
eugeosyncline, 211
Euglena, 289
eukaryotes, 264, 648, 649; earliest, 284–286; early evolution of, 286–289, 294
Euramerian flora, 420, 421, 461, 466, 469
Eurasia, 46, 212, 213, 214, 216, 217, 589; became separate continent, 508, 545–546; collision with Africa, 608, 609; collision with Australia, 614; formation of, 297, 317–320; tundra, 43
Eurasian plate, 213, 214, 215
Europe, 231, 368, 381, 478–480, 555; biogeographic provinces, 465–466; in breakup of Pangaea, 463; Chalk Seas, 520–521; climate, 421–422, 575; coal deposits, 423, 424–427; in

Europe (cont.)
  continental drift theory, 181–182; continental movement and evaporites, 436–438; fossiliferous deposits, 544, 552–555; Jurassic Period, 466, 476–480; Neogene Period, 609–612; soil formation, 63; Triassic Period, 466, 471–476
eurypterid arthropods, 364, 370, 389
evaporation, 37, 55, 58, 63, 86, 385, 627
evaporites, 69, 70, 71, 73, 198, 258, 423, 632, 633; in breakup of Pangaea, 463–465; Cretaceous Period, 516; deep-water, 124; Delaware Basin, 435; early Mesozoic, 461, 472–473, 475, 507; late Paleozoic, 397, 418, 422, 432, 434, 436–438; Mediterranean Sea, 609, 610–611; middle Paleozoic, 381; North America, 383–385, 432, 434, 436; Old Red Sandstone continent, 387; in sabkas, 103–104
Everglades swamp (Fla.), 428
evergreen coniferous forests, 43
evolution, 8, 9, 13, 24, 29, 135, 137, 616–617; anaerobic, 279; anatomical evidence of, 141–142; of aquatic life, 58; Archean, 266–270; early, 262, 264, 269; of eukaryotes, 286–289; and fossil record, 135–165; human, 156–157, 559, 565, 578–589, 616–617; irreversibility of, 162; mechanisms or, 163–164; in Proterozoic Eon, 273, 283–294; rate of, 157–160, 163; sequence of, 18; soil, 64
evolutionary "experimentation," 334, 404
evolutionary opportunism 367; lung as example of, 378
evolutionary recovery(ies), 147–148, 360, 361–370, 398
evolutionary trends: large-scale, 157, 162, 163, 164; long-term, 152–164
exposures, 3
extinction(s), 23, 137, 147–151; of dinosaurs, 2, 147, 148, 150, 364, 459, 466, 401–411; of large mammals, 588–589; latitudinal patterns of, 511–512; pulses of, 607; rates of, 148–151, 164; see also mass extinctions
extrusive (volcanic) igneous rock, 3, 623
Exuma Sound, 96

F

facies, 124, 254
facies change, 124, 125, 131
failed rifts, 195, 199; North American craton, 302–303, 304–305, 309, 310, 313; Siberian platform, 311; Wopmay orogen, 276
Famennian Stage, 381
Fammenian Age, 383, 395
family(ies), 5, 148; and body size, 154; evolution of, 150

fault(s), 14, 18, 198, 201, 643, 645; western North America, 593, 599–600
fault blocks, 196
faulting, 14, 17, 645–647; Coastal Region (Calif.), 593, 594, 596; western North America, 594, 599
fauna(s), 31, 58, 83; fossil, 86, 554; shifting of, by glaciers, 568–569; similarity of, in disparate locations, 170–171; soft-bodied, 292–294; see also Animalia (kingdom) (animals)
Fayum deposits (Egypt), 556
fecal pellets, 98, 99
feldspar, 15, 121, 431, 622, 627, 628, 629, 638
felsic minerals, 627
felsic rocks, 15, 251, 254, 622, 623, 624, 627; corridors of, 170, 174
felsic roots: uplifting in western North America, 598–599
fermentation, 267, 279
fern(s), 31, 343, 371, 373, 374, 404; early Mesozoic, 452–453, 454, 466; life cycle, 372
fertilization: double, 497; flowering plants, 497–498; seed plants, 375
Fig Tree Group (Africa), 263
finback reptiles, 412–413
finches, 139, 147, 150, 151
fishes, 50, 52, 55, 58, 389, 395, 471; adaptive radiation, 150; Cambrian, 332; Cambro-Ordovician, 366; Cretaceous Period, 483–484; early Mesozoic, 443; fins, 366, 367, 368, 369, 486; fossils, 65; freshwater, 158, 366–368, 386; jawed, 359, 367–368, 370; jawless, 339–340, 366; late Paleozoic, 398; leaving water for land, 378–379; middle Paleozoic, 364–370; similarity across regions, 386; see also bony fishes; ray-finned fishes; sharks; teleost fishes
fissile rocks, 629
fissility, 637
fission-track dating, 121, 609
flint, 633; see also chert
flood basalts, 196, 623
floods, 2, 77, 84, 85, 101; Cambrian, 343, 344, 345; see also transgressions
flora, 31; late Paleozoic, 420–422; Pangaea, 461; shifting of, by glaciers, 568–569; see also plants
Florida Keys, 99, 577
flowering plants (angiosperms), 44, 376, 453, 454; Cretaceous Period, 483, 484, 495–498, 504; fossil, 44, 554; herbaceous (term), 497; life cycle of, 375; Neogene Period, 562–564; Paleogene time, 527, 530–531; as thermometers of the past, 539–540
fluid movement, 37–38
flysch deposition, 207; Alps, 212, 214, 555; Appalachians, 219, 221, 222, 223, 227; Himalayas, 216; Wopmay orogenic sequence, 274, 276
fold-and-thrust belt(s), 206, 207, 643; Alps, 209, 212; Andes, 219;

Appalachians, 219, 220, 229, 235; Himalayas, 216; Laramide orogeny, 546; Ouachita Mountains, 430; western North America, 514; Wopmay orogen, 274–275, 276
folding, 201, 639–640, 645–647; in mountain building, 207–208, 214; in origin of Appalachians, 220, 230, 235
folds, 14; overturned, 643
foliation, 636, 637, 638
Folkstone beds, 520
Folsom projectile point, 589
food, 29, 30; competition for, 31
food chain(s), 31–32, 447
food web(s), 32, 58, 378, 452; fishes, 367–368; marine, 50–55; pelagic, 484–488
foraminifer skeletons, 571–572
foraminifers, 49–50, 288, 398, 521, 555; bottom-dwelling, 517, 545; Cretaceous Period, 489; in Cretaceous mass extinction, 510, 511, 512; see also benthic foraminifers; fusilid foraminifers; planktonic foraminifers
foredeep(s), 207–208, 426; Alps, 214, 555; Appalachians, 221, 222, 226, 227, 229, 230; Cordilleran mountain chain, 513, 514, 515, 516; eastern Europe, 436–438; western North America, 471; Wopmay orogen sequence, 276
forereef slope: Delaware Basin, 435
forests, 43, 64, 359, 376, 542, 562; early Mesozoic, 454; Eocene, 550; see also rain forests, tropical
formations (stratigraphic), 5, 130–131; type section, 131
formic acid, 264
fossil: term, 9
fossil fuels, 13, 36, 578; see also coal; petroleum
fossil groups: classification of, 648–655
fossil plants: as indicators of climate, 44
fossils, 9–13, 18, 617; accuracy of correlation vs. radioactivity, 121–123, and continental drift theory, 17, 174–75, 178–179, 180–181; of early life, 262–264; evolution and, 135–136; as lake deposits, 65; as system boundaries, 23; used to determine relative ages of rock strata, 115; vertical ordering of, 18, 19; see also index fossils
Fountain Arkose, 431–432
fragmentation, 627
Franciscan sequence, 469–470, 513–514, 515, 593
Frasnian Stage, 381, 383
Frasnian Age, 395; reef-building in, 390–391
freshwater, 84; in lagoons, 86; locked up in glacial ice, 570
freshwater environments, 58
freshwater habitats, 30; late Paleozoic, 408–415
freshwater life, 325

fringing reefs, 93
frogs, 378, 411, 457, 499, 536; fossil, 549, 554; Neogene, 559, 564
Front Range uplift, 431, 432
Fungi (kingdom), 5, 9, 325
fusulinid foraminifers, 401–402, 420, 435; and mass extinction, 415, 423, 444

G

gabbro, 623
Galápagos Islands, 139, 147, 150, 151, 269
galaxies, 242, 246
Galilei, Galileo, 252
gamete(s), 144, 147
gamete-bearing plant, 372, 374, 375
Ganges Plain, 215
Ganges River, 216, 612
Garlock Fault, 593
gases, 242; inert, 121; volcanic, 626
gastropod mollusks, 10, 361, 383, 401; carnivorous, 493–494; Cretaceous Period, 483, 490; early Mesozoic, 444, 446; in mass extinctions, 415
Gawler Belt, 315
Geisel Valley (Germany), 554
Geiseltal deposit, 12
gene pool, 147
genes, 143–147
genetic change, 157, 162
genetic engineering, 617
genetics, 143–145
genus (genera), 5, 148
geobiology, 23
geochemistry, 23
geologic events: prediction of future, 616
geologic history, 647
geologic intervals: dating of, 123, 131–132
geologic periods, 19–20
geologic science: fields of, 23
geologic systems, 22–23; formal definition of, 305
geologic time, 2, 18–22, 23; boundaries in, 19, 23, 240
geologic time scale, 20; absolute, 119–121, 123; growth of, 22–23; for magnetic reversals, 186; relative, 119
geology: fundamental principles of, 17–18
geophysics, 23
geosyncline(s), 209, 210–211
Germany, 389, 471–475
geysers, 191, 549, 626
Giant's Causeway, 553
gibbon family, 578
Gibraltor, 612
Gigantopithecus, 579
ginkgos, 453, 454
giraffes, 170, 564, 579
gizzard stones, 459
glacial deposits: Gondwana sequence, 176, 177–178; Proterozoic, 276–279
glacial environments, 65–69

glacial episodes, 349–351, 418, 427; nature of, 573–576; Pleistocene, 604, 606–607, 616, 565–568, 569–576, 616; nature of, 573–576; Proterozoic, 273, 277–279, 302
glacial "pavement," 66
glaciation, 66; causes of, 576–578; chronology of, 572–573; Ordovician Period, 349–351; Pleistocene, 565–571; Pliocene, 606–607; Proterozoic, 276–277, 309, 316, 421; pulses in, 571, 572; waxing and waning, 66, 427; western North America, 598; worldwide, 286; *see also* continental glaciation; glaciers
glaciers, 34, 42–43, 121, 125–126; Antarctica, 564, 565; continental, 66, 417, 420, 426–427 (*see also* continental glaciation); expansion and contraction of (unstable process), 576; Gondwanaland, 349–351; gravity in movement of, 208; late Paleozoic, 397; melting, 77, 381, 541; movement of, 177–178; Pleistocene, 239, 320, 573, 588
glass sponges, 383
glauconite, 121
global cooling; *see* cooling; Pliocene, 606–607
global oxygen cycle, 36
global refrigeration; *see* cooling
global warming, 616; Paleogene Period, 554; *see also* climate(s)
globigerina ooze, 108
globigerinacean foraminifers, 484–485, 562
globigerines, 152
Glossopteris flora, 171, 173, 175, 176, 177, 404, 420, 421
gneiss, 307, 636–637
Goat Seep Formation, 435
goats, 533, 564, 588
gold deposits, 260
Gondwana (India), 173
Gondwana flora, 461, 466
Gondwana sequence, 175–178, 180
Gondwanaland, 171–175, 178, 209, 297, 343, 344, 383, 395, 443, 454, 461, 607; breakup of, 216, 317, 369, 463, 483, 506, 575; coal deposits, 420; early history, 311–317; effect of movement of, on climate, 347, 349; glaciers, 349–351, 397, 417, 427; limestone deposition, 418; marginal growth, 356; mountain building, 325; plant life, 404, 421; position of, 230, 231, 234, 381; sutured to Old Red Sandstone continent, 230, 231, 397, 419–420, 430
gorillas, 6, 578
Gowganda Formation (Canada), 276–277
grabens, 186–187, 191, 196; three-armed, 194, 195
grade (term), 637
graded bed(s), 630
Grand Banks (Newfoundland), 106
Grand Canyon of the Colorado River, 21, 401, 590, 596, 598, 610, 647

granite(s), 3, 199, 251, 254, 315, 317, 622, 623
graules, 629
granulites, 253, 638
graphite, 262, 619
graptolites, 335–336, 348, 350–352, 361, 381; as index fossils, 116, 117–118, 122–123
grasses, 44, 454, 495, 562, 564, 565; origin of, 530–531; spread of, 559
grasslands, 41–42, 564, 574, 613
gravel, 76, 77, 629
gravity, 4, 45; law of, 17; in mountain building, 208, 217; of planets, 243, 244
graywackes, 107, 199, 629; in Archean rocks, 254–257, 259
grazers, 52
Great Artesian Basin, 521–522
Great Bahama Bank, 96, 99
Great Barrier Reef (Australia), 394, 613
Great Basin, 41, 574, 591, 599, 645–657
Great Britain, 46, 230, 426; paleogeography, 351; as part of Laurentia, 298–311; Triassic Period, 475; *see also* British Isles; England; Ireland; Scotland; Wales
Great Estuarine Series, 480
Great Glen Fault, 645
Great Lakes (N. Amer.), 568
Great Meteorite Shower, 251–253
Great Oolite Series, 480
Great Plains (U.S.), 41–42, 300, 550, 598
Great Smokey Mountains, 220
Great Valley (Calif.), 547–548, 593, 594, 596–597
Great Valley Ophiolite, 470
Great Valley Sequence, 593
Greater Antilles, 603
Greece, 609
Green Mountains, 220
Green River basin, 548–549, 550
Green River Formation (Wyo.), 71–74, 549
greenhouse effect, 36, 578, 616
Greenland, 230, 251, 368, 386, 423, 438, 471, 480, 507; became separate continent, 508; formation of, 311; glacial center, 66, 575; ice cap, 29, 42, 570; mafic dikes, 259; as part of Laurentia, 298–311, 345; position of, 229; Precambrian basement, 308; separation of Europe from, 545–546, 552–553, 554–555; vertebrate fossils, 378
Greenland Shield, 306
Greensands, 520
greenstone belts, 253–257, 259, 260, 270, 273, 300, 315, 317, 319
Grenville Belt, 304, 305, 307, 343
Grenville orogenic episode, 300, 304–305, 307, 311
Grenville Province (N. Amer. craton), 299–300, 304, 305, 320
groundwater, 70, 598
groups (stratigraphic), 5, 130–131, 152
Guadalupian Age, 435

guanine, 145, 146
Gubbio, Italy, 511
guide fossils; *see* index fossils
Gulf Coast (U.S.), 1, 85, 486, 513, 515–518; deposition along, 544, 551–552
Gulf of Aden, 191, 195, 607
Gulf of Gaeta (Italy), 90
Gulf of Mexico, 84, 86, 231, 234, 463, 507, 508, 515–516, 518, 552, 609; example of geosyncline, 211
Gulf Stream, 46, 575, 577, 606, 607
Gunflint Chert, 284
Guyana Shield, 313
Guyot, Arnold, 182
guyots, 182, 183, 184, 198
gymnosperms, 406–408, 421, 423, 484, 496, 649; Cretaceous Period, 495, 497; early Mesozoic, 443, 452–454
gypsum, 70, 103, 632
gyre(s), 39, 46, 54, 508, 607; northern Atlantic, 576

H

habitats, 30–31, 32, 42; moist, 372, 373–374, 375; plants as, 39
Hadrian's Wall, 386
half-life (lives), 119, 120, 121
halite, 70, 103, 385, 432, 619–621, 632; in Mediterranean, 609; Zechstein basin, 438
Hall, James, 210, 211
Halloy, d'Omalius D', 483
"hard parts," 9–10, 13
hardness, 619
hardwood trees, 44, 453, 454, 495
Hawaiian Islands, 191, 350; glaciers, 568
heat production, 259
Helderberg Group, 226
helium, 242, 248, 251
Hellenide range, 609
Helvetic Alps (zone), 209, 212
hematite (Fe$_2$O$_3$), 280, 281, 621
hemichordates, 118
Hemicyclaspis, 366
Henslow, J. S., 137
herbivores, 31, 34, 42, 43, 330; benthic, 52; dinosaurs, 457, 459, 498, 510; large, galloping, 152; mammals, 533, 535; Neogene Period, 564–565; in ocean, 50; reptile, 413–415
herbs, 562–564
Hercynian mountain chain, 425–426
Hercynian (or Vascarian) orogeny, 231, 419–420, 425, 430, 478
Hercynian suture, 463
Hercynides mountains, 419–420
herring, 52, 368
Hesperornis, 488
Hess, Harry H.: "History of Ocean Basins," 182–185, 189
hexacorals, 148, 445–446, 475, 489, 491; adaptive breakthrough of, 148, 150
High Atlas, 555
Himalayan Mountains, 21, 317, 555, 559, 607; future of, 616; history of,

209, 215–217, 612, 613; structure of, 215–216
hippopotamus, 568
"History of Ocean Basins," (Hess), 182–185, 189
H.M.S. *Challenger* expedition, 173
Holmes, Arthur, 180, 187–188
Hominoidea (hominoids), 565, 578–579; fossil, 609
Homo (genus), 582–583, 584
Homo erectus, 583–584, 585, 586
"*Homo*" *habilis,* 584
Homo neanderthalensis, 584
Homo sapiens, 584, 585, 588, 617; expansion of, 588–589
Homo sapiens neanderthalensis, 584
homogeneous accretion, 248–250
homology, 141
horned dinosaurs, 498, 510
hornfels, 637
horse(s), 135, 533, 537, 554–555, 564, 604; adaptive radiation, 148; evolution of, 157, 162, 163; large, 588
hot spots, 191, 194, 195, 555
hot springs, 263, 269, 549
Hovey Channel, 435
Hudson Bay, 48, 568
Hudsonian (orogenic) interval, 300–302, 304, 311
human culture, 121
human evolution, 156–157, 559, 565, 578–589, 616–617
humans, 6, 583–584, 585; ancestors of, 370; ecologically disruptive, 565; entering New World, 570; evolutionary future of, 616–617; expansion of, 588–589; and extinction of large mammals, 589; fossil, 589, 609; influence on climate, 578
humus, 63, 64, 69, 370
Hunsrück Shale, 389
Huronian sequence, 280, 302
Huronian Supergroup (Canada), 276
Hutton, James, 2–3, 20–21, 306, 385
hybodonts, 449–450
hydrogen, 242, 246, 248, 251, 264; in atmosphere, 250
hydrogen chloride, 250
hydrogen sulfide H$_2$S, 267–268, 269
hydrological cycle, global, 36, 251
hyena family, 564, 565
hypersalinity, 55–58, 86
hypsometric curve, 29, 48, 83
Hyracotherium, 554–555
hyraxes, 155

I

Iapetus Ocean, 229, 230, 231, 305, 308; closure of, 381, 383; narrowing of, 351–352, 368
Iberian Peninsula, 213
ice age(s), 23, 66
Ice Age, modern (recent), 121, 125–126, 541, 545, 564, 589, 606, 616; hypotheses about, 565

ice caps, 42, 559, 570, 576, 577; Antarctica, 602
ice sheets, 39, 419, 541, 568, 571, 573; size of, and oxygen-isotope ratio in sea water, 572
icebergs, 67–68, 421, 565, 570
Iceland, 186, 565, 571, 606
ichthyosaurs, 451, 466, 486
Ichthyostega (genus), 378, 379
Idaho, 514, 596
igneous activity, 217, 218, 270, 305, 553, 593; Canadian Shield, 299, 300; in formation of earth's crust, 251, 253; Gondwanaland, 311–312; Laramide orogeny, 546, 547; mechanisms of, 598–600; in origin of Appalachians, 220, 227, 230; in regional metamorphism, 636; sediments destroyed by, 258; western North America, 513–514, 549, 590, 596; in Wopmay orogen, 276
igneous arc, 206, 207, 208, 229
igneous petrology, 23
igneous rocks, 3, 4, 14, 621–626, 627, 635, 638, 639; Alps, 214; anatomy of mountain chain, 206, 208; clues to subduction zone, 199; continental accretion, 297; dating, 120, 121; intrusive and extrusive, 3, 17–18, 274, 276, 469; silicates in, 621; Wopmay orogen, 274, 276
Illinois, 66, 427
impressions, 11
index fossils, 122–123, 151; ammonoids, 447, 486; Neogene globigerinaceans, 562; Ordovician, 335–336; Pleistocene, 568–569; Precambrian, 239; trilobites, 327; and zones, 115–119
India, 216, 217, 311, 312, 317, 612, 616; became separate continent, 506; in continental drift theory, 174, 175, 177; fauna, 170–171; glacial activity, 66; Precambrian record, 316–317
Indian craton, 216–217
Indian Ocean, 46, 171, 174, 185, 198, 216, 608
Indochina, 216, 217
Indo-Pacific region, 609
Indus River, 216, 612
industrial melanism, 143
inheritance, 143; blending, 145
inhomogeneous accretion, 250
inoceramid bivalves, 510, 513
insectivores, 531
insects, 30, 31, 39, 64, 294, 325, 564; ecological role of, 408–410; evolution of, 398; flightless, 376; fossil, 549; immature growth stages of, 58; middle Paleozoic, 359; Paleogene, 536–537; as reproductive mechanism for flowering plants, 497–498
interglacial intervals, 571, 573, 577, 616; nature of, 573–576
Interior Oolite Series, 480
intertidal zone(s), 48, 49, 99
intrusion(s), 623; Antarctica, 316
intrusive relationships, 17–18; principle of, 259

invertebrate paleontology, 23
invertebrates: Neogene, 561
ionic bond, 618, 619
ions, 618, 619, 621
Ireland, 306, 308, 553
iridium anomaly, 512
iron (Fe), 14, 15, 129–130, 250, 304, 619, 621, 623; as oxygen sink, 282; in Precambrian rocks, 279–280
iron formations, 633
iron ore, 281, 301; deposits of, 2, 633
iron oxide(s), 63, 281, 627, 631
iron sulfide, 281
island arc(s), 190, 216, 218, 231; in Andes, 219; Caribbean, 603; western North America, 391
islands: with dunes, 103; life forms on, 138–139; as sites of adaptive radiation, 150–151; see also barrier islands
isolation of species, 150; during glaciation, 569–570, 606
isostasy: principle of, 15–16, 206
isotopes: oxygen, 55; radioactive, 119–120, 121
Isthmus of Panama, 577, 589, 603–606
Isua (Greenland), 257, 262
Italy, 213
iterative evolution, 152

J

Japan, 217
Java, 582
"Java man," 582
jaws: dinosaurs, 459; early human, 582, 583; fish, 449, 486; mammals, 502; reptile, 412, 415
jellyfishes, 56, 292, 332
Jet Rock, 478
Joggins Formation (Nova Scotia), 412
joints, 643
Jovian planets, 242–243
jungles, 542
Juniata formation (Appalachians), 223, 351
Jupiter, 242–243
Jura Mountains (Alps), 209–212
Jurassic Period, 157, 443, 465–466, 483, 484; carbonate platforms, 96; Europe, 476–80; flora, 408; Gondwana sequence, 177; hexacorals, 150; igneous activity, 514; land plants, 452, 454; mammals, 500; marine life, 446, 447, 448, 451, 489; mass extinctions, 466; mountain building, 212–213; paleogeography, 467, 469, 470, 471, 476, 477; Pangaea, 463; plate movement since, 191; seas, 508; terrestrial animals, 454, 457, 460; vertebrates, 499
Jurassic System, 444, 461; informal divisions, 478–480
juvenile form: shifted to adult, 155–157

K

kangaroos, 152, 503
Kansas, 427, 486

Karroo Basin (S. Africa), 454
Karroo sequence, 175–176, 177, 415
Kayenta Formation, 469
Kelvin, Lord, 21, 119, 189
Kenoran orogenic interval, 300
Kenyan Dome, 607, 608
Keuper deposits, 472
Keweenawan Supergroup, 303, 304
key beds, 123–124
kingdoms, 5, 9
Klamath Arc, 391, 435, 436
Klamath Mountains, 391
Kuenen, Philip, 105

L

Labrador, 386
Labrador Trough, 301
Lachapelle-aux-Saints, France, 585
lacy bryozoans, 362, 401–402, 435, 444
ladybugs, 31
lagoons, 48, 61, 62, 86–90; Cretaceous Interior Seaway, 516–517; lime mud deposits, 98; salinity, 55, 58; see also barrier island–lagoon complex
Laguna Madre (Texas), 86
Lake Tanganyika, 196
Lake Victoria (Uganda), 150, 151
Lakes, 58, 61, 65, 73, 83, 198, 568; glacial, 66–67; as sites of adaptive radiation, 150–151; turbidity currents, 104–105; western North America, 548–549, 574; see also playa lakes
Lamar Valley (Wyo.), 550
lamination, 4, 630
land: animals move onto, 376–380; life forms, 359, 370–376, 402–408, 443, 484, 495, 562–565; limiting factors on, 31; and sea, 29–30
land barriers, 34
land bridges (corridors), 34, 170–173, 174, 554–555, 570, 609; Isthmus of Panama, 603–604
land plants: Cretaceous Period, 342–343, 402–408, 495, 511; early, 386; early Mesozoic, 452–454, 461; evolution of: Paleogene Period, 530–531; middle Paleozoic, 372
landmasses, ancient: connected, 170–173, 174–175, 177–178, 181–182, 191
Lapworth, Charles, 22–23
Laramide orogeny, 518, 544, 546–548, 550, 590, 598
Late Neolithic stone culture, 586
laterites, 63
Laurasia, 178, 386; see also Old Red Sandstone continent
Laurentia, 229, 230, 298–311, 320, 343, 346, 347, 351, 384; geography of, 344–345; marginal accretion, 311; marine fauna, 348; movements of sea, 354–356; sutured to Baltica, 359, 381, 385, 386
lava, 187, 211, 259, 623–625, 627; multiple flows, 130
lead, 120

leaf margins, 44, 504, 539, 564
leaves, 135, 141, 370, 373; desert plants, 41; flowering plants, 496, 497; tips, 135; waxy, 44, 135
Lepidodendron, 404, 420
Lesser Antilles, 603
levees, natural, 77–79, 83, 99, 101
Lewisian sequence, 307, 308
Libya, 555
life, 5–13; Archean, 262–270; Cambrian, 22; Cenozoic, 560; common biological heritage of, 137, 141–142; Cretaceous, 483–501; definition of, 5; early Mesozoic, 443, 444–461; early Paleozoic, 325–343; environments and, 24, 29–59; expansion of, 616; history of, 23–24, 617; interpreting past through sedimentary environments, 61; late Paleozoic, 398–417; middle Paleozoic, 360–380; Neogene, 559, 560–589; Paleogene, 527, 528–543; Proterozoic Eon, 283–294; requirements for, 29; see also marine life; terrestrial life
life habit(s), 30; marine, 50–55
light, 52; as limiting factor, 49
lignites (brown coal), 554, 602
limb structure: early apes, 579
limestones, 13, 633–634, 635, 639; biogeography of, 55; in British Isles deposits, 308; in continental accretion, 297; in coral reefs, 56; Cretaceous Period, 483, 486, 520; early Mesozoic, 466, 475, 477, 478–480; in Himalayas, 216; late Paleozoic, 397, 398, 401, 418, 423–424; Mediterranean region, 555; in mountain building, 211; in North American craton, 309; Ordovician, 348–349; see also carbonate sediments
limestone reefs, 61, 90–92, 93
limiting factors, 30–31; mode of reproduction as, 373–374; on populations, 147; salinity as, 55–58; temperature as, 34; water depth as, 49
lions, 31, 42, 170, 565
lithification, 4, 631, 635, 636
lithified deposits: Cretaceous, 483; Paleogene, 527
lithology, 131
lithosphere, 16–17, 189, 190, 209, 639; divided into plates, 190–191; see also asthenosphere-lithosphere boundary
Little Bahama Bank, 96
Little Dal Group (Canada), 290
Little Ice Age, 578, 616
living fossils, 160, 369, 406, 453, 503
lizards, 484, 486–488, 499
Llano uplift, 304
lobe-finned fishes, 368, 369–370, 378, 383
lobsters, 294, 446
Loch Ness, 645
modes of, early apes, 579
loess, 573

London Basin, 553–554
London Clay, 553–554
longitude, 181
longshore currents, 86
longshore drift, 47
Louann Salt, 463–465
"Lucy" (skeleton), 579
lung fishes, 368, 369, 383
lungs, 369, 370, 378, 449
lycopods, 373, 404, 406, 452, 453
Lyell, Charles: *Principles of Geology,* 3, 137, 528, 559
Lystrosaurus, 180–181, 454, 463, 510–511

M

Maastrichtian Age, 509, 510, 511, 512, 539
mackeral sharks, 450
Madagascar, 170–171
mafic crust, 251
mafic dikes, 196, 198, 259
mafic minerals, 627–628
mafic rocks, 15, 170, 622–623, 627
mafic volcanoes, 196
magma, 3, 190, 199, 251, 623–624, 627, 635; in anatomy of mountain chain, 206, 218; of greenstone belts, 254
magnesium (Mg), 15, 304, 619, 621, 623
magnetic anomalies, 185–186, 189, 302
magnetic field, 150; age of, 250; reversal in polarity of, 129–130, 181–182, 185–186, 572
magnetic polar wander paths, 181–182
magnetic stratigraphy, 129–130
magnetic "striping," 185–186, 195
magnetite (Fe₃O₄), 281, 621
magnolia tree, 575
Malay Archipelago, 613
Malay Peninsula, 554
Malm (Upper Jurassic), 480
Malvinokaffric Province, 381, 383
Mammalia (class) (mammals), 6, 398, 411, 564, 616; adaptive radiation, 148, 151; Africa, 608; common ground plan, 136; Cenozoic Era, 501; Cretaceous Period, 484, 498–501; early, 502–503; early Mesozoic, 443, 455; entered New World, 570; extinctions, 151, 510, 540, 588–589; feeding and locomotory adaptations, 531; fossil, 550; geographic distribution, 180; Indian peninsula, 216; limitations of body plan, 152; Neogene Period, 564–565; on ocean islands, 138–139; Oligocene Epoch, 537–539; Paleogene Period, 527, 531–533, 535, 554–555; rate of evolution, 158, 160; spread from Africa to Eurasia, 579
mammoths, 34, 147, 588
Mammuthus (genus), 34, 588
manatees (sea cows), 154–155
mantle (earth), 14, 15, 16, 184, 185, 187, 240, 619, 623; in concentric

layering, 248–251; convective motion, 178, 191, 195
mantle plume(s); *see* hot spots
marble, 637
marginal marine sabkas, 103–104
maria (moon), 252
marine deposition: Africa, 555–556; Atlantic Coast (U.S.), 600–602; Cretaceous time, 483; Eocene Epoch, 555, 556; Europe, 471–476; middle Paleozoic, 359; Pliocene Epoch, 565; western North America, 469–471
marine ecosystem, 325, 340–341, 443
marine fauna, 326–334; early Mesozoic, 466; evolutionary recovery, 361–370 (*see also* adaptive radiation); multicellular, 326
marine fossils: Delaware Basin, 434–435
marine habitats, 30
marine life, 45–58, 137–138; Archean, 262–270; Atlantic Ocean, 606; biogeographic provinces of, 465–466; early Paleozoic, 324–326; and Cretaceous mass extinction, 509–510, 511–512; Cretaceous Period, 484–488; diversification of, 325; early Mesozoic, 444–451, 463; extinction rates, 151; late Paleozoic, 398–402, 415; Neogene Period, 561–562; Paleogene Period, 527, 529–530; Proterozoic, 283–294; and water depth, 49
marine sedimentary environments, 83–111
Maritime Alps, 215
marker beds; *see* key beds
Mars, 242, 243, 244–245, 262
marshes, 48, 62, 83, 89, 90, 373
marsupials, 151–152, 500–501, 503, 531, 604, 614
Martinsburg Formation (Appalachians), 222
mass extinctions, 147–148, 391, 454; acritarchs, 286; Cambrian, 325, 334, 335, 343–346; climatic change and, 539–543; Cretaceous Period, 509–513, 529; Devonian, 359, 446; early Mesozoic, 466; Eocene, 562; evolutionary recovery from, 360, 361–370; Frasnian time, 395; Jurassic, 471; Late Devonian, 381–383, 398, 512; late Paleozoic, 397, 415–416, 422–23; Mesozoic Era, 529; Ordovician, 343, 349–351, 360, 361; Paleozoic Era, 452, 489; Permian, 443, 444; periodic, 345–346
massif(s), 311, 313, 478, 520
Matterhorn, 209
Matthews, Drummond, 185, 189
Mauch Chunk formation (Appalachians), 227
Mauritanide mountains (Africa), 231, 419
mayflies, 408
meandering rivers, 61, 77–79, 131, 376; cycles of, 426, 428; deposits of, 107, 227

Mediterranean climates, 43–44
Mediterranean region, 196, 231, 471, 606; mountain chains in, 209, 212, 215; Paleogene Period, 555–556
Mediterranean Sea, 212, 234, 423, 461, 559, 565, 609–612
mélange(s), 200; Appalachians, 231; western North America, 470
meltwater, 66–67, 573, 627
members (rocks), 5, 130–131
Mendel, Gregor, 143–145
Mercury, 242, 243
Mesohippus, 537
Mesolimulus, 480
mesonychids, 535
Mesosaurus, 175, 176, 177
Mesozoic-Cenozoic boundary, 512
Mesozoic Eon: Gulf Coast in, 551
Mesozoic Era, 19, 110, 364, 375, 499, 515, 547, 604; adaptive radiation, 148; animal life, 531; apparent-polar-wonder path, 182; and size of North America, 299; configuration of Pangaea, 234; continental fragmentation, 209, 231, 311, 312, 316; continental rifting, 213, 216, 506; disappearance of dinosaurs, 150–151, 510; early, 443–481; fossils, 151; index fossils, 115, 119; magnetic reversals, 130; mammals, 148; marine life, 423, 562; mountain building, 212, 218, 219, 235; Pangaea, 179, 228, 231; position of continents, 417; sedimentation, 126; vegetation, 408
Messinian Event, 565, 602, 612
Messinian Stage, 612
metamorphic belt, 206; Alps, 209; Wopmay orogen sequence, 275, 276
metamorphic petrology, 23
metamorphic rocks, 3, 4–5, 13, 635–638, 639; dating of, 121; silicates in, 621
metamorphism, 4–5, 13, 195, 253, 259, 297, 635–638; Antarctica, 316; Appalachians, 219, 220; Canadian Shield, 299; British Isles, 307; Canadian Shield, 299; in continental accretion, 297; Cretaceous System, 501; in formation of earth's crust, 251; 253; Gondwanaland, 311–312; India, 317; North American craton, 299–300, 301, 302, 305; Precambrian rocks, 239; resets radioactive clock, 120; sedimentary deposits destroyed by, 258; South America, 313; upper limit of, 638; Wopmay orogen, 274, 276; *see also* regional metamorphism
metamorphosis: juvenile to adult form, 155–156, 157
meteorites, 2, 241, 246, 248, 251–253, 264
methane (CH₄), 242, 264, 269
Mexico, 76
mica(s), 3, 121, 627, 637
mice, 559, 564
Michigan Basin, 384, 385, 425, 640
Michigan River delta, 423–424
microcontinents, 297, 515

microfossils, 23
micropaleontology, 23
microplates, 209, 213, 231, 297, 317, 430
Mid-Atlantic Ridge, 173, 183, 186, 191, 195
Mid-Atlantic Rift, 546, 606
Midcontinent Province, 420
Midcontinental Gravity High, 304
Midland Basin, 433
midocean ridges, 173, 183–184, 185, 189, 196; created by crustal rifting, 205–206; hot springs along, 195, 269; spreading rate, 186–187, 190–191
Milky Way, 242
minerals, 2, 64; defined, 3; felsic, mafic, 251; oxidation state of, 279–282; properties of, 618–621; radioactive, 120, 121; rock-forming, 621, 627–628, 629, 632, 634–635, 637, 639
mining, 3
Minnesota, 300
Miocene Epoch, 527, 559, 607, 609–610; animal life, 561–562, 564; development of American west in, 593–596, 597, 598; early, 564; late, 214–215, 216; marine life, 562; soils, 64
Miocene-Pliocene boundary, 565
Miocene Series, 579
Miocene time: Atlantic Ocean and environs, 600–602; climate, 542, 562, 565; Hominoidea fossil record, 579; plants, 531; rift valleys, 198
miogeosyncline, 211
missing link(s), 461; *Ichthyostega,* 378
Mississippi Embayment, 518, 540, 551
Mississippi River, 85, 195, 552
Mississippi River delta, 84–85, 423–424
Mississippian Period, 398, 401, 423, 436; early, 391; *see also* carboniferous
mitochondrion(a), 286–287, 648
mobile belts: Gondwanaland, 312, 317
mode of life, 30
Moeritherium (genus), 533–534
Mohorovičić discontinuity (Moho), 15, 178, 187, 189
Moinian sediments, 307, 308
Moinian Thrust Fault, 307
Moine sequence, 351
Molasse Basin (Alps), 209
molasse deposition, 207–208, 471; Alps, 212, 214–215, 555; Appalachians, 219, 221, 223, 227, 230; Himalayas, 216, 217; late Paleozoic, 424, 425, 426; Middle Paleozoic, 386; Wopmay orogen sequence, 274, 276
Molasse Plateau (Alps), 212
mold(s), 10, 13
mollusks, 327, 389, 606; bottom-dwelling, 527, 559; early Mesozoic, 443, 444, 466; extinction of, 540, 607; freshwater, 408; shelled, 58; *see also* bivalve mollusks; gastropod mollusks

Monera (kingdom), 5, 263, 648
monkeys, 539, 565, 604, 614
Montana, 498, 512
Monterey Formation (Calif.), 109
Moodies Group, 260
moon: age of, 241, 246; and great meteorite shower, 251–253; origin of, 248
moraines, 66, 77; Pleistocene, 568, 573, 578
Morley, L. W., 189
Morrison Formation, 457, 471
mosasaurs, 488, 510
mosses, 43, 158, 343
Mount Everest, 215
Mount Kilimanjaro, 608
Mount Monadnock (N. H.), 568
Mount Rushmore, 300
Mount St. Helens, 592, 596, 624, 625
mountain belts, 21, 169, 206, 420, 616; alignment of, in continental drift theory, 178; anatomy of, 206–207; axis of, 208; Eurasian, 209; ideal, 206, 209; origin of, 14, 24; Precambrian, 239–40; as species refuge during Ice Age, 569–570, 606; submarine, 205–206
mountain building, 17, 24, 200, 205–235, 297, 381; Australia, 356; as cause of ice Age, 576–577; Cordilleran region, 589–590; Europe, 397; Gondwanaland, 325; late Paleozoic, 427–432; Mediterranean region, 555; Middle Paleozoic, 385–386; North America, 301, 397, 469, 513–515, 518, 527, 546–550; Proterozoic Eon, 273–275; *see also* orogenesis
Mousterian stone culture, 585–586
Mowry Formation, 515
Mowry Sea, 515, 516
mud(s), 108, 199, 227, 257, 629; in deltas, 83, 85, 86, 90; lime, 98; pelleted, 99; in river systems, 77, 79
mud-cracked deposits, 71, 73
mud cracks, 71, 77, 309; Appalachians, 222; carbonate platforms, 102; Wopmay orogen sequence, 276
mud flats, 73
mudstones, 631; Archaen rocks, 254, 257
multituberculates, 502–503, 531, 534
Murchison, Roderick, 22, 359, 398
Murchison meteorite, 264
Muschelkalk Sea, 474–475
Muschelkalk sequence, 472, 474–475
Musgrave Belt, 315
muskrat, 568
mutations, 145–147
Mya arenaria, 606

N

Nain Province (N. Amer. craton), 300
natural gas, 1, 61
natural law(s), 1, 2, 29
natural resources, 61
natural selection, 136–137, 142–143, 147, 162

nautiloids, 332, 339, 364, 370, 395, 398; mass extinctions, 335
Navajo Sandstone, 469
Nazca oceanic plate, 206
Neander Valley (Germany), 584
Neanderthal, 584–588
Nebraska, 64
nebula, 246
nektonic animals, 50, 52, 364–370, 383, 398
Neoceratodus, 369
Neogastropods, 490
Neogene Period, 215, 316, 527–528, 546, 548, 559–615; animal life, 539; continental glaciation, 66; epochs of, 559; regional events, 589–614; uplifts, 527, 550; worldwide events, 560–589
Neptune, 242–243
Netherlands, 86
Neumayr, Melchior, 170, 171
Nevada, 391, 436, 514
Nevadan orogeny, 470
New Brunswick, 232, 233
New England, 232, 233; in Little Ice Age, 578
New Guinea, 577, 613, 614
New Mexico, 76
New Red Sandstone, 475
New World Monkeys, 565
New York State, 227, 361, 383, 385
New Zealand, 416, 613
Newark Group, 458, 467
Newark rift, 199
Newark Series, 304, 518
nickel, 250
Niger River (Africa), 85
Nile delta, 556, 609–610
Nile River, 85, 609
Niobrara Chalk, 517–518
nitrogen, 54, 264; in atmosphere, 35, 36, 250, 251
nitrogen *14,* 121
noble gases, 251
nodules, 103, 633; *see also* caliche nodules (calcrete)
nonconformity, 22; *see also* unconformity
nonmarine clastic sediments, 198
nonmarine deposition, 61–81; rarity of, in Archean rocks, 257–259
Noonday Dolimite, 309
normal fault(s), 643
North Africa, 381
North America, 46, 319, 320, 368, 381, 471, 480, 575, 589; became separate continent, 509, 545–546, 555; in breakup of Pangaea, 463; central, 302–304; climate change, 421–422; coal deposits, 423, 424–427; in continental drift theory, 181–182; divided by Cretaceous Interior Seaway, 509, 514, 515–518, 521; eastern, 302–305, 383–385, 518–520; failed rifts, 302–303, 304–305, 309, 310, 313; flooding, 344; formation of, 297, 311; growth by accretion, 299; mountain building, 301, 397, 469, 513–515, 518, 527,

North America (*cont.*)
546–550, 594; Ordovician mass
extinction, 343; paleogeography, 351,
352–354, 383–385; as part of
Laurentia, 229, 298–311, 345;
stabilization, 352–356; welded to
Gondwanaland, 420
North America, western, 308–311;
development of, 589–600; early
Mesozoic, 469–471; late Paleozoic,
435–438; mountain building,
513–515; provinces of, 590–593; reef
building and orogeny, 390–393;
tectonics, 546–550
North American craton, 220, 231–235
North Atlantic Ocean, 46, 508, 551, 571
North Dakota, 512
North Pole, 42, 508, 544, 545–546, 576
North Pole (Western Australia), 263
North Sea, 438, 553, 554, 555, 602
North Sea basin, 553
Northern Hemisphere, 43, 45, 47, 508,
576, 606; climates, 421, 545, 565,
574; glaciation, 559, 568, 575, 616
Northwest Foreland, 307
Norway, 229
nothosaurs, 450–451, 474
nucleic acids, 266
nucleotide bases, 145, 146, 266
nummulitids, 555
nutrition, 30; chemosynthetic, 270;
dinosaurs, 458–459; single-celled
organisms, 266, 267

O

oak trees, 43, 44, 496
ocean(s), 29, 30, 31, 201; currents,
44–46; new, 616; origin of, 251;
partial draining of, in glacial intervals,
571; solar radiation and, 36–37
ocean basins: age of, 169, 183
oceanic crust, 15, 17, 29, 48, 96, 178,
623; in continental accretion, 297;
formation of, 183–184, 185,
186–187, 189; inhomogeneities of,
186; movement of, 182–185; origin
of, 251
oceanic realm, 48
Ogallala Formation, 598
Ohio basin, 384, 385
oil shales, 71, 549
Old Red Sandstone continent, 230,
231, 368, 378, 381, 383, 385–390,
418, 423; collision with
Gondwanaland, 397, 419–420, 430
Old Red Sandstone Formation, 230
Old World Monkeys, 565, 579
Oldowan stone tools, 584
Olduvai Gorge, 584, 609
Olduvai Event, 572
Oligocene Epoch, 64, 527, 531, 535,
550, 555, 564, 592; animal life,
536–537, 565; climate, 562, 564,
565; Gulf Coast, 551–552; mammals,
537–539; mountain building, 209,
214; sea level, 542
Oligocene Series, 528
olivine, 619, 623
Olympic Mountains (Wash.), 548, 593

*On the Origin of Continents and
Oceans* (Wegener), 173
*On the Origin of Species by Natural
Selection* (Darwin), 137, 143, 210,
570, 584, 613
Ontario, 384
ontogeny, 155–157
Onverwacht Group, 260
ooid shoals, 103
ooids, 62, 98–99
oolites, 98–99, 476, 480
ooze, 108–109; calcareous, 485–486;
carbonate, 564; siliceous, 484, 564
Ophiceras (genus), 444
ophiolites, 201, 205; Alps, 209, 214;
Appalachians, 231; Himalayas, 216;
Klamath Arc, 391
opossum family, 503, 604
opportunistic species, 34, 346
orangutan, 578
order(s), 5, 650, 653, 654; and body
size, 154
Ordovician Period, 384; adaptive
radiation, 341; Antarctica in, 316;
continental glaciation, 66; dating of
rocks, 119; ice caps, 576; index
fossils, 117, 118; Lower, 229–230;
marine fauna, 326, 327, 335–343;
mass extinctions, 360, 361; Middle,
222–223; mountain building,
222–223, 229–230; paleogeography,
346–357; sea level, 325, 381
Ordovician System, 22–23, 115, 131,
325, 371
ore deposits, 1
Oregon, 514, 548, 550, 596
organic skeletons, 627, 633, 635
organs: common "ground plan" of, 136,
141–142
original horizonality, principle of, 17
original lateral continuity, principle of,
17
origination (taxa): rate of, 148–151
Oriskany Formation, 226
orogenesis, 212–215; Appalachians,
220, 228–235; Australia, 315;
Gondwanaland, 311–312, 316, 344;
late Paleozoic, 423; plate tectonics
and, 205–206, 228–235; western
North America, 436, 469–471
orogeny, 205, 426; Africa, 312;
Gondwanaland, 356; middle
Paleozoic, 359, 383, 385, 386,
390–393; Proterozoic Eon, 273–276;
pulses of, 604–606; western North
America, 390–393
ornithischian dinosaurs, 457
orthids, 361
ostracoderms, 366–367, 370
ostracods, 327, 338, 389, 521, 540, 611
ostriches, 137, 608
Ouachita Mountains, 231, 299, 420,
427–430, 433
outcrops, 3
outwash (glacial), 66
overturned fold, 640, 643
Oxfordian Stage, 466
oxidation, 279, 282–283, 387
oxidation state: minerals, 279–282
oxides, 621

oxygen, 47, 262; in atmosphere, 35–36,
64, 250, 279–283, 294; in deep-sea
deposition, 504; from photosynthesis,
268; in weathering, 627
oxygen *16,* 55
oxygen *18,* 55
oxygen in sea: Chalk Seas, 520–521
oxygen-isotope cycle, 540–541,
571–572, 577
oxygen sinks, 282
Ozark Dome, 518

P

Pacific Coast (U.S.): tectonism,
599–600
Pacific Ocean, 217, 463, 508, 546, 602;
clays in, 108; deep-sea trenches,
189; subduction zones, 190
Pacific plate, 191, 599–600, 603
pack ice, 570, 575, 607
Padre Island (Tex.), 86
paleobiology, 23
paleobotany, 23
Paleocene Epoch, 453, 512, 527, 528,
563; animal life, 531, 535; climate,
539–540; Gulf Coast, 551; tectonic
activity, 546
Paleocene Series, 528
Paleocene-Neogene boundary, 559
Paleogene Period, 527–557; animal
life, 564; regional events, 544–556;
sea level, 600; worldwide events,
528–543
Paleogene System, 528
paleogeography, 61, 334; Cambrian,
343–346; Cretaceous Period,
501–522; early Mesozoic, 461–480;
early Paleozoic, 351–356; late
Paleozoic, 417–438; middle
Paleozoic, 381–395; Neogene, 560;
Ordovician, 346–351; Paleogene,
529
paleomagnetic data: Appalachians,
233–235; continental drift, 381;
Gondwanaland, 234; North American
craton, 304, 305; Proterozoic cratons,
297
paleomagnetic dating, 606, 609, 613
paleomagnetism, 181–182, 185–189,
219, 228–229
paleontology, 23
Paleozoic Era, 19, 22, 23, 283, 299,
312, 317; apparent-polar-wander
paths, 182; coal deposits, 171; flora,
452; Gondwanaland, 311; index
fossils, 117–118, 119; late
continental glaciation, 66, ice caps,
576; reefs, 95; marine life, 325, 444,
446, 494; mountain building, 218,
219; Pangaea formed, 173, 178;
plate positions, 229; South America,
314–315
Palisades, 468
Pangaea, 173–174, 178–179, 191, 212,
213, 228, 343, 397, 417, 454, 467;
configuration, 234, 443; creation of,
228, 231; fragmentation of, 173, 174,
231, 235, 311, 443, 461, 463–465,

Pangaea (*cont.*)
231, 235, 311, 443, 461, 463–465, 506; rifting, 179, 191; Triassic Period, 461–463
Paradox Basin, 432
Paradoxides (genus), 233, 345
Paraná Basin, 314, 381
parasites, 32
Paratethys, 609, 611, 612
Paris Basin, 528, 553–554
Parnaiba Basin, 314
particulate inheritance, 143–145
passerine birds; *see* songbirds
patch reefs, 93
pea plants, 143–145
peat, 13, 48, 89
pebbles, 99, 102, 309, 629
"Peking man," 582
pelagic life, 50, 383; Cretaceous Period, 484–488; early Mesozoic, 446–451; Neogene Period, 562
pelagic sediments, 108–110, 475, 477–478
Pelomyxa palustris, 287
pelycosaurs, 412–413, 415
penguins, 529
Peninsular range, 593
Pennales, 562
Pennine zone (Alps), 209
Penninic Ocean, 213, 214
Pennsylvania, 79, 385, 426
Pennsylvanian Period, 398, 411, 424, 426, 428–429, 430, 433; *see also* Carboniferous Period
Pennsylvanian System, 23
Permian mass extinction, 443, 466, 509, 510–511
Permian Period, 397, 417, 461, 463; animal life, 408–415; climate, 420–422, 469; early; Age of Amphibians, 379; Gondwana sequence, 176, 177; marine life, 402, 415; mountain building, 227; orogenic activity, 436; plant life, 406–408, 420–422, 423, 452, 453; plate tectonic movements, 430, 431, 436–438; reef banks, 432–435; terrestrial animals, 454
Permian System, 423, 436; founding of, 398; west Texas, 432–435
Permian-Triassic boundary, 415–416, 454, 461
permineralization, 10, 64
Persian Gulf, 103
Petrified Forest (Ariz.), 469
petroleum, 23, 56, 61, 85, 390, 424, 544, 552; in lake deposits, 549; in limestone reefs, 91, 95; in oil shales, 71
petroleum reservoirs, 1, 423, 433, 465, 597
petrology, 23
Phanerozoic Eon, 19, 20, 39, 49, 126, 129, 148, 217, 238, 260, 273, 317, 551; Antarctica, 316; atmosphere, 280; floras, 420; plate movements, 230; Precambrian, 313
Phanerozoic orogenies, 240
Phanerozoic rocks, 22, 257; Canadian Shield, 298–299

phosphate rocks, 633
phosphate structures, 266
phosphorus, 54
photic zone, 49, 52, 55
photosynthesis, 36, 43, 250, 263, 268, 484; in ocean, 49, 50, 52, 56; in production of oxygen, 279, 282; in protoctists, 289
photosynthetic bacteria, 268, 279, 282
phylogeny, 155–157, 160, 163
phylum(a), 5
physical and chemical conditions, 30–31
physical laws, 17
physical weathering, 627
phytoplankton, 50, 52, 54, 55, 108, 109, 330–332, 386; calcareous, 529; Cretaceous Period, 484; early Palezoic, 330–332; in freshwater environments, 58; late Paleozoic, 398; middle Paleozoic, 361; Proterozoic, 284
Piedmont Province (Appalachians), 219, 220, 222–223, 227, 235, 304, 305
pig family, 533, 539, 564, 604, 614
pigmy chimpanzee, 578
Pilbara Shield, 263, 315
pillow basalt, 187, 308, 436, 623–624; Archean rocks, 254, 259
pillow lavas; *see* pillow basalt
pine trees, 43, 453, 495
pinnacle reefs, 93
pinnipeds, 529–530
Pithecanthropus, 583
placentals, 500–501, 502–503, 531, 533, 604
placoderms, 368, 383, 398
placodont reptiles, 450, 466, 474
planets, 241, 242–245, 577; formation of, 246–248; and great meteorite shower, 251–252, 253
plankton, 50, 383; Cretaceous Period, 484, 485; distribution of, 55, 465; in rivers and streams, 58
planktonic algae: *see* phytoplankton
planktonic ecosystems, 510, 511
planktonic foraminifers, 108, 109, 484–485, 540; Cretaceous mass extinctions, 510, 511, 512; fossil, 573–574; as index fossils, 115–116; iterative evolution, 152
planktonic species: early Mesozoic, 446–447; of graptolites, 117, 118
plant fossils: in ancient-soil identification, 64
Plantae (kingdom) (plants), 5, 370, 373, 614, 649–650; adaptive forms and features, 135–136; bottom-dwelling, 52; climate and, 39–44; colonized land, 359, 361, 370–376 (*see also* land plants); desert, 41; distribution of, 170; early Mesozoic, 466; in fossil record, 10, 44; habitat, 31, 32; large, woody, 64; in levees and backswamps, 79; living fossils, 160; Neogene Period, 560; new, 397; in ocean, 52 (*see also* marine life), predation on, 33–34; similarities among widely separated, 174, 175; soils as medium for growth of, 63; in tundras, 43; *see also* flowering

plants; seed plants; spore-bearing plants
plate boundaries, 190, 202; triple junction of, 191–195
plate convergence, 199, 200, 201, 202
plate movements, 17, 21, 24, 205; clues to ancient, 195–202; Cretaceous Period, 501; northern Atlantic, 546; in origin of Appalachians, 219, 228–235; relative and absolute, 190–191
plate-tectonic theory, 349, 614; and mountain building, 205, 211
plate tectonics, 23, 169–203, 639; Archean, 240, 259; continental movement in, 191; Middle Paleozoic, 359; mountain building in, 208–209; and Old Red Sandstone continent, 386; in origin of Alps, 212–215; and orogenesis, 205–206; Proterozoic time, 300–301; rise of, 182–195; western North America, 599; Wopmay orogen, 274–276
plates, 17; descending, 189–190, 205
playa lakes, 69
Pleistocene Epoch, 34, 427, 527, 528, 559, 560, 574; animal life, 588; climate, 565; climatic change, 573–576; dating of materials from, 121; glaciation, 559, 565–571, 572–580, 604, 606–607, 616
Pleistocene Ice Age, 66
plesiosaurs, 451, 466, 486, 510
Pliocene Epoch, 34, 527, 528, 559, 565; animal life, 564, 588; climate, 564; development of American west in, 594, 596–598; early human fossil record, 579; glaciation, 559; Hominoidea fossil record, 579; plate movement since, 191
Plio-Pleistocene time: Atlantic Ocean and environs, 602–607; development of American west in, 596–598
plunging fold, 640
Pluto, 243
plutonism, 218; Gondwanaland, 317; Wopmay orogen sequence, 276
plutonium *244* (isotope), 246
plutons, 206, 219, 592, 593, 623
Po River, 609
Po River Valley (Italy), 212
point bar(s), 77, 79
point mutations, 145
polar ice caps, 29; *see also* ice caps
polar regions, 420; cooling of, 381, 527, 564
polar wander, 304; *see also* magnetic polar-wander-paths
pole-fleeing force, 178
poles, 38, 45; ocean currents, 46–47; temperature, 39; *see also* North Pole; South Pole
pollen, 374, 375, 376, 573, 600; Cretaceous Period, 496, 497; as index fossil, 568–569
pollen analysis, 565, 600, 613
pollination: by insects, 497–498; by wind, 531
polychaete worms, 329–330
polypeptides, 264

polyps, 56
Pongidae, 578–579
Pongola Supergroup, 260, 263
poor sorting, 629
populations, 147; geographic isolation of, 147, 569–570, 606
porcupine, 604, 614
Port Jackson shark, 449
Posidonia Shale, 478
potassium, 120, 250
potassium *40,* 120, 121
potassium-argon dating system, 121, 122, 241
potassium salts, 438
Pottsville formation (Appalachian), 227
Pound Quartzite, 316
Precambrian-Cambrian boundary, 343
Precambrian interval, 19, 23
Precambrian provinces: and continental accretion, 299–300
Precambrian rocks, 239–240, 253, 633; of British Isles, 305–308; dating of, 123, 124; Gondwanaland, 312–317; western North America, 308–309
Precambrian shields, 19, 23, 239–240, 298; Antarctica, 316; Eurasia, 319–320
Precambrian time, 19, 23, 64; Archean Eon, 239–271; continental glaciation, 66; dating of rocks from, 123, 124; mountain building, 221; Proterozoic Eon, 239
precession of the equinoxes, 578
precipitation of minerals, 2, 4
predation, 31, 32–33, 334, 337, 339, 411, 531; as limiting factor, 147, 148, 150
predators, 154, 332, 359, 370, 444, 535; dinosaurs as, 498; ectotherms, 459; late Paleozoic, 398; marine, 341, 493–495, 529; mobile, 398; swimming, 360, 364–370, 443, 447
primary waves (*P* waves), 14, 15
Primates (order), 6, 531, 539, 565
prodelta deposits, 83, 84, 90
producers, 31, 50, 52, 58, 287, 446–447; early forms of cell life as, 266, 268–270; photosynthetic, 291; planktonic algae, 330
productids, 398–401
progradation, 84, 85, 89, 90, 92, 104, 125, 556
prokaryotic cell(s), 263, 264, 286–287; fossil, 284
prokaryotes, 270, 648
proteins, 34, 264, 266
Proterozoic Eon (Precambrian time), 19, 238, 240, 273; adaptive radiation, 326; cratons, 297–321; diversity of marine life, 325; fossils, 264; global events, 273–295; late Precambrian, 76; magma intruded into cratonic rocks, 259; middle, 302–304
protists, 291, 325
protocontinents, 259
Protoctista (kingdom), 5, 287–288, 648–649
protons, 618
protozoans, 287–288, 294

provinces: American west, 590–593; Devonian Period, 381
pseudoextinction, 148
psychrosphere, 49, 541, 544–545, 546
Pteraspis, 366
pterobranchs, 118
pterosaurs, 460
Puerto Rico, 603
Purcell Supergroup, 309
Pyrenees, 574–575
pyrite (FeS$_2$), 280
pyroxenes, 637

Q

Quarternary Period, 528
quartz, 3, 251, 622, 627, 629, 637, 638; Archean rocks, 257; as cementing agent, 631; crystalline, 109
quartz sand, 74
quartz sandstone, 629, 637; Archean rocks, 257, 258, 259; Wopmay orogen sequence, 275
quartzite, 637
Queenston formation, 351

R

radioactive clock: reset, 259
radioactive decay, 19, 21, 184, 241; rate of, 119, 120, 259
radioactive elements, 250
radioactive dating, 19, 21, 23, 115
radioactivity, 189; and absolute ages, 119–121
radioactivity vs. fossils: accuracy of correlation, 121–123
radiocarbon dating, 121, 123, 585
radiolarians, 109, 395, 434–435, 478, 510
radiolarites, 478
radiometric clocks: reset, 120, 251, 252, 297
radiometric dating, 119, 132, 191, 606, 609; accuracy of, 121–123; Canadian Shield, 299, 300; meteorites, 241; moon craters, 252–253; North American craton, 302; Precambrian rocks, 239, 312, 313
rain forests, tropical, 39, 42, 542, 574, 604–606, 608, 613; plants in, 135–136
rain shadow(s), 41, 69, 386, 418; Ancestral Rocky Mountains, 432; Andes, 604; Kenyan Dome, 608; Rocky Mountains, 598; Sierra Nevada, 598
rainfall, 39, 41, 55–58, 251, 627; in arid basins, 70, 71, 73; high, 74–75; middle Paleozoic, 387; Neogene Period, 563
Ramapithecidae, 579
Ranidae, 536
rats, 559, 564, 604
ray-finned fishes, 368, 398, 408, 448–449
Recent (Holocene) Epoch, 527, 529, 537, 559
recumbent fold, 640
Red Bedford delta, 423, 424
red beds, 198, 280–281, 294, 631;

Appalachians, 221, 223, 227; Old Red Sandstone continent, 389
Red Sea, 191, 195, 196, 198, 607
red shift, 242
reef builders, 360, 361–363, 383, 398
reef-building corals, 55, 56, 398, 491, 492; Caribbean Sea, 606; Cretaceous period, 507; early Mesozoic, 443, 445–446; Paleogene Period, 529
reef flats, 91, 92
reefs, organic, 90–95, 390, 562; Cretaceous Period, 491–493; Devonian, 381; early, 332–334; early Mesozoic, 475, 477; Gulf Coast (U.S.), 511–513, 515, 516; late Paleozoic, 398, 432–435; middle Paleozoic, 359, 361–363, 381, 383; North American, 383–385; Ordovician, 341, 361; Tethyan realm, 515–516; western North America, 390–393
regional metamorphism, 206, 297–98, 311–312, 635, 636–637
regional strike(s), 639
regression(s), 125, 129, 389, 422, 436, 570; as agent of mass extinction, 511; Cretaceous Period, 517–518, 522; middle Paleozoic, 381; Paleogene, 553–554, 555; Pliocene, 596–597; rapid, 427, 428–429; *see also* sea level
relict distribution, 34
religion, 588
remobilization, 297–298, 300
reproduction: plant, 373–374, 495–496, 497–498, 531; reptile, 412
reproductive barriers, 147
reproductive isolation, 8, 498, 617; early human, 585
reproductive rate, 164; large animals, 153–154; and natural selection, 142
reptiles, 44, 398, 411–415, 536; Cretaceous Period, 484, 486–488, 494, 499; dinosaurs as, 459, 502; early Mesozoic, 443, 450–451, 460; flying, 480, 499; mammal-like, 397, 415, 454–455, 457, 502, 510–511; marine, 510, 527
respiration, 279, 287
reverse fault(s), 643
rheas, 137
Rhine Graben, 555
Rhine River, 602
rhinocerous family, 170, 533, 537–539, 564, 604
Rhone River, 609
Rhynia (genus), 370, 373
Rhynie Chert (Scotland), 376–378
rhyolite, 623; ashfall deposits, 594
rift basins: Africa, 609
rift systems: Africa, 313; Himalayas, 217
rift valleys, 258; Africa, 174, 191, 196–198, 589, 608, 616
rift zones, 205–206; Gondwanaland, 312
rifting, 174, 185, 191–195, 205–206, 616; Africa, 195, 196–198, 607–609, 616; ancient, 196–199; Cenozoic Era, 544; Cretaceous Period, 508;

rifting (*cont.*)
   early Mesozoic, 461, 467, 469, 476;
   effect on continental size, 298;
   Laurentia, 299; Mesozoic, 518; North
   American craton, 297, 302–305,
   309–310; and origin of Alps,
   212–213; Paleogene Period, 546;
   Pangaea, 463–465; sedimentary
   sequence in, 198–199; Tethyan, 476;
   three-armed, 194, 195, 555; western
   North America, 599; Wopmay orogen
   sequence, 276; *see also* failed rifts
"Ring of Fire" (Pacific Ocean), 190,
   217
ripples, 630
Riss-Würm interval; *see* Wisconsin
   (Riss-Würm) interval
river cycles, 85
river deltas, 423–424
river-dominated delta, 84–85
river systems, 74–79
rivers: currents, 65; eastern Asia, 612;
   faunas of, 58; flow into sea, 75, 79,
   83, 86; western North America, 596,
   598; *see also* meandering rivers
RNA (ribonucleic acid), 146, 264–266,
   287
Roberts Mountains Thrust, 391, 393
rock ages, 17–18; absolute, 123, 132,
   651; relative, 119
rock dating; *see* dating of rock record
rock flour, 69
rock formation, 2–5; principle of
   actualism and, 2
rock salt, 385, 632; *see* halite
rock-stratigraphic units, 130–132
rocks, 618; bending and flowing of,
   639–642; breaking of, 642–645;
   composition of, 3; determination of
   ages of, 18–19 (*see also* dating of
   rock record); deformation structures
   in, 639–647; types of, 3–5, 130–132,
   621–638; uniformitarian view of, 2–5;
   see also under individual types of
   rocks, e.g., sedimentary rocks
Rocky Mountains, 21, 71, 209, 327,
   343, 386, 401, 431–432, 527, 544,
   548; elevation of, 589, 590; Front
   Range uplift, 548; future of, 616;
   modern formation of, 550, 559, 590,
   595–596, 598; *see also* Ancestral
   Rocky Mountains
rodents, 41, 43, 64, 148, 534–535, 537,
   564
root casts, 428
root tubes, 64
roots: mountain, 598–599, 601; plant,
   370, 373, 376
Ross orogen, 316
Royal Gorge, 598
rubidium, 250
rubidium *87*, 120
rubidium-strontium dating system, 121,
   241
rudists, 491–493, 507, 510, 511, 513,
   527, 529
rugose corals (horn corals), 337, 362,
   381, 383, 394; extinction of, 415,
   423, 444, 445
Russian platform, 317, 320

S

sabkas, 103–104
Sahara Desert, 41, 349, 555, 574
St. Peter Sandstone, 355
salamander(s), 157, 378, 411, 484, 499
salinity(ies), 55–58, 86, 251, 422, 471;
   as limiting factor, 55–58; *see also*
   hypersalinity
Salisbury Embayment, 600
salt, 1, 251, 309
salt deposits, 422, 467
salt domes, 1, 465, 609
salt flats, 104
Salt Range (Pakistan), 415–416
Samfrau, 178
Samfrau mountain belt, 311, 315, 316
San Andreas Fault (Calif.), 187, 593,
   597, 599–600, 616
San Francisco earthquake, 593
sand(s), 627, 629; in barrier islands, 86,
   87–89, 90; in deltas, 83, 85; in river
   systems, 76, 77, 79; in turbidite
   deposits, 107; *see also* oolites
sand dollars, 17, 73–74, 529
sand grains, glacial, 68
sandstone, 2, 309, 424, 629, 631, 639
Sangamon (Eem) interval, 577
Santa Barbara Basin, 354
São Francisco Basin, 314
Saturn, 242–243
saurischian dinosaurs, 457
sauropods, 457, 466, 471, 498
savannahs, 41–42, 542, 564–565, 574,
   588, 608
scales (fish), 449, 486
scallops, 491
Scandinavia, 230, 298, 305, 348, 568;
   glacial center, 575; Little Ice Age, 578
scavengers, 32
Schimper, W. P., 528
schist, 637
Schooley Peneplain, 601
Scotia Arc, 218
Scotland, 305–306, 307, 308, 351, 368,
   418, 553, 645; late Proterozoic tillites,
   278; origin of rocks in, 3; as part of
   Laurentia, 345; position of, 230;
   unconformities, 385–386
scorpions, 364, 376
sea: depth of, 48–50; land and, 29–30;
   see also ocean(s); seas
sea floor, 30, 45, 49, 52–54, 201;
   geographic distribution of species on,
   55; sediments: Cretaceous Period,
   501–504, 511; *see also* ophiolites
sea-floor life, 50–52, 545; Cretaceous,
   488–493, 494, 495; Cretaceous mass
   extinction, 510; early Mesozoic,
   444–446, 463; late Paleozoic, 398,
   401–402; Neogene, 562
sea-floor spreading, 182–185
sea grass, 490–491
sea level, 29, 34, 39, 90, 123, 325, 359;
   Cambrian time, 343, 344, 345, 346;
   change in, 125–129, 426–427,
   428–429, 510, 511, 577–578;
   Cretaceous Period, 483, 504–505,
   508–509, 513, 517; early Mesozoic,
   443, 461, 465, 471; during glacial

episodes, 571; and mass extinctions,
   422, 510, 511; middle Paleozoic, 381;
   Miocene Epoch, 565, 602, 610;
   Neogene Period, 600; Ordovician,
   349–350, 355, 356; Paleogene
   Period, 540, 541, 542, 555; Pliocene
   time, 606, 607
sea pens, 292
sea urchins, 10, 31, 327, 334, 338, 489,
   529; early Mesozoic, 444–445, 446
seas: epicontinental seas; ocean(s);
   shallow seas
seaways, interior, 106
secondary waves (*S* waves), 14, 15
sedentary bivalves, 483
Sedgwick, Adam, 22, 137, 359
sedimentary deposition, 61;
   Appalachians, 210; Archean, 273;
   Australia, 315–316; Europe,
   471–480; Great Valley (Calif.), 593;
   Wopmay orogen sequence, 274–276;
   see also deposition
sedimentary environments, nonmarine,
   61–81; marine, 83–111; *see also*
   depositional environments
sedimentary petrology (sedimentology),
   23
sedimentary rocks, 3–4, 13, 18,
   626–635; dating of, 19; Precambrian
   shields, 239; silicates in, 621
sedimentary structures, 90, 630
sediments, 3–4, 24, 626–635; ancient,
   253–257; vulnerability to erosion,
   258–259
seed(s), 374–376, 412, 496; exposed,
   453; nutrition for, 497
seed ferns, 452
seed plants, 158, 374–376, 404, 406;
   late Paleozoic, 397, 421; life cycle,
   375
seismic analysis, 219
seismic stratigraphy, 125–129
seismic waves, 14–15, 16, 183, 619
seismographs, 14
selection pressures: modern society,
   617
selective breeding, 142, 617
self-regulation, capacity for, 5
self-replication, capacity for, 5, 264–266
Serengeti Plain, 608, 609
series (rocks), 131–132
Seuss, Eduard, 170, 171–173
Sevier orogenic belt, 516
Sevier orogeny, 515, 546–547
sexual maturation, 155
sexual recombination, 147
shale, 629, 639
Shansi (rift system), 217
sharks, 52, 368, 398, 494, 529, 600;
   Cretaceous Period, 486; early
   Mesozoic, 449–450; freshwater, 408
sheep, 533, 564, 588
sheet silicas, 627
shelled animals, 325, 326
shield volcanoes, 623
shorelines: relocation of, 123, 125, 126,
   129
Siberia, 34, 298, 343, 509, 546, 570;
   flora, 420, 461, 466
Siberian platform, 310–311, 317, 319

Sierra Nevada Mountains, 41, 391, 469, 470, 527, 592–593, 594, 598; Neogene uplift, 593; uplift, 589, 597–598
Sierra Nevada region, 514, 548
Sigillaria, 404, 420
silica, 108, 109, 633
silicates, 4, 621
siliceous ooze, 484, 564
siliceous sediments, 109, 276, 513, 565
siliciclastic rocks, 4, 628–631
siliciclastic sediments, 90, 626–627, 635; Appalachians, 221–222, 227; Atlantic coast (U.S.), 519, 600; Great Valley (Calif.), 594; Green River Formation, 71; Gulf Coast (U.S.), 516, 518; Mediterranean region, 556
silicon, 15, 622, 623
sills, 198, 468, 623
silt, 77, 83, 85
Silurian Period, 226, 230, 341, 349, 351, 359; dating of, 122–123; index fossils, 117; land animals, 376–380; marine fauna, 335, 361–370; paleogeography, 381–385; plants, 342, 370–376
Silurian rocks: dating of, 119
Silurian System, 115, 132, 398; formally designated, 22, 359; zones, 117
Siluro-Devonian boundary, 381
single-celled organisms, 283; animal-like, 266, 267–268; plantlike, 266, 267, 268–270; new, 484, 485
Simpson Group (Okla.), 340
Siwalik beds (Pakistan/India), 612
skeletons, 10, 11, 13, 148, 334; evolution of, 327; fish, 366, 449; in fossil record, 325, 398; opaline, 109; organic, 627, 633, 635
skulls: dinosaurs, 455; early human, 580–581, 582, 583, 584, 585
slate, 637
Slave Province, 254, 274, 276, 300, 302, 311
sloths, 137, 604
Smith, William "Strata," 18, 115, 119
snails, 52, 334, 338, 434, 490, 495, 510, 512, 529
Snake River Plain, 596
Snake River Plateau, 591–592
snakes, 484, 499, 500, 559, 564
snapping turtle, 160
Snider-Pellegrini, Antonio, 169
snowshoe hares, 43
sodium, 619
sodium chloride, 632
soft-bodied organisms, 326; fossil record of, 327–330
"soft parts," 11–13
soil(s), 62–63, 64; ancient, 63–64, 387; formation of, 63
soil environments, 62–64
soil horizons, 63, 64
solar energy, 36, 37
solar nebula theory, 246, 253
solar radiation, 36–37; and glaciation, 578
solar system, 253; origin of, 241–242, 245–248

solar wind, 248
Solnhöfen Limestone, 460, 480
songbirds, 536, 559, 564
Sonoma orogeny, 436, 469
South Africa, 171, 175–176, 177, 180
South America, 39, 171, 259, 311, 420, 467, 604; animal life, 137, 369; became separate continent, 506, 508; in breakup of Pangaea, 463; in continental drift theory, 169, 175, 177, 178, 180; glacial episodes, 66, 427, 575; mountain building, 218; position, 234, 312; Precambrian rock record, 313–315
South Atlantic Ocean, 174, 195, 463, 465, 608; opening of, 507
South Pacific Ocean, 174
South Pole, 42, 66, 350, 381, 417, 544, 575, 576, 612–613
Southeast Asia, 420–421
Southern Alps (zone), 209, 212
Southern Hemisphere, 43, 45, 46, 508, 565, 589
Southern Highlands (Scotland), 343
Southern Province (N. Amer. craton), 300
Soviet Union, 415, 454
Space Age, 617
Spain, 231
Speciation, 147, 155, 498; evolution and, 157, 160–162, 163; rapid, 157, 162; rate of, 164
species, 5, 8–9, 139, 147; disappearance of, 162; distribution of, 29, 34, 39, 55; extinct, 116; in marine realm, 49–50, marooned during Ice Age, 569–570, 575, 606; new, 8–9, 137, 147, 148–151, 155, 157, 158; position in environment, 30–31; restricted to discrete regions, 137–139; transition from one to another, 157
species selection, 164
spherical cell-like bodies, 266, 269
sperm (plants), 372, 373, 374–376
sphenopsids, 404, 406, 452, 453
Sphinx, 555
spiders, 294
sponge reefs, 480
sponges, 327, 334, 337–338, 402, 434; Devonian Great Barrier Reef, 395; reef-building, 435, 475
spore(s), 370–371, 374–376
spore-bearing plants, 371, 372, 373, 374, 404, 412, 421, 496; extinction of, 397
spreading zone(s), 191–194, 196
stabilization, 297; Antarctica, 316; Archean crust, 300; earth against erosion, 375; Eurasian Precambrian shields, 320; late Proterozoic sediments, 316; in North America, 352–356
stage boundaries, 651
stages (rocks), 131–132
stalked crinoids, 495
stalked echinoderms, 415
starfishes, 31, 52, 327, 334, 339
stars, 242; origin of, 245–246, 248

stegosaurian dinosaurs, 466
stems (plant), 370, 373, 376
Steno, Nicholas, 17
Stoer Group, 307, 308
stone cultures, 584, 585–586, 609
stony-iron meteorites, 241
stony meteorites, 241
storm waves, 34, 101
Straits of Florida, 96
stratification, 4
stratified rocks: axioms for interpreting, 17
stratigraphic record, 115–119
stratigraphic section(s), 115
stratigraphic stages, 651–653
stratigraphy, 23; units of, 130–132
stratum(a), 4
streams, 65, 74–75, 76; temporary, 69, 70; see also braided steams
Streptococcus (bacteria), 266
strike(s), 639
strike-slip faults, 187, 202, 217, 599, 643
stromatolites, 99–101, 102, 309, 315–316, 354, 395; algal, 327, 330; Appalachians, 222; Archean, 263; decline of, 341–342; and oxygen, 282; Proterozoic Eon, 283–284; Wopmay orogen sequence, 275–276
stromatoporoids, 337, 341, 361–362, 381, 383, 390, 394, 398
strontium, 121, 619, 621
structural basin, 207
structural cross sections, 645–647
structural gene(s), 146
structural geology, 23
subduction, 195, 351, 391; Adriatic plate, 555; angle of, western North America, 514, 515; deep-sea trenches, 189–190; Pacific Ocean, 217; resistance to, 205; western North America, 469–470, 515, 593, 594, 599
subduction zones, 190, 191–194, 206, 231, 345, 501; ancient, 199–201; Caribbean, 603; continental accretion, 297; continental margin of Africa, 312; formation of North American craton, 305; Laramide orogeny, 548; mountain building, 205–206; Pacific Ocean, 217, 218; suturing along, 297; Taconic orogeny, 229; western North America, 469
subfossil, 9
submarine lavas, 211
subsidence, 14
substratum(a), 50–52
subtidal sea floor, 49
subungulates, 155
subzones, 116
Sulawesi, 613–614
sulfate(s), 267, 621
sulfides, 267, 621
sulfur, 1, 268, 282
sun: origin of, 246, 248
Sundance Sea, 469–471
supercontinent; see Pangaea
supergroups (rocks), 5, 131
Superior Province (N. Amer. craton), 254, 299, 300, 302, 311

supernova, 246, 510
superposition, principle of, 17, 119, 123, 253
supratidal zone, 48
surf zone, 47, 49
surface waves, 47
suspension feeders, 52, 327, 332, 335, 339, 401
suturing, 205, 216, 230, 297, 312, 317, 359; creation of Pangaea, 230–231; Devonian Period, 385, 386; Gondwanaland/Old Red Sandstone continent, 397
swamps, 62, 83, 397, 398, 404, 413, 516, 554
swim bladder, 449
Switzerland, 214
symbiotic algae, 148, 446, 493
symbiotic relationships, 148, 498
synclines, 254, 639, 642
system boundaries, 23
system(s); see geologic systems

T

T Tauri stage, 248
tabulate-strome reefs, 361–362, 383, 398, 384, 390–391, 393, 395, 512; demise of, 383
tabulates, 337–338, 341, 361–362, 381, 383, 394, 398; extinction of, 415, 423
Taconic Mountains, 226, 351
Taconic Orogeny, 223, 226, 228, 229–230, 232, 351
tails: fish, 449, 486; prehensile, 565
talus, 92
talus slope, 435
Tasman orogenic belt, 316, 351, 356, 393
Taurus Mountains, 609
taxonomic groups (taxa), 5–8; rates of origination and extinction, 148–151; tropical, 383
taxonomy, 5
Taylor, Frank B., 173, 174
tectonic cycles: Appalachians, 220–228
tectonic events (episodes); see tectonism
tectonics, 23, 169, 639
tectonism: American southwest, 423–432; Africa, 607–608; Atlantic Ocean region, 600, 603; effect on climate, 562–563; Gondwanaland, 311–312, 317; Laramide orogeny, 546–547; western North America, 391, 466, 469–471, 546–550; western U.S., 590, 599–600
teeth, 136, 141, 148, 564; amphibians, 378; canine (eye), 579, 581; cheek, 135, 136, 502, 579, 582; early human, 581, 582, 583–584, 585; fish, 449; reptile, 412, 415
teleost fishes, 486, 493–494, 495, 527, 529
temperature, 29, 34, 250, 251, 262, 276, 279; average annual, 578; and circulation in atmosphere, 36–39; contrasts in, late Carboniferous, 418–419; decline in global, 512 (see

also cooling); effect on rocks, 635, 636; greenhouse effect on, 36; leaf margins as measure of, 44; as limiting factor, 49; marine, 55, 56, 350, 544–545, 607; tropical climates, 41; see also climate(s)
temperature gradients, 39; Cretaceous Period, 507; early Mesozoic, 466; north-south, 574, 575–576; Ordovician, 347; Pliocene Epoch, 565
terrestrial habitats, 30, 31
terrestrial life: Cretaceous Period, 485–504; early Mesozoic, 452–461, 463; late Paleozoic, 402–417; middle Paleozoic, 370–380; Neogene Period, 562–565; Paleogene Period, 530–539
terrestrial realm, 39–44
Tertiary Period, 528
Tertiary rocks, 593
Tethyan Seaway, 212–213, 461, 463, 466, 467, 471, 473, 475–478, 480, 507–508, 511–512, 520, 554, 589, 606, 609; closure of, 609–612
Tethys (sea); see Tethyan Seaway
Texas, 76, 86, 95, 422; Permian System, 432–435
thecodonts, 455, 457, 463
therapsids, 415, 416, 422, 463, 466
thermal plume, 191, 195
thorium, 250
thorium dating, 577
Thornton reef (Ill.), 385
thrust faults, 206, 207, 208, 643; Alps, 209; Appalachians, 219; western North America, 391
thrust sheets, 208, 643; Appalachians, 219; Canadian Rockies, 546; Himalayas, 217
thrusting: Alps, 214; Appalachians, 235
thymine, 145, 146
Tibetan plateau, 215, 217
tidal currents, 86, 87, 90, 99
tidal delta, 87
tidal inlets, 86, 87
tidal range, 86, 87
tides, 47–48
tiger shark, 450
till, 66, 68, 77, 124
tillite(s), 66, 124, 309, 316, 606; Gowganda Formation, 177, 276–277; late Proterozoic, 278
time-parallel boundaries, 131, 132
time-parallel surfaces, 123–130, 131
time-stratigraphic units, 132, 651
titanotheres, 539
toads, 378, 411, 564
toes, 141
Tommotian Stage, 326–327, 334
Tongue of the Ocean, 96, 98
tools; see stone cultures
topsoil, 63
Torridonian sequence, 308
tortoises, 139, 150
trace fossils, 11, 291, 209, 327
tracks, 11
trade winds, 39, 41, 45–46, 98, 386, 418, 507, 575–576, 607, 608, 613, 616

trails, 11
Transcontinental Arch, 354–355, 356, 386, 389
transform fault(s), 187, 190, 191, 194; in rock record, 201–202
transgression(s), 125, 129, 389; Cretaceous Period, 515, 517–518, 520; early Mesozoic, 469, 478; Jurassic Period, 466; Paleogene, 553–554, 555; Pliocene time, 596–597; rapid, 427, 428–429; see also sea level
transpiration, 37
Transverse range, 593
trees, 42, 43, 135, 376; Carboniferous Period, 404–406; early Mesozoic, 452–454, 469; seed-bearing, 398; spore-bearing, 397, 398; see also forests
Triassic-Jurassic boundary, 478
Triassic Period, 398, 423, 443; continental rifting, 212–213; early, 436; Europe in, 471–476; flora, 408; Lower, 181; mammals, 500; marine ecosystem, 415–416; marine life, 444, 445, 446, 448, 463; mass extinctions, 466; Pangaea during, 234; 461–464; paleogeography, 467, 469; pelagic life, 450–451; terrestrial animals, 454, 455, 457, 460; terrestrial life, 451, 453, 454
Triassic System, 443; Gondwana sequence, 176, 177
trilobites, 327, 332, 334, 338, 340, 348, 361, 389; diversity, 370; extinction, 334, 335, 345–346, 383, 415, 444
triple junction, 191–194, 195
tropical climates, 41
trough cross-bedding, 73, 74
true breeding, 143–145
tuft, 625
tundra, 43, 575
turbidite deposition: Appalachians, 222, 227; Archean rocks, 254; British Isles, 308, Himalayas, 216; mountain building, 207, 211, 212; Wopmay orogen sequence, 275
turbidites, 104–108, 201; Late Paleozoic, 436
turbidity currents, 104–107, 108; Archean, 254, 257; submarine, 571
Turgai Strait, 554, 555
turtles, 457, 471, 484, 499
Turkey, 609
Tuscarora Formation (Appalachians), 223
type section (rocks), 131
Tyrannosaurus, 498

U

Uinta basin, 548–559
Ukrainian massif, 319–320
ultramafic rocks, 15, 623
Uncompahgre uplift, 431
unconformity(ies), 21–22, 63–64, 385–386; as time-parallel surfaces, 125–129
underclay, 428

uniformitarianism, principle of, 1–2, 20–21, 29, 137, 528

ungulates, even-toed, 533, 539, 569; odd-toed, 533, 537–538, 569

United States, 351; depositional environments, 389; development of western, 589–600; earth movements in southwest, 423, 427–432; soil formation, 63; *see also* North America

U.S. Geological Survey, 398

universe: ages of, 241–242

uplift, orogenic, 14, 218, 219, 235; Alps, 607; Coastal Ranges (Calif.), 593, 594; Gondwanaland, 316, 317, 344; Himalayan region, 612; mechanisms of, 598–600; North American craton, 300, 304; South America, 313; southwestern U.S., 430–432, 436; western North America, 548, 550, 595–598; Wopmay orogenic sequence, 276

upright posture, 415, 580, 585

upwelling, 54–55, 109, 185, 282, 565

Ural Mountains, 205, 231, 317, 398, 423, 436, 454, 554

uraninite (UO_2), 280

uranium, 119, 250, 279–280

uranium *238*, 120, 121

uranium-thorium decay system, 241

Uranus, 242–243

Utah, 548, 549

V

Valley and Ridge Province (Appalachians), 219, 220, 235, 304, 430

valleys: faults as, 645; U-shaped, 568, 598

Variscan orogeny, 231

varves, 67, 276, 316, 573

vascular plants, 370, 371, 373, 378, 383

vegetation: climates and, 39–44; Pleistocene, 574; stabilization of land through, 376

vent(s), 623; deep-sea, 269, 270

Venus, 242, 243–244, 262

Venus's-flytrap, 135–136

vertebrate paleontology, 23, 554

vertebrates, 325, 359, 378, 463, 600; Cretaceous Period, 484, 498, 499–501; flying, 499; late Paleozoic, 410–415; Neogene Period, 559; Paleogene Period, 536; Permian, 422; terrestrial, 370, 564–565

Vine, Fred, 185, 189

viruses, 5

volcanic arc, 594

volcanic ash, 123–124, 625

volcanic dust, 625

volcanic emissions, 250

volcanic islands, 190, 422

volcanic rocks, 549–550, 623, 625; Archean, 254, 257

volcanic seamounts, 182, 183, 184

volcanism, 217, 218, 351, 476; Africa, 607, 608–609; as agent of mass extinction, 510, 512; Australia, 316, 356; in origin of Alps, 212–213; in origin of banded iron formations, 281; and oxygen, 282–283; western North America, 548, 549–550, 591, 592, 594

volcanoes, 108, 169, 191, 218, 623–626; in mountain chain, 206, 211; oceanic, 108, 183, 184, 189–190

Von Alberti, Friedrich August, 443

W

Waddenzee (Holland), 86

Walchia, 406, 422

Walcott, Charles, 352–353, 354

Wales, 18, 325, 343, 359

Wallace, Alfred Russel, 613–614

Wallace's line, 613–614

walnut family, 496

walruses, 529

Walther's law, 79, 84, 89, 90, 104

Wasatch Formation, 548, 549

Washington, 548, 550, 596

Washington, George, 578

water, 35, 241, 262; in erosion, 627–628

water barriers, 34

water column, 108, 504, 510

water depth: as limiting factor, 49

water movement(s), 44–48, 56, 58; cycle of, 37

water vapor (H_2O), 36, 37, 39; in atmosphere, 250, 251

wave-dominated deltas, 85

waves, 47, 49, 65, 99

Wealden beds, 498

weasel family, 31, 535

weathering, 3–4, 63, 64, 69, 108, 279, 627; in formation of earth's crust, 251, 253, 254

weedy plants, 34, 495, 559, 562–563, 565

Wegener, Alfred, 169, 173–175, 177, 178–179, 180, 181, 191; *On the Origin of Continents and Oceans,* 173

well drilling, 3

well sorted, 629

Werner, Abraham Gottlob, 2

West Germany, 12

west-wind drifts, 46

westerlies, 39

Western orogen (Antarctica), 317

whales, 52, 148, 486, 529, 531, 600; evolution of, 159; Neogene Period, 561–562, vestigial bones, 142

White Cliffs of Dover, 520

Wilkins Peck Member (Green River Formation), 71, 549

wind barriers, 608

wind shadow(s), 73, 98, 389

wind systems, 575–576

Windermere Formation, 344

Windermere Group, 309, 310

winds, 385; direction of ancient, 74; and formation of sand dunes, 73, 74; and ocean currents, 45–46, 47

Wingate Sandstone, 469

Wisconsin (Riss-Würm) interval, 573–574, 585, 607

Witwatersrand sequence, 260, 280

wood: fossil, 10, 83, 121

woodlands, 42, 564, 565, 588

Woodmont Shale, 227

Wopmay orogen (Canada), 273–276, 301

worms, 52, 101

Wyoming, 548, 549; dinosaur fossils, 498

Wyoming Province (N. Amer. craton), 254, 300, 302, 311

X

xenon *129* (isotope), 246

Xenodiscus, 444

Y

Yellow Sea, 423

Yellowstone National Park, 64, 191, 263, 549, 550, 626

Yilgarn Shield, 315

Yorkshire, England, 478

Yucatán Peninsula (Mexico), 96, 603

Yugoslavia, 609

Z

Zagros Mountains (Iraq), 588

Zechstein deposits, 438

Zechstein Sea, 436–438, 474

Zechstein sequence, 472–473

Zion National Park, 469

zircon, 120

zonation: in reefs, 93

zone(s): index fossils and, 116–119; lower boundary, 116; upper boundary, 116, 131, 132

zooplankton, 50, 52, 55, 56, 58, 562

Zumaya, Spain, 511

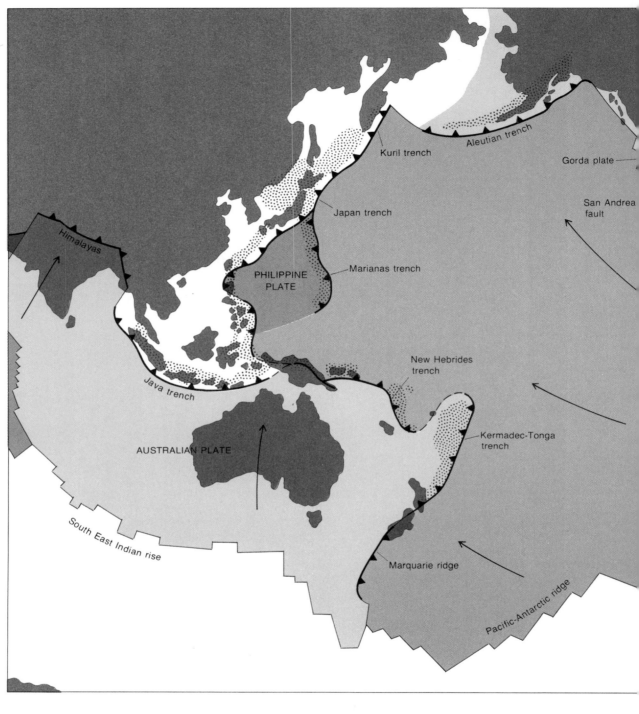

Kuril trench

Aleutian trench

Gorda plate

San Andrea
fault

Japan trench

Marianas trench

PHILIPPINE
PLATE

New Hebrides
trench

Himalayas

Java trench

Kermadec-Tonga
trench

AUSTRALIAN PLATE

South East Indian rise

Marquarie ridge

Pacific-Antarctic ridge

Subduction zone   Ridge axis   Direction of plate motion

Transform   Uncertain plate zone   Areas of deep-focus earthquakes